An Introduction to Bayesian Inference and Decision

SERIES IN QUANTITATIVE METHODS FOR DECISION MAKING

ROBERT L. WINKLER, Consulting Editor

Probability and Statistics for Decision Making, Y. Chou
Statistics: Probability, Inference, and Decision, Volumes I and II and Combined ed., W. L. Hays and R. L. Winkler
Decision Making and the Theory of the Firm, I. Horowitz
An Introduction to Probability, Decision, and Inference, I. H. LaValle
Elements of Probability and Statistics, S. A. Lippman
Modern Mathematical Methods for Economics and Business, R. E. Miller
Applied Multivariate Analysis, S. J. Press
Linear Optimization, W. A. Spivey and R. M. Thrall
An Introduction to Bayesian Inference and Decision, R. L. Winkler

INTERNATIONAL SERIES IN DECISION PROCESSES

INGRAM OLKIN, Consulting Editor

A Basic Course in Statistics with Sociological Applications, 2d ed., T. R. Anderson and M. Zelditch, Jr.
Probability Theory and Elements of Measure Theory, H. Bauer
Probability and Statistics for Decision Making, Y. Chou
Statistics: Probability, Inference, and Decision, Volumes I and II and Combined ed., W. L. Hays and R. L. Winkler
Decision Making and the Theory of the Firm, I. Horowitz
Introduction to Statistics, R. A. Hultquist
Reliability Handbook, B. A. Kozlov and I. A. Ushakov (edited by J. T. Rosenblatt and L. H. Koopmans)
An Introduction to Probability, Decision, and Inference, I. H. LaValle
Elements of Probability and Statistics, S. A. Lippman
Modern Mathematical Methods for Economics and Business, R. E. Miller
Applied Multivariate Analysis, S. J. Press
Fundamental Research Statistics for the Behavioral Sciences, J. T. Roscoe
Linear Optimization, W. A. Spivey and R. M. Thrall
Applied Probability, W. A. Thompson, Jr.
Elementary Statistical Methods, 3d ed., H. M. Walker and J. Lev
An Introduction to Bayesian Inference and Decision, R. L. Winkler

An Introduction to Bayesian Inference and Decision

Robert L. Winkler

Indiana University

HOLT, RINEHART AND WINSTON, INC.

New York · Chicago · San Francisco · Atlanta
Dallas · Montreal · Toronto · London · Sydney

Library of Congress Catalog Card Number: 70-166109
ISBN: 0-03-081327-1
Printed in the United States of America
2 3 4 5 038 1 2 3 4 5 6 7 8 9

To my mother and to the memory of my father

PREFACE

This book is intended as an introduction to statistical inference and decision from a Bayesian viewpoint. The emphasis is largely conceptual; although specific classes of situations are considered, the primary objective is to present a general framework for handling problems of statistical inference and decision and to develop an appreciation for the basic concepts and the theory underlying this framework. Furthermore, although a considerable amount of space is devoted to the solution of problems once they have been expressed in terms of the general framework presented here, the expression of real-world situations in terms of the desired framework (that is, the modeling process) and the determination of the necessary inputs are also discussed at some length.

The mathematical prerequisite is college algebra; calculus is used in a few places, but the reader unfamiliar with calculus can easily skip over the technical details without loss of continuity. Furthermore, no previous knowledge of probability and statistics is assumed. However, those who have been exposed to more "traditional" statistics courses may find this book of interest, and their previous training may enable them to proceed at a faster rate, particularly in the early chapters, than those with no exposure to probability and statistics. In general, considerable flexibility in the choice and emphasis of topics is possible, and an instructor can take advantage of this flexibility to "fit" the book to a particular course and to a particular group of students. For example, some courses might be primarily inference oriented, whereas others might be primarily decision oriented; with regard to different groups of students, there may be variability in general interests as well as in mathematical and statistical backgrounds.

The book is introductory in that it requires no previous exposure to statistics, and it is self-contained in that it can be read without referring to other sources. However, references are given at the end of each chapter for readers interested in further discussions of any topic (at an elementary, intermediate, or advanced level), in more detailed and more concrete applications, in more complete theoretical developments, in extensions to models not considered here, in discussions of related topics, in documentation of historical developments, and so on. These references increase the flexibility of the book, for a judicious choice of readings to accompany the book can vary the emphasis and level considerably. As noted above, however, the book is self-contained and can be read without consulting any of the references.

Numerous problems are included at the end of each chapter, ranging from straightforward applications of the textual material to problems that require considerably more thought. The problems are an integral part of the book, serving to reinforce the reader's grasp of the concepts presented in the text and to point out possible applications and extensions of these concepts. Answers to selected problems are presented at the end of the book, and a solutions manual containing reasonably detailed solutions of the problems is available to instructors from the publisher.

With the usual disclaimer, I would like to acknowledge the assistance of numerous individuals. During the course of this project I was fortunate to receive extensive comments from David P. Baron, William A. Ericson, Ingram Olkin, Richard A. Olshen, Richard E. Trueman, and Roger L. Wright. Their comments led to some changes in the general orientation of the book as well as to modifications of specific discussions and examples. Other helpful comments and assistance were received from Timothy J. Heintz, Gero Meyersiek, Swaminathan Sankaran, and Robert H. Taylor. For performing the tedious task of typing the manuscript, I am grateful to Donna Britton, Maggie Heintz, and Karen Hower. In addition to these specific acknowledgments, there are many people who contributed in an indirect manner (probably more than they realize) to this book, including contributors to the literature in probability and statistics (and in related areas as well), my colleagues, and my students; I acknowledge their contributions collectively rather than individually. However, my wife, Dorth, deserves my greatest thanks for her assistance in making the final product more understandable and in the unenviable chore of proofreading, but most of all for her seemingly endless supply of love and encouragement during the writing of this book. Finally, my love and thanks go to Kevin and Kristin, who, in sacrificing many hours with their father, demonstrated patience beyond their years.

Bloomington, Indiana
January 1972

R. L. W.

CONTENTS

1

INTRODUCTION

Applications of statistics occur in virtually all fields of endeavor—business, the social sciences, the physical sciences, the biological sciences, the health sciences, education, and so on, almost without end. Although the specific details differ somewhat in the different fields, the problems can all be treated with the same general theory of statistics. To begin, it is convenient to identify three major branches of statistics: descriptive statistics, inferential statistics, and statistical decision theory. *Descriptive statistics* is a body of techniques for the effective organization, summarization, and communication of data. When the "man on the street" speaks of "statistics," he usually means data organized by the methods of descriptive statistics. *Inferential statistics* is a body of methods for arriving at conclusions extending beyond the immediate data. For example, given some information regarding a small subset of a given population, what can be said about the entire population? Inferential statistics, then, refers to the process of drawing conclusions or making predictions on the basis of limited information. Finally, *statistical decision theory* goes one step further; instead of just making inferential statements, the decision maker uses the available information to choose among a number of alternative actions.

This book is concerned with the Bayesian approach to problems of statistical inference and decision. There is a common denominator to all such problems—uncertainty. Given some information regarding a small subset of a population, one can be certain about the subset of the population that has been observed, but not about the rest of the population. Thus, in statistical inference, the statistician, given incomplete information, attempts to draw inferences under uncertainty. In statistical decision theory, the decision maker must choose an action in the face of uncertainty (statistical decision making is often referred to as decision making under uncertainty).

In the Bayesian approach to statistics, an attempt is made to utilize all available information in order to reduce the amount of uncertainty present in an inferential or decision-making problem. As new information is obtained, it is combined with any previous information to form the basis for statistical procedures. The formal mechanism used to combine the new information with the previously available information is known as *Bayes' theorem;* this explains why the term *"Bayesian"* is often used to describe this general approach to statistics. Bayes' theorem involves the use of probabilities, which is only natural, since probability can be thought of as the mathematical language of uncertainty. At any given point in time, the statistician's (or the decision maker's) state of information about some uncertain quantity can be represented by a set of probabilities. When new information is obtained, these probabilities are revised in order that they may represent all of the available information.

This book presents a general framework for handling problems of statistical inference and decision. In any application, therefore, it is necessary to express the real-world situation of interest in terms of this framework; in other words, it is necessary to construct a *model* of the real-world situation. In constructing such a model, it is generally impossible to take into account *all* of the relevant factors in the real-world situation; doing so would make the model too complex and too difficult to solve. In a sense, then, a mathematical model is an abstraction of reality, and the results obtained by using such a model are reasonable only if the model itself is a reasonable representation of reality. In this book considerable emphasis is on the solution of problems once they have been expressed in terms of the Bayesian framework for statistical inference and decision. In addition, a substantial amount of space is devoted to the process of expressing problems in terms of the desired framework and to the process of determining the necessary inputs, and it must be emphasized that in practice, these processes are just as important as the methodology for dealing with the problem once it has been expressed in the Bayesian framework and all of the inputs have been determined.

Although many of the examples and the exercises in this book involve computations, the problem of "efficient" computational methods is not discussed. On the one hand, working through numerical examples and exercises by hand or on a desk calculator is most useful in the sense that it helps the reader to understand the methods being illustrated by enabling him to see firsthand "what happens to the numbers." On the other hand, real-world applications often necessitate the use of a computer. Because of differences among computers (especially in light of the rapid development of bigger and better computers) and because some readers may not have access to a computer, none of the examples or exercises requires the use of a computer. It should be noted, however, that problems of statistical inference and decision are generally quite amenable to solution by computer.

This book is introductory in that it requires no previous exposure to statistics, and it is self-contained in that it can be read without reference to other sources. However, references are cited at the end of each chapter for readers interested in further discussions of any topic (at an elementary, intermediate, or advanced level), more detailed and more concrete applications, more complete theoretical developments, extensions to models not considered here, discussions of related topics, documentation of historical developments, and so on. The complete list of references is presented at the end of the book. While these references are by no means exhaustive, they are quite extensive, covering virtually all aspects of statistical inference and decision.

The organization of this book can be summarized briefly as follows: Chapter 2 deals with probability, Chapters 3 and 4 with inferential procedures, Chapters 5 and 6 with decision theory, and Chapter 7 with the relation between the Bayesian approach and other approaches to statistical inference and decision.

2

PROBABILITY: MEASURING UNCERTAINTY

A key concept in the theory of statistical inference and decision is that of *uncertainty*. In a decision-making problem, one is often uncertain about the outcomes of various situations which have an important bearing on the problem. For example, when you contemplate the purchase of shares of a particular common stock, you are uncertain about the future price performance of that stock. When a firm introduces a new product, there is uncertainty about the sales of the product. When a horse-racing fan bets on a race, he is uncertain about the outcome of the race. In these and countless other situations, it is necessary to make decisions in the face of uncertainty.

The uncertainty in a situation is often reflected in such expressions as "it is likely," "it is probable," "the chances are," "probably," and so on. When such expressions are quantified, one is dealing with *probabilities*. For instance, your local weather bureau may report that the probability of rain today is ten percent ($\frac{1}{10}$). If a coin is to be tossed, you might say that the probability is one-half that the coin will come up heads. Since the theory of probability enables a person to express uncertainty in quantitative terms, it is very important in problems of inference and decision. *Probability, then, can be thought of as the mathematical language of uncertainty.*

2.1 THE MATHEMATICAL THEORY OF PROBABILITY

Although this book is concerned with probability only to the extent that it is a useful way to measure uncertainty in problems of statistical inference and decision, it is possible to study probability solely as a branch of mathematics. Given that certain basic axioms are satisfied, probability can be developed as an abstract mathematical system. Fortunately, this abstract mathematical system is applicable to real-world systems involving uncertainty. Thus, some knowledge of the mathematics of probability is most useful in the study of problems of statistical inference and decision. This section contains a very brief discussion of elementary probability theory, and the following sections are concerned primarily with the *interpretation* of probability.

Before the axioms of probability are presented, the notion of an event must be introduced. In any problem involving uncertainty, the "true situation" is, of course, not known for certain. However, if the problem is well-defined, it should be possible to envision the various possible outcomes. Any set, or class, of possible outcomes is called an *event*. If an event cannot be decomposed into a number of "smaller" events, then it is called an *elementary event*; otherwise, it is called a *compound event*. For example, if you are interested in the price per share of a particular stock one year from now, the event "price is 20" is an elementary event, whereas the event "price is greater than 20" is a compound event rather than an elementary event because it can be decomposed into "price is 21," "price is $22\frac{1}{2}$," and many other elementary events. In any problem involving uncertainty, one is interested in the *sample space*, or *event space*, which is defined as the set of all possible elementary events. In the stock example, the sample space consists of all possible prices of the stock one year from now. If the stock in question is traded on a stock exchange where prices are quoted in combinations of dollars and eighths of dollars, then the sample space consists of all such combinations, excluding negative prices (and possibly excluding prices greater than some large number, such as $500).

For another example, suppose that you are interested in the number of customers entering a certain store during a particular hour. The events "two customers enter the store" and "five customers enter the store" are elementary events, whereas "at least two customers enter the store" and "no more than five customers enter the store" are compound events. The sample space consists of all possible elementary events, which can be represented by zero and the positive integers. It should be pointed out that the sample space could be defined differently in this example. If you want to differentiate between male customers and female customers, the sample space might be defined to include all possible combinations of male and female customers. Then the events "four males and three females enter the store" and "five males and two females enter the store" are two different elementary events. Whether the sample space is defined in terms of the number of customers, the combinations of male and female customers, or otherwise depends upon the particular problem of interest.

The concepts of "event" and "sample space" are basic to the development of probability theory, and in any application of probability theory it is important to define the sample space carefully. If you are familiar with set theory, you should recognize that events are *sets* of possible outcomes and that relationships among different events can be expressed in terms of set theory. A set is a well-defined collection of objects (well-defined means that given any object, you can tell whether or not it is a member of the set in question). In probability theory, the sets of interest happen to be events, which are sets of possible outcomes of an uncertain situation. The "uncertain situation" is traditionally referred to as an "experiment," but this term is not always used in this book because it connotes a carefully controlled laboratory experiment. The "uncertain situations" encountered in real-world problems of inference and decision seldom involve such ideal circumstances. The sample space for an uncertain situation or experiment corresponds to the notion of the "universal set" (the set consisting of all possible objects of interest) in set theory, and any event is a subset of this universal set.

Events, like other types of sets, can be illustrated on Venn diagrams. In Figure 2.1.1, for instance, the rectangle represents the sample space, denoted by S, and the circle represents the event E. A set may also be represented by a listing, enclosed in brackets, of the members of the set. In the example of customers entering a store, the set $\{0,1,2,3,4\}$ represents the event "less than five customers enter the store" and the sample space is simply $\{0,1,2,3,\ldots\}$, provided that the customers are not differentiated by sex. Note that an elementary event is a set consisting of just one member. For instance, the elementary event "two customers enter the store" can be represented by $\{2\}$ in set-theoretic notation.

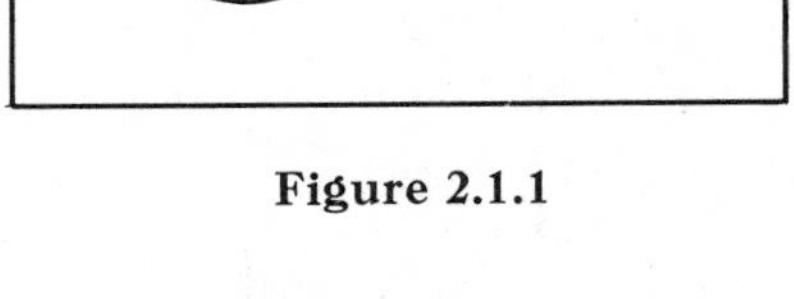

Figure 2.1.1

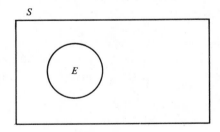

Given that the sample space is carefully defined in any problem involving uncertainty, the next step is to consider the *probabilities* associated with the various events comprising the sample space. As noted above, probability can be developed as an abstract mathematical system on the basis of certain basic axioms. One version of the axioms of probability can be stated informally as follows:

1. The probability of an event E, written $P(E)$, is nonnegative.
2. If S denotes the set of all possible events (the sample space), then the probability of S, $P(S)$, is equal to one.
3. If two events E_1 and E_2 are *mutually exclusive* (that is, they cannot *both* occur), then the probability that at least one of the two events will occur is the sum of the individual probabilities $P(E_1)$ and $P(E_2)$.

In terms of a Venn diagram in which probability is represented by area, the first axiom reflects the fact that the area enclosed by the geometric figure representing an event E cannot be negative; the second axiom states that the area of the rectangle representing S is equal to one; and the third axiom states that the total area enclosed by two non-overlapping figures is equal to the sum of their individual areas (Figure 2.1.2).

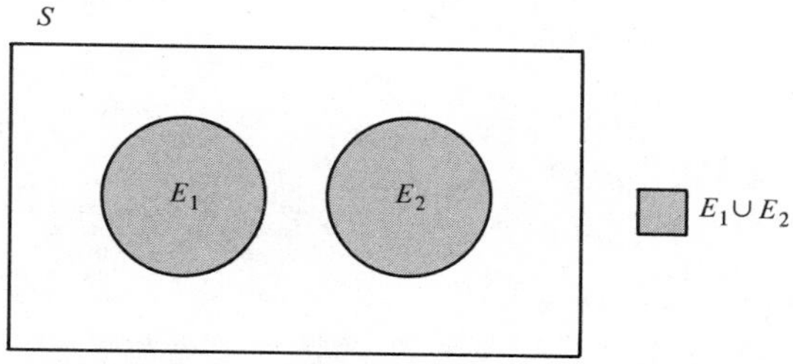

Figure 2.1.2

In set-theoretic terminology, the event "E_1 *or* E_2 (or both) will occur" is called the *union* of E_1 and E_2, and the event "both E_1 *and* E_2 will occur" is called the *intersection* of E_1 and E_2. In the stock example, suppose that E_1 is the event "price is greater than 20 but less than 25" and E_2 is the event "price is greater than 22." The union of E_1 and E_2, written $E_1 \cup E_2$, is "price is greater than 20"; the intersection of E_1 and E_2, written $E_1 \cap E_2$, is "price is greater than 22 but less than 25." Axiom 3, then, states that if E_1 and E_2 are mutually exclusive, then $P(E_1 \cup E_2) = P(E_1) + P(E_2)$.

From the three basic axioms, various theorems of probability can be easily deduced. For instance, the probability of an event is a number which is between zero and one, inclusive:

$$0 \leq P(E) \leq 1.$$

If E_1 is a *subset* of E_2 (written $E_1 \subset E_2$), then

$$P(E_1) \leq P(E_2).$$

(E_1 is a subset of E_2 if each elementary event in E_1 is also in E_2—see Figure 2.1.3). For instance, the probability that a football team wins a game by at least 14 points is less than the probability that the team wins the game. Also, if the *complement* of E, denoted by $\bar{E}$, is defined to be the event "E does *not* occur," then

$$P(\bar{E}) = 1 - P(E).$$

This is illustrated in Figure 2.1.4. For example, if the probability of rain on a particular day is $\frac{2}{3}$, then the probability of no rain must be $\frac{1}{3}$.

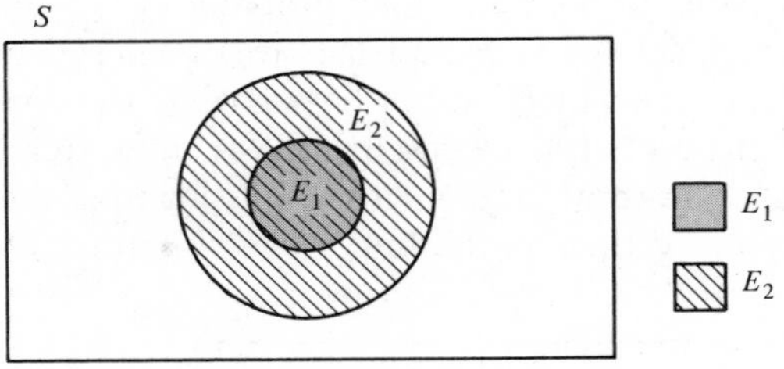

Figure 2.1.3

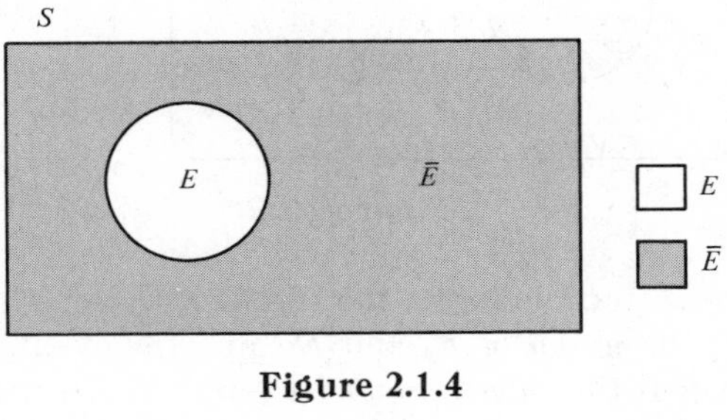

Figure 2.1.4

If two events E_1 and E_2 are not necessarily mutually exclusive, then the probability that at least one of them occurs (in set-theoretic terms, the probability of their union) is equal to the sum of their individual probabilities minus the probability that *both* will occur (in set-theoretic terms, the probability of their intersection):

$$P(E_1 \text{ or } E_2 \text{ or both}) = P(E_1) + P(E_2) - P(\text{both } E_1 \text{ and } E_2),$$

or

$$P(E_1 \cup E_2) = P(E_1) + P(E_2) - P(E_1 \cap E_2).$$

This rule is illustrated in Figure 2.1.5. Note that $P(E_1 \cap E_2)$, the portion where the two sets overlap, must be subtracted to avoid "double counting," since this portion is added twice, once with $P(E_1)$ and once with

$P(E_2)$. If E_1 and E_2 are mutually exclusive, there is no overlap, or intersection; thus, $P(E_1 \cap E_2) = 0$, and the rule reduces to the third axiom of probability. For an example of the rule, suppose that a single card is to be dealt from a well-shuffled, standard 52-card deck of playing cards. Let E_1 be the event that the card is a club, and let E_2 be the event that the card is an ace. The probability of dealing either a club or an ace (or both) is

$$P(E_1) + P(E_2) - P(E_1 \cap E_2) = \frac{13}{52} + \frac{4}{52} - \frac{1}{52} = \frac{16}{52}$$

(assuming that all cards are equally likely to be dealt). Note that 1/52 is subtracted to avoid double counting of the ace of clubs.

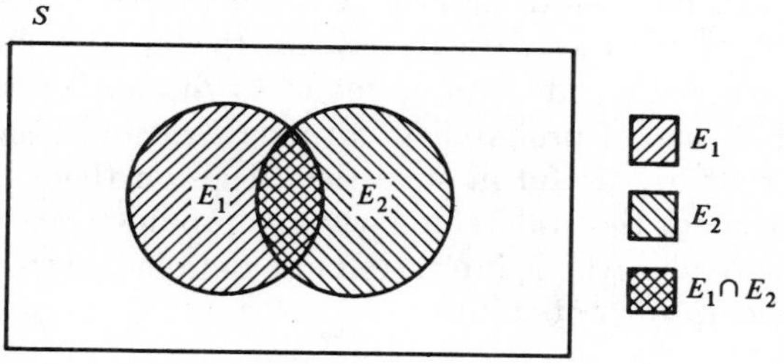

Figure 2.1.5

It is easy to extend these elementary rules of probability to the situation in which more than two events are of interest. For example, you might be interested in the probability that E_1 or E_2 or E_3 (or any combination of the three events) will occur; this is the probability of the union of the three events E_1, E_2, and E_3, and it is written $P(E_1 \cup E_2 \cup E_3)$. The general rule for the calculation of this probability is

$$\begin{aligned} P(E_1 \cup E_2 \cup E_3) = {} & P(E_1) + P(E_2) + P(E_3) - P(E_1 \cap E_2) \\ & - P(E_1 \cap E_3) - P(E_2 \cap E_3) + P(E_1 \cap E_2 \cap E_3), \end{aligned}$$

where $P(E_1 \cap E_2 \cap E_3)$ is the probability of the intersection of the three events (that is, the probability that all three will occur). This rule is illustrated in Figure 2.1.6. Note that the probabilities of the intersections $E_1 \cap E_2$, $E_1 \cap E_3$, and $E_2 \cap E_3$ are subtracted to avoid "double counting." As a result, the intersection of all three events is added three times (once with E_1, once with E_2, and once with E_3) and subtracted three times (once with each of the pairwise intersections). Thus, it must be added again, and that is why the last term in the above equation must be included.

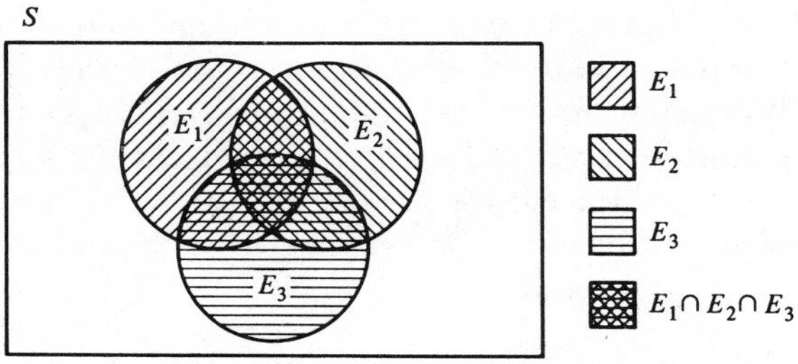

Figure 2.1.6

Elementary rules of probability such as these are extremely useful in the calculation of probabilities. Often, certain probabilities are known and the rules can be used to determine other probabilities. Because this book is concerned with statistical decision theory and not with probability theory per se, no attempt is made to discuss systematically the mathematical theory of probability, although various facts from probability theory that are useful in statistical decision theory are presented as they are needed. Several references are provided at the end of the chapter for the reader who is interested in a detailed, systematic development of the theory of probability.

2.2 THE FREQUENCY INTERPRETATION OF PROBABILITY

In the previous section, probability was considered as a mathematical system; given certain probabilities, the theory of probability can be used to determine certain other probabilities. If the probability of obtaining a "head" on a single toss of a coin is $\frac{1}{2}$, then the probability of obtaining a "tail" is also $\frac{1}{2}$ (assuming that other outcomes, such as the coin landing on edge or the coin disintegrating in mid-air, are not considered as possible events). The mathematical theory can thus be extremely useful in calculating probabilities. However, an important question is still unanswered: What do these probabilities mean? What does it mean to say that the probability of obtaining a "head" on a single toss of a coin is $\frac{1}{2}$? If a marble is to be drawn from a box containing 30 red marbles and 70 blue marbles, what does it mean to say that the probability is $\frac{3}{10}$ that the marble drawn is red? What does it mean to say that the probability of the event "odd number" in one throw of a die is $\frac{1}{2}$? What does it mean to say that the probability of obtaining a "full house" in a five-card poker hand is .0014? In this section, one possible interpretation of probability is considered: the *frequency* interpretation. In the next section, a second possible interpretation is discussed: the *subjective,* or *personal,* interpretation. The frequency and subjective interpretations are not the only interpretations of probability, but they are by far the most commonly encountered interpretations; thus, other interpretations are not discussed in this book. References concerning other interpretations are given at the end of this chapter and at the end of Chapter 7.

In the first of the examples mentioned in the preceding paragraph, if the coin is tossed repeatedly, then how often would one expect to obtain "heads"? Over a long series of tosses, it seems reasonable to expect to observe heads about 50 percent of the time. Similarly, if one keeps drawing marbles out of the box with 30 red and 70 blue marbles and if after each draw the marble drawn is replaced in the box before the next drawing occurs, then in the *long run*, after numerous observations, what proportion of the marbles drawn would be expected to be red? It seems reasonable to expect that about three out of 10 observations will be red. In the same way, in the third example, if one throws the die repeatedly, then it seems reasonable that in the long run, over a large number of such observations, about $\frac{1}{2}$, or one in two, of the throws should result in an odd number. Finally, in the poker example, one should expect a full house about 14 times in 10,000 five-card poker hands in the long run. In these examples, the *probability* of an event has been interpreted as the *relative frequency* of occurrence of that event to be expected in the long run.

Simple examples of this sort usually involve the assumption that all of the elementary events of interest are equally likely. In many applications of probability theory, especially to games of chance, this assumption is made. For example, the statement "the probability of a head on a single toss of a coin is $\frac{1}{2}$" implies that heads and tails are equally likely. If the elementary events are all equally likely, then the probability of an event E can be determined by taking the relative frequency of occurrence of E in the sample space S:

$$P(E) = \frac{\text{number of elementary events comprising } E}{\text{total number of elementary events in } S}. \tag{2.2.1}$$

This interpretation of probability is sometimes called the *classical* interpretation. Note that it deals with the relative frequency of occurrence of an event *in a sample space* rather than "in the long run." In the marble example, if each marble is thought of as an elementary event, then the probability of drawing a red marble is equal to the number of red marbles, 30, divided by the total number of marbles, 100.

The applicability of Equation (2.2.1) is based on symmetry arguments, which may be reasonable in games of chance and similar situations. In most uncertain situations, however, it is *not* reasonable to assume that all elementary events are equally likely. Although Equation (2.2.1) is no longer relevant, the relative-frequency interpretation of probability applies even when the various events, or outcomes, are *not* assumed to be equally likely. An important theorem of probability theory, called the *law of large numbers,* states essentially that if an experiment (such as tossing a die or dealing a five-card poker hand) is repeated many times under identical conditions, the relative frequency of occurrence of any event (such as an even number on a die or a full house in a poker hand) is likely to be close to the probability of that event. Furthermore, the theorem states that the relative frequency and the probability are *more* likely to be close as the number of repetitions of the experiment increases. Formally, if the experiment is repeated n times and if r denotes the number of times that the event E occurs in these n repetitions, then

$$P\left[\left|\frac{r}{n} - P(E)\right| \geq \epsilon\right] \rightarrow 0 \qquad \text{as } n \rightarrow \infty, \tag{2.2.2}$$

where ϵ is any arbitrarily small positive number. As the number of repetitions, n, increases, the relative frequency of occurrence of the event E, r/n, tends to get closer and closer to the probability, $P(E)$. For instance, if a symmetric, six-sided die is tossed repeatedly, the relative frequency of occurrence of the event "six comes up" is more likely to be close to $\frac{1}{6}$ as the number of tosses, n, increases.

The law of large numbers formally expresses in mathematical terms what might be thought of as "statistical regularity." Such statistical regularity is very important in, for example, the gambling and insurance businesses. Suppose that the death rate for a certain age group in the United States has remained reasonably stable over the past decade at five deaths per 100 persons per year. Barring special conditions such as epidemics or new "wonder drugs" (which would violate the requirement of "identical conditions"), an insurance company with a large number of policies among members of the given age group can expect that the relative frequency of death for these policyholders in the coming year will be very close to .05. If only a few policyholders are considered, variations from .05 are not surprising. As more and more policyholders are considered, statistical regularity is observed, as illustrated in Figure 2.2.1; the relative frequency tends to stabilize around .05. If a new drug that cures cancer is introduced, this would tend to reduce the death rate, and the relative frequency might stabilize around, say, .046 instead of .05. The important point is that if repeated groups of, for example, 1000 policyholders are considered, the relative frequencies of death will be very close for the different groups.

It is useful to note that relative frequencies satisfy the three axioms of probability. Clearly a relative frequency cannot be negative, and the relative frequency of the entire sample space is one (if an elementary event not included in the sample space occurs, then the sample space has been incorrectly specified, since it is defined as the set of *all* possible elementary events). Finally, if two events are mutually exclusive, the relative frequency of the union of the two events is equal to the sum of their individual relative frequencies. In a particular series of tosses of a die, if the event "2 or 4 comes up" occurs with relative frequency .30 and the event "6 comes up" occurs with relative frequency .18, then the event "an even number comes up" occurs with relative frequency .48.

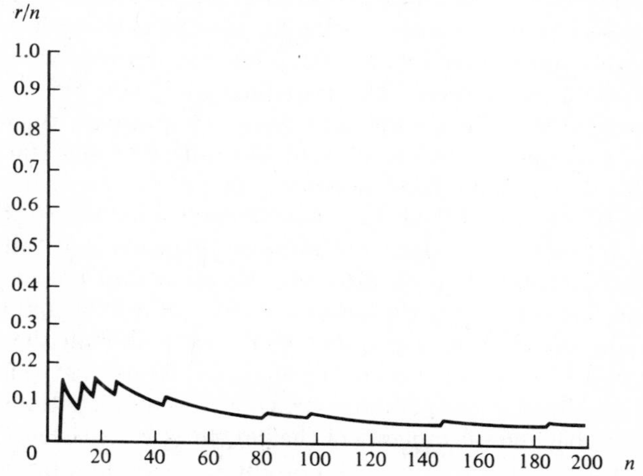

Figure 2.2.1

The relative-frequency interpretation of probability, then, seems intuitively reasonable and empirically sound. If an uncertain situation is repeated a large number of times, the frequency of an event's occurrence should somehow reflect the probability of the event. Most people think of this phenomenon as a "law of averages." It is important to note, however, that no matter how many repetitions of the uncertain situation are considered, the actual relative frequency of an event need not *exactly* equal the probability of the event. If a fair coin is tossed 1,000,000 times, the number of occurrences of heads is expected to be close to, but not necessarily equal to, 500,000. The law of large numbers says that the relative frequency and the probability need be equal only for an infinite number of repetitions, which is a theoretical ideal clearly unattainable in practice. Indeed, if n is small, then r/n may be quite different from $P(E)$, as illustrated in Figure 2.2.1. For any large but finite number of repetitions, you may expect r/n to be very close to $P(E)$, but you can never be guaranteed that they will be equal.

A common misinterpretation of the "law of averages" is the notion that a particular event is "due to occur" because it has not occurred for some time. In the law of large numbers, it is assumed that the trials (repetitions of the experiment) are independent, which essentially means that the outcome on any single trial has no bearing on the outcomes on any other trials. In terms of tossing a fair coin, the fact that it comes up heads on one toss bears no apparent relationship to the outcome of the previous toss, the next toss, or any other toss. Thus, an event cannot be thought of as "due" to occur. But if this is the case, how does the law of large numbers allow for unusual but possible occurrences such as 100 heads in a row? It does so not by compensating for the 100 heads in a row with 100 tails in a row, but rather by a "swamping" procedure, which works as follows. If the first 100 tosses of a sequence are heads, this is an unusual event (provided that the coin is fair). Suppose, however, that in the next 1000 trials, you observe 500 heads and 500 tails. For all 1100 tosses, you have 600 heads and 500 tails, which does not seem as unusual as 100 heads and 0 tails. If you tossed the coin 1,000,000 additional times, with 500,000 heads and 500,000 tails, the total number of heads and tails would be 500,600 and 500,500. This result does not seem unusual at all. As the sample size has increased, the original "run" of 100 straight heads has been "swamped" by the subsequent trials, even though the subsequent trials resulted in exactly 50 percent heads and 50 percent tails.

It is interesting to note at this point that in some situations you might have cause to believe the *opposite* of the "event is due to occur" argument. If you meet a stranger in a bar who produces a coin and proceeds to toss 100 straight heads, you would *not* think that a tail was "due." More likely, you would suspect that the coin was not fair and that the probability of a head on a single toss was greater than $\frac{1}{2}$ (if the coin were two-headed, the probability of a head would be 1). This amounts to questioning the assumption of a fair coin and demonstrates an important use of the law of large numbers; if you have no idea what the probability of a particular event is and if you observe a large number of independent trials, you can claim with some confidence that the probability of the event is close to the relative frequency of the event in the sequence of trials. Of course, what constitutes a "large" number of trials depends on the situation and on the desired accuracy. Reasonable accuracy to four decimal places requires a much larger value of n than accuracy to only two decimal places, and thus a large number of trials is needed in the case of a "rare event" with, for instance, probability .0005.

One serious problem with the relative-frequency interpretation of probability is that technically it requires a long series of trials *under identical conditions* to determine probabilities with reasonable accuracy. Very often one is interested in the probability of an event but is unable to observe repeated trials, as you shall see in the next three sections. In this sense, the relative-frequency interpretation is a *conceptual* interpretation, not an *operational* interpretation, since a long series (formally, an infinitely long series) of trials is needed to determine probabilities. Even a *finite* series of trials is at best costly and time-consuming and at worst impossible. Thus, for most practical purposes, the relative-frequency interpretation does not provide a realistic procedure for determining probabilities.

2.3 THE SUBJECTIVE INTERPRETATION OF PROBABILITY

Relative frequency is but one interpretation that can be given to the notion of probability. It is important to remember that this is an *interpretation* of the abstract model and that there is no universally agreed upon interpretation. The model per se is a system of relations among numbers that happen to be called probabilities and rules for calculating with these numbers.

The probability concept acquired an interpretation in terms of relative frequency because it was originally developed to describe certain games of chance where plays (such as spinning a roulette wheel or tossing dice or dealing cards) are indeed repeated for a large number of trials and where it is reasonable to assume that the elementary events of interest are equally likely. Similarly, there are numerous situations in which statisticians make many observations under essentially the same conditions, and the mathematical theory of probability can be given a relative-frequency interpretation in these situations as well. For instance, a quality-control statistician may observe thousands of items produced by a certain production process, and he may record the weight of each item or simply whether each item is defective or nondefective. An actuarial statistician may observe the records of millions of persons, recording for each the number of claims with regard to health insurance or automobile insurance. A medical statistician may observe thousands of persons with a certain disease, recording for each the drug or drugs used and whether or not the person is cured of the disease.

On the other hand, there are many events that can be thought about in a probabilistic sense but that cannot have a probability in terms of the relative-frequency interpretation. Indeed, an everyday use of the probability concept does not have this relative-frequency connotation at all. You say, "It will probably rain tomorrow," or "The Packers will probably not win the championship," or "I am unlikely to pass this test." Extending this notion, you might say, "The chances are two in three that it will rain tomorrow," or "The odds are nine to one that the Packers will not win the championship," or "I only have about one chance in 10 of passing this test." These statements appear to be probability statements, and their meanings should be clear to most listeners; but it is very difficult to see how they could describe long-run relative frequencies of outcomes of simple experiments repeated over and over again. The problem is that these events are unique (that is, the situations cannot be duplicated). Although some information may be available regarding past occurrences in *similar* situations, no information in the form of observed frequencies is available regarding repeated trials under *identical* conditions. You may have some information available regarding the past occurrence of rain in a certain location, at a certain time of year, under certain atmospheric conditions (for example, temperature, wind velocity and direction, and humidity), and so on, but it is doubtful that any of this information represents situations exactly identical to the current situation. Similarly, you may have past information about the performance of a football team such as the Packers; but each year new players enter the league, injuries occur, key players retire, and so on. Surely no previous season was played under conditions identical to those of the current season. The same idea holds for the statement involving the passing of a test. Each of the three probability statements given above describes the speaker's *degree of belief* about a situation that will occur once and once only. In each case it will not be possible to observe repetitive trials of the uncertain situation, so the probability statements cannot be explained in terms of the long-run frequency interpretation of probability.

An interpretation of probability that enables one to explain the probability statements in the preceding paragraph is the *subjective,* or *personal,* interpretation. In this approach a probability is interpreted as a measure of degree of belief, or as the quantified judgment of a particular individual. The probability statements given above thus represent the judgments of the individual making the statement. Since a probability, in this interpretation, is a measure of a degree of belief rather than a long-run frequency, it is perfectly reasonable to assign a probability to an event that involves a nonrepetitive situation. As a result, you can think of a probability as representing the individual's judgment concerning what will happen in a single trial of the uncertain situation in question rather than a statement about what will happen in the long run.

It is not necessary, of course, for an experiment to be nonrepetitive for the subjective interpretation of probability to be applicable. Consider once more the four examples that were discussed from a long-run frequency viewpoint in the preceding section: tossing a coin, drawing marbles from a box, throwing a die, and dealing a poker hand. In each of these examples, the probability of any particular event could be interpreted as a degree of belief, although the experiments are all capable of being repeated. For instance, an individual might have a certain degree of belief that a full house will occur *on a particular deal* in a game of poker. In the die example, an individual who is willing to accept certain judgmental assumptions (the die is "fair," the method of tossing the die does not favor any side or sides of the die, and so on) would feel that the probability of the die coming up odd *on a particular throw* is $\frac{1}{2}$. Notice that this is a probability statement about a particular throw, not about a long sequence of throws. The point is simply that the subjective interpretation of probability makes sense whether the experiments in question are repetitive or nonrepetitive. Furthermore, the subjective interpretation is operationally more useful than the long-run frequency interpretation because it allows a person to consider individual situations instead of appealing to the "long run" and to the concept of statistical regularity. In this respect, the subjective interpretation is both conceptual *and* operational, whereas the frequency approach is conceptual but *not* operational.

In some respects, subjective probability can be thought of as an *extension* of the frequency interpretation of probability. As you have seen, the justification of long-run frequencies is based on certain *assumptions* that are necessary for the proof of the law of large numbers. One important assumption is that the trials comprising the "long run" are independent. As noted in Section 2.2, this essentially means that the outcome on any one trial will in no way affect the outcome on any other trial. Another assumption is that the trials are conducted under identical conditions. If probability is defined in terms of relative frequencies by Equation (2.2.1), then it must be assumed that the elementary events of interest are equally likely. Although techniques are available for statistically investigating assumptions such as these, it should be pointed out that the decision as to whether the assumptions seem reasonable in any given situation is ultimately a subjective decision. Thus, there is an element of subjectivity in the relative-frequency interpretation of probability. If a person feels that the assumptions are reasonable, it is perfectly acceptable for him to make his subjective probability for a given event equal to the probability determined by the frequency approach, whether this latter probability is based on symmetry arguments or on actual observed relative frequencies. If he does not feel that the assumptions are reasonable or if he has some other information (other than frequencies) about the event in question, his subjective probability may differ from the frequency probability. In this respect (and in the respect that the subjective interpretation allows one to make probability statements about nonrepetitive events), subjective probability can be thought of as an *extension* of frequency probability.

For an example involving the assumption of equally likely elementary events, consider a six-sided die that is not perfectly symmetric. After examining the asymmetries of the die, you might feel that the six sides are equally likely to appear face up if the die is tossed, in which case Equation (2.2.1) is applicable for determining probabilities of various events such as "an even number comes up." On the other hand, the asymmetry of the die may lead you to believe that the side marked "1" is considerably more likely to occur than any other side, in which case the elementary events should not be considered to be equally likely and Equation (2.2.1) would not be applicable. In either case, the assignment of probabilities to the events of interest is subjective.

For an example involving observed frequencies, suppose that you are interested in the probability that a given person will be involved in an automobile accident (as the driver of a car, not as a passenger) during the next year. You know from recent data that among drivers in the United States of the same age as the person of interest, the frequency of drivers involved in one or more accidents in a year is 12/100. If you feel that the given person is representative of drivers in his age group (and this is a subjective judgment on your part), you might be willing to assign the probability .12 to the event that the person will be involved in an accident in the next year. On the other hand, you might feel that the person is a somewhat reckless driver, so that the probability should be, say, .20; or you might want to look up the historical frequencies for drivers in the same state (as opposed to all drivers in the United States), or for drivers of a particular type of car, and so on. The point is that symmetry assumptions or observed relative frequencies may be quite useful in the assessment of probabilities, but the ultimate assessment is subjective; thus, the subjective interpretation of probability can be thought of as an extension of the frequency interpretation. Furthermore, the subjective interpretation is applicable in *any* uncertain situation, whether or not the situation is repeatable. Because many uncertain situations of interest in problems of statistical inference and decision are unique, nonrepetitive situations, the subjective interpretation is extremely useful. As noted earlier in this section, the subjective interpretation is operational as well as conceptual.

2.4 PROBABILITIES, LOTTERIES, AND BETTING ODDS

Because they represent degrees of belief, subjective probabilities may vary from person to person for the same event. Different individuals may have different degrees of belief, or judgments. Such differences may be due to differences in background, knowledge, and/or information about the event in question. For example, you might state that the probability of rain tomorrow is $\frac{1}{10}$, and the meteorologist at the weather bureau might state that the probability of rain tomorrow is $\frac{1}{2}$. According to the subjective interpretation of probability, both of these values can properly be considered as probabilities and subjected to the usual mathematical treatment. Such a difference in probabilities is permissible, provided that you and the meteorologist both feel that your respective probabilities accurately reflect your respective judgments concerning the chance of rain tomorrow. In addition, the quantification of these judgments is not as arbitrary as it might first appear. It is true that a person may be somewhat vague when attempting to convert his judgments into probabilities, and the subjective interpretation of probability is sometimes criticized on the grounds that the inclusion of subjective judgment results in probabilities that are not "objective," but are in some respects arbitrary. For instance, you may feel that the probability of rain tomorrow is between .10 and .20, but you find it difficult to come up with a single number representing your probability. There are "objective" procedures, such as lotteries and betting odds, that can be extremely helpful in reducing vagueness and in accurately quantifying judgments.

One possible way of formally defining subjective probability is in terms of *lotteries*. For instance, suppose that you are offered a choice between Lottery A and Lottery B.

Lottery A: You win \$100 with probability $\frac{1}{2}$.
You win \$0 with probability $\frac{1}{2}$.

Lottery B: You win \$100 if it rains tomorrow.
You win \$0 if it does not rain tomorrow.

It is assumed that since the "prize" is the same in both lotteries (\$100), you should prefer the lottery that gives you the greater chance of winning the prize. Thus, if you choose Lottery B, then you must feel that the probability of rain tomorrow is greater than $\frac{1}{2}$; if you choose Lottery A, then you must feel that this probability is less than $\frac{1}{2}$; if you are indifferent between the two lotteries, then you must feel that the probability of rain tomorrow is equal to $\frac{1}{2}$. Now consider the same lotteries, except that the probabilities in Lottery A are changed to $\frac{1}{4}$ and $\frac{3}{4}$. If you still prefer Lottery A, implying that you feel that you have a greater chance of winning in Lottery A than in Lottery B, then your subjective probability of rain is less than $\frac{1}{4}$. Presumably you could keep changing the probabilities in Lottery A until you are just indifferent between Lotteries A and B; if this happens when the probabilities are .1 and .9, then your subjective probability of rain is .1. In a similar manner, you can assess your probability for any event.

A formal definition of subjective probability can be given in terms of lotteries as follows. Your subjective probability $P(E)$ of an event E is the number $P(E)$ that makes you indifferent between the following two lotteries.

Lottery A: You obtain X with probability $P(E)$.
You obtain Y with probability $1 - P(E)$.

Lottery B: You obtain X if E occurs.
You obtain Y if E does not occur.

Here X and Y are two "prizes." The only restriction on X and Y is that one must be preferred to the other; if you are indifferent between X and Y, then you are indifferent between the two lotteries regardless of the choice of $P(E)$. In the above example, X = \$100, Y = \$0, and E = "rain tomorrow." It should be noted that X and/or Y could be undesirable prizes, such as negative amounts of money. For instance, X and Y could be −\$10 and −\$20.

Lottery A: You lose \$10 with probability $\frac{1}{2}$.
You lose \$20 with probability $\frac{1}{2}$.

Lottery B: You lose \$10 if it rains tomorrow.
You lose \$20 if it does not rain tomorrow.

In order to use the lottery procedure operationally to determine a person's subjective probability for an event E, it is necessary to specify how the outcome in Lottery A is determined. In the first illustration of lotteries given above, suppose that you choose Lottery A; do you win \$100 or \$0? The probabilities in Lottery A are called *canonical* probabilities; they are probabilities that everyone would agree upon. For instance, if the probabilities are $\frac{1}{2}$ and $\frac{1}{2}$, then the outcome can be decided by the selection of a ball from an urn containing 50 red balls and 50 green balls. If the balls are all of the same size and if they are thoroughly mixed before the drawing, then the chance of drawing a red ball is the same as the chance of drawing a green ball. If the probabilities are $\frac{1}{10}$ and $\frac{9}{10}$ in Lottery A, then an urn with 10 red balls and 90 green balls could be used. To assess your subjective probability of rain tomorrow to the nearest .01, consider the following two lotteries.

Lottery A: You win \$100 if a red ball is drawn from an urn containing R red balls and $100 - R$ green balls.
You win \$0 if a green ball is drawn.

Lottery B: You win \$100 if it rains tomorrow.
You win \$0 if it does not rain tomorrow.

If you can determine a number R that makes you indifferent between

Lottery A and Lottery B, then your subjective probability of rain tomorrow is $R/100$. For example, you may feel that if there are fewer than 16 red balls in the urn, you prefer Lottery B; if there are more than 16 red balls, you prefer Lottery A; and if there are exactly 16 red balls (and hence 84 green balls), you are indifferent between the two lotteries. Then your subjective probability of rain tomorrow must be .16.

It is not necessary to use the idea of balls in an urn to determine canonical probabilities. If you feel unsure about the reliability of such a device (perhaps because such devices have been used without proper mixing of balls and have produced decidedly nonrandom results in military draft lotteries, for instance), other devices are available. The most convenient (and perhaps the most reliable) such device is the use of a table of random digits. Such tables are constructed so that at any point in the table, each digit is equally likely to appear (other details, such as independence and equal probabilities for the various possible combinations of digits at successive points in the table, are also required of a table of random digits). For Lottery A, a two-digit number could be chosen from a table of random digits. If the number is between 00 and 15, you win \$100; and if the number is between 16 and 99, you win \$0. This is equivalent to the urn with 16 red balls and 84 green balls. The point is that the probabilities in Lottery A must be reasonable to an individual attempting to assess his subjective probabilities; thus, it is convenient if they are probabilities about which different persons would agree. Note, by the way, that to assess probabilities to the nearest .001 instead of just to the nearest .01, you could consider an urn with 1000 balls or a three-digit random number. Of course, it is only possible to go so far in combating vagueness. For some events, it may even be difficult to obtain a subjective probability to the nearest .01. In other words, you may find it difficult to determine a single probability that makes you indifferent between two lotteries. The consideration of lotteries, whether real or hypothetical, should help to reduce vagueness, but it is generally not possible to eliminate it entirely.

Another procedure that is useful in the assessment of subjective probabilities involves *betting odds*. A probability may be thought of in terms of the odds at which a person would be willing to bet. For example, if you say, "The chances are one in 10 that Green Bay will win the championship of the National Football League," this implies that

$$P(\text{Green Bay wins championship}) = \tfrac{1}{10}$$

and

$$P(\text{Green Bay does not win championship}) = \tfrac{9}{10}.$$

Thus, Green Bay is nine times as likely to lose the championship as it is to win the championship, in your opinion. This means that you think that nine to one odds against Green Bay winning would be fair odds. The formal relationship between probabilities and odds is as follows: if the *probability* of an event is equal to p, then the *odds in favor of that event* are *p to (1 − p)*. If the chances are two in three that it will rain tomorrow, the probability of rain is $\frac{2}{3}$ and the odds in favor of rain are $\frac{2}{3}$ to $\frac{1}{3}$, or 2 to 1. If you only have about one chance in 10 of passing a particular test, the probability of your passing the test is $\frac{1}{10}$ and the odds in favor of your passing the test are $\frac{1}{10}$ to $\frac{9}{10}$, or 1 to 9.

It is also possible to convert odds into probabilities: if the *odds in favor of an event* are *a to b*, then the *probability* of that event is equal to $a/(a + b)$. If the odds are even that the Dow–Jones Average will rise tomorrow, the odds are a to a, so the probability that the Dow–Jones Average will rise tomorrow is $a/(a + a)$, or $\frac{1}{2}$. If you know the probability of an event, you can calculate the odds in favor of the event, and vice versa. This is important with regard to the assessment of an individual's subjective probabilities, for some persons find it easier and more convenient to think in terms of odds rather than in terms of probability.

It is possible to go one step further and consider actual or hypothetical bets. Suppose that two persons, A and B, make the following bet.

> A bets \$3 that the price of IBM common stock goes up tomorrow.
>
> B bets \$2 that the price of IBM common stock does not go up tomorrow.

If you think that this is a "fair" bet (that is, if you are indifferent between A's side of the bet and B's side of the bet), then you apparently feel that 3 to 2 odds in favor of IBM stock going up tomorrow are "fair" odds. This implies that your probability that IBM stock goes up tomorrow is $3/(3 + 2)$, or .60. Although it is not possible at this point to show formally why this is true, a brief intuitive explanation can be given. If you take A's side of the bet, you feel that you have a probability of .60 of winning \$2 (the amount your opponent, B, is wagering), and you have a probability of .40 of losing \$3 (the amount you, as A, are wagering). But (.60)(\$2) is equal to (.40)(\$3), so the amount you "expect" to gain from the bet is \$0. It turns out that if you take B's side of the bet, the amount you "expect" to gain is still \$0. As a result, you are indifferent between the two sides of the bet, and thus you consider the bet to be "fair." If your probability that IBM stock goes up tomorrow is greater than .60, the bet is no longer "fair" in your estimation; you prefer A's side of the bet. On the other hand, if your probability is less then .60, then you prefer B's side of the bet. You only feel that the bet is "fair" if your probability that IBM stock goes up tomorrow is .60. If subjective probabilities are thought of in terms of betting odds and the corresponding actual or hypothetical bets, then it can be seen that they are not really arbitrary, even though they may differ from person to person.

It should be noted that the use of betting situations to assess probabilities introduces a problem that is not encountered in the use of lotteries. The problem concerns the amounts of the stakes involved in the bets. In the lottery situation, the amounts involved are identical in Lottery A and Lottery B. In the betting situation, the amounts are different in A's side of the bet than they are in B's side of the bet. In the above example, A is risking \$3 in order to possibly gain \$2, whereas B is risking \$2 in order to possibly gain \$3. For amounts as small as these, there hopefully will be little difficulty. If the amounts considered are \$300,000 and \$200,000, however, or if the odds are very high in favor of one of the events, then the person's attitude toward risk becomes a crucial factor in the choice of one side of the bet. A full discussion of this problem is given in Chapter 5 when utility is considered. At that time you will see that it is necessary to consider the "expected gain" from a bet in terms of utility rather than in terms of money. Provided that the stakes for the bets are not too large, it can be assumed that the utility problem does not arise and that the betting situation can be used in the assessment of subjective probabilities.

A device similar to the betting situation involves bidding for a *reference contract,* which is a prize contingent upon the event of interest. For instance, suppose that I offer you a contract that will pay you \$10 if the maximum official temperature at San Diego, California, tomorrow is greater than 80°F. What is the largest amount that you would be willing to pay for this contract? If it is \$3.50, say, then your probability for the event in question is \$3.50/\$10, or .35. The argument used to arrive at this conclusion is similar to the argument used in the betting situation. The "expected gain" of the contract is equal to the product of the amount involved (\$10) and your probability for the event, less the cost of buying the contract. You should be willing to buy the contract as long as this "expected gain" is nonnegative [that is, when $(\$10)P(E) - C \geq 0$, where C is the cost of the contract]. It is zero when the cost is equal to the product of \$10 and your probability, negative for larger costs, and positive for smaller costs. Thus, the largest amount you should be willing to pay is the amount that makes the "expected gain" zero, implying that your probability for the event "the maximum official temperature at San Diego tomorrow is greater than 80°F" is $C/\$10$, or $\$3.50/\$10 = .35$. Unfortunately, this argument relies once again on the assumption that the amounts involved are small enough that attitude toward risk is not a relevant factor in the choice of a "bid" for the reference contract.

It should be noted that subjective probabilities assessed through the use of betting situations, lotteries, or reference contracts must obey the three basic axioms of probability if the person making the assessments wishes to avoid the possibility of having a "Dutch book" set up against him. That is, if he does not obey the rules of probability presented in Section 2.1, then it is possible to set up a series of bets or lotteries such that he is sure to lose no matter what happens. For example, suppose that you claim that the probability of rain tomorrow is $\frac{2}{3}$ and that the probability of no rain tomorrow is also $\frac{2}{3}$. These probabilities clearly violate the rule that the probability of the complement of an event E must be equal to one minus the probability of the event E. Since you think the probability of rain is $\frac{2}{3}$, this implies that you consider the following bet to be a "fair" bet:

A bets \$2 that it will rain tomorrow.

B bets \$1 that it will not rain tomorrow.

Furthermore, since you claim that the probability of no rain is $\frac{2}{3}$, you also consider the following bet to be "fair":

C bets \$2 that it will not rain tomorrow.

D bets \$1 that it will rain tomorrow.

Since you consider these two bets to be fair, you will take either side of the bet in each case. Suppose that I choose sides B and D, leaving you with sides A and C. If it rains, you win \$1 on the first bet and lose \$2 on the second bet; if it does not rain, you lose \$2 on the first bet and win \$1 on the second bet. Thus, no matter what happens, you end up with a net loss of \$1; in other words, I have made a "Dutch book" against you.

The same idea can be demonstrated with reference contracts. Your probabilities imply that you would be willing to pay \$2 for a \$3 reference contract contingent on rain and that you would pay \$2 for a \$3 reference contract contingent on no rain. If I sell you both contracts, you will win \$3 in exactly one of the two cases (since it will either rain or not rain). However, you are paying a total of \$4 for the two contracts, so you are sure to lose \$1 whether or not it actually rains. Your violation of the rules of probability is being exploited and you are being used as a "money pump."

An alternative approach to the idea of subjective probability is to begin with certain reasonable "axioms of coherence," or "axioms of consistent behavior," and to show that if a person obeys these axioms, then the existence of subjective probabilities obeying the usual rules of mathematical probability is implied. If one or more of the coherence axioms is violated, a "Dutch book" can be set up. All of the axioms of coherence are not presented at this point, but two examples of such axioms are given. The axiom of *transitivity* states that preferences are transitive. That is, if you prefer Lottery A to Lottery B and Lottery B to Lottery C, then you must prefer Lottery A to Lottery C. This seems intuitively reasonable. The axiom of *substitutability* states that if you are indifferent between X and Y, then X can be substituted for Y as a prize in a lottery or as a stake in a bet without changing your preferences with regard to the bets or lotteries in question. For example, suppose that you are indifferent between the following two lotteries.

Lottery A: You win \$2 for certain.

Lottery B: You win a concert ticket with probability $\frac{1}{2}$.
You win \$0 with probability $\frac{1}{2}$.

Then you should be willing to substitute Lottery B for the \$2 stake in the bet considered earlier regarding the price of IBM common stock. Moreover, this substitution should not affect your indifference between the two sides of the bet. The "axioms of coherence" are discussed in more detail in the development of utility theory in Chapter 5.

Devices such as lotteries, betting situations, and reference contracts have the common property of resulting in statements that can be converted into probabilities. They represent the "behavioral" approach to the quantification of judgment; that is, they force the person who is attempting to determine his subjective probabilities to state what *action* he would take in a particular decision-making situation rather than just to state directly what the probability is. His probabilities can then be inferred from his behavior (that is, from his decisions). In this regard, the subjective interpretation of probability should have great applicability in decision-making situations. The subjective interpretation is also valuable in making decisions because it allows probability statements for nonrepetitive situations as well as for repetitive situations. Important decisions often depend on the outcomes of nonrepetitive experiments. For example, an investment decision may depend on the probability that Congress will raise taxes or the probability that the stock market will go up in the next year. A decision concerning a price increase for a certain product may depend on the probability that other firms producing the product will refuse to go along with the increase. These examples refer to nonrepetitive situations. In the next section, some simple examples of the use of probabilities in decision making are presented, examples in which the subjective interpretation of probability is most useful.

2.5 PROBABILITY AND DECISION MAKING

In real-world problems, one is often faced with the task of making inferences or decisions under uncertainty. As you have seen, the uncertainty about any event may be represented quantitatively in the form of a probability, and some procedures for arriving at such probabilities have been discussed. One advantage of expressing uncertainty in terms of probability is that it makes it much easier to communicate the nature of the uncertainty. Instead of presenting long verbal arguments about whether the price of IBM common stock will or will not go up in the coming month, it is simpler to summarize your judgments by saying, "I think that the probability that the price of IBM common stock will go up in the coming month is .65." In business decision-making, an important decision may involve uncertainty about a large number of events and may be of concern to quite a few people. For instance, some of the events of interest may involve the stock market, in which case a financial analyst should be consulted; some of the events may involve pending legislation, in which case a person familiar with the political scene should be consulted; some of the events may involve potential sales of a certain product, in which case the sales and advertising managers should be consulted; and so on. The point is that a large-scale problem may involve many uncertainties, and numerous experts may be consulted regarding these uncertainties. The use of probabilities to represent the uncertainties provides a convenient and useful medium for the communication of judgments. Furthermore, in order to apply the formal decision theory that is presented in Chapters 5 and 6, it is necessary to express uncertainty in terms of probability. To illustrate this informally, some very simple decision-making problems are presented in this section.

Consider the following problem: you are preparing to leave the house in the morning, and you must decide whether or not to carry an umbrella. You are wearing your new sport coat, which would be ruined if it got wet, and you will be outside all day. On the other hand, it is a nuisance to have to carry an umbrella all day. Should you take the umbrella or not?

This problem is a very trivial one; most of us make numerous decisions of this nature every day, and we do so on an intuitive, informal basis. It should be instructive, however, to consider the problem in a more formal manner in order to illustrate the statistical theory of decision-making. Many important decisions are similar in nature to the problem of whether or not to carry an umbrella, and the same formal theory can be used for important decisions and for less-important decisions. When the decision is a minor one, such as the umbrella example, the formal theory is seldom applied because of the time and effort required to do so. For important decisions, however, the time and effort required to implement the theory may be minor considerations in comparison with the potential gains or losses to be realized from the decision.

In order to apply the formal theory to the umbrella example, it is necessary to specify a few more details of the problem. Suppose that the sport coat is worth $60 and that it would be completely worthless if it were rained upon. Furthermore, suppose that you feel that the cost of carrying the umbrella, in terms of the inconvenience and the possibility of leaving the umbrella somewhere and thus losing it, is equivalent to $1. This implies that you would be willing to pay $1 to avoid carrying the umbrella, and it demonstrates the consideration of nonmonetary factors in the analysis. You have attempted to express the nonmonetary factors in terms of an equivalent amount of money. The consequences of your decision, which depend on whether or not you carry an umbrella and on whether or not it rains, can be expressed in a *payoff table*, as follows:

		EVENT	
		Rain	*No rain*
YOUR DECISION	*Carry umbrella*	−$1	−$1
	Do not carry umbrella	−$60	$0

The cost of carrying an umbrella is $1 whether it rains or not (a cost of $1 is the same as a payoff of −$1). If you do not carry an umbrella, the cost is $60 if it rains and $0 if it does not rain.

The payoffs in the above table are important factors in your decision. Another important factor is the chance of rain. If you were sure that it would rain, you would certainly carry your umbrella. On the other hand, if you were positive that it would not rain, you would not carry the umbrella. These cases are of little interest because the decision is obvious, so consider the case in which you are uncertain as to whether or not it will rain. Your uncertainty can be expressed in terms of probabilities, P(rain) and P(no rain). Possible sources of information in this situation include a weather report and data regarding past frequencies of rain for the particular location and time of year. Suppose that you do not have access to such information but that you look out the window and see a few scattered dark clouds. On the basis of this, you assess the probability of rain to be $\frac{1}{10}$. This is a subjective, or degree-of-belief, probability, and it implies that you feel that the event "no rain" is nine times as likely ($\frac{9}{10}$ to $\frac{1}{10}$) as the event "rain." It also implies that you would be indifferent between the first two lotteries in the preceding section if the probabilities in Lottery A were $\frac{1}{10}$ and $\frac{9}{10}$.

Formal decision theory prescribes certain criteria for making decisions based on inputs regarding the possible consequences, or payoffs, and inputs regarding the probabilities for the various possible events. These criteria are discussed in Chapter 5. At this point, however, an informal, intuitive discussion of how you might make your decision in the umbrella example can be offered. If you carry the umbrella, you are certain to incur a cost equivalent to \$1. If you do not carry the umbrella, you will incur a cost of \$60 with probability $\frac{1}{10}$ and a cost of \$0 with probability $\frac{9}{10}$. If you do not carry the umbrella, then, your payoff is uncertain, depending on whether or not it rains. To evaluate this payoff, the concept of mathematical expectation, which is discussed in Sections 3.1 and 4.1, is used to compute the "expected value" of the second action, "Do not carry the umbrella." Obtaining a payoff of $-\$60$ with probability $\frac{1}{10}$ and a payoff of \$0 with probability $\frac{9}{10}$ corresponds to an "expected payoff" of $-\$60\ (\frac{1}{10}) + \$0\ (\frac{9}{10})$, or $-\$6$. Comparing this with the payoff which you will obtain if you carry the umbrella, $-\$1$, you can see that your decision should be to carry the umbrella.

In order to see how the procedure used in the umbrella example can be applied to a problem that is not quite so trivial, consider a decision-making problem involving the drilling of an oil well. The decision maker must decide whether or not to drill for oil at a particular location. To simplify the problem, assume that there are two possible events, "oil" and "no oil" (in an actual drilling situation, one might obtain varying amounts of oil). If the decision is to drill and the event "oil" occurs, the oil obtained will be worth \$130,000. The cost of drilling is \$30,000, so the net payoff obtained by drilling and striking oil is \$100,000. If the decision is not to drill, the decision maker will exercise an option to sell the drilling rights to an oil wildcatter for a fixed sum of \$10,000 plus another \$10,000 contingent upon the presence of oil. The payoff table can thus be represented as follows:

		EVENT	
		Oil	*No oil*
DECISION	*Drill*	\$100,000	−\$30,000
	Do not drill	\$20,000	\$10,000

The other input to the problem involves the decision maker's uncertainty about the presence of oil. Although the decision maker may have some experience with similar drilling opportunities, this is a unique situation, not a situation involving repetitive trials. To represent his uncertainty, the decision maker should determine his subjective probability that there is oil at the particular site in question. Suppose that he decides to hire a geologist to assess the probability of oil. On the basis of similar oil-drilling ventures and geological information regarding the site, the geologist assesses the probability of oil to be $\frac{3}{10}$. That is, the odds against oil are 7 to 3. Therefore, the decision maker feels that if he drills, he will gain \$100,000 with probability $\frac{3}{10}$ and lose \$30,000 with probability $\frac{7}{10}$. This corresponds to an expected payoff of \$100,000 $(\frac{3}{10})$ − \$30,000 $(\frac{7}{10})$, or \$9,000. If he does not drill, he will gain \$20,000 with probability $\frac{3}{10}$ and \$10,000 with probability $\frac{7}{10}$, so the expected payoff is \$20,000 $(\frac{3}{10})$ + \$10,000 $(\frac{7}{10})$, or \$13,000. Thus, the decision should be not to drill [note that if $P(\text{oil})$ were $\frac{4}{10}$ instead of $\frac{3}{10}$, the best decision would be to go ahead and drill].

For a final example, consider the decision by a firm as to whether or not to initiate a special advertising campaign for a certain product. The firm feels that a competitor might introduce a competing product, and that if such a new product is introduced, sales of the firm's own product would greatly decrease unless an advertising campaign for the product was in progress. The three actions under consideration are "no advertising," "minor advertising campaign," and "major advertising campaign." Taking into account the cost of advertising and its anticipated effect on sales, the decision maker determines the following payoff table, where the numbers represent net profits, or net payoffs, to the firm.

		EVENT: *Competitor introduces new product*	EVENT: *Competitor does not introduce new product*
DECISION	*No advertising*	100,000	700,000
DECISION	*Minor ad campaign*	300,000	600,000
DECISION	*Major ad campaign*	400,000	500,000

On the basis of currently available information concerning the actions of the competing firm, the decision maker decides that the probability is .60 that the new product will be introduced. The "expected payoffs" for the three actions are \$340,000 for "no advertising," \$420,000 for "minor advertising campaign," and $440,000 for "major advertising campaign." Thus, the decision should be to launch a major advertising campaign.

It should be stressed that in all of the decision-making examples in this section, the possible effects of attitude toward risk have been ignored. The amounts involved are large enough in the second and third examples to cast serious doubt on the wisdom of ignoring the utility problem. The purpose of the examples, however, is to demonstrate the importance of probabilities in decision-making problems, and complicating the examples by bringing in utility considerations would not further this purpose. Such considerations are explored in detail in Chapter 5.

The umbrella example, the oil-drilling example, and the advertising example demonstrate the use of probabilities in decision making. In actual decision-making situations such as these, the situations of interest are often nonrepetitive. For additional examples, consider decisions regarding a bet on a football game, the purchase of some common stock, the creation of a small business firm, or the medical treatment of someone who is ill. In each of these cases, it is not possible to observe repeated trials of the situation at hand, although it may be possible to observe trials involving similar situations, such as the outcomes of previous football games, the past performance of stocks, the success of other business ventures, or the reaction of other patients with similar symptoms to various medical treatments. Because of the nonrepetitive nature of the situations, the probabilities used as inputs to the decision-making procedure must be subjective probabilities.

Of course, not *all* decision-making situations involve nonrepetitive events. Nevertheless, the subjective interpretation of probability can be thought of as an extension of the frequency interpretation, as noted in Section 2.3, and any information regarding observed relative frequencies of events can be considered in the determination of subjective probabilities. Often the decision maker has both information regarding observed frequencies *and* other information of a subjective nature. In Section 2.7 and in Chapters 3 and 4, the possibility of revising probabilities on the basis of additional information is discussed. In this way, it is possible to utilize formally all of the available information that is relevant to the decision-making situation.

2.6 THE CONDITIONAL NATURE OF PROBABILITY

An important concept in the theory of probability is that of *conditional probability*. Often one is interested in the probability that one event will occur, given that a particular second event has occurred or will occur. For instance, you might be interested in the probability that a student will earn at least a "B" average at a particular university, given that the student is a male. You might be interested in the probability that the price of a certain stock will go up, given that taxes remain the same. You might be interested in the probability that sales of a certain firm's product will go down, given that a rival firm introduces a competing product. In each of these cases you are interested in the *conditional probability* of one event, given the occurrence of a second event. The conditional probability of event E_2, given event E_1, is denoted by $P(E_2 \mid E_1)$.

It is possible to develop the concept of conditional probability directly by using devices discussed in Section 2.4, such as lotteries, bets, and reference contracts; in each case, however, the device must be modified to allow for the "given" conditions. Suppose that you are offered a choice between Lottery A and Lottery B:

Lottery A: If it rains today, you win \$100 with probability $\frac{1}{2}$ and you win \$0 with probability $\frac{1}{2}$.
If it does not rain today, you win \$25.

Lottery B: If it rains today, you win \$100 if it rains tomorrow and you win \$0 if it does not rain tomorrow.
If it does not rain today, you win \$25.

If it does not rain today, then Lottery A and Lottery B give you the same prize, \$25. If it does rain today, however, then you must choose between winning \$100 with a fixed probability of $\frac{1}{2}$ and winning \$100 if it rains tomorrow. Since this portion of each lottery only applies if it rains today, "rain today" is a "given" condition. Thus, if you choose Lottery B, you must feel that $P(\text{rain tomorrow} \mid \text{rain today})$ is greater than $\frac{1}{2}$; if you choose Lottery A, then you must feel that this conditional probability is less than $\frac{1}{2}$; and if you are indifferent between the two lotteries, then you must feel that the probability is equal to $\frac{1}{2}$. By changing the canonical probabilities in Lottery A until you are just indifferent between the two lotteries, you can determine your conditional probability of rain tomorrow, given rain today.

Conditional lotteries of this nature can be used as a basis for a formal definition of conditional probability. Your conditional probability $P(E_2 \mid E_1)$ is the number $P(E_2 \mid E_1)$ that makes you indifferent between the following two lotteries.

Lottery A: If E_1 occurs, you obtain X with probability $P(E_2 \mid E_1)$ and you obtain Y with probability $1 - P(E_2 \mid E_1)$.
If E_1 does not occur, you obtain Z for certain.

Lottery B: If E_1 occurs, you obtain X if E_2 occurs and you obtain Y if E_2 does not occur.
If E_1 does not occur, you obtain Z for certain.

Here, X, Y, and Z are three "prizes," and it is necessary that you not be indifferent between X and Y (if you are indifferent between X and Y, you will be indifferent between the two lotteries regardless of the choice of canonical probabilities in Lottery A). Since the situation in which E_1 does not occur is irrelevant in your choice between the two lotteries, it is convenient to set Z equal to zero, thereby essentially eliminating the event "E_1 does not occur" from consideration in the lotteries. If this is done, the two lotteries are identical to those considered in Section 2.4 except that they are conditional on the occurrence of E_1; the lotteries are called off if E_1 does not occur. Note that since the prizes (X, Y, and Z) are the same in the two lotteries, the problem of your attitude toward risk should not arise in the choice between the lotteries.

It is also possible to consider conditional probability in terms of conditional bets or conditional reference contracts. Suppose that two persons, A and B, make the following bet:

A bets \$5 that the San Francisco Giants baseball team wins over half of its games in the coming season.

B bets \$2 that the team does not win over half of its games, with the proviso that the bet is called off if any of a stipulated list of key Giant players is injured during the course of the season.

This bet is a conditional bet which is called off if any key players are injured. Thus, if you think that this is a "fair bet," then you apparently feel that 5 to 2 odds in favor of the team winning over half of its games are "fair" odds, given no injuries to key players. This implies that P(Giants win over half of their games | no injuries to key players) is equal to $5/(5 + 2) = \frac{5}{7}$. Relating this to the notion of a conditional reference contract, you should be willing to pay \$5 for a reference contract that will pay you \$7 if the Giants win over half of their games, where the contract will be canceled and your \$5 returned to you if any injuries to key players occur. It should be noted that the utility problem, involving attitude toward risk, may affect the results of conditional bets and conditional reference contracts; to use these devices, it is necessary to assume that the stakes involved are small enough so that the utility problem will not arise.

If the only information available concerning the events of interest is in the form of relative frequencies, the concept of conditional probability can be considered in terms of these relative frequencies. Suppose that an experiment is repeated many times under identical conditions, and either (or both) of the events E_1 and E_2 can occur on any single trial. To investigate the probability $P(E_2 \mid E_1)$, one could consider the relative frequency of occurrence of E_2, not in the entire sequence of trials, but just in those trials on which E_1 occurs. For instance, suppose that the records of an insurance company indicate that of the one million persons holding health insurance policies with the company, 250,000 were over 50 years old at the beginning of the past year. Of these 250,000 policyholders, 100,000 submitted at least one claim during the year. Thus, among policyholders over 50, the relative frequency of claims was 100,000/250,000, or .4. If you feel that this relative frequency satisfactorily represents your information, then for a given period of one year,

$$P(\text{submit at least one claim} \mid \text{over 50 years old}) = .4.$$

Note that the denominator in the relative frequency is 250,000 (the number of policyholders over 50 years old) rather than one million (the total number of policyholders).

It is sometimes convenient to represent conditional probability in terms of a *tree diagram*. For instance, consider the tree diagram in Figure 2.6.1, which concerns the occurrence or nonoccurrence of rain. Starting at the left side of the diagram, there is a "chance fork," which is a situation in which there are a number of possible outcomes, of which exactly one will occur. The "chance" indicates that there is uncertainty about which outcome will occur. At the first chance fork in this example, either E_1 (rain today) or $\bar{E}_1$ (no rain today) will occur. In either case, there is a second chance fork at which either E_2 (rain tomorrow) or $\bar{E}_2$ (no rain tomorrow) will occur. On the right-hand side of the diagram, the four possible events in this example are listed: rain today and tomorrow, rain today but not tomorrow, rain tomorrow but not today, and no rain on either day. Events such as these can be defined in set-theoretic terms as the intersection of two or more events; for instance, "rain today *and* rain tomorrow" is the intersection of "rain today" and "rain tomorrow."

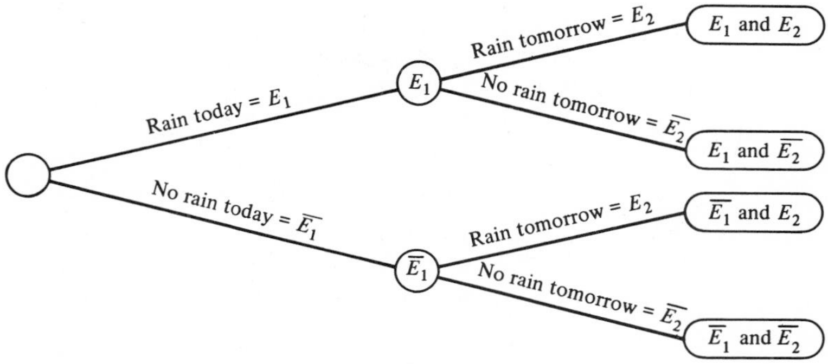

Figure 2.6.1

The uncertainty at any chance fork of a tree diagram can be represented in terms of probabilities for the possible outcomes at that fork. At the first fork in the illustration, the probabilities of interest are P(rain today) and P(no rain today). If it rains today, the probabilities at the second fork are P(rain tomorrow | rain today) and P(no rain tomorrow | rain today). If it does not rain today, the probabilities at the second fork are P(rain tomorrow | no rain today) and P(no rain tomorrow | no rain today). Thus, at any given fork in the tree diagram, the probabilities are conditional upon the outcomes at previous forks.

From any of the four endpoints in the tree diagram, it is possible to work backward to see the sequence of occurrences at the chance forks. For instance, "rain today and rain tomorrow" is clearly the result of "rain today" followed by "rain tomorrow given rain today." As a result, the probability of the event "rain today *and* rain tomorrow" (this is called a *joint probability*) is the product of P(rain today) and P(rain tomorrow | rain today). In general, for any two events E_1 and E_2,

$$P(E_1,E_2) = P(E_1)P(E_2 \mid E_1), \tag{2.6.1}$$

where $P(E_1,E_2)$ represents the joint probability of the two events. In terms of set theory, as discussed in Section 2.1, the joint probability $P(E_1,E_2)$ is simply the probability of the intersection of E_1 and E_2, $P(E_1 \cap E_2)$.

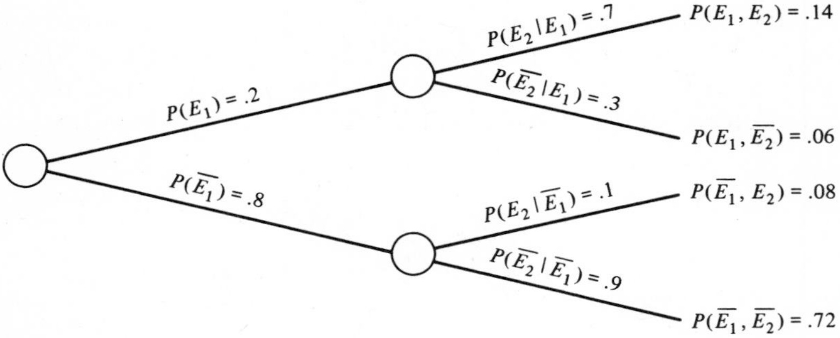

Figure 2.6.2

For example, suppose that P(rain today) $= .2$, P(rain tomorrow | rain today) $= .7$, and P(rain tomorrow | no rain today) $= .1$. This situation is illustrated in terms of a tree diagram in Figure 2.6.2. An alternative way to present joint probabilities is in terms of a table, as follows:

	E_2	$\bar{E}_2$	
E_1	$P(E_1,E_2)$	$P(E_1,\bar{E}_2)$	$P(E_1)$
$\bar{E}_1$	$P(\bar{E}_1,E_2)$	$P(\bar{E}_1,\bar{E}_2)$	$P(\bar{E}_1)$
	$P(E_2)$	$P(\bar{E}_2)$	

For the rain example, this table is:

	Rain tomorrow	*No rain tomorrow*	
Rain today	.14	.06	.20
No rain today	.08	.72	.80
	.22	.78	

The entries in the body of the table are joint probabilities, whereas the entries on the margins of the table are the individual probabilities, P(rain today) = .20, P(no rain today) = .80, P(rain tomorrow) = .22, and P(no rain tomorrow) = .78. Each individual probability is obtained by summing the values in the relevant row or column. For instance, since "rain tomorrow" can occur either in conjunction with "rain today" or with "no rain today,"

$$\begin{aligned} &P(\text{rain tomorrow}) \\ &\quad = P(\text{rain today, rain tomorrow}) + P(\text{no rain today, rain tomorrow}) \\ &\quad = .14 + .08 = .22. \end{aligned}$$

Equation (2.6.1) can be used to determine conditional probabilities. If both sides of Equation (2.6.1) are divided by $P(E_1)$, which can be done only if $P(E_1) > 0$, the result is

$$P(E_2 \mid E_1) = \frac{P(E_1,E_2)}{P(E_1)}. \tag{2.6.2}$$

That is, the conditional probability of event E_2 given event E_1 is equal to the joint probability of the two events divided by the probability of the given event, E_1. Thus, if the joint probability and the probability of the given event are known, a conditional probability can be determined from Equation (2.6.2) instead of being determined directly through the consideration of conditional lotteries, bets, reference contracts, relative frequencies, or any other such devices.

As a simple example, suppose that 60 percent of the students at a particular university are male and that 10 percent of the students are male *and* have at least a "B" average. The probability that a student chosen randomly from the student body has a "B" average, given that the student is a male, is therefore .10/.60, or $\frac{1}{6}$. The terminology "chosen randomly" means that all of the members of the student body have the same chance, or probability, of being chosen (for instance, the students could be numbered and a table of random digits could be used to choose a student). For another example, if a fair die is tossed,

$$\begin{aligned} P(\text{even number comes up} \mid 6 \text{ does not come up}) &= \frac{P(\text{even, not } 6)}{P(\text{not } 6)} \\ &= \frac{P(2 \text{ or } 4)}{P(\text{not } 6)} \\ &= \frac{\frac{2}{6}}{\frac{5}{6}} = \frac{2}{5}. \end{aligned}$$

It should be emphasized that Equation (2.6.2) can be proven formally either in terms of subjective probabilities or in terms of relative frequencies. In terms of subjective probabilities, three different lotteries (or bets or reference contracts) could be considered: one involving just E_1, one involving the intersection of the events E_1 and E_2, and one conditional lottery for E_2 given E_1. If the probabilities determined via the three lotteries, bets, or reference contracts do not satisfy Equation (2.6.2), then a "Dutch book" can be set up against the person assessing the probabilities. That is, the payoffs for the three lotteries, bets, or reference contracts can be determined so as to guarantee that the assessor will lose no matter which events occur.

For example, suppose that you assess $P(E_1) = .50$, $P(E_1,E_2) = .25$, and $P(E_2 \mid E_1) = .75$. This is not consistent with Equation (2.6.2), since $P(E_1,E_2)/P(E_1) = .50$. Suppose that you are offered three reference contracts that will pay, respectively, \$6 if E_1 occurs; \$8 if E_1 and E_2 do not both occur; and \$8 if E_2 occurs, given that E_1 occurs (the third contract is canceled and the money returned if E_1 does not occur). From your probabilities, you should be willing to pay \$3, \$6, and \$6, respectively, for the three contracts. Thus, you are paying a total of \$15. If

E_1 *and* E_2 occur, you "cash in" on the first and third contracts, winning \$14. If E_1 occurs but E_2 does not, you receive the prizes in the first and second contracts, winning \$14. If E_2 occurs but E_1 does not, you win the \$8 prize in the second lottery and the third lottery is canceled, so your \$6 is refunded, for a total of \$14. Thus, no matter what happens, you win \$14. But you originally paid \$15 for the three contracts, so you are sure to lose \$1; a "Dutch book" has been made against you.

In terms of relative frequencies, the "conditional" relative frequency must equal the relative frequency of the joint event divided by the relative frequency of the "given" event. In the insurance example presented earlier in this section, the relative frequency of the joint event "over 50 years old and at least one claim" is 100,000/1,000,000, or .10, and the relative frequency of the "given" event "over 50 years old" is 250,000/1,000,000, or .25. This yields a "conditional" relative frequency of $.10/.25 = .40$, which agrees with the conditional relative frequency determined directly earlier in the section.

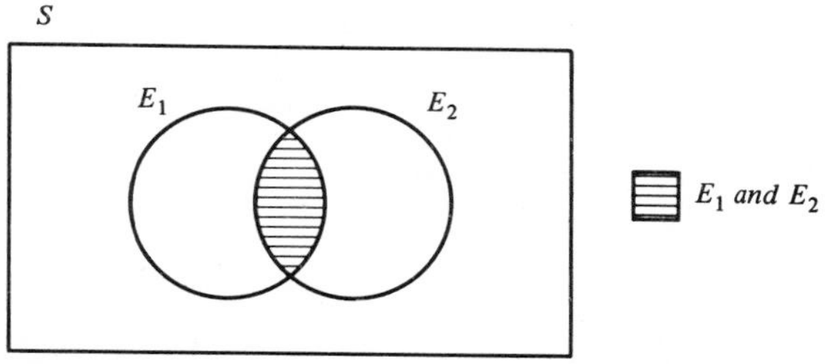

Figure 2.6.3

A final justification of Equation (2.6.2) does not involve the interpretation of probability at all. The events E_1 and E_2 for any uncertain situation are contained in S, the sample space for that uncertain situation. Their probabilities can be illustrated in terms of a Venn diagram, as discussed in Section 2.1. In Figure 2.6.3, consider what happens if it is known that E_1 has occurred or will occur. Any points of the sample space outside of E_1 are no longer relevant; in other words, the sample space has been reduced, and E_1 now comprises the entire "new" sample space. The new sample space might be thought of as the sample space conditional upon the occurrence of E_1. Conditional upon E_1, then, E_2 can occur only in conjunction with E_1; this means that E_2 can occur only via the event "E_1 *and* E_2" (the intersection of E_1 and E_2). This event is represented in Figure 2.6.3 by the portion of E_2 that intersects E_1. Thus, the conditional probability of E_2 given E_1 is represented on the Venn diagram by the area of the intersection divided by the entire area of E_1

(this division is necessary because of the reduction of the sample space due to the "given" event E_1).

In the previous sections, probability was discussed with no mention of conditional probability. In a sense, however, *all probabilities are conditional.* That is, they may be conditional upon some assumptions, upon the details of an experiment, upon some action, or upon numerous similar factors. For example, if you say that the probability of rain is $\frac{1}{3}$, you should write something like this:

$$P(\text{rain} \mid \text{current atmospheric conditions}) = \tfrac{1}{3}.$$

When you say that the probability of getting a full house in a five-card poker hand is .0014, you should write

$$P(\text{full house} \mid \text{standard, well-shuffled deck of 52 cards}) = .0014.$$

Similarly, considering the probability that a person is involved in an automobile accident in a one-year period, you might be interested in

$$P(\text{accident} \mid \text{age of the person}),$$

or

$$P(\text{accident} \mid \text{occupation of the person}),$$

or

$$P(\text{accident} \mid \text{number of miles the person drives per year}),$$

and so on.

Of course, the notion of conditional probability is not restricted to the case in which there is only one "given" event. For example, consider the conditional probability

$$P(\text{accident} \mid \text{age of the person, occupation of the person, and number of miles the person drives per year}),$$

which might be written in the form $P(A \mid B,C,D)$, the probability of event A, given events B, C, and D. This can be thought of in terms of a tree diagram with four successive chance forks, where $P(A \mid B,C,D)$ is the probability of A at the fourth chance fork, given the results B, C, and D at the previous chance forks.

Because it is bothersome to write out long expressions such as this, the "given" conditions often are not written down in probability statements. Often these "given" conditions are fairly obvious, and it is assumed that they are understood by the reader. In some situations, however, particularly when there are competing hypotheses or sets of assumptions, probabilities are written in conditional form. At any rate, you should keep in mind that all probabilities are conditional upon some factors, even if, for the sake of convenience, these factors are not explicitly noted in the probability statement.

Conditional probabilities provide information about the relationships among events. In particular, the independence or nonindependence of events is of much interest in problems of statistical inference and decision. Two events A and B are said to be *independent* if their joint probability is equal to the product of their individual probabilities (which are also called *marginal probabilities*):

$$P(A,B) = P(A)P(B). \tag{2.6.3}$$

In terms of conditional probabilities, this can be written as follows, using Equation (2.6.2):

$$P(A \mid B) = P(A)$$

or

$$P(B \mid A) = P(B).$$

Intuitively, two events are independent when knowledge about one does not change the probability of the other. If A and B are independent events, the conditional probabilities $P(A \mid B)$ and $P(B \mid A)$ are the same as the respective unconditional probabilities, $P(A)$ and $P(B)$. In terms of lotteries, if A and B are independent, then the choice of a lottery involving A alone is equivalent to the choice of a lottery involving A but conditional on the occurrence of B. Similarly, knowledge about the occurrence of one of the two events does not affect the betting odds or the choice of a bid for a reference contract concerning the other event.

For a simple example of two independent events, consider two tosses of a fair coin, and let H_1 and H_2 represent "heads comes up on the first toss" and "heads comes up on the second toss," respectively. Clearly, the result of the first toss should have no effect on the second toss, so

$$P(H_2 \mid H_1) = P(H_2) = \tfrac{1}{2}.$$

The rain example presented earlier in this section is an example involving events that are not independent. From the table of joint and marginal probabilities,

$$P(\text{rain today, rain tomorrow}) = .14,$$

while

$$P(\text{rain today})P(\text{rain tomorrow}) = (.20)(.22) = .044.$$

In terms of conditional probabilities, which are given in Figure 2.6.2,

$$P(\text{rain tomorrow} \mid \text{rain today}) = .7,$$

whereas

$$P(\text{rain tomorrow} \mid \text{no rain today}) = .1.$$

According to these probabilities, tomorrow's weather is clearly not independent of today's weather.

Sometimes the independence of two events A and B is conditional upon a third event (or assumption, or hypothesis) C. That is,

$$P(A,B \mid C) = P(A \mid C)P(B \mid C),$$

but

$$P(A,B \mid \bar{C}) \neq P(A \mid \bar{C})P(B \mid \bar{C}).$$

For example, suppose that you are a marketing manager and that you are interested in C, the event that a rival firm will not introduce a new product to compete with a certain product your firm produces. You consider the possibility that the rival firm is about to conduct an extensive market survey (event A) and about to hire a large number of new employees (event B). If C is true (the rival firm will *not* introduce a competing product), then you are willing to assume that A and B are independent. If C is not true, then you feel that A and B are *not* independent.

The concept of independence can be extended to more than two events. The K events $A_1, A_2, \ldots, A_K$ are said to be *mutually independent* if for all possible combinations of events chosen from the set of K events, the joint probability of the combination is equal to the product of the respective individual, or marginal, probabilities. For example, the three events A_1, A_2, and A_3 are mutually independent if

$$P(A_1,A_2) = P(A_1)P(A_2),$$
$$P(A_1,A_3) = P(A_1)P(A_3),$$
$$P(A_2,A_3) = P(A_2)P(A_3),$$

and

$$P(A_1,A_2,A_3) = P(A_1)P(A_2)P(A_3).$$

If the first three of these equations are satisfied but the fourth is not, then the three events A_1, A_2, and A_3 are said to be pairwise independent but not mutually independent. In this book, the "independence" of a number of events should be interpreted as *mutual* independence unless otherwise stated.

The independence of events may greatly simplify probability calculations in statistical problems, especially when repeated trials of an uncertain situation are being observed. Recall that independence of trials is required in the law of large numbers, which was discussed in Section 2.2. When it is reasonable, this is a most useful assumption, since it implies that the joint probability of the n specific occurrences in a series of n trials is equal to the product of the n individual probabilities.

2.7 BAYES' THEOREM

Sometimes it is not easy to compute conditional probabilities directly from the conditional probability formula given in the preceding section.

A convenient formula that gives the relationship among various conditional probabilities is *Bayes' theorem*, named after an English clergyman who did early work in probability theory. The simplest version of this theorem can be stated as follows:

$$P(A \mid B) = \frac{P(B \mid A)P(A)}{P(B \mid A)P(A) + P(B \mid \bar{A})P(\bar{A})}, \tag{2.7.1}$$

where $\bar{A}$ represents the complement of the event A (that is, "not A").

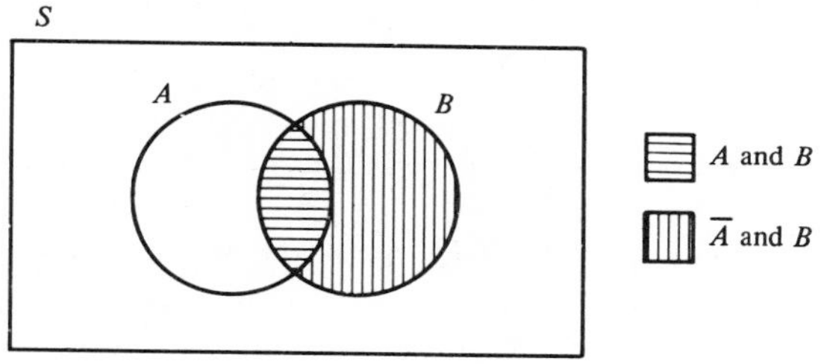

Figure 2.7.1

It is not difficult to derive Bayes' theorem from the conditional probablity formulas in the previous section. From Equation (2.6.2),

$$P(A \mid B) = \frac{P(A,B)}{P(B)}.$$

Similarly,

$$P(B \mid A) = \frac{P(A,B)}{P(A)} \quad \text{and} \quad P(B \mid \bar{A}) = \frac{P(\bar{A},B)}{P(\bar{A})},$$

so that

$$P(A,B) = P(B \mid A)P(A)$$

and

$$P(\bar{A},B) = P(B \mid \bar{A})P(\bar{A}).$$

From Figure 2.7.1, it can be seen that the events "A and B" and "$\bar{A}$ and B" are mutually exclusive (they do not "overlap" on the Venn diagram). Furthermore, the union of these two events (that is, the event that at least one of the two events will occur) is equal to B. Thus, by the third axiom of probability in Section 2.1,

$$P(B) = P(A,B) + P(\bar{A},B).$$

That is, B can occur in conjunction with A or in conjunction with $\bar{A}$, so $P(B)$ is equal to the sum of the joint probabilities of the events "A and B" and "$\bar{A}$ and B." In terms of a table such as those presented in the previous section, a marginal probability is the sum of the joint probabilities in the relevant row or column. Thus, using the above formulas for $P(A,B)$ and $P(\bar{A},B)$, the marginal probability $P(B)$ is

$$P(B) = P(B \mid A)P(A) + P(B \mid \bar{A})P(\bar{A}), \tag{2.7.2}$$

so that

$$P(A \mid B) = \frac{P(A,B)}{P(B)} = \frac{P(B \mid A)P(A)}{P(B \mid A)P(A) + P(B \mid \bar{A})P(\bar{A})}.$$

This completes the derivation of Bayes' theorem.

To illustrate Bayes' theorem, consider a simple example. Suppose that the probability that a student passes an examination is .80 if he studies for the examination and .50 if he does not study. Furthermore, suppose that 60 percent of the students in a particular class study for a particular examination. If a student chosen randomly from the class passes the examination, what is the probability that he studied?

This is obviously a problem involving conditional probability. Let B stand for the event "passes the examination" and let A stand for the event "studied for the examination." Of course, $\bar{A}$ then stands for the event "did not study for the examination." The probability that is wanted is the probability that a student studied for the examination, given that he passed. This is $P(A \mid B)$. Since 60 percent of the students in the class studied and the student of interest has been randomly chosen from the class, the *unconditional* probability that he studied for the examination, $P(A)$, is .60. Bayes' theorem can be used to determine the *conditional* probability that he studied, given that he passed the examination:

$$P(A \mid B) = \frac{P(B \mid A)P(A)}{P(B \mid A)P(A) + P(B \mid \bar{A})P(\bar{A})} = \frac{.80(.60)}{.80(.60) + .50(.40)}$$

$$= \frac{.48}{.48 + .20} = \frac{.48}{.68} = .706.$$

Note that .48/.68 is not *exactly* equal to .706; to three decimal places, it is .706. In this book, fractions are often expressed in decimal form and "rounded off" if necessary, so you should be aware that certain numerical "equalities" are only approximate.

As a second example of the use of Bayes' theorem, consider a medical situation. Suppose that a person chosen at random from the population of a certain city is given a tuberculin skin test and that the reading on the test is positive. Given the results of the skin test, what is the probability that the person has tuberculosis? In order to calculate this probability, you obtain the following information from a medical expert.

1. The probability that a person with tuberculosis will have a positive reading on a tuberculin skin test is 0.98. That is,

$$P(\text{positive reading} \mid \text{tuberculosis}) = .98.$$

2. The probability that a person without tuberculosis will have a positive reading on a tuberculin skin test is .05. That is,

$$P(\text{positive reading} \mid \text{no tuberculosis}) = .05.$$

3. Of the members of the population in the city of interest, 1 percent have tuberculosis. That is, for a randomly selected person from this population, $P(\text{tuberculosis}) = .01$.

In order to calculate the desired probability, let B represent the event "positive reading on a tuberculin skin test," and let A represent the event "has tuberculosis." The probability $P(A \mid B)$ is then

$$\begin{aligned} P(A \mid B) &= \frac{P(B \mid A)P(A)}{P(B \mid A)P(A) + P(B \mid \bar{A})P(\bar{A})} = \frac{.98(.01)}{.98(.01) + .05(.99)} \\ &= \frac{.0098}{.0098 + .0495} = \frac{.0098}{.0593} = .165. \end{aligned}$$

Before the tuberculin skin test, the probability that the person has tuberculosis is .01, since the only information available is that 1 percent of the members of the population of the city have tuberculosis and the person was chosen at random from this population. After a positive reading on the test, the probability that the person has tuberculosis increases to .165. It follows that after a positive reading, the probability of "no tuberculosis" is $1 - .165$, or .835.

Suppose that the reading on the skin test is negative rather than positive. Since $P(\text{positive reading} \mid \text{tuberculosis}) = .98$, $P(\text{negative reading} \mid \text{tuberculosis}) = .02$. Similarly, $P(\text{negative reading} \mid \text{no tuberculosis}) = 1 - P(\text{positive reading} \mid \text{no tuberculosis}) = 1 - .05 = .95$. The new information is now $\bar{B}$ (negative reading) instead of B, and

$$\begin{aligned} P(A \mid \bar{B}) &= \frac{P(\bar{B} \mid A)P(A)}{P(\bar{B} \mid A)P(A) + P(\bar{B} \mid \bar{A})P(\bar{A})} = \frac{.02(.01)}{.02(.01) + .95(.99)} \\ &= \frac{.0002}{.9407} = .0002. \end{aligned}$$

Thus, if the reading on the skin test is negative, the probability of tuberculosis drops from .01 to .0002; if the reading is positive, this probability increases from .01 to .165.

Bayes' theorem can be put into much more general form. If the J events $A_1, A_2, \ldots, A_J$ are mutually exclusive (that is, no two of the events can both occur; if one of them occurs, none of the other $J - 1$ can occur) and collectively exhaustive (that is, one of the events *must* occur; the J events exhaust all of the possible results), and B is another event (the "given" event), then

$$P(A_j \mid B) = \frac{P(B \mid A_j)P(A_j)}{P(B \mid A_1)P(A_1) + P(B \mid A_2)P(A_2) + \cdots + P(B \mid A_J)P(A_J)} \tag{2.7.3}$$

for any event A_j, where j is an integer between 1 and J, inclusive.

To illustrate the general form of Bayes' theorem, consider another medical example. A person chosen at random from the population of a given city is given a chest X-ray, and the results turn out to be "positive" (that is, a "spot" shows up on the X-ray). For the sake of simplicity, assume that the person could have lung cancer, tuberculosis, or neither (it is assumed that he cannot have *both* cancer and tuberculosis). The following information is available.

1. The probability that a person with lung cancer will have a positive chest X-ray is .90.
2. The probability that a person with tuberculosis will have a positive chest X-ray is .95.
3. The probability that a person without lung cancer or tuberculosis will have a positive chest X-ray is .07.
4. In the city of interest 2 percent of the members of the population have lung cancer and 1 percent have tuberculosis (no one has both diseases).

Using the notation of Equation (2.7.3), let A_1, A_2, and A_3 represent the events "has lung cancer," "has tuberculosis," and "has neither lung cancer nor tuberculosis," respectively; and let B represent the event "positive chest X-ray." The probabilities $P(A_1 \mid B)$, $P(A_2 \mid B)$, and $P(A_3 \mid B)$ are

$$P(A_1 \mid B) = \frac{P(B \mid A_1)P(A_1)}{P(B \mid A_1)P(A_1) + P(B \mid A_2)P(A_2) + P(B \mid A_3)P(A_3)}$$
$$= \frac{(.90)(.02)}{(.90)(.02) + (.95)(.01) + (.07)(.97)} = \frac{.0180}{.0954} = .1887,$$

$$P(A_2 \mid B) = \frac{P(B \mid A_2)P(A_2)}{P(B \mid A_1)P(A_1) + P(B \mid A_2)P(A_2) + P(B \mid A_3)P(A_3)}$$
$$= \frac{(.95)(.01)}{(.90)(.02) + (.95)(.01) + (.07)(.97)} = \frac{.0095}{.0954} = .0996,$$

and

$$P(A_3 \mid B) = \frac{P(B \mid A_3)P(A_3)}{P(B \mid A_1)P(A_1) + P(B \mid A_2)P(A_2) + P(B \mid A_3)P(A_3)}$$
$$= \frac{(.07)(.97)}{(.90)(.02) + (.95)(.01) + (.07)(.97)} = \frac{.0679}{.0954} = .7117.$$

Before the chest X-ray, the probabilities for lung cancer, tuberculosis, and neither disease are .02, .01, and .97, since the person was chosen at random from the population of the city and the only information available is that 2 percent of the members of the population have lung cancer, 1 percent have tuberculosis, and 97 percent have neither disease. After a positive chest X-ray, the probability of cancer increases to .1887, the probability of tuberculosis increases to .0996, and the probability of neither disease decreases to .7117 (the probability of having one disease or the other increases from .03 to .2883).

For another example, suppose that the chairman of a pollution control board for a particular area is concerned about the possibility that a certain firm is polluting a river. The chairman considers three events: the firm is a heavy polluter (A_1), a light polluter (A_2), or neither (A_3). Furthermore, he feels that the odds are 7 to 3 against the firm's being a heavy polluter, so that $P(A_1) = .3$; the odds are even that the firm is a light polluter, so that $P(A_2) = .5$; and the odds are 4 to 1 in favor of the firm's being a polluter, so that $P(A_3) = .2$. In order to obtain more information, the chairman takes a sample of river water 10 miles downstream from the firm in question. The water is visibly polluted and has a terrible odor. Since there are a few other possible sources of pollution for this river, the new information is not conclusive evidence that the firm in question is a polluter. However, the chairman of the pollution control board judges that if the firm is a heavy polluter, the probability that the water would be so polluted 10 miles downstream is .9; if the firm is a light polluter, this probability is .6; and if the firm is not a polluter, the probability is only .3. Thus, $P(B \mid A_1) = .9$, $P(B \mid A_2) = .6$, and $P(B \mid A_3) = .3$, where B represents the information from the sample of river water. Applying Bayes' theorem,

$$P(A_1 \mid B) = \frac{P(B \mid A_1)P(A_1)}{P(B \mid A_1)P(A_1) + P(B \mid A_2)P(A_2) + P(B \mid A_3)P(A_3)}$$
$$= \frac{.9(.3)}{.9(.3) + .6(.5) + .3(.2)} = \frac{.27}{.63} = .429,$$

$$P(A_2 \mid B) = \frac{P(B \mid A_2)P(A_2)}{P(B \mid A_1)P(A_1) + P(B \mid A_2)P(A_2) + P(B \mid A_3)P(A_3)}$$

$$= \frac{.6(.5)}{.9(.3) + .6(.5) + .3(.2)} = \frac{.30}{.63} = .476,$$

and
$$P(A_3 \mid B) = \frac{P(B \mid A_3)P(A_3)}{P(B \mid A_1)P(A_1) + P(B \mid A_2)P(A_2) + P(B \mid A_3)P(A_3)}$$

$$= \frac{.3(.2)}{.9(.3) + .6(.5) + .3(.2)} = \frac{.06}{.63} = .095.$$

Before the sample of river water is observed, the probabilities for "heavy polluter," "light polluter," and "no polluter" are .3, .5, and .2. After seeing the river water, the probability of "heavy polluter" increases to .429, the probability of "light polluter" decreases slightly to .476, and the probability of "no polluter" decreases to .095.

Bayes' theorem is a key feature of the decision-theoretic approach to statistics, and the following terminology is used to describe the various probabilities appearing in Bayes' theorem: $P(A_j)$ represents a *prior probability;* $P(B \mid A_j)$ represents a *likelihood,* which involves the additional information B; and $P(A_j \mid B)$ represents a *posterior probability.* The "likelihood" can be interpreted as the likelihood, or probability, of the observed information B, given a particular event A_j. Of course, the likelihood will generally be different for different events. The likelihood of the observed information (a positive reading) given tuberculosis in the first medical example is .98, whereas the likelihood of the observed information given no tuberculosis is .05. The terms "prior" and "posterior" are relative to the observed information.

In terms of the first medical example, the prior probabilities of tuberculosis and no tuberculosis are .01 and .99, respectively. After seeing some additional information (the positive reading on the skin test), the posterior probabilities given this new information are .165 and .835. In terms of the second medical example, the prior probabilities of lung cancer, tuberculosis, and neither are .02, .01, and .97, respectively. After the new information (the positive chest X-ray), the posterior probabilities determined from Bayes' theorem are .1887, .0996, and .7117. The likelihood of the observed information (the positive chest X-ray) given lung cancer is .90, the likelihood given tuberculosis is .95, and the likelihood given neither disease is .07. In terms of the pollution example, the prior probabilities of "heavy polluter," "light polluter," and "no polluter" are .2, .5, and .3, respectively. After the new information (the polluted river water), the posterior probabilities are .429, .476, and .095. The likelihood of the observed information given that the firm is a heavy polluter is .9, the likelihood given that the firm is a light polluter is .6, and the likelihood given that the firm is not a polluter is .3.

Note that since exactly one of the events $A_1, A_2, \ldots, A_J$ must occur, the union of the J events is equal to the sample space S, and the sum of the probabilities for the J events must equal one. Thus, the prior probabilities $P(A_1), P(A_2), \ldots, P(A_J)$ must sum to one; and the posterior probabilities $P(A_1 \mid B), P(A_2 \mid B), \ldots, P(A_J \mid B)$ must sum to one. This is easily verified for the two medical examples and the pollution example. There are no restrictions on the sum of the likelihoods (moreover, it is meaningless to sum likelihoods); in the three examples, this sum happens to be 1.03, 1.92, and 1.80, respectively.

In the model that is presented in Chapters 5 and 6 for decision making under uncertainty, uncertainty is formally represented in terms of probability. In making a decision, it is desirable to utilize all available information. Thus, since Bayes' theorem provides a means of revising probabilities as new information is obtained, it is an extremely valuable tool in decision theory. In the next two chapters, the use of Bayes' theorem to revise probabilities on the basis of new information is discussed in further detail; the resulting probabilities are used in decision-making problems in Chapters 5 and 6.

2.8 REFERENCES AND SUGGESTIONS FOR FURTHER READING

For elementary to intermediate level discussions of the mathematical theory of probability, see Feller (1968); Goldberg (1960); Hays and Winkler (1970, Chapters 1 and 2); Mosteller, Rourke, and Thomas (1970, Chapters 1–3); and Parzen (1960). An important historical reference at an advanced level is Kolmogorov (1933). With regard to the interpretation of probability, key historical references are Ramsey (1931), de Finetti (1937), and Savage (1954); the Savage reference contains an extensive annotated bibliography. In Kyburg and Smokler (1964), some important articles, including the above-mentioned Ramsey article and an English translation of the de Finetti article, are reprinted, and a bibliography is included. In addition, see de Finetti (1968, 1970); Fellner (1965); Fishburn (1964); Good (1950; 1959; 1965, Chapter 2); and Savage *et al.* (1962) for discussions regarding the interpretation of probability, with emphasis on subjective interpretations. Other references relating to subjective probability are given in later chapters. Proponents of other interpretations, with emphasis on symmetry arguments, are Carnap (1950); Jeffreys (1961); Keynes (1921); and Laplace (1820). The primary reference regarding Bayes' theorem is Bayes (1763), and further references are given in Chapters 3 and 4 as Bayesian inferential procedures are discussed.

The above references concern specific aspects of probability. In addition, some general references at a level roughly comparable to this chapter are Aitchison (1970*a*, Chapters 2 and 3); Hadley (1967, Chapter 1); LaValle (1970, Chapters 1, 5, and 6); Lindley (1965, Chapter 1; 1971, Chapters 2 and 3); Morris (1968, Chapter 3); Pratt, Raiffa, and Schlaifer (1965, Chapter 2); Raiffa (1968, Chapter 5); Roberts (1966*b*, Chapters 2–4); Schlaifer (1959, Chapters 1 and 9; 1969, Chapter 6); and Schmitt (1969, Chapter 1).

EXERCISES

1. Explain the statement that probability can be thought of as the mathematical language of uncertainty.
2. Discuss the importance of carefully defining the sample space in any application of probability theory. In the oil-drilling example presented in Section 2.5, the sample space consists of just two events, "oil" and "no oil," in an attempt to simplify the problem. Can you think of some more realistic ways to define the sample space in this example?
3. A coin is tossed n times, and each time either a head or a tail occurs. If the elementary events for this experiment are viewed as the possible *sequences* of results in n tosses, how many different elementary events are there? If $n = 4$, is the event "no heads in four tosses" an elementary event? Is the event "two heads in four tosses" an elementary event?
4. Suppose that a politician is uncertain about the outcome of a particular election. If there are just two candidates in the election, give at least two possible ways of defining the sample space. If the politician has held public office for some time and plans to retire from politics if he loses the election, might he define the sample space differently than would a younger man who is a candidate for the first time?
5. Briefly explain the connection between set theory and probability theory and discuss the usefulness of Venn diagrams to illustrate elementary rules of probability.
6. Suppose that a red die and a green die are each tossed and that r and g represent the numbers appearing on the red and green die, respectively. Assume that the dice are perfectly symmetric.
   (a) Specify the sample space completely. How many elementary events are there?
   (b) Find the probability that $r = g$.
   (c) Find the probability that $r \leq g$.

(d) Find the probability that $r < 2$ *and* $g \geq 3$.
(e) Find the probability that $r < 2$ *or* $g \geq 3$ (including the possibility that *both* will occur).
(f) Find the probability that $r + g = 7$.
(g) Find the probability that $r + g = 8$ *and* $r - g \geq 0$.

7. A letter is selected at random from the English alphabet (that is, all possible letters are equally likely to be chosen). Find the probability that
(a) the letter is a vowel,
(b) the letter is a consonant,
(c) the letter occurs in the last 10 positions of the alphabet (given the conventional ordering of the alphabet),
(d) the letter is a consonant falling between the two vowels "a" and "i" in the conventional ordering.

8. Two letters are drawn at random from the alphabet, and the first is replaced before the second is drawn. What is the sample space for this experiment? Find the probability that the first letter drawn precedes the second letter in the conventional ordering of the alphabet. If the first letter is *not* replaced before the second is drawn, so that the two letters must be different, find the sample space and determine the probability that the first letter precedes the second letter in the conventional ordering of the alphabet.

9. Suppose that a fair die is thrown until a six appears. What is the probability that this will take more than three throws?

10. Illustrate on a Venn diagram that $P(A \cap B) \leq P(A) \leq P(A \cup B)$ for any events A and B. Also, give a brief intuitive explanation of this rule of probability.

11. A famous eighteenth-century mathematician, d'Alembert, argued that in two tosses of a coin, heads could appear once, twice, or not at all, and thus each of these three events should be assigned a probability of one-third. Do you agree with him? Comment on the issues involved and on the implications of the situation with regard to the use of symmetry assumptions to determine probabilities.

12. It is claimed by some that defining probability in terms of Equation (2.2.1) is a circular definition because it involves "equally likely" events, thereby using the notion of probability in the definition of probability. Discuss this claim.

13. The so-called "gambler's fallacy" goes something like this: in a dice game, for example, a seven has not turned up in quite a few rolls of a pair of honest dice; now a seven is said to be "due" to come up. Why is this a fallacy?

14. The law of large numbers has occasionally been called "the link between the mathematical concept of probability and the real world about us." Discuss this proposition.

15. Take a fair die and throw it 300 times, recording the number that appears on each throw. Find the frequency of each of the six possible events after 2, 5, 10, 20, 50, 100, and 300 throws. Repeat this experiment several times and comment on the results in light of the law of large numbers and the concept of statistical regularity.

16. In what sense is the relative-frequency interpretation of probability a conceptual interpretation rather than an operational interpretation?

17. Find the probability of the event E if the odds in favor of E are
(a) 2 to 1,
(b) 1 to 2,
(c) 3 to 7,
(d) 9 to 2.

18. Find the odds in favor of the event E if the probability of E is
(a) .50,
(b) .20,
(c) .875,
(d) .001.

19. Explain why the subjective interpretation of probability can be thought of as an extension of the frequency interpretation. Are there any restrictions on the types of situations for which subjective probabilities can be used?

20. Discuss the following statement: "If a person feels subjectively that an event will occur with relative frequency p in a long series of identical, independent trials, then his subjective probability of the event occurring on any single trial should be p."

21. Explain why the basic axioms of probability presented in Section 2.1 must be obeyed under both (a) the relative-frequency interpretation of probability and (b) the subjective interpretation of probability.

22. (a) What is your subjective probability that it will rain tomorrow?
(b) What, in your opinion, are the odds in favor of rain tomorrow?
(c) Are your answers to (a) and (b) consistent? If not, why not?
(d) When "precipitation probabilities" are given out to the public, how do you interpret them? How do you think the "average person" interprets them?

23. Three football teams are fighting for the league championship. A sportswriter claims that the odds are even, or 1 to 1, that team A will win the championship, the odds are even that team B will win the championship, and the odds are even that team C will win the championship. Is this reasonable? Explain your answer.

24. In Exercise 23, the sportswriter states that he is willing to bet for or against team A at even odds, and likewise for teams B and C. Even without knowing anything about the teams, would you like to bet with him? Explain.

25. An investment analyst thinks that the following is a "fair" bet:

A bets \$10 that the price of XYZ stock will increase in the next year.

B bets \$7 that the price of XYZ stock will *not* increase in the next year.

What is the analyst's subjective probability that the price of XYZ stock will increase in the coming year?

26. What is your personal probability that the recorded maximum temperature on next January 1 in San Francisco is greater than 50°F? Would you be willing to bet at the odds implied by this probability?

27. How would your answer to Exercise 26 change as a result of obtaining the additional information that for the 30-year period from 1931 to 1960, the average maximum daily temperature during January in San Francisco was 55°? Would it change if you also were told that the average minimum daily temperature during January in San Francisco was 42° during the same period?

28. What is your subjective probability that a member of the Democratic party will win the next presidential election in the United States? How much would you be willing to pay for a reference contract that would give you $5 if the winner of the next presidential election is a Democrat? If you already held such a reference contract, for how much would you be willing to *sell* it? Discuss the relationship between the problem of buying the reference contract and the problem of selling it.

29. Explain how lotteries can be used to help a person express his judgments in terms of probability. What is the "behavioral approach" to the assessment of probabilities?

30. Discuss the concept of canonical probabilities and their use in the assessment of probabilities via devices such as lotteries. What devices might be used to determine canonical probabilities?

31. Suppose that you are offered a choice between the following two lotteries.

    Lottery A: A box contains 50 blue marbles and 50 yellow marbles. A marble will be drawn at random from the box, and you will win $50 if the marble is blue and $0 if the marble is yellow.

    Lottery B: A box contains 100 marbles, each of which is either blue or yellow, but you do not know how many of each color are in the box. A marble will be drawn at random from the box, and you will win $50 if the marble is blue and $0 if the marble is yellow.

    (a) Do you prefer Lottery A, do you prefer Lottery B, or are you indifferent between the two lotteries? Explain your answer.
    (b) Although psychological experiments have indicated that many individuals are *not* indifferent between the two lotteries, it is claimed by some that they *should* be indifferent. Discuss.

32. Discuss each of the following situations with regard to the different interpretations of probability.
    (a) A coin has been tossed 1000 times and exactly 500 heads have been observed. What is the probability of heads on the next toss?

(b) A brand-new coin is obtained at a bank. What is the probability of heads on the next toss?
(c) A coin was tossed at the beginning of the 1970 Stanford–Washington football game to determine choice of kicking versus receiving or choice of goal. What is the probability that the coin came up heads?
(d) A coin is tossed, placed on a table, and covered with a sheet of paper. What is the probability that "heads" is up?
(e) A stranger produces a coin and tosses it 10 times. Each time it comes up heads. What is the probability of heads on the next toss?
(f) You take a brand new coin and toss it 10 times. Each time it comes up heads. What is the probability of heads on the next toss?

33. What, in your opinion, is the probability that Ohio State will win the next football game they play against Michigan? What is the probability that Ohio State beat Michigan in their football game in 1960?

34. Suppose that you are contemplating a picnic on the coming Fourth of July. If you are concerned about the weather on that day, how might you define the relevant sample space, taking into account any factors about the weather that might have some effect on your decision concerning the picnic?

35. If $P(B) = 0$, does it make sense to consider the conditional probability $P(A \mid B)$? Explain.

36. Discuss the statement "In a sense, all probabilities are conditional."

37. Explain the difference between mutually exclusive events and independent events. Is it possible for two events A and B to be both mutually exclusive *and* independent? Explain.

38. Suppose that you are offered the following choice of lotteries:

Lottery A: If the winner of the next presidential election in the United States is a member of the Democratic party, then you win \$10 with probability $\frac{1}{2}$, and you lose \$10 with probability $\frac{1}{2}$.
If the winner of the election is not a Democrat, you win \$0.

Lottery B: If the winner of the next presidential election in the United States is a member of the Democratic party, then you win \$10 if the Senate has more Democrats than Republicans at the beginning of the president's term of office, and you lose \$10 if the Senate does not have more Democrats than Republicans at that time.
If the winner of the election is not a Democrat, you win \$0.

Which lottery would you select? What canonical probabilities in Lottery A would make you indifferent between the two lotteries? What subjective probability of yours is being assessed here?

39. Combining the results of Exercises 28 and 38, what is your joint probability for the event that the next presidential election is won by a Democrat *and*

the Senate has more Democrats than Republicans at the beginning of this president's term of office? Assuming a strict two-party system, represent the possible events in this situation in terms of a tree diagram.

40. Suppose that someone offers to sell you a reference contract that will pay you \$5 if the price of IBM stock goes up tomorrow, with the condition that the contract is called off and you get your money back if it rains in Seattle tomorrow. How much would you be willing to pay for this contract? If the contract is changed so that it is called off only if it does *not* rain in Seattle tomorrow, how much would you pay for it? What do your answers imply about the relationship between price changes of IBM stock and rain in Seattle?

41. Suppose that you are interested in whether Congress raises taxes in the next year and in whether the sales of a particular product are higher in the next year than they were in the past year. Furthermore, your joint probabilities are P(taxes raised, sales higher) = .15, P(taxes raised, sales not higher) = .20, P(taxes not raised, sales higher) = .60, and P(taxes not raised, sales not higher) = .05.
    (a) Find the individual probabilities P(taxes raised), P(taxes not raised), P(sales higher), and P(sales not higher).
    (b) Find the conditional probabilities P(sales higher | taxes raised), P(sales higher | taxes not raised), P(sales not higher | taxes raised), and P(sales not higher | taxes not raised).
    (c) Represent this situation in terms of a tree diagram.
    (d) Represent this situation in terms of a table.
    (e) Are the events "taxes raised" and "sales higher" independent events?

42. Give an example of three events which are pairwise independent but not mutually independent. [*Hint:* Consider a simple experiment, such as the tossing of two dice.]

43. Determine a formula similar to Equation (2.6.1) to express the joint probability $P(E_1,E_2,E_3)$ as a product of probabilities, and do the same for $P(E_1,E_2, \ldots, E_K)$.

44. Imagine three identical wallets; one contains two \$5 bills, one contains two \$10 bills, and one contains one \$5 bill and one \$10 bill. You choose a wallet randomly, take a bill from it randomly, and find that it is a \$10 bill. What is the probability that the other bill in the chosen wallet is also a \$10 bill?

45. As a contractor, you have submitted a bid for a certain large project (Project A). If your bid is not successful, you plan to submit a bid for a second large project (Project B), but if your bid for Project A is successful, your additional resources will only be sufficient to allow you to bid for a smaller project (Project C). Since Project B is even larger than Project A, bidding for Project B precludes a bid for Project C, and the deadline for bids for Project C falls before the announcement of the low bidder for Project B. You feel that the probabilities of successful bids are .40 for A, .30 for B, and .60 for C. Furthermore, you judge the success or failure of

your bids to be independent for the different projects. Represent this situation on a tree diagram. What is the joint probability that you will successfully obtain contracts for Projects A and B? What is the joint probability that you will successfully obtain contracts for Projects A and C? Given that you are successful in your bid for Project A, what is the probability that you will *not* be successful in your bid for Project C?

46. Explain why Equation (2.6.2) is applicable (a) under the relative-frequency interpretation of probability and (b) under the subjective interpretation of probability.

47. In the first medical example in Section 2.7, the new information, which consists of a positive reading on the tuberculin skin test, increases the probability of tuberculosis from .01 to .165. This result is surprising to some people, who claim that they expected a greater increase in the probability of tuberculosis. Explain why the increase is no greater, despite the fact that the tuberculin skin test results in only a .02 chance of a "false negative" reading for a person with tuberculosis and a .05 chance of a "false positive" reading for a person without tuberculosis.

48. The probability that 1 percent of the items produced by a certain process are defective is .80, the probability that 5 percent of the items are defective is .10, and the probability that 10 percent of the items are defective is .10. An item is randomly chosen, and it is defective. *Now* what is the probability that 1 percent of the items are defective? That 5 percent are defective? That 10 percent are defective? Suppose that a *second* item is randomly chosen from the output of the process, and it too is defective. Following this second observation, what are the probabilities that 1, 5, and 10 percent, respectively, of the items produced by the process are defective?

49. In a football game, suppose that a team has time for one more play and that they need to score a touchdown on this play to win the game. As a fan, you feel that there are only three possible plays that can be used, and that the three plays are equally likely to be used. The plays are a long pass, a screen pass, and an end run. Based on past experience in similar situations, you feel that the probabilities of getting a touchdown with these plays are as follows:

Play	$P(touchdown \mid play)$
Long pass	.50
Screen pass	.30
End run	.10

The team scores a touchdown on the play. What is the probability that they used the long pass? The screen pass? The end run?

50. Suppose that Urn A is filled with 700 red balls and 300 green balls and that Urn B is filled with 700 green balls and 300 red balls. One of the two urns is selected at random (that is, the two urns are equally likely to be selected). The experiment consists of selecting a ball at random from the chosen urn, recording its color, and then replacing it and thoroughly mixing the balls again. This experiment is conducted three times, and each time the ball chosen is red. What is the probability that the urn chosen is Urn A?

51. You are interested in the stock of the XYZ corporation. You feel that if the stock market goes up in the next year, the probability is .9 that the price of XYZ stock will go up. If the market goes down, the probability is .4 that XYZ will go up. Finally, if the market remains steady, the probability is .7 that XYZ will go up. Furthermore, you think that the probabilities are .5, .3, and .2 for the market to go up, go down, or remain steady. At the end of the year, the price of XYZ stock has *not* gone up. What is the probability that the stock market as a whole went up?

3

BAYESIAN INFERENCE FOR DISCRETE PROBABILITY MODELS

As indicated in the last chapter, probability can be thought of as a measure of uncertainty. At any given point in time, you may be uncertain about which of a number of events will occur. The information you have concerning the situation at hand can be represented by a set of probabilities for the possible events, and these probabilities can be used to help you make decisions. In many situations, it may be possible to obtain additional information before making a decision.

For instance, you may be uncertain about the proportion of items that turn out to be defective from a particular production process. You have some information about this proportion, and it may be possible to gain additional information by observing some items from the process. In this case, you might be interested in making a decision regarding the process (such as a decision as to whether or not to overhaul a particular machine), and you would like to utilize both the old and the new information in making your decision. For another example, suppose that the manager of a chain of retail stores must decide whether or not to open a new store in a particular area. He is uncertain about the potential sales of the proposed store, and he would like to obtain additional information before he makes his decision. To obtain this information, he decides to conduct a survey among the consumers in the area. For a third example, suppose that you are interested in the proportion of voters who will vote for a certain candidate in a forthcoming election. You have some judgments concerning this proportion, but you are quite uncertain, so you obtain additional information in the form of a survey of, say, 1 percent of the registered voters. In order to make an inferential statement about the proportion of votes the candidate of interest will obtain, you would like to combine the information from this survey with your prior judgments.

In each of these situations, what is needed is a method for *combining* the new information with the previously available information. The resulting decision or inferential statement can then be based on *all* available information. The mechanism used to combine information is called Bayes' theorem, which was introduced in Section 2.7 and which is studied in detail in this chapter and in Chapter 4. In this chapter, Bayes' theorem is applied to discrete probability models (continuous probability models are considered in Chapter 4). First, the concepts of discrete random variables and discrete probability models are presented in Section 3.1.

3.1 DISCRETE PROBABILITY MODELS

In Section 2.7, Bayes' theorem was presented in terms of probabilities of various events. These events were expressed in a qualitative fashion, however, and it is often convenient to quantify them by assigning numbers to the various events. When such numbers are assigned, the "function" that assigns a number to each elementary event in the sample space is called a *random variable.* Formally, a random variable is a real-valued function that is defined on a sample space. (Recall from Chapter 2 that a sample space is the set of all possible outcomes, or elementary events, of interest.) Thus, a random variable is a function that associates a single real number with each elementary event in the sample space.

For instance, if a coin is to be tossed once, the two elementary events are heads and tails, so one possible assignment of numbers to these events would be "1" for heads and "0" for tails. If you are interested in whether or not it will rain tomorrow, the two elementary events are "rain tomorrow" and "no rain tomorrow," so you could assign "1" to "rain tomorrow" and "0" to "no rain tomorrow." When there are only two events of interest, the random variable that assigns "1" to one event and "0" to the other is called an *indicator* random variable. Of course, this is not the only possible assignment; for example, you could assign "−5" to "rain tomorrow" and "3" to "no rain tomorrow" if you so desired. Indicator random variables often are useful, however, and they help to clarify the relationship between events and random variables.

Sometimes the events in question already have some numerical values, in which case it is convenient to take these values as the values of a random variable. In tossing a single die, the elementary events are "1 comes up," "2 comes up," "3 comes up," "4 comes up," "5 comes up," and "6 comes up." The most convenient assignment of numbers to these events is to assign "1" to "1 comes up," and so on, and the random variable can then be defined in words as "the number of points coming up, or showing, on the toss of the die." Other examples of random variables with numerical values equal to the numerical values associated with the corresponding events are the sales of a particular product, the total number of points scored in a football game, the temperature (in degrees Fahrenheit) at a particular time and location, and the number of vehicles passing a certain point during some given period of time.

It is often useful to think of a random variable simply as an *uncertain quantity*, since probability can be regarded as a measure of uncertainty. The terms "random variable" and "uncertain quantity" are used interchangeably in this book. It is important to note that the "uncertainty" involved in an uncertain quantity or a random variable refers to uncertainty on the part of the statistician or decision maker in any given situation. For example, you might consider making a bet regarding some "trivia" item such as a football game conducted 20 years ago. The outcome of the game is clearly not uncertain, but because your memory of the game is somewhat vague, it is uncertain to you, so you can consider it a random variable. For another example, a production manager is interested in the proportion of parts that are defective in a newly received shipment. Clearly this proportion is fixed and could be determined with certainty by inspecting the entire shipment. However, such an inspection may be infeasible because of cost considerations or because the testing is destructive (that is, it may not be possible to determine whether or not an item is defective without destroying the item). Thus, the proportion of defective parts should be treated by the production manager as a random variable. A third illustration of the relative nature of the uncertainty in a random variable involves a firm attempting to determine the price of a new product. A rival firm has developed a similar product and has decided upon its price. The first firm, however, is not able to obtain information regarding this price, so it must treat the rival firm's price as a random variable.

In general, a random variable or an uncertain quantity is denoted in this book by a letter with a tilde (~) over it, and a specific *value* of the random variable is denoted by the same letter without the tilde. For instance, if $\tilde{x}$ is the random variable that assigns 1 to heads and 0 to tails, as in the first paragraph of this section, then the probability of heads can be written $P(\tilde{x} = 1)$. If the random variable $\tilde{y}$ refers to the price of a particular stock on a certain day in the future, you might be interested in $P(\tilde{y} = y)$, where y is some specific number, such as 20. Often a probability statement such as $P(\tilde{y} = y)$ is written in abbreviated form as $P(y)$. When the particular random variable of interest is obvious, the probability $P(\tilde{y} = 20)$ may be written $P(20)$, although considerable care is taken not to abuse this notation. A final note with regard to the use of tildes is that in some contexts there is ambiguity concerning the inclusion or the exclusion of a tilde, although every attempt is made to be consistent in the distinction among random variables, values of random variables, and other variables (variables about which no uncertainty exists).

This chapter involves *discrete* random variables, which are random variables with only a finite or countably infinite number of possible values. The random variable $\tilde{x}$ discussed above has only two possible values, so it is discrete. For an example involving a countably infinite number of values, consider an experiment in which a coin is tossed repeatedly until heads occurs. If $\tilde{z}$ represents the number of tosses until the first occurrence of heads, then the possible values of $\tilde{z}$ are the positive integers 1, 2, 3, . . . , which form a countably infinite set. In general, a set is countably infinite if the members of the set can be put in one-to-one correspondence with the positive integers (that is, if the members of the set can be "indexed" by the positive integers). Thus, $\tilde{z}$ has a countably infinite number of possible values. If $\tilde{t}$ represents the total number of points scored in a football game, then $\tilde{t}$ has a countably infinite number of values and is a discrete random variable. Note that although $\tilde{t}$ technically has a countably infinite number of values, for all practical purposes $\tilde{t}$ will not be larger than, say, 300, so that the values of $\tilde{t}$ might be limited to 0 and the positive integers less than 300.

When a random variable takes on an infinite (but uncountable) number of values (for instance, the random variable corresponding to the weight of a particular item, which could conceptually have any positive real number as a value), it is called a *continuous* random variable. Continuous random variables are studied in Chapter 4. In the remainder of this section some useful facts about discrete probability models are briefly reviewed.

Since probabilities can be determined for the various possible events in any situation, these probabilities also refer to any random variable that is a quantification of the events. If the discrete random variable $\tilde{x}$ can assume the J values $x_1, x_2, x_3, \ldots, x_J$, then the set of J possible values, together with their probabilities $P(\tilde{x} = x_1)$, $P(\tilde{x} = x_2)$, $P(\tilde{x} = x_3)$, . . . , $P(\tilde{x} = x_J)$, comprises the *probability distribution* of the random variable $\tilde{x}$. For instance, if $\tilde{x}$ represents the number of times heads occurs in two tosses of a coin, the sample space consists of the elementary events (H,H), (H,T), (T,H), and (T,T), where the first letter in each pair stands for the result on the first toss and the second letter stands for the result on the second toss. If the coin is "fair," each of the four elementary events has probability $\frac{1}{4}$. These probabilities can be used to determine probabilities for the three possible values of $\tilde{x}$:

$$P(\tilde{x} = 0) = P[(T,T)] = \tfrac{1}{4},$$

$$P(\tilde{x} = 1) = P[(H,T)] + P[(T,H)] = \tfrac{1}{2},$$

and

$$P(\tilde{x} = 2) = P[(H,H)] = \tfrac{1}{4}.$$

The probability distribution of $\tilde{x}$ can be represented in tabular form as follows.

x	$P(\tilde{x} = x)$
0	$\frac{1}{4}$
1	$\frac{1}{2}$
2	$\frac{1}{4}$

In this example, the probabilities are all nonnegative, and they sum to 1. In general, for a random variable with J possible values it is necessary for all J probabilities to be nonnegative and for their sum to equal 1:

$$P(\tilde{x} = x_j) \geq 0 \quad \text{for } j = 1, 2, 3, \ldots, J, \tag{3.1.1}$$

and

$$\sum_{j=1}^{J} P(\tilde{x} = x_j) = 1, \tag{3.1.2}$$

where $\sum_{j=1}^{J}$ stands for "sum over the values $j = 1, j = 2$, and so on up to $j = J$." If these two conditions are not satisfied, the basic axioms of probability presented in Section 2.1 are violated, and therefore it is not possible to call the numbers probabilities.

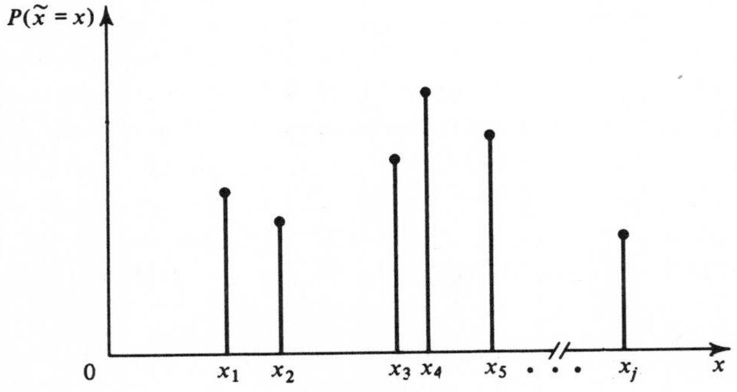

Figure 3.1.1

The probability distribution of a random variable $\tilde{x}$ can be represented in graphical form as in Figure 3.1.1. The possible values of $\tilde{x}$ are on the horizontal axis, and the vertical axis represents probability. Such a graph is often referred to as the graph of the *probability mass function* (abbreviated PMF) of $\tilde{x}$. Another way of representing a probability distribution graphically is by means of a *cumulative distribution function* (CDF), which is defined as the function $P(\tilde{x} \leq x)$. Here the vertical axis represents *cumulative* probability. To illustrate this, let $\tilde{x}$ represent the number that comes up on a single toss of a "fair" die. The PMF and CDF of $\tilde{x}$ are presented in Figures 3.1.2 and 3.1.3. The CDF of a discrete random variable is always a "step function," as in Figure 3.1.3; the "jumps" in the function occur at the possible values of the random variable, and the amount of the "jump" at any particular value x_j is equal to the probability of that value, $P(x_j)$. For each horizontal segment of the CDF function, the left-hand end point of the segment is included and the right-hand end point is not. For example, consider the segment from $x = 1$ to $x = 2$ in Figure 3.1.3, which is at "height" $\frac{1}{6}$. Since $P(\tilde{x} \leq 1) = P(\tilde{x} = 1) = \frac{1}{6}$, the left-hand end point, 1, is included in this segment. However, $P(\tilde{x} \leq 2) = P(\tilde{x} = 1) + P(\tilde{x} = 2) = \frac{2}{6}$, so the right-hand end point, 2, is included in the *next* segment, the segment from 2 to 3. Thus, the value of the CDF at a "jump" is read from the higher segment [for example, $P(\tilde{x} \leq 5)$ is equal to $\frac{5}{6}$, not $\frac{4}{6}$]. Note that for values of $\tilde{x}$ greater than 6, $P(\tilde{x} \leq x) = 1$, so the CDF is equal to 1.

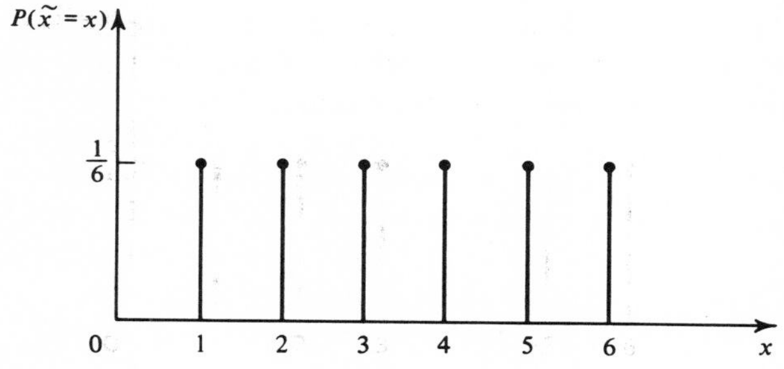

Figure 3.1.2

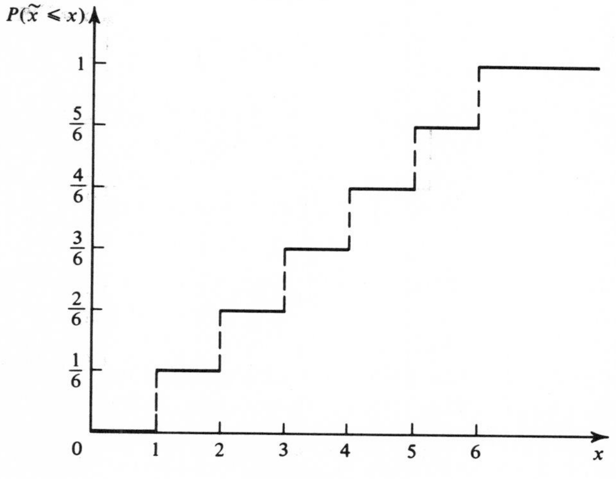

Figure 3.1.3

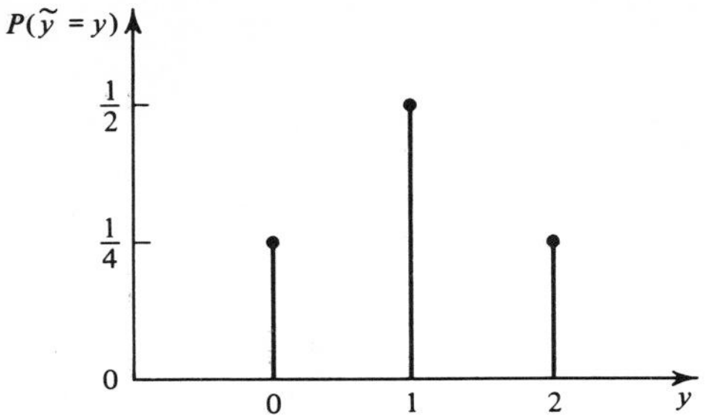

Figure 3.1.4

A second example of the PMF and CDF of a discrete random variable is illustrated in Figures 3.1.4 and 3.1.5, which represent the probability distribution of $\tilde{y}$, the number of heads in two tosses of a "fair" coin. For a final example, let $\tilde{z}$ represent the number of rainy days (days with measurable precipitation) in Chicago next week. Suppose that after obtaining weather reports and any other information that you consider relevant, you assess the following probability distribution for $\tilde{z}$.

z	$P(\tilde{z} = z)$
0	.13
1	.19
2	.29
3	.18
4	.11
5	.06
6	.03
7	.01

The PMF and CDF of $\tilde{z}$ are presented in Figures 3.1.6 and 3.1.7.

Often certain *summary measures* of a probability distribution are of particular interest. One such measure is the *expectation* of the random variable, which is defined as follows for a discrete random variable $\tilde{x}$:

$$E(\tilde{x}) = \sum_{j=1}^{J} x_j P(\tilde{x} = x_j). \tag{3.1.3}$$

This expectation, which is also denoted by $\mu_{\tilde{x}}$, or simply μ (Greek mu) when the variable of interest is clear, is a weighted average of the possible values of $\tilde{x}$. The weights are simply the probabilities of these values. For instance, if $\tilde{x}$ represents the number that comes up on a single toss

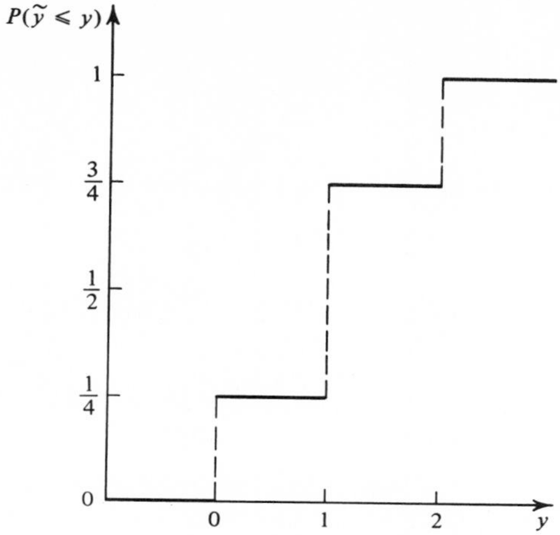

Figure 3.1.5

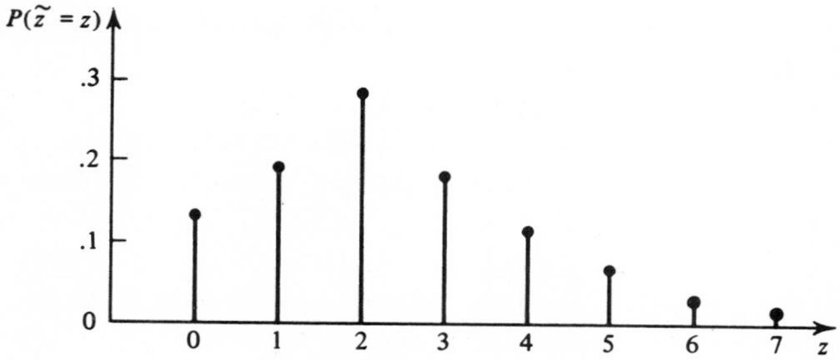

Figure 3.1.6

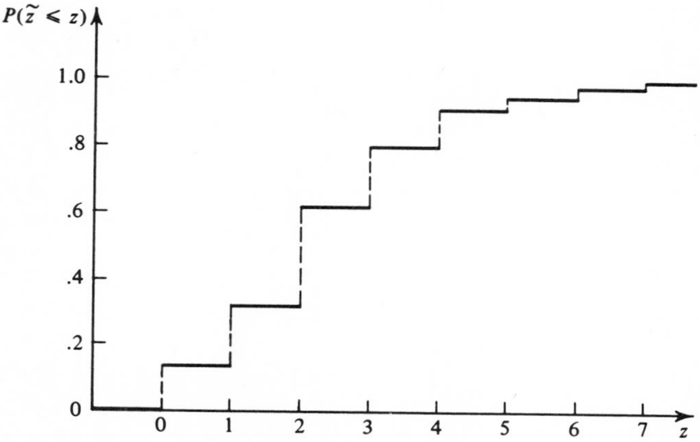

Figure 3.1.7

of a "fair" die,

$$\begin{aligned} E(\tilde{x}) &= 1(\tfrac{1}{6}) + 2(\tfrac{1}{6}) + 3(\tfrac{1}{6}) + 4(\tfrac{1}{6}) + 5(\tfrac{1}{6}) + 6(\tfrac{1}{6}) \\ &= \tfrac{21}{6} = \tfrac{7}{2}. \end{aligned}$$

If $\tilde{y}$ represents the number of heads in two tosses of a "fair" coin,

$$E(\tilde{y}) = 0(\tfrac{1}{4}) + 1(\tfrac{1}{2}) + 2(\tfrac{1}{4}) = 1.$$

If $\tilde{z}$ represents the number of rainy days in Chicago next week, and the distribution of $\tilde{z}$ is as given above, then

$$\begin{aligned} E(\tilde{z}) = 0(.13) &+ 1(.19) + 2(.29) + 3(.18) + 4(.11) + 5(.06) \\ &+ 6(.03) + 7(.01) = 2.30. \end{aligned}$$

If $\tilde{t}$ is an indicator random variable that equals 1 if some event E occurs and 0 if E does not occur, then

$$E(\tilde{t}) = 1 \cdot P(E) + 0 \cdot P(\bar{E}) = P(E).$$

Thus, by using indicator random variables, probabilities can be interpreted in terms of expectations.

$E(\tilde{x})$ is also called the "*expected value* of $\tilde{x}$," or the "*mean* of $\tilde{x}$," and it can be thought of as an "average value." Note that the expectation of a random variable can be considered as a measure of location, or central tendency, for that random variable. That is, $E(\tilde{x})$ can be used as a summary measure of the probability distribution of $\tilde{x}$, representing an "average value" and thus roughly indicating the location of the "center" of the distribution. In terms of the PMF of a random variable, the expectation of the random variable can be interpreted as a center of gravity, as illustrated in Figure 3.1.8. Suppose that the horizontal axis is thought of as a thin board and that the vertical lines, or "masses," are thought of as weights proportional to the heights of the lines. If you move a fulcrum along (and below) the board, the point at which the board exactly balances corresponds to the expected value of the random variable.

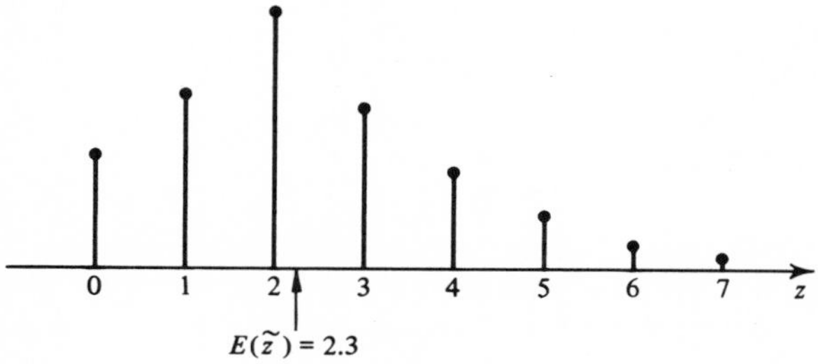

Figure 3.1.8

Recall that in Chapter 2 "expected payoffs" were used as decision-making criteria in several examples. Sometimes one is interested in the expectation of a *function* of a random variable $\tilde{x}$:

$$E[t(\tilde{x})] = \sum_{j=1}^{J} t(x_j)P(\tilde{x} = x_j), \tag{3.1.4}$$

where $t(\tilde{x})$ is some function of $\tilde{x}$. For example, suppose that you are offered the following bet: you win \$10 if heads does not occur in two tosses of a "fair" coin, and you lose \$5 if heads occurs at least once. Letting $\tilde{x}$ represent the number of heads, you win \$10 if $\tilde{x} = 0$ and you lose \$5 if $\tilde{x} = 1$ or $\tilde{x} = 2$. Thinking of your payoff as a *function* t of $\tilde{x}$, you could write

$$t(\tilde{x}) = \begin{cases} \$10 & \text{if } \tilde{x} = 0, \\ -\$5 & \text{if } \tilde{x} = 1 \text{ or } \tilde{x} = 2. \end{cases}$$

Thus,

$$E[t(\tilde{x})] = (\$10)(\tfrac{1}{4}) + (-\$5)(\tfrac{1}{2}) + (-\$5)(\tfrac{1}{4}) = -\$1.25.$$

Your expected payoff is −\$1.25, so the bet does not look advantageous. If the amounts in a wager are small enough that utility considerations can be ignored (this point is discussed in Chapter 5), it is said that a "fair" bet is a bet with an expected payoff of \$0. To illustrate this, consider a bet in which you will win \$5 if it rains tomorrow, and suppose that you feel that $P(\text{rain tomorrow}) = .6$. For the bet to be a "fair" bet, how much should you have to pay if it does not rain tomorrow? Your expected payoff is $(\$5)(.6) + (-L)(.4)$, where L is the amount you lose if it does not rain. For this to equal zero, L must be \$7.50. These illustrations demonstrate the fact that expectations are very important in decision theory.

In working with expectations, certain types of functions of random variables are encountered very frequently in problems of statistical inference and decision. Consequently, the following mathematical rules ("laws of expectation"), most of which are presented here without proof, are often quite useful:

1. For any constant c, $E(c) = c$.
2. For any constant c and any random variable $\tilde{x}$, $E(c\tilde{x}) = cE(\tilde{x})$.
3. For any constant c and any random variable $\tilde{x}$, $E(\tilde{x} + c) = E(\tilde{x}) + c$.
4. For any two random variables $\tilde{x}$ and $\tilde{y}$, $E(\tilde{x} + \tilde{y}) = E(\tilde{x}) + E(\tilde{y})$.
5. For any two random variables $\tilde{x}$ and $\tilde{y}$ and any two constants c and d,

$$E(c\tilde{x} + d\tilde{y}) = cE(\tilde{x}) + dE(\tilde{y}).$$

6. For any M random variables $\tilde{x}_1, \tilde{x}_2, \ldots, \tilde{x}_M$ and any M constants $c_1, c_2, \ldots, c_M$,

$$E\left(\sum_{i=1}^{M} c_i\tilde{x}_i\right) = \sum_{i=1}^{M} c_i E(\tilde{x}_i).$$

The proofs of these laws of expectation are not difficult. For instance,

$$E(c) = \sum_{j=1}^{J} cP(\tilde{x} = x_j) = c \sum_{j=1}^{J} P(\tilde{x} = x_j).$$

But, from Equation (3.1.2), $\sum_{j=1}^{J} P(\tilde{x} = x_j) = 1$, so $E(c) = c$. Also,

$$E(c\tilde{x}) = \sum_{j=1}^{J} cx_jP(\tilde{x} = x_j) = c \sum_{j=1}^{J} x_jP(\tilde{x} = x_j) = cE(\tilde{x}).$$

Once the first four rules are proved, the fifth rule follows easily. From the fourth rule,

$$E(c\tilde{x} + d\tilde{y}) = E(c\tilde{x}) + E(d\tilde{y}).$$

But from the second rule, $E(c\tilde{x}) = cE(\tilde{x})$ and $E(d\tilde{y}) = dE(\tilde{y})$. Thus,

$$E(c\tilde{x} + d\tilde{y}) = cE(\tilde{x}) + dE(\tilde{y}).$$

One expectation of particular interest is the *variance* of the random variable $\tilde{x}$, which is defined as the expectation of the squared deviation from the mean:

$$V(\tilde{x}) = E[\tilde{x} - E(\tilde{x})]^2 = E[(\tilde{x} - \mu)^2]. \qquad (3.1.5)$$

The variance of $\tilde{x}$ may be denoted by $V(\tilde{x})$, var $(\tilde{x})$, $\sigma_{\tilde{x}}^2$, or even just σ^2 (Greek sigma, squared) if it is understood that $\tilde{x}$ is the variable of interest. The variance is a measure of the dispersion, or variability, of a random variable. The more "spread out" the probability distribution, the larger the variance. In Figure 3.1.9, $\tilde{x}$ has a larger variance than $\tilde{y}$, although their means are equal. If you are acquainted with elementary physics, you might note that just as $E(\tilde{x})$ can be interpreted as a center of gravity, $V(\tilde{x})$ can be interpreted as a moment of inertia. Furthermore, it gives the minimum squared error in the sense that $E(\tilde{x} - c)^2$ is smallest when $c = E(\tilde{x})$. A final note regarding variance is that because the variance is expressed in squared units, it is sometimes more convenient to work with its square root, which is called the *standard deviation*. The standard deviation of a random variable is expressed in the same units as the random variable itself.

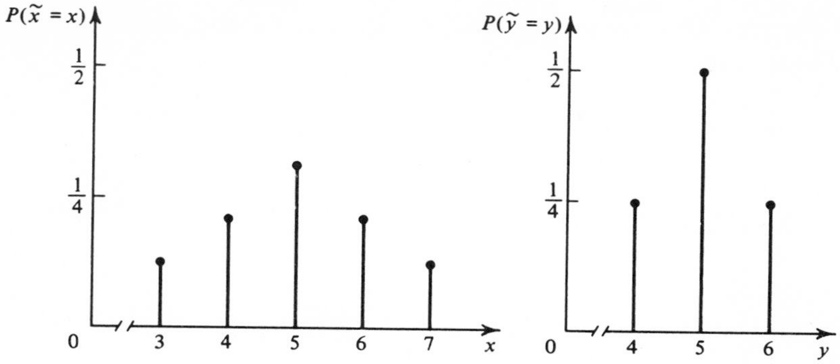

Figure 3.1.9

If $\tilde{x}$ is the number that comes up on a single toss of a "fair" die, $E(\tilde{x}) = \frac{7}{2} = 3.5$, and

$$V(\tilde{x}) = E(\tilde{x} - 3.5)^2 = (1 - 3.5)^2(\tfrac{1}{6}) + (2 - 3.5)^2(\tfrac{1}{6}) + (3 - 3.5)^2(\tfrac{1}{6}) + (4 - 3.5)^2(\tfrac{1}{6}) + (5 - 3.5)^2(\tfrac{1}{6}) + (6 - 3.5)^2(\tfrac{1}{6}) = 35/12.$$

The standard deviation of $\tilde{x}$ is $\sqrt{35/12}$. If $\tilde{y}$ is the number of heads on two tosses of a "fair" coin, $E(\tilde{y}) = 1$, and

$$V(\tilde{y}) = E(\tilde{y} - 1)^2 = (0 - 1)^2(\tfrac{1}{4}) + (1 - 1)^2(\tfrac{1}{2}) + (2 - 1)^2(\tfrac{1}{4}) = \tfrac{1}{2}.$$

The standard deviation of $\tilde{y}$ is $\sqrt{\frac{1}{2}}$. If z is the number of rainy days in Chicago next week, with the probability distribution given earlier in this section, then $E(\tilde{z}) = 2.3$, and

$$V(\tilde{z}) = E(\tilde{z} - 2.3)^2 = (0 - 2.3)^2(.13) + (1 - 2.3)^2(.19) + (2 - 2.3)^2(.29) + (3 - 2.3)^2(.18) + (4 - 2.3)^2(.11) + (5 - 2.3)^2(.06) - (6 - 2.3)^2(.03) + (7 - 2.3)^2(.01) = 2.51.$$

The standard deviation of $\tilde{z}$ is $\sqrt{2.51} = 1.58$. For the bet in which you win \$10 if no heads occur and you lose \$5 otherwise,

$$V[t(\tilde{x})] = E\{t(\tilde{x}) - E[t(\tilde{x})]\}^2 = E[t(\tilde{x}) - (-\$1.25)]^2 = (\$11.25)^2(\tfrac{1}{4}) + (-\$3.75)^2(\tfrac{1}{2}) + (-\$3.75)^2(\tfrac{1}{4}) = 42.1875.$$

The standard deviation of your payoff is $\sqrt{42.1875} = \$6.50$.

At this point, a few more useful rules for dealing with expectations (in particular, with variances, which are a special type of expectation) can be presented.

7. For any random variable $\tilde{x}$, $V(\tilde{x}) = E(\tilde{x}^2) - [E(\tilde{x})]^2$.
8. For any random variable $\tilde{x}$ and any constant c, $V(c\tilde{x}) = c^2V(\tilde{x})$.
9. For any constant c, $V(c) = 0$.
10. For any random variable $\tilde{x}$ and any constant c, $V(\tilde{x} + c) = V(\tilde{x})$.

Note that the mean and the variance (or standard deviation) are summary measures of a probability distribution. The mean is a measure of location, or central tendency, and the variance (standard deviation) is a measure of dispersion, or variability. The mean and the variance are not the only summary measures of a probability distribution. For example, two measures of location (other than the mean) are the mode (the most likely value of the random variable) and the median, which is discussed in Chapter 4. There are many other summary measures (for instance, fractiles of a probability distribution, which are discussed in Chapter 4), but the mean and the variance are often the most useful and mathematically the most convenient summary measures with which to work.

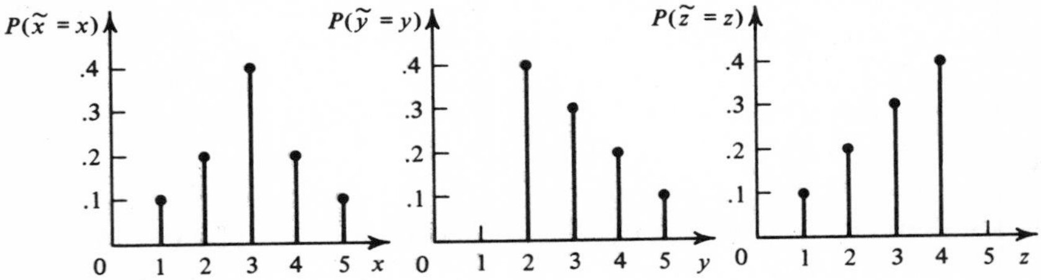

Figure 3.1.10

The mean and the variance of a random variable do not, of course, completely describe the probability distribution of the random variable. In Figure 3.1.10, for instance, the three random variables $\tilde{x}$, $\tilde{y}$, and $\tilde{z}$ have the same mean and similar variances $[V(\tilde{x}) = 1.2,\ V(\tilde{y}) = V(\tilde{z}) = 1.0]$, but their probability distributions are not identical. The distribution of $\tilde{x}$ is symmetric about the mean, whereas the distributions of $\tilde{y}$ and $\tilde{z}$ are positively skewed and negatively skewed, respectively. If a distribution is asymmetric with the "long tail" of the distribution to the right, it is said to be positively skewed; if the long tail is to the left, it is said to be negatively skewed. Measures of skewness and of characteristics of probability distributions other than location and dispersion have been developed, but such measures are not discussed here. Many such measures are based on the *moments* of a random variable; the kth moment of $\tilde{x}$

about the origin is $E(\tilde{x}^k)$, and the kth moment of $\tilde{x}$ about the mean is $E[\tilde{x} - E(\tilde{x})]^k$.

To investigate the position of a particular value of $\tilde{x}$ relative to other values of $\tilde{x}$, one can consider the corresponding *standardized* value,

$$\tilde{z} = \frac{\tilde{x} - E(\tilde{x})}{\sqrt{V(\tilde{x})}} = \frac{\tilde{x} - \mu}{\sigma}. \tag{3.1.6}$$

It is possible to transform any random variable $\tilde{x}$ into a *standardized random variable* $\tilde{z} = [\tilde{x} - E(\tilde{x})]/\sqrt{V(\tilde{x})}$ by subtracting the mean of $\tilde{x}$ from each value and dividing the result by the standard deviation. If $\tilde{x}$ represents the number of heads in two tosses of a "fair" coin, $E(\tilde{x}) = 1$ and $\sqrt{V(\tilde{x})} = .707$. The corresponding standardized random variable $\tilde{z}$ takes on the values $(0 - 1)/.707 = -1.414$, $(1 - 1)/.707 = 0$, and $(2 - 1)/.707 = 1.414$ with probabilities $\frac{1}{4}$, $\frac{1}{2}$, and $\frac{1}{4}$, respectively. Furthermore, $E(\tilde{z}) = 0$ and $V(\tilde{z}) = 1$; the mean and the variance of a standardized random variable are always 0 and 1, respectively.

Just as joint probabilities were considered in the previous chapter, it is possible to consider *joint probability distributions* of more than one random variable. If $\tilde{x}$ takes on the values $x_1, x_2, \ldots, x_J$, and $\tilde{y}$ takes on the values $y_1, y_2, \ldots, y_K$, then the joint probability distribution of $\tilde{x}$ and $\tilde{y}$ is represented by pairs of values (x_j, y_k) and their probabilities $P(\tilde{x} = x_j, \tilde{y} = y_k)$ for $j = 1, 2, \ldots, J$ and $k = 1, 2, \ldots, K$. $P(\tilde{x} = x_j, \tilde{y} = y_k)$ is the probability that $\tilde{x} = x_j$ *and* $\tilde{y} = y_k$. To obey the basic axioms of probability, the probabilities comprising the joint distribution must be nonnegative and must sum to 1:

$$P(\tilde{x} = x_j, \tilde{y} = y_k) \geq 0 \quad \text{for all } j = 1, \ldots, J \text{ and } k = 1, \ldots, K,$$

and

$$\sum_{j=1}^{J} \sum_{k=1}^{K} P(\tilde{x} = x_j, \tilde{y} = y_k) = 1.$$

For a simple example of a joint probability distribution, let $\tilde{y} = 1$ if it rains today and $\tilde{y} = 0$ if it does not rain today, and let $\tilde{x} = 1$ if it rains tomorrow and $\tilde{x} = 0$ if it does not rain tomorrow. Suppose that you feel that the joint probabilities are

$$P(\tilde{x} = 1, \tilde{y} = 1) = .20,$$
$$P(\tilde{x} = 1, \tilde{y} = 0) = .08,$$
$$P(\tilde{x} = 0, \tilde{y} = 1) = .02,$$

and

$$P(\tilde{x} = 0, \tilde{y} = 0) = .70.$$

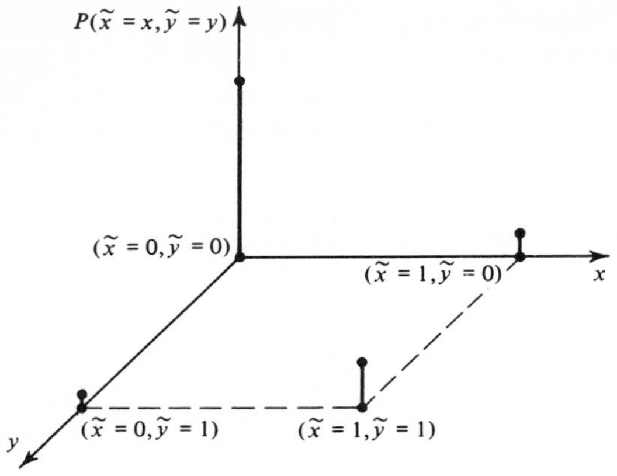

Figure 3.1.11

These probabilities can be represented on a three-dimensional graph, as in Figure 3.1.11. A more convenient way to represent them is in terms of a table, as follows.

		$\tilde{y}$ = 1	$\tilde{y}$ = 0
$\tilde{x}$	1	.20	.08
	0	.02	.70

The entries within the table are the joint probabilities, and they sum to 1.

Given a joint probability distribution of $\tilde{x}$ and $\tilde{y}$, it is possible to find the distributions of $\tilde{x}$ and $\tilde{y}$ individually. These distributions, which may be found by summing the joint distribution over one of the variables, are often referred to as the *marginal distributions* of $\tilde{x}$ and $\tilde{y}$. To find the marginal probabilities of x_j and y_k, respectively, sum as follows:

$$P(\tilde{x} = x_j) = \sum_{k=1}^{K} P(\tilde{x} = x_j, \tilde{y} = y_k) \tag{3.1.7}$$

and

$$P(\tilde{y} = y_k) = \sum_{j=1}^{J} P(\tilde{x} = x_j, \tilde{y} = y_k). \tag{3.1.8}$$

In terms of a table, the marginal probabilities are found on the margins of the table. For the rain example, the marginal probabilities for $\tilde{x}$ are

$$P(\tilde{x} = 1) = P(\tilde{x} = 1, \tilde{y} = 1) + P(\tilde{x} = 1, \tilde{y} = 0) = .28$$

and

$$P(\tilde{x} = 0) = P(\tilde{x} = 0, \tilde{y} = 1) + P(\tilde{x} = 0, \tilde{y} = 0) = .72.$$

The marginal probabilities for $\tilde{y}$ are

$$P(\tilde{y} = 1) = P(\tilde{x} = 1, \tilde{y} = 1) + P(\tilde{x} = 0, \tilde{y} = 1) = .22$$

and

$$P(\tilde{y} = 0) = P(\tilde{x} = 1, \tilde{y} = 0) + P(\tilde{x} = 0, \tilde{y} = 0) = .78.$$

The completed table is as follows.

		$\tilde{y}$		
		1	0	
$\tilde{x}$	1	.20	.08	.28
	0	.02	.70	.72
		.22	.78	

As Section 2.6 indicated, all probabilities can be thought of as conditional probabilities, conditional upon certain assumptions, certain information, and so on. Consequently, the *conditional probability distribution* of one random variable $\tilde{x}$, given a value of a second random variable $\tilde{y}$, often is of interest. The *conditional distribution of* $\tilde{x}$ *given that* $\tilde{y} = y_k$ is simply the set of values of $\tilde{x}$ together with the conditional probabilities $P(\tilde{x} = x_1 \mid \tilde{y} = y_k)$, $P(\tilde{x} = x_2 \mid \tilde{y} = y_k)$, $\ldots$, $P(\tilde{x} = x_J \mid \tilde{y} = y_k)$, where

$$P(\tilde{x} = x_j \mid \tilde{y} = y_k) = \frac{P(\tilde{x} = x_j, \tilde{y} = y_k)}{P(\tilde{y} = y_k)}. \qquad (3.1.9)$$

(Of course, only values of $\tilde{y}$ for which $P(\tilde{y} = y_k) > 0$ are of interest.) Note that there are K different conditional distributions of $\tilde{x}$ given $\tilde{y}$, corresponding to the K values of $\tilde{y}$.

For the rain example, the conditional distribution of $\tilde{x}$ given that $\tilde{y} = 1$ can be represented by

$$P(\tilde{x} = 1 \mid \tilde{y} = 1) = \frac{P(\tilde{x} = 1, \tilde{y} = 1)}{P(\tilde{y} = 1)} = \frac{.20}{.22} = \frac{10}{11}$$

and

$$P(\tilde{x} = 0 \mid \tilde{y} = 1) = \frac{P(\tilde{x} = 0, \tilde{y} = 1)}{P(\tilde{y} = 1)} = \frac{.02}{.22} = \frac{1}{11}.$$

The conditional distribution of $\tilde{x}$ given that $\tilde{y} = 0$ can be represented by

$$P(\tilde{x} = 1 \mid \tilde{y} = 0) = \frac{P(\tilde{x} = 1, \tilde{y} = 0)}{P(\tilde{y} = 0)} = \frac{.08}{.78} = \frac{4}{39}$$

and

$$P(\tilde{x} = 0 \mid \tilde{y} = 0) = \frac{P(\tilde{x} = 0, \tilde{y} = 0)}{P(\tilde{y} = 0)} = \frac{.70}{.78} = \frac{35}{39}.$$

If it rains today ($\tilde{y} = 1$), the probability of rain tomorrow is 10/11, or approximately .91, whereas if it does not rain today, this probability is 4/39, or approximately .10. If you do not yet know if $\tilde{y} = 1$ or $\tilde{y} = 0$, then the *marginal* distribution of $\tilde{x}$ is relevant, in which case the probability of rain tomorrow is $P(\tilde{x} = 1) = .28$.

As you shall see in the next section, the application of Bayes' theorem to discrete probability models results in a conditional distribution, where $\tilde{y}$ represents the new information and $\tilde{x}$ is the variable of interest. Expectations can be calculated from conditional distributions in the usual manner. The *conditional expectation* of $\tilde{x}$, given that $\tilde{y} = y$, is

$$E(\tilde{x} \mid \tilde{y} = y) = \sum_{j=1}^{J} x_j P(\tilde{x} = x_j \mid \tilde{y} = y), \tag{3.1.10}$$

which may be denoted by $E(\tilde{x} \mid y)$ or $E_{x|y}(\tilde{x})$. For the rain example,

$$\begin{aligned} E(\tilde{x} \mid \tilde{y} = 1) &= 1\,P(\tilde{x} = 1 \mid \tilde{y} = 1) + 0\,P(\tilde{x} = 0 \mid \tilde{y} = 1) \\ &= 1(10/11) + 0(1/11) = 10/11 \end{aligned}$$

and

$$\begin{aligned} E(\tilde{x} \mid \tilde{y} = 0) &= 1\,P(\tilde{x} = 1 \mid \tilde{y} = 0) + 0\,P(\tilde{x} = 0 \mid \tilde{y} = 0) \\ &= 1(4/39) + 0(35/39) = 4/39. \end{aligned}$$

The *conditional variance* of $\tilde{x}$, given that $\tilde{y} = y$, is

$$V(\tilde{x} \mid \tilde{y} = y) = \sum_{j=1}^{J} [x_j - E(\tilde{x} \mid \tilde{y} = y)]^2\, P(\tilde{x} = x_j \mid \tilde{y} = y), \tag{3.1.11}$$

which can also be expressed in the form

$$V(\tilde{x} \mid \tilde{y} = y) = E(\tilde{x}^2 \mid \tilde{y} = y) - [E(\tilde{x} \mid \tilde{y} = y)]^2.$$

For the rain example,

$$V(\tilde{x} \mid \tilde{y} = 1) = \left(1 - \frac{10}{11}\right)^2 \left(\frac{10}{11}\right) + \left(0 - \frac{10}{11}\right)^2 \left(\frac{1}{11}\right) = \frac{10}{121}$$

and

$$V(\tilde{x} \mid \tilde{y} = 0) = \left(1 - \frac{4}{39}\right)^2 \left(\frac{4}{39}\right) + \left(0 - \frac{4}{39}\right)^2 \left(\frac{35}{39}\right) = \frac{140}{1521}.$$

Thus, if it rains today, the conditional mean and the conditional variance of $\tilde{x}$ are 10/11 and 10/121. If it does not rain today, the conditional mean and the conditional variance of $\tilde{x}$ are 4/39 and 140/1521.

In Section 2.6, the relationship among events was briefly discussed. Often one is interested in the relationship between two random variables. In particular, the random variables $\tilde{x}$ and $\tilde{y}$ are said to be *independent* if, for all j and k, $P(\tilde{x} = x_j, \tilde{y} = y_k) = P(\tilde{x} = x_j)P(\tilde{y} = y_k)$. That is, $\tilde{x}$ and $\tilde{y}$ are independent if their joint probabilities equal the products of their marginal probabilities. From this definition of independence and from Equation (3.1.9), $\tilde{x}$ and $\tilde{y}$ are independent if

$$P(\tilde{x} = x_j \mid \tilde{y} = y_k) = P(\tilde{x} = x_j) \quad \text{for all } j \text{ and } k \tag{3.1.12}$$

or if

$$P(\tilde{y} = y_k \mid \tilde{x} = x_j) = P(\tilde{y} = y_k) \quad \text{for all } j \text{ and } k. \tag{3.1.13}$$

Two variables are independent if the conditional distribution of either variable, given a value of the other, is identical to the marginal distribution of the first variable. That is, knowledge about one of the variables does not affect the probabilities for the other variable. In the example concerning rain today and rain tomorrow, $\tilde{x}$ and $\tilde{y}$ clearly are *not* independent. It should be noted that just as in Section 2.6 (in terms of events), the concepts of *conditional independence* ($\tilde{x}$ and $\tilde{y}$ independent conditional on some event E, but not independent if E does not occur) and *mutual independence* ($\tilde{x}_1, \tilde{x}_2, \ldots, \tilde{x}_K$ independent) can be developed in terms of random variables.

The concept of independence is very important and useful in statistical inference and decision, as mentioned in Section 2.6. Utilizing this concept, a few more rules for dealing with expectations can be presented.

11. If $\tilde{x}$ and $\tilde{y}$ are independent random variables, $E(\tilde{x}\tilde{y}) = E(\tilde{x})E(\tilde{y})$.
12. If $\tilde{x}$ and $\tilde{y}$ are independent random variables,

$$V(\tilde{x} + \tilde{y}) = V(\tilde{x}) + V(\tilde{y}).$$

13. If $\tilde{x}$ and $\tilde{y}$ are independent random variables and if c and d are any constants, $V(c\tilde{x} + d\tilde{y}) = c^2V(\tilde{x}) + d^2V(\tilde{y})$.
14. If $\tilde{x}_1, \tilde{x}_2, \ldots, \tilde{x}_M$ are independent random variables and if $c_1, c_2, \ldots, c_M$ are any constants,

$$V\left(\sum_{i=1}^{M} c_i\tilde{x}_i\right) = \sum_{i=1}^{M} c_i^2 V(\tilde{x}_i).$$

It should be noted that rules 1 through 14 also apply to *conditional* expectations and variances (with conditional independence substituted for independence).

This section was not intended to be either detailed or comprehensive; its purpose was to provide a brief discussion of some of the important concepts involving discrete probability models, concepts that are used in the remainder of this chapter and elsewhere in the book. References are provided at the end of the chapter for the reader interested in more detailed discussions of these concepts.

3.2 BAYES' THEOREM FOR DISCRETE PROBABILITY MODELS

In Section 2.7, *Bayes' theorem* was discussed in terms of events. It is also possible to interpret Bayes' theorem in terms of the conditional distribution of a random variable, discrete or continuous. In this section Bayes' theorem is presented in terms of a discrete probability model.

In the form given by Equation (2.7.3), Bayes' theorem provides a convenient way to determine conditional probabilities in some situations. If the conditional probabilities $P(B \mid A_j)$ and the probabilities $P(A_j)$ are known, it is a simple matter to apply the formula to determine the conditional probabilities $P(A_j \mid B)$. Suppose, however, that the events A_j, $j = 1, \ldots, J$, represent the J possible values $x_1, x_2, \ldots, x_J$ of a discrete random variable $\tilde{x}$, and B represents a particular value y of a second random variable $\tilde{y}$. Then Bayes' theorem can be rewritten in the following form:

$$P(\tilde{x} = x_j \mid \tilde{y} = y) = \frac{P(\tilde{y} = y \mid \tilde{x} = x_j)P(\tilde{x} = x_j)}{\begin{array}{l} P(\tilde{y} = y \mid \tilde{x} = x_1)P(\tilde{x} = x_1) \\ \quad + P(\tilde{y} = y \mid \tilde{x} = x_2)P(\tilde{x} = x_2) + \cdots \\ \qquad + P(\tilde{y} = y \mid \tilde{x} = x_J)P(\tilde{x} = x_J) \end{array}},$$

or

$$P(\tilde{x} = x_j \mid \tilde{y} = y) = \frac{P(\tilde{y} = y \mid \tilde{x} = x_j)P(\tilde{x} = x_j)}{\sum_{i=1}^{J} P(\tilde{y} = y \mid \tilde{x} = x_i)P(\tilde{x} = x_i)}. \tag{3.2.1}$$

The set of probabilities $P(\tilde{x} = x_j \mid \tilde{y} = y)$ for $j = 1, \ldots, J$ corresponds to the conditional probability distribution of the random variable $\tilde{x}$, given that the random variable $\tilde{y}$ takes on the value y. Thus, Bayes' theorem can be used to determine the conditional distribution of a discrete random variable.

How, then, can Bayes' theorem be used to revise probabilities (or entire probability distributions) in the light of new information? A digression to discuss statistical terminology is necessary at this point. Statistical problems are generally formulated by defining some *population* of interest, such as the population of items produced by a particular manufacturing process, the population of licensed drivers in a particular state, or the population of students at a particular university. Alternatively, the statistician might envision a *process*, sometimes called a *stochastic process* or a *data-generating process*, which generates the items of interest. For instance, if he is interested in the incoming telephone calls at a switchboard, he can consider these calls to be "generated" by some underlying process. The items produced by a manufacturing process can also be thought of as being "generated" in this sense. Whether the statistician thinks of his problem in terms of a population or in terms of a process (the terms are essentially interchangeable), he is usually interested in certain summary measures, or *parameters*, of the population or process. Examples of such parameters are the proportion of defective items produced by a manufacturing process, the average weight and length of items produced by another manufacturing process, the proportion of drivers over 50 years old in a particular state, the mean and the variance of the grade-point averages of students at a particular university, and the mean rate at which incoming calls "arrive" at a switchboard.

In order to make inferential statements or decisions concerning the parameters of a given process or population, information is obtained from the process or population. Statisticians think of the process of obtaining new information as *sampling*, so the new information is generally referred to as *sample information*. In statistics, the term "sampling" often implies a carefully designed procedure for selecting items from a population or process. In this book, however, the terms "sampling" and "sample information" are used to indicate the process of obtaining new information and the new information itself, regardless of how the information is obtained.

Inferential statements in statistics, then, generally concern one or more unknown parameters of a given population or data-generating process. Since Greek letters are usually used to represent such parameters and since the uncertain quantity of interest in statistical inference and decision may often be thought of as a population (or process) parameter, it is convenient to replace $\tilde{x}$ with $\tilde{\theta}$ (Greek theta) and x_j with θ_j in Equation

(3.2.1), thus arriving at the following expression:

$$P(\tilde{\theta} = \theta_j \mid \tilde{y} = y) = \frac{P(\tilde{y} = y \mid \tilde{\theta} = \theta_j)P(\tilde{\theta} = \theta_j)}{\sum_{i=1}^{J} P(\tilde{y} = y \mid \tilde{\theta} = \theta_i)P(\tilde{\theta} = \theta_i)}. \tag{3.2.2}$$

When the uncertain quantities of interest are obvious, this is written in shorter form:

$$P(\theta_j \mid y) = \frac{P(y \mid \theta_j)P(\theta_j)}{\sum_{i=1}^{J} P(y \mid \theta_i)P(\theta_i)}. \tag{3.2.3}$$

This equation makes it possible to revise probabilities concerning an uncertain quantity $\tilde{\theta}$ on the basis of new sample information, which may be represented by the sample statistic $\tilde{y}$.

At this point some new terminology, which is related to the terminology briefly discussed in Section 2.7, should be introduced. Suppose that the random variable $\tilde{\theta}$ is of interest to you and that you would like to make some inferences about $\tilde{\theta}$ or to make some decision related to $\tilde{\theta}$. Suppose further that your information concerning $\tilde{\theta}$ is summarized by the set of values $\theta_1, \ldots, \theta_J$ and by the set of probabilities $P(\tilde{\theta} = \theta_j)$, $j = 1, \ldots, J$, which represent the *prior distribution* of $\tilde{\theta}$. You then observe a sample, and the outcome of the sample can be summarized by y, the observed value of a random variable $\tilde{y}$ (since the value y is computed from the sample, the random variable $\tilde{y}$ is called a *sample statistic*). The probabilities $P(\tilde{y} = y \mid \tilde{\theta} = \theta_j)$, called *likelihoods*, are determined from the conditional distributions of $\tilde{y}$, given different values $\theta_1, \ldots, \theta_J$ of $\tilde{\theta}$. Once a sample result, y, is observed, the likelihoods (as a function of $\tilde{\theta}$, with y fixed) can be taken to be any positive multiple of the conditional probabilities $P(\tilde{y} = y \mid \tilde{\theta} = \theta_j)$. After $\tilde{y}$ is known, the uncertainty in the problem involves just $\tilde{\theta}$, and the likelihoods will generally be different for different values of $\tilde{\theta}$. Using Bayes' theorem, the prior information (represented by the prior distribution) and the sample information (represented by the likelihoods) are combined to form the *posterior distribution* of $\tilde{\theta}$, the set of probabilities $P(\tilde{\theta} = \theta_j \mid \tilde{y} = y)$. The posterior distribution, then, summarizes your information concerning $\tilde{\theta}$, given the sample outcome y. The adjectives "prior" and "posterior" are relative terms relating to the observed sample. If after observing a particular sample and computing the posterior distribution of $\tilde{\theta}$, you decide to take another sample, the distribution just computed would now be the *prior distribution relative to the new sample.*

To illustrate Equation (3.2.2) and the above terminology, it might be helpful to restate the third example of Section 2.7 in terms of random variables. Consider an individual chosen at random from the city of interest, and let $\tilde{\theta} = 2$ if he has lung cancer, $\tilde{\theta} = 1$ if he has tuberculosis, and $\tilde{\theta} = 0$ if he has neither disease. From the information given in Section 2.7,

$$P(\tilde{\theta} = 0) = .97,$$
$$P(\tilde{\theta} = 1) = .01,$$
and
$$P(\tilde{\theta} = 2) = .02.$$

These probabilities constitute the *prior distribution* of $\tilde{\theta}$, the probability distribution of $\tilde{\theta}$ prior to obtaining any information about the individual other than the knowledge that he was randomly selected from the population of the city of interest.

In order to obtain more information about $\tilde{\theta}$, you take a chest X-ray of the individual. In statistical terminology, the result of the chest X-ray is *sample information*. This new information can be summarized by the *sample statistic* $\tilde{y}$, where $\tilde{y} = 1$ if the X-ray is "positive" (that is, if a "spot" shows up on the X-ray) and $\tilde{y} = 0$ if the X-ray is not "positive." Suppose that the X-ray turns out positive, so that $\tilde{y} = 1$. From Section 2.7,

$$P(\tilde{y} = 1 \mid \tilde{\theta} = 0) = .07,$$
$$P(\tilde{y} = 1 \mid \tilde{\theta} = 1) = .95,$$
and
$$P(\tilde{y} = 1 \mid \tilde{\theta} = 2) = .90.$$

These are the conditional probabilities of the *observed* value of the sample statistic, given the three different values of $\tilde{\theta}$. These conditional probabilities may be taken as the *likelihoods* in this problem. Note that $\tilde{y} = 1$ in each of the three probabilities, whereas the three different values of $\tilde{\theta}$ are all considered. In other words, the likelihood is a function of $\tilde{\theta}$, with $\tilde{y} = 1$ fixed. Since the X-ray was positive, $\tilde{y} = 0$ is totally irrelevant and thus of no interest to the statistician.

Applying Bayes' theorem to combine the prior information (as represented by the prior distribution) and the sample information (as represented by the likelihoods),

$$P(\tilde{\theta} = 0 \mid \tilde{y} = 1) = \frac{P(\tilde{y} = 1 \mid \tilde{\theta} = 0)P(\tilde{\theta} = 0)}{P(\tilde{y} = 1 \mid \tilde{\theta} = 0)P(\tilde{\theta} = 0) + P(\tilde{y} = 1 \mid \tilde{\theta} = 1)P(\tilde{\theta} = 1) + P(\tilde{y} = 1 \mid \tilde{\theta} = 2)P(\tilde{\theta} = 2)}$$

$$= \frac{(.07)(.97)}{(.07)(.97) + (.95)(.01) + (.90)(.02)} = .7117,$$

$$P(\tilde\theta = 1 \mid \tilde y = 1) = \frac{P(\tilde y = 1 \mid \tilde\theta = 1)P(\tilde\theta = 1)}{P(\tilde y = 1 \mid \tilde\theta = 0)P(\tilde\theta = 0) + P(\tilde y = 1 \mid \tilde\theta = 1)P(\tilde\theta = 1) + P(\tilde y = 1 \mid \tilde\theta = 2)P(\tilde\theta = 2)}$$

$$= \frac{(.95)(.01)}{(.07)(.97) + (.95)(.01) + (.90)(.02)} = .0996,$$

and

$$P(\tilde\theta = 2 \mid \tilde y = 1) = \frac{P(\tilde y = 1 \mid \tilde\theta = 2)P(\tilde\theta = 2)}{P(\tilde y = 1 \mid \tilde\theta = 0)P(\tilde\theta = 0) + P(\tilde y = 1 \mid \tilde\theta = 1)P(\tilde\theta = 1) + P(\tilde y = 1 \mid \tilde\theta = 2)P(\tilde\theta = 2)}$$

$$= \frac{(.90)(.02)}{(.07)(.97) + (.95)(.01) + (.90)(.02)} = .1887.$$

These probabilities constitute the *posterior distribution* of $\tilde\theta$, the probability distribution of $\tilde\theta$ after seeing the results of the chest X-ray. If some additional information concerning the individual could be obtained (for instance, the results of some medical test), then this posterior distribution (posterior to seeing the X-ray) would be the prior distribution (prior to seeing the additional information), and Bayes' theorem could be applied a second time.

To emphasize the point that the likelihoods can be taken to be any positive multiple of the conditional probabilities $P(\tilde y = 1 \mid \tilde\theta = 0)$, $P(\tilde y = 1 \mid \tilde\theta = 1)$, and $P(\tilde y = 1 \mid \tilde\theta = 2)$, suppose that the likelihoods are

$$100P(\tilde y = 1 \mid \tilde\theta = 0) = 7,$$
$$100P(\tilde y = 1 \mid \tilde\theta = 1) = 95,$$
and
$$100P(\tilde y = 1 \mid \tilde\theta = 2) = 90.$$

Using these likelihoods in Bayes' theorem,

$$P(\tilde\theta = 0 \mid \tilde y = 1) = \frac{7(.97)}{7(.97) + 95(.01) + 90(.02)} = .7117,$$

$$P(\tilde\theta = 1 \mid \tilde y = 1) = \frac{95(.01)}{7(.97) + 95(.01) + 90(.02)} = .0996,$$

and
$$P(\tilde\theta = 2 \mid \tilde y = 1) = \frac{90(.02)}{7(.97) + 95(.01) + 90(.02)} = .1887.$$

These results agree with the results obtained without multiplying the conditional probabilities $P(\tilde y = 1 \mid \tilde\theta = \theta_j)$ by 100 because the factor of 100 cancels out in the application of Bayes' theorem (it occurs in both the numerator and the denominator). The point is that only *ratios* of likelihoods, or relative values of likelihoods, are needed to apply Bayes' theorem. In general, unless it is convenient to do otherwise, the conditional probabilities $P(\tilde{y} = y \mid \tilde{\theta} = \theta_j)$ are taken as likelihoods in this book. You should keep in mind, however, that any constant multiple of these conditional probabilities could also be used, and this point is discussed in further detail in Chapter 4.

For another example, suppose that you have two urns, one containing 70 red balls and 30 green balls and the other containing 70 green balls and 30 red balls. You choose one of the two urns at random, but since the urns look alike, you do not know which urn has been chosen. Let $\tilde{\theta}$ represent the number of red balls in the chosen urn. Since the urn is selected at random, the two urns are equally likely, so $P(\tilde{\theta} = 70) = P(\tilde{\theta} = 30) = \frac{1}{2}$; these two probabilities constitute the prior distribution of the uncertain quantity $\tilde{\theta}$ (there are two possible values of $\tilde{\theta}$, $\theta_1 = 70$ and $\theta_2 = 30$). To help you identify the urn, you draw a single ball from the urn and observe its color. How does this sample information modify your probability distribution of $\tilde{\theta}$? Using the notation in Equation (3.2.2), let $\tilde{y}$ equal 1 if the ball drawn is red and let $\tilde{y}$ equal 2 if the ball is green. Suppose that the ball is red, so that $\tilde{y} = 1$. The likelihoods are proportional to $P(\tilde{y} = 1 \mid \tilde{\theta} = 70)$ and $P(\tilde{y} = 1 \mid \tilde{\theta} = 30)$. If $\tilde{\theta} = 70$, then the urn contains 70 red balls and 30 green balls, and thus the likelihood of drawing a red ball is 70/100, or 7/10, provided that you feel that the 100 balls were equally likely to be chosen. Similarly, if $\tilde{\theta} = 30$, the likelihood of drawing a red ball is 30/100, or 3/10. Thus, you know the prior probabilities and the likelihoods, so you can apply Bayes' theorem [in the form of Equation (3.2.2)] to determine your posterior distribution:

$$P(\tilde{\theta} = 70 \mid \tilde{y} = 1) = \frac{P(\tilde{y} = 1 \mid \tilde{\theta} = 70)P(\tilde{\theta} = 70)}{P(\tilde{y} = 1 \mid \tilde{\theta} = 70)P(\tilde{\theta} = 70) + P(\tilde{y} = 1 \mid \tilde{\theta} = 30)P(\tilde{\theta} = 30)}$$
$$= \frac{(7/10)(1/2)}{(7/10)(1/2) + (3/10)(1/2)} = \frac{7/20}{10/20} = \frac{7}{10}$$

and

$$P(\tilde{\theta} = 30 \mid \tilde{y} = 1) = \frac{P(\tilde{y} = 1 \mid \tilde{\theta} = 30)P(\tilde{\theta} = 30)}{P(\tilde{y} = 1 \mid \tilde{\theta} = 70)P(\tilde{\theta} = 70) + P(\tilde{y} = 1 \mid \tilde{\theta} = 30)P(\tilde{\theta} = 30)}$$
$$= \frac{(3/10)(1/2)}{(7/10)(1/2) + (3/10)(1/2)} = \frac{3/20}{10/20} = \frac{3}{10}.$$

These two posterior probabilities (posterior to observing the sample information consisting of one red ball in a single draw from the urn) constitute the posterior distribution of the uncertain quantity $\tilde{\theta}$. The two posterior probabilities must add up to 1 so that the posterior distribution is a proper probability distribution (a distribution is proper if the probabilities are nonnegative and sum to 1). Therefore, it is not really necessary to compute the second posterior probability from Equation (3.2.2); it could be obtained by simply subtracting the first probability from 1. It is useful, however, to calculate the second probability as a computational check.

3.3 DISCRETE PRIOR DISTRIBUTIONS: A BERNOULLI EXAMPLE

The example considered in this section comes from the area of quality control. Suppose that a production manager is concerned about the items produced by a certain manufacturing process. More specificially, he is concerned about the proportion of these items that are defective. From past experience with the process, he feels that the proportion defective, an uncertain quantity denoted by $\tilde{p}$, takes on four possible values: .01, .05, .10, and .25. These are the four possible values of the population parameter $\tilde{p}$, or, in decision-theoretic terminology, the four possible *states of nature*. The manager feels that one of these four "states" will actually occur, but he does not know which one. The four values of $\tilde{p}$ might be explained in terms of malfunctions in the production process. Either (1) there are no malfunctions, in which case $\tilde{p} = .01$ and the process is "in control"; (2) there is a "type x" malfunction, in which case $\tilde{p} = .05$; (3) there are *two* malfunctions, "type x" and "type y," in which case $\tilde{p} = .10$; or (4) there are *three* malfunctions, "type x," "type y," and "type z," in which case $\tilde{p} = .25$. This is obviously a simplification of the real situation. It is highly unlikely that $\tilde{p}$ would take on only the four given values and no intermediate values. It might be more realistic to think of $\tilde{p}$ as a continuous random variable rather than as a discrete random variable (and this is done in Chapter 4), but for the purposes of this example $\tilde{p}$ is considered to be discrete.

Although the production manager has never formally determined the proportion of defectives in any "batch" from the manufacturing process, he has observed the process and he has some information concerning $\tilde{p}$. Suppose that he feels that this information can be summarized in terms of the following degree-of-belief, or subjective, probabilities for the four possible values of $\tilde{p}$:

$$P(\tilde{p} = .01) = .60,$$
$$P(\tilde{p} = .05) = .30,$$
$$P(\tilde{p} = .10) = .08,$$
and
$$P(\tilde{p} = .25) = .02.$$

In words, the probability is .60 that 1 percent of the items are defective, the probability is .30 that 5 percent of the items are defective, and so on. For instance, he might arrive at these probabilities by considering various lotteries or bets concerning $\tilde{p}$. (The assessment of prior probabilities is considered in Section 3.5.) These four probabilities constitute the production manager's prior distribution of $\tilde{p}$.

In addition to this prior information, the production manager decides to obtain some sample information from the process. In doing so, he assumes that the process can be regarded as a *Bernoulli process*, with the assumptions of *stationarity* and *independence* appearing reasonable. A Bernoulli process is a series of trials, each of which has two possible outcomes, which can be labeled "success" and "failure." In this case the production manager observes a number of items from the process, and each item is either defective or not defective. Let "success" represent the outcome "defective," and let "failure" represent "nondefective." The assumptions of stationarity and independence for a Bernoulli process mean (1) that $\tilde{p}$, the probability that any given item is defective, remains constant for all items produced (stationarity of the process) and (2) that conditional upon $\tilde{p}$, the condition of any given item (defective or nondefective) is statistically independent of the condition of the other items from the process (independence of the process). The Bernoulli process is a data-generating process with a single parameter, $\tilde{p}$, the probability of "success" on a single trial. Given the assumptions of a Bernoulli process, specifying a value of $\tilde{p}$ amounts to completely specifying the process.

An alternative way to consider the Bernoulli process is in terms of *exchangeability* rather than stationarity and independence. Exchangeability means that different sequences of Bernoulli trials are considered to be exchangeable, or equally likely, if these sequences consist of the same number of successes and the same number of failures. That is, the different sequences are equally likely regardless of the patterns of successes and failures within the sequences, and this is true for sequences of any number of trials with any number of successes. Let D represent defective and N represent nondefective. If the production manager thinks that the manufacturing process can be regarded as a Bernoulli process, then the sequences (D,D,N,N,N), (D,N,D,N,N), (D,N,N,D,N), (D,N,N,N,D), (N,D,D,N,N), (N,D,N,D,N), (N,D,N,N,D), (N,N,D,D,N), (N,N,D,N,D), and (N,N,N,D,D), which are the 10 possible ways of obtaining two defective items and three nondefective items, are considered to be equally likely. Similarly, all sequences with, say, one defective and nine nondefectives must be equally likely, and so on for any combination of defectives and nondefectives.

It should be emphasized that whether the process is thought of in terms of stationarity and independence or in terms of exchangeability, the assumption that the process behaves like a Bernoulli process is a subjective assumption. Other sets of assumptions lead to different "processes," or "models," as you shall see in Sections 3.4 and 3.6. Given such assumptions, it is easy to express the likelihood function (representing new sample information from the process) in relatively simple form. Keep in mind, however, that this simplification, like most simplifications in problems of statistical inference and decision, depends on a subjective assumption.

A sample of five items is taken from the production process, and one of the five is found to be defective. How can this information be combined with the prior information? Under the assumptions the production manager has made, the probability distribution of $\tilde{r}$, the number of defectives in five trials, given any particular value of $\tilde{p}$, is a *binomial distribution*, which is of the following form:

$$P(\tilde{r} = r \mid n, p) = \binom{n}{r} p^r(1 - p)^{n-r}. \tag{3.3.1}$$

Here n is the number of trials, p is the probability of success at any trial, r is the number of defectives in the n trials, and $\binom{n}{r}$ is a combinatorial term that can be written in terms of factorials:

$$\binom{n}{r} = \frac{n!}{r!(n - r)!}. \tag{3.3.2}$$

[$n!$ is read "n factorial" and is equal to the product $n(n - 1)(n - 2) \cdots 2 \cdot 1$; the other factorials are computed in similar fashion.] For the binomial distribution, $E(\tilde{r} \mid n,p) = np$ and $V(\tilde{r} \mid n,p) = np(1 - p)$. Note that the mean and the variance of the binomial distribution are expressed in conditional form, since the binomial distribution is the *conditional* distribution of $\tilde{r}$, given n and p.

The development of the binomial distribution is relatively straightforward. The probability of "success" on any individual trial is p and the probability of "failure" is $(1 - p)$, so that by the independence of trials the probability of any sequence of r successes and $n - r$ failures is the product of the individual probabilities, $p^r(1 - p)^{n-r}$. Independence allows the multiplication of the probabilities for the individual trials to arrive at the joint probability for the entire sequence, as indicated in Section 3.1. Exactly r of these probabilities equal p (corresponding to the r successes) and the remaining $n - r$ probabilities equal $1 - p$ (corresponding to the $n - r$ failures), so the probability of the entire sequence is $p^r(1 - p)^{n-r}$. However, there are $\binom{n}{r}$ distinct sequences with r successes and $n - r$ failures, since the combinatorial term $\binom{n}{r}$

represents the number of distinct choices of r trials out of n (that is, the choices of the r trials on which the successes occur). Thus, the probability of r successes, regardless of the exact sequence, is $\binom{n}{r} p^r(1-p)^{n-r}$.

The mean and the variance of the binomial distribution are easy to derive. Let $\tilde{x}_i = 1$ if the result of the ith trial is a success, and let $\tilde{x}_i = 0$ if the result of the ith trial is a failure. This is an indicator random variable, with

$$E(\tilde{x}_i \mid p) = 1 \cdot p + 0 \cdot (1-p) = p,$$
$$E(\tilde{x}_i^2 \mid p) = 1^2 \cdot p + 0^2 \cdot (1-p) = p,$$

and

$$V(\tilde{x}_i \mid p) = E(\tilde{x}_i^2 \mid p) - [E(\tilde{x}_i \mid p)]^2 = p - p^2 = p(1-p).$$

But a series of n independent Bernoulli trials can be represented by n such indicator random variables $\tilde{x}_1, \tilde{x}_2, \ldots, \tilde{x}_n$. Furthermore, $\tilde{r} = \sum_{i=1}^{n} \tilde{x}_i$, since $\tilde{x}_i$ merely indicates if the ith trial is a success or a failure. Thus, using rules 6 and 14 from Section 3.1,

$$E(\tilde{r} \mid n,p) = E\left(\sum_{i=1}^{n} \tilde{x}_i \mid n,p\right) = \sum_{i=1}^{n} E(\tilde{x}_i \mid p) = \sum_{i=1}^{n} p = np$$

and

$$V(\tilde{r} \mid n,p) = V\left(\sum_{i=1}^{n} \tilde{x}_i \mid n,p\right) = \sum_{i=1}^{n} V(\tilde{x}_i \mid p) = \sum_{i=1}^{n} p(1-p)$$
$$= np(1-p).$$

For the production example, $\tilde{p}$ is the uncertain quantity the production manager is interested in, and $\tilde{r}$ and n represent the sample information. (Since n is known, it is not a random variable and hence is written without a tilde.) Hence, the likelihoods for the application of Bayes' theorem can be determined from the binomial distribution. A table giving binomial probabilities for various values of $\tilde{r}$, n, and $\tilde{p}$ is presented at the end of the book. For the production example, $n = 5$, $\tilde{r} = 1$, and the four possible values of $\tilde{p}$ are .01, .05, .10, and .25, so the likelihoods are as follows:

$$P(\tilde{r} = 1 \mid n = 5, \tilde{p} = .01) = \binom{5}{1}(.01)(.99)^4 = .0480,$$

$$P(\tilde{r} = 1 \mid n = 5, \tilde{p} = .05) = \binom{5}{1}(.05)(.95)^4 = .2036,$$

$$P(\tilde{r} = 1 \mid n = 5, \tilde{p} = .10) = \binom{5}{1}(.10)(.90)^4 = .3280,$$

and

$$P(\tilde{r} = 1 \mid n = 5, \tilde{p} = .25) = \binom{5}{1}(.25)(.75)^4 = .3955.$$

In this example, Bayes' theorem can be written in the form

$$P(p_j \mid y) = \frac{P(y \mid p_j)P(p_j)}{\sum_i P(y \mid p_i)P(p_i)},$$

where y represents the sample result, one defective in five trials. Notice that the numerator is the product of a likelihood and a prior probability and that the denominator is just the sum of all of the possible numerators. To determine the posterior distribution, it is convenient to set up the following table and apply Bayes' theorem in tabular form:

(1)	(2)	(3)	(4)	(5)
$\tilde{p}$	*Prior probability*	*Likelihood*	*(Prior probability) × (likelihood)*	*Posterior probability*
.01	.60	.0480	.02880	.232
.05	.30	.2036	.06108	.492
.10	.08	.3280	.02624	.212
.25	.02	.3955	.00791	.064
	1.00		.12403	1.000

The first column in the table lists the possible values of $\tilde{p}$ (that is, the values of $\tilde{p}$ having a nonzero prior probability). The second and third columns give the prior probabilities and the likelihoods, and the fourth column is the product of these two columns. Finally, the fifth column is obtained from the fourth column by dividing each individual entry by the sum of the four entries. This determines the posterior probabilities, since Bayes' theorem can be written as

$$\text{posterior probability} = \frac{\text{(prior probability)(likelihood)}}{\Sigma\text{(prior probability)(likelihood)}}.$$

This example illustrates some features of Bayes' theorem. For one thing, the prior probabilities must sum to one (since they form the prior distribution, which must be a proper probability distribution), whereas there is no such restriction on the sum of the likelihoods. The likelihoods used in the example come from different binomial distributions (one from the binomial distribution with $n = 5$ and $\tilde{p} = .01$, one from the binomial distribution with $n = 5$ and $\tilde{p} = .05$, and so on), so there is no restriction on their sum. Furthermore, since the likelihoods can be taken to be any positive multiple of the conditional probabilities, the sum of the likelihoods can be varied simply by using different multiplying factors.

The posterior probabilities, however, must sum to 1 (since they form the posterior distribution, which must be a proper probability distribution), and this is why it is necessary to divide by Σ(prior probability) (likelihood) in Bayes' theorem. The sum of the entries in column four of the above table is not equal to 1, but when each entry is divided by this sum, the resulting values (column five) do add to 1. This process of making the probabilities sum to 1 is sometimes referred to as "*normalizing*" the probabilities. Another way of describing this procedure is to note that the posterior probability of any value of $\tilde{p}$ is *proportional* to the product of the corresponding prior probability and likelihood. The constant of proportionality is simply $1/\Sigma$ (prior probability) (likelihood).

Another feature of Bayes' theorem that is illustrated here is that if the prior probability of any value of $\tilde{p}$ is 0, the posterior probability of that value will also be 0, since the posterior probability is a multiple of the prior probability (the multiplier being the likelihood divided by the normalizing sum). Since the prior probability of $\tilde{p} = .20$ is 0, for example, the posterior probability of $\tilde{p} = .20$ is 0. The moral is obvious: when determining prior probabilities for discrete probability models, make sure that any value that you consider to be even remotely possible has a nonzero prior probability. As noted above, this production example is highly simplified and unrealistic because values such as .02, .09, .20, and so on, are assigned prior probabilities of 0 (and thus designated as essentially "impossible" events). A related point is that if the likelihood corresponding to any value of $\tilde{p}$ is 0, the posterior probability of that value will be 0. In the production example, for instance, $P(\tilde{r} = 1 \mid n = 5, \tilde{p} = 0) = 0$ (it is impossible to observe any defective items if $\tilde{p} = 0$), so even if $\tilde{p} = 0$ had a nonzero prior probability, its posterior probability would be 0.

Returning to the production manager, suppose that he decides to take yet another sample in order to obtain more information about the production process. In particular, he takes another sample of size five and obtains two defectives. His prior probabilities are now the posterior probabilities calculated after the first sample, and the likelihoods are once again determined via the binomial distribution, with the following results.

$\tilde{p}$	*Prior probability*	*Likelihood*	*(Prior probability) × (likelihood)*	*Posterior probability*
.01	.232	.0010	.00023	.005
.05	.492	.0214	.01053	.244
.10	.212	.0729	.01545	.359
.25	.064	.2637	.01688	.392
	1.000		.04309	1.000

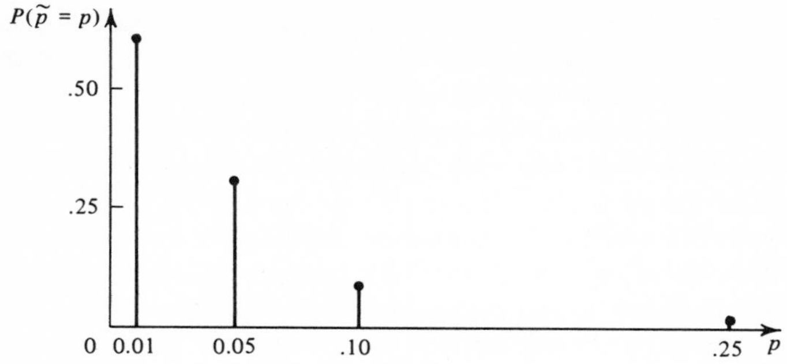

Figure 3.3.1

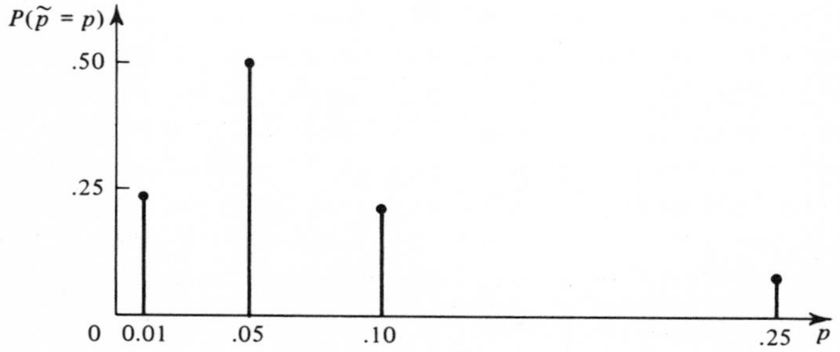

Figure 3.3.2

It is interesting to observe the changes in the probabilities as new sample information is obtained. The original prior distribution, which is presented in graphical form in Figure 3.3.1, indicates that the production manager thought that the low values of $\tilde{p}$ (.01 and .05) were more likely than the higher values. The mean, $E(\tilde{p})$, was .034. The first sample of size five, with 1 (that is, 20 percent) defective, shifts the probabilities so that .05 appears to be the most likely value of $\tilde{p}$ (.01 is the most likely value according to the production manager's original prior distribution). After the first sample, the mean is .064, as compared with the prior mean of .034 (the distribution is illustrated in Figure 3.3.2). The second sample of size five, with 2 (that is, 40 percent) defective, shifts the probabilities so that the highest possible value, .25, becomes the most likely, while the lowest value now has a probability of only .005, compared with the original prior probability of .60 assessed by the production manager. After the second sample, the mean of the posterior distribution is .146 (this posterior distribution is illustrated in Figure 3.3.3). In other words, the sample results (a total of three defectives in 10 items) change the probabilities considerably, making the high value of $\tilde{p}$, .25, much more likely than it seems under the original prior distribution. You should be careful not to interpret this as meaning that the sample will always change the probabilities so drastically. If the statistician had observed no defectives or one defective instead of three defectives in the sample of 10 items, the posterior probabilities would be much closer to the prior probabilities. You can verify this for yourself by performing the calculations.

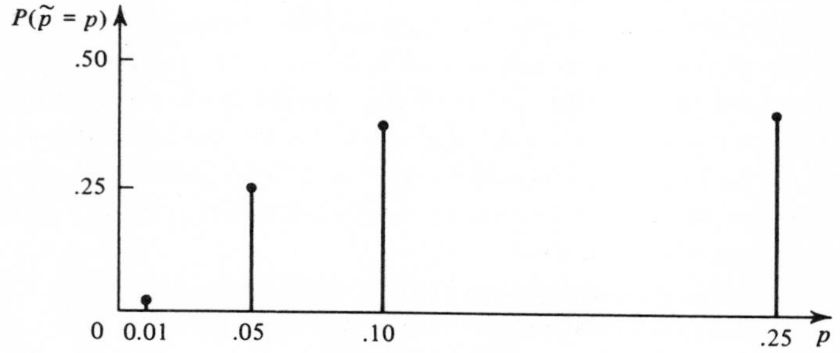

Figure 3.3.3

In this example, the production manager took two samples and revised his probabilities after each sample. A useful feature of Bayes' theorem is that he could have waited and revised his probabilities just once, using the combined results of the two samples (that is, three defectives in 10 trials). When applying Bayes' theorem for independent trials, the same result is obtained whether the calculations are made after each trial or just once at the end of all of the trials. Starting with the original prior probabilities and using all 10 trials to determine the likelihoods, the results are as follows.

$\tilde{p}$	*Prior probability*	*Likelihood*	*(Prior probability)* × *(likelihood)*	*Posterior probability*
.01	.60	.0001	.00006	.005
.05	.30	.0105	.00315	.245
.10	.08	.0574	.00459	.359
.25	.02	.2503	.00501	.391
	1.00		.01281	1.000

These probabilities are identical to the probabilities obtained from revising twice, once after each sample. The only differences (at the third decimal place) in the posterior probabilities actually presented in the two tables are due to rounding errors in the calculations. This feature of

Bayes' theorem is very important, since it reduces the number of applications of the theorem required to revise probabilities on the basis of several samples. Of course, the production manager may want to revise after each sample, primarily because the resulting probabilities may help him to decide whether or not to take another sample, as you shall see in Chapter 6.

Now that the posterior probabilities have been calculated, consider the following question: Why might the production manager be interested in $\tilde{p}$ in the first place? There are a number of explanations, but they all relate to the fact that a high value of $\tilde{p}$ could be quite costly to the firm. It could be quite costly to repair defective items, perhaps so costly that they are scrapped entirely. In addition, if any defective items are shipped to the firm's customers, the customers might be displeased and might take their business elsewhere.

Often the quality-control statistician (or his boss, the production manager) simply wants an estimate of $\tilde{p}$. He can determine such an estimate from the posterior distribution by taking some summary measure, such as the mean of the posterior distribution. In this manner, the statistician bases his estimate on all available information, as reflected in the posterior distribution.

It may be that the production manager must make a decision concering the process instead of just estimating $\tilde{p}$. Suppose that he has three possible actions: (1) he can leave the process as is; (2) he can have a mechanic make a minor adjustment in the process that will guarantee that $\tilde{p}$ will be no higher than .05 (specifically, if $\tilde{p}$ is .10 or .25, the adjustment reduces it to .05; if $\tilde{p}$ is .05 or .01, the adjustment has no effect); or (3) he can have the mechanic make a major adjustment in the process that will guarantee that $\tilde{p}$ will be .01. In the first paragraph of this section, the four possible values of $\tilde{p}$ were related to malfunctions in the production process. In terms of that explanation of the values of $\tilde{p}$, a major adjustment corrects *all* malfunctions while a minor adjustment corrects only "type y" and "type z" malfunctions.

Suppose that the production manager has just received an order for 1000 items and that he must choose one of the above three actions before the production run is started. There are certain costs that are relevant to his decision. First, the profit to the firm is \$0.50 per item sold, and the cost of replacing defectives is a fixed amount, \$2.00 per item. The other relevant costs are the costs of making adjustments in the process: a minor adjustment costs \$25, while a major adjustment costs \$100. From these figures, it is clear that the profit to be realized from the run of 1000 items is 1000 (\$0.50) = \$500. From this amount the cost of replacing defectives and the cost of any adjustments must be subtracted. If $\tilde{p}$ is .01, the expected number of defectives is .01(1000) = 10, and the cost of replacing these is \$2 per item, or a total of \$20. Similarly, for the other three possible values of $\tilde{p}$, the expected costs of replacing defectives are \$100, \$200, and \$500, respectively. From this, the production manager can determine a "payoff table," or "reward table," which shows the net profit to the firm for each combination of an action and a possible event. Since there are three actions and four events, the payoff table has a total of 12 entries. For example, if $\tilde{p} = .10$ and a minor adjustment is made, the adjustment changes $\tilde{p}$ to .05, so the expected cost of replacing defectives is \$100. The cost of the minor adjustment is \$25, so the payoff is \$500 − \$100 − \$25 = \$375.

		EVENT (VALUE OF $\tilde{p}$)			
		.01	.05	.10	.25
	Major adjustment	\$380	\$380	\$380	\$380
ACTION	*Minor adjustment*	\$455	\$375	\$375	\$375
	No adjustment	\$480	\$400	\$300	\$0

In Chapter 5, various criteria for making decisions are discussed. Suffice it to say at this point that a reasonable criterion is to choose the act for which the expected payoff or reward (expected net profit to the firm in this example) is largest. The expectations should be taken with respect to the production manager's posterior distribution of $\tilde{p}$, since this represents all of the information available to him concerning $\tilde{p}$, the state of nature. The posterior probabilities of the four states of nature are .005, .245, .359, and .391. The expected rewards (denoted by ER) are then

$$ER(\text{Major adjustment}) = \$380,$$
$$ER(\text{Minor adjustment}) = .005(\$455) + .995(\$375) = \$375.40,$$
and
$$ER(\text{No adjustment}) = .005(\$480) + .245(\$400) + .359(\$300) = \$208.10.$$

This indicates that the production manager should have the mechanic make a major adjustment. This is reasonable, since the probability of a high proportion of defectives (.10 or .25) is quite high. It is interesting to note that if the production manager had obtained no sample information, he would base his decision on the prior distribution, in which case no adjustment would be made. You can verify this by calculating the expected rewards with respect to the prior distribution of $\tilde{p}$.

This quality-control example illustrates the use of Bayes' theorem to revise probabilities on the basis of sample information and the use of the resulting posterior probabilities as inputs in a decision-making situation. It should be emphasized again that the example was purposely simplified and that a somewhat more realistic approach to the problem is discussed in the next chapter.

3.4 DISCRETE PRIOR DISTRIBUTIONS: A POISSON EXAMPLE

For an example involving a process other than the Bernoulli process, consider a problem involving the owner of an automobile dealership. The owner is concerned about the performance of his salesmen; in particular, he is concerned with their success or lack of success in selling cars. From past experience with the dealership, he feels that salesmen can be divided roughly into three categories, which he calls "great," "good," and "poor." A "great" salesman sells cars at a rate of one every other day. It should be noted that this does not mean that such a salesman will sell one car exactly every other day; a "great" salesman may have periods of several days with no sales or he may have several sales in a single day. On the average, however, he sells about one car every two days. A "good" salesman sells cars at a rate of one every fourth day, and a "poor" salesman sells cars at a rate of one every eighth day. The owner decides to express this information in terms of the uncertain quantity $\tilde{\lambda}$ (Greek lambda), which represents the average sales rate per day. Therefore, the three possible values of $\tilde{\lambda}$ that he is considering are $\frac{1}{2}$ (the "great" salesman), $\frac{1}{4}$ (the "good" salesman), and $\frac{1}{8}$ (the "poor" salesman). As in the example of the previous section, it is highly unlikely that $\tilde{\lambda}$ would take on only these three values. For some salesmen, $\tilde{\lambda}$ might be .4, .28, and so on. To allow for all of these intermediate values, it could be assumed that $\tilde{\lambda}$ is a continuous random variable rather than a discrete random variable, and this is considered in Chapter 4. In this section, however, assume that in order to simplify the model, the owner decides to consider only the three values $\frac{1}{2}$, $\frac{1}{4}$, and $\frac{1}{8}$.

The owner of the dealership has just hired a new salesman. Based on past experience with salesmen and on a personal interview with the new employee, the owner judges that the odds are 4 to 1 against his being a great salesman but that the odds are even that he could be a good salesman. The owner feels that such odds would be fair odds for bets concerning the performance of the salesman. This implies that the probability that he is a great salesman is $1/(1 + 4) = .2$ and the probability that he is a good salesman is $1/(1 + 1) = .5$. It follows that since the only other alternative is being a poor salesman, the probability of this alternative is .3. The owner's prior distribution for $\tilde{\lambda}$, then, may be represented by the following probabilities:

$$P(\tilde{\lambda} = \tfrac{1}{2}) = .2,$$
$$P(\tilde{\lambda} = \tfrac{1}{4}) = .5,$$
and
$$P(\tilde{\lambda} = \tfrac{1}{8}) = .3.$$

In the first 24 days with the dealership, the new salesman sells 10 cars. How can the owner use this new sample information to revise his probabilities concerning $\tilde{\lambda}$? He could judgmentally assess the conditional probability of this sample result given that $\tilde{\lambda} = \frac{1}{2}$, the conditional probability of the sample result given that $\tilde{\lambda} = \frac{1}{4}$, and the conditional probability of the sample result given that $\tilde{\lambda} = \frac{1}{8}$. These conditional probabilities can be taken as his likelihoods. If he is willing to make certain assumptions about the "process" of selling cars, he can use a mathematical model called a *Poisson* model and he can determine the likelihoods through this model.

A *Poisson process* is similar to a Bernoulli process with the exception that events occur over some continuum, such as time, rather than occurring on fixed "trials." For instance, consider the arrival of cars at a toll booth or the arrival of incoming calls at a telephone switchboard. One could approximate such processes by using the Bernoulli model. For example, each minute could be considered a "trial." This creates difficulties, however, because it is possible for more than one car or incoming call to arrive in any given minute, and the Bernoulli model does not allow for multiple occurrences on any single trial. In a sense, a Poisson process can be thought of as a limiting form of a Bernoulli process as the time interval corresponding to a single trial is made shorter and shorter, approaching a length of 0. As in the Bernoulli process, the assumptions of stationarity and independence are required for the Poisson process. In the Poisson model, stationarity means that $\tilde{\lambda}$, which is called the *intensity* of the process per unit of time, remains constant over time; the Poisson process has one parameter, $\tilde{\lambda}$. Independence means that conditional upon $\tilde{\lambda}$, the number of occurrences, or events, in a particular period of time is independent of occurrences in other time periods. It is also possible to think of the Poisson process in terms of exchangeability. This essentially means that given any fixed length of time and any fixed number of occurrences, the various possible distributions of the occurrences over the fixed amount of time are regarded as roughly exchangeable.

It should be emphasized, as it was in the last section, that the assumption that a real-world process behaves like a certain model, such as a

Poisson process, is strictly a subjective assumption. Generally the assumptions of stationarity and independence are not met exactly by real-world situations, but they may be met approximately, so that the model is a good approximation of reality. In the example, sales might tend to be higher on certain days of the week than on other days, thus violating the stationarity assumption. However, suppose that the owner of the automobile dealership feels that the Poisson model is a reasonable approximation and that he is willing to determine his likelihoods by using this model. If the intensity of a Poisson process per unit of time is $\tilde{\lambda}$, then the probability distribution for $\tilde{r}$, the number of occurrences in t units of time, is a *Poisson distribution,* which is of the form

$$P(\tilde{r} = r \mid t, \lambda) = \frac{e^{-\lambda t}(\lambda t)^r}{r!}, \tag{3.4.1}$$

where $e = 2.71828 \ldots$ is the base of the Naperian or natural logarithm system. The mean and the variance of the Poisson distribution are $E(\tilde{r} \mid \lambda,t) = \lambda t$ and $V(\tilde{r} \mid \lambda,t) = \lambda t$. This distribution can be developed formally directly from the assumptions of a Poisson process or it can be developed as a limiting form of the binomial distribution. For those interested in such developments, references are given at the end of the chapter.

In the automobile-salesman example, $\tilde{\lambda}$ is the uncertain quantity of interest to the owner and $\tilde{r}$ represents the sample information. Hence, the likelihoods for the application of Bayes' theorem can be determined from the Poisson distribution. A table giving Poisson probabilities for various values of $\tilde{r}$ and $\tilde{\lambda}t$ is presented at the end of the book. In the example, the unit of time is a day ($\tilde{\lambda}$ is expressed in sales per day). The sample information consists of 10 sales in 24 days, so that $\tilde{r} = 10$, $t = 24$, and the three possible values of $\tilde{\lambda}$ are $\frac{1}{2}$, $\frac{1}{4}$, and $\frac{1}{8}$. The likelihoods are

$$P(\tilde{r} = 10 \mid t = 24, \tilde{\lambda} = \tfrac{1}{2}) = \frac{e^{-12}(12)^{10}}{10!} = .1048,$$

$$P(\tilde{r} = 10 \mid t = 24, \tilde{\lambda} = \tfrac{1}{4}) = \frac{e^{-6}(6)^{10}}{10!} = .0413,$$

and

$$P(\tilde{r} = 10 \mid t = 24, \tilde{\lambda} = \tfrac{1}{8}) = \frac{e^{-3}(3)^{10}}{10!} = .0008.$$

Bayes' theorem can be written in the form

$$P(\lambda_j \mid y) = \frac{P(y \mid \lambda_j)P(\lambda_j)}{\sum_i P(y \mid \lambda_i)P(\lambda_i)},$$

where y represents the sample result, 10 sales in 24 days. In tabular form, the posterior probabilities are determined as follows.

$\tilde{\lambda}$	*Prior probability*	*Likelihood*	*(Prior probability)* × *(likelihood)*	*Posterior probability*
$\frac{1}{2}$	.2	.1048	.02096	.501
$\frac{1}{4}$	.5	.0413	.02065	.493
$\frac{1}{8}$	.3	.0008	.00024	.006
	1.0		.04185	1.000

The prior probabilities for "great salesman," "good salesman," and "poor salesman" are .2, .5, and .3. After seeing the sample information, the revised, or posterior, probabilities for these three situations are .501, .493, and .006, respectively. Informally, this means that the new salesman sold enough cars in the 24-day period to virtually convince the owner that he is not a poor salesman and to greatly increase the odds, in the owner's estimation, in favor of his being a great salesman. Using the prior distribution, the odds are 4 to 1 *against* his being a great salesman; using the posterior distribution, the odds are approximately even. The prior and posterior expectations of $\tilde{\lambda}$ are .26 and .37. Prior to seeing the sample information, the owner would guess that, on the average, the salesman would sell about .26 cars per day, or about one car every 3.8 days. After the salesman sells 10 cars in 24 days, the owner feels that, on the average, the salesman will sell about .37 cars per day, or about one car every 2.7 days. The prior and the posterior distributions of $\tilde{\lambda}$ are illustrated in Figures 3.4.1 and 3.4.2, respectively.

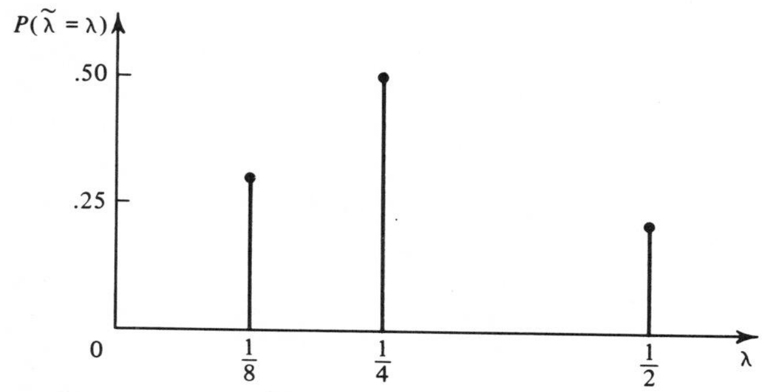

Figure 3.4.1

Note that the owner's best guess about $\tilde{\lambda}$ based on the sample information alone is the average number of sales per day during the 24 days, or $10/24 = .42$. The posterior mean (.37) is between the prior mean (.26) and the average during the sample (.42), which is called the sample mean. When Bayes' theorem is applied, the mean of the posterior distribution is generally between the mean of the prior distribution and the sample mean (a few special cases are exceptions to this rule, but they are relatively unimportant for practical purposes). This is an intuitively reasonable result, since the posterior distribution represents a combination of the prior information, as represented by the prior distribution, and the sample information, as represented by the likelihood function.

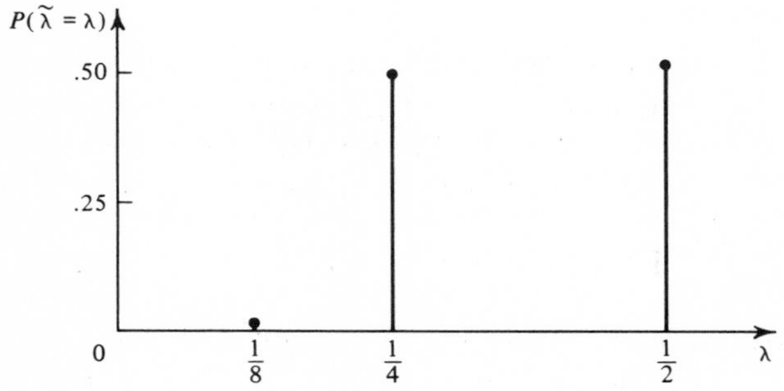

Figure 3.4.2

The owner of the automobile dealership could use the posterior distribution of the uncertain quantity, $\tilde{\lambda}$, in a formal decision-theoretic framework. Suppose that each salesman is paid \$10 per day plus a commission on each car that he sells. The owner has determined that his marginal profit per car (after subtracting commissions and similar marginal costs) is about \$50. Thus, the payoff to the owner per day that is attributable to a particular salesman who sells $\tilde{\lambda}$ cars per day is simply $\tilde{\lambda}(\$50) - \$10$. Given a distribution for $\tilde{\lambda}$, the expected payoff to the owner per day is $E(\tilde{\lambda})(\$50) - \$10$. Considering the posterior distribution determined above, the expected payoff per day due to the new salesman is $(.37)(\$50) - \$10$, or \$8.50. Note that if the expected value of $\tilde{\lambda}$ were $\frac{1}{5}$, the expected payoff per day would be 0. Thus, the owner certainly would like to avoid hiring any salesman who, in his opinion, will not average better than one car every five days. The "poor salesman" averages only one car every eight days, so he falls into this category. The "good salesman" averages one car every four days, so he results in a positive expected payoff to the owner; the "great salesman," of course, even surpasses that.

This example illustrates the use of Bayes' theorem to revise probabilities when the sample information is assumed to be generated by a Poisson process. It should be stressed that in general the Poisson process does not represent any real-world situation perfectly, although it may be a very close approximation. The use of the Poisson model requires the assumption that the model provides a satisfactory approximation of reality, and this assumption involves such considerations as the stationarity and independence of the process. If it is felt that the Poisson model is not realistic enough in any given situation, then the likelihoods should be determined by using some other model or they should be assessed directly by using such devices as the conditional lotteries, conditional bets, and so on, discussed in Section 2.7. The assessment of likelihoods is considered in Section 3.6.

3.5 THE ASSESSMENT OF PRIOR PROBABILITIES

An important problem is the determination of the prior distribution and the likelihoods, two of the primary inputs to a formal Bayesian analysis (when decision theory is discussed in Chapters 5 and 6, a third important input is included, involving payoffs or losses). First, the decision maker must decide which uncertain quantities, or random variables, are of interest to him, and then he must express his information about these random variables in probabilistic form. Once all of the inputs are determined, the application of Bayes' theorem is a simple arithmetic task. In this section, the assessment of prior probabilities is discussed; the assessment of likelihoods is considered in Section 3.6.

The situation in which there is just one uncertain quantity of interest is the easiest to handle. Often the choice of the set of possible values of the uncertain quantity is obvious, but in other cases it is a subjective choice. If there is a large number of possible values, the decision maker may want to simplify the analysis by considering fewer values. Sometimes a variable that could be treated as a continuous variable is instead assumed to be discrete. In the Bernoulli and Poisson examples in Sections 3.3 and 3.4, a discrete probability model was used as a convenient approximation. Clearly the random variables $\tilde{p}$ and $\tilde{\lambda}$ in these examples could take on more than just three or four values. Consideration of a large number of values does not make the analysis conceptually more difficult, but it does increase the computational burden (if the problem is to be handled on a computer, however, the computational burden becomes relatively unimportant).

In the discrete case, the prior distribution consists of a set of values of the uncertain quantity, together with their probabilities. The only restrictions are that the probabilities must be nonnegative and must sum to 1. Subject to these restrictions, how can the decision maker determine a prior distribution?

The prior probabilities should reflect the decision maker's prior information about the uncertain quantity in question. If this information is primarily in the form of sample results, then the prior probabilities should be close to the observed relative frequencies. In the automobile-salesman example in Section 3.4, suppose that the owner of the dealership knows virtually nothing about his new salesman. He does know, however, that in the past, 10 percent of his salesmen turned out to be "great," 40 percent turned out to be "good," and 50 percent turned out to be "poor." The owner has a high turnover of salesmen because he refuses to retain a salesman once he feels quite strongly that the salesman is "poor" [this could be expressed in terms of the statistical model by saying tht a salesman will be fired if, say, $P(\text{"poor"}) > .9$]. If the owner has no special information about a new salesman that would differentiate the salesman from previous employees, he might be willing to use the past observed relative frequencies (.1, .4, and .5) as his prior probabilities of "great salesman," "good salesman," and "poor salesman." For the new salesman considered in Section 3.4 the owner apparently had some information, perhaps from a personal interview, that indicated that the salesman in question appeared more likely to be successful than the past relative frequencies would indicate. Using both the past frequencies *and* the additional information from the interview, the owner assessed prior probabilities of .2, .5, and .3 for "great," "good," and "poor."

If there is little or no sample information, then the probabilities should be based on whatever other relevant information is available. The ultimate choice of a prior distribution by the statistician is a subjective choice, no matter what form the prior information takes, so it will be assumed that the prior probabilities are subjective probabilities. As noted in Chapter 2, a subjective probability reflects the degree of belief of a given person (the decision maker) about a given proposition (say, for example, the proposition that $\tilde{p} = .10$). But if probabilities are based on the judgments of a person, what is needed is a way for the person to quantify his judgments (that is, to express them in probabilistic terms).

Numerous techniques are available for the quantification of judgment; some of these were described in Sections 2.4 and 2.6. A person who understands the concept of probability might simply assign probabilities directly to the various possible values of the uncertain quantity of interest. In the example in Section 3.3, the production manager might have said, "I think the probability is .60 that the proportion of defectives produced by this production process is .01." Alternatively, the production manager might have chosen to think in terms of betting odds rather than probabilities: "I think the odds are 3 to 2 in favor of the proportion defective being .01." In Section 2.4 the relationship between probabilities and betting odds was discussed; if the odds in favor of an event are a to b, then the probability of that event is equal to $a/(a + b)$. In the example, $a = 3$ and $b = 2$, so the required probability is $3/(3 + 2)$, or .60. Another possibility is the consideration of lotteries, as discussed in Section 2.4. For instance, if the production manager feels that $P(\tilde{p} = .01) = .60$, then presumably he would be indifferent between receiving (1) a lottery ticket that would pay him \$1 if a ball drawn at random from an urn with 60 red balls and 40 blue balls turns out to be red, and (2) a lottery ticket that would pay him \$1 if the proportion of defectives from the production process turns out to be .01. Alternatively, he should be willing to pay up to \$.60 as an insurance premium against a loss of \$1.00 that is contingent upon $p = .01$, since his expected payoff is $(-\$1.00)(.60) = -\$.60$. There are numerous such devices (lotteries, betting odds, reference contracts, insurance premiums, and so on) to aid the decision maker in expressing his judgments in probabilistic form.

Of course, there is an element of vagueness in the determination of subjective prior probabilities. You may feel certain that the probability of rain tomorrow is between .10 and .30, for instance, but you find it difficult to pick a single number to represent your probability. In situations in which vagueness is present, devices such as those discussed above may prove useful. Would you bet for or against rain tomorrow at odds of 4 to 1 in favor of "no rain"? If you would bet for rain at these odds, then your probability of rain must be greater than .20. If you would prefer a lottery ticket paying \$5 if it rains tomorrow to a second lottery ticket paying \$5 if no heads are obtained in two tosses of a fair coin, then your probability of rain must be greater than .25, since the probability of obtaining no heads in two tosses of a fair coin is .25. Considering betting situations and lotteries of this nature helps to combat the problem of vagueness. If the problem at hand is quite important, then it should be well worth your while to take your time in carefully attempting to express your prior judgments in terms of probabilities.

It might be noted in passing that vagueness can be represented formally by using "second-order probabilities." That is, you can assess a probability distribution for the probability of interest. Unless the inferential or decision-making problem at hand is highly sensitive to slight variations in the prior probabilities, however, the consideration of second-order probabilities is an unnecessary and time-consuming procedure. Furthermore, the mean of the second-order probability distribution can usually be used to represent the first-order probability with no loss of precision. The questions of the sensitivity of statistical problems to changes in the inputs and of the possible use of a single summary measure such as the mean to represent an entire probability distribution are discussed in Chapters 4, 5, and 6.

When there are more than two possible events or more than two possible values of the uncertain quantity of interest, it is often convenient to determine a prior distribution by considering the *relative* chances of the various events or values of the uncertain quantity. If $\tilde{\theta}$ has three values, 1, 2, and 3, it may be difficult to assess probabilities directly for these values but relatively easy to compare the values. Suppose that you are interested in tomorrow's *change* in price of a given common stock; let $\tilde{\theta} = 1$ correspond to the event that the change is positive (the stock price goes up), let $\tilde{\theta} = 2$ correspond to the event that there is no change, and let $\tilde{\theta} = 3$ correspond to the event that the change is negative. You feel that the stock price is twice as likely to go up as it is to go down and that it is three times as likely to go down as it is to remain the same. This implies that

$$\frac{P(\tilde{\theta} = 1)}{P(\tilde{\theta} = 3)} = 2 \qquad \text{and} \qquad \frac{P(\tilde{\theta} = 3)}{P(\tilde{\theta} = 2)} = 3.$$

Thus, $P(\tilde{\theta} = 1) = 2P(\tilde{\theta} = 3)$ and $P(\tilde{\theta} = 2) = P(\tilde{\theta} = 3)/3$. But the prior probabilities must sum to 1, so

$$P(\tilde{\theta} = 1) + P(\tilde{\theta} = 2) + P(\tilde{\theta} = 3) = 1,$$

which implies that

$$2P(\tilde{\theta} = 3) + \frac{P(\tilde{\theta} = 3)}{3} + P(\tilde{\theta} = 3) = 1,$$

or

$$\frac{10}{3} P(\tilde{\theta} = 3) = 1,$$

or

$$P(\tilde{\theta} = 3) = \frac{3}{10}.$$

From the equations relating the various values of $\tilde{\theta}$,

$$P(\tilde{\theta} = 1) = 2P(\tilde{\theta} = 3) = \frac{6}{10}$$

and

$$P(\tilde{\theta} = 2) = \frac{P(\tilde{\theta} = 3)}{3} = \frac{1}{10}.$$

Thus, the prior distribution has been determined. Note that in the determination of the prior distribution, it is not necessary to consider all possible values of the random variable at once. The values can be considered in pairs, odds can be determined for each pair, and the probabilities can be found from these odds.

It should be emphasized that whatever the technique or techniques used to assess the prior probabilities, the probabilities must be nonnegative and must sum to 1. If these requirements are not satisfied, then the probabilities are said to be *inconsistent.* An assessor who obeys certain reasonable *axioms of coherent assessment* (such as, if he considers A to be more likely than B, and B to be more likely than C, then he should consider A to be more likely than C—this is an axiom of transitivity) will always be consistent in a decision-making sense. Such an assessor (sometimes referred to as the "rational man") will have no inconsistencies in his probability assessments. The axioms of coherence are discussed in Section 5.7. Some experiments have indicated that individuals do not always assess probabilities in a "rational" manner. For example, they frequently assess probabilities that do not sum to 1, perhaps because of carelessness, vagueness, or a misunderstanding of the assessment procedures. At any rate, it should not present a major problem, since inconsistencies can be removed if the assessor is made aware of their existence. If someone pointed out an arithmetic error to you, you would no doubt correct the error. Similarly, if someone pointed out that your prior probabilities summed to, say, 1.20, you could change the probabilities so that they would sum to 1. In this regard, it is sometimes useful to attempt to assess the same set of probabilities in more than one way in order to "check" your assessments. For instance, a probability distribution can be assessed once by using lotteries and a second time by considering odds. If the resulting two distributions differ in any respect, you should reconcile the differences. Once again, this is analogous to performing "checks" on arithmetic calculations, although the arithmetic calculations do have a single correct answer, which is not the case with probability assessments.

Since all probabilities are ultimately subjective under the subjective interpretation of probability, why is Bayes' theorem needed at all? Why doesn't the decision maker simply observe the sample information and subjectively determine his posterior probabilities, thereby saving himself a lot of trouble? An individual who obeys the above-mentioned axioms of coherence must revise probabilities via Bayes' theorem. Otherwise, one or more of the axioms of coherence is being violated. Numerous psychological studies have indicated that people do not always intuitively revise probabilities in the manner specified by Bayes' theorem. That is, if the revision of probabilities on the basis of new sample information is done intuitively rather than by formally applying Bayes' theorem, people tend not to give enough "weight" to the sample information. In general, this means that they do not change their probabilities as much as Bayes' theorem tells them that they should. For instance, consider once again the two urns discussed in Section 3.2—one with 70 red balls and 30 green balls and one with 70 green balls and 30 red balls. Suppose that you initially feel that the two urns are equally likely to be chosen and that you take a sample of nine balls from the chosen urn. Assume that the sample is taken with replacement; that is, after a ball is drawn and its color observed, it is replaced in the urn and the contents of the urn are well-mixed before the next drawing. Thus, the contents of the urn (and hence the probabilities of obtaining a red or green ball) remain constant from drawing to drawing. If, under these conditions, you draw seven red balls and two green balls, what is your posterior probability that the urn selected is the predominately red urn (that is, the urn with 70 red and 30 green balls)? Try to guess intuitively what this probability is, and then compute it using Bayes' theorem. If you are like the average subject in experiments of this nature, your intuitive probability will be much lower than the calculated probability. Consequently, it would seem that the revision of probabilities in the face of new sample evidence should be done formally via Bayes' theorem rather than intuitively. After all, adding machines, mechanical calculators, and high-speed computers are used because they are much faster and more accurate than people at performing arithmetic operations. In a similar manner, the formal application of Bayes' theorem is more accurate than the judgmental revision of probabilities.

In many real-world problems, there are several uncertain quantities of interest, and a *joint* prior distribution for these uncertain quantities is needed. If the uncertain quantities of interest are judged to be independent, then the joint prior distribution is simply the product of a number of individual, or marginal, distributions, and the procedures discussed above for the assessment of a distribution of a single uncertain quantity are applicable. Generally the assumption of independence is quite unrealistic, however, and the assessment process becomes more difficult. The most straightforward approach would simply be to assess probabilities directly for all possible joint events. This procedure was illustrated in Section 3.1, where joint probabilities were assessed directly for the events "rain today, rain tomorrow," "rain today, no rain tomorrow," "no rain today, rain tomorrow," and "no rain today, no rain tomorrow." Note that the uncertain quantities $\tilde{x}$ (corresponding to rain or no rain tomorrow) and $\tilde{y}$ (corresponding to rain or no rain today) are not independent, so it is not possible to assess their marginal distributions separately and simply to multiply these distributions in order to arrive at the joint distribution.

Unfortunately, it is often difficult to think in terms of probabilities of joint events. For instance, in making an investment decision you might be interested in the future price of a particular common stock *and* in whether or not Congress will raise taxes. You might, with little difficulty, be able to assess prior probabilities for various possible future prices of the stock in question, and you might be able to assess the probability that Congress will raise taxes. If you feel that the two variables in question are not independent, however, you may find it very difficult to assess probabilities directly for situations such as "the stock price one year from now is 40 *and* Congress raises taxes this year."

A convenient way to avoid this problem involves breaking up the joint probability distribution of the two variables into the product of the marginal distribution of one and the conditional distribution of the second, given the first. For instance, in the investment example, you might first assess the probability that Congress will raise taxes this year and then assess two conditional distributions: the conditional distribution of the year-end price of the stock under consideration, given that Congress raises taxes this year, and the conditional distribution of the year-end price of the stock, given that Congress does not raise taxes this year. Suppose that $\tilde{\theta}$ represents the year-end price of the stock and $\tilde{\phi}$ (Greek phi) is a random variable that equals 1 if taxes are raised and equals 0 if taxes are not raised. Then, from Equation (3.1.9),

$$P(\tilde{\theta} = \theta_j \mid \tilde{\phi} = \phi_k) = \frac{P(\tilde{\theta} = \theta_j, \tilde{\phi} = \phi_k)}{P(\tilde{\phi} = \phi_k)},$$

so that

$$P(\tilde{\theta} = \theta_j, \tilde{\phi} = \phi_k) = P(\tilde{\phi} = \phi_k)P(\tilde{\theta} = \theta_j \mid \tilde{\phi} = \phi_k). \qquad (3.5.1)$$

The joint distribution is a product of a marginal distribution and a conditional distribution.

In order to illustrate this problem conveniently, assume that $\tilde{\theta}$ has five possible values: 30, 35, 40, 45, and 50. Obviously, this is a gross simplification. It is useful to graph Equation (3.5.1) in the form of a tree diagram, as discussed in Section 2.6. A tree diagram for the investment problem is presented in Figure 3.5.1. The first chance fork involves whether or not Congress raises taxes, and the second chance fork involves the year-end price of the stock. The probability that taxes will be raised is .3, and given that taxes are raised, the conditional probabilities for the possible prices (50, 45, 40, 35, and 30) are .1, .1, .2, .4, and .2, respectively. Given that taxes are *not* raised, the conditional probabilities for the respective prices are .2, .3, .2, .2, and .1. The probabilities for the various possible results at a chance fork are given in parentheses. The joint probabilities, which are products of the probabilities along different paths of the tree diagram, are given in parentheses to the right of the corresponding joint events.

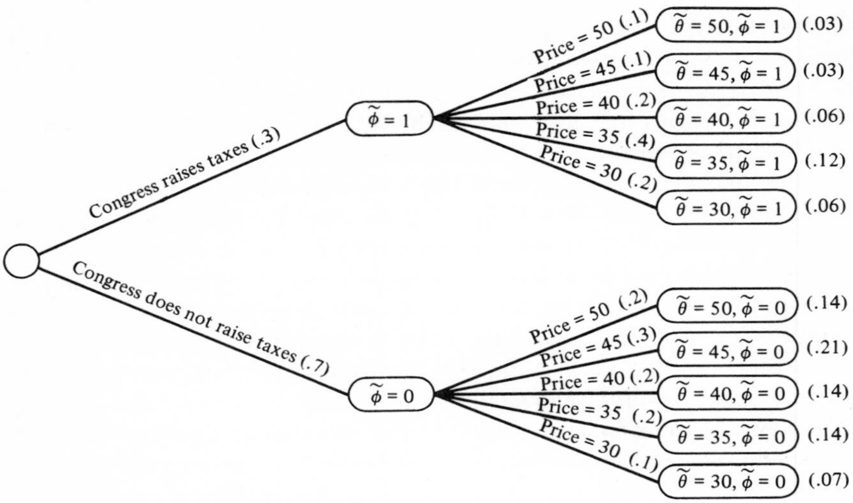

Figure 3.5.1

Note that the conditional distribution of $\tilde{\theta}$, given that taxes are raised, places higher probabilities on the lower prices for the stock than does the conditional distribution of $\tilde{\theta}$, given that taxes are not raised. In assessing such conditional distributions, it might be helpful to consider such devices as conditional lotteries or bets, which were discussed in Section 2.6. In any event, it seems that it is easier to assess joint probabilities in terms of a product of marginal probabilities and conditional probabilities instead of assessing the joint probabilities directly. If there are three random variables of interest, $\tilde{\theta}$, $\tilde{\phi}$, and $\tilde{\psi}$ (Greek psi), a third set of chance forks is added to the tree diagram (see Figure 3.5.2), and the joint probabilities are determined as follows:

$$P(\theta_j,\phi_k,\psi_h) = P(\psi_h)P(\phi_k \mid \psi_h)P(\theta_j \mid \phi_k,\psi_h). \tag{3.5.2}$$

The extension to more than three variables is straightforward but increasingly tedious, involving additional sets of chance forks on the tree diagram.

The question of whose prior probabilities are being assessed has not been raised in this section. Generally, the person faced with the inferential or decision-making problem of interest (that is, the statistician or the decision maker) is the person assessing the prior probabilities. In many situations, however, particularly for important problems, it may be worthwhile to hire an "expert" to assess prior probabilities. For instance, you might hire a meteorologist to assess the probability of rain, a stock-market analyst to assess probabilities for future stock prices, a political analyst to assess probabilities involving potential legislation by Congress, and so on. This raises the question of the evaluation of probability assessments. If an expert frequently assesses probabilities for events within his sphere of expertise, it is possible to evaluate these probabilities in light of the events that actually occur. Carrying this one step further, the statistician or decision maker might be able to calibrate the expert by investigating his past performance at assessing probabilities. For instance, if the decision maker feels that a particular meteorologist consistently underestimates the probability of rain by approximately .1, then he might take the probabilities he receives from the meteorologist and increase them by .1. This is a crude form of calibration, of course, and more sophisticated techniques are available. The point is that although an expert may be brought in to make probability assessments, the ultimate choice of prior probabilities rests with the statistician or decision maker.

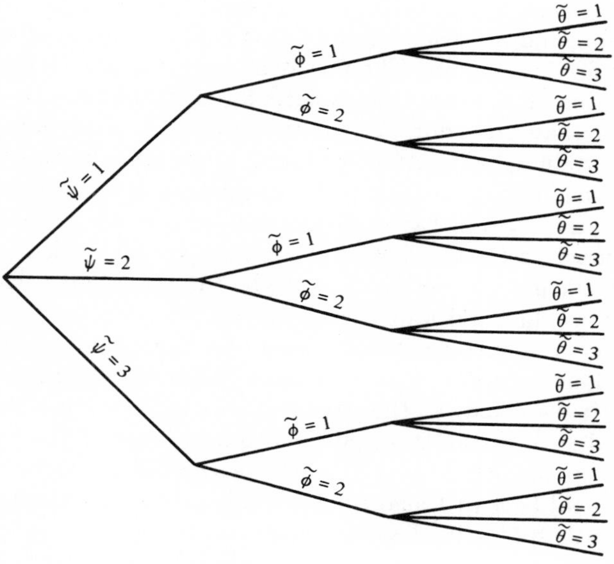

Figure 3.5.2

In summary, then, anyone who understands the basic concept of probability should be able to assess a prior distribution for a discrete random variable or a number of discrete random variables, although this may not be an easy task, especially when there are several random variables of interest. When dealing with continuous random variables, the assessment of a prior distribution may be more difficult, as you shall see in the next chapter. The main thing to keep in mind is that the prior distribution should reflect all of the relevant information available to the decision maker prior to observing a particular sample.

3.6 THE ASSESSMENT OF LIKELIHOODS

The assessment of prior probabilities is an important problem in Bayesian inference and decision. The determination of another input to the Bayesian model, the likelihoods, is also quite important. Since a likelihood is proportional to a conditional probability, many of the techniques discussed in Sections 2.6 and 3.5 for the assessment of probabilities in general and conditional probabilities in particular are relevant. That is, likelihoods can be assessed directly, using such devices as conditional bets, conditional lotteries, and so on.

For a simple example involving the assessment of likelihoods, suppose that the president of a firm is concerned that a competing firm might be planning to introduce a new product that could seriously affect the sales of the first firm's products. The president feels that the chances are only one in five that the competitor is actually planning to introduce a new product, so that his prior probabilities are $P(\text{new product}) = .2$ and $P(\text{no new product}) = .8$. After assessing these probabilities, he receives the additional information that the competing firm is building a new plant. In order to use Bayes' theorem to revise his probabilities, the president must assess the likelihoods associated with this new information. These likelihoods can be taken to be the probability that the competing firm would build a new plant, given that it plans to introduce a new product, and the probability that the new plant would be built, given no plans regarding a new product. After careful consideration of the competitor's situation, the president decides that if a new product is to be introduced, the odds would be 3 to 2 in favor of the building of a new plant; on the other hand, if a new product is not to be introduced, the odds would be 4 to 1 against the building of a new plant. Therefore, the president's subjective likelihoods are

$$P(\text{new plant} \mid \text{new product}) = \frac{3}{(3+2)} = .6$$

and

$$P(\text{new plant} \mid \text{no new product}) = \frac{1}{(1+4)} = .2.$$

Using the notation of Section 3.2, let $\tilde{\theta} = 1$ if the new product is to be introduced and let $\tilde{\theta} = 0$ if the new product is not to be introduced; the sample information y consists of the knowledge that the new plant is being built. Bayes' theorem can then be applied as follows:

$\tilde{\theta}$	*Prior probability*	*Likelihood*	*(Prior probability) × (likelihood)*	*Posterior probability*
1	.2	.6	.12	.429
0	.8	.2	.16	.571
	1.0		.28	1.000

The new information concerning the building of the plant increases the probability that $\tilde{\theta} = 1$ (that is, the probability that the competitor is planning to introduce a new product) from .2 to .429. Note that since likelihoods can be taken to be any positive multiple of the relevant conditional probabilities, it is only necessary to give the *relative* likelihoods. In other words, instead of assessing $P(\text{new plant} \mid \text{new product}) = .6$ and $P(\text{new plant} \mid \text{no new product}) = .2$, it is sufficient to say that the new plant is three times as likely given the new product as it is given no new product. In general, it should be easier to assess relative likelihoods than to assess all of the conditional probabilities of interest.

Often several new "pieces" of information are obtained over a period of time, and likelihoods are determined for each new piece of information as it is obtained so that Bayes' theorem can be applied to update the probabilities for the uncertain quantity of interest. An important point is that the likelihoods for any piece of information are conditional upon the previously obtained information as well as upon $\tilde{\theta}$. This point has been ignored so far in this chapter because in all of the examples it has been assumed that the information is generated by an *independent* process. That is, it has been assumed that the successive trials, or pieces of information, are conditionally independent, given any value θ of $\tilde{\theta}$. Such *conditional independence* is explicitly required of the Bernoulli and Poisson processes, for example. *Given* a value of $\tilde{p}$, the probability of heads, successive tosses of a particular coin are independent. If it is not realistic to assume that successive pieces of information are conditionally independent, then it is necessary to take account of their *conditional dependence* formally by assessing likelihoods that are conditional on all previous information.

To illustrate conditional dependence of successive pieces of information, consider the preceding example and suppose that a short time after learning that his competitor is building a new plant, the president of the firm in question learns that the competing firm has just hired a marketing research firm to conduct an extensive market survey. Although no details are available, this information seems quite relevant to the president's concern about $\tilde{\theta}$. After some thought, he assesses the following conditional probabilities:

$$P(\text{market survey} \mid \text{new plant, new product}) = .8,$$
$$P(\text{market survey} \mid \text{new plant, no new product}) = .5,$$
$$P(\text{market survey} \mid \text{no new plant, new product}) = .7,$$
and
$$P(\text{market survey} \mid \text{no new plant, no new product}) = .2.$$

In this situation, only the former two probabilities are needed to apply Bayes' theorem, but the latter two probabilities are given here to illustrate the conditional dependence of the information regarding the new plant and the information regarding the market survey. Given the previous information that the competitor is building a new plant, the market survey is $.8/.5 = 1.6$ times as likely given the new product as it is given no new product. On the other hand, if the previous information had been that *no* new plant is being built, the market survey is $.7/.2 = 3.5$ times as likely given the new product as it is given no new product. The difference between these figures is easy to explain: information that a new plant is being built favors $\tilde{\theta} = 1$ to the extent that the additional information that a market survey is to be conducted is somewhat redundant, whereas information that no new plant is being built might favor $\tilde{\theta} = 0$, in which case the additional information that a market survey is to be conducted has a considerable amount of impact. Because the pieces of information are not conditionally independent, the relative likelihood for the second piece of information depends on the previously obtained information.

If no information regarding a new plant had been obtained, the likelihoods of interest upon receiving the information about the market survey would be

$$P(\text{market survey} \mid \text{new product})$$

and

$$P(\text{market survey} \mid \text{no new product}).$$

Given the previous information about the new plant and the conditional dependence of the information sources, however, the likelihoods of interest are

$$P(\text{market survey} \mid \text{new plant, new product}) = .8$$

and

$$P(\text{market survey} \mid \text{new plant, no new product}) = .5.$$

Bayes' theorem can then be applied as follows.

$\tilde{\theta}$	*Prior probability*	*Likelihood*	*(Prior probability) × (likelihood)*	*Posterior probability*
1	.429	.8	.3432	.546
0	.571	.5	.2855	.454
	1.000		.6287	1.000

Note that if the previous information had been that no new plant is to be built, both the prior probabilities and the likelihoods would be changed; the prior probabilities for this application of Bayes' theorem would be the posterior probabilities following "no new plant" instead of the posterior probabilities following "new plant," and the likelihoods would be .7 and .2 instead of .8 and .5. In successive applications of Bayes' theorem, it is very important to keep in mind that the likelihoods at any one trial are conditional upon the results of the previous trials. If the nth piece of information (the result on the nth trial) is represented by $\tilde{y}_n$, where n is a positive integer, then the application of Bayes' theorem at the nth trial is of the form

$$P(x_j \mid y_1, \ldots, y_n) = \frac{P(x_j \mid y_1, \ldots, y_{n-1})P(y_n \mid x_j, y_1, \ldots, y_{n-1})}{\sum_{i=1}^{J} P(x_i \mid y_1, \ldots, y_{n-1})P(y_n \mid x_i, y_1, \ldots, y_{n-1})}. \tag{3.6.1}$$

Here $P(x_j \mid y_1, \ldots, y_{n-1})$ and $P(x_j \mid y_1, \ldots, y_n)$ represent the prior probability of x_j and the posterior probability of x_j with respect to the nth piece of information, and the likelihoods are $P(y_n \mid x_j, y_1, \ldots, y_{n-1})$ for $j = 1, \ldots, J$. The previous trials can be ignored in the assessment of likelihoods only if it can be assumed that the data-generating process is conditionally independent [that is, if

$$P(y_n \mid x_j, y_1, \ldots, y_{n-1}) = P(y_n \mid x_j) \tag{3.6.2}$$

for all values of n and for $j = 1, \ldots, J$].

In many situations it is possible to represent the data-generating process that generates new sample information in terms of a well-known statistical model. For instance, the examples in Sections 3.3 and 3.4 utilized the Bernoulli and Poisson models, both of which are widely applied to statistical problems. In each case, it was emphasized that certain simplifying assumptions were necessary in order to use the model. The Bernoulli model requires a dichotomous situation, and the Poisson model requires that events occur over a continuum, such as time. Both models involve assumptions of stationarity and independence (alternatively, the models can be thought of in terms of exchangeability, as pointed out in Sections 3.3 and 3.4). Of course, real-world data-generating processes are unlikely to obey *exactly* the assumptions of any statistical model. Thus, it must be recognized that a statistical model, like any type of mathematical model, is an idealization that one cannot expect to realize exactly in practice. However, in a great many situations, such a model may be a close approximation to reality and its use may be justified. Certainly models like the Bernoulli and Poisson models greatly simplify the determination of likelihoods. Given the assumptions of the models, the likelihoods can be expressed in terms of certain probability distributions, many of which are well known and extensively tabled.

The Bernoulli and Poisson models are by no means the only statistical models that are frequently encountered. For instance, in a dichotomous situation, suppose that sample information is generated by sampling *without replacement* from a finite population. Since each item is not replaced when it is sampled, the composition of the population changes from trial to trial. Thus, the data-generating process is not stationary. Also, the probabilities of the various possible outcomes on any trial depend on the results of previous trials, so the trials are not independent. For example, in sampling without replacement from a well-shuffled, standard deck of 52 cards, the conditional probability of obtaining a heart on the fifth draw clearly depends on the results of the first four draws. If the first four draws all result in hearts, the probability of a heart on the fifth draw is 9/48, since there are 48 cards remaining in the deck and 9 of them are hearts. If the first four draws yield no hearts, the probability of a heart on the fifth draw is 13/48. Clearly the Bernoulli model is inapplicable. The Bernoulli process requires sampling with replacement, or, alternatively, sampling from an infinite population or process (if the population or process is infinite, the distinction between sampling with replacement and sampling without replacement becomes irrelevant). The assumptions of stationarity and independence imply that the Bernoulli process "has no memory"; what happens on future trials does not depend on what happened on any past trials. In sampling without replacement from a finite population, the process does indeed have a "memory"; every time a heart is drawn, the probability of a heart on a future draw goes down. In an extreme case, if the first 13 cards drawn are hearts, then the probability of drawing a heart is zero thereafter.

The statistical model used to represent this situation is called the *hypergeometric* model. Let $\tilde{r}$, n, $\tilde{R}$, and $\tilde{N}$ represent the number of "successes" in the sample, the number of trials (the sample size), the number of "successes" in the entire finite population, and the number of items in the entire population (the population size), respectively. Then the conditional distribution of $\tilde{r}$ given n, R, and N is the *hypergeometric distribution,* which is of the following form:

$$P(\tilde{r} = r \mid n, R, N) = \frac{\binom{R}{r}\binom{N-R}{n-r}}{\binom{N}{n}}, \tag{3.6.3}$$

where $\binom{R}{r}$, $\binom{N-R}{n-r}$, and $\binom{N}{n}$ are combinatorial terms, as defined in Equation (3.3.2). The term in the denominator represents the number of different possible samples of size n that can be drawn without replacement from a population of size N. The numerator represents the number of these possible samples that include exactly r "successes" and $n - r$ "failures"; the first term is the number of different ways of selecting r "successes" from the R "successes" in the population, and the second term is the number of different ways of selecting $n - r$ "failures" from the $N - R$ "failures" in the population.

In sampling without replacement from a well-shuffled, standard deck of 52 cards, let $\tilde{r}$ represent the number of hearts in a five-card hand. Using the hypergeometric distribution, $\tilde{R} = 13$ (there are 13 hearts in the 52-card deck), $\tilde{N} = 52$ and $n = 5$, so the distribution of $\tilde{r}$ is as follows:

$$P(\tilde{r} = 0 \mid n = 5, \tilde{R} = 13, \tilde{N} = 52) = \binom{13}{0}\binom{39}{5}\Big/\binom{52}{5} = .2215,$$

$$P(\tilde{r} = 1 \mid n = 5, \tilde{R} = 13, \tilde{N} = 52) = \binom{13}{1}\binom{39}{4}\Big/\binom{52}{5} = .4114,$$

$$P(\tilde{r} = 2 \mid n = 5, \tilde{R} = 13, \tilde{N} = 52) = \binom{13}{2}\binom{39}{3}\Big/\binom{52}{5} = .2743,$$

$$P(\tilde{r} = 3 \mid n = 5, \tilde{R} = 13, \tilde{N} = 52) = \binom{13}{3}\binom{39}{2}\Big/\binom{52}{5} = .0815,$$

$$P(\tilde{r} = 4 \mid n = 5, \tilde{R} = 13, \tilde{N} = 52) = \binom{13}{4}\binom{39}{1}\Big/\binom{52}{5} = .0107,$$

and $$P(\tilde{r} = 5 \mid n = 5, \tilde{R} = 13, \tilde{N} = 52) = \binom{13}{5}\binom{39}{0}\Big/\binom{52}{5} = .0005.$$

For another example, if a neighborhood consists of 40 families with yearly incomes of at least \$12,000 and 60 families with yearly incomes below \$12,000, what is the probability of observing exactly five families with incomes of at least \$12,000 in a random sample (without replacement) of 10 families from the neighborhood? From the hypergeometric distribution, this probability is

$$P(\tilde{r} = 5 \mid n = 10, \tilde{R} = 40, \tilde{N} = 100) = \frac{\binom{40}{5}\binom{60}{5}}{\binom{100}{10}} = .21.$$

Suppose that the sample of 10 families is to be taken *with* replacement rather than *without* replacement. Then it is possible for a family to be included in the sample more than once. The composition of the population remains the same on each trial and the trials are independent, so the Bernoulli model applies. On any single trial, the probability that a randomly chosen family has an income of at least \$12,000 is $\tilde{p} = \tilde{R}/\tilde{N} = 40/100 = .4$. For a sample of $n = 10$ families with replacement, then, the binomial distribution can be used to find the probability of observing exactly five trials in which the family income is at least \$12,000:

$$P(\tilde{r} = 5 \mid n = 10, \tilde{p} = .4) = \binom{10}{5}(.4)^5(.6)^5 = .2007.$$

Notice that this is very close to the probability calculated from the hypergeometric distribution under the assumption of sampling without replacement. When the sample size n is small relative to the population size $\tilde{N}$ (in this case, $n = 10$ and $\tilde{N} = 100$), the binomial distribution with $\tilde{p} = \tilde{R}/\tilde{N}$ is a good *approximation* to the hypergeometric distribution. Such an approximation often is convenient because extensive tables of the binomial distribution are available.

The binomial approximation of the hypergeometric distribution illustrates the possibility of using one statistical model as an approximation of a second model. Often a particular model is used to represent a real-world situation because the model seems to provide a good abstraction of reality. It may turn out, however, that the probabilities of interest are very difficult to compute from the chosen model, perhaps because adequate tables are not available and because the computation of the probabilities from the relevant formula is an arduous task. Frequently it is possible to find a second statistical model which, for the situation at hand, provides a very close approximation of the first model and possesses the advantage of being easier to work with computationally. The binomial approximation to the hypergeometric model is one example.

Another example is the use of the Poisson model to approximate the binomial. When n is large in the Bernoulli process, binomial probabilities may not be tabled, and their direct computation may be difficult. The Poisson distribution provides a good approximation to the binomial distribution when n is large and $\tilde{p}$ is close to 0 or 1. If $\tilde{p}$ is close to 0, the approximation takes $\tilde{\lambda}t = n\tilde{p}$, whereas if $\tilde{p}$ is close to 1, $\tilde{\lambda}t = n(1 - \tilde{p})$ and "success" and "failure" are interchanged. For example, suppose that $n = 50$, $\tilde{r} = 1$, and $\tilde{p} = .01$. The binomial probability is

$$P(\tilde{r} = 1 \mid n = 50, \tilde{p} = .01) = \binom{50}{1} (.01)^1(.99)^{49} = .3056,$$

while the Poisson approximation is

$$P(\tilde{r} = 1 \mid \tilde{\lambda}t = n\tilde{p} = .5) = e^{-.5}(.5)^1/1! = .3033.$$

The "closeness" of the approximation improves as n increases and as $\tilde{p}$ gets closer to 0 (or, for high values of $\tilde{p}$, as $\tilde{p}$ gets closer to 1). If n is large and $\tilde{p}$ is close to $\frac{1}{2}$, a different approximation to the binomial distribution can be used; this approximation involves the normal distribution, which is discussed in Chapter 4.

Another model of some interest is the *multinomial* model, which is similar to the Bernoulli model except that the number of possible events at each trial is greater than two. Suppose that at each trial, exactly one of the K events $E_1, E_2, \ldots, E_K$ will occur, and that the probabilities of these events are represented by $\tilde{p}_1 = p_1, \tilde{p}_2 = p_2, \ldots, \tilde{p}_K = p_K$, where $\sum_{i=1}^{K} p_i = 1$. Furthermore, assume that the process is stationary and independent (the p_i's remain constant from trial to trial and the trials are independent). If a sample of size n is taken, the probability that E_1 occurs r_1 times, E_2 occurs r_2 times, and so on, where $\sum_{i=1}^{K} r_i = n$, is given by the *multinomial distribution:*

$$P(\tilde{r}_1 = r_1, \tilde{r}_2 = r_2, \ldots, \tilde{r}_K = r_K \mid n, p_1, \ldots, p_K)$$
$$= \frac{n!}{r_1!r_2!\cdots r_K!} p_1^{r_1}, p_2^{r_2} \cdots p_K^{r_K}. \quad (3.6.4)$$

Clearly this is just an extension of the binomial distribution; the binomial distribution is a special case of the multinomial distribution with $K = 2$.

To illustrate the multinomial distribution, suppose that a defective item from a production process can have one, two, three, or four defects. Suppose further that the probabilities for the different numbers of defects (given that an item is defective) are $\tilde{p}_1 = .80$, $\tilde{p}_2 = .18$, $\tilde{p}_3 = .01$, and $\tilde{p}_4 = .01$. In addition, assume that these probabilities are the same for all defective items and that the trials are independent. This can be considered as a multinomial process, and the probability of obtaining exactly five items with one defect ($\tilde{r}_1 = 5$), three items with two defects ($\tilde{r}_2 = 3$), two items with three defects ($\tilde{r}_3 = 2$), and one item with four defects

$(\tilde{r}_4 = 1)$ in a sample of 11 defective items is

$$P(\tilde{r}_1 = 5, \tilde{r}_2 = 3, \tilde{r}_3 = 2, \tilde{r}_4 = 1 \mid n = 11, \tilde{p}_1 = .80, \tilde{p}_2 = .18, \tilde{p}_3 = .01, \tilde{p}_4 = .01) = \frac{11!}{5!3!2!1!} (.80)^5(.18)^3(.01)^2(.01)^1.$$

In a similar fashion, the hypergeometric distribution can be extended to the case in which there are K possible outcomes at each trial:

$$P(\tilde{r}_1 = r_1, \tilde{r}_2 = r_2, \ldots, \tilde{r}_K = r_K \mid n, R_1, R_2, \ldots, R_K) = \frac{\binom{R_1}{r_1}\binom{R_2}{r_2}\cdots\binom{R_K}{r_K}}{\binom{N}{n}}, \qquad (3.6.5)$$

where $\sum_{i=1}^{K} r_i = n$ and $\sum_{i=1}^{K} R_i = N$. In sampling without replacement from a well-shuffled, standard deck of 52 cards, the probability of obtaining one heart, three spades, five diamonds, and four clubs in a 13-card hand is

$$P(\tilde{r}_1 = 1, \tilde{r}_2 = 3, \tilde{r}_3 = 5, \tilde{r}_4 = 4 \mid n = 13, \tilde{R}_1 = \tilde{R}_2 = \tilde{R}_3 = \tilde{R}_4 = 13) = \frac{\binom{13}{1}\binom{13}{3}\binom{13}{5}\binom{13}{4}}{\binom{52}{13}}.$$

As noted earlier, real-world data-generating models are unlikely to obey *exactly* the assumptions of any statistical model, so that statistical models are idealizations that are not realized in practice. In many instances, however, the model is a close enough approximation of reality for the problem at hand. The primary advantage of using a statistical model is that it generally leads to the use of well-known and extensively tabled probability distributions, thus simplifying the determination of likelihoods. Of course, this advantage is relevant only when the model is realistic. Various procedures are available for statistically investigating the applicability of models, including tests of the underlying assumptions (such as stationarity and independence) and techniques for comparing observed data with the predictions of the model. This process of "fitting" models is often very important in statistical problems. It should be mentioned that the desired degree of realism of the model varies from situation to situation, depending on the precision needed for an inferential or decision-making problem. For some problems, the final inferential statement or decision may be quite sensitive to variations in the model (and hence in the likelihoods), while in other situations the result may be very insensitive to such variations. This question of sensitivity, which relates to prior probabilities as well as to likelihoods (for instance, the presence of vagueness in the assessment of prior probabilities may not be of serious concern if the problem at hand is reasonably insensitive to variations in the prior probabilities), is discussed at greater length in Chapters 4 and 5.

Statistical models such as those presented in this section and in Sections 3.3 and 3.4 play an important role in problems of inference and decision. They greatly simplify the determination of likelihoods, and, to the extent that different persons would agree about the use of a particular model to represent a particular real-world situation, they result in likelihoods about which many people might be in agreement. That is, given that different persons are willing to accept the assumptions of a certain model, they should agree about the likelihoods determined from that model. As a result, the determination of likelihoods is considered by many to be less controversial than the determination of prior probabilities. However, the willingness to accept the assumptions of a model and to consider that model as a satisfactory representation of reality is a subjective choice. Whether judgments are used to select a model or whether they are used to assess probabilities directly without using a model, the element of subjectivity is still present. If a statistical model seems to be realistic in any given situation, it may simplify the assessment of likelihoods, but it should be emphasized that the decision regarding the use of such a model remains a subjective decision.

3.7 THE PREDICTIVE DISTRIBUTION

Bayes' theorem is used to revise probabilities concerning a random variable $\tilde{\theta}$ on the basis of new sample information that is represented by $\tilde{y}$. Inferences or decisions concerning $\tilde{\theta}$ may then be based on the posterior distribution of $\tilde{\theta}$ determined via Bayes' theorem. In some situations, you may also be interested in making *predictions* about the sample result before it is observed. Before the new sample information is obtained, you are uncertain about which sample outcome will occur. As you have seen, probability is the language of uncertainty, so that in order to make predictions you would like to express your uncertainty about the sample outcome $\tilde{y}$ in the form of a probability distribution, which for obvious reasons is called a *predictive distribution*. The set of possible sample outcomes $y_1, y_2, \ldots, y_K$, together with their probabilities

$P(\tilde{y} = y_k)$, $k = 1, \ldots, K$, comprises the predictive distribution for $\tilde{y}$. For example, if you are taking a sample of size 10 from a Bernoulli process, the uncertain sample result is the number of "successes," which is generally denoted by $\tilde{r}$. The possible values of $\tilde{r}$ are 0, 1, 2, 3, 4, 5, 6, 7, 8, 9, and 10, and the marginal, or unconditional, probabilities of these values are called predictive probabilities.

Why might you be interested in predictive probabilities? Often a simple betting situation depends on a sample outcome. For example, consider a bet on a single toss of a coin. The bet concerns the outcome on that toss—heads or tails. In the quality-control example of Section 3.3, the production manager might be interested in $\tilde{r}$, the number of defectives in a lot of, say, 20 items (as opposed to being interested in $\tilde{p}$, the underlying proportion of defective items produced by the machine). Perhaps he has an order for 20 items that must be shipped tomorrow. The variable $\tilde{r}$ can be thought of as the number of defectives in a sample of size 20 from the process. The owner of the auto dealership in the example in Section 3.4 may be interested not in the general rate at which a particular salesman sells cars, but rather in the exact number of cars sold by that salesman in a particular month. Perhaps the owner must order some cars from the factory in advance, and in order to decide how many cars to order he needs to determine his predictive probabilities for the number of cars that will be sold in the coming month. Another use of the predictive distribution is in deciding whether or not to sample and, more generally, how large a sample to take. In a decision-making problem, additional sample information is of value in the sense that it may change your probabilities in such a way that your decision is changed (that is, the decision that appears to be best under the prior distribution may not appear best under the posterior distribution). The predictive distribution is used in calculating the expected value of sample information in Chapter 6.

Given that you are interested in predictive probabilities, how can they be determined? Like any other probabilities, they can be determined directly, perhaps by using devices such as bets and lotteries. They also can be determined, however, in terms of the distributions that have already been considered in some detail in this chapter. Before this is demonstrated, it should be useful to give a brief review of these distributions and the relationships among them. From Equation (3.2.3), Bayes' theorem can be written in the form

$$P(\theta_j \mid y_k) = \frac{P(y_k \mid \theta_j)P(\theta_j)}{\sum_{i=1}^{J} P(y_k \mid \theta_i)P(\theta_i)},$$

where θ_j represents a value of the random variable of interest, $\tilde{\theta}$, and y_k represents new sample information. In the terminology of Section 3.1, the values of $\tilde{\theta}$ together with their probabilities $P(\theta_j)$ constitute the *marginal distribution of* $\tilde{\theta}$, which is called the *prior distribution* because it represents the state of uncertainty about $\tilde{\theta}$ prior to a particular sample. The values of $\tilde{\theta}$ together with the conditional probabilities $P(\theta_j \mid y_k)$ constitute the *conditional distribution of* $\tilde{\theta}$ given the sample result y_k, which is called the *posterior distribution* because it represents the state of uncertainty about $\tilde{\theta}$ after observing the sample result y_k. In applying Bayes' theorem, the *likelihoods* are determined from the *conditional distribution* of y_k given the different values of $\tilde{\theta}$.

The prior and posterior distributions may be used to make inferences or decisions concerning $\tilde{\theta}$ before and after observing the sample result y_k. Likelihoods may be used to determine probabilities for sample results y_k, *conditional* on $\tilde{\theta}$. From the binomial distribution, for example, one can find the probability of the sample result "4 successes in 10 trials," *given* a particular value of $\tilde{p}$. In general, however, $\tilde{p}$ is *not* known, so it is impossible to predict a sample outcome by using the binomial distribution. The problem arises because the binomial distribution is conditional on $\tilde{p}$. Since $\tilde{p}$ is not known, what is needed is the *unconditional,* or marginal, probability for each possible sample outcome. But such a probability is simply the denominator of Bayes' theorem (provided that the conditional probabilities $P(y_k \mid \theta_j)$, not multiples of these probabilities, are used as likelihoods). From Equation (3.1.9),

$$P(y_k \mid \theta_i) = \frac{P(y_k,\theta_i)}{P(\theta_i)},$$

so that

$$P(y_k \mid \theta_i)P(\theta_i) = P(y_k,\theta_i).$$

Hence, the denominator of Bayes' theorem is

$$\sum_{i=1}^{J} P(y_k \mid \theta_i)P(\theta_i) = \sum_{i=1}^{J} P(y_k,\theta_i),$$

which, from Equation (3.1.8), is $P(y_k)$, the marginal probability of the sample outcome y_k. Therefore,

$$P(y_k) = \sum_{i=1}^{J} P(y_k \mid \theta_i)P(\theta_i). \tag{3.7.1}$$

Thus, a predictive probability for a sample result y_k is the denominator in Bayes' theorem, which is the sum of products of prior probabilities of the form $P(\theta_i)$ and conditional probabilities of the form $P(y_k \mid \theta_i)$.

For the second medical example in Section 2.7, the predictive probability of a positive result on the chest X-ray is, from Equation (3.7.1),

$$\begin{aligned}P(\tilde{y} = 1) &= P(\tilde{y} = 1 \mid \tilde{\theta} = 0)P(\tilde{\theta} = 0) + P(\tilde{y} = 1 \mid \tilde{\theta} = 1)P(\tilde{\theta} = 1) \\ &\qquad + P(\tilde{y} = 1 \mid \tilde{\theta} = 2)P(\tilde{\theta} = 2) \\ &= (.07)(.97) + (.95)(.01) + (.90)(.02) \\ &= .0954.\end{aligned}$$

Note that this is the same as the denominator of Bayes' theorem, which is given for this example in Section 3.2. The probability of a negative result is

$$P(\tilde{y} = 0) = 1 - P(\tilde{y} = 1) = .9046,$$

and this can be verified by using Equation (3.7.1):

$$\begin{aligned}P(\tilde{y} = 0) &= P(\tilde{y} = 0 \mid \tilde{\theta} = 0)P(\tilde{\theta} = 0) + P(\tilde{y} = 0 \mid \tilde{\theta} = 1)P(\tilde{\theta} = 1) \\ &\qquad + P(\tilde{y} = 0 \mid \tilde{\theta} = 2)P(\tilde{\theta} = 2) \\ &= (.93)(.97) + (.05)(.01) + (.10)(.02) \\ &= .9046.\end{aligned}$$

When Bayes' theorem is expressed in tabular form, the predictive probability is simply the sum of the column labeled (*Prior probability*) × (*likelihood*). Recall that in all examples in this chapter, the likelihoods have been taken as the conditional probabilities $P(y_k \mid \theta_i)$, not as multiples of these conditional probabilities. In the quality-control example in Section 3.3, the first table represents an application of Bayes' theorem following a sample of size five with one defective item. The sum of the (*Prior probability*) × (*likelihood*) column is .12403. This means that before seeing this sample, the marginal, or predictive, probability of obtaining one defective item in a sample of size five is .12403. By considering similar tables for the other values of $\tilde{r}$(0,2,3,4,5), an entire predictive distribution can be determined. These tables are presented in Section 6.3 when the value of sample information is discussed; the predictive distribution is as follows:

$$\begin{aligned}P(\tilde{r} = 0) &= .854726, \\ P(\tilde{r} = 1) &= .124030, \\ P(\tilde{r} = 2) &= .018126, \\ P(\tilde{r} = 3) &= .002736, \\ P(\tilde{r} = 4) &= .000324,\end{aligned}$$

and

$$P(\tilde{r} = 5) = .000020.$$

These probabilities do not quite sum to 1 because of rounding error in the determination of the likelihoods from the binomial distribution.

In the automobile-dealership example in Section 3.4, the sample result is 10 sales in 24 days, and the predictive probability for this sample result is .042. This means that at the time the salesman is hired, the probability that he will sell exactly 10 cars in his first 24 days is .042. In the example involving a new product in Section 3.6, the new information is that the competitor is building a new plant, and $P(\text{new plant}) = .28$. This means that before this new information is obtained, the predictive probabilities are $P(\text{new plant}) = .28$ and $P(\text{no new plant}) = .72$.

In predicting a sample outcome, there are two types of uncertainty involved. First, even if it is known for certain that $\tilde{\theta} = \theta_i$, there is some uncertainty about $\tilde{y}$. This uncertainty is reflected in the conditional distribution $P(y_k \mid \theta_i)$. Sometimes this sort of uncertainty is referred to as "chance variation." Even if $\tilde{p}$ is known in a Bernoulli process, for instance, we are uncertain about $\tilde{r}$, and our uncertainty is represented by a binomial distribution that gives $P(\tilde{r} = r \mid n, p)$. The second type of uncertainty in the prediction problem is uncertainty about $\tilde{\theta}$, and this uncertainty is reflected in the prior distribution $P(\theta_i)$. Since $\tilde{\theta}$ is not known for certain, there is more uncertainty in the prediction problem than simply chance variation. The predictive distribution, as expressed in Equation (3.7.1), involves both types of uncertainty. In essence, a predictive probability $P(y_k)$ is a weighted average of conditional probabilities $P(y_k \mid \theta_i)$, where the weights are the prior probabilities $P(\theta_i)$. The predictive distribution depends both on the likelihoods and on the prior probabilities.

For an example of a prediction problem, consider once again the automobile-dealership example discussed in Section 3.4. After the new salesman's first 24 days, the posterior probabilities of his being a "great salesman" ($\tilde{\lambda} = \frac{1}{2}$), a "good salesman" ($\tilde{\lambda} = \frac{1}{4}$), and a "poor salesman" ($\tilde{\lambda} = \frac{1}{8}$), respectively, are .501, .493, and .006. What is the probability that in the *next* 24 days, he will again sell exactly 10 cars? Taking the above posterior probabilities as the new prior probabilities and noting that the likelihoods determined from the Poisson distribution are

$$P(\tilde{r} = 10 \mid t = 24, \tilde{\lambda} = \tfrac{1}{2}) = .1048,$$
$$P(\tilde{r} = 10 \mid t = 24, \tilde{\lambda} = \tfrac{1}{4}) = .0413,$$

and

$$P(\tilde{r} = 10 \mid t = 24, \tilde{\lambda} = \tfrac{1}{8}) = .0008,$$

the predictive probability that $\tilde{r} = 10$ is

$$\begin{aligned}
P(\tilde{r} = 10) &= \sum_{j=1}^{3} P(\tilde{r} = 10 \mid t = 24, \tilde{\lambda} = \lambda_j)P(\tilde{\lambda} = \lambda_j) \\
&= P(\tilde{r} = 10 \mid t = 24, \tilde{\lambda} = \tfrac{1}{2})P(\tilde{\lambda} = \tfrac{1}{2}) \\
&\qquad + P(\tilde{r} = 10 \mid t = 24, \tilde{\lambda} = \tfrac{1}{4})P(\tilde{\lambda} = \tfrac{1}{4}) \\
&\qquad\qquad + P(\tilde{r} = 10 \mid t = 24, \tilde{\lambda} = \tfrac{1}{8})P(\tilde{\lambda} = \tfrac{1}{8}) \\
&= .1048(.501) + .0413(.493) + .0008(.006) \\
&= .073.
\end{aligned}$$

Note that at the time the salesman was first hired, the predictive probability that he would sell exactly 10 cars in his first 24 days was .042. The entire predictive distribution for $\tilde{r}$, the number of cars sold during his first 24 days, is as follows:

r	$P(\tilde{r} = r)$	r	$P(\tilde{r} = r)$	r	$P(\tilde{r} = r)$	r	$P(\tilde{r} = r)$
0	.016	7	.084	13	.024	19	.003
1	.052	8	.067	14	.019	20	.002
2	.090	9	.053	15	.015	21	.001
3	.112	10	.042	16	.011	22	.001
4	.118	11	.034	17	.008	23	.000
5	.113	12	.028	18	.005	24	.000
6	.102						

This distribution is illustrated in Figure 3.7.1. After the new salesman sells 10 cars in his first 24 days, the predictive probability that he will sell exactly 10 more cars in the *next* 24 days is .073. After the first 24 days, the owner of the dealership places higher probability on the salesman's being "great" and less probability on his being "poor," so the predictive probability of 10 sales in 24 days is higher than it was when the salesman was first hired. The entire predictive distribution for $\tilde{r}$, the number of cars sold in the next 24 days, is as follows:

r	$P(\tilde{r} = r)$	r	$P(\tilde{r} = r)$	r	$P(\tilde{r} = r)$	r	$P(\tilde{r} = r)$
0	.002	7	.090	13	.055	19	.008
1	.008	8	.084	14	.046	20	.005
2	.024	9	.078	15	.037	21	.003
3	.046	10	.073	16	.027	22	.002
4	.070	11	.068	17	.019	23	.001
5	.086	12	.063	18	.013	24	.000
6	.092						

This distribution is illustrated in Figure 3.7.2. As compared with the previous predictive distribution (the distribution of $\tilde{r}$ for the first 24 days), low values of $\tilde{r}$ are less likely and high values of $\tilde{r}$ are more likely. This demonstrates the dependence of the predictive distribution on the prior distribution, for the likelihoods are identical in the two cases. The only differences involve the prior distribution.

The introduction of the concept of a predictive distribution completes the development of the Bayesian framework in the discrete case (discrete $\tilde{\theta}$ and discrete $\tilde{y}$). This framework is used in decision-making situations in Chapters 5 and 6. In Chapter 4, the Bayesian framework for continuous probability models (continuous $\tilde{\theta}$ and/or continuous $\tilde{y}$) is presented.

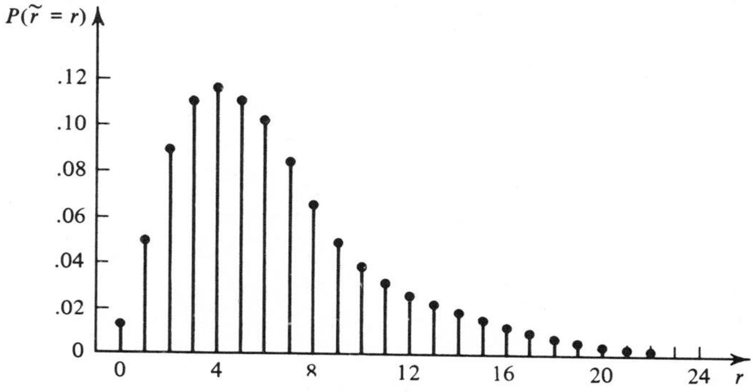

Figure 3.7.1

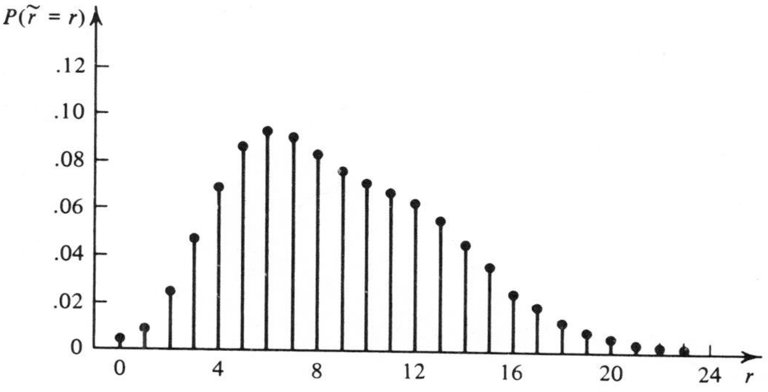

Figure 3.7.2

3.8 REFERENCES AND SUGGESTIONS FOR FURTHER READING

For discussions of discrete probability models, see Feller (1968); Hays and Winkler (1970, Chapters 3 and 4); Mosteller, Rourke, and Thomas (1970, Chapters 4–6); and Parzen (1960). In addition, numerous discrete models are catalogued in Johnson and Kotz (1969) and Patil and Joshi (1968). For a discussion of exchangeability, which is not included in the above references, see de Finetti (1937).

As indicated in Chapter 2, the original reference regarding Bayes' theorem is Bayes (1763); another early reference of interest is Pearson (1925).

General references concerning Bayesian inference for discrete probability models at a level roughly comparable to this chapter are Aitchison (1970*a*, Chapters 3 and 5); Brown (1969); Hadley (1967, Chapters 1, 5, 6, and 10); Pratt, Raiffa, and Schlaifer (1965, Chapters 5, 7, 9, and 10); Roberts (1966*b*, Chapters 4–6 and 8); Schlaifer (1959, Chapters 2, 3, 10–13, 16, 19, 21, and 23; 1969, Chapters 7–10); and Schmitt (1969, Chapters 2 and 3). Of these references, Schlaifer (1959) deserves special mention as the first introductory statistics book written from the Bayesian viewpoint [see Roberts (1960)]. Two references regarding predictive distributions are Roberts (1965; 1966*b*, Chapter 8).

As noted in Section 3.5, numerous psychological studies regarding the assessment and revision of probabilities have been conducted. For bibliographies of work in this area, see Edwards (1954, 1961, 1969); in addition, two recent general references are Peterson and Beach (1967) and Shelly and Bryan (1964), and a related review article and bibliography is Becker and McClintock (1967). Specific references involving the intuitive revision of probabilities include Du Charme (1970); Edwards (1968); Green, Halbert, and Robinson (1965); Phillips and Edwards (1966); and Phillips, Hays, and Edwards (1966). This research has led to increased interest in "clinical versus statistical prediction" and "man-machine systems"; for instance, see Edwards (1962, 1966); Edwards, Phillips, Hays, and Goodman (1968); Howell, Gettys, and Martin (1971); Kaplan and Newman (1966); Meehl (1954); Pankoff and Roberts (1968); and Sawyer (1966). Finally, a topic mentioned briefly in Section 3.5 is the evaluation of probability assessors and assessments, which is discussed in de Finetti (1962, 1965); Murphy and Winkler (1970); Savage (1971); Staël von Holstein (1970*a*, 1970*b*); Winkler (1967*b*, 1969, 1971); and Winkler and Murphy (1968*a*, 1968*b*).

EXERCISES

1. The random variable $\tilde{x}$ has the following probability distribution.

x	$P(\tilde{x} = x)$
-1	.2
0	.3
3	.2
4	.2
6	.1

(a) Graph the PMF and the CDF of $\tilde{x}$.
(b) Explain the relationship between the PMF and the CDF.
(c) Find $P(\tilde{x} \geq 2.5)$.
(d) Find $P(-1 < \tilde{x} < 4)$.
(e) Find $P(-1 \leq \tilde{x} \leq 4)$.
(f) Find $P(\tilde{x} < -3)$.
(g) Find $P(\tilde{x} = 1)$.

2. The cumulative distribution function of $\tilde{x}$ is given by the rule

$$F(x) = \begin{cases} 1 & \text{if } x \geq 2, \\ \frac{1}{4} & \text{if } 1 \leq x < 2, \\ 0 & \text{if } x < 1. \end{cases}$$

(a) Find the corresponding PMF and graph it.
(b) Find $P(1 < \tilde{x} < 2)$.
(c) Find the mean and the variance of $\tilde{x}$.

3. Suppose that the random variable $\tilde{x}$ represents the number of heads occurring in three independent tosses of a fair coin. Represent the distribution of $\tilde{x}$
(a) by a listing,
(b) by a graph of the PMF,
(c) by a graph of the CDF.
Do the same for $\tilde{y}$, the number of heads occurring in *four* independent tosses of a fair coin.

4. Suppose that the face value of a playing card is regarded as a random variable $\tilde{x}$, with an ace counting as 1 and any face card (jack, queen, or king) counting as 10. You draw one card at random from a standard, well-shuffled deck of 52 cards. Construct a PMF showing the probability distribution for this random variable, calculate $E(\tilde{x})$ and $V(\tilde{x})$, and find the probabilities of the following events:
(a) $P(\tilde{x} \leq 6)$,
(b) $P(\tilde{x} > 4)$,
(c) $P(\tilde{x} = 10)$,
(d) $P(\tilde{x}$ is an even number$)$.

5. For the random variable $\tilde{x}$ in Exercise 1,
(a) find $E(\tilde{x})$,
(b) find $E(\tilde{x}^2)$,
(c) find $V(\tilde{x})$.
For the random variable $\tilde{z} = [\tilde{x} - E(\tilde{x})]/\sqrt{V(\tilde{x})}$,
(d) represent the distribution of $\tilde{z}$ by a listing,
(e) represent the distribution of $\tilde{z}$ by a graph of the PMF,
(f) find $E(\tilde{z})$ and $V(\tilde{z})$.

6. Suppose that $\tilde{x}$ represents the daily sales of a particular product and that the probability distribution of $\tilde{x}$ is as follows:

x	$P(\tilde{x} = x)$
7000	.05
7500	.20
8000	.35
8500	.19
9000	.12
9500	.08
10,000	.01

(a) Find the expectation and the variance of daily sales. [*Hint:* to find the variance, it is easiest to find first the variance of a different variable, such as $\tilde{y} = (\tilde{x} - 7000)/1000$.]

(b) If the net profit resulting from sales of $\tilde{x}$ items can be given by $\tilde{z} = 5\tilde{x} - 38{,}000$, find the expectation and the variance of the net profit, $\tilde{z}$.

7. Is the distribution of the random variable $\tilde{x}$ in Exercise 6 symmetric, positively skewed, or negatively skewed?

8. Give examples of some random variables that might be expected to have roughly symmetric distributions, some that might be expected to have positively skewed distributions, and some that might be expected to have negatively skewed distributions. For instance, if $\tilde{x}$ represents the yearly income of a randomly chosen family in the United States, how would you expect the distribution of $\tilde{x}$ to be shaped?

9. Prove the first 10 "laws of expectation" presented in Section 3.1.

10. Explain intuitively the 14 "laws of expectation" presented in Section 3.1, assuming that the random variables represent uncertain payoffs in decision-making situations (for instance, the first rule simply states that if your payoff is a constant c dollars regardless of which event occurs, then your expected payoff is obviously c dollars).

11. For the random variable $\tilde{x}$ in Exercise 6, find the corresponding standardized random variable and represent the distribution of this standardized random variable
    (a) by a listing,
    (b) by a graph of the PMF,
    (c) by a graph of the CDF.

12. Suppose that a person agrees to pay you \$10 if you throw at least one six in four tosses of a fair die. How much would you have to pay him for this opportunity in order to make it a "fair" gamble? (You pay him in advance and then receive either \$10 or \$0; your payment is not returned.)

13. Discuss the statement, "All observed numerical events represent values of discrete variables, and continuous variables are only an idealization."

14. Suppose that the joint probability distribution of $\tilde{x}$ and $\tilde{y}$ is represented by the following table:

$\tilde{x}$ \ $\tilde{y}$	1	2	3	4
1	.12	.18	.24	.06
2	.06	.09	.12	.03
3	.02	.03	.04	.01

(a) Graph the joint distribution of $\tilde{x}$ and $\tilde{y}$ on a three-dimensional graph.
(b) Are $\tilde{x}$ and $\tilde{y}$ independent? Explain your answer.
(c) Determine the marginal distributions of $\tilde{x}$ and $\tilde{y}$.
(d) Determine the conditional distribution of $\tilde{x}$ given that $\tilde{y} = 2$. What does this tell you about the conditional distribution of $\tilde{x}$ given *any* particular value of $\tilde{y}$?
(e) Find $E(\tilde{x})$, $E(\tilde{y})$, $E(\tilde{x} + \tilde{y})$, and $E(4\tilde{x} - 2\tilde{y})$.
(f) Find $V(\tilde{x})$, $V(\tilde{y})$, $V(\tilde{x} + \tilde{y})$, and $V(4\tilde{x} - 2\tilde{y})$.
(g) Using the distribution obtained in (d), find $E(\tilde{x} \mid \tilde{y} = 2)$.

15. Complete the following table, given that $P(\tilde{x} = 1 \mid \tilde{y} = 2) = \frac{1}{3}$ and $P(\tilde{y} = 3 \mid \tilde{x} = 2) = \frac{1}{2}$.

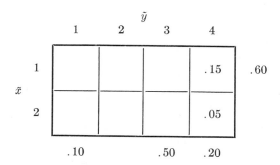

$\tilde{x}$ \ $\tilde{y}$	1	2	3	4	
1				.15	.60
2				.05	
	.10		.50	.20	

16. Complete the following table of probabilities, given that $\tilde{x}$ and $\tilde{y}$ are independent.

$\tilde{x}$ \ $\tilde{y}$	10	20	30	
-1	.05			
0			.20	.50
1				
	.20			

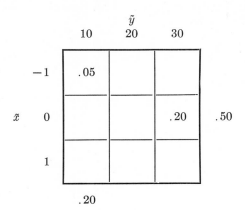

17. From the joint distribution of $\tilde{x}$ and $\tilde{y}$ presented in Exercise 15,
    (a) find the conditional distribution of $\tilde{x}$, given that $\tilde{y} = 1$,
    (b) find the conditional distribution of $\tilde{x}$, given that $\tilde{y} = 3$,
    (c) find the conditional distribution of $\tilde{y}$, given that $\tilde{x} = 1$.
    (d) find the conditional distribution of $\tilde{y}$, given that $\tilde{x} = 2$,
    (e) find $E(\tilde{x} \mid \tilde{y} = 1)$, $E(\tilde{x} \mid \tilde{y} = 3)$, $E(\tilde{y} \mid \tilde{x} = 1)$, and $E(\tilde{y} \mid \tilde{x} = 2)$,
    (f) find $V(\tilde{x} \mid \tilde{y} = 1)$, $V(\tilde{x} \mid \tilde{y} = 3)$, $V(\tilde{y} \mid \tilde{x} = 1)$, and $V(\tilde{y} \mid \tilde{x} = 2)$,
    (g) given the marginal distributions of $\tilde{x}$ and $\tilde{y}$, what would the joint distribution be if $\tilde{x}$ and $\tilde{y}$ were independent?

18. With respect to Bayes' theorem, what do the prior distribution, the likelihoods, and the posterior distribution represent? Explain the statement, "The terms *prior* and *posterior*, when applied to probabilities, are relative terms, relative to a given sample." Is it possible for a set of probabilities to be both prior *and* posterior probabilities? Explain.

19. Explain the statement, "Likelihoods are only relative (to each other), so they can be taken to be the appropriate conditional probabilities $P(y \mid \theta)$ for fixed $\tilde{y}$ and different values of $\tilde{\theta}$ *or* they can be taken to be any positive multiple of these conditional probabilities."

20. A Bernoulli process is a series of dichotomous trials that are stationary and independent. Explain the terms "stationary" and "independent" with regard to a Bernoulli process. Give some examples of situations in which a series of dichotomous trials is observed but in which the assumptions of stationarity and/or independence seem unrealistic.

21. A multiple choice examination is constructed so that, in principle, the probability of a correct choice for any item by guessing alone is $\frac{1}{4}$. If the test consists of 10 items, what is the probability that a student will have

exactly five correct answers if he is just guessing? What is the probability that he will have *at least* five correct answers if he is just guessing?

22. The probability that an item produced by a certain production process is defective is .05. Assuming stationarity and independence, what is the probability that a lot of 10 items from the process contains no defectives? If $\tilde{r}$ represents the number of defectives in a lot of 10 items, find $E(\tilde{r} \mid n,p)$ and $V(\tilde{r} \mid n,p)$.

23. Suppose that it is believed that $\tilde{p}$, the proportion of consumers (in a large population) who will purchase a certain product is either .2, .3, .4, or .5. Furthermore, the prior probabilities for these four values are $P(.2) = .2$, $P(.3) = .3$, $P(.4) = .3$, and $P(.5) = .2$. A sample of size 10 is taken from the population, and of the 10 consumers, 3 state that they will buy the product. What are the posterior probabilities? In calculating the posterior probabilities, what assumptions did you need to make?

24. You feel that $\tilde{p}$, the probability of heads on a toss of a particular coin, is either .4, .5, or .6. Your prior probabilities are $P(.4) = .1$, $P(.5) = .7$, and $P(.6) = .2$. You toss the coin three times and obtain heads once and tails twice. What are the posterior probabilities? If you then toss the coin three *more* times and once again obtain heads once and tails twice, what are the posterior probabilities? Also, compute the posterior probabilities by pooling the two samples and revising the original probabilities just once; compare with your previous answer.

25. In Exercise 24, suppose that you feel that in comparing $\tilde{p} = .6$ and $\tilde{p} = .4$, the odds are 3 to 2 in favor of $\tilde{p} = .6$. Furthermore, in comparing $\tilde{p} = .4$ and $\tilde{p} = .5$, the odds are 5 to 1 in favor of $\tilde{p} = .5$. On the basis of these odds, find your prior distribution for $\tilde{p}$.

26. A medical researcher feels that $\tilde{p}$, the probability that a person who is exposed to a certain contagious disease will catch it, is either .10, .12, .14, .16, .18, or .20. He assesses his prior probabilities to be $P(\tilde{p} = .10) = .05$, $P(\tilde{p} = .12) = .08$, $P(\tilde{p} = .14) = .13$, $P(\tilde{p} = .16) = .30$, $P(\tilde{p} = .18) = .34$, and $P(\tilde{p} = .20) = .10$. From this prior distribution, find $E(\tilde{p})$ and $V(\tilde{p})$. The researcher conducts an experiment in which 20 persons are exposed to the disease and 2 catch it. On the basis of this new information, find the researcher's posterior distribution for $\tilde{p}$ and find the mean and the variance of this posterior distribution.

27. A veteran public official is up for reelection and is concerned about $\tilde{p}$, the proportion of the votes cast that will be for him. In the past, he has run for office 20 times; he received 48 percent of the vote four times, 50 percent of the vote four times, 52 percent of the vote five times, 54 percent of the vote four times, 56 percent of the vote twice, and 58 percent of the vote once. In the upcoming election, however, his opponent is young, vigorous, and has a large campaign budget. As a result, the veteran official feels that the past frequencies slightly overstate his chances. He feels that this opponent

will obtain about 2 percent more of the votes than his past opponents, so that his prior distribution consists of the probabilities $P(\tilde{p} = .46) = \frac{4}{20}$, $P(\tilde{p} = .48) = \frac{4}{20}$, $P(\tilde{p} = .50) = \frac{5}{20}$, and so on. He then conducts a small survey of 100 voters, 49 of whom state that they will vote for him and 51 of whom state that they will vote for his young opponent. Find the posterior distribution of $\tilde{p}$ and use this distribution to calculate the expected proportion of votes the veteran public official will obtain.

28. In sampling from a Bernoulli process, it is possible to sample with a fixed r and a variable $\tilde{n}$ instead of with a fixed n and a variable $\tilde{r}$. The former sampling procedure is called Pascal sampling, while the latter procedure is called binomial sampling. For an example of Pascal sampling, a quality-control statistician might decide to observe items from a production process until he finds five defectives, so that $r = 5$ and $\tilde{n}$ is an uncertain quantity (it may only take five trials to find five defectives, or it may take, say, 1000 trials). The conditional probability distribution of $\tilde{n}$, given r and $\tilde{p}$, which is known as the Pascal distribution, is of the form

$$P(\tilde{n} = n \mid r, p) = \binom{n-1}{r-1} p^r(1-p)^{n-r} = \frac{(n-1)!}{(r-1)!(n-r)!} p^r(1-p)^{n-r},$$

where $\tilde{n}$ can take on the values $r, r+1, r+2, \ldots$. For the example presented in Section 3.3, calculate the posterior probabilities under the assumption that the sampling was conducted with r fixed rather than with n fixed (that is, for the first sample, assume Pascal sampling with $r = 1$; for the second sample, assume Pascal sampling with $r = 2$). Compare your results with the posterior probabilities calculated in Section 3.3 and explain the relationship between the two sets of results.

29. Explain the similarities and differences between the Bernoulli process and the Poisson process. Give some examples of realistic situations in which the Bernoulli model might be applicable, and do the same for the Poisson model. Explain the terms "stationary" and "independent" with regard to a Poisson process.

30. If $\tilde{p}$ is small and n is large, binomial probabilities may be approximated by Poisson probabilities with $\tilde{\lambda}t = n\tilde{p}$. For $n = 100$ and $\tilde{p} = .05$, compare the binomial distribution with the Poisson approximation.

31. Suppose that cars arrive at a toll booth according to a Poisson process with intensity five per minute. What is the probability that no cars will arrive in a particular minute? What is the probability that more than eight cars will arrive in a particular minute? What is the probability that exactly nine cars will arrive in a given two-minute period, and what is the expected number of arrivals during that period? Comment on the applicability of the Poisson process for this example.

32. Suppose that incoming telephone calls behave according to a Poisson process with intensity 12 per hour. What is the probability that more than 15 calls will occur in any given one-hour period? If the person receiving the calls

takes a 15-minute coffee break, what is the probability that no calls will come in during the break? What is the expected number of incoming calls during an eight-hour work day?

33. Suppose that you feel that accidents along a particular stretch of highway occur roughly according to a Poisson process and that the intensity of the process is either 2, 3, or 4 accidents per week. Your prior probabilities for these three possible intensities are .25, .45, and .30, respectively. If you observe the highway for a period of three weeks and if 10 accidents occur, what are your posterior probabilities?

34. In Exercise 32, suppose that the switchboard operator is unsure whether the intensity of incoming calls is 10, 11, 12, or 13 per hour. She feels that 12 is the most likely value and that 12 is twice as likely as 13, twice as likely as 11, and five times as likely as 10. Find her prior distribution for $\tilde{\lambda}$. On the basis of a sample of 30 minutes with seven calls, revise this distribution of $\tilde{\lambda}$ by using Bayes' theorem.

35. As the owner of the automobile dealership discussed in Section 3.4, you have just hired another new salesman. You know little about your new salesman, so you feel that the past data (10 percent "great," 40 percent "good," and 50 percent "poor") accurately reflect your judgments about him. In the first 12 days on the job, the new salesman sells five cars. What is your posterior distribution? You decide to keep him on the payroll if the expected number of cars sold per day (according to the posterior distribution) is at least .32. Should the new salesman be kept on the payroll?

36. Given the limitations of the Poisson table in the back of the book, how would you determine the posterior distribution in Exercise 35 by using this table if the sample consists of 50 cars sold in 120 days?

37. A bank official is concerned about the rate at which the bank's tellers provide service for their customers. He feels that all of the tellers work at about the same speed, which is either 30, 40, or 50 customers per hour. Furthermore, 40 customers per hour is twice as likely as each of the other two values, which are assumed to be equally likely. In order to obtain more information, the official observes all five tellers for a two-hour period, noting that 380 customers are served during that period. Use this new information to revise the official's probability distribution of the rate at which the tellers provide service.

38. Suppose that you are concerned about the degree of pollution of a particular body of water. In particular, you decide to measure the pollution in terms of the number of microorganisms (within a certain class of microorganisms) in the body of water. You assume that the "process" of the occurrence of microorganisms is stationary and independent; that is, the expected number of microorganisms per cubic millimeter of water is the same everywhere in the body of water, and the number of microorganisms found in one cubic millimeter is independent of the number found in any other cubic millimeter.

You feel that the "intensity of occurrence" of these microorganisms is either 50 per cubic millimeter, in which case the body of water is not polluted; 75 per cubic millimeter, in which case it is mildly polluted; or 100 per cubic millimeter, in which case it is highly polluted. Furthermore, on the basis of your current information about the body of water, you think that (a) the odds are 3 to 1 that it is polluted and (b) given that it is polluted, the odds are even that it is highly polluted. You obtain sample information in the form of two cubic millimeters of water from the body of water, and these two cubic millimeters contain 180 microorganisms. What is your posterior distribution for the intensity of occurrence of the microorganisms?

39. Suppose that $\tilde{p}$ represents the proportion (rounded to the nearest tenth) of males among the full-time students at the University of Chicago. Thus, the possible values of $\tilde{p}$ are 0, .1, .2, .3, .4, .5, .6, .7, .8, .9, and 1.0. Assess your subjective probability distribution for $\tilde{p}$. If you take a random sample of 10 students and observe 6 males, use this sample to revise your distribution.

40. Do the first part of Exercise 39, letting $\tilde{p}$ represent (a) the proportion of students at the University of Chicago who are undergraduates, (b) the proportion of students at the University of Chicago who are United States citizens, and (c) the proportion of 21-year-olds in the United States who are *full-time* students at colleges or universities.

41. In Exercise 31, suppose that after seeing a sample in which eight cars arrive at the toll booth in a two-minute period, your posterior distribution for $\tilde{\lambda}$, intensity per minute, is $P(\tilde{\lambda} = 4) = .40$, $P(\tilde{\lambda} = 5) = .50$, and $P(\tilde{\lambda} = 6) = .10$. What was your prior distribution for $\tilde{\lambda}$ before seeing this sample?

42. Suppose that $\tilde{\lambda}$ represents the number of planes per minute landing at O'Hare Field in Chicago during the hours of 5 P.M. to 7 P.M. on a Friday night in June. Assuming for simplicity that $\tilde{\lambda}$ can only take on integer values, assess your prior distribution for $\tilde{\lambda}$. If you are then told that in a randomly chosen 30-second period during the above-stated hours, exactly four planes landed, revise your distribution for $\tilde{\lambda}$ intuitively without using any formulas. Then, assuming a Poisson process, revise your distribution formally according to Bayes' theorem. Explain any differences in the two revised distributions; what are the possible implications of these differences with respect to this example and with respect to Bayesian inference in general?

43. Suppose that you are interested in $\tilde{\theta}$, the sales (in thousands) of a particular product, and that $\tilde{\theta}$ depends on whether or not a competing firm introduces a new product. Let $\tilde{\phi} = 1$ if the new product is introduced and let $\tilde{\phi} = 0$ otherwise. From your knowledge of the competing firm, you assess $P(\tilde{\phi} = 1) = .30$. Furthermore, suppose that the possible values of $\tilde{\theta}$ are taken to be 200, 250, 300, and 350. You feel that if the competitor introduces a new product, your probabilities for the four possible values of $\tilde{\theta}$ are .30, .50, .15, and .05, respectively. If the new product is not introduced, your probabilities for $\tilde{\theta}$ are .10, .10, .40, and .40, respectively. Express these probability assessments on a tree diagram and calculate the joint probability

distribution of $\tilde{\phi}$ and $\tilde{\theta}$. From this joint distribution, find the marginal distribution of $\tilde{\theta}$.

44. Suppose that $\tilde{\theta} = 1$ if a member of the Democratic party is elected president of the United States in the next presidential election, $\tilde{\theta} = 2$ if a member of the Republican party is elected president, and $\tilde{\theta} = 3$ if a person who is a member of neither party is elected president. Furthermore, let $\tilde{\phi}$ represent the number of senators from the Democratic party after the next presidential election. Assess your subjective probability distribution of $\tilde{\theta}$, and assess your subjective conditional probability distribution of $\tilde{\phi}$, given that $\tilde{\theta} = 1$, given that $\tilde{\theta} = 2$, and given that $\tilde{\theta} = 3$. Use these distributions to find the joint distribution of $\tilde{\theta}$ and $\tilde{\phi}$ and the marginal distribution of $\tilde{\phi}$.

45. Why is the assessment of a joint probability distribution of two random variables greatly simplified if it can be assumed that the two variables are independent?

46. Explain the differences among the binomial, multinomial, and hypergeometric distributions, and for each, give examples of realistic situations in which the model might be applicable. When might the binomial distribution serve as a good approximation to the hypergeometric distribution? Explain your answer.

47. A university committee consists of six faculty members and four students. If a subcommittee of four persons is to be chosen randomly from the committee, what is the probability that it will consist of *at least* two faculty members? Compute this probability from the hypergeometric distribution *and* from the binomial approximation with $\tilde{p} = \tilde{R}/\tilde{N}$. How good is the approximation?

48. If 12 cards are to be drawn at random without replacement from a standard, 52-card deck, what is the probability that there will be exactly three cards of each suit? What is this probability if the 12 cards are drawn *with* replacement? (Assume that the deck is thoroughly shuffled before each draw.)

49. For any given match, the probabilities that a soccer team will win, tie, or lose are .3, .4, and .3. If the team plays 10 matches and if the outcomes of the different matches are assumed to be independent, what is the probability that they will win five, tie three, and lose two? What is the probability that they will win exactly eight matches?

50. What is the probability that in 24 tosses of a fair die, each face will occur exactly four times? What is the probability that in six tosses, each face will occur exactly once?

51. An instructor in an introductory statistics class knows that out of 50 students, 12 are freshmen, 25 are sophomores, 11 are juniors, and only 2 are seniors. He assigns these students to 5 extra review sessions randomly, 10 to a session. What is the probability that the first session consists of 1 freshman, 1 sophomore, and 8 juniors? What is the probability that the first session

consists only of sophomores *and* that the second session consists of 8 juniors and 2 seniors?

52. Why are statistical models such as the Bernoulli and the Poisson models useful with regard to the assessment of likelihoods? What are the advantages and disadvantages of using such models?

53. Suppose that a judge is presiding over an antitrust case and that he is concerned with the effect of a particular merger on competition in the brewing industry. Let $\tilde{\theta} = 2$ if the merger has a severe effect on competition, $\tilde{\theta} = 1$ if it has a minor effect, and $\tilde{\theta} = 0$ if it has no effect. The judge feels that $\tilde{\theta} = 0$ and $\tilde{\theta} = 1$ are equally likely and that each is three times as likely as $\tilde{\theta} = 2$. An economist is called in as a witness to provide additional information, and the gist of his testimony is that the merger is not harmful to competition in the brewing industry. The judge considers this to be important evidence, but he does not accept it completely at face value because he knows that the economist is being retained by the main firm involved in the merger. In particular, he feels that the testimony given by the economist is most likely if, in fact, $\tilde{\theta} = 0$ and least likely if, in fact, $\tilde{\theta} = 2$. Furthermore, he views the new information as four times as likely if $\tilde{\theta} = 0$ than if $\tilde{\theta} = 2$ and twice as likely if $\tilde{\theta} = 0$ than if $\tilde{\theta} = 1$. After the economist testifies, what is the judge's probability distribution for $\tilde{\theta}$?

54. In Exercise 53, suppose that a second economist testifies that the merger is not harmful to the brewing industry. Like the first economist, he is being retained by the main firm involved in the merger. The judge feels that this new testimony is twice as likely if $\tilde{\theta} = 0$ than if $\tilde{\theta} = 2$ and 1.5 times as likely if $\tilde{\theta} = 0$ than if $\tilde{\theta} = 1$. After the second economist testifies, what is the judge's probability distribution for $\tilde{\theta}$? Compare the likelihoods in Exercise 53 with the likelihoods in this exercise, and comment on any differences. Might the likelihoods assessed by the judge for the second economist's testimony be different if the second economist were not being retained by any of the firms involved in the merger or if the *first* economist had stated that the merger is harmful to the brewing industry? Explain your answer.

55. Discuss the implications of the conditional dependence of pieces of information with regard to the assessment of likelihoods and to successive applications of Bayes' theorem. Give some examples of situations in which it might be assumed that the data-generating process of interest is conditionally independent and some examples of situations in which this assumption would not be reasonable.

56. Predictive probabilities and likelihoods both involve the probability of a future sample outcome $\tilde{y}$. What, then, is the difference between the two types of probabilities?

57. In Exercise 23, prior to taking the sample of size 10, find the predictive distribution of $\tilde{r}$, the number of consumers in the sample who state that they will buy the product.

58. In Exercise 24, prior to taking the first sample of size three, find the predictive distribution of $\tilde{r}$, the number of heads in the three tosses. Then, once the first sample is observed (heads once in three tosses), find the predictive distribution of the number of heads in the next three tosses. Why is this second predictive distribution different from the first predictive distribution?

59. In Exercise 33, after seeing the sample of three weeks and revising your distribution of $\tilde{\lambda}$ accordingly, find the predictive distribution for the number of accidents in the next week.

60. In Exercise 35, after seeing the sample of five cars sold in 12 days, what is your predictive distribution for the number of cars sold by the salesman in the next four days?

4

BAYESIAN INFERENCE FOR CONTINUOUS PROBABILITY MODELS

4.1 CONTINUOUS PROBABILITY MODELS

The previous chapter dealt with *discrete* random variables, which take on only a finite or countably infinite number of values. There are many variables that, conceptually at least, could take on an infinite (but not countably infinite) number of values. For example, consider the weight of an object or the temperature at a particular time and place. The underlying sample space in these examples consists not just of integers or decimal expressions given to the nearest one-hundredth; conceptually, variables such as weight and temperature can take on any real number as a value. In practice, limitations of measurement devices imply that the *observed* weight or temperature is discrete. For instance, a balance may only be accurate to the nearest tenth of an ounce. It is still convenient, however, to use *continuous* probability models as approximate representations in such situations. In this section, the concept of continuous probability models is briefly discussed, and the remainder of the chapter is concerned with the application of Bayes' theorem to such models.

Because a continuous random variable can take on an infinite number of possible values (such as the entire real line), it is not worthwhile to talk of probabilities of the form $P(\tilde{x} = x)$. If $\tilde{x}$ is continuous, $P(\tilde{x} = x) = 0$ for all possible values x of $\tilde{x}$; no single point can have positive probability in a continuous probability model. Instead of considering probabilities for single values, then, it is necessary to consider probabilities for intervals of values, such as $P(a \leq \tilde{x} \leq b)$. Thus, if $\tilde{x}$ represents the maximum official temperature in degrees Fahrenheit at Chicago on December 25, 1990, and if it is assumed that $\tilde{x}$ is continuous, then it is worthwhile to consider probabilities such as $P(20 \leq \tilde{x} \leq 40)$, $P(25 \leq \tilde{x} \leq 35)$, $P(29 \leq$

$\tilde{x} \leq 31)$, $P(29.9 \leq \tilde{x} \leq 30.1)$, $P(29.99 \leq \tilde{x} \leq 30.01)$, and so on, but it is *not* worthwhile to consider $P(\tilde{x} = 30)$.

The probability distribution of a continuous random variable $\tilde{x}$ is represented by a *probability density function*, or PDF, rather than by a probability mass function. The density function is developed by considering the probability of an interval divided by the length of the interval:

$$\frac{P(a \leq \tilde{x} \leq b)}{b - a}.$$

If the left-hand end point of the interval, a, is fixed while the right-hand end point, b, moves closer to a, then the probability associated with the interval and the length of the interval both decrease. In the limit, as the length of the interval approaches 0, the ratio $P(a \leq \tilde{x} \leq b)/(b - a)$ approaches the probability *density* of the random variable $\tilde{x}$ at the value a (provided, of course, that a limit exists). In general, a probability density function will be denoted by f, so that the value of the density function of $\tilde{x}$ at the value x is simply $f(x)$.

The above development of a probability density function becomes clearer when it is presented in graphical form. Consider a histogram (bar graph) in which the probability of an interval is represented by the area of the "bar" over the interval (Figure 4.1.1). This area is, of course, given by the product of the height of the bar and the length of the interval; thus, the height of the bar equals the probability of the interval divided by the length of the interval. If each of the intervals is divided into two subintervals, a new histogram could be drawn using the new subintervals, as in Figure 4.1.2. If this process of dividing intervals is repeated over and over, the length of each interval becomes smaller and the histogram begins to look more like a smooth curve. In the limit, it *is* a smooth curve, the probability density function (Figure 4.1.3). Since the height of a bar of the histogram equals the probability of the interval divided by the length of the interval, the limit of this ratio is the height of the density function, as indicated in the previous paragraph.

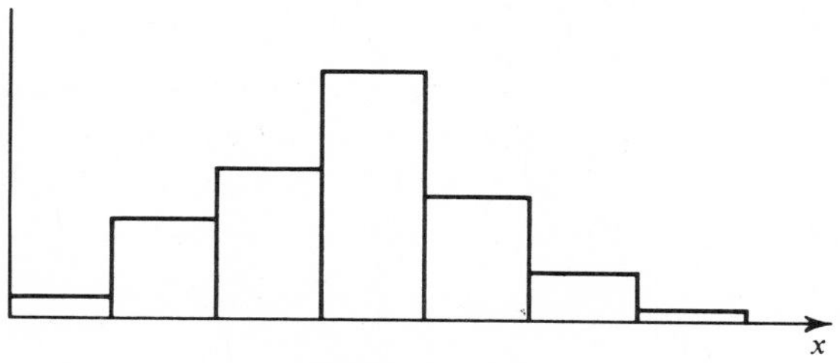

Figure 4.1.1

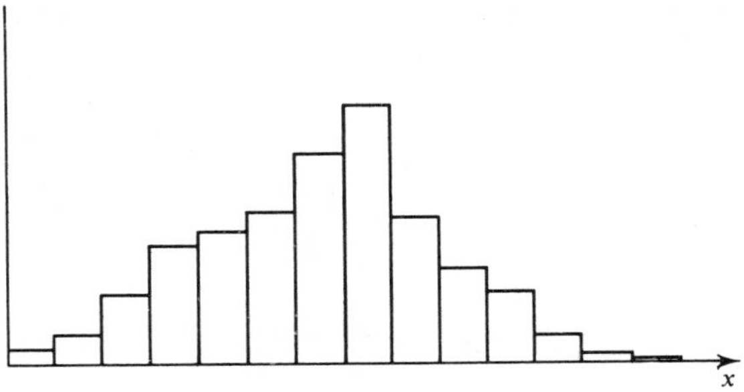

Figure 4.1.2

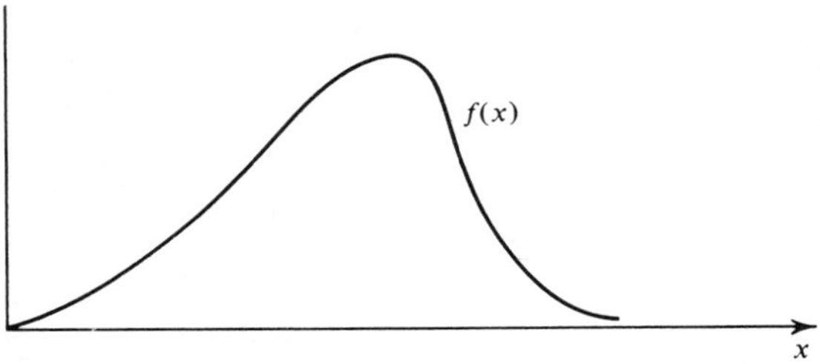

Figure 4.1.3

To illustrate this process, suppose that $\tilde{x}$ represents the weight (in pounds) of a person randomly chosen from the student body of a particular university. First, the weights might be grouped into 50-pound intervals, yielding a histogram comparable to Figure 4.1.1. Next, to obtain greater precision, the weights might be grouped into 25-pound intervals, yielding a histogram comparable to Figure 4.1.2. This process of reducing the length of each interval can be repeated, yielding histograms with 12.5-pound intervals, 6.25-pound intervals, and so on. As the interval length decreases (and correspondingly, the number of intervals increases), the histogram looks more and more like the smooth curve illustrated in Figure 4.1.3.

For the histogram, probability is represented by the areas of the "bars." But the PDF is simply a limiting form of the histogram, so probability in a continuous probability model is represented by area under the PDF. Mathematically, the area under a curve over some interval of values is represented by a definite integral, so that the probability of an interval can be written as follows (see Figure 4.1.4):

$$P(a \leq \tilde{x} \leq b) = \int_a^b f(x)\, dx. \tag{4.1.1}$$

If you are unfamiliar or uncomfortable with calculus, you can think of a definite integral as the limiting form of a sum or as the area under a curve. The notation $\int_a^b f(x)\,dx$ is merely a mathematical notation for the area under the graph of the function $f(x)$ between the points a and b on the horizontal axis. This, of course, relates to the above discussion (and to Figures 4.1.1 through 4.1.3) concerning the concept of a density function. If the smooth curve in Figure 4.1.3 is divided into a large number of vertical segments and approximated by a histogram (such as the one in Figure 4.1.2), then the sum of the areas of the bars under the histogram between two points a and b on the horizontal axis should be approximately equal to the area under the smooth curve between a and b, as determined by Equation (4.1.1).

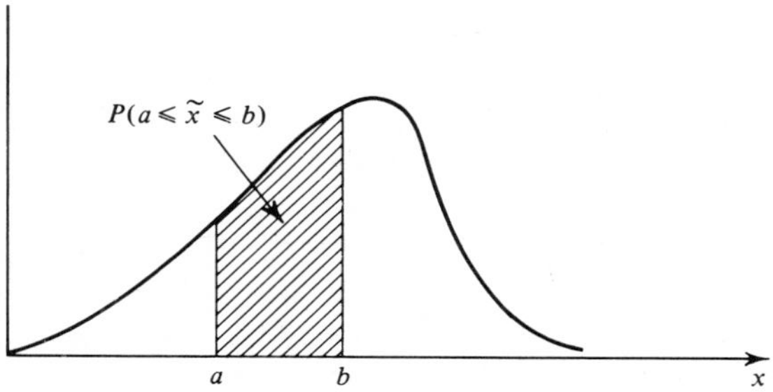

Figure 4.1.4

It should be strongly emphasized that the *height* of the density function does *not* represent probability; indeed, the probability of any single point is 0. It is the *area* under the curve that represents probability. In order that probabilities such as this will always satisfy the basic axioms of probability theory, a probability density function $f(x)$ must satisfy two properties,

$$f(x) \geq 0 \quad \text{for all } x$$

and

$$\int_{-\infty}^{\infty} f(x)\,dx = 1.$$

That is, $f(x)$ cannot be negative, and the total area under the curve must equal 1. These properties are analogous to the two properties required of discrete probability models (the probabilities must be nonnegative and they must sum to 1). Incidentally, one interesting feature of continuous probability models is that since the probability of any single point is always 0, the probability of an interval is the same whether or not the end points of the interval are included in the interval. In other words, $P(a \leq \tilde{x} \leq b) = P(a < \tilde{x} < b)$. Of course, this is not true for discrete probability models [see Exercise 1, parts (d) and (e), in Chapter 3].

The cumulative probability $P(\tilde{x} \leq x)$ is the probability of the interval from minus infinity to x. Thus, the *cumulative distribution function* (CDF) for a continuous random variable is given by

$$P(\tilde{x} \leq x) = \int_{-\infty}^{x} f(x)\, dx. \tag{4.1.2}$$

Sometimes the CDF is denoted by $F(x)$ in both the discrete case and the continuous case.

For continuous random variables, the above equation provides a means of determining $F(x)$ from the density function, $f(x)$. To find the density function from the cumulative function, simply differentiate:

$$f(x) = \frac{d}{dx} F(x). \tag{4.1.3}$$

Recall that $f(x)$ was defined to be the limit, as b approaches x, of $P(x \leq \tilde{x} \leq b)/(b - x)$, which is equal to $[F(b) - F(x)]/(b - x)$. But the limit of this last expression as b approaches x is nothing other than the derivative of $F(x)$. Graphically, the derivative of a function at a point x corresponds to the slope of the line tangent to the function at that point. Once again, if you are unfamiliar with calculus, you should not concern yourself with the mathematical definition of a derivative; simply interpret the notation $\frac{d}{dx} F(x)$ as the slope of the function $F(x)$ at the value x. An example of a CDF for a continuous probability model is illustrated in Figure 4.1.5; note that whereas a CDF in the discrete case is always a step function, a CDF in the continuous case is a continuous, nondecreasing function. The graph of a CDF in the continuous case is a reasonably "smooth" curve.

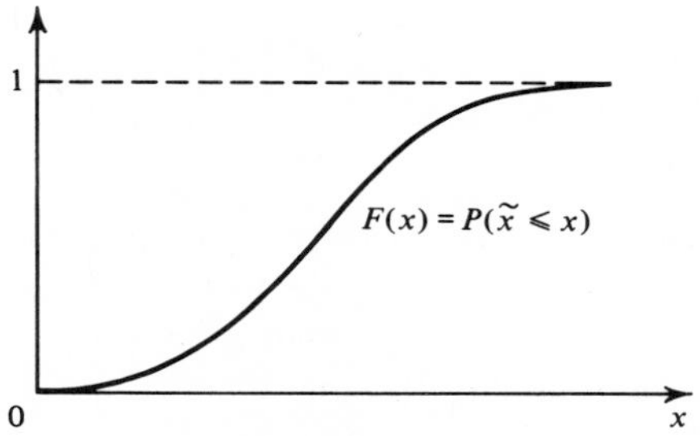

Figure 4.1.5

For a simple example of a continuous random variable, consider a variable $\tilde{x}$ with a "uniform" density function on the interval from a

to b. The density function, which is graphed in Figure 4.1.6, is of the form

$$f(x) = \begin{cases} 1/(b-a) & \text{if } a \le x \le b, \\ 0 & \text{elsewhere.} \end{cases}$$

Note that the two requirements of a density function are satisfied, since $f(x) \ge 0$ for all x and the total area under the density function is simply the area of a rectangle of length $(b - a)$ and height $1/(b - a)$; this area is equal to 1. The CDF of $\tilde{x}$, which is graphed in Figure 4.1.7, is of the form

$$F(x) = \begin{cases} 1 \quad \text{if } x > b, \\ \int_a^x [1/(b-a)]\, dx = (x-a)/(b-a) \quad \text{if } a \le x \le b, \\ 0 \quad \text{if } x < a. \end{cases}$$

Note that the derivative of $F(x)$ is equal to $f(x)$. Whenever integration or differentiation is used to obtain a result in this book, as is the case in this example, the reader unfamiliar with calculus can safely ignore the technical details and concentrate on the final result. A basic understanding of what the result means is much more important than an understanding of the technical details of its derivation, and graphs such as Figures 4.1.6 and 4.1.7 are extremely useful in this regard.

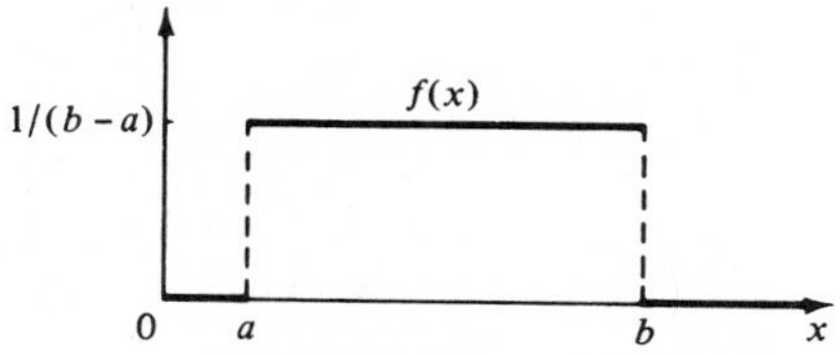

Figure 4.1.6

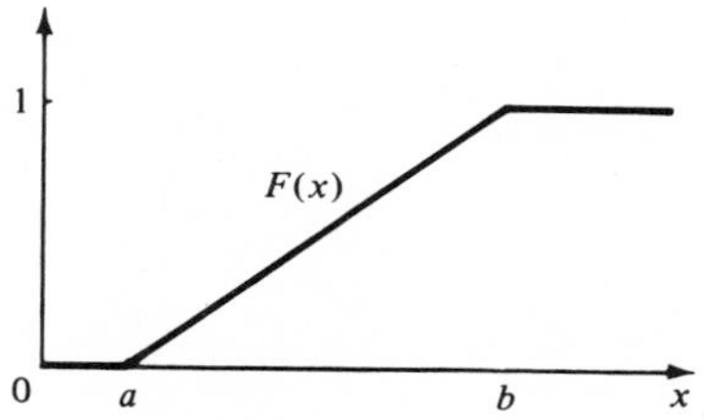

Figure 4.1.7

Under what conditions might a uniform density be reasonable? Suppose that you feel that the temperature (in degrees Fahrenheit) at noon tomorrow will be between 50 and 70. In addition, you think that any interval of width one degree within the limits of 50 and 70 is just as likely as any other interval of width one degree within these limits. That is, any interval of width one degree within the limits has probability $\frac{1}{20}$. This means that for any x such that $50 \leq x \leq 69$ (x does not have to be an integer), you feel that 19 to 1 odds are fair odds against the temperature being between x and $x + 1$. Equivalently, you are indifferent between two lotteries, one offering you \$100 if the temperature is between x and $x + 1$ and \$0 otherwise, and a second offering you \$100 with probability 1/20 and \$0 with probability 19/20. Furthermore, suppose that you feel that all intervals of *any* specified width are equally likely, not just intervals of width one. If this assumption is met in the temperature example, the distribution of temperature is a uniform distribution with $a = 50$ and $b = 70$.

For a second example of a continuous random variable, consider a random variable $\tilde{x}$ with a "triangular" density function of the form

$$f(x) = \begin{cases} 2x/b^2 & \text{if } 0 \leq x \leq b, \\ 0 & \text{elsewhere.} \end{cases}$$

The corresponding CDF is

$$F(x) = \begin{cases} 1 & \text{if } x > b, \\ \int_0^x (2x/b^2)\, dx = x^2/b^2 & \text{if } 0 \leq x \leq b, \\ 0 & \text{if } x < 0. \end{cases}$$

The PDF and CDF of $\tilde{x}$ are graphed in Figures 4.1.8 and 4.1.9. Note that for this random variable, intervals of a given width are more probable as the limits are increased. If $b = 4$, for instance, $P(3 \leq \tilde{x} \leq 4) = F(4) - F(3) = 7/16$, whereas $P(0 \leq \tilde{x} \leq 1) = F(1) - F(0) = 1/16$.

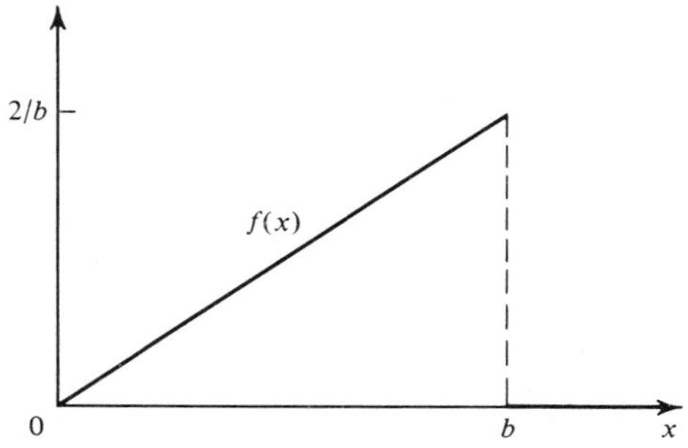

Figure 4.1.8

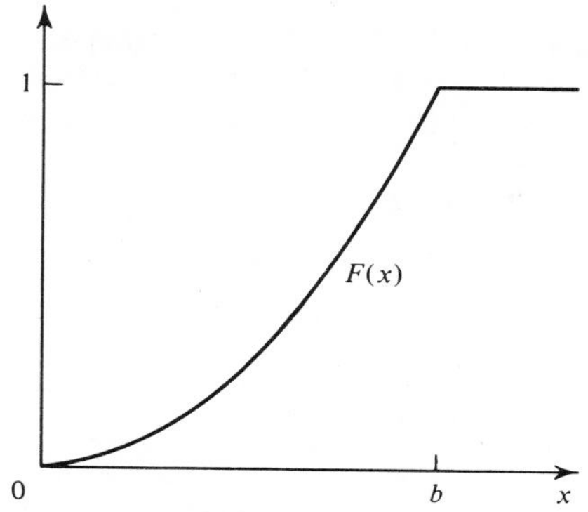

Figure 4.1.9

As in the discrete case, certain summary measures of a probability distribution are often of interest. The *expectation* of a continuous random variable (or a function of the variable) is defined analogously to the expectation of a discrete random variable [Equations (3.1.3) and (3.1.4)], with the density function $f(x)$ replacing the individual probabilities $P(\tilde{x} = x_j)$ and an integral sign replacing the summation sign:

$$E(\tilde{x}) = \int_{-\infty}^{\infty} xf(x)\,dx, \tag{4.1.4}$$

and

$$E[t(\tilde{x})] = \int_{-\infty}^{\infty} t(x)f(x)\,dx. \tag{4.1.5}$$

For instance, for the uniform distribution discussed above,

$$E(\tilde{x}) = \int_{-\infty}^{\infty} xf(x)\,dx = \int_a^b \frac{x}{(b-a)}\,dx = \left[\frac{x^2}{2(b-a)}\right]\Bigg|_a^b$$
$$= \frac{b^2 - a^2}{2(b-a)} = \frac{b+a}{2}.$$

This is the midpoint of the interval from a to b, which is an intuitively reasonable result, given the "uniform" nature of the distribution. For the triangular distribution,

$$E(\tilde{x}) = \int_{-\infty}^{\infty} xf(x)\,dx = \int_0^b \frac{2x^2}{b^2}\,dx = \left[\frac{2x^3}{3b^2}\right]\Bigg|_0^b$$
$$= \frac{2b^3}{3b^2} = \frac{2b}{3}.$$

Note that this is greater than the midpoint of the interval from 0 to b, since the density function is higher for the larger values of $\tilde{x}$.

To illustrate the use of Equation (4.1.5), suppose that someone offers you a bet in which you will "win" $t(\tilde{x}) = 5 - \tilde{x}^2$, where $\tilde{x}$ has a triangular distribution with $b = 4$ (if the result is negative, your "win" will actually be a loss). The expected amount you will "win" is then

$$E[t(\tilde{x})] = E(5 - \tilde{x}^2) = \int_{-\infty}^{\infty} (5 - x^2)f(x)\,dx$$

$$= \int_0^4 (5 - x^2)\frac{x}{8}\,dx = \int_0^4 \frac{5x - x^3}{8}\,dx$$

$$= \left[\frac{5x^2}{16} - \frac{x^4}{32}\right]\Bigg|_0^4 = \frac{5(16)}{16} - \frac{256}{32} = 5 - 8 = -3.$$

Since your expected winnings are negative, this is not an advantageous bet for you. The intuitive interpretation of expectation is the same in the continuous and discrete cases; it can be thought of as an "average value" or, in terms of a graph, as a center of gravity.

The definitions of *variance* and of *standard deviation* and all 14 of the rules presented in Section 3.1 for dealing with expectations and variances apply in the continuous case as well as in the discrete case. For the uniform and triangular distributions, respectively,

$$V(\tilde{x}) = E(\tilde{x}^2) - [E(\tilde{x})]^2$$

$$= \left[\int_a^b \frac{x^2}{(b-a)}\,dx\right] - \left[\frac{(b+a)^2}{4}\right]$$

$$= \left[\frac{x^3}{3(b-a)}\right]\Bigg|_a^b - \left[\frac{(b+a)^2}{4}\right]$$

$$= \frac{b^3 - a^3}{3(b-a)} - \left[\frac{(b+a)^2}{4}\right]$$

$$= \frac{b^2 + ab + a^2}{3} - \frac{b^2 + 2ab + a^2}{4} = \frac{(b-a)^2}{12}$$

and

$$V(\tilde{x}) = E(\tilde{x}^2) - [E(\tilde{x})]^2$$

$$= \int_0^b \frac{2x^3}{b^2}\,dx - \frac{4b^2}{9}$$

$$= \left[\frac{x^4}{2b^2}\right]\Bigg|_0^b - \frac{4b^2}{9}$$

$$= \frac{b^4}{2b^2} - \frac{4b^2}{9}$$

$$= \frac{b^2}{2} - \frac{4b^2}{9} = \frac{b^2}{18}.$$

The concept of a standardized random variable is also applicable in the continuous case, as are the concepts of skewness and moments of a random variable (see Section 3.1).

In continuous probability models, joint probability distributions of more than one random variable can be represented by *joint density functions*. If $\tilde{x}$ and $\tilde{y}$ are continuous random variables, then $f(x,y)$ represents their joint density function. As in continuous models of just one random variable, the probability of any *single* point, which is a pair of values ($\tilde{x} = x$, $\tilde{y} = y$), is 0; it is necessary to consider joint *intervals* of values. The function $f(x,y)$ is a surface in three-dimensional space, with probability represented by *volume* under this surface (between the surface and the xy-plane). An example of a joint density function is presented in Figure 4.1.10.

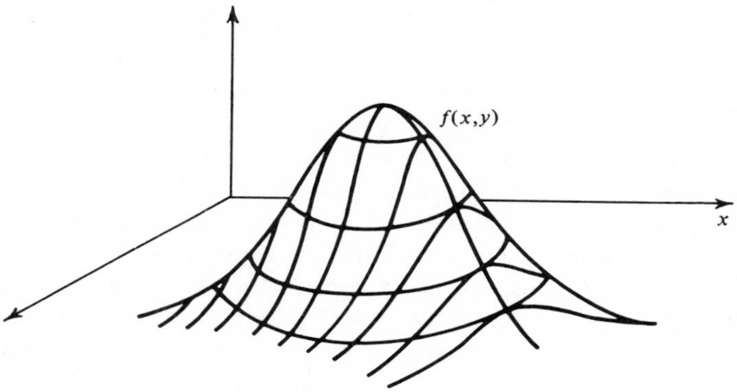

Figure 4.1.10

Sometimes the joint density is given, but one of the individual densities (called a *marginal density function*) is wanted. This can be obtained by integrating the joint density with respect to the *other* variable:

$$f(x) = \int_{-\infty}^{\infty} f(x,y)\, dy \tag{4.1.6}$$

and

$$f(y) = \int_{-\infty}^{\infty} f(x,y)\, dx. \tag{4.1.7}$$

This is analogous to the discrete case, in which the marginal distribution is obtained by summing the joint distribution across the other variable (Section 3.1). The continuous variables $\tilde{x}$ and $\tilde{y}$ are said to be *independent* if their joint density function equals the product of their individual density functions, that is, if $f(x,y) = f(x)f(y)$.

As indicated in the previous chapter, the *conditional probability distribution* of one random variable $\tilde{x}$, given a value of a second random variable

$\tilde{y}$, is often of interest. In the continuous case, the *conditional density function of $\tilde{x}$ given that $\tilde{y} = y$* is defined as follows:

$$f(x \mid \tilde{y} = y) = \frac{f(x,y)}{f(y)}. \tag{4.1.8}$$

Of course, the value y of $\tilde{y}$ must be such that $f(y) > 0$. Note that $\tilde{x}$ and $\tilde{y}$ are independent if $f(x \mid \tilde{y} = y) = f(x)$ for each possible value of $\tilde{y}$ or if $f(y \mid \tilde{x} = x) = f(y)$ for each possible value of $\tilde{x}$. That is, knowledge about one of the variables does not affect the density function of the other variable.

As you shall see in the next section, the application of Bayes' theorem to continuous probability models results in a conditional density function, where $\tilde{y}$ represents sample information and $\tilde{x}$ is the random variable (uncertain quantity) of interest. The concepts of conditional expectation and conditional variance, defined for discrete random variables in Equations (3.1.10) and (3.1.11), are defined analogously for continuous random variables.

As in the previous chapter, it is assumed that all probability distributions are ultimately of a subjective nature, whether they are directly assessed subjectively or whether they are determined from some statistical model, in which case the choice of a model (that is, the acceptance of the assumptions of a given model) is subjective. In general, the assessment of probability distributions may be more difficult in the continuous case than in the discrete case, although techniques such as bets and lotteries may still be valuable in the assessment process. The assessment of probability distributions for continuous random variables is discussed in Section 4.8.

This section has involved continuous random variables, whereas Chapter 3 covered discrete random variables. It should be mentioned that it is possible for a random variable to be both discrete and continuous. That is, part of the distribution can be represented by a density function and the remainder of the distribution can be represented by a mass function. Such random variables, which are sometimes called "mixed" random variables, can be handled by straightforward combinations of the methods presented in Section 3.1 and in this section. Mixed random variables are utilized in Chapter 7.

This section was similar to Section 3.1 in that it was not meant to be either detailed or comprehensive; its purpose was to provide a brief discussion of some of the important concepts involving continuous probability models. Moreover, references are given at the end of the chapter for the reader who is interested in more detailed discussions of these concepts. Continuous probability models can be thought of as limiting forms of discrete probability models; the usual operations with the model are similar, with an integral replacing a sum. Of course, continuous models represent an idealization that is never realized in practice. The limitations of available measuring instruments and computing devices essentially imply that the uncertain quantities actually observed in the real world are discrete (one cannot obtain measurements to an infinite number of decimal places, and even if one could, computers could not handle the numbers). But if this is true, why are continuous models even studied? They are studied because (1) they are often mathematically easier to work with than discrete models, and (2) they provide an excellent approximation to numerous discrete models. A discrete random variable with a large number of possible values may prove very cumbersome to handle, and it may be much easier to use a continuous model as an approximation. In the remainder of this chapter, the application of Bayes' theorem to such models is discussed.

4.2 BAYES' THEOREM FOR CONTINUOUS PROBABILITY MODELS

In many problems involving statistical inference and decision, it is realistic and convenient to assume that the random variable (uncertain quantity) of interest is continuous. It should be emphasized that measurement procedures are such that actual observed random variables are never truly continuous. In many situations, however, the measurement may be precise enough so that for all practical purposes it can be assumed that a random variable is continuous. Indeed, this assumption is often quite realistic. For example, in the quality-control illustration used in Section 3.3, it would be much more realistic to assume that $\tilde{p}$, the proportion of items that are defective, is a continuous random variable than to assume that it only takes on the four values .01, .05, .10, and .25. In Section 4.4, the quality-control illustration is modified so that $\tilde{p}$ is taken to be continuous. In the automobile-salesman example presented in Section 3.4, it was assumed that $\tilde{\lambda}$, the average number of cars sold per day, was either $\frac{1}{2}$, $\frac{1}{4}$, or $\frac{1}{8}$; it would be more realistic to consider $\tilde{\lambda}$ as a continuous random variable.

Suppose that the uncertain quantity of interest in some inferential or decision-making problem is continuous, and call this uncertain quantity $\tilde{\theta}$. Furthermore, suppose that sample information involving $\tilde{\theta}$ can be summarized by the sample statistic $\tilde{y}$. If $\tilde{y}$ contains all of the information from the sample that is relevant with regard to the uncertainty about $\tilde{\theta}$, then $\tilde{y}$ is called a *sufficient* statistic. For instance, under the assumptions of the Bernoulli model presented in Section 3.3, the information in a sample may be summarized by n (the sample size) and r (the number of "successes"). Knowledge of the actual *sequence* of "successes" and "failures" provides no more information about $\tilde{p}$ than does knowledge of n and r, so $y = (n,r)$ is a sufficient statistic. In terms of Bayes' theorem, this means that knowledge of n and r is sufficient to determine the likelihoods, so that the posterior distribution of $\tilde{p}$ given r and n is exactly the same as the posterior distribution of $\tilde{p}$ given the entire sequence of results.

If $\tilde{\theta}$ is continuous, the prior and posterior distributions of $\tilde{\theta}$ can be represented by density functions. The posterior distribution is the conditional density of $\tilde{\theta}$, given the observed value, y, of the sample statistic $\tilde{y}$. But, from Equation (4.1.8), this can be written as

$$f(\theta \mid \tilde{y} = y) = \frac{f(\theta,y)}{f(y)}.$$

The conditional density of one random variable given a value of a second random variable is simply the joint density of the two random variables divided by the marginal density of the second random variable.

But the joint density $f(\theta,y)$ and the marginal density $f(y)$ are not generally known; usually, only the prior distribution and the likelihood function are assessed. Fortunately, the required joint and marginal densities can be written in terms of the prior distribution and the likelihood function, as in the discrete case. From Equation (4.1.8),

$$f(\theta,y) = f(\theta)f(y \mid \theta). \tag{4.2.1}$$

Then, from Equations (4.1.7) and (4.2.1),

$$f(y) = \int_{-\infty}^{\infty} f(\theta,y)\, d\theta = \int_{-\infty}^{\infty} f(\theta)f(y \mid \theta)\, d\theta. \tag{4.2.2}$$

The posterior density can thus be written as follows:

$$f(\theta \mid y) = \frac{f(\theta)f(y \mid \theta)}{\int_{-\infty}^{\infty} f(\theta)f(y \mid \theta)\, d\theta}. \tag{4.2.3}$$

This is Bayes' theorem for continuous random variables. The densities $f(\theta \mid y)$ and $f(\theta)$ represent the posterior distribution and the prior distribution, respectively, and $f(y \mid \theta)$ represents the likelihood function. These terms have the same interpretation for continuous random variables as they have for discrete random variables. The prior and posterior distributions must be proper density functions. That is, they must be nonnegative, and the total area under the curve, determined by integrating the density function over its entire domain, must be equal to 1. The integral in the denominator of Equation (4.2.3) serves the same purpose as the sum in the denominator of Equation (3.2.3); it makes the posterior distribution a proper probability distribution. As in the discrete case, the likelihood function is a function of θ, with y fixed (equal to the observed value of $\tilde{y}$). It can be seen that Bayes' theorem for continuous probability models is analogous to Bayes' theorem for discrete probability models. In the discrete case, Bayes' theorem can be expressed in words as

$$\text{posterior probability} = \frac{(\text{prior probability})(\text{likelihood})}{\Sigma(\text{prior probability})(\text{likelihood})},$$

and the corresponding statement for continuous random variables is

$$\text{posterior density} = \frac{(\text{prior density})(\text{likelihood})}{\int(\text{prior density})(\text{likelihood})}.$$

It should be noted that in Equation (4.2.3), it is assumed that $\tilde{\theta}$ is a continuous random variable. The sample statistic $\tilde{y}$, on the other hand, may be discrete or continuous. If the random variable of interest is $\tilde{p}$ from a Bernoulli process and if the sample information consists of the number of "successes" $\tilde{r}$ in a fixed number of trials, then $\tilde{r}$ is discrete, whereas the prior distribution of $\tilde{p}$ may either be discrete (as discussed in Section 3.3) or continuous (as discussed in Section 4.3). The terminology "Bayes' theorem for discrete probability models" and "Bayes' theorem for continuous probability models" refers to the form of the prior and the posterior distributions (that is, it refers to whether the random variable of interest, $\tilde{\theta}$, is assumed to be discrete or continuous).

To illustrate Equation (4.2.3), suppose that $\tilde{\theta}$ represents the market share (expressed as a decimal, not as a percentage) of a new brand of a certain product. The new brand is considerably different from the other brands of the product, so you are quite uncertain about the share of the market that it will attract. You think that it might, if it "catches on," attract virtually the entire market for the product (that is, $\tilde{\theta}$ might be close to 1), it might not be successful at all (that is, $\tilde{\theta}$ might be close to 0), or it might be moderately successful. It seems reasonable to assume that $\tilde{\theta}$ is continuous. Furthermore, you think that low values of $\tilde{\theta}$ are more likely than high values, and you assess a prior distribution for $\tilde{\theta}$ that is a variant of the "triangular" distribution discussed in Section 4.1:

$$f(\theta) = \begin{cases} 2(1-\theta) & \text{if } 0 \leq \theta \leq 1, \\ 0 & \text{elsewhere.} \end{cases}$$

This prior distribution is illustrated in Figure 4.2.1.

In order to obtain more information about $\tilde{\theta}$, a sample of five consumers of the product in question is taken; one purchases the new brand and the other four purchase other brands. You are willing to assume that the process of purchasing this product can be satisfactorily represented by a Bernoulli process, where $\tilde{p}$, the probability that a randomly selected consumer purchases the new brand, is equal to $\tilde{\theta}$, the market share. From Section 3.3, the sample information (one "success" in five trials) can be represented in terms of a likelihood function by the binomial distribution. In the notation of Equation (4.2.3),

$$f(y \mid \theta) = P(\tilde{r} = 1 \mid n = 5, \tilde{\theta} = \theta) = \binom{5}{1}\theta^1(1-\theta)^4$$
$$= 5\theta(1-\theta)^4.$$

This likelihood function is graphed as a function of θ in Figure 4.2.2.

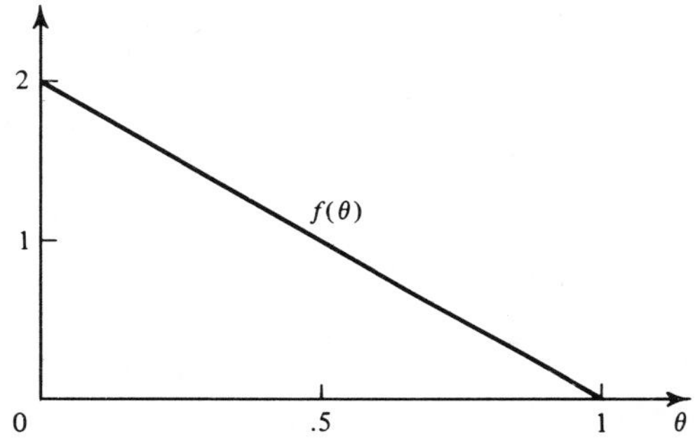

Figure 4.2.1

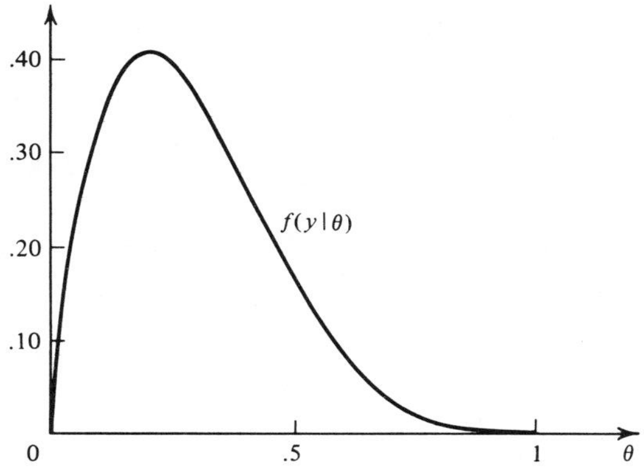

Figure 4.2.2

Applying Equation (4.2.3),

$$f(\theta \mid y) = \frac{f(\theta)f(y \mid \theta)}{\int_{-\infty}^{\infty} f(\theta)f(y \mid \theta)\,d\theta} = \frac{[2(1-\theta)][5\theta(1-\theta)^4]}{\int_0^1 [2(1-\theta)][5\theta(1-\theta)^4]\,d\theta}$$
$$= \frac{10\theta(1-\theta)^5}{10\int_0^1 \theta(1-\theta)^5\,d\theta} = \frac{\theta(1-\theta)^5}{\int_0^1 \theta(1-\theta)^5\,d\theta}$$

if $0 \leq \theta \leq 1$, and $f(\theta \mid y) = 0$ elsewhere. The integral in the denominator is equal to $(1!)(5!)/(7!) = \frac{1}{42}$; thus, the posterior density function of $\tilde{\theta}$ is

$$f(\theta \mid y) = \begin{cases} 42\theta(1-\theta)^5 & \text{if } 0 \leq \theta \leq 1, \\ 0 & \text{elsewhere.} \end{cases}$$

This posterior distribution is illustrated in Figure 4.2.3.

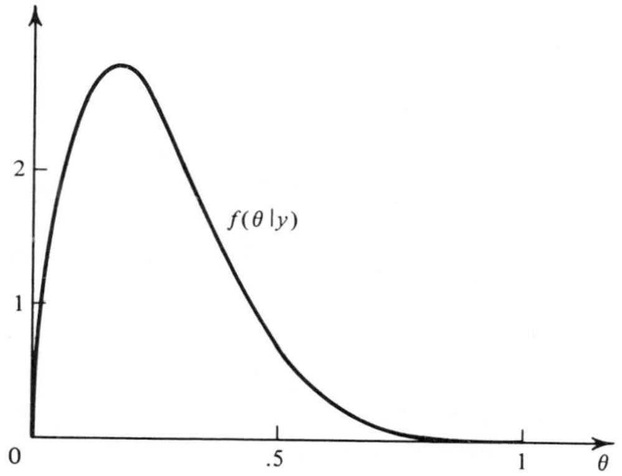

Figure 4.2.3

Conceptually, Equation (4.2.3) provides a convenient way to revise density functions in the light of sample information. In practice, however, it may prove quite difficult to apply this formula. If $f(\theta)$ and $f(y \mid \theta)$ are not fairly simple mathematical functions, it may be a hard task to carry out the integration in the denominator of Equation (4.2.3). In some situations it may be necessary to resort to advanced mathematical techniques to determine the exact posterior distribution. One way to avoid this difficulty is to use discrete approximations of continuous probability models; this approach, which falls under the general heading of numerical methods, is considered in Section 4.9. A second approach, involving the restriction of the prior distribution to a certain family of distributions that depends on the form of the likelihood function, is discussed in the remainder of this section and in the following sections.

Because of potential difficulties in the application of Equation (4.2.3), Bayesian statisticians have developed the concept of *conjugate prior distributions*, which essentially are families of distributions that ease the computational burden when used as prior distributions. Of course, the posterior distribution depends on the likelihood function as well as on the prior distribution. Usually, once certain assumptions are made about the population or process that is being sampled (for example, the assumption that the population is normally distributed or the assumption that the process is a stationary and independent Bernoulli process), the likelihood function is uniquely determined according to the chosen statistical model. In other words, if a particular data-generating model is specified, the likelihood function is known. For any particular model and likelihood function, the Bayesian statistician attempts to determine a conjugate family of prior distributions. This is a set of distributions, each of which can be combined with the given likelihood function without too much difficulty. It should be emphasized that the "conjugate" nature of a family of distributions in any given situation depends on the form of the likelihood function, which in turn depends on the choice of a statistical model. In the market-share example presented earlier in this section, the data-generating process is assumed to follow a Bernoulli model, in which case the likelihood function can be represented by the binomial distribution. The prior and the posterior distributions in that example are members of the family of beta distributions, and the beta family is "conjugate" with respect to the Bernoulli process, as you shall see in Sections 4.3 and 4.4.

Three properties that have been put forth as desirable properties for conjugate families of distributions are (1) mathematical tractability, (2) richness, and (3) ease of interpretation. The first property, mathematical tractability, is the property that motivated the development of conjugate prior distributions. A prior distribution is mathematically tractable if (1) it is reasonably easy to specify the posterior distribution given the prior distribution and the likelihood function; (2) it results in a posterior distribution that is also a member of the same conjugate family, so that successive applications of Bayes' theorem are not difficult; and (3) it is feasible to calculate expectations (such as those required in decision theory) from the prior distribution. The property of mathematical tractability, then, refers to the computational burdens involved in Bayesian inference and decision.

The prior distribution should, of course, reflect the statistician's prior information. Since different people may (and no doubt *will*) have different prior information, a conjugate family of distributions should include, insofar as possible, distributions with different locations, dispersions, shapes, and so on, so as to be able to represent a wide variety of states of prior information. This property is called richness. If a family of distributions is not rich in this sense, there may be no member of the family that accurately reflects a particular individual's prior information. If this is the case, then the statistician will be unable to take advantage of the mathematical tractability of the family.

Finally, it should be possible to specify the conjugate family in such a way that it could be readily interpreted by the person whose prior information is of interest. Prior information often consists at least partially of previous sample results. Thus, the easiest way to interpret a prior distribution might be in terms of previous sample results (whether they are actual or hypothetical). In general, conjugate prior distributions can be interpreted in this manner.

It is important to remember that a conjugate prior distribution is "conjugate" only with respect to a given likelihood function (that is, a given statistical model). Conjugate families of distributions corresponding to numerous likelihood functions have been developed. Only two situations are discussed in detail here. The first situation involves sampling from a stationary and independent Bernoulli process, in which case the conjugate family is the family of beta distributions. The second situation involves sampling from a normally distributed population with known variance, in which case the conjugate family is the family of normal distributions. These two cases are discussed in the following sections. The formal mathematical derivations of these conjugate families of distributions from the given likelihood functions are not presented, but an attempt is made to demonstrate (informally if not formally) how they satisfy the three properties listed in this section.

4.3 CONJUGATE PRIOR DISTRIBUTIONS FOR THE BERNOULLI PROCESS

Recall, from Section 3.3, that a Bernoulli process is a data-generating process with two possible outcomes on each trial, such that the probabilities for these outcomes remain constant from trial to trial (stationarity) and the outcomes of the trials are independent. Alternatively, a Bernoulli process may be thought of in terms of exchangeability, as noted in Section 3.3. In that section, an example of Bayesian techniques involving a Bernoulli process was presented under the assumption that the prior distribution was discrete. In most situations it is unrealistic, however, to limit $\tilde{p}$, the probability of success on a Bernoulli trial, to a finite number of values. Conceptually, $\tilde{p}$ can assume any real value from 0 to 1, so that the prior distribution should be continuous rather than discrete. Using Equation (4.2.3), the statistician can assess a continuous prior distribution and then find the posterior distribution following an observed sample. This could be a difficult task unless the prior distribution is a member of the family of distributions that is conjugate with respect to the Bernoulli process. The conjugate family in this case is the family of beta distributions.

The density function of the beta distribution is very similar to the probability mass function of the binomial distribution. If the probability distribution of $\tilde{p}$ is a *beta distribution* with parameters r and n, where

$n > r > 0$, then

$$f(p) = \begin{cases} \dfrac{(n-1)!}{(r-1)!(n-r-1)!} p^{r-1}(1-p)^{n-r-1} & \text{if } 0 \leq p \leq 1, \\ 0 & \text{elsewhere.} \end{cases} \tag{4.3.1}$$

This expression corresponds to the binomial probability function (see Section 3.3) with $n - 1$ substituted for n, $r - 1$ substituted for r, and $n - r - 1$ substituted for $n - r$. But the beta distribution is continuous, whereas the binomial distribution is discrete. The explanation for this difference is quite simple; in the binomial distribution, the uncertain quantity is $\tilde{r}$, the number of "successes," and since $\tilde{r}$ obviously can take on only integer values, it is discrete. In the beta distribution, the uncertain quantity is $\tilde{p}$, which in terms of a Bernoulli process is the probability of success on any single Bernoulli trial. But the only limitation on probabilities is that they must be between 0 and 1, inclusive. Thus, it is reasonable to assume that $\tilde{p}$ can take on an infinite number of values (all real numbers between 0 and 1) and to use a continuous distribution (such as the beta distribution) as the distribution of $\tilde{p}$.

Note that the beta distribution has two parameters, r and n. It is necessary that $n > r > 0$, but n and r need not be integers. It should be mentioned that if n and r are not integers, the factorial terms $(n - 1)!$, $(r - 1)!$, and $(n - r - 1)!$ in Equation (4.3.1) must be replaced by mathematical functions known as gamma functions, denoted by $\Gamma(n)$, $\Gamma(r)$, and $\Gamma(n - r)$. Formally,

$$\Gamma(t) = \int_0^\infty x^{t-1}e^{-x}\,dx$$

for $t > 0$. For integer r and n, however, these gamma functions are equal to the corresponding factorial terms [that is, $\Gamma(t) = (t - 1)!$ if t is an integer], so for most purposes you need not be concerned with the mathematical concept of a gamma function.

The mean and the variance of a beta distribution with parameters r and n are

$$E(\tilde{p} \mid r,n) = \frac{r}{n} \tag{4.3.2}$$

and

$$V(\tilde{p} \mid r,n) = \frac{r(n-r)}{n^2(n+1)}. \tag{4.3.3}$$

The shape of the beta distribution depends on r and n. If $r = n/2$, the distribution is symmetric. If $r < n/2$, the distribution is positively skewed (the "long tail" of the density function is to the right), while if $r > n/2$, the distribution is negatively skewed (the "long tail" is to the left). If

$r > 1$ and $n - r > 1$, the distribution is unimodal (that is, it has a single mode) with the mode [the value p for which $f(p)$ is largest] equal to $(r - 1)/(n - 2)$. If $r \leq 1$ *or* $n - r \leq 1$, the distribution is either unimodal with mode at 0 or 1, U-shaped with modes at 0 *and* 1, or simply the uniform distribution on the unit interval [that is, $f(p)$ is constant; this occurs when $r = 1$ and $n = 2$]. Thus, the family of beta distributions can take on a wide variety of shapes as r and n vary, so the property of richness, which was discussed in the previous section, should be satisfied. Some examples of beta distributions with r/n equal to 1/2, 1/20, and 19/20 are presented in Figures 4.3.1 through 4.3.3.

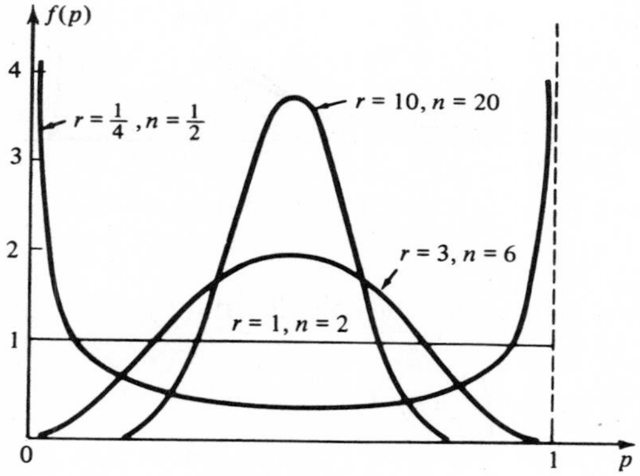

Figure 4.3.1

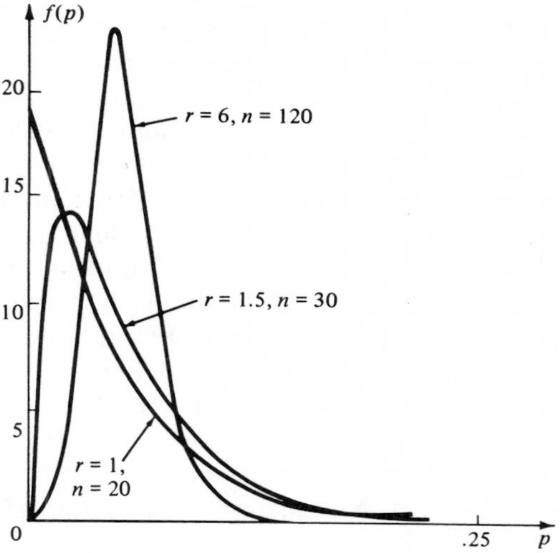

Figure 4.3.2

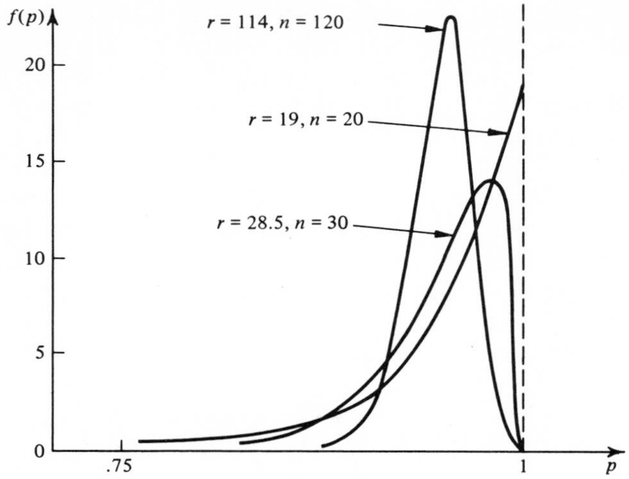

Figure 4.3.3

Of course, probabilities can be calculated for any interval of values of $\tilde{p}$:

$$\begin{aligned} P(a \leq \tilde{p} \leq b) &= \int_a^b f(p)\, dp \\ &= \int_a^b \frac{(n-1)!}{(r-1)!(n-r-1)!} p^{r-1}(1-p)^{n-r-1}\, dp, \end{aligned}$$

provided that $0 \leq a \leq b \leq 1$. Such computations could become quite tedious, so a table that gives various *fractiles* of the beta distribution for numerous values of r and n is presented in the back of the book. An f fractile of the distribution of a continuous random variable $\tilde{x}$ is a point x_f such that $P(\tilde{x} \leq x_f) = f$. That is, it is the point x_f such that the value of the CDF at x_f is exactly f. The .50 fractile, for example, is the *median* of the distribution, which for a continuous random variable divides the area under the density function into two equal portions, each with area $\frac{1}{2}$. To find the f fractile from a CDF of a continuous random variable, simply find f on the vertical axis, see where the CDF curve reaches that height, and read the corresponding value from the horizontal axis. In the table presented in the back of the book, the .001, .01, .05, .10, .25, .50, .75, .90, .95, .99, and .999 fractiles are presented for numerous beta distributions. For example, the CDF for the beta distribution with $r = 3$ and $n = 10$ is presented in Figure 4.3.4. Note, from this graph *and* from the table of fractiles of the beta distribution, that the .75 fractile of this distribution is .3905. It should also be noted that cumulative beta probabilities and values of the beta density function $f(p)$ can be determined from tables of the binomial distribution provided that r and n are integers:

$$P_\beta(\tilde{p} \leq p \mid r, n) = \begin{cases} P_{\text{bin}}(\tilde{r} \geq r \mid n-1, p) & \text{if } p \leq \frac{1}{2}, \quad (4.3.4) \\ P_{\text{bin}}(\tilde{r} < n - r \mid n-1, 1-p) & \text{if } p \geq \frac{1}{2}, \quad (4.3.5) \end{cases}$$

and

$$f_\beta(p \mid r,n) = (n-1)P_{\text{bin}}(\tilde{r} = r-1 \mid n-2, p). \qquad (4.3.6)$$

A word on notation: prior distributions and parameters of prior distributions (such as the mean of a prior distribution) are denoted with single primes, whereas posterior distributions and parameters of posterior distributions are denoted with double primes. For example, $f'(\theta)$ and $f''(\theta \mid y)$ represent the prior and posterior distributions of $\tilde{\theta}$; $\mu_{\tilde{\theta}}'$ and $\mu_{\tilde{\theta}}''$ [or $E'(\tilde{\theta})$ and $E''(\tilde{\theta})$] represent the means of the prior and posterior distributions, respectively. Of course, you should remember that the terms "prior" and "posterior" are relative terms, relating to a particular sample result.

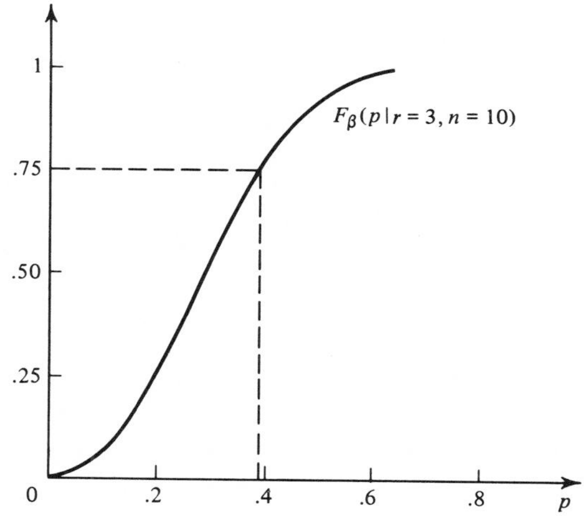

Figure 4.3.4

Suppose that you are sampling from a dichotomous process and that you are willing to assume that the process behaves like a Bernoulli process (that is, you are willing to accept the assumptions of stationarity and independence). Then if your prior distribution for $\tilde{p}$, the probability of "success" on any single trial, is a beta distribution, your posterior distribution will also be a beta distribution. Using the formula for the beta density function, suppose that the prior density is of the form

$$f'(p) = \frac{(n'-1)!}{(r'-1)!(n'-r'-1)!} p^{r'-1}(1-p)^{n'-r'-1} \quad \text{for } 0 \leq p \leq 1. \qquad (4.3.7)$$

Furthermore, suppose that the sample results in r "successes" in n trials. Then the posterior density is also a member of the beta family (this is demonstrated formally later in this section):

$$f''(p \mid y) = \frac{(n''-1)!}{(r''-1)!(n''-r''-1)!} p^{r''-1}(1-p)^{n''-r''-1} \quad \text{for } 0 \leq p \leq 1, \tag{4.3.8}$$

where

$$n'' = n' + n, \tag{4.3.9}$$

$$r'' = r' + r, \tag{4.3.10}$$

and y represents the sample results. To determine the posterior parameters of the beta distribution, n'' and r'', the statistician just adds the prior parameters, n' and r', to the sample statistics, n and r. It is not necessary to go to the trouble of directly applying Equation (4.2.3).

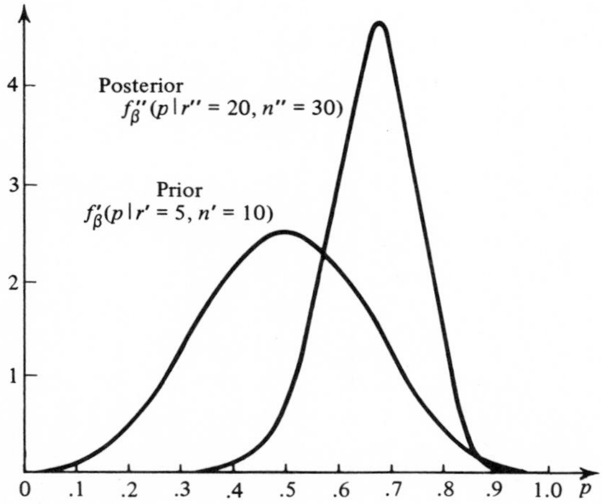

Figure 4.3.5

In the market-share example presented in Section 4.2, the prior distribution is a beta distribution with $r' = 1$ and $n' = 3$; the sample results are $r = 1$ and $n = 5$; and the posterior distribution is a beta distribution with $r'' = r' + r = 2$ and $n'' = n' + n = 8$. For another simple example, suppose that the statistician's prior distribution is a beta distribution with $r' = 5$ and $n' = 10$ and that he observes $r = 15$ successes in $n = 20$ Bernoulli trials. His posterior distribution is then a beta distribution with $r'' = 20$ and $n'' = 30$, as illustrated in Figure 4.3.5. How does the sample information change his distribution? First of all, since the mean of the beta distribution with parameters r and n

is just r/n, the mean of the statistician's prior distribution is $\frac{5}{10} = \frac{1}{2}$ and the mean of his posterior distribution is $\frac{20}{30} = \frac{2}{3}$. The sample results shift the mean upward, making higher values of $\tilde{p}$ appear more likely. Note that the sample mean is $r/n = 15/20 = 3/4$. The posterior mean always lies between the prior mean and the sample mean when the prior distribution is a beta distribution and the sample is taken from a Bernoulli process.

The variance of the beta distribution with parameters r and n is given by Equation (4.3.3). Applying this to the example, the prior variance is .0227 and the posterior variance is .0072. In most, but not all, cases, the posterior variance will be smaller than the prior variance. This is intuitively reasonable, since the sample provides new information, and increased information should reduce the uncertainty about the uncertain quantity of interest, which might be measured by the variance. Thus, it might be expected that additional information would reduce the variance, just as the variance of a sample mean is smaller as the sample size becomes larger. In some situations, notably when the sample information greatly shifts the mean of the posterior distribution toward $\frac{1}{2}$, the variance of the posterior beta distribution may be larger than the variance of the prior beta distribution.

For sampling from a Bernoulli process, the beta family of distributions clearly satisfies the requirement that a conjugate family of prior distributions be mathematically tractable. Given the prior distribution and its parameters r' and n', it is very easy to determine the posterior distribution after a sample is observed, using Equations (4.3.9) and (4.3.10). Also, if the prior distribution is a member of the beta family, so is the posterior distribution. Finally, since formulas for the mean and the variance (and other summary measures as well) of a beta distribution are known, it should not prove too difficult in most situations to calculate expectations of importance for decision theory. From Figures 4.3.1 through 4.3.3, it appears that the beta family satisfies the requirement of richness. The beta family includes symmetric, skewed, unimodal, U-shaped, J-shaped, and other distributions. By varying the parameters r' and n', the statistician has available a wide variety of shapes of distributions within the beta family.

What about the final requirement of a conjugate family, the requirement that the prior distribution can be interpreted in some manner? As might be suspected from the similarity between the prior parameters r' and n' and the sample statistics r and n, the beta distribution can be interpreted in terms of "equivalent sample information." A beta distribution with parameters r' and n' can be interpreted as follows: the prior information is roughly equivalent to the information contained in a sample of size n' with r' successes from the Bernoulli process of interest.

The beta distribution can be thought of as representing a person's knowledge about $\tilde{p}$, given that his knowledge consists solely of having observed a sample from the process. This does not necessarily mean that he has observed an *actual* sample from the process, but simply that he considers his information to be roughly *equivalent* to the information from such a sample. It should be noted that this interpretation should not be taken to imply that the parameters r' and n' must be integers. Since the interpretation is in terms of a "hypothetical" sample, not necessarily an "actual" sample, r' and n' need not be integers. They can take on any real values subject to the restriction that $n' > r' > 0$.

So far, nothing has been said in this section about the observed sample that is used to revise the prior distribution other than that the sample results are r successes in n trials. At least two ways of arriving at such sample results could be considered. First, the sample size n could be fixed, in which case $\tilde{r}$ is a random variable and the distribution of $\tilde{r}$ given p and n is a *binomial distribution*, as discussed in Section 3.3:

$$P(r \mid n,p) = \binom{n}{r} p^r(1 - p)^{n-r}. \tag{4.3.11}$$

A second possibility is that the number of successes r could be fixed (sample until r successes are obtained), in which case $\tilde{n}$ is a random variable and the distribution of $\tilde{n}$ given p and r is a *Pascal distribution*, which is of the following form:

$$P(n \mid r,p) = \binom{n-1}{r-1} p^r(1 - p)^{n-r}. \tag{4.3.12}$$

Notice that these two sampling distributions differ only in the first terms, the combinatorial terms. The reason for this is that the last trial must be a success in Pascal sampling (since Pascal sampling is defined as sampling until the rth success), while there is no such restriction in binomial sampling.

It turns out that in the application of Bayes' theorem, the combinatorial terms are irrelevant since they do not involve $\tilde{p}$, the variable of interest. In Equation (4.2.3), the combinatorial term can be moved outside of the integral sign in the denominator, since it does not involve $\tilde{p}$. But then the term appears in both the numerator and the denominator, so that it can be canceled out. Consequently, Equations (4.3.9) and (4.3.10) are applicable whether the sample is taken with n fixed (binomial sampling) or with r fixed (Pascal sampling). The only important part of the sampling distribution for Bayesian purposes is the part involving $\tilde{p}$. Therefore, the likelihood function for the Bernoulli process can be taken to be equal to $p^r(1 - p)^{n-r}$, so that it is not even necessary to know whether the sampling is done with n fixed or with r fixed. To determine the posterior distribution, the only information needed about the sample consists of the values of the sample statistics, r and n. Hence, r and n are sufficient statistics. The procedure used to tell the statistician when to stop sampling is called a stopping rule, and if the stopping rule has no effect on the posterior distribution, then it is said to be *noninformative.* The two stopping rules discussed above (sample until you have n trials and sample until you have r successes) are both noninformative.

To demonstrate this formally, consider an application of Equation (4.2.3) following binomial sampling (sampling with fixed sample size n). If the prior distribution is a beta distribution with parameters r' and n', then the posterior density is

$$f''(p \mid y) = \frac{f'(p)f(y \mid p)}{\int_0^1 f'(p)f(y \mid p)\,dp} = \frac{f'(p)P(r \mid n,p)}{\int_0^1 f'(p)P(r \mid n,p)\,dp}$$

$$= \frac{\dfrac{(n'-1)!}{(r'-1)!(n'-r'-1)!}\,p^{r'-1}(1-p)^{n'-r'-1}\dbinom{n}{r}p^r(1-p)^{n-r}}{\displaystyle\int_0^1 \dfrac{(n'-1)!}{(r'-1)!(n'-r'-1)!}\,p^{r'-1}(1-p)^{n'-r'-1}\dbinom{n}{r}p^r(1-p)^{n-r}\,dp}. \tag{4.3.13}$$

The terms not involving p can be moved outside of the integral in the denominator and canceled with the like terms in the numerator, leaving

$$f''(p \mid y) = \frac{p^{r'-1}(1-p)^{n'-r'-1}p^r(1-p)^{n-r}}{\int_0^1 p^{r'-1}(1-p)^{n'-r'-1}p^r(1-p)^{n-r}\,dp}. \tag{4.3.14}$$

As in all applications of Bayes' theorem, the denominator serves merely to assure that the posterior distribution will be a proper probability distribution; the denominator of Equation (4.3.14) is a constant with regard to p (since it involves an expression integrated over p), so denote it by k:

$$f''(p \mid y) = \frac{1}{k}\,p^{r'-1}(1-p)^{n'-r'-1}p^r(1-p)^{n-r}.$$

Combining terms in this expression,

$$f''(p \mid y) = \frac{1}{k}\,p^{r'+r-1}(1-p)^{n'+n-r'-r-1}. \tag{4.3.15}$$

Now, define n'' and r'' as follows: $n'' = n' + n$ and $r'' = r' + r$. Then the posterior density can be written in the form

$$f''(p \mid y) = \frac{1}{k} p^{r''-1}(1 - p)^{n''-r''-1} \tag{4.3.16}$$

But Equation (4.3.16) is of the same form as Equation (4.3.8) with

$$k = \frac{(r'' - 1)!(n'' - r'' - 1)!}{(n'' - 1)!}, \tag{4.3.17}$$

so the posterior distribution is simply a beta distribution with parameters r'' and n''.

Notice that if the sampling is done with a fixed value of r (Pascal sampling) rather than with a fixed value of n (binomial sampling), the above analysis is not affected at all, since the only difference is the replacement of $\binom{n}{r}$ in Equation (4.3.13) with $\binom{n-1}{r-1}$. In either case, the term in question cancels out, so the final result is the same.

The above analysis also serves to prove an assertion made earlier in this section, namely that if the prior distribution is a beta distribution and if the sample is taken from a Bernoulli process, then the posterior distribution is also a beta distribution. Incidentally, it is much more important that you gain an *understanding* of how Bayes' theorem works than that you concern yourself unduly over mathematical details. Now that it has been shown that the beta family is conjugate to the Bernoulli process, it is not necessary for you to go through the steps of Equation (4.3.13) through Equation (4.3.16) every time you wish to revise a beta prior distribution on the basis of a sample from a Bernoulli process. Given the prior distribution and the sample results, you need simply note that $r'' = r' + r$ and $n'' = n' + n$ are the parameters of the posterior beta density. This is a most useful feature of the concept of conjugate prior distributions.

It should be stressed that the applicability of conjugate prior distributions in any given situation depends in part on the applicability of a particular statistical model, such as the Bernoulli model, because the conjugate family of distributions depends on the form of the likelihood function, which in turn depends on assumptions concerning a statistical model. In some situations, no statistical model is applicable, and it is necessary to assess likelihoods directly (see Section 3.6), in which case no conjugate family can be found. Furthermore, even if a certain model is applicable to the data-generating process and if the corresponding conjugate family is known, it may be that no member of the family adequately represents the assessor's prior judgments. In sampling from a Bernoulli process, it may not be possible to find a beta distribution that is a "good fit" to the prior judgments. Hopefully, the beta distribution is rich enough that this will be the exception rather than the rule.

4.4 THE USE OF BETA PRIOR DISTRIBUTIONS: AN EXAMPLE

You are now prepared to consider a modification of the quality-control example presented in Section 3.3 to make it somewhat more realistic. Suppose that the production manager feels that his prior information concerning $\tilde{p}$, the proportion of defective items produced by the process, can be well-represented by a beta distribution with parameters $r' = 1$ and $n' = 20$ (procedures for assessing such distributions are discussed in Section 4.8). This distribution has a mean of $r'/n' = .05$ and a variance of $r'(n' - r')/n'^2(n' + 1) = .0023$ (the standard deviation is thus .048). The graph of the density function is presented in Figure 4.4.1. This prior distribution seems more realistic than the discrete distribution used in Section 3.3, in which only four possible values of $\tilde{p}$ were considered. Notice that the density function reaches its highest value at $\tilde{p} = 0$ and decreases as $\tilde{p}$ increases.

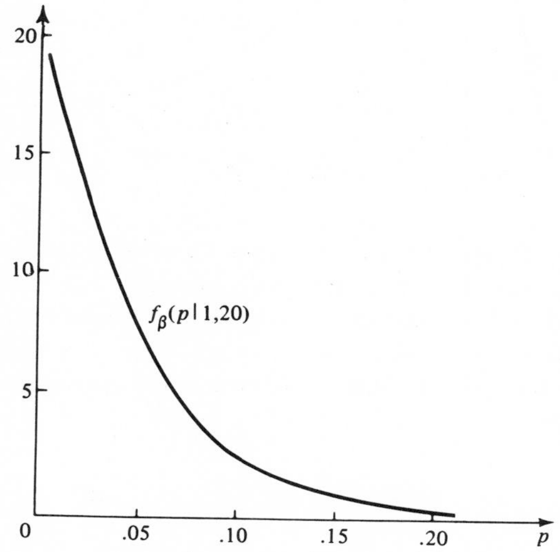

Figure 4.4.1

Following the assessment of his prior distribution, the production manager takes a sample of five items from the production process, observing one defective item. Under the assumption that the production process behaves as a stationary and independent Bernoulli process, with $\tilde{p}$ representing the probability of a defective item on any trial, the posterior distribution is a beta distribution with parameters

$$r'' = r' + r = 1 + 1 = 2$$

and

$$n'' = n' + n = 20 + 5 = 25.$$

This distribution is shown in Figure 4.4.2. The mean of the distribution is $2/25 = .08$, and the variance is $2(23)/25^2(26) = .0028$. Note that the variance of the posterior distribution is slightly higher than the variance of the prior distribution; this happens because of the shift in the mean. The posterior mean is .03 units to the right of the prior mean. Notice also that the mode of the posterior distribution is no longer 0; instead, it is equal to 1/24, or about .042. In general, if $r > 1$ and $n - r > 1$, the mode of the beta distribution with parameters r and n is $(r - 1)/(n - 2)$.

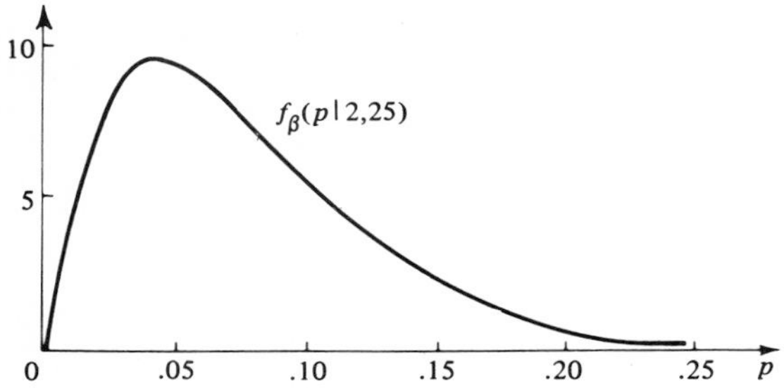

Figure 4.4.2

Suppose that the production manager decides to take a second sample of size five and observes two defective items. The posterior distribution calculated after the first sample is now the prior distribution with respect to the new sample, so $r' = 2$ and $n' = 25$. After the second sample, the posterior distribution is a beta distribution with parameters

$$r'' = r' + r = 2 + 2 = 4$$

and

$$n'' = n' + n = 25 + 5 = 30.$$

This distribution is shown in Figure 4.4.3. The mean of the distribution is $4/30 = .133$, and the variance is $4(26)/(30)^2(31) = .0037$. Once again the mean is shifted to the right and the variance increases. The mode is now $3/29 = .103$.

Just as in the discrete case, the production manager could wait and revise his original prior distribution once after observing both samples. If he does this, the samples can be summarized by $r = 3$ and $n = 10$, and the parameters of the posterior beta distribution are

$$r'' = r' + r = 1 + 3 = 4$$

and

$$n'' = n' + n = 20 + 10 = 30.$$

This is identical to the posterior distribution determined from two successive applications of Bayes' theorem.

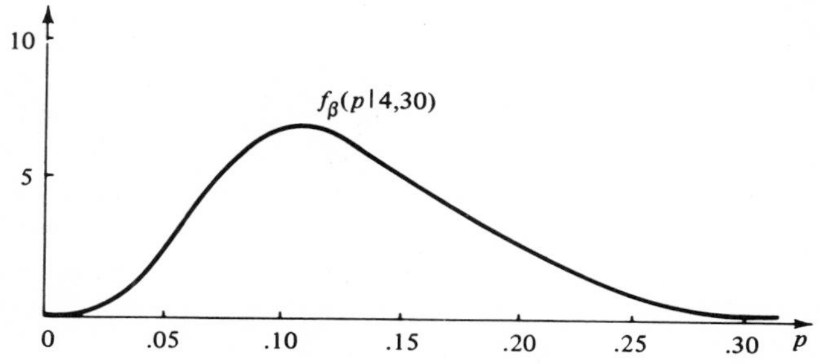

Figure 4.4.3

If the production manager wants an estimate of $\tilde{p}$, he should base his estimate on the posterior distribution. For example, he might want to use the mean of the posterior distribution, which in this case is .133. Suppose that instead of just estimating $\tilde{p}$, he must make a decision concerning the process, a decision that depends on the value of $\tilde{p}$. It is convenient to modify the example from Section 3.3 so that the costs are the same but the actions available to the manager are slightly different. The two actions are as follows: (1) he can leave the process as is, or (2) he can have a mechanic make an adjustment in the process that will halve the proportion of defectives, $\tilde{p}$—this adjustment costs \$100.

The profit from a run of 1000 items is 1000(\$0.50) = \$500, less the costs of replacing defectives and adjusting the process (if the second action is chosen). The expected number of defectives in a run of 1000 is $1000\tilde{p}$, and the cost of replacing a single defective is \$2, so the total expected cost of replacing defectives is $\$2000\tilde{p}$ if the process is not adjusted. If the adjustment is made, the proportion of defectives is cut in half, so the total expected cost of replacing defectives is $\$2000\tilde{p}/2 = \$1000\tilde{p}$. Thus, the profits to the firm associated with the two actions are

$$\text{Profit(no adjustment)} = \$500 - \$2000\tilde{p}$$

and

$$\text{Profit(adjustment)} = \$500 - \$1000\tilde{p} - \$100.$$

The corresponding expected profits, or expected rewards, for the two acts are

$$ER(\text{no adjustment}) = \$500 - \$2000E(\tilde{p})$$

and

$$ER(\text{adjustment}) = \$400 - \$1000E(\tilde{p}).$$

From the posterior distribution, $E(\tilde{p}) = .133$, so the respective expected payoffs are \$234 and \$267. In order to maximize expected profit, the production manager should have the process adjusted.

It is interesting to note that if the production manager had not obtained any sample information, the decision would have been based on the prior distribution. The prior mean is $E(\tilde{p}) = .05$, and the expected payoffs are \$400 and \$350. Solely on the basis of the prior information, then, the production manager should leave the process as is. As in the decision-making example of Section 3.3, the sample information revises the prior distribution in such a way as to make large values of $\tilde{p}$ more likely and small values less likely. Since large values of $\tilde{p}$ are more costly due to the costs of replacing defectives, the expected payoffs calculated from the posterior distribution are considerably smaller (for both actions) than the expected payoffs calculated from the prior distribution.

This example illustrates the use of Bayes' theorem to revise a continuous probability distribution and the use of the resulting posterior distribution in a decision-making problem. Decision theory is considered in more detail in Chapter 5. In the next section, Bayesian techniques for a normally distributed population or process are discussed.

4.5 CONJUGATE PRIOR DISTRIBUTIONS FOR THE NORMAL PROCESS

While it may often prove useful, the discussion of Bayesian inference for a Bernoulli process is limited in application to cases in which the assumptions of a Bernoulli process are satisfied. There must be two possible outcomes at each trial (or if there are more than two, the outcomes must be divided into two sets), the probabilities of these two outcomes must remain constant from trial to trial, and the trials must be independent. In many situations the data-generating process is such that there are more than two possible outcomes at each trial; in particular, any real number within some specified interval (which may be the entire real line) may be a potential outcome. An example is sampling from a *normal population,* or *normal process.* This section deals with the procedure by which prior information can be easily combined with sample information from a normal process if the prior information can be expressed in the form of a normal distribution.

If the probability distribution of $\tilde{x}$ is a *normal distribution* with mean μ and variance σ^2, then the density function is of the form

$$f(x) = \frac{1}{\sigma\sqrt{2\pi}} e^{-(x-\mu)^2/2\sigma^2}, \qquad (4.5.1)$$

where $e = 2.71828 \ldots$ (the base of the natural logarithm system) and $\pi = 3.1415 \ldots$ are constants. The graph of this density is a symmetric, bell-shaped curve (Figure 4.5.1). Increasing (decreasing) the mean merely shifts the entire distribution to the right (left) along the horizontal axis; increasing (decreasing) the variance makes the distribution more (less) "spread out" and lowers (raises) the "peak" of the distribution. Unlike the beta distribution, the normal distribution is always symmetric about its mean, μ.

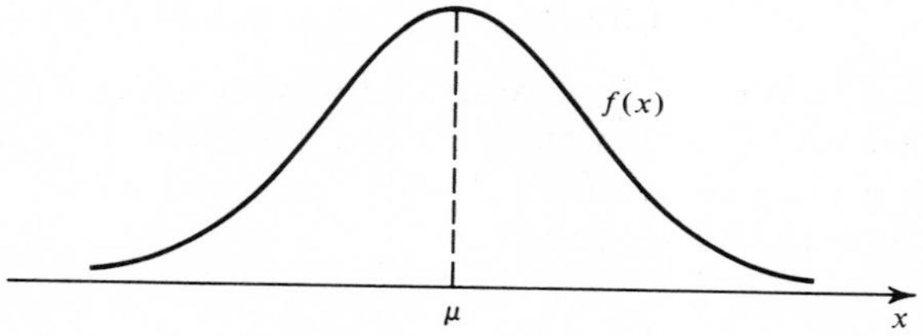

Figure 4.5.1

The normal distribution has several attractive mathematical properties that greatly enhance its usefulness in statistics. For instance, any linear function of a normally distributed random variable is also normally distributed. In particular, if $\tilde{x}$ is normally distributed with mean μ and variance σ^2, then the standardized random variable corresponding to $\tilde{x}$ is normally distributed with mean 0 and variance 1 (it is said to have the *standard normal distribution*). That is, if

$$\tilde{z} = \frac{\tilde{x} - \mu}{\sigma} = \left(\frac{1}{\sigma}\right)\tilde{x} - \frac{\mu}{\sigma},$$

then

$$f(z) = \frac{1}{\sqrt{2\pi}} e^{-z^2/2}. \qquad (4.5.2)$$

Probabilities for intervals of values of $\tilde{z}$ can be determined from the table of cumulative standard normal probabilities that is presented at the back of the book. From this table, for instance, $P(\tilde{z} \leq 0) = F(0) =$

.50, $P(\tilde{z} \leq 1) = F(1) = .84$, and $P(\tilde{z} \leq 1.96) = F(1.96) = .975$. Thus,

$$P(0 \leq \tilde{z} \leq 1) = F(1) - F(0) = .34$$

and

$$P(0 \leq \tilde{z} \leq 1.96) = F(1.96) - F(0) = .475.$$

By the symmetry of the standard normal distribution about $\tilde{z} = 0$, $F(-z) = 1 - F(z)$, so that

$$P(-1 \leq \tilde{z} \leq 1) = 2(.34) = .68$$

and

$$P(-1.96 \leq \tilde{z} \leq 1.96) = 2(.475) = .95.$$

These probabilities can be converted into the following probabilities for intervals of values of $\tilde{x}$ by noting that $\tilde{x} = \mu + \sigma\tilde{z}$:

$$P(\mu - \sigma \leq \tilde{x} \leq \mu + \sigma) = .68$$

and

$$P(\mu - 1.96\sigma \leq \tilde{x} \leq \mu + 1.96\sigma) = .95.$$

In words, approximately 68 percent of the area under a normal density function is contained within one standard deviation from the mean, and approximately 95 percent of the area is contained within 1.96 standard deviations from the mean. In general,

$$\begin{aligned} P(a \leq \tilde{x} \leq b) &= P\left(\frac{a - \mu}{\sigma} \leq \frac{\tilde{x} - \mu}{\sigma} \leq \frac{b - \mu}{\sigma}\right) \\ &= P\left(\frac{a - \mu}{\sigma} \leq \tilde{z} \leq \frac{b - \mu}{\sigma}\right) \end{aligned} \tag{4.5.3}$$

and

$$\begin{aligned} P(c \leq \tilde{z} \leq d) &= P(\mu + \sigma c \leq \mu + \sigma\tilde{z} \leq \mu + \sigma d) \\ &= P(\mu + \sigma c \leq \tilde{x} \leq \mu + \sigma d). \end{aligned} \tag{4.5.4}$$

Thus, probabilities for intervals of values of $\tilde{x}$ can be found from probabilities for intervals of values of $\tilde{z}$, and vice versa.

A table of values of the standard normal density function, $f(z)$, is also presented at the back of the book. These values are useful in determining likelihoods, as you will see in the next section. By the symmetry of the standard normal distribution about $z = 0$, $f(-z) = f(z)$. The table can be used to find values for any normal density. Since $\tilde{z} = (\tilde{x} - \mu)/\sigma$, the terms in the exponents of Equations (4.5.1) and (4.5.2) are equal. Therefore, the only difference between the two equations is the appearance of σ in the denominator of Equation (4.5.1), so the relationship between $f(x)$ and $f(z)$ is simply

$$f(x \mid \mu,\sigma^2) = \frac{f(z \mid 0,1)}{\sigma}. \tag{4.4.5}$$

For instance, if $\tilde{x}$ is normally distributed with mean 50 and variance 100, then the height of the density function of $\tilde{x}$ at $x = 55$ can be found from the table of values of the standard normal density function by taking

$$f(x \mid 50{,}100) = \frac{f(z \mid 0{,}1)}{\sigma} = \frac{f\left(\frac{55-50}{10} \middle| 0{,}1\right)}{10} = \frac{f(.5 \mid 0{,}1)}{10}$$
$$= \frac{.3521}{10} = .03521.$$

At this point a digression regarding sample statistics is needed. Just as μ and σ^2 are summary measures of probability distributions, summary measures of samples can also be determined. For instance, in sampling from a Bernoulli process (see Sections 3.3 and 4.3), r and n are summary measures of the sample information. In general, suppose that a sample consists of a series of n independent trials and that the sample information on the ith trial is represented by the random variable $\tilde{x}_i$. Sample statistics analogous to μ and σ^2 are the *sample mean,*

$$\tilde{m} = \frac{\sum_{i=1}^{n} \tilde{x}_i}{n}$$

and the *sample variance,*

$$\tilde{s}^2 = \frac{\sum_{i=1}^{n} (\tilde{x}_i - \tilde{m})^2}{n-1}.$$

The "tildes" in $\tilde{m}$ and $\tilde{s}^2$ reflect the fact that before the sample is taken, $\tilde{m}$ and $\tilde{s}^2$ are random variables. Also, to coordinate the notation in the Bayesian model, $\tilde{m}$ is used instead of the more common $\tilde{\bar{x}}$ to represent the sample mean.

To illustrate the sample mean and the sample variance, suppose that a sample consists of the weight (in grams) of four randomly-selected items from the output of a manufacturing process. These weights are 15, 10, 16, and 15. The sample mean and the sample variance are

$$m = \frac{15 + 10 + 16 + 15}{4} = 14$$

and

$$s^2 = \frac{(15-14)^2 + (10-14)^2 + (16-14)^2 + (15-14)^2}{3} = 7.33.$$

The sample mean and the sample variance can be related to the Bernoulli model. In sampling from a Bernoulli process with fixed n, let $\tilde{x}_i$ equal 1 if the ith trial yields a "success" and 0 if the ith trial yields a "failure." Then $\tilde{m} = \tilde{r}/n$ and $\tilde{s}^2 = \tilde{r}(n - \tilde{r})/n(n - 1)$.

The sample mean and the sample variance may be regarded as measures of location and dispersion, respectively, of the sample information. Other descriptive measures are available to investigate location and dispersion as well as additional properties, such as skewness. An investigation of such measures, together with the use of various types of graphs to represent sample information, falls within the realm of descriptive statistics. This book is concerned with problems of statistical inference and decision rather than with problems of descriptive statistics, and the discussion of sample statistics is limited to their use in inferential and decision-making situations. Thus, if $\tilde{\theta}$ is the uncertain quantity of interest, a sample statistic $\tilde{y}$ is useful in an inferential sense only to the extent that it provides information about $\tilde{\theta}$.

Returning to the discussion of the normal distribution, another useful feature of this distribution is that if a number of random variables are independent and normally distributed, then any linear combination of them is also normally distributed. In particular, if the n random variables $\tilde{x}_1, \tilde{x}_2, \ldots, \tilde{x}_n$ represent a random sample of size n from a normally distributed population with mean μ and variance σ^2, then the sample mean $\tilde{m}$ is normally distributed with conditional mean $E(\tilde{m} \mid \mu,\sigma^2) = \mu$ and conditional variance $V(\tilde{m} \mid \mu,\sigma^2) = \sigma^2/n$. The conditional standard deviation of a sample statistic is often called a standard error, so $\sigma/\sqrt{n}$ is the standard error of the sample mean $\tilde{m}$. In addition, since $\tilde{m}$ is normally distributed, $(\tilde{m} - \mu)/(\sigma/\sqrt{n})$, the corresponding standardized random variable, has a standard normal distribution.

The above results require that the population of interest be normally distributed. However, even if the population is *not* normally distributed, an important theorem known as the *central limit theorem* states that as the sample size gets larger (that is, as $n \to \infty$), the distribution of $(\tilde{m} - \mu)/(\sigma/\sqrt{n})$ gets closer and closer to a normal distribution. This result holds for *any* population or process, with the only requirement being that the population or process must have finite mean and variance. The central limit theorem provides the justification for much of the widespread use of the normal distribution in statistics.

The normal model serves as a good approximation to many other models. For instance, if n is large and $\tilde{p}$ is near $\frac{1}{2}$, the normal distribution is a good approximation of the binomial distribution. For $n = 50$ and $\tilde{p} = .5$,

$$P(\tilde{r} \geq 23 \mid n = 50, \tilde{p} = .5) = .7603.$$

Standardizing by subtracting $E(\tilde{r} \mid n,p) = np = 25$ and dividing by $\sqrt{V(\tilde{r} \mid n,p)} = \sqrt{np(1-p)} = \sqrt{12.5} = 3.536$ yields

$$\tilde{z} = \frac{\tilde{r} - np}{\sqrt{np(1-p)}} = \frac{r - 25}{3.536}.$$

The normal approximation, then, is

$$P(\tilde{r} \geq 23 \mid n = 50, \tilde{p} = .5) \approx P\left(\tilde{z} \geq \frac{23 - 25}{3.536}\right)$$
$$= P(\tilde{z} \geq -.57) = .7157.$$

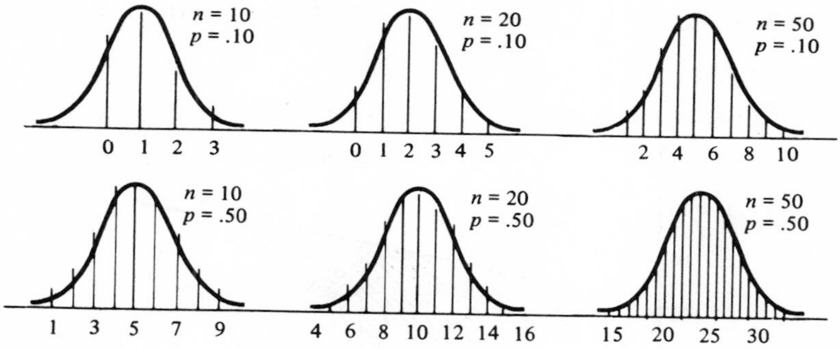

Figure 4.5.2

A better approximation is obtained by using a *continuity correction*, which allows for the fact that a continuous distribution (the normal distribution) is being used to approximate a discrete distribution (the binomial distribution). Note that since the normal distribution is continuous, the interval from $\tilde{r} = 22$ to $\tilde{r} = 23$ has positive probability, $P(22 < \tilde{r} < 23)$. The continuity correction divides the probability of this interval into two parts, $P(22 < \tilde{r} < 22.5)$ and $P(22.5 < \tilde{r} < 23)$, and associates these probabilities with $\tilde{r} = 22$ and $\tilde{r} = 23$, respectively (see Figure 4.5.2). Thus, using this continuity correction of $\frac{1}{2}$,

$$P(\tilde{r} \geq 23 \mid n = 50, \tilde{p} = .5) = P(\tilde{r} \geq 22.5 \mid n = 50, \tilde{p} = .5)$$
$$\approx P\left(\tilde{z} \geq \frac{22.5 - 25}{3.536}\right) = P(\tilde{z} \geq -.71)$$
$$= .7611.$$

The continuity correction improves the approximation and also provides a means of approximating the probability of a single value of $\tilde{r}$ (as opposed to an interval of values of $\tilde{r}$). From the binomial distribution,

$$P(\tilde{r} = 45 \mid n = 100, \tilde{p} = .4) = .0478;$$

from the normal approximation,

$$\begin{aligned} P(\tilde{r} = 45 \mid n = 100, \tilde{p} = .4) &= P(44.5 \leq \tilde{r} \leq 45.5) \mid n = 100, \tilde{p} = .4) \\ &\approx P\left(\frac{44.5 - 40}{\sqrt{24}} \leq \tilde{z} \leq \frac{45.5 - 40}{\sqrt{24}}\right) \\ &= P(.92 \leq \tilde{z} \leq 1.12) = .0474. \end{aligned}$$

This illustrates the use of the normal distribution to approximate the binomial distribution, and the approximation improves as n increases and as $\tilde{p}$ approaches $\frac{1}{2}$, as indicated in Figure 4.5.3. Furthermore, the normal distribution is also used as an approximation to many other distributions.

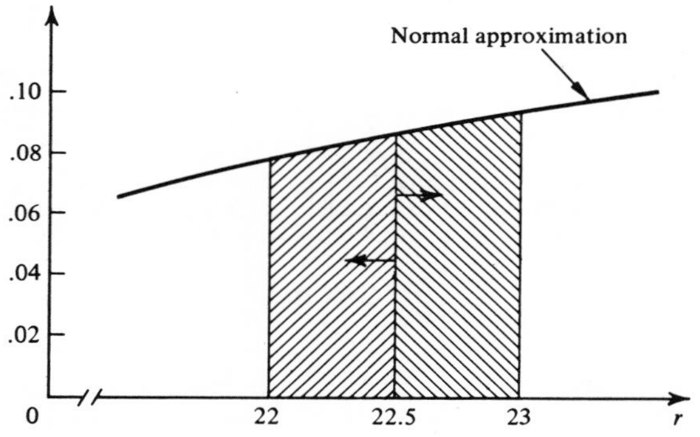

Figure 4.5.3

It should be strongly emphasized that the normal distribution is by no means a "universal nature's rule," although it does have wide applicability. Whenever a normal distribution is used in statistical inference or decision, its applicability should be carefully investigated. Various procedures are available for such investigations. For instance, if a graph of a CDF is drawn on a special type of graph paper called "normal probability paper," then it should result in a straight line if the random variable in question is normally distributed. Another way to see if a particular normal distribution is a "good fit" to some given distribution is to calculate probabilities for various intervals from both distributions and to compare these probabilities. If there are no "big" differences, then the normal distribution is a reasonable "fit." Of course, the term "big" depends on the problem; for some problems, more accuracy is necessary than for other problems. It should be noted that this procedure of comparing probabilities can also be used to check the "fit" of other types of distributions, such as beta distributions, to some given distribution. The point of this general discussion is that the assumption of normality, like many other assumptions used in statistical inference and decision, is ultimately a subjective assumption, and the statistician should take care to see that it can be justified.

Suppose that a statistician is sampling from a normal process or population and that he knows σ^2, the variance of the process. He does not know the mean, $\tilde{\mu}$, however, although he has some prior information concerning $\tilde{\mu}$. If his prior distribution of the uncertain quantity $\tilde{\mu}$ is a *normal distribution*, his posterior distribution will also be a normal distribution. From Equation (4.5.1), suppose that the prior density is of the form

$$f'(\mu) = \frac{1}{\sqrt{2\pi\sigma'^2}} e^{-(\mu-m')^2/2\sigma'^2}. \tag{4.5.6}$$

This is a normal density function with mean m' and variance σ'^2. If the statistician takes a sample of size n and observes a sample mean of m, his posterior density is a normal density,

$$f''(\mu \mid y) = \frac{1}{\sqrt{2\pi\sigma''^2}} e^{-(\mu-m'')^2/2\sigma''^2}, \tag{4.5.7}$$

where y represents the sample results and the posterior parameters m'' and σ''^2 can be determined from the formulas

$$\frac{1}{\sigma''^2} = \frac{1}{\sigma'^2} + \frac{n}{\sigma^2} \tag{4.5.8}$$

and

$$m'' = \frac{(1/\sigma'^2)m' + (n/\sigma^2)m}{(1/\sigma'^2) + (n/\sigma^2)}. \tag{4.5.9}$$

Note that n and m, the sample size and the sample mean, are the only sample statistics needed to determine the posterior distribution, so they are sufficient statistics in this situation.

The reciprocal of the posterior variance is equal to the sum of the reciprocal of the prior variance and the reciprocal of the variance of $\tilde{m}$, σ^2/n. The posterior mean is a weighted average of the prior mean and the sample mean, the weights being the reciprocals of the respective variances. As σ'^2, the variance of the prior distribution, decreases, the amount of prior uncertainty decreases, and the prior information is given more weight in the determination of the posterior distribution. The same is true for the sample information with respect to σ^2/n, the variance of the sample mean. Be sure not to confuse σ^2 and σ'^2; the former is the variance of the data-generating process or the population in question, whereas the latter is the prior variance of $\tilde{\mu}$, the *mean* of the process or population. This distinction is explained in more detail later in this section.

For a numerical example, suppose that the process variance is known to be $\sigma^2 = 400$, but $\tilde{\mu}$ is unknown. The prior distribution for $\tilde{\mu}$ is a normal distribution with mean 225 and variance 25. A sample of size four is taken, and the sample mean is 200. Thus, $m' = 225$, $\sigma'^2 = 25$, $m = 200$, and $n = 4$. Using Equations (4.5.8) and (4.5.9),

$$\frac{1}{\sigma''^2} = \frac{1}{25} + \frac{4}{400} = \frac{5}{100} = \frac{1}{20}$$

and

$$m'' = \frac{(1/25)225 + (4/400)200}{(1/25) + (4/400)} = \frac{9 + 2}{(1/20)} = 220.$$

The posterior parameters are $m'' = 220$ and $\sigma''^2 = 20$. As in the Bernoulli case, it is always true that the posterior mean lies between the prior mean and the sample mean. Unlike the Bernoulli case, the posterior variance is *always* smaller than the prior variance. This can be seen from Equation (4.5.8). As long as $n > 0$, the second term on the right-hand side of the equation is greater than 0, and thus the reciprocal of the posterior variance is larger than the reciprocal of the prior variance. This implies that the posterior variance is smaller than the prior variance.

Does the family of normal distributions possess the desirable properties of a conjugate family in this case? Mathematical tractability appears to be satisfied. Although the computations necessary to determine the posterior parameters are more complicated than the simple addition used for the beta distribution and the Bernoulli process, they are still quite simple and should pose no computational problems. The posterior distribution is a member of the same family (the normal family) as the prior distribution, so repeated applications of Bayes' theorem create no difficulties. Finally, because of its importance in statistics, the normal distribution has been studied extensively. Tables are available showing values of both the cumulative distribution function and the density function, and the moments of the distribution (including higher order moments as well as the mean and the variance) are known.

The normal family is reasonably rich, for the mean and the variance can assume any values. All members of the family are unimodal and symmetric, though, so no variation in the basic shape of the density function is possible. As a result, there are situations in which a person's prior distribution is definitely nonnormal, particularly when the distribution is highly skewed. While it is hoped that conjugate families of prior distributions are rich enough to be widely applicable, clearly there are times when no member of the conjugate family adequately represents the prior information. In such cases, it is necessary to apply Equation (4.2.3) directly rather than to use the short cuts provided by conjugate priors. This point is discussed further in Section 4.9.

The question of the interpretation of a normal prior distribution has been avoided up to this point. A normal prior distribution can be interpreted in terms of "equivalent sample information" from the process of interest, just as a beta prior distribution is interpreted in terms of an equivalent sample from a Bernoulli process. To see this in a clear fashion, however, it is necessary to consider a different parametrization of the distribution. The distribution will not be changed, but one of the parameters used to define it will be transformed to a new parameter.

Consider the parameter n', defined by the formula

$$n' = \frac{\sigma^2}{\sigma'^2}. \tag{4.5.10}$$

Therefore,

$$\sigma'^2 = \frac{\sigma^2}{n'}, \tag{4.5.11}$$

so that the prior variance can be written in terms of n' and the process variance σ^2. The prior distribution is thus a normal distribution with mean m' and variance σ^2/n'. Now, if n'' is defined as

$$n'' = \frac{\sigma^2}{\sigma''^2}, \tag{4.5.12}$$

then Equation (4.5.8) becomes

$$\frac{n''}{\sigma^2} = \frac{n'}{\sigma^2} + \frac{n}{\sigma^2},$$

or simply

$$n'' = n' + n. \tag{4.5.13}$$

Furthermore, Equation (4.5.9) becomes

$$m'' = \frac{(n'/\sigma^2)m' + (n/\sigma^2)m}{(n'/\sigma^2) + (n/\sigma^2)},$$

or

$$m'' = \frac{n'm' + nm}{n' + n}. \tag{4.5.14}$$

Comparing Equations (4.5.13) and (4.5.14) with Equations (4.5.8) and (4.5.9), notice that the prior distribution is expressed in terms of m' and n' rather than m' and σ'^2 ,and the posterior distribution is expressed in terms of m'' and n'' rather than m'' and σ''^2. How can n' (and n'') be interpreted? From Equation (4.5.11), n' appears to be the sample size required to produce a variance of σ'^2 for a sample mean, because the variance of the sample mean from a sample of size n' is equal to σ^2/n'. This suggests an interpretation for the prior distribution: the prior distribution is roughly equivalent to the information contained in a sample of size n' with a sample mean of m'. Under this interpretation, Equations (4.5.13) and (4.5.14) can be thought of as formulas for pooling the information from two samples. The pooled (posterior) sample size is equal to the sum of the two individual sample sizes (one from the prior distribution, one from the sample), and the pooled (posterior) sample mean is equal to a weighted average of the two individual sample means (one from the prior, one from the sample). This does not mean that the prior information must consist solely of a prior sample from the process; it merely indicates that the prior information can be thought of as being roughly equivalent to such sample information. It should be noted that n' need not be an integer (recall that for a beta prior distribution, r' and n' do not have to be integers). For instance, if the variance of the data-generating process is $\sigma^2 = 110$ and the variance of the prior distribution is $\sigma'^2 = 4$, then $n' = \sigma^2/\sigma'^2 = 27.5$.

Using the new parameter introduced in Equation (4.5.10), the relative weights of the prior information and the sample information can be investigated. If you want to estimate $\tilde{\mu}$, for example, a reasonable estimate is the posterior mean, m'', which is always between the prior mean m' and the sample mean m. Under what conditions is m'' closer to m' than to m, and under what conditions is m'' closer to m? From Equation (4.5.14), m'' is a weighted average of m' and m, with the weights being equal to $n'/(n' + n)$ and $n/(n' + n)$, respectively. Therefore, if $n' > n$, the prior mean is given more weight, and the posterior mean m'' is closer to m' than to m. If $n' < n$, the sample mean is given more weight, and m'' is closer to m than to m'. If $n' = n$, m'' is exactly midway between m' and m.

In terms of the above interpretation of the prior distribution, these results appear obvious; in pooling two samples, the one with the larger sample size automatically receives more weight in the determination of a pooled mean. Recalling that the variance of the prior distribution is equal to σ^2/n' and that the conditional variance of the sample mean $\tilde{m}$ is equal to σ^2/n, notice that the prior information receives more (less) weight than the sample information if the prior variance is less than (greater than) the variance of $\tilde{m}$. If this is viewed in terms of information, a smaller variance implies more information. In determining the posterior distribution, then, the Bayesian statistician is pooling information. If there is more prior information than sample information (where information in this context is inversely related to variance), then the posterior distribution is affected more by the prior information (as expressed in terms of a prior distribution) than by the sample information (as expressed in terms of a likelihood function).

Consider the example presented earlier in this section. In this example

$$n' = \frac{\sigma^2}{\sigma'^2} = \frac{400}{25} = 16$$

and $n = 4$, so that

$$n'' = n' + n = 16 + 4 = 20$$

and

$$m'' = \frac{n'm' + nm}{n' + n} = \frac{16(225) + 4(200)}{16 + 4} = 220.$$

Notice that the posterior mean, 220, is closer to the prior mean, 225, than to the sample mean, 200. This is because $n' > n$; the prior information is roughly equivalent to the information contained in a sample of size 16 from the process with sample mean 225. Also notice that the results are identical to those found by using the original parametrization. The posterior mean is the same in both cases, and

$$n'' = \frac{\sigma^2}{\sigma''^2} = \frac{400}{20} = 20.$$

It should be noted that sometimes a yet different parametrization is used for the normal model. In this parametrization, the "amount of information" contained in a probability distribution, or the "precision" of the distribution, is represented by the reciprocal of the variance of the distribution. Thus, the amount of prior information (the prior precision) is simply

$$I' = \frac{1}{\sigma'^2};$$

the amount of information contained in the sample (the precision of the sample) is

$$I = \frac{n}{\sigma^2};$$

and the amount of posterior information (the posterior precision) is

$$I'' = \frac{1}{\sigma''^2}.$$

In terms of these measures of information, then, Equations (4.5.8) and (4.5.9) can be written as follows:

$$I'' = I' + I \tag{4.5.15}$$

and

$$m'' = \frac{(I'm' + Im)}{(I' + I)}. \tag{4.5.16}$$

For the example given earlier in this section, $I' = 1/25$, $I = 4/400$, and $I'' = 1/20$. The three different parametrizations, of course, simply provide three different ways of looking at the model; the numerical results are equivalent.

4.6 THE USE OF NORMAL PRIOR DISTRIBUTIONS: AN EXAMPLE

Suppose that a retailer is interested in the distribution of weekly sales at one of his stores. In particular, he is willing to assume that the random variable $\tilde{x}$, weekly sales (expressed in terms of dollars), is normally distributed with unknown mean $\tilde{\mu}$ and known variance $\sigma^2 = 90{,}000$. It should be stressed that this is a subjective assumption. Perhaps the retailer feels that the normality assumption is justified because he has found that for other stores that he has investigated, $\tilde{x}$ has tended to be approximately normally distributed. Furthermore, at various other stores that are quite similar to the store in question, mean weekly sales have differed from store to store, but the variance has been reasonably stable at about 90,000. Hence, the retailer is willing to make the simplifying assumption that σ^2 is known. Finally, in assuming an independent normal data-generating model, he is implicitly assuming that there is no trend or seasonal effect to worry about. That is, he feels that for the type of store in question, sales generally do not change greatly over time and are not affected by such things as the weather, holidays, and so on. These are the usual types of assumptions that must be considered in the selection of a statistical model.

In order to illustrate the use of discrete prior distributions in conjunction with a continuous data-generating process, suppose that the retailer decides to consider only five potential values of $\tilde{\mu}$: 1100, 1150, 1200, 1250, and 1300. Based on experience with similar stores, he assesses the following prior probabilities:

$$P(\tilde{\mu} = 1100) = .15,$$
$$P(\tilde{\mu} = 1150) = .20,$$
$$P(\tilde{\mu} = 1200) = .30,$$
$$P(\tilde{\mu} = 1250) = .20,$$

and

$$P(\tilde{\mu} = 1300) = .15.$$

In assessing the prior probabilities, perhaps the assessor used one of the techniques discussed in Section 3.5. For instance, he might have decided that the odds in favor of $\tilde{\mu} = 1200$ are 2 to 1 when $\tilde{\mu} = 1200$ is compared with either $\tilde{\mu} = 1100$ or $\tilde{\mu} = 1300$ and 1.5 to 1 when $\tilde{\mu} = 1200$ is compared with either $\tilde{\mu} = 1150$ or $\tilde{\mu} = 1250$.

The retailer decides that he would like to obtain more information about the sales of the store, so he takes a sample from the past sales records of the store, assuming that weekly sales from different weeks can be regarded as independent. For a sample of 60 weeks, he finds the average weekly sales (the sample mean) to be $m = 1240$. Prior to seeing this sample, the conditional distribution of $\tilde{m}$, given $\tilde{\mu} = \mu$, was a normal distribution with mean μ and variance $\sigma^2/n = 90{,}000/60 = 1500$. Using Equation (4.5.5) and the table of values of the standard normal density function, the likelihoods are

$$f(1240 \mid \mu = 1100, \sigma/\sqrt{n} = 38.73) = f\left(\frac{1240 - 1100}{38.73} \,\middle|\, 0,1\right)\Big/ 38.73 = \frac{f(3.61 \mid 0,1)}{38.73} = \frac{.0006}{38.73},$$

$$f(1240 \mid \mu = 1150, \sigma/\sqrt{n} = 38.73) = f\left(\frac{1240 - 1150}{38.73} \,\middle|\, 0,1\right)\Big/ 38.73 = \frac{f(2.32 \mid 0,1)}{38.73} = \frac{.0270}{38.73},$$

$$f(1240 \mid \mu = 1200, \sigma/\sqrt{n} = 38.73) = f\left(\frac{1240 - 1200}{38.73} \,\middle|\, 0,1\right)\Big/ 38.73 = \frac{f(1.03 \mid 0,1)}{38.73} = \frac{.2347}{38.73},$$

$$f(1240 \mid \mu = 1250, \sigma/\sqrt{n} = 38.73) = f\left(\frac{1240 - 1250}{38.73} \,\middle|\, 0,1\right)\Big/ 38.73 = \frac{f(-0.26 \mid 0,1)}{38.73} = \frac{.3857}{38.73},$$

and $$f(1240 \mid \mu = 1300, \sigma/\sqrt{n} = 38.73) = f\left(\frac{1240 - 1300}{38.73} \,\middle|\, 0,1\right)\Big/ 38.73 = \frac{f(-1.55 \mid 0,1)}{38.73} = \frac{.1200}{38.73}.$$

Note that since $\sigma/\sqrt{n} = 38.73$ is the same for all values of $\tilde{\mu}$, it is not actually necessary to divide each term by 38.73 to determine the likelihoods. The random variable of interest is $\tilde{\mu}$, and since $\sigma/\sqrt{n}$ bears no relationship to $\tilde{\mu}$, the 38.73 in the denominator of the likelihood simply cancels out in the application of Bayes' theorem for the same reason that the combinatorial terms in the binomial and Pascal distributions cancel out in the application of Bayes' theorem when the data-generating process is assumed to follow a Bernoulli model (see Section 4.3). Thus, since likelihoods can be multiplied by a positive constant, it is convenient to multiply each of the above likelihoods by 38.73.

The application of Bayes' theorem can be represented in tabular form when the prior distribution is discrete.

$\tilde{\mu}$	*Prior probability*	*Likelihood*	*(Prior probability)* × *(likelihood)*	*Posterior probability*
1100	.15	.0006	.00009	.001
1150	.20	.0270	.00540	.032
1200	.30	.2347	.07041	.412
1250	.20	.3857	.07714	.450
1300	.15	.1200	.01800	.105
	1.00		.17104	1.000

The prior and the posterior distributions of $\tilde{\mu}$ are graphed in Figures 4.6.1 and 4.6.2. The sample information has increased the probabilities for the higher values of $\tilde{\mu}$ and has decreased the probabilities for the lower values of $\tilde{\mu}$.

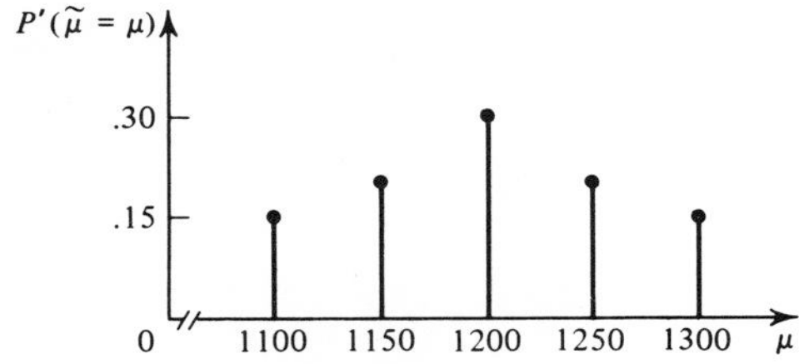

Figure 4.6.1

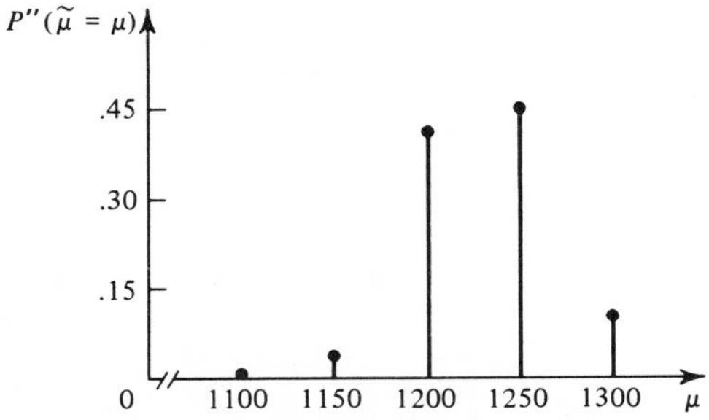

Figure 4.6.2

Although this illustrates the use of discrete prior distributions in conjunction with a continuous data-generating process, it would be more realistic to assume that $\tilde{\mu}$ is continuous. Suppose that the retailer assumes that $\tilde{\mu}$ is continuous. From informal conversations with the manager of the store and from knowledge about sales at similar stores, he decides that the mean of the prior distribution, m', is 1200 and that the standard deviation, σ', is 50. Furthermore, he feels that his prior distribution is symmetric about the mean and is shaped roughly like a normal density function, so he assumes that the distribution is normal. Be sure to distinguish between $\sigma^2 = 90{,}000$, which represents the variance of weekly sales, and $\sigma'^2 = 2500$, which represents the variance of the retailer's prior distribution of $\tilde{\mu}$, *mean* weekly sales.

The retailer may be interested in the distribution of sales for several reasons. Perhaps he must decide whether to keep the store or to sell it, whether to open more stores of a similar nature or in a similar neighborhood, or whether to bring in a new manager for the store. Such decisions necessitate the consideration of a loss function and of relevant variables other than sales. To keep the example from becoming too complicated, assume that the retailer does not have to make a decision at this particular moment. Instead, he just wants some idea about the mean of the distribution of weekly sales, and he decides to look at an interval of values of $\tilde{\mu}$.

Given the prior distribution, an interval that is centered at m' and that has a probability of .95 can be determined. Recall, from Section 4.5, that 95 percent of the area under a normal density function is contained within 1.96 standard deviations from the mean, so that this interval is simply

$$(m' - 1.96\sigma', m' + 1.96\sigma'),$$

or

$$(1200 - 1.96(50), 1200 + 1.96(50)),$$

or

$$(1102{,}1298).$$

This is called a "*credible interval.*" Based on the prior distribution, then, the interval from 1102 to 1298 is a 95 percent credible interval for $\tilde{\mu}$.

On the basis of the sample of 60 weeks with sample mean $m = 1240$, the retailer can revise his prior distribution. Following this sample, the posterior distribution for $\tilde{\mu}$ is a normal distribution with mean m'' and variance σ''^2, where

$$\frac{1}{\sigma''^2} = \frac{1}{\sigma'^2} + \frac{n}{\sigma^2} = \frac{1}{2500} + \frac{60}{90{,}000} = \frac{96}{90{,}000}$$

$$m'' = \frac{(1/\sigma'^2)m' + (n/\sigma^2)m}{(1/\sigma'^2) + (n/\sigma^2)} = \frac{(1/2500)(1200) + (60/90{,}000)(1240)}{(1/2500) + (60/90{,}000)}$$

$$= 1225.$$

The mean and the variance of the posterior distribution are 1225 and 90,000/96 = 937.5, and the prior and posterior distributions of $\tilde{\mu}$ are graphed in Figure 4.6.3. The limits for a 95 percent credible interval based on the posterior distribution are

$$m'' - 1.96\sigma'' \quad \text{and} \quad m'' + 1.96\sigma'',$$

or

$$1225 - 1.96(30.6) \quad \text{and} \quad 1225 + 1.96(30.6),$$

or

$$1165 \quad \text{and} \quad 1285.$$

Notice that the width of the 95 percent credible interval based on the prior distribution is 1298 − 1102 = 196, whereas the width of the interval based on the posterior distribution is 1285 − 1165 = 120. This reflects the fact that the introduction of the sample information reduces the standard deviation from 50 to 30.6. Incidentally, a corresponding interval based solely on the sample information has limits of approximately 1164 and 1316.

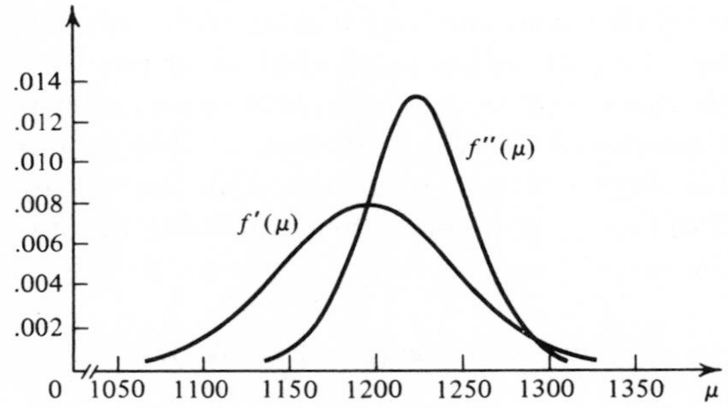

Figure 4.6.3

In terms of the other parametrizations presented in Section 4.5,

$$n' = \sigma^2/\sigma'^2 = 90{,}000/2500 = 36,$$
$$I' = 1/\sigma'^2 = 1/2500 = .0004,$$

and

$$I = n/\sigma^2 = 60/90{,}000 = .00067.$$

Applying Equations (4.5.13) and (4.5.15),

$$n'' = n' + n = 36 + 60 = 96$$

and

$$I'' = I' + I = .0004 + .00067 = .00107.$$

Thus, the "equivalent prior sample size" is 36, while the size of the actual sample is 60. The "amount of sample information," as measured by I, is greater than the "amount of prior information," as measured by I'.

This explains why the mean of the posterior distribution is closer to m than to m'.

One aspect of this example seems particularly unrealistic—the assumption that the variance of the process is known. The retailer may have some information about the variance, but it is highly unlikely that he knows it for certain. Unfortunately, a formal Bayesian approach to inference for a normal process is somewhat more complex if the variance is not assumed to be known. If the variance is not known, the retailer would have to assess a *joint* prior distribution for the variance and the mean and modify this distribution on the basis of the sample information. In this case, the sample variance would be of interest as well as the sample mean, since it provides information concerning the process variance. At any rate, assessing and revising a *joint* prior distribution is more complicated than assessing and revising a univariate prior distribution; this problem is not taken up in detail, although it is discussed very briefly in Section 4.7.

Does this mean that you will be unable to carry out a Bayesian analysis for the normal process with unknown variance without using a joint prior distribution? In some cases, yes. In other cases, however, an approximation may provide sufficient accuracy for the problem at hand. In Equations (4.5.8) and (4.5.9), σ^2 only appears in connection with the sample information, not the prior information. If you replace σ^2 by the *sample variance* $\tilde{s}^2 = \sum_{i=1}^{n} (\tilde{x}_i - \tilde{m})^2/(n-1)$, which can be used as an *estimator* of the population variance σ^2, the distribution of $\tilde{z} = (\tilde{m} - \mu)/(\tilde{s}/\sqrt{n})$ is not a standard normal distribution. If the sample size n is large enough, though, this distribution, which is called a t distribution, is virtually identical to the standard normal distribution. For large n, substituting s^2 for σ^2 in Equations (4.5.8) and (4.5.9) and proceeding as though σ^2 were known provides a reasonable approximation. For example, if the retailer does not know σ^2 but observes $s^2 = 96{,}000$, you can verify that the approximate posterior mean and variance of $\tilde{\mu}$ are 1224 and 975.6, respectively.

4.7 CONJUGATE PRIOR DISTRIBUTIONS FOR OTHER PROCESSES

In the past four sections, a discussion of the use of conjugate prior distributions has been presented for two particular statistical models, the Bernoulli model and the normal model with known variance. Similar procedures have been developed for numerous other models. For instance, consider the Poisson model, which was discussed in Section 3.4. The random variable of interest is $\tilde{\lambda}$, the intensity of the process, and the conjugate family for the Poisson process is the family of *gamma distributions*, which have density functions of the form

$$f(\lambda) = \frac{e^{-\lambda t'}(\lambda t')^{r'-1}t'}{(r'-1)!}, \qquad (4.7.1)$$

where $r' > 0$ and $t' > 0$. Just as in the beta distribution, the factorial term $(r-1)!$ technically should be a gamma function, $\Gamma(r)$, to allow for noninteger values of r, but you need not be concerned with this technicality. The mean and the variance of the gamma distribution are $E(\tilde{\lambda} \mid r',t') = r'/t'$ and $V(\tilde{\lambda} \mid r',t') = r'/t'^2$.

For instance, in the automobile-salesman example presented in Section 3.4, suppose that the owner of the dealership feels that his prior information about the average number of sales per day for the salesman in question can be represented by a gamma distribution with $r' = 4$ and $t' = 16$. The mean and the variance of the distribution are $E(\tilde{\lambda} \mid r',t') = .25$ and $V(\tilde{\lambda} \mid r',t') = .0156$, and the distribution may be interpreted in terms of equivalent prior information. The owner's prior information is roughly equivalent to a sample of 16 days with four sales (note that r' and t' in this case are analogous to r' and n' for the beta prior distribution in the Bernoulli case). On the basis of this prior distribution and the sample of 10 sales in 24 days, the owner's posterior distribution for $\tilde{\lambda}$ is a gamma distribution with $r'' = r' + r = 4 + 10 = 14$ and $t'' = t' + t = 16 + 24 = 40$. Thus, the posterior mean and variance are $r''/t'' = .35$ and $r''/t''^2 = .00875$.

This Poisson example demonstrates the use of conjugate prior distributions for a model other than the Bernoulli model or the normal model with known variance. For any data-generating process that can be described by a general statistical model, the idea of a conjugate family of prior distributions is the same. The likelihood function is determined from the statistical model, and a family of conjugate distributions is found. (In this book the process of actually finding such a family is not formally considered.) In general, the conjugate family and the likelihood function have similar functional forms; essentially, the conjugate family is determined by taking the general functional form of the likelihood function (not including factors that do not involve the variable of interest) and including additional factors to normalize the expression (that is, to make it a proper density function). If the prior distribution is a member of the conjugate family, the posterior distribution is a member of the same family, and the prior distribution and the likelihood function can be combined through a few reasonably simple formulas such as Equations (4.3.9) and (4.3.10) or (4.5.8) and (4.5.9).

As the process or population of interest becomes more complex, so that the mathematical form of the likelihood function becomes more complicated, it may be more difficult to apply Bayesian methods unless approximations (such as the normal distribution) can be used. The difficulties are increased when the statistician is interested in several variables at once. The prior distribution in such situations is a joint distribution of several variables.

For instance, if you are interested in a normal data-generating process with unknown mean $\tilde{\mu}$ *and* unknown variance $\tilde{\sigma}^2$, a complete Bayesian approach requires the assessment of a joint prior distribution for $\tilde{\mu}$ and $\tilde{\sigma}^2$. The conjugate family in this case is the family of *normal-gamma distributions*, which have density functions that can be expressed as products of a normal distribution and a gamma distribution. The normal distribution is the *conditional* distribution of $\tilde{\mu}$, given $\tilde{\sigma}^2 = \sigma^2$, and it can be expressed in terms of a mean m' and a variance $\sigma'^2 = \sigma^2/n'$, just as in the case where the variance of the process is known. The gamma distribution is the *marginal* distribution of the reciprocal of the process variance, $1/\tilde{\sigma}^2$, and its density function is of the form

$$f\left(\frac{1}{\tilde{\sigma}^2}\right) = \frac{e^{-v'd'/2\sigma^2}(v'd'/2\sigma^2)^{(d'/2)-1}(d'v'/2)}{[(d'/2) - 1]!},$$

where $v' > 0$ and $d' > 0$. The mean and the variance of this distribution are $E(1/\tilde{\sigma}^2 \mid v',d') = 1/v'$ and $V(1/\tilde{\sigma}^2 \mid v',d') = 2/d'v'^2$.

Thus, the conjugate normal-gamma distribution for $\tilde{\mu}$ and $\tilde{\sigma}^2$ from a normal data-generating process is of the form

$$f(\tilde{\mu},\tilde{\sigma}^2 \mid m',n',v',d') = f(\tilde{\mu} \mid m',n',\sigma^2)f(1/\tilde{\sigma}^2 \mid v',d').$$

Suppose that this represents the prior distribution of $\tilde{\mu}$ and $\tilde{\sigma}^2$. If a sample of size n is taken from the process, with sample mean m and sample variance s^2, then the posterior distribution of $\tilde{\mu}$ and $\tilde{\sigma}^2$ is a normal-gamma distribution with parameters m'', n'', v'', and d'',

$$f(\tilde{\mu},\tilde{\sigma}^2 \mid m'',n'',v'',d'') = f(\tilde{\mu} \mid m'',n'',\sigma^2)f(1/\tilde{\sigma}^2 \mid v'',d''),$$

where

$$m'' = \frac{n'm' + nm}{n' + n},$$

$$n'' = n' + n,$$

$$v'' = \frac{(d'v' + n'm'^2) + [(n-1)s^2 + nm^2] - n''m''^2}{d' + n},$$

and

$$d'' = d' + n.$$

In the example from the previous section involving the retailer, suppose that $\tilde{\mu}$ and $\tilde{\sigma}^2$ are both unknown and that the retailer's prior distribution for $\tilde{\mu}$ and $\tilde{\sigma}^2$ is a normal-gamma distribution with $m' = 1200$, $n' = 36$, $v' = 90{,}000$, and $d' = 90$. The sample statistics are $m = 1240$, $n = 60$, and $s^2 = 96{,}000$. Thus, the posterior distribution for $\tilde{\mu}$ and $\tilde{\sigma}^2$ is a normal-gamma distribution with

$$m'' = \frac{36(1200) + 60(1240)}{36 + 60} = 1225,$$

$$n'' = 36 + 60 = 96,$$

$$v'' = \frac{[90(90{,}000) + 36(1200)^2] + [59(96{,}000) + 60(1240)^2] - 96(1225)^2}{90 + 60}$$

$$= 92{,}000,$$

and

$$d'' = 90 + 60 = 150.$$

From this posterior distribution, the conditional expectation of $\tilde{\mu}$, given σ^2, is $E(\tilde{\mu} \mid \sigma^2, m'', n'', v'', d'') = m'' = 1225$; the expectation of $1/\tilde{\sigma}^2$ is $E(1/\tilde{\sigma}^2 \mid m'', n'', v'', d'') = 1/v'' = 1/92{,}000$.

In general, Bayesian inference is more difficult when there are two or more variables of interest than when there is only one variable. The problem lies not in the mathematics, although the difficulties that arise when the prior distribution is not a member of the conjugate family may be more severe in the multivariate case. Instead, the problem lies in the determination of the necessary inputs to the problem. The assessment of both a prior distribution and a likelihood function may become more complex in the multivariate case. For instance, even if the statistician is willing to assume that the process of interest is a normal process with unknown mean $\tilde{\mu}$ and variance $\tilde{\sigma}^2$ (this assumption determines the likelihood function), it may not be easy to assess a prior distribution. In the conjugate family of normal-gamma distributions, $\tilde{\mu}$ and $\tilde{\sigma}^2$ are dependent, so that it is not possible to determine the marginal distributions of $\tilde{\mu}$ and $\tilde{\sigma}^2$ separately and simply to multiply these marginal distributions to arrive at the joint prior distribution. The assessment of prior distributions is discussed in the following section.

4.8 THE ASSESSMENT OF PRIOR DISTRIBUTIONS

In all of the examples, it has been assumed that the prior distribution is given (that is, that the statistician has already determined his prior distribution). Suppose that this is not true. Instead, suppose that the statistician has certain information and wants to express this information in terms of a prior distribution. How can he quantify his judgments and express them as a probability distribution? In Section 3.5 the assessment of individual probabilities was discussed. If the random variable in question is discrete, the statistician's distribution will consist of a number of such probabilities. If the random variable is continuous, the task facing the assessor is somewhat different. Several techniques have been proposed to aid the assessor in this task.

The most obvious way to attempt to assess a continuous probability distribution is simply to specify the density function or the cumulative distribution function. One way to do this is to specify the functional form of the density function. The relationship between subjective judgments and a mathematical function, if such a relationship does exist, is usually not at all obvious. Attempting to specify directly the functional form of the distribution, then, might not be such a good approach. A second approach involves the graph of the density function and/or the cumulative distribution function. The assessor may have some idea of the general shape of the distribution, even if he has no idea about its mathematical form. Perhaps he can draw at least a rough graph of the PDF or CDF. Of these two, the PDF is probably more meaningful to most assessors when expressed in graphical form; that is, they can translate their prior information into a graph of a PDF more easily than into a graph of a CDF.

One way to determine a PDF is to assess probabilities for certain intervals, draw a histogram (a bar graph) corresponding to these probabilities, and attempt to fit a smooth density function to the histogram. Recall, from Section 4.1, that a density function can be viewed as the limiting form of a histogram as the number of intervals increases and the length of each interval decreases. Figure 4.8.1 illustrates the assessment of a PDF by this "grouping and smoothing" technique. For instance, if you want to assess a probability distribution for $\tilde{x}$, the sales (in dollars) of a particular item, you might assess $P(\tilde{x} < 10{,}000)$, $P(10{,}000 \leq \tilde{x} \leq 12{,}000)$, $P(12{,}000 \leq \tilde{x} \leq 14{,}000)$, and so on, construct a histogram, and then try to find a continuous distribution that is a good approximation to the histogram. The probabilities for the intervals of interest can be assessed by using techniques such as those discussed in Section 3.5.

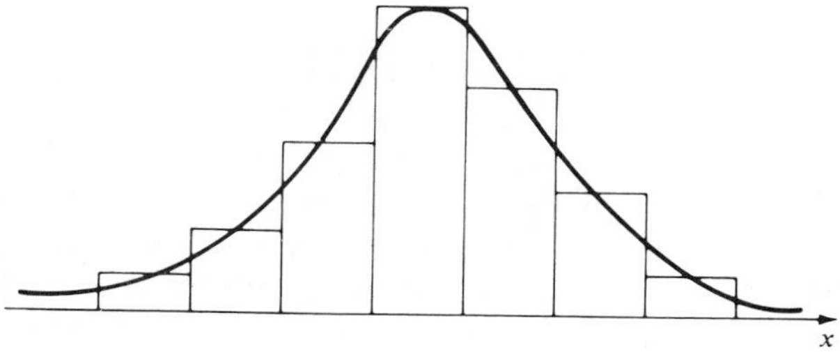

Figure 4.8.1

In some situations, much of the prior information may consist of historical data. For instance, if the random variable of interest is $\tilde{p}$, the proportion of defective items produced by a certain production process, past data may be available regarding the number of defectives in various lots of, say, 1000 items each. This data may be expressed in the form of a histogram, which can be used as a starting point for the assessment of a continuous prior distribution. Note that unlike the case in which the histogram is determined subjectively, the prior distribution need not be a particularly "good fit" to the histogram when past data is used to construct the histogram. Certain intervals may, simply by chance, occur very seldom in the past data despite the fact that neighboring intervals occur much more frequently. If there appears to be no explanation for such results, then the chance variations should be "smoothed out" in the assessment of a continuous prior distribution. For example, consider the histogram in Figure 4.8.2, representing historical data regarding $\tilde{\mu}$, average weekly sales at a particular store. If there is no obvious explanation for the "jumps" that occur around $\tilde{\mu} = 150{,}000$ and $\tilde{\mu} = 220{,}000$, then the assessed continuous distribution for $\tilde{\mu}$ need not take these "jumps" into account. Another reason that the assessed distribution may differ from the histogram is that other information may be available in addition to the past data. As indicated in Chapter 2, historical frequencies are helpful in the assessment of probabilities, but other prior information may cause the assessed probabilities to differ somewhat from the historical data.

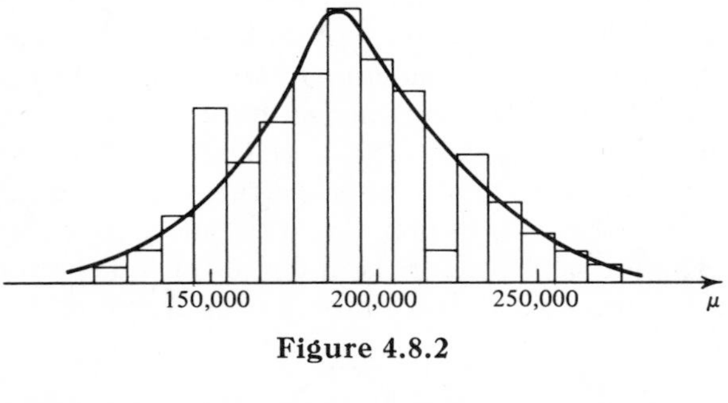

Figure 4.8.2

In determining a PDF or CDF, it might be helpful to consider certain summary measures of the distribution. In particular, some measure of location, such as the mean, the median, or the mode of the prior distribution, would be useful. This would give the assessor some idea of where the "center" of the distribution is. Next, a measure of dispersion might be considered. The variance is probably not a good measure in this context, because it is expressed in terms of squared units and because the relationship between a certain variance and an amount of dispersion is not intuitively obvious. The standard deviation has the advantage of being expressed in the same units as the random variable, but unless the assessor has enough experience with probability and statistics to gain some "feel" for the standard deviation, it might not be too meaningful to him. Measures such as credible intervals are better suited for the assessment of prior distributions. Recall from the previous section that a credible interval is simply an interval of values with some given probability under a prior or a posterior distribution. The assessor might ask himself, "Can I find an interval centered at the mean that includes 50 percent, or 75 percent, or 95 percent, of the prior probability?"

The use of credible intervals suggests the assessment of fractiles of the prior distribution. Recall, from Section 4.3, that an f fractile of the distribution of a continuous random variable $\tilde{x}$ is a point x_f such that $P(\tilde{x} \leq x_f) = f$. The .50 fractile, for example, is the median of the distribution, which for a continuous random variable divides the area under the density function into two equal parts, each with area $\frac{1}{2}$. Intuitively, the median of the assessor's prior distribution is the value that the assessor feels is equally likely to be exceeded or not exceeded. Once the median is assessed, the .25 and .75 fractiles are the values that divide the two halves of the distribution in half again, and so on. This process is illustrated in Figure 4.8.3. The advantage of this technique is that it merely requires the assessor to determine a series of fractiles, and the assessment of any single fractile is similar to the assessment of an individual probability. After a number of fractiles are assessed and plotted on a CDF graph, the assessor can then draw a rough curve through them and thus determine the entire distribution.

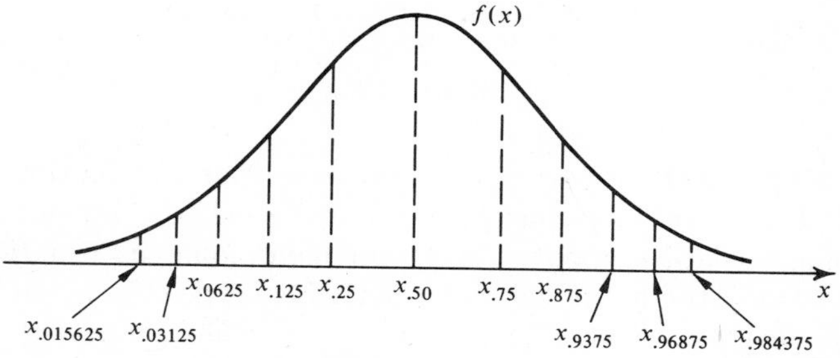

Figure 4.8.3

For example, suppose that a statistician wants to assess his prior distribution for the proportion of defectives produced by a production process. First of all, he thinks that the proportion, $\tilde{p}$, is just as likely to be above .05 as below .05. In other words, he is indifferent between receiving a lottery ticket that pays $100 if $\tilde{p}$ is less than .05 and receiving a lottery ticket that pays \$100 if $\tilde{p}$ is greater than .05. For him, the odds in favor of either of these events are even. This implies that .05 is the median of his distribution. Then, looking just at the interval below .05 (that is, the interval from 0 to .05), he decides that .035 divides this interval into two equally likely subintervals. That is, he is indifferent between a lottery that pays \$100 if $\tilde{p}$ is less than .035 (and pays nothing otherwise) and a lottery that pays \$100 if $.035 < \tilde{p} < .05$ (nothing otherwise), so that $P(\tilde{p} < .035) = P(.035 < \tilde{p} < .05)$. Thus, .035 is the .25 fractile of the distribution. Similarly, he assesses the .75, .125, .375, .625, and .875 fractiles. In addition, to get some idea about the extreme tails of the distribution, he assesses the .01 and .99 fractiles. He then plots these assessed points on a graph and draws a rough CDF through them, as illustrated in Figure 4.8.4. To check his assessments, he might look at the curve to see if it reflects his prior information reasonably well. As noted in Section 3.5, there may be some vagueness on the part of the assessor, but devices such as hypothetical lotteries or bets should help to reduce this vagueness.

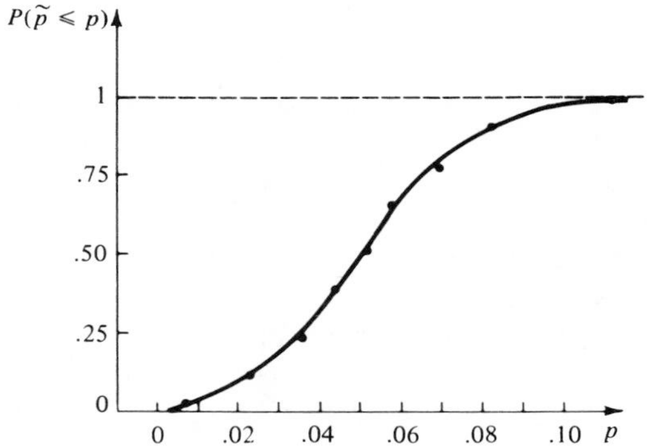

Figure 4.8.4

It should be clear by now that assessing a prior distribution is not an easy task if a careful assessment is desired. To see how difficult it can be, select some variables about which you have some information (some ideas: the proportion of vehicles of a particular type on a given highway; the price, one year from now, of a certain common stock; the total amount of precipitation in the next year in a given city; and so on) and *carefully* attempt to assess a prior distribution in each case. To encourage careful assessment, pretend that your prior distribution in each case will be used as an input for a very important decision-making problem. It might be interesting to select several variables about which you have considerable knowledge and several about which you have very little information and then to compare the resulting distributions. It is also interesting to assess a prior distribution for a variable that you will be able to observe shortly. You can then evaluate your distribution after observing the actual event. For example, assess a distribution for the point spread in an upcoming football game.

At this point, the discussion of the assessment of prior distributions is interrupted to ask this question: why do you want a prior distribution in the first place? Presumably, you want to represent your prior uncertainty in probabilistic terms in order to make an inferential statement or to make a decision, and you want to base the inferential statement or decision on all of the information that happens to be available. You may intend to take a sample and use Bayes' theorem to combine the prior distribution and the likelihood function, forming a posterior distribution. But if that is the case, then it is convenient if the prior distribution is a member of a conjugate family of distributions. If you are sampling from a Bernoulli process, for example, and if you are interested in $\tilde{p}$, the probability of a success on any single trial, the analysis is greatly simplified if the prior distribution happens to be a beta distribution. For sampling from a normal process with known variance, the conjugate family is the normal family of distributions.

A subjectively assessed prior distribution is not necessarily a member of a conjugate family of distributions. In general, it does not follow any mathematical function exactly. However, it may be possible to find a member of the conjugate family that is a "good fit" to the assessed distribution. Thus, the concept of conjugate distributions provides a convenient model that may be realistic in many situations. This is where the property of richness of conjugate families is important. Unless the distribution is quite irregular, there is a good chance that a member of the conjugate family will approximate it reasonably well. The assessor could try a number of conjugate distributions to see if any prove satisfactory. A more precise notion of what is meant by "satisfactory" and "good fit" is given later in this section.

It is possible for the assessor to consider the conjugate family of distributions before assessing his entire distribution. If he feels that his prior information is roughly equivalent to sample information from the process or population of interest, he may be able to determine the prior parameters directly, using the interpretation of conjugate families presented earlier. If he can do this, the assessment procedure is greatly simplified. Suppose that the production manager interested in the proportion of defective items produced by a certain process feels that his prior information is roughly equivalent to a sample of 40 items with three defectives. His prior distribution is then a beta distribution with parameters $r' = 3$ and $n' = 40$.

Suppose that the production manager in the above example is not able to translate his prior information into equivalent sample evidence, but that he does think that a beta distribution would adequately represent his information. How can he choose a single beta distribution? If he can determine the mean and the variance of the distribution, he can equate these to the formulas for the mean and the variance of a beta distribution and solve for r' and n'. Suppose that the production manager's mean and standard deviation are .08 and .04, respectively. Then from Equations (4.3.2) and (4.3.3),

$$\frac{r'}{n'} = .08$$

and

$$\frac{r'(n' - r')}{n'^2(n' + 1)} = (.04)^2 = .0016.$$

Solving these two equations simultaneously yields $r' = 3.6$ and $n' = 45$. Recall that r' and n' need not be integers.

In the normal case, of course, the mean and the variance are themselves the prior parameters; thus, there are no equations to solve. However, the assessor might wish to assess a pair of fractiles and use the tables of the normal distribution to find m' and σ'. For a very simple example, suppose that you are interested in $\tilde{x}$, the maximum temperature next July 4 in Chicago. You feel that the .50 fractile, or the median, is 86 and that the .75 fractile is 92. Furthermore, you decide that your distribution of $\tilde{x}$ is approximately normal. Since the normal distribution is symmetric, the mean and the median are equal, so that the mean is 86. From the table of the cumulative standard normal distribution, the .75 fractile of the standard normal distribution is approximately .67. This implies that the .75 fractile of any normal distribution is .67 standard deviations, or $.67\sigma'$, to the right of the mean. But 92 is six units to the right of the mean, 86, so

$$6 = .67\sigma'.$$

Therefore,

$$\sigma' = 8.96.$$

Your distribution for $\tilde{x}$ is thus a normal distribution with mean 86 and standard deviation 8.96.

It is also possible to assess a beta distribution by using fractiles. It is very difficult to assess the variance of a distribution intuitively, so the procedure for fitting a beta distribution that involves both the mean and the variance might not be too practical. Instead, the assessor could asesss a pair of fractiles, or, if he prefers, the mean and a single fractile. For instance, suppose that the assessor feels that his prior distribution for some uncertain quantity $\tilde{p}$ is a beta distribution with a mean of .25 and a .25

fractile of .20. Using the formula for the mean of a beta distribution yields $r'/n' = .25$, or $n' = 4r'$. There is no formula for the .25 fractile of a beta distribution, but this fractile is given for numerous beta distributions in a table in the back of the book. Since $n' = 4r'$, it is necessary only to look at the .25 fractiles for the beta distributions with $r' = 1$, $n' = 4$; $r' = 2$, $n' = 8$; and so on. For $r' = 1$, $n' = 4$, the .25 fractile is .0914. As r' and n' increase with r'/n' held constant, the .25 fractile increases; for $r' = 9$, $n' = 36$, the .25 fractile is .1990, and for $r' = 10$, $n' = 40$, it is .2017. If the assessor wants to use integer values for r' and n', then the distribution that comes closest to his assessments is the beta distribution with $r' = 9$ and $n' = 36$, since .1990 is closer to .2 than is .2017. It is not necessary to use integer values for r' and n', although it is more convenient because the tables only consider integer values, and it should be close enough for all practical purposes, since the assessment of .20 for the .25 fractile is probably just a rough approximation anyway. After the assessor goes through the above procedure and determines r' and n', he might want to assess another fractile judgmentally and see if this agrees with the corresponding fractile given in the table for r' and n'. This serves as a check on his assessments and on the applicability of the beta distribution as his judgmental probability distribution. If the assessed fractile does not agree reasonably well with the tabled fractile for r' and n', then the assessor should re-evaluate his assessments. If he then still feels that his assessments accurately reflect his judgments, perhaps his distribution is of such a form that it cannot be fit by a beta distribution. Informal checks such as this are extremely useful in the assessment of prior distributions.

Another technique that may be useful in the assessment of prior distributions involves the consideration of hypothetical sample information. For instance, suppose that you are interested in $\tilde{p}$, the proportion of consumers who will purchase a certain product, and that you are willing to assume that your prior information can be adequately represented by a beta distribution. Your expected value of $\tilde{p}$ is .20. Furthermore, you claim that if you observed a random sample of 100 consumers and only 10 bought (or stated that they would buy) the product, then your expected value of $\tilde{p}$ would shift to .18. That is, your mean is currently $r'/n' = .20$, but if you observed this particular sample, your mean would be $r''/n'' = (r' + 10)/(n' + 100) = .18$. Solving these two equations for r' and n' yields $r' = 80$ and $n' = 400$. By considering the effect of hypothetical samples such as this on the parameters of your prior distribution, you can determine the values of these parameters. Note that this procedure involves a "backward" application of Bayes' theorem; given a particular sample and a particular posterior distribution following this sample, what can be said about the original prior distribution?

The assessment procedures discussed in this section can be extended to the assessment of conditional distributions and joint distributions. For instance, it is possible to assess the distribution of average sales of a particular product, conditional upon the price of the product. This can be done for various possible prices, and it might be extremely useful in making a pricing decision for the product in question. To carry this one step further, suppose that the price of the product is beyond the control of the firm that manufactures the product (for instance, the price might be determined by a governmental agency). Then the firm would be interested in a joint distribution of $\tilde{x}$, mean sales, and $\tilde{y}$, price. If a prior distribution is assessed for $\tilde{y}$, then the joint distribution of $\tilde{x}$ and $\tilde{y}$ may be found by combining $f(y)$ and $f(x \mid y)$ through the use of Equation (4.1.8), which implies that

$$f(x,y) = f(y)f(x \mid y).$$

An example of this procedure is presented for discrete random variables in Section 3.5, and the extension to continuous random variables is straightforward, so the procedure is not discussed further here.

In this section several techniques for assessing prior distributions have been discussed. The technique used in any specific situation should depend on the form of the prior information. If the prior information consists entirely of sample information or can be thought of as equivalent to sample information, a conjugate prior distribution can be assessed directly. In other cases, it may be desirable to determine a few summary measures, or perhaps the entire subjective distribution, before attempting to fit a conjugate distribution. In yet other cases, the decision maker might not be able to find a conjugate distribution that adequately reflects his prior information. Just as a statistical model such as the Bernoulli model is an approximation to reality that (1) may be realistic in many situations and (2) may greatly simplify the determination of likelihoods, a conjugate prior distribution is an approximation that may often be realistic and that should greatly simplify the application of Bayes' theorem to determine a posterior distribution. Thus, conjugate prior distributions are convenient models that may provide good approximations in some (but not all) situations.

Because of the subjective nature of the assessment process, different techniques of assessment are likely to lead to different distributions. If two such distributions are assessed, the decision maker may be interested in how similar they are to each other as well as how similar they are to members of the conjugate family of distributions. In particular, he may be concerned with a decision-theoretic approach to "goodness-of-fit."

In most applications of Bayes' theorem, the ultimate aim of the decision maker is to make a decision on the basis of the posterior distribution. Variations in the prior distribution will cause variations in the posterior distribution. The decision maker is interested in this question: how will these variations affect the ultimate decision? Or, in other words, how sensitive is the decision-making procedure to variations in the prior distribution? An investigation of this question is called a *sensitivity analysis.* If large variations in the prior distribution tend not to affect the decision, then the decision-making procedure is said to be *insensitive* to variations in the prior distribution. If very slight variations in the prior distribution are likely to cause the decision to be changed, the decision-making procedure is said to be highly *sensitive* to such variations.

It should be pointed out that sensitivity depends on the particular decision-making procedure, including the sample that is taken, the actions available to the decision maker, and the potential payoffs or losses. Technically, then, it is not possible to speak of sensitivity except in connection with a particular decision-making situation. It is possible, however, to make some generalizations concerning sensitivity. The more sample information the decision maker has, the less sensitive the procedure usually is with regard to the prior distribution. As more sample information is gathered, the prior distribution is given less weight in the computation of the posterior distribution (as in Section 4.5). As a result, variations in the prior distribution are of less importance. Any time there is a great deal of sample information relative to the prior information, insensitivity can be expected to hold. This fact is of great importance in the discussion of diffuse prior distributions in Section 4.10.

Some research concerning sensitivity analysis has indicated that in a wide variety of situations, the decision-making procedure is reasonably insensitive to moderate variations in the prior distribution. This implies that a conjugate prior distribution, while it may not be a perfect fit to a subjectively assessed distribution, will often be a "satisfactory" fit. Furthermore, insensitivity implies that the choice of a particular distribution from the family of conjugate distributions may not be crucial. For instance, the ultimate decision may be the same for a wide variety of prior distributions. If there is any doubt in a particular application, however, it is prudent to investigate the sensitivity of the decision-making procedure. If it is not too burdensome computationally (and it should not be, particularly if a high-speed computer is available), several posterior distributions can be calculated. In this manner, it is possible to see how variations in the prior distribution are reflected in the posterior distribution and in the decision-making problem. Since decision theory is presented in Chapters 5 and 6, further discussions of sensitivity are deferred until those chapters.

This section has involved the assessment of prior distributions, an important problem in Bayesian inference and decision. The assessment of another input to the Bayesian model, likelihood functions, is also quite important. The discussion in Section 3.6 applies in the continuous case as well. If a particular statistical model seems to be a reasonable approximation to a real-world situation, then likelihoods can be determined from the model; if the situation at hand cannot be represented by such a model, then it is necessary to assess the likelihoods directly. (Since the likelihoods are proportional to conditional probabilities, this amounts to the assessment of conditional probability distributions.) For certain models, such as the Bernoulli and Poisson models, the applicability of the model can be investigated in terms of certain assumptions, such as stationarity and independence, concerning the data-generating process of interest. For other models, such as the normal model, this approach is not possible, since the models are not developed in terms of a few simple assumptions. As noted in Section 4.5, there are various devices, such as "normal probability paper," for investigating the applicability of the normal model to an actual data-generating process. It should be emphasized again that the assessment of likelihoods, whether done directly or in terms of a certain statistical model, is ultimately of a subjective nature. Furthermore, another point of interest is that sensitivity analysis can be used for likelihoods as well as for prior distributions. If the decision-making problem at hand is relatively insensitive to variations in the likelihoods, then the choice of a statistical model is not crucial. On the other hand, if the problem is highly sensitive, then it is important to investigate carefully the applicability of any model that is being considered.

4.9 DISCRETE APPROXIMATIONS OF CONTINUOUS PROBABILITY MODELS

Whenever a prior distribution can be approximated closely by a conjugate distribution, where "closely" is to be interpreted in terms of the discussion of sensitivity analysis in the previous section, then it is convenient to use this conjugate distribution as the prior distribution. This is true regardless of whether the actual prior distribution is discrete or continuous (as noted in Section 4.1, continuous probability models are an idealization). Thus, it often may be convenient to approximate a discrete prior distribution by a continuous prior distribution.

If, however, the prior distribution cannot be approximated by a member of the conjugate family in any given situation with a continuous probability model, the application of Bayes' theorem to determine the posterior distribution is considerably more difficult, involving the application of Equation (4.2.3):

$$f(\theta \mid y) = \frac{f(\theta)f(y \mid \theta)}{\int_{-\infty}^{\infty} f(\theta)f(y \mid \theta)\, d\theta}.$$

As pointed out in Section 4.2, potential difficulties in evaluating the integral in the denominator of this equation provide much of the motivation for the development of conjugate families of distributions. Very often this integral cannot be evaluated by elementary techniques of integration. Even if it can be evaluated, the result of the application of Equation (4.3.2) may be quite complicated, and if *further* information is obtained, the second application of the equation is likely to be even more troublesome than the first.

Whenever an integral is so complicated that it is difficult to evaluate analytically, it is generally evaluated numerically, utilizing a computer. In the case of Bayes' theorem for continuous probability models, this amounts to approximating the prior density function, $f(\theta)$, by a discrete mass function, $P(\theta)$, and applying Bayes' theorem for discrete probability models. Recall the discussion in Section 4.1 in which continuous probability models were developed as a limiting form of discrete probability models. The probability of any interval is represented by a definite integral in the continuous case, but this is just the limiting form of a sum. To approximate a continuous density function, $f(\theta)$, by a discrete mass function, simply divide the set of possible values of the uncertain quantity $\tilde{\theta}$ into a number of intervals and determine the probability of each interval:

$$P(a \leq \tilde{\theta} \leq b) = \int_a^b f(\theta)\, d\theta.$$

To form the discrete approximation, simply assume that this probability $P(a \leq \tilde{\theta} \leq b)$, instead of being distributed continuously over the interval, is a mass that is concentrated at the midpoint of the interval, $(a + b)/2$. The discrete approximation then consists of a number of masses equal to the number of intervals originally chosen, each mass being located at the midpoint of an interval. Bayes' theorem for discrete probability models, presented in Chapter 3, can then be applied to determine the posterior distribution, which of course will also be discrete.

The accuracy of this numerical approximation increases as the number of intervals increases and the width of the intervals decreases. (It should be noted that the intervals do not have to be of equal width; generally they should be narrower where the density function is high and wider where the density function is low.) By increasing the number of intervals appropriately, it is possible to obtain any desired degree of accuracy. This should not create any computational problems, for numerical procedures such as this usually involve the use of a high-speed computer.

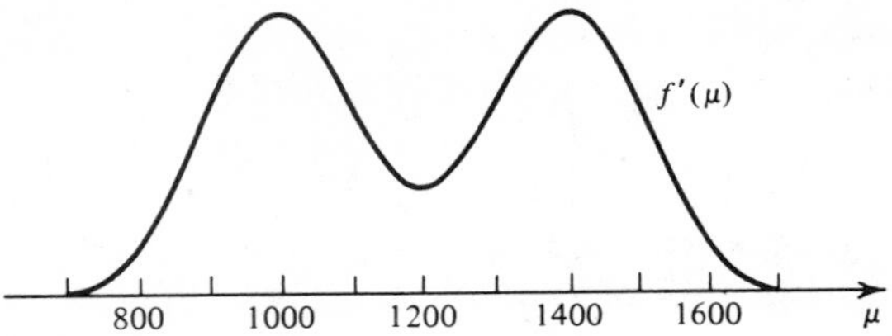

Figure 4.9.1

For an illustration, consider a modification of the example presented in Section 4.6 in which a retailer assessed a normal prior distribution for the uncertain quantity $\tilde{\mu}$, mean weekly sales at a particular store. Instead of a normal prior distribution for $\tilde{\mu}$, suppose that the retailer's prior distribution is represented by the density function shown in Figure 4.9.1. This form of density function is called *bimodal* because it has two modes, or high points of the density function. A distribution such as this might be reasonable if some extraneous consideration is expected to affect mean sales. For instance, the retailer's best guess about mean sales might be 1400 provided that consumers expect that Congress will reduce personal income taxes, and his best guess might be 1000 provided that consumers do not expect that taxes will be reduced. The bimodal distribution results because he is uncertain as to whether or not consumers expect that Congress will reduce taxes; the two "halves" of the distribution are equal (the distribution is symmetric), so apparently he feels that the probability that consumers expect a tax reduction is about $\frac{1}{2}$. Given the assumptions presented in Section 4.6 (sales are normally distributed with unknown mean $\tilde{\mu}$ and known variance $\sigma^2 = 90{,}000$) and given some sample information, Bayes' theorem can be applied to revise the prior distribution, which is shown in Figure 4.9.1. Suppose that a sample of size 27 is taken, with sample mean $m = 1100$. In that case, the posterior density function for $\tilde{\mu}$ is shown in Figure 4.9.2. Note that the horizontal scale in Figure 4.9.2 is not the same as that in Figure 4.9.1; in relation to the prior distribution, the posterior distribution is even more "concentrated" than the graphs imply at first glance. Furthermore, the posterior distribution is unimodal (it has only one mode). It appears that the sample mean, 1100, is far enough from 1400 to strongly discount the right-hand half of the prior distribution, which corresponds to the possibility of a tax reduction.

Even though the prior distribution in this example is not a conjugate distribution (the conjugate family is the family of normal distributions), it is a *mixture* of conjugate distributions. That is, the distribution is a combination of two normal distributions, one with mean $m' = 1000$ and standard deviation $\sigma' = 100$ and the other with mean $m' = 1400$ and standard deviation $\sigma' = 100$. A weighted average of these two distributions is taken as the prior distribution, where the weights are $\frac{1}{2}$ and $\frac{1}{2}$. Given a mixture of conjugate distributions such as this, it is possible to find the posterior distribution, which is also a mixture of conjugate distributions, without using a discrete approximation. This is, however, a convenient example in which to illustrate the use of a discrete approximation to a continuous probability model (such an approximation may be more convenient in many situations anyway).

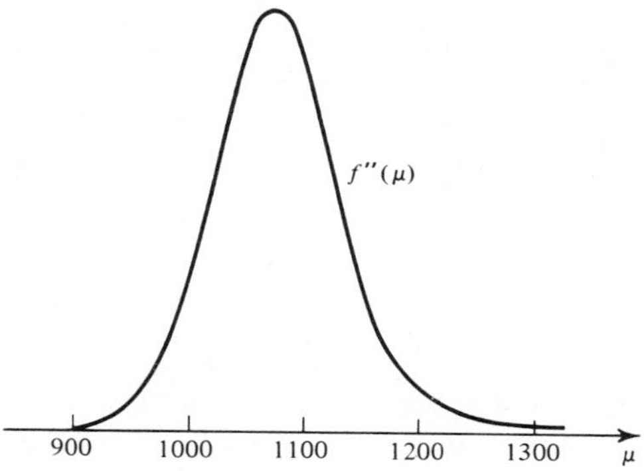

Figure 4.9.2

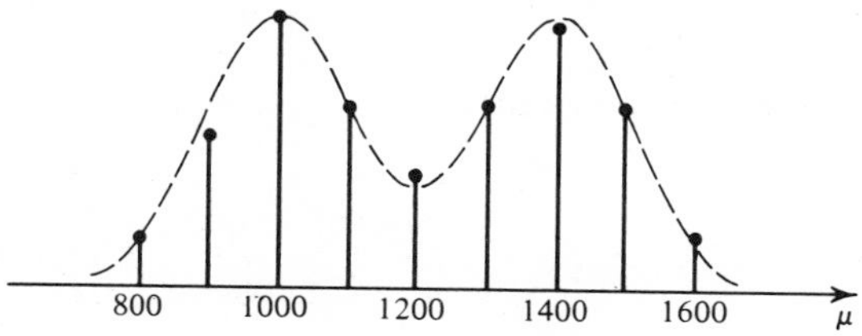

Figure 4.9.3

For a very rough discrete approximation to the prior distribution, consider intervals of width 100 centered at 800, 900, and so on, up to 1600. The prior probabilities of these intervals can be determined by integrating the prior density function over each interval. The functional form of the prior density function has not been given, however, although it was pointed out that it is a mixture of normal distributions. Very often the prior distribution is presented only in graphical form, and the functional form of the density function is not known. If this is the case, the probabilities of the chosen intervals can be determined approximately by drawing the prior density function on graph paper and by counting the number of "squares" under the curve in each interval. The resulting discrete approximation for the example is then as follows (the mass function is graphed in Figure 4.9.3 together with the prior density function that is being approximated, with the vertical scales adjusted to facilitate comparisons of the two functions):

Interval	*Midpoint*	*Prior probability*
750–850	800	.03
850–950	900	.12
950–1050	1000	.19
1050–1150	1100	.12
1150–1250	1200	.08
1250–1350	1300	.12
1350–1450	1400	.19
1450–1550	1500	.12
1550–1650	1600	.03

In order to apply Bayes' theorem to this discrete approximation, the likelihoods must be determined. For any given value μ of $\tilde{\mu}$, the distribution of the sample mean $\tilde{m}$ is a normal distribution with mean μ and variance $\sigma^2/n = 90{,}000/27 = 3333$, as indicated in Section 4.7. But the observed sample mean is $\tilde{m} = 1100$, so the likelihood of any value of $\tilde{\mu}$ is simply the height of the density function of $\tilde{m}$ given the value of $\tilde{\mu}$, evaluated at $\tilde{m} = 1100$. For instance, if $\tilde{\mu} = 1000$, the likelihood can be determined by using Equation (4.5.5) and the table of the standard normal density function:

$$f(1100 \mid 1000{,}3333) = f\left(\frac{1100 - 1000}{\sqrt{3333}} \,\middle|\, 0{,}1\right) \Big/ \sqrt{3333}$$

$$= \frac{f(1.73 \mid 0{,}1)}{57.7} = \frac{.0893}{57.7}$$

$$= .001548.$$

The entire set of likelihoods for the example is as follows:

$$f(1100 \mid 800{,}3333) = f(5.20 \mid 0{,}1)/57.7 = 0,$$
$$f(1100 \mid 900{,}3333) = f(3.47 \mid 0{,}1)/57.7 = .000017,$$
$$f(1100 \mid 1000{,}3333) = f(1.73 \mid 0{,}1)/57.7 = .001548,$$
$$f(1100 \mid 1100{,}3333) = f(0 \mid 0{,}1)/57.7 = .006913,$$
$$f(1100 \mid 1200{,}3333) = f(-1.73 \mid 0{,}1)/57.7 = .001548,$$
$$f(1100 \mid 1300{,}3333) = f(-3.47 \mid 0{,}1)/57.7 = .000017,$$
$$f(1100 \mid 1400{,}3333) = f(-5.20 \mid 0{,}1)/57.7 = 0,$$
$$f(1100 \mid 1500{,}3333) = f(-6.93 \mid 0{,}1)/57.7 = 0,$$

and $$f(1100 \mid 1600{,}3333) = f(-8.67 \mid 0{,}1)/57.7 = 0.$$

The values given as 0 are not *exactly* 0, but they are 0 for all practical purposes. Bayes' theorem for discrete probability models can now be applied as follows.

$\tilde{\mu}$	*Prior probability*	*Likelihood*	*(Prior probability)* $\times$ *(likelihood)*	*Posterior probability*
800	.03	0	0	0
900	.12	.000017	.00000204	.002
1000	.19	.001548	.00029412	.235
1100	.12	.006912	.00082944	.662
1200	.08	.001548	.00012384	.099
1300	.12	.000017	.00000204	.002
1400	.19	0	0	0
1500	.12	0	0	0
1600	.03	0	0	0
	1.00		.00125148	1.000

It should be noted that in the calculation of the likelihoods for this example, it would have been possible not to divide each standard normal density value by the term $\sigma/\sqrt{n} = 57.7$. This term does not involve $\tilde{\mu}$, so it is the same for all values of $\tilde{\mu}$. In this sense it is similar to the terms $\binom{n}{r}$ and $\binom{n-1}{r-1}$, which were discussed in Section 4.3 in relation to sampling from a Bernoulli process (also, see the example with the discrete prior distribution in Section 4.6).

The posterior probability mass function is graphed in Figure 4.9.4, together with the density function that is being approximated. (As in Figure 4.9.3, the vertical scales are adjusted to facilitate comparisons.) Despite the fact that the discrete prior approximation was quite rough (in practice the decision maker would probably want to choose more intervals and to make the intervals narrower in the regions in which the prior density function is high), the discrete posterior distribution is a very good approximation to the actual posterior density function. This illustrates the use of discrete approximations when the prior distribution is not a member of the relevant conjugate family of distributions. Discrete approximations may also be quite useful when the data-generating process cannot be satisfactorily represented by a well-known statistical model, in which case a conjugate distribution cannot be found. Moreover, with the aid of a computer it is convenient to use discrete approximations and to investigate the sensitivity of the results to such things as changes in the choice of intervals.

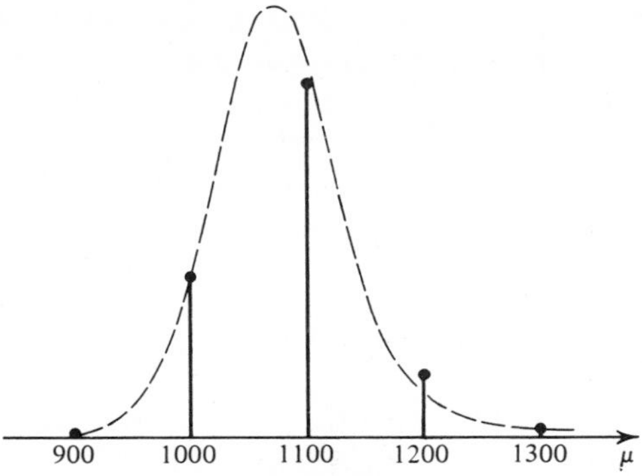

Figure 4.9.4

4.10 REPRESENTING A DIFFUSE PRIOR STATE

Suppose that a decision maker wants to assess a prior distribution in a situation in which he has very little or no prior information. More specifically, his prior information is such that it is "overwhelmed" by the sample information. Then it is said that the decision maker has a *diffuse* state of prior information. The situation described is not necessarily an informationless state in the usual meaning of the word; it is informationless in a relative sense. When it is said that someone's prior distribution is diffuse, this means only that it is *diffuse relative to the sample information.*

A good example of diffuseness concerns the determination of the weight of a potato. If you were given a potato and asked to assess a distribution of its weight, you would clearly have some information about the weight. On the other hand, your information would probably be of a rather vague nature. You could probably specify some limits within which you were sure the weight would lie, and your distribution might have a peak, or mode, somewhere within these limits. It is doubtful, however, that your distribution would have a sharp "spike" anywhere.

If the potato were weighed on a balance of known precision (say, a standard deviation, or standard error of measurement, of $\frac{1}{2}$ gram), how would your posterior distribution look following a single weighing? Since your distribution is probably quite "spread out," or diffuse relative to the likelihood function, the sample information receives much more weight than the prior distribution. In fact, the posterior distribution depends almost solely on the sample information. Notice that the sample size here is only 1, but the precision of the balance is so high that the likelihood function is much more precise than the prior distribution. It should be noted that increased precision in a sample can result from a larger sample size or from more precise measurement.

Graphically, the situation is represented in Figure 4.10.1. Let $\tilde{\theta}$ represent the weight of the potato and let y represent the sample result, the weight obtained from one reading of the balance. Relative to the likelihood function, the prior distribution $f'(\theta)$ is "flat." It is not strictly a uniform distribution, but it *is* virtually uniform *relative* to $f(y \mid \theta)$. Note that as a likelihood function, $f(y \mid \theta)$ is considered as a function of θ, with y fixed (see Section 3.2). From Section 4.2, the posterior density function is proportional to the product of the prior density and the likelihood function,

$$f''(\theta \mid y) \propto f'(\theta)f(y \mid \theta), \tag{4.10.1}$$

where the $\propto$ sign is read "is proportional to." The constant of proportionality is the reciprocal of the normalizing integral,

$$\int_{-\infty}^{\infty} f'(\theta)f(y \mid \theta)\, d\theta.$$

But the prior distribution is essentially "flat," so it can be approximated by a constant function, $f'(\theta) = k$. The posterior density is then proportional to the likelihood function:

$$f''(\theta \mid y) \propto f(y \mid \theta). \tag{4.10.2}$$

This expresses formally what was previously explained in a heuristic manner; if the prior distribution is diffuse relative to the likelihood function, the posterior distribution depends almost solely on the likelihood function. The proportionality in Equation (4.10.2) is not a *strict* proportionality because $f'(\theta)$ is not *strictly* uniform. Unless the decision-making problem at hand is quite sensitive to slight variations in the prior distribution, though, this approximation can safely be used.

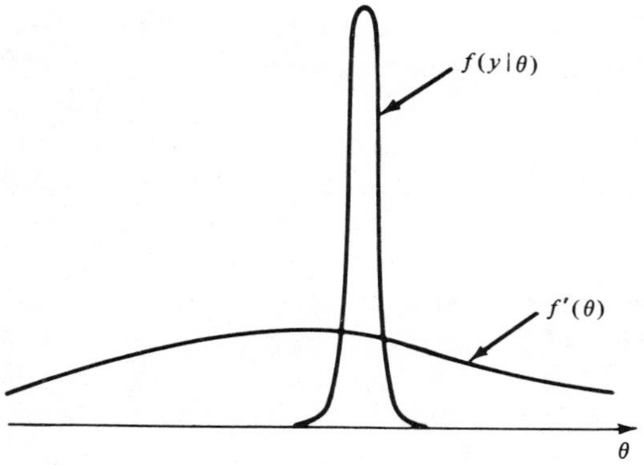

Figure 4.10.1

It is important to remember that diffuseness is a relative term. It is not necessary for the prior distribution to be perfectly "flat" to be diffuse, although it should not have any sharp "spikes," or "peaks." It is also not necessary for the prior distribution to be diffuse over the entire range of values of the random variable (all positive real numbers in the potato example). It *should* be diffuse in the neighborhood of the likelihood function (that is, where the likelihood function is nonnegligible). Thus, it need only be locally diffuse. The notion of local diffuseness does not imply that the prior distribution should be assessed *after* the sample information is observed (until then, it is not known where the likelihood function is nonnegligible). If the assessed prior distribution turns out to be relatively flat where the likelihood function is nonnegligible, however, it may be convenient to use a diffuse distribution to approximate the assessed distribution.

The discussion of diffuseness has involved prior distributions of continuous random variables. The notion of diffuseness does not seem as useful in the discrete case, but if $\tilde{\theta}$ can take on K possible values θ_1, $\theta_2, \ldots, \theta_K$, then one way to represent diffuseness is simply to let

$$P(\tilde{\theta} = \theta_i) = \frac{1}{K} \quad \text{for } i = 1, 2, \ldots, K.$$

In an application of Bayes' theorem for discrete random variables, the posterior probabilities will then be proportional to their respective likelihoods. This is by no means the only way to represent diffuseness in the discrete case, but it is convenient and useful *if* it is a reasonably good representation of the actual prior distribution.

If the decision maker's prior distribution is diffuse relative to the sample information, it should not make too much difference exactly how he specifies the mathematical form of the density function, provided of course that the function chosen is relatively "flat." In order to simplify the process of revising the distribution on the basis of sample information, he would prefer his distribution to be a member of the conjugate family of distributions. Conjugate distributions have been discussed in some detail for two situations. First, the beta family of distributions is conjugate when the sample is from a Bernoulli process. Second, the normal family of distributions is conjugate when the sample is from a normal process with known variance. It would therefore be most convenient to have members of these two conjugate families that could be used as diffuse prior distributions. The *exact* form of the prior distribution is not too important because of the diffuseness, so the statistician might as well ease the computational burden by selecting a diffuse conjugate distribution.

An investigation of various beta distributions reveals that when $r' = 1$ and $n' = 2$, the beta distribution is a uniform distribution over the unit interval. This can be demonstrated by substituting $r' = 1$ and $n' = 2$ in the formula for the beta distribution,

$$f'(p) = \frac{(n'-1)!}{(r'-1)!(n'-r'-1)!} p^{r'-1}(1-p)^{n'-r'-1} \quad \text{for } 0 \leq p \leq 1.$$

Since $r' - 1 = 0$ and $n' - r' - 1 = 0$,

$$f'(p) = \frac{1!}{0!0!} p^0(1-p)^0 \quad \text{for } 0 \leq p \leq 1.$$

By convention, $0! = 1$, and anything raised to the 0 power is equal to 1, so the distribution is simply

$$f'(p) = 1 \quad \text{for } 0 \leq p \leq 1, \tag{4.10.3}$$

which is the uniform distribution defined on the unit interval.

In terms of its density function, then, the beta distribution with $r' = 1$ and $n' = 2$ seems to be a good representation of a diffuse state of prior knowledge. In terms of an interpretation of the distribution, however, some questions arise. According to the "equivalent prior sample information" interpretation, the distribution represents information roughly equivalent to a sample of size two with one "success" from the process in question. While this is not a great deal of information, it may be enough information relative to some samples (in particular, small samples) to cast some doubt on the applicability of the distribution as a diffuse distribution. For example, suppose that you have no information whatsoever regarding a Bernoulli process. You decide to let the beta distribution with $r' = 1$ and $n' = 2$ represent this diffuse state of affairs. A sample of size three is taken, and one success is observed. Using Bayes' theorem, your posterior distribution is a beta distribution with parameters

$$r'' = r' + r = 1 + 1 = 2$$

and

$$n'' = n' + n = 2 + 3 = 5.$$

The mean of your posterior distribution is $r''/n'' = \frac{2}{5} = .40$. But how can this be, if you start out with no information and you then observe one success in three trials? Intuitively, your posterior mean should equal $\frac{1}{3}$, or .33. Here the choice of a prior distribution evidently *does* have some effect on the posterior distribution, in which case it could not be called diffuse.

Difficulties such as this do not arise too often. The sample must be small (in which case a diffuse prior distribution might not be applicable anyway) and preferably should consist of almost all successes or almost all failures. The fact that it does arise at all is bothersome because it means that the uniform beta distribution seems unreasonable as a diffuse prior distribution in some situations. It should be noted that if the mode, $(r'' - 1)/(n'' - 2)$, is used instead of the mean as a summary measure of the posterior distribution, the above difficulty does not arise. In addition, a uniform distribution for $\tilde{p}$ implies a nonuniform distribution for variables such as $\tilde{p}^2$. That is, the uniform nature of the distribution is not invariant with respect to transformations of the random variable of interest. But diffuse prior information concerning $\tilde{p}$ should imply, at least roughly, diffuse prior information concerning $\tilde{p}^2$. Because of the invariance with respect to transformations, then, it seems unreasonable to limit the representation of diffuse prior information to strictly uniform distributions.

Another approach is to look not at the density function, but at the interpretation of the prior parameters. If the prior distribution is interpreted in terms of equivalent sample information, the obvious choice is $r' = 0$ and $n' = 0$. Unfortunately, there is a drawback to this choice; the beta distribution is a "proper" distribution only for r' and n' such that $n' > r' > 0$. When $n' = r' = 0$, the total area under the curve is not equal to 1; the integral representing this area does not converge. You need not concern yourself with the mathematical details; just remember that the area under the density function is not equal to 1.

If you take $r' = 0$ and $n' = 0$ and apply the standard formulas for revising a beta distribution, $r'' = r' + r$ and $n'' = n' + n$, you can see that the posterior parameters are equal to the observed sample results. This result is so intuitively appealing in a truly "diffuse" situation that many Bayesian statisticians take $r' = 0$ and $n' = 0$ as their prior parameters in a diffuse situation, even though the implied prior distribution is "improper." If you look only at the terms involving $\tilde{p}$ in this prior distribution, you can see that the density is proportional to $p^{r'-1}(1 - p)^{n'-r'-1}$, or $p^{-1}(1 - p)^{-1}$. A graph of this function looks U-shaped, reaching a minimum at $p = \frac{1}{2}$. At $p = 0$ and $p = 1$, the function is not defined; as p approaches either 0 or 1, the function increases without bound, becoming infinite. This sort of a function does not exactly agree with a mental picture of how a diffuse prior distribution should look. But the real test of a diffuse prior distribution in Bayesian inference and decision should be whether or not it affects the posterior distribution (and, ultimately, the decision in a decision-making problem). As noted in the previous section, the ultimate aim of the Bayesian approach is to use the posterior distribution in an inferential or decision-making situation. Therefore, in all examples calling for a diffuse beta distribution in this book, the improper beta distribution with $r' = 0$ and $n' = 0$ is used.

With regard to the normal process, how can a diffuse state of affairs be represented by a conjugate normal prior distribution? The density function of a normal distribution is not nearly as flexible as that of members of the beta family. For any choice of m' and σ'^2, the distribution is symmetric, unimodal, and roughly "bell-shaped." For any given m', the distribution is more "spread out" as the variance, σ'^2, is increased. When σ'^2 is very large, the distribution is almost, but not quite, uniform. For a diffuse normal distribution, then, it is necessary to make σ'^2 large.

How large should σ'^2 be? The answer depends on the variance of the process. The prior variance should be larger than the variance of the sampling distribution of $\tilde{m}$, which is σ^2/n. This is still not too satisfactory, for what is needed is a single number that would be applicable in most, if not all, cases. Consider the interpretation of the normal prior distribution in terms of equivalent sample information. To do this, it is necessary to reparametrize and look at n' instead of σ'^2, where, from Equation (4.5.10),

$$n' = \frac{\sigma^2}{\sigma'^2}.$$

For the prior variance to be large relative to the variance σ^2/n for fixed n, n' should be small. But n' can be interpreted as the sample size of an equivalent prior sample. If no information is available regarding the process, the obvious choice is $n' = 0$. Note that in terms of the third parametrization discussed in Section 4.5, this corresponds to setting $I' = 0$, where I' is the "amount of prior information." This certainly seems like a reasonable choice for a diffuse prior state.

When $n' = 0$, the same problem arises as in the case of the beta distribution: the implied distribution is not a proper distribution. The process variance is known and is presumably greater than 0 (if it is equal to 0, then a sample of size one will determine $\tilde{\mu}$ for certain). From Equation (4.5.10), then, n' can only be 0 when the prior variance is infinite. In the mathematical expression for the density of the normal distribution, though, both parameters, the mean and the variance, must be finite.

Since the Bayesian is primarily interested in the posterior distribution, what happens to the posterior parameters when $n' = 0$? Using Equations (4.5.13) and (4.5.14),

$$n'' = n' + n = 0 + n = n$$

and

$$m'' = \frac{n'm' + nm}{n' + n} = \frac{0m' + nm}{0 + n} = \frac{nm}{n} = m.$$

The posterior parameters depend solely on the sample size n and the sample mean m. Notice that nothing has been said about the prior mean

m'. When $n' = 0$, m' becomes irrelevant; it receives no weight in the calculation of m'' because it is weighted by n', which is 0. Even though it is an improper density, then, the normal prior density with $n' = 0$ has no effect on the posterior distribution. As a result, this density is used to represent a diffuse prior state when the sample is from a normal distribution with known variance.

It should be stressed that in general, there is no such thing as a "totally informationless" situation and that the use of particular distributions to represent diffuse prior states of information is a convenient approximation that is applicable only if the prior information is "overwhelmed" by the sample information. In most real-world situations, nonnegligible prior information (nonnegligible relative to the sample information) is available, and the concept of a diffuse prior distribution is not applicable. A further discussion of diffuse prior distributions is presented in Chapter 7 in relation to a comparison of Bayesian and non-Bayesian inferential and decision-making procedures.

4.11 PREDICTIVE DISTRIBUTIONS

In Section 3.7, the determination of predictive distributions for future sample outcomes under a discrete prior distribution was discussed. The predictive distribution under a continuous prior distribution is analogous. In the discrete case, a predictive probability is a denominator in Bayes' theorem. In the continuous case, Bayes' theorem can be expressed in the following form [from Equation (4.2.3)]:

$$f(\theta \mid y) = \frac{f(\theta)f(y \mid \theta)}{\int_{-\infty}^{\infty} f(\theta)f(y \mid \theta)\, d\theta}.$$

From Equation (4.1.8),

$$f(y \mid \theta) = \frac{f(y,\theta)}{f(\theta)},$$

so that the denominator of Bayes' theorem is

$$\int_{-\infty}^{\infty} f(\theta)f(y \mid \theta)\, d\theta = \int_{-\infty}^{\infty} f(y,\theta)\, d\theta,$$

which is simply the marginal distribution of $\tilde{y}$ [Equation (4.1.7)], $f(y)$. But the random variable $\tilde{y}$ represents the sample results, so the marginal distribution of $\tilde{y}$ is the predictive distribution. Thus, the *predictive distribution* can be written in the form

$$f(y) = \int_{-\infty}^{\infty} f(\theta)f(y \mid \theta)\, d\theta. \tag{4.11.1}$$

In general, it may be very difficult to apply Equation (4.11.1) to find the predictive distribution. If the likelihood function $f(y \mid \theta)$ can be related to a given statistical model and if the prior distribution $f(\theta)$ is a member of the appropriate conjugate family of distributions, however, it should be possible to relate $f(y)$ to the sample statistics and to the parameters of the prior distribution. For instance, suppose that the data-generating process is assumed to be a normal process with unknown mean $\tilde{\mu}$ and with known variance σ^2 and that the prior distribution for $\tilde{\mu}$ is a normal distribution with mean m' and variance $\sigma'^2 = \sigma^2/n'$, as discussed in Section 4.5. The predictive distribution for the sample mean $\tilde{m}$ from a sample of size n can be shown to be a normal distribution with mean

$$E(\tilde{m}) = m' \tag{4.11.2}$$

and variance

$$V(\tilde{m}) = \frac{\sigma^2}{n'} + \frac{\sigma^2}{n} = \sigma^2 \left(\frac{1}{n'} + \frac{1}{n}\right). \tag{4.11.3}$$

Note that the mean of the predictive distribution is equal to the mean of the prior distribution. Prior to seeing the sample, the expected value of the unknown process (or population) mean, $\tilde{\mu}$, is m', and this is also the expected value of the sample mean, $\tilde{m}$. The variance of the predictive distribution is equal to σ^2/n', the variance of the prior distribution, plus σ^2/n, the variance of the conditional distribution of $\tilde{m}$, given $\tilde{\mu}$.

As in the discrete case, it is important to distinguish carefully between the *predictive* distribution of the sample result, which is a marginal, or unconditional, distribution, and the *conditional* distribution of the sample result, given a value of the unknown random variable. Here, the predictive distribution for $\tilde{m}$ is a normal distribution with mean m' and variance $\sigma^2 \left(\frac{1}{n'} + \frac{1}{n}\right)$; this distribution reflects uncertainty about $\tilde{\mu}$ *and* uncertainty about $\tilde{m}$, given $\tilde{\mu}$. The conditional distribution of $\tilde{m}$, given $\tilde{\mu} = \mu$, is a normal distribution with mean μ and variance σ^2/n; this distribution reflects uncertainty only about $\tilde{m}$, given $\tilde{\mu}$. Since the predictive distribution takes into account both types of uncertainty (uncertainty about $\tilde{\mu}$ and uncertainty about $\tilde{m}$, given $\tilde{\mu}$), the variance of the predictive distribution is greater than the variance of the conditional distribution of $\tilde{m}$.

To illustrate the determination of a predictive distribution for a sample mean, consider the example in Section 4.6, in which a retailer was interested in the distribution of weekly sales at one of his stores. He assumed that $\tilde{x}$, weekly sales, was normally distributed and that the variance of $\tilde{x}$ was 90,000. Furthermore, his prior distribution for $\tilde{\mu}$, the mean of $\tilde{x}$, was a normal distribution with mean $m' = 1200$ and variance

$\sigma'^2 = 2500$. From Equation (4.5.10), $n' = \sigma^2/\sigma'^2 = 90{,}000/2500 = 36$. Prior to observing the sample of 60 weeks, the retailer could determine the predictive distribution for $\tilde{m}$, the sample mean. From Equations (4.11.2) and (4.11.3), this predictive distribution is a normal distribution with

$$E(\tilde{m}) = m' = 1200$$

and

$$\begin{aligned} V(\tilde{m}) &= \sigma^2\left(\frac{1}{n'} + \frac{1}{n}\right) = 90{,}000\left(\frac{1}{36} + \frac{1}{60}\right) \\ &= 90{,}000\left(\frac{96}{2160}\right) = 4000. \end{aligned}$$

After the sample of size 60 is observed and the prior distribution is revised as in Section 4.6, the posterior distribution is a normal distribution with mean $m'' = 1387.5$ and variance $\sigma''^2 = 937.5$. If the retailer then wants to predict sales at the store in question for the *next* 10 weeks, he should determine a predictive distribution for $\tilde{m}$, the sample mean over the next 10 weeks. This distribution is a normal distribution with

$$E(\tilde{m}) = m'' = 1387.5$$

and

$$\begin{aligned} V(\tilde{m}) &= \sigma^2\left(\frac{1}{n''} + \frac{1}{n}\right) = 90{,}000\left(\frac{1}{96} + \frac{1}{10}\right) \\ &= 90{,}000\left(\frac{106}{960}\right) = 9937.5. \end{aligned}$$

The standard deviation of $\tilde{m}$ is 99.7, so that the limits for a 68 percent credible interval (or predictive interval) for $\tilde{m}$ are

$$1387.5 - 99.7 \quad \text{and} \quad 1387.5 + 99.7,$$

or

$$1287.8 \quad \text{and} \quad 1487.2.$$

That is, the retailer's predictive distribution implies that the probability is .68 that average sales over the next 10 weeks will be between 1287.8 and 1487.2. Since total sales over the next 10 weeks will equal 10 times average sales, the probability is .68 that total sales will be between 12,878 and 14,872.

For another example of the determination of a predictive distribution when the prior distribution is continuous, suppose that the data-generating process is assumed to be a Bernoulli process and that the prior distribution of $\tilde{p}$ is a beta distribution with parameters r' and n', as discussed in Section 4.3. For binomial sampling (sampling with fixed sample size n), the predictive distribution for $\tilde{r}$, the number of "successes," can be determined from Equation (4.11.1):

$$P(r \mid n,r',n') = \int_0^1 f_\beta(p \mid r',n')P_{\text{bin}}(r \mid n,p)\, dp.$$

This expression, which is called a *beta-binomial distribution*, can be simplified to the form

$$P(r \mid n,r',n') = \frac{(r + r' - 1)!(n + n' - r - r' - 1)!n!(n' - 1)!}{r!(r' - 1)!(n - r)!(n' - r' - 1)!(n + n' - 1)!}. \tag{4.11.4}$$

Of course, $\tilde{r}$ can take on only the values 0, 1, 2, . . . , n. The mean and the variance of this beta-binomial predictive distribution are

$$E(\tilde{r}) = \frac{nr'}{n'} \tag{4.11.5}$$

and

$$V(\tilde{r}) = \frac{n(n + n')r'(n' - r')}{n'^2(n' + 1)}. \tag{4.11.6}$$

Note that $E(\tilde{r})$ is equal to n times the prior mean, r'/n'. That is, the expected value of $\tilde{p}$ from the prior distribution is r'/n', so the expected number of "successes" in a sample of size n is simply $E(\tilde{r}) = nE(\tilde{p})$. Furthermore, the variance of $\tilde{r}$ is equal to $n(n + n')V(\tilde{p})$.

To illustrate the use of the beta-binomial distribution, consider the quality-control example presented in Section 4.4. The prior distribution of $\tilde{p}$, the proportion of defective items produced by the production process, is a beta distribution with $r' = 1$ and $n' = 20$. If the production manager contemplates taking a sample of size $n = 5$ from the process, the predictive distribution of the number of defective items in this sample is a beta-binomial distribution with $n = 5$, $r' = 1$, and $n' = 20$. The expected number of defective items is $E(\tilde{r}) = nr'/n' = 5(1)/20 = 1/4$, and the variance of $\tilde{r}$ is

$$V(\tilde{r}) = n(n + n')r'(n' - r')/n'^2(n' + 1) = 5(25)(1)(19)/400(21) = .283.$$

The probabilities for the six possible values of $\tilde{r}$ can be determined from Equation (4.11.4):

$$\begin{aligned} P(\tilde{r} = r \mid n = 5,\ r' = 1,\ n' = 20) \\ &= \frac{(r + 1 - 1)!(5 + 20 - r - 1 - 1)!5!(20 - 1)!}{r!(1 - 1)!(5 - r)!(20 - 1 - 1)!(5 + 20 - 1)!} \\ &= \frac{r!(23 - r)!5!19!}{r!0!(5 - r)!18!24!} = \frac{(23 - r)!5!(19)}{24!(5 - r)!}. \end{aligned}$$

These probabilities are

$$P(\tilde{r} = 0 \mid n,r',n') = \frac{19}{24} = .7917,$$

$$P(\tilde{r} = 1 \mid n,r',n') = \frac{19(5)}{24(23)} = .1721,$$

$$P(\tilde{r} = 2 \mid n,r',n') = \frac{19(5)(4)}{24(23)(22)} = .0313,$$

$$P(\tilde{r} = 3 \mid n,r',n') = \frac{19(5)(4)(3)}{24(23)(22)(21)} = .0045,$$

$$P(\tilde{r} = 4 \mid n,r',n') = \frac{19(5)(4)(3)(2)}{24(23)(22)(21)(20)} = .0004,$$

and $$P(\tilde{r} = 5 \mid n,r',n') = \frac{19(5)(4)(3)(2)(1)}{24(23)(22)(21)(20)(19)} = .0000.$$

Suppose that the production manager decides not to take a sample of size five. Instead, he will produce 1000 items for a given customer's purchase order. He would like to know the distribution of the number of defective items in this lot of 1000. This predictive distribution is a beta-binomial distribution with $n = 1000$, $r' = 1$, and $n' = 20$. The expected number of defective items is $E(\tilde{r}) = 1000(1)/20 = 50$, and the variance of $\tilde{r}$ is $V(\tilde{r}) = 1000(1020)(1)(19)/400(21) = 2307$.

This discussion of predictive distributions completes the development of the Bayesian framework in the continuous case. This framework, together with the discrete framework presented in Chapter 3 (the discrete and continuous cases are analogous), is used in decision-making situations in Chapters 5 and 6.

4.12 THE POSTERIOR DISTRIBUTION AND DECISION THEORY

This chapter and the preceding chapter have not been decision-oriented, although the potential application of Bayesian techniques in decision-making situations has been suggested in some of the examples. In the following chapters the theory of making decisions under conditions of uncertainty is discussed. In the example of Sections 3.3 and 4.4, a decision has to be made about a production process. The uncertainty involves the proportion of defectives, $\tilde{p}$, produced by the process. If the production manager knew $\tilde{p}$ for certain, the choice of an action would be easy. He does not know $\tilde{p}$ for certain, but he *does* have some information regarding $\tilde{p}$. This information is represented by his posterior distribution of $\tilde{p}$. As a result, the posterior distribution is an important input to the decision-making problem.

The posterior distribution, then, is one input to a problem of decision-making under uncertainty, and this is why so much space has been devoted to the revision of prior probabilities and the determination of posterior probabilities. A second important input, one that has not yet been discussed in any detail, is a payoff function or a loss function. Such a function tells the decision maker what the consequences to him will be if he takes a particular action and if at the same time a particular value of $\tilde{\theta}$ occurs, where $\tilde{\theta}$ is the uncertain quantity of interest (the posterior distribution will then be a distribution of $\tilde{\theta}$).

Given a set of possible actions, a set of payoffs or losses, and a posterior distribution, the decision maker can use decision theory to choose one of the possible actions as the "best" action. Of course, if no sample information is available, then it is the prior distribution that is an input to the decision-making problem. It also should be noted that some decision-making problems are directly related to a sample outcome, in which case the predictive distribution is the distribution of interest. Prior, posterior, or predictive, the distribution represents the decision maker's state of uncertainty, or state of information, in a situation in which he must make a decision under uncertainty.

4.13 REFERENCES AND SUGGESTIONS FOR FURTHER READING

For discussions of continuous probability models, see Freund (1962, Chapters 5 and 6); Hadley (1967, Chapter 7); Hays and Winkler (1970, Chapters 3 and 4); and LaValle (1970, Chapters 7–13). In addition, numerous probability models, both discrete and continuous, are summarized in Raiffa and Schlaifer (1961, Chapters 7 and 8), and continuous models are catalogued in Johnson and Kotz (1970). An introductory book on stochastic processes is Parzen (1962); more advanced references are Doob (1953) and Karlin (1966); and an applications-oriented book is Massy, Montgomery, and Morrison (1970).

General references concerning Bayesian inference for continuous probability models at a level roughly comparable to this chapter (or in some cases, slightly more advanced) are Lindley (1965, Chapters 2–5); LaValle (1970, Chapter 14); Morris (1968, Chapter 6); Pratt, Raiffa, and Schlaifer (1965, Chapters 7–11, 13, 16, and 18); Roberts (1966*b*, Chapters 7–9, 13); Schlaifer (1959, Chapters 6, 14, 16–18, 24–31; 1969, Chapters 6, 8, 11–13); and Schmitt (1969, Chapters 4–7, 10). The concept of conjugate families of distributions is developed in Chapters 2 and 3 of Raiffa and Schlaifer (1961), and conjugate distributions for several statistical models are presented in Chapters 9–13; for another presentation (also at an intermediate to advanced level), see De Groot (1970, Chapters 3–6, 9–10).

Some of the above references include discussions of processes other than the Bernoulli, Poisson, and normal processes; other references of interest (primarily at an intermediate to advanced level), including references to problems involving two or more variables, are Altham (1969); Ando and Kaufman (1965); Bhattacharya (1967); Blackman (1971); Box and Tiao (1962, 1964); Chetty (1968); Ericson (1969*a*, 1970); Good (1965); Hill (1963, 1967, 1968, 1970); Hoadley (1969); Lindley (1964); Martin (1967); Novick (1969); Novick and Grizzle (1965); Novick and Hall (1965); Press (1971); and Silver (1965). One important topic not covered in this book is regression analysis, and Bayesian approaches to regression analysis are discussed in Hartigan (1969); Hill (1969); Hoadley (1970); Hurst (1968); Lindley (1968); Lindley and El-Sayyad (1968); Raiffa and Schlaifer (1961, Chapter 13); Roberts (1966*b*, Chapters 16 and 17); Tiao and Zellner (1964); Zellner (1970); Zellner and Chetty (1965); and Zellner and Tiao (1964).

With regard to the specific topics covered in the latter part of the chapter, the assessment of prior distributions for continuous models is discussed in Winkler (1967*a*); a related topic of interest, the combination of a number of prior distributions from different assessors, is discussed in Winkler (1968) [also, see Eisenberg and Gale (1959) and Helmer (1966)]. The concept of diffuse distributions is developed in Savage *et al.* (1962) and Edwards, Lindman, and Savage (1963). Finally, predictive distributions are discussed in Roberts (1965), which also includes a formal Bayesian model for handling multimodal distributions such as the prior distribution encountered in Section 4.9.

Many of the above references require a much greater facility with calculus than does this book. For the reader with limited exposure to calculus (or with no exposure at all), a good introductory text that is particularly appropriate for self-study because it is written in programmed form is Martin (1969). This book also includes material on linear algebra, which is useful in problems involving several variables (including regression problems). Another calculus book is Bers (1969), and other introductory references regarding linear algebra include Hadley (1961) and Searle and Hausman (1970).

EXERCISES

1. The density function of $\tilde{x}$ is given by

$$f(x) = \begin{cases} kx(1-x) & \text{for } 0 \leq x \leq 1, \\ 0 & \text{elsewhere.} \end{cases}$$

(a) Find k and graph the density function.
(b) Find $P(\frac{1}{4} < \tilde{x} < \frac{1}{2})$.
(c) Find $P(-\frac{1}{2} \leq \tilde{x} \leq \frac{1}{4})$.
(d) Find the CDF and graph it.
(e) Find $E(\tilde{x})$, $E(\tilde{x}^2)$, and $V(\tilde{x})$.

2. The CDF of $\tilde{x}$ is given by

$$F(x) = \begin{cases} 1 & \text{for } x \geq 2, \\ x^2/4 & \text{for } 0 \leq x < 2, \\ 0 & \text{for } x < 0. \end{cases}$$

(a) Find $f(x)$, the density function, and show that it satisfies the two requirements for a density function.
(b) Graph $f(x)$ and $F(x)$.
(c) Find $E(\tilde{x})$ and $V(\tilde{x})$.
(d) Find $E(3\tilde{x} - 5)$ and $V(3\tilde{x} - 5)$.

3. The density function of $\tilde{x}$ is given by

$$f(x) = \begin{cases} 2(3 - x)/9 & \text{for } 0 \leq x \leq 3, \\ 0 & \text{elsewhere.} \end{cases}$$

Without using integration, show that the area under the curve is equal to 1, and find $P(1 < \tilde{x} < 1.5)$ and $P(\tilde{x} > 2)$.

4. Why is it necessary to deal with probability densities rather than probabilities such as $P(\tilde{x} = a)$ when the variable under consideration is continuous?

5. Does the function

$$f(x) = \begin{cases} 2x/3 & \text{for } -1 \leq x \leq 2, \\ 0 & \text{elsewhere,} \end{cases}$$

satisfy the two requirements for a density function?

6. Suppose that $\tilde{x}$ and $\tilde{y}$ are continuous random variables with joint density function given by the rule

$$f(x,y) = \begin{cases} k(x + y) & \text{for } 0 \leq x \leq 2,\ 0 \leq y \leq 2, \\ 0 & \text{elsewhere.} \end{cases}$$

(a) Find k.
(b) Find the marginal density functions of $\tilde{x}$ and $\tilde{y}$.
(c) Find the conditional density function of $\tilde{x}$, given that $\tilde{y} = 1$.
(d) Find the conditional density function of $\tilde{x}$, given that $\tilde{y} = \frac{1}{2}$.
(e) Are $\tilde{x}$ and $\tilde{y}$ independent?
(f) Find $E(\tilde{x})$ and $E(\tilde{y})$.

7. In Bayes' theorem, why is it necessary to divide by Σ(prior probability) $\times$ (likelihood) in the discrete case and by $\int$(prior density) $\times$ (likelihood) in the continuous case?

8. Suppose that $\tilde{\theta}$ represents the rate of return (expressed in decimal form, not in percentage form) for a particular investment and that your uncertainty about $\tilde{\theta}$ can be expressed in terms of the following probability distribution:

$$f(\theta) = \begin{cases} 100(\theta + .10)/3 & \text{if } -.10 \le \theta \le .10, \\ 200(.20 - \theta)/3 & \text{if } \quad .10 \le \theta \le .20, \\ 0 & \text{elsewhere.} \end{cases}$$

   (a) Graph this prior distribution and discuss what it implies about your judgments concerning $\tilde{\theta}$.
   (b) An investment analyst is trying to convince you that this is a good investment (he will receive a commission if you make the investment), and he claims that the return on the investment will be .15. Treating this claim as new information and denoting it by y, you decide that your likelihood function (as a function of θ) is

$$f(y \mid \theta) = 5 \quad \text{if } -.10 \le \theta \le .20.$$

   Graph this likelihood function and comment on its implications concerning the new information.
   (c) On the basis of the new information y, revise your distribution of $\tilde{\theta}$.
   (d) The posterior distribution in part (c) applies only to this specific problem. However, can you generalize this result? Explain.

9. In Exercise 8, suppose that your likelihood function is

$$f(y \mid \theta) = \begin{cases} 8(\theta + .10) & \text{if } -.10 \le \theta \le .12, \\ 22(.20 - \theta) & \text{if } .12 \le \theta \le .20, \\ 0 & \text{elsewhere.} \end{cases}$$

   (a) Graph this likelihood function and comment on its implications concerning the new information.
   (b) On the basis of the new information, revise your distribution of $\tilde{\theta}$.

10. Suppose that you are interested in $\tilde{p}$, the proportion of station wagons among the registered vehicles in a particular state. Your prior distribution for $\tilde{p}$ is a normal distribution with mean .05 and variance .0004. To obtain more information, a random sample of 50 registered vehicles is taken, and three are station wagons.
   (a) In using Equation (4.2.3) to revise your distribution of $\tilde{p}$, what difficulties are encountered?
   (b) How might you avoid such difficulties in this situation? Can they always be avoided?

11. Discuss the importance of conjugate families of distributions in Bayesian statistics.

12. Find the mean and the variance of the beta distribution with parameters $r' = 2$ and $n' = 6$, and graph the density function. Do the same for the following beta distributions:

$$r' = 4,\ n' = 6; \qquad r' = 4,\ n' = 12; \qquad r' = 8,\ n' = 12.$$

Explain how the different values of r' and n' affect the shape and the location of these four distributions.

13. If the mean and the variance of a beta distribution with parameters r' and n' are 2/3 and 1/72, respectively, find r' and n'.

14. If the mean and the .05 fractile of a beta distribution with parameters r' and n' are .20 and .13, respectively, find r' and n'.

15. In Exercise 23 in Chapter 3, suppose that the prior distribution could be represented by a beta distribution with $r' = 4$ and $n' = 10$. Find the posterior distribution. Also, find the posterior distribution corresponding to the following beta prior distributions:

$$r' = 2,\ n' = 5; \qquad r' = 8,\ n' = 20; \qquad r' = 6,\ n' = 15.$$

In each of the four prior distributions considered in this exercise, the mean of the prior distribution is .40. How, then, do you explain the differences in the means of the posterior distributions?

16. In Exercise 24 in Chapter 3, suppose that you feel that the mean of your prior distribution is $\frac{1}{2}$ and that the variance of the distribution is $\frac{1}{20}$. If your prior distribution is a member of the beta family, find r' and n' and determine the posterior distribution following the sample of size six. Graph the density functions and find the mean and the variance of the posterior distribution.

17. In sampling from a Bernoulli process, the posterior distribution is the same whether one samples with n fixed (binomial sampling) or with r fixed (Pascal sampling). Explain why this is true. Suppose that a statistician merely samples until he is tired and decides to go home. Would the posterior distribution still be the same (that is, is the stopping rule noninformative)?

18. Try to assess a subjective distribution of $\tilde{p}$, the probability of rain tomorrow. Can you find a beta distribution that expresses your judgments reasonably well?

19. In Exercise 26 in Chapter 3, suppose that the medical researcher decides to treat $\tilde{p}$ as a continuous random variable. He subjectively assesses the mean, the .25 fractile, and the .75 fractile of his distribution to be .167, .118, and .187, respectively. Can you find a beta distribution satisfying these assessments? If not, explain (in terms of the general "shape" of the distribution) why not.

20. Suppose that $\tilde{x}$ is normally distributed with mean 3 and variance 16.
    (a) Find $P(1 \leq \tilde{x} \leq 7)$, $P(\tilde{x} \leq 5)$, $P(-2 \leq \tilde{x} \leq 1.5)$.
    (b) Find a number c such that $P[(\tilde{x} - 2) < c] = .95$.
    (c) Find the .05, .25, .67, and .99 fractiles of the distribution of $\tilde{x}$.
    (d) Graph the density function of $\tilde{x}$.

21. If the .35 fractile of a normal distribution is 105 and the .85 fractile is 120, find the mean and the standard deviation of the distribution.

22. The normal model of a data-generating process cannot be conveniently explained in terms of assumptions such as stationarity and independence, as can the Bernoulli and Poisson models. How, then, can the use of the normal model be justified in any specific application? Give some examples of realistic situations in which the normal model might be a suitable representation of a data-generating process.

23. For a random sample of size n from a data-generating process with mean μ and variance σ^2, show that

$$E(\tilde{m} \mid \mu,\sigma^2) = \mu$$

and

$$V(\tilde{m} \mid \mu,\sigma^2) = \frac{\sigma^2}{n},$$

where $\tilde{m}$ is the sample mean. If $n = 5$ and the sample results are 12, 18, 13, 15, and 19, find the sample mean and the sample variance.

24. A sample of size 500 is taken from a Bernoulli process with $\tilde{p} = .4$. Using the normal approximation to the binomial distribution, find the probability of observing at least 180 "successes" and the probability of observing no more than 210 "successes."

25. For each of the following Bernoulli situations, determine and graph the appropriate binomial distribution, the normal approximation to the binomial, and the Poisson approximation to the binomial.
    (a) $n = 10$, $\tilde{p} = .05$.
    (b) $n = 20$, $\tilde{p} = .05$.
    (c) $n = 10$, $\tilde{p} = .40$.
    (d) $n = 20$, $\tilde{p} = .40$.
    Compare the two approximations in each case.

26. In Exercise 24 in Chapter 3, suppose that the sample information consists of 85 heads in 200 tosses of the coin. Revise your distribution of $\tilde{p}$, using the normal approximation to the binomial distribution to determine the likelihoods.

27. A production manager is interested in the mean weight of items turned out by a particular process. He feels that the weight of items from the process is normally distributed with mean $\tilde{\mu}$ and that $\tilde{\mu}$ is either 109.4, 109.7, 110.0, 110.3, or 110.6. The production manager assesses prior probabilities of $P(\tilde{\mu} = 109.4) = .05$, $P(\tilde{\mu} = 109.7) = .20$, $P(\tilde{\mu} = 110.0) = .50$, $P(\tilde{\mu} = 110.3) = .20$, and $P(\tilde{\mu} = 110.6) = .05$. From past experience, he is willing

to assume that the process variance is $\sigma^2 = 4$. He randomly selects five items from the process and weighs them, with the following results: 108, 109, 107.4, 109.6, and 112. Find the production manager's posterior distribution and compute the means and the variances of the prior and posterior distributions.

28. In Exercise 27, if $\tilde{\mu}$ is assumed to be continuous and if the prior distribution for $\tilde{\mu}$ is a normal distribution with mean 110 and variance .4, find the posterior distribution.

29. You are attempting to assess a prior distribution for the mean of a process, and you decide that the .25 fractile of your distribution is 160 and the .60 fractile is 180. If your prior distribution is normal, determine the mean and the variance.

30. In reporting the results of a statistical investigation, a statistician reports that his posterior distribution for $\tilde{\mu}$ is a normal distribution with mean 52 and variance 10 and that his sample of size four with sample mean 55 was taken from a normal population with variance 100. On the basis of this information, determine the statistician's *prior* distribution.

31. Explain the parametrization of a normal prior distribution in terms of n' and m', as given in Section 4.5. How does this parametrization make it easier to see the relative weights of the prior and the sample information in computing the mean of the posterior distribution? For the prior and posterior distributions in Exercise 28, express the distributions in terms of n', m', n'', and m''. How could you interpret this prior distribution in terms of an equivalent sample? Also, express the distributions in Exercise 28 in terms of the parametrization involving the measures of information I', I, and I''.

32. In assessing a distribution for the mean height of a certain population of college students, a physical-education instructor decides that his distribution is normal, the median is 70 inches, the .20 fractile is 67 inches, and the .80 fractile is 72 inches. Can you find a normal distribution with these fractiles? Comment on the ways in which the instructor could make his assessments more consistent.

33. Suppose that a data-generating process is a normal process with unknown mean $\tilde{\mu}$ and with known variance $\sigma^2 = 225$. A sample of size $n = 9$ is taken from this process, with the sample results 42, 56, 68, 56, 48, 36, 45, 71, and 64. If your prior judgments about $\tilde{\mu}$ can be represented by a normal distribution with mean 50 and variance 14, what is your posterior distribution for $\tilde{\mu}$? From this distribution, find $P(\tilde{\mu} \geq 50)$ and $P(\tilde{\mu} \geq 55)$.

34. The number of customers entering a certain store on a given day is assumed to be normally distributed with unknown mean $\tilde{\mu}$ and unknown variance $\tilde{\sigma}^2$. As the store manager, you feel that your prior distribution for $\tilde{\mu}$ is a normal distribution with mean 1000 and that $P(900 \leq \tilde{\mu} \leq 1100) = .95$. You then take a random sample of 10 days, observing a sample mean of $m = 900$

customers and a sample variance of $s^2 = 50{,}000$. Find your approximate posterior distribution for $\tilde{\mu}$. Aside from the usual argument concerning the applicability of the normal model, why is your posterior distribution only approximate?

35. Using Equation (4.2.3), prove that if the data-generating process of interest is a normal process with unknown mean $\tilde{\mu}$ and known variance σ^2; if the prior distribution of $\tilde{\mu}$ is a normal distribution with mean m' and variance σ^2/n'; and if the sample information consists of a sample of size n from the process with sample mean m; then the posterior distribution of $\tilde{\mu}$ is a normal distribution with mean m'' and variance σ^2/n'', where n'' and m'' are given by Equations (4.5.13) and (4.5.14).

36. In Exercise 34, suppose that your joint prior distribution for $\tilde{\mu}$ and $\tilde{\sigma}^2$ is a normal-gamma distribution with $m' = 1000$, $n' = 20$, $v' = 60{,}000$, and $d' = 11$.
    (a) Find the conditional prior distribution of $\tilde{\mu}$, given that $\tilde{\sigma}^2 = 60{,}000$.
    (b) Find the marginal prior distribution of $1/\tilde{\sigma}^2$.
    (c) Find the joint posterior distribution of $\tilde{\mu}$ and $\tilde{\sigma}^2$.

37. In Exercise 27, suppose that the production manager is unwilling to assume that $\tilde{\sigma}^2$ is known. Instead, he assesses a normal-gamma prior distribution for $\tilde{\mu}$ and $\tilde{\sigma}^2$ with parameters $m' = 110$, $n' = 10$, $v' = 4$, and $d' = 6$. Find the posterior distribution of $\tilde{\mu}$ and $\tilde{\sigma}^2$ and compute $E(\tilde{\mu} \mid \sigma^2)$ and $E(1/\tilde{\sigma}^2)$ from this distribution.

38. In Exercises 38 and 39 in Chapter 3, assume that $\tilde{p}$ is continuous and assess a continuous distribution for $\tilde{p}$ in each case. Try to fit beta distributions to the subjectively assessed distributions.

39. Try to assess a probability distribution for $\tilde{T}$, the maximum temperature tomorrow in the city where you live. In assessing the distribution, use two or three of the methods proposed in the text and compare the results. Save the distribution and look at it again after you find out the true value of $\tilde{T}$. Do this for three or four consecutive days and comment on any difficulties that you encounter in attempting to express your subjective judgments in probabilistic form.

40. Follow the procedure in Exercise 39 for the variable $\tilde{D}$, the daily *change* in the price of one share of IBM common stock.

41. Suppose that a date is to be chosen randomly from next year's calendar, and let $\tilde{\theta}$ represent the maximum official temperature (in degrees Fahrenheit) on that date in Chicago. Assuming that $\tilde{\theta}$ is continuous, assess a probability distribution for $\tilde{\theta}$. Furthermore, let $\tilde{\phi}$ represent the maximum official temperature in Chicago on the day after the chosen date. Assuming that $\tilde{\phi}$ is continuous, assess a conditional distribution for $\tilde{\phi}$, given that $\tilde{\theta} = 60$. Repeat this process for $\tilde{\theta} = 80$, $\tilde{\theta} = 40$, and $\tilde{\theta} = 20$.

42. In Exercise 41, can you approximate your subjectively assessed probability distributions by members of any of the statistical models that have been

discussed in this book (or by any other statistical models)? Also, although your conditional distribution of $\tilde{\phi}$, given $\tilde{\theta}$, no doubt varies considerably for different values of $\tilde{\theta}$, can you express your conditional distribution of $\tilde{\phi}$, given $\tilde{\theta}$, as a function of $\tilde{\theta}$ [for example, can you express $E(\tilde{\phi} \mid \tilde{\theta} = \theta)$ as a function of θ, and so on]? Discuss the use of approximations such as this and the use of statistical models in inferential problems.

43. In assessing a prior distribution for $\tilde{p}$, the proportion of votes that will be cast for a particular candidate in a statewide election, a political analyst feels that the mean of his prior distribution is .45. Furthermore, he feels that if he observes a random sample of 2000 voters, 960 of whom state that they will definitely vote for the candidate and 1040 of whom state that they will *not* vote for the candidate (the sample includes no undecided voters), the mean of his posterior distribution would be .47. Assuming that the process behaves approximately as a Bernoulli process, that stated voting intentions are representative of actual voting behavior, and that the political analyst's prior distribution is a member of the beta family of distributions, find the exact form of the prior distribution.

44. Suppose that a statistician decides that his prior distribution for an uncertain quantity $\tilde{\theta}$ is an exponential distribution and that the .67 fractile of this distribution is 2. If the density function for the exponential distribution is given by

$$f(\theta) = \begin{cases} \lambda e^{-\lambda\theta} & \text{for } \theta \geq 0, \\ 0 & \text{for } \theta < 0, \end{cases}$$

find the exact form of the statistician's prior distribution for $\tilde{\theta}$.

45. In Exercise 33 in Chapter 3, suppose that your prior distribution of $\tilde{\lambda}$, the intensity of accidents per week, is a gamma distribution with mean 3.5 and variance .5. Find the posterior distribution of $\tilde{\lambda}$ and determine the mean and the variance of this posterior distribution.

46. Comment on the following statement: "One cannot speak of sensitivity except in connection with a particular decision-making situation." Can you think of an example in which the decision-making procedure would be quite insensitive to changes in the prior distribution? Can you think of an example in which it would be quite sensitive?

47. Discuss the importance of discrete approximations to continuous prior distributions in Bayesian analysis. Since the concept of conjugate prior distributions greatly simplifies the analysis, why is it ever necessary to use discrete approximations?

48. In Exercise 26 in Chapter 3, suppose that the medical researcher assesses a normal prior distribution for $\tilde{p}$ with mean .16 and variance .0009. Using a discrete approximation to this prior distribution with intervals of width .01, find the approximate posterior distribution of $\tilde{p}$ following the sample of

size 20 with $\tilde{r} = 2$. Repeat the procedure with intervals of width .05 and comment on the differences in the approximate posterior distributions.

49. In Exercise 28, use a discrete approximation to the prior distribution with intervals of width .2 and compare the resulting posterior distribution with the posterior distribution found in Exercise 28.

50. In Exercise 10,
   (a) try to find a beta distribution to approximate the normal prior distribution (for example, you might find the beta distribution with the same mean and variance as this normal distribution), discuss the "closeness" of the approximation, and find the posterior distribution;
   (b) use a discrete approximation to the normal prior distribution with intervals of width .01, discuss the "closeness" of the approximation, and find the posterior distribution.

51. Comment on the following statement: "A diffuse prior state of information is not informationless in the usual meaning of the word, but rather in a relative sense."

52. Give a few examples of situations in which your prior distribution would effectively be diffuse and a few examples in which it would definitely *not* be diffuse relative to a given sample.

53. The beta distribution with $r' = n' = 0$ and the beta distribution with $r' = 1$ and $n' = 2$ have both been used by statisticians as "diffuse" beta distributions. Discuss the advantages and disadvantages of each.

54. In Exercise 28, find predictive distributions (based on the prior distribution) for the sample mean $\tilde{m}$ for samples of sizes 1, 2, 5, 10, 50, and 100.

55. In Exercise 33, find the predictive distribution (based on the prior distribution) for the sample mean $\tilde{m}$ if $n = 9$. Find $P(\tilde{m} \geq 50)$ and $P(\tilde{m} \geq 55)$.

56. After revising your distribution in Exercise 16, you contemplate the possibility of taking another sample of size six. What is your predictive distribution for $\tilde{r}$? From this distribution, find $E(\tilde{r})$ and $V(\tilde{r})$.

57. If a data-generating process is assumed to follow the Bernoulli model and if the prior distribution of $\tilde{p}$ is a beta distribution with $r' = 1$ and $n' = 2$, find the predictive distribution for $\tilde{r}$, the number of successes in a sample of size n. For $n = 5$, graph the PMF of $\tilde{r}$.

5

DECISION THEORY

Why should you be interested in decision theory? Life is a constant sequence of decision-making situations. Every action you take, with the exception of a few involuntary physiological actions, such as breathing, can be thought of as a decision. Of course, most of these decisions are quite minor because the consequences involved are not very important. For example, consider the decision whether to walk up a flight of stairs or to take the elevator. Unless you have a heart condition or some other physical reason for taking the elevator, this is not an important decision. The relevant factors entering into the decision include the exertion required to walk up the stairs, the availability of the elevator, whether or not you are in a hurry, and so on. Minor decisions such as this are usually made intuitively without conscious thought. Other examples are the decision to smoke a cigarette, to drive home via a certain route, to have a particular beverage with your evening meal, and so on. For some individuals, these decisions may require some thought. A person trying to break the smoking habit may find the decision as to whether or not to smoke a cigarette to be a difficult decision. For most people, though, such decisions are usually made out of habit. They do not consciously think about the various possible actions because they have faced the situation (or similar situations) many times in the past.

Other decisions require some thought but can still be made intuitively. Examples are a choice of brands of a product, a choice of movies on a Saturday night, and so on. Often you face a decision-making problem that requires some serious thought. Consider a major purchasing decision, such as the decision to buy a new car. In such a situation, you evaluate such factors as the condition of your present car, the cost of a new car, the enjoyment of owning a new car, and so on. You probably would make this decision subjectively, but it is the type of decision for which the formal decision theory to be discussed in this chapter might prove useful. Examples of similar problems are investment decisions (either by individuals or by corporations), medical decisions (whether or not to undergo surgery), and job decisions (for the individual, whether or not to accept a job offer; for the corporation, whether or not to make such an offer). It may not always be easy to specify the problem in quantitative form and to apply formal decision-making procedures. Whether to proceed at an informal level or at a formal level is a decision in itself. Whenever the consequences of making a wrong decision could be quite serious, however, decision theory should prove worthwhile.

5.1 CERTAINTY VERSUS UNCERTAINTY

This chapter and the next chapter deal with decision making under the condition of *uncertainty*. This condition refers to uncertainty about the true value of a variable related to the decision, or, in simpler terms, uncertainty about the actual state of the world. Formally, a consequence of a decision, which may be expressed in terms of a payoff or a loss to the decision maker, is the result of the interaction of two factors: (1) the *decision*, or the *action*, selected by the decision maker; and (2) the *event*, or *state of the world*, that actually occurs.

In general, the decision-making problem is easier to solve if the decision maker knows which event will occur. For a trivial example, suppose that you have to decide whether or not to carry an umbrella when you leave home in the morning. There are two possible actions: carry the umbrella or leave it at home. Also, there are two relevant events, or states of the world: rain or no rain. If you know for certain that it will rain, you will definitely carry your umbrella. If you know for certain that it will not rain, you will leave the umbrella at home. In this situation, you are making a decision under the condition of *certainty*.

For another example, suppose that you are faced with a decision concerning the purchase of common stock. For the sake of simplicity, assume that you intend to invest exactly \$1000 in a single common stock and to hold the stock for exactly one year, at which time you will sell it at the market price. Furthermore, assume that you are considering only three stocks, A, B, and C, each of which currently sells for \$50 a share. Thus, you intend to buy 20 shares of one of the three stocks. Your selection of a single stock from the three constitutes your *decision*, or *action*. The prices of the three stocks one year from now constitute the *state of the world*. The combination of your action and the state of the world determines your payoff. Suppose that at the end of one year, the prices of stocks A, B, and C are \$60, \$40, and \$50. The payoffs for the three possible actions are then +\$200, −\$200, and \$0, respectively. For this state of the world, the best decision is to buy stock A, for this results in the highest payoff. If you know the state of the world with certainty, the decision can be made in this manner, and this is once again decision making *under certainty.* In this example, if you know what the prices of the stocks will be in one year, then you will simply buy the stock that will give you the maximum payoff—that is, the stock that will increase in value the most during the coming year.

In general, the actual state of the world is *not* known with certainty. On many mornings, you may not be at all certain whether or not it will rain that day (technically, you could *never* be absolutely certain about the weather, although you may often feel certain for all practical purposes). You listen to the weather report on the morning newscast and you look out the window at the sky, so that you have some information, which may be reflected in probabilistic terms. Subjectively, you can determine your personal probabilities for the two events "rain" and "no rain." Or, if your local weather bureau issues "precipitation probabilities" and if you have confidence in the weather bureau, you may decide to use these probabilities. At any rate, you are operating under the condition of *uncertainty,* and your uncertainty can be represented by your probabilities for the two possible states of the world. Recall, from Chapter 2, that probability can be thought of as the language of uncertainty.

It should be noted that some authors differentiate between what they call "decision making under risk" (decision making when the state of the world is not known but probabilities for the various possible states are known) and "decision making under uncertainty" (decision making when the state of the world is not known and probabilities for the various possible states are not known). Under the subjective interpretation of probability, as discussed in Chapter 2, it is always possible to assess probabilities for the possible events, or states of the world. Hence, the "risk versus uncertainty" dichotomy is extremely artificial (in fact, it is nonexistent according to the subjective interpretation of probability), and in this book any decision-making problem in which the state of the world is not known for certain is called decision making under uncertainty. The term "risk" is reserved for discussions relating to utility, which is covered in Sections 5.5 through 5.8.

How about the stock example? Unless you are fortunate enough to receive copies of the *Wall Street Journal* one year in advance or unless you possess extrasensory perception, you could not know for certain what the prices of the three stocks would be one year hence. Thus, treating the purchase of common stock as a decision under certainty is not at all realistic. You no doubt have some ideas regarding the prices, ideas that may be based on your impressions of the economy in general, of particular industries, and of specific firms within the industries. Your knowledge may stem from years and years of careful study of the stock market, or it may consist of a hot tip from a "friend." At any rate, you have some knowledge, but not perfect knowledge. Once again you are faced with a problem of decision making *under uncertainty*, and you can express your uncertainty in probabilistic terms. There are three variables of interest here: the prices one year hence of the three stocks. What is needed is a *joint* distribution of the three random variables. If the variables are assumed to be independent, you could assess three individual marginal distributions and multiply them together to get the joint distribution. For these three variables, the assumption of independence is, to say the least, an unrealistic assumption.

There are other more realistic assumptions that provide some shortcuts in the assessment procedure, but that is not the point of this discussion. The point is this: in general, the problem of making decisions is more complex and more difficult under uncertainty than it is under certainty. As a result, it is important when formally specifying a decision problem to keep it as simple as possible, including only the most relevant variables. Otherwise it may prove too difficult to apply the techniques of this chapter. This problem occurs whenever one attempts to build a mathematical model of a real-world situation; the model must include enough variables and must be detailed enough to be realistic, yet it should not be so complex that it becomes difficult to handle. This is a drawback that limits the application of decision theory somewhat, although modern computers permit the solution of complex decision problems that would be impossible (or at least infeasible) to solve manually or on a desk calculator. As in the stock example, the difficulty may lie in determining the inputs to the decision. One important set of inputs consists of the decision maker's probabilities concerning the events or random variables of interest. In Chapters 2 through 4, methods for assessing and manipulating probabilities were discussed in some detail. Another important set of inputs consists of the potential payoffs or losses in the decision-making problem; this set of inputs is discussed in the following section.

There are numerous examples of decision making under uncertainty. In Section 2.5, examples were presented concerning an oil-drilling decision under uncertainty about the presence of oil and concerning an advertising decision under uncertainty about the actions of a competing firm. Most investment decisions are made under uncertainty about the eventual return on the investment. Production decisions are made under uncertainty about the demand for a product. Numerous medical decisions are made under uncertainty (for instance, uncertainty about the outcome of surgery or the effect of a drug). The area of decision making under uncertainty clearly includes many important decision-making problems.

It should be stressed that decision making under *certainty* is not always as trivial as it seems to be in the examples presented in this section. For example, suppose that a manufacturer must ship a certain product from a number of factories to a number of warehouses. Each factory produces a certain number of units of the product, and each warehouse requires a certain number of units. Furthermore, the cost of shipping from any given factory to any given warehouse depends on the amount shipped, the particular factory, and the particular warehouse. The decision-making problem is this: what shipping pattern minimizes the total transportation costs? That is, what is the least expensive way to transport the product from the factories to the warehouses? In this problem there is no uncertainty; the amounts produced at the various factories, the amounts needed at the various warehouses, and the costs of shipping are all known. Even under certainty, this decision-making problem is clearly not trivial to solve (although it becomes much *more* complex if the amounts produced, the amounts needed, or the costs are uncertain). It may be solved by a technique known as linear programming, a special (though very important) case of mathematical programming, which is a general approach to certain types of decision-making problems. Other decision-making problems under certainty require the use of different types of mathematical optimization procedures. Since this book is concerned with the case of uncertainty rather than certainty, such procedures are not discussed, although you should realize that decision making under certainty includes many important and by no means mathematically trivial problems.

5.2 PAYOFFS AND LOSSES

It has been noted that the consequences of a decision can be expressed in terms of either payoffs or losses. It is important to define explicitly what is meant by these two terms. A *payoff*, or *reward*, can be interpreted in the usual manner; if it is expressed in monetary units, it represents the net change in your total wealth as a result of your decision and the actual state of the world (the actual value of some uncertain quantity). This can be either positive or negative. It is important to remember that payoffs are always expressed in net terms rather than gross terms, so that all relevant factors can be taken into consideration. For instance, when making a decision about the price of a particular product and the quantity to be produced, a firm should take into account all relevant costs as well as the gross profits attributable to the item of interest.

In the umbrella example, consider the *payoff table* presented in Section 2.5.

		STATE OF THE WORLD	
		Rain	*No rain*
ACTION	*Carry umbrella*	−\$1	−\$1
	Do not carry umbrella	−\$60	\$0

A payoff table consists of the set of payoffs for all possible combinations of actions and states of the world. If there are m actions and n states of the world, the payoff table will have $n \times m$ entries. In the umbrella example, there are four entries, all of which happen to be negative or zero. In this book, payoffs are denoted by R (for "reward") because P is used to represent probability. For instance,

$$R(\text{carry umbrella, rain}) = -\$1.$$

Note that all of the payoffs in the above table are expressed in terms of dollars, even though the potential consequences include nonmonetary considerations (such as the possibility of getting wet and perhaps becoming ill, and so on). In this example, the possible consequences are converted into *cash equivalents.* It is assumed that you feel that the cost of carrying the umbrella, in terms of the inconvenience and the possibility of leaving the umbrella somewhere and thus losing it, is equivalent to \$1. Therefore, −\$1 is a cash equivalent for the consequence of carrying the umbrella, whether or not it actually rains. Similarly, the consequence corresponding to not carrying the umbrella when it does, in fact, rain includes the possible harm to your clothing, the possible danger to your health, the possible inconvenience of waiting inside until the rain stops, and so on. The cash equivalent of this consequence is −\$60. In many situations it may be possible to express nonmonetary outcomes in terms of cash equivalents, although this may cause difficulties in other situations, as you shall see when utility theory is discussed later in this chapter.

How can the above payoff table be converted to an equivalent *loss table?* "Loss" in this context refers to "*opportunity loss.*" For any combination of an action and a state of the world, the question is: could you have obtained a higher payoff, given that particular state of the world? If the answer is no, then your loss is zero. If the answer is yes, then your loss is the positive difference between the given payoff and the *highest possible payoff under that state of the world.* Keep in mind that in this book, the term "loss" is always to be interpreted in terms of opportunity loss. Negative values in a payoff table are to be regarded as negative payoffs, negative rewards, or costs; in terms of opportunity loss, they may correspond to losses of zero.

Consider the first entry in the first column of the above payoff table. It is higher than the other entry in that column, so it is the best you could do given that it rains, and the loss is zero. For the second entry in the first column, the payoff is \$59 lower than the first entry, so the loss is \$59. Similarly, in the second column, the first entry is \$1 lower than the second entry, so the corresponding losses are \$1 and \$0. The *loss table* follows.

		STATE OF THE WORLD	
		Rain	*No rain*
ACTION	*Carry umbrella*	\$0	\$1
	Do not carry umbrella	\$59	\$0

This is sometimes called a *regret table,* since the entries reflect the decision maker's regret at not having made the decision that turns out to be optimal under the actual state of the world.

For another example, consider the stock-purchasing decision from the previous section. Furthermore, for the sake of simplicity assume that there are only four possible states of the world. In the first (state I), the prices of the stocks A, B, and C at the end of one year are, respectively, \$60, \$50, and \$40. For state II, the prices are \$80, \$60, and \$40. For state III, the prices are \$20, \$40, and \$10. Finally, for state IV, the prices are \$60, \$60, and \$60. This results in the following payoff table, where the payoffs are expressed in dollars.

		STATE OF THE WORLD			
		I	II	III	IV
	Buy A	200	600	−600	200
ACTION	*Buy B*	0	200	−200	200
	Buy C	−200	−200	−800	200

For example, if stock A is bought and if state II occurs, the price goes up from \$50 to \$80, or \$30 per share. Since 20 shares will be bought, the payoff is \$600. This payoff table corresponds to the following loss table.

		STATE OF THE WORLD I	II	III	IV
ACTION	*Buy A*	0	0	400	0
	Buy B	200	400	0	0
	Buy C	400	800	600	0

For a third example, consider the advertising example presented in Section 2.5. The possible actions are not to advertise, to conduct a minor advertising campaign, and to conduct a major advertising campaign; the states of the world involve the action of a competing firm (whether or not the firm will introduce a competing product). Taking into account the cost of advertising and its anticipated effect on sales under both states of the world, the decision maker arrives at the following payoff table (the payoffs are given in dollars).

		STATE OF THE WORLD *Competitor introduces new product*	*Competitor does not introduce new product*
ACTION	*No advertising*	100,000	700,000
	Minor ad campaign	300,000	600,000
	Major ad campaign	400,000	500,000

Note that this example differs slightly from the previous examples in that the uncertainty involves a competing firm's action. Such situations are often classified under the heading of "game theory." In game theory, the state of the world is determined by some "opponent" who is interested in making the best decision possible, as opposed to having the state of the world being determined by a disinterested opponent, such as nature. (In decisions involving the occurrence of rain or the presence of oil, the state of the world may be thought of as being determined by nature.) Some examples of realistic game-theoretic situations are a bidding situation, in which competing firms bid for a contract and the firm with the lowest bid is awarded the contract; a bargaining situation, in which two parties (for example, union and management) attempt to reach an agreement; and a military situation, in which each side must consider the possible actions that could be taken by the opponents. Since subjective probabilities can be assessed for the various possible actions available to an opponent as well as for the various possible states of the world that could be determined by nature, game-theory problems are no different from other problems of decision making under uncertainty. In assessing subjective probabilities for an "opponent's" action in a game-theoretic situation, you must, of course, take into consideration the fact that your "opponent" is trying to make a "good" decision and that he is probably concerned about the potential actions that you might take. In the advertising example, by considering past actions of the competing firm in similar situations, its general willingness or lack of willingness to introduce new products, the capabilities of its key personnel, information concerning its present situation, and so on, the decision maker should be able to represent his uncertainty about the competing firm's action in terms of subjective probabilities.

Returning to the relationship between payoffs and losses, the loss table for the advertising example is as follows.

		STATE OF THE WORLD	
		Competitor introduces new product	*Competitor does not introduce new product*
ACTION	*No advertising*	300,000	0
	Minor ad campaign	100,000	100,000
	Major ad campaign	0	200,000

The procedure for converting a payoff table to a loss table should be clear by now. First, work with each column of the table separately. In each column find the highest payoff (which may be negative, zero, or positive). The opportunity loss corresponding to this payoff is zero, and the loss corresponding to any other payoff in the same column is obtained by subtracting that payoff from the highest payoff in the column. Formally, if R(action i, state j) and L(action i, state j) represent the payoff and the loss, respectively, corresponding to the combination of the ith row (ith action) and the jth column (jth state of the world) in the table, then

$$L(\text{action } i, \text{ state } j) = \max_k R(\text{action } k, \text{ state } j) - R(\text{action } i, \text{ state } j). \quad (5.2.1)$$

Here, $\max_k R(\text{action } k, \text{ state } j)$ is the largest payoff in the jth column of the payoff table.

The loss tables presented above demonstrate some interesting characteristics of losses. First, losses cannot be negative. This can be seen from the way in which they are defined; each loss is either zero or the (positive) difference between two payoffs [in Equation (5.2.1), $\max_k R(\text{action } k,$ state $j)$ is, by definition, at least as large as R(action i, state j)]. Payoffs can be negative, zero, or positive; losses, on the other hand, can only be zero or positive (there is no such thing as a negative regret!). In addition, if all the payoffs in a column are equal, the corresponding losses will all be zero. This is because no payoff is higher than another, so if that particular state of nature occurs, it makes no difference which action is chosen.

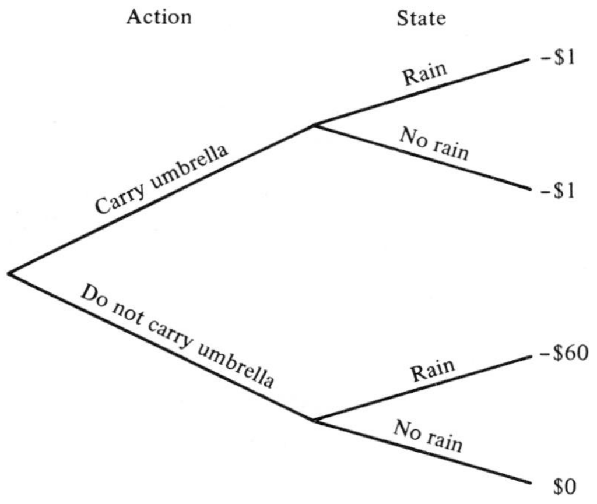

Figure 5.2.1

An alternative way to present the payoffs or losses in a decision-making problem is in terms of a *tree diagram*. For example, tree diagrams are presented for the umbrella, stock, and advertising examples in Figures 5.2.1 through 5.2.3. In each case, the first fork corresponds to the action chosen by the decision maker and the second fork corresponds to the state of the world. Thus, each terminal branch at the right-hand side of the tree diagram corresponds to a combination of a particular action and a particular state of the world. The numbers at the end of these terminal branches are the corresponding payoffs (alternatively, losses could be used in place of payoffs). Tree diagrams are particularly useful in representing complex decision-making problems with sequences of actions and events over time, as you shall see in Chapter 6, for it is convenient to add more branches and forks to the diagram; such complex problems are very difficult to represent clearly in tabular form.

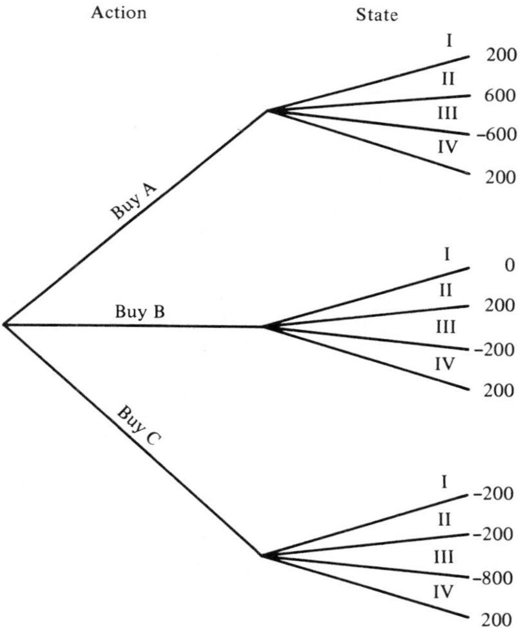

Figure 5.2.2

One other concept of importance is the concept of *admissible actions*, or *admissible decisions*. An action is said to *dominate* a second action if for *each possible state of the world*, the first action leads to at least as high a payoff (or at least as small a loss) as the second action, and if for at least one state of the world, the first action leads to a higher payoff (or smaller loss) than the second action. If one action dominates another, then it would never be reasonable to choose the second action (you could always do at least as well with the first action, and for at least one state of the world, you could do better with the first action). Therefore, the second action is said to be *inadmissible.* Notice that the last requirement in the definition of dominance (that the first action leads to a *higher* payoff for at least one state of the world) is necessary to prevent actions with identical payoffs from being classified as "dominated by each other."

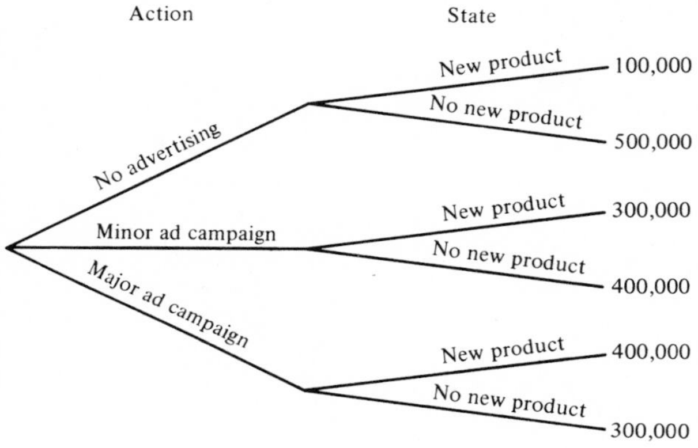

Figure 5.2.3

In the stock-purchasing example, the third action (buy C) is dominated by both of the other two actions, as you can see by looking at either the payoff table or the loss table. The first action is better than the third action for states I, II, and III, and it is just as good as the third action for state IV. The same is true of the second action as compared with the third action. Hence, the third action is *inadmissible;* there is no need to consider it further. Be careful to note that in order to be inadmissible, an action must be dominated by another *single* action, not just by a combination of other actions. In the advertising example, the minor advertising campaign has a smaller payoff than the major advertising campaign if the competing firm introduces a new product, and it has a smaller payoff than no advertising if the competing firm does not introduce a new product. However, "minor ad campaign" is not dominated by "major ad campaign" alone or by "no advertising" alone. Thus, it is *not* inadmissible.

In making a decision, it is necessary to consider only *admissible acts,* acts that are not dominated by any other act. The practical importance of this is that the elimination of inadmissible acts may greatly reduce the "dimension" of the decision-making problem by enabling the decision maker to confine his attention to the admissible acts. For instance, in setting up the problem, the decision maker may consider, say, 15 actions. Upon formally determining the relevant payoffs (or losses), he may find that 10 of the actions are dominated, or inadmissible. Thus, he can immediately eliminate these 10 actions from further consideration, leaving a much simpler (at least computationally) problem involving a choice among only five actions.

A final point of interest with regard to payoffs and losses is that it has been assumed that the payoffs and losses are known with certainty. In other words, for any action and any state of the world, it is assumed that you can determine for certain what the payoff will be. This is not always the case, however. In the umbrella example, you cannot be certain what the consequences will be if you do not carry your umbrella and if it rains. If you are outside when it rains, you will get wet and your new sport coat might be ruined. On the other hand, you might be inside when it starts raining, and you may choose to stay inside until the rain ends, in which case the consequence is simply the inconvenience of being delayed while it rains. In the advertising example, the decision maker cannot be certain about the effect of the rival firm's action or the effect of advertising on sales and profits. There are many situations in which some payoffs are uncertain. This possibility could formally be included in the decision-making model presented in this chapter. It amounts to introducing a second type of uncertainty, uncertainty about payoffs (uncertainty about the state of the world is already included in the model). In terms of a decision tree, this can be taken into account by adding another set of branches. For instance, in Figure 5.2.1, you may be uncertain about the payoff that will result if you do not carry the umbrella and if it rains. Instead of a known payoff of $-\$60$, then, you might consider a number of branches with various possible payoffs. Whether or not to include these new branches is a question of how complex (and how realistic) you want the model to be and is one aspect of the general question of how "bushy" the decision tree should be or where the tree should be truncated.

In situations in which the payoff is uncertain, the *expected* payoff often is used (in other words, the expected payoff is treated as a payoff that is known for certain). In the advertising example, the decision maker cannot be certain what the payoff will be if there is no advertising and if the rival firm introduces a competing product, but perhaps he can say that under these conditions, the expected payoff to the firm is $100,000. Note that the expectation here is taken with regard to the uncertainty about the payoff, not the uncertainty about the state of the world. The use of the expected payoff in this fashion is a reasonable procedure, for in many cases a formal treatment of the uncertain payoffs, which requires a probability distribution for the payoffs, would lead to the same (or virtually the same) result. If the amounts involved are large enough to make the decision maker's attitude toward risk an important factor, however, the entire probability distribution for the payoffs (as opposed to just the mean of this distribution) should be considered.

5.3 NONPROBABILISTIC CRITERIA FOR DECISION MAKING UNDER UNCERTAINTY

If a table or a tree diagram of payoffs or losses is known in a decision problem and if the actual state of the world is known for certain, the decision is obvious: choose the action giving the highest payoff (smallest loss) for that particular state of the world. In terms of the table, choose the action corresponding to the largest entry in the relevant column of the payoff table or the smallest entry in the relevant column of the loss table. Because of the way in which losses were defined in Section 5.2, the same action is selected in both cases.

Under uncertainty, however, it is not possible for the decision maker to confine his attention to a single column of the table. Because there is uncertainty as to which state of nature will occur, he must consider the entire table. But then it is not obvious that one particular action is a better choice than all of the other actions. It might be the best action given one state of the world, but the worst given another state of the world. The only way to formally introduce the nature of your uncertainty about the state of the world is to assess a probability distribution. Most of the available information is likely to be subjective, so that the probability distribution would be interpreted in terms of degrees-of-belief. Several decision-making criteria that are based solely on the payoff (or loss) table have been developed, at least partially in an attempt to avoid the assessment of probabilities. These decision-making rules ignore the probabilistic nature of decision-making under uncertainty. Some of these rules are discussed briefly in this section, and probabilistic decision-making criteria are covered in the next section.

Consider the stock-purchasing example in Section 5.2, and change the payoff table slightly so that the third action is admissible. Under state III, assume that the price of C is \$50 rather than \$10. The payoff is then \$0 rather than −\$800, and C is no longer dominated by A or B. Also, under state II, assume that the price of C is \$30 rather than \$40, which changes the payoff from −\$200 to −\$400. The new payoff table is as follows.

		STATE OF THE WORLD			
		I	II	III	IV
	Buy A	200	600	−600	200
ACTION	*Buy B*	0	200	−200	200
	Buy C	−200	−400	0	200

One decision rule, called the *maximin* rule, says: for each action, find the smallest possible payoff, and then choose the action for which this smallest payoff is largest. The smallest payoffs for the three actions are −600, −200, and −400. According to this rule, the decision maker should choose the second action. The maximin rule amounts to maximizing the minimum payoff. In effect, it tells the decision maker to assume that for each action, the worst possible state of the world will occur. This seems to be an extremely conservative approach, since it considers only the lowest payoff in each row and ignores the magnitudes of the other payoffs. It is easy to construct payoff tables for which the maximin rule gives ridiculous results.

		STATE OF THE WORLD	
		X	*Y*
ACTION	*1*	100,000	−1
	2	0	0

In this payoff table, the smallest payoffs for the two actions are −1 and 0, respectively, so the second action is selected by the maximin criterion, since 0 is greater than −1. The possible gain of $100,000 is completely ignored. These two actions could be thought of as "do something" and "stay as is." According to the maximin rule, the decision maker should never take an action if there is a possibility of a negative payoff, since he could always do better in a maximin sense by doing nothing and obtaining a payoff of zero. This is most unrealistic!

A second decision rule, called the *maximax* rule, assumes that the best will happen, not the worst. According to this rule, the decision maker should find the *largest* possible payoff for each action and then choose the action for which this largest payoff is largest. This maximizes the maximum payoff. In the stock-purchasing example, the maximum payoffs are 600, 200, and 200, so the decision maker should buy stock A if he follows the maximax rule. Just as the maximin rule ignores the high payoffs, the maximax rule ignores the low payoffs, thereby implicitly assuming that for each action, the best possible state of the world will occur. As with the maximin rule, it is possible to construct a payoff table for which the maximax rule gives ridiculous results.

		STATE OF THE WORLD *X*	*Y*
ACTION	*1*	100,000	99,999
	2	100,001	0

In this payoff table, the highest payoffs for the two actions are 100,000 and 100,001; the maximax criterion chooses the second action. Thus, in order to gain possibly one extra dollar (from 100,000 to 100,001 dollars), the decision maker is taking the chance that he could end up with nothing, whereas he is assured of at least $99,999 if he takes the first action. In this example, then, the maximax rule is unreasonable.

The maximin and maximax rules deal with the payoff matrix. A third rule, called the *minimax loss*, or *minimax regret*, criterion, says: for each action, find the largest possible loss, and then choose the action for which this largest loss is smallest. For the stock-purchasing example, the loss table is as follows.

		STATE OF THE WORLD I	II	III	IV
	Buy A	0	0	600	0
ACTION	*Buy B*	200	400	200	0
	Buy C	400	1000	0	0

The largest losses for the three actions are 600, 400, and 1000. The smallest of these three numbers is 400, and thus the second action is chosen by the minimax loss rule. As the name implies, this rule minimizes the maximum possible loss. The minimax loss rule is neither as conservative as the maximin rule nor as risky as the maximax rule. An illustration of this is provided by considering the loss tables corresponding to the payoff tables (other than the stock example) given on pages 233 and 234.

		STATE OF THE WORLD *X*	*Y*
ACTION	*1*	0	1
	2	100,000	0

		STATE OF THE WORLD *X*	*Y*
ACTION	*1*	1	0
	2	0	99,999

In each case the minimax loss rule chooses the first action.

In the example presented in Section 3.3 involving the proportion of defectives produced by a given production process, the following payoff table was given.

		STATE OF THE WORLD (VALUE OF $\tilde{p}$) .01	.05	.10	.25
	Major adjustment	380	380	380	380
ACTION	*Minor adjustment*	455	375	375	375
	No adjustment	480	400	300	0

The minimum payoffs for the three actions are 380, 375, and 0. Thus, the maximin rule would advise the decision maker to make a major adjustment in the production process. This is a conservative decision: "Let's make the major adjustment so we won't have to worry about what could happen if we didn't make the adjustment and if $\tilde{p}$ happened to be .10 or .25." The maximum payoffs for the three actions are 380, 455, and 480, and the maximax rule chooses the third action, "no adjustment." This is a risky decision: "Let's go for the highest payoff, $480, and hope that $\tilde{p}$ is not .10 or .25." Notice, by the way, that the maximax rule always chooses the action corresponding to the highest payoff in the entire table.

The loss table in the production example follows.

		STATE OF THE WORLD			
		.01	.05	.10	.25
ACTION	*Major adjustment*	100	20	0	0
	Minor adjustment	25	25	5	5
	No adjustment	0	0	80	380

The maximum losses for the three actions are 100, 25, and 380. According to the minimax loss rule, a minor adjustment should be made in the production process. In this example, the three decision rules result in three different decisions.

Other nonprobabilistic decision criteria, some of which may be better than the above rules, have been proposed. For instance, one rule is a combination of the maximin and maximax rules, using an "optimism-pessimism" index. These rules can all be criticized on the grounds that they ignore the probabilistic nature of uncertainty. As a result, each nonprobabilistic criterion leads to the same decision regardless of how likely the various states of the world are. Looking at the revised payoff table in the stock-purchasing example, for instance, the first action (buy A) looks best if $P(\text{state II}) = .99$, and the third action (buy C) looks best if $P(\text{state III}) = .99$. The rules in the next section take into account the *probabilistic* nature of the problem of decision making under uncertainty.

5.4 PROBABILISTIC CRITERIA FOR DECISION MAKING UNDER UNCERTAINTY

If some information, but not perfect information, is available regarding the states of the world, then the decision maker is operating under uncertainty. If the decision maker expresses his uncertainty in terms of probabilities, then these probabilities can be used as inputs to the decision-making process (recall that probability was called the mathematical language of uncertainty in Chapter 2). In particular, they can be used to calculate expected payoffs or expected losses of the potential actions. The *expected payoff* (ER) criterion says to choose the act with the highest expected payoff. This criterion has been used in decision-making examples in previous chapters. Correspondingly, the *expected loss* (EL) criterion says to choose the act with the smallest expected loss. For any decision-making problem, the ER and EL criteria always yield identical decisions. As you shall see in Sections 5.5 through 5.8 when utility theory is discussed, the primary justification for the use of ER (or EL) as a decision-making criterion is the fact that under certain assumptions about an individual's utility function, maximizing ER or minimizing EL is equivalent to maximizing expected utility.

In the umbrella example in Section 5.2, suppose that you assess $P(\text{rain}) = .20$ and $P(\text{no rain}) = .80$ on the basis of a weather report and a look at the sky. Then the expected payoffs of the two acts are

$$ER(\text{carry umbrella}) = .20(-\$1) + .80(-\$1) = -\$1$$

and

$$ER(\text{do not carry umbrella}) = .20(-\$60) + .80(\$0) = -\$12.$$

Similarly, from the loss table,

$$EL(\text{carry umbrella}) = .20(\$0) + .80(\$1) = \$0.80$$

and

$$EL(\text{do not carry umbrella}) = .20(\$59) + .80(\$0) = \$11.80.$$

The act "carry umbrella" has the larger expected payoff and the smaller expected loss. Furthermore, its ER is \$11 higher than the ER of the other action, and its EL is \$11 lower than the EL of the other action. This demonstrates a relationship between expected payoffs and expected losses; the action with the highest expected payoff will always have the lowest expected loss, and vice versa. Furthermore, if the expected payoff of one action is A units *higher* than the ER of a second action, then the EL of the first action will be exactly A units *lower* than the EL of the second action. That is, if

$$ER(\text{action 1}) - ER(\text{action 2}) = A,$$

then

$$EL(\text{action 2}) - EL(\text{action 1}) = A.$$

Notice that there is a direct relationship between the *differences* in expected payoffs and the *differences* in the corresponding expected losses.

The relationship between the *magnitudes* of expected payoffs and the *magnitudes* of expected losses, however, is not quite so simple. From Equation (5.2.1),

$$L(\text{action } i, \text{ state } j) = \max_k R(\text{action } k, \text{ state } j) - R(\text{action } i, \text{ state } j),$$

where R(action i, state j) and L(action i, state j) represent the payoff and loss, respectively, corresponding to the combination of the ith action and the jth state of the world. Hence, if the uncertain quantity of interest is $\tilde{\theta}$ and if there are J possible states of the world (values of $\tilde{\theta}$) $\theta_1, \theta_2, \ldots, \theta_J$, then

$$\begin{aligned} EL(\text{action } i) &= \sum_{j=1}^{J} P(\theta = \theta_j) L(\text{action } i, \text{state } j) \\ &= \sum_{j=1}^{J} P(\theta = \theta_j) \, [\max_k . \, R(\text{action } k, \text{state } j) \\ &\qquad\qquad - R(\text{action } i, \text{state } j)] \\ &= \sum_{j=1}^{J} P(\theta = \theta_j) \max_k . \, R(\text{action } k, \text{state } j) \\ &\qquad\qquad - \sum_{j=1}^{J} P(\theta = \theta_j) R(\text{action } i, \text{state } j) \\ &= \sum_{j=1}^{J} P(\theta = \theta_j) \max_k . \, R(\text{action } k, \text{state } j) - ER(\text{action } i). \end{aligned}$$

But since $\sum_{j=1}^{J} P(\theta = \theta_j) \max_k . \, R(\text{action } k, \text{state } j)$ does not depend on the index i, it is the same for all actions. Thus, letting

$$T = \sum_{j=1}^{J} P(\theta = \theta_j) \max_k . \, R(\text{action } k, \text{state } j), \tag{5.4.1}$$

$$EL(\text{action } i) = T - ER(\text{action } i). \tag{5.4.2}$$

For the umbrella example, "carry umbrella" is the best action corresponding to the state "rain," and "do not carry umbrella" is the best action corresponding to the state "no rain." Therefore,

$$\begin{aligned} T &= P(\text{rain}) R(\text{carry umbrella, rain}) \\ &\qquad + P(\text{no rain}) R(\text{do not carry umbrella, no rain}) \\ &= .2(-\$1) + .8(\$0) = -\$0.20. \end{aligned}$$

From Equation (5.4.2),

$$\begin{aligned} EL(\text{carry umbrella}) &= T - ER(\text{carry umbrella}) \\ &= -\$0.20 - (-\$1) = \$0.80 \end{aligned}$$

and

$$\begin{aligned} EL(\text{do not carry umbrella}) &= T - ER(\text{do not carry umbrella}) \\ &= -\$0.20 - (-\$12) = \$11.80. \end{aligned}$$

From Equation (5.4.2), it can be seen formally that the ER and EL criteria always yield identical decisions. T is a constant with respect to the different actions; as you shall see in Chapter 6, T is the "expected payoff under perfect information." Thus, the action with the largest ER must also have the smallest EL. If another action existed with a smaller EL, then it would also have a larger ER, since $ER = T - EL$. In some decision-making problems, it is convenient to work with payoffs, while in others it is more convenient to work with losses. Because ER and EL always yield the same result, it makes no difference whether a problem is expressed in terms of payoffs or losses.

For another example, consider the stock-purchasing example, using the payoff table and the loss table presented in Section 5.3. Suppose that the probabilities of states I, II, III, and IV are .1, .4, .3, and .2, respectively. The expected payoffs are then

$$\begin{aligned} ER(\text{buy A}) &= .1(200) + .4(600) + .3(-600) + .2(200) = 120, \\ ER(\text{buy B}) &= .1(0) + .4(200) + .3(-200) + .2(200) = 60, \\ \text{and } ER(\text{buy C}) &= .1(-200) + .4(-400) + .3(0) + .2(200) = -140. \end{aligned}$$

The expected losses are

$$\begin{aligned} EL(\text{buy A}) &= .1(0) + .4(0) + .3(600) + .2(0) = 180, \\ EL(\text{buy B}) &= .1(200) + .4(400) + .3(200) + .2(0) = 240, \\ \text{and} \quad EL(\text{buy C}) &= .1(400) + .4(1000) + .3(0) + .2(0) = 440. \end{aligned}$$

From Equation (5.4.1),

$$T = .1(200) + .4(600) + .3(0) + .2(200) = 300.$$

The decision should be to buy A on the basis of ER and EL. Of course, the expected payoffs and losses (and hence the resulting decision) depend on the values of the probabilities. Had the probabilities been .1, .1, .7, and .1, the decision would be to buy C, since the EL's would be 420, 200, and 140. If the probabilities were .4, .1, .4, and .1, the decision would be to buy B, since the EL's would be 240, 200, and 260. The different sets of probabilities correspond to different information about the state of the world, and different information may lead to a different decision.

How are the probabilities of the states of the world determined? In general, they are posterior probabilities. As such, they represent all relevant information, sample information or otherwise. If no sample information is available, of course, then prior probabilities may be used to calculate expected payoffs and losses.

Consider once again the example involving the proportion of defective items, $\tilde{p}$, produced by a certain process. In Section 3.3, the statistician's prior probabilities are given as

$$P(\tilde{p} = .01) = .60,$$
$$P(\tilde{p} = .05) = .30,$$
$$P(\tilde{p} = .10) = .08,$$

and

$$P(\tilde{p} = .25) = .02.$$

He then observes a sample of size 10 with three defectives. Applying Bayes' theorem, his posterior probabilities are

$$P(\tilde{p} = .01) = .005,$$
$$P(\tilde{p} = .05) = .245,$$
$$P(\tilde{p} = .10) = .359,$$

and

$$P(\tilde{p} = .25) = .391.$$

Using the prior probabilities, the expected payoffs of the three actions are

$$ER(\text{major adjustment}) = 380,$$
$$ER(\text{minor adjustment}) = 423,$$

and

$$ER(\text{no adjustment}) = 432.$$

Using the posterior probabilities, the expected payoffs are

$$ER(\text{major adjustment}) = 380,$$
$$ER(\text{minor adjustment}) = 375.5,$$

and

$$ER(\text{no adjustment}) = 208.1.$$

The optimal act under the prior distribution is "no adjustment," since there is a high probability that $\tilde{p}$ is small (a probability of .90 that $\tilde{p}$ is either .01 or .05). The particular sample observed makes the higher values of $\tilde{p}$ seem more likely (the posterior probability is only .25 that $\tilde{p}$ is either .01 or .05), and the optimal act under the posterior distribution is "major adjustment."

Expected payoffs and expected losses are applicable when the random variable of interest is continuous as well as when it is discrete. In the modification of the quality-control example, presented in Section 4.4, the two available actions are "no adjustment" and "adjustment," and the corresponding payoffs are $\$500 - \$2000\tilde{p}$ and $\$400 - \$1000\tilde{p}$. The expected payoffs are thus

$$ER(\text{no adjustment}) = \$500 - \$2000E(\tilde{p})$$

and

$$ER(\text{adjustment}) = \$400 - \$1000E(\tilde{p}).$$

The posterior distribution of $\tilde{p}$ is a beta distribution with $r'' = 4$ and $n'' = 30$. Therefore, $E(\tilde{p}) = r''/n'' = 4/30 = .133$, and the expected payoffs are

$$ER(\text{no adjustment}) = \$500 - \$2000(.133) = \$234$$

and

$$ER(\text{adjustment}) = \$400 - \$1000(.133) = \$267.$$

The optimal act under the posterior distribution according to the ER criterion is to have the process adjusted.

The quality-control example illustrates the possibility of a continuous random variable representing the state of the world. Often the payoff or loss in a decision-making problem may be expressed as a function of some continuous random variable. Although most of the examples presented up to this point have involved discrete random variables, the framework presented in this chapter is equally applicable to decision-making problems involving continuous random variables, and some problems of this nature are discussed in Chapter 6.

It should be mentioned that the random variable in a decision-making problem may be an unknown sample statistic from a future sample. If this is the case, the probabilities of interest are predictive probabilities, which were discussed in Sections 3.7 and 4.11. In the automobile-salesman example presented in Section 3.4, the owner of the dealership assesses a prior distribution for $\tilde{\lambda}$, the mean number of sales per day by the salesman. In making a decision, the owner might be particularly concerned with $\tilde{r}$, the number of cars the salesman will sell in the next six months. The distribution of $\tilde{r}$ is a predictive distribution that can be determined from the owner's prior distribution and likelihoods.

Because the ER and EL criteria take into account the probabilistic nature of the problem of decision making under uncertainty, they are more useful than the nonprobabilistic decision-making criteria discussed in Section 5.3. The uncertainty in the decision-making situation is expressed in terms of probabilities, and the probabilities are used to calculate expectations of payoffs and losses. Even the ER and EL criteria are not without their drawbacks, however, as you shall see in the next section.

5.5 UTILITY

In all of the examples involving payoff and loss tables, the payoffs and losses have been expressed in monetary terms. This is not always the case; it is easy to think of examples in which the consequences of a decision are nonmonetary. The consequences may involve quantities of a good or a service. If the good or the service has a known monetary value, then the payoffs can be expressed in dollars, pounds, francs, marks, rupees, or whatever. In the production-process example, the consequence involving the services of a mechanic was expressed in terms of the cost of these services. In most decisions, however, there are additional factors to consider, such as time, inconvenience, social approval, and so on. In the umbrella example, it was said that the consequence of leaving the umbrella home and being caught in the rain is a ruined sport coat with a value of \$60. To simplify the problem, other potential consequences, such as the danger to your health and the discomfort of being caught in the rain with no umbrella or raincoat, were ignored. If you carry the umbrella and if it does not rain, the only cost you suffer is the inconvenience of carrying the umbrella. In the example, this inconvenience was expressed in monetary terms as a cash equivalent, the amount of money that you would be willing to pay not to have to carry the umbrella. It may not always be easy to express consequences in this manner.

There is another problem concerning the use of money to describe consequences and the related use of *ER* and *EL* as decision-making criteria. Suppose that you were offered the following bet on one toss of a coin (you have the opportunity to bet only once—repeated bets are not allowed): you win \$1 if the coin comes up heads, and you lose \$0.75 if the coin comes up tails. If you are convinced that the coin is a fair coin, so that $P(\text{heads}) = P(\text{tails}) = \frac{1}{2}$, your expected payoff is $\frac{1}{2}(\$1) + \frac{1}{2}(-\$0.75) = \$0.125$ if you take the bet and \$0 if you do not take the bet. According to the *ER* criterion, you should take the bet. This is reasonable, since the bet looks intuitively advantageous. Unless you are opposed in principle to gambling, you probably would take the bet. Now suppose that the amounts involved are \$10,000 and \$7500 rather than \$1 and \$0.75. The expected payoff is now \$1250 if you take the bet and \$0 if you do not take the bet. According to the *ER* rule, you should take the bet. Would you do so? Probably not, unless you are wealthy enough so that a loss of \$7500 would not seriously affect your financial position. The possible gain of \$10,000 is tempting, but there is still a 50–50 chance of a loss of \$7500. By the way, if you still would take the bet, change the values to \$100,000 and \$75,000 and make the choice again.

If you choose not to take the bet in any of these examples, you are violating the *ER* criterion. Why? The monetary payoffs are unambiguous, so there must be factors involved in this example other than these payoffs. If you lose \$7500, the consequence is not just that loss, but the possible accompanying factors, such as the impact of a sharp reduction in your savings account, which is supposed to tide you over in your old age, or the embarrassment when you tell your spouse or your friends that you lost \$7500 on the toss of a coin.

For another example, consider a gamble in which a fair coin is to be tossed until heads first appears and you will receive 2^{k-1} dollars if this occurs on the kth toss of the coin. That is, you will receive $2^0 = 1$ dollar if heads appears on the first toss, $2^1 = 2$ dollars if tails appears on the first toss and heads on the second toss, $2^2 = 4$ dollars if tails appears on the first two tosses and heads on the third toss, and so on. How much would you be willing to pay in order to be able to participate in this gamble? In order to compute the expected payoff of the gamble, it is necessary to determine the probabilities of the various possible outcomes. If heads first occurs on the kth toss, then the sequence of observations consists of $(k - 1)$ tails followed by heads. But $P(\text{tails}) = P(\text{heads}) = \frac{1}{2}$ on any single toss, since the coin is fair. Furthermore, the tosses are assumed to be independent, so that the probability of a sequence of $(k - 1)$ tails followed by heads is simply $(\frac{1}{2})^{k-1}(\frac{1}{2})$, or $(\frac{1}{2})^k$. The expected payoff of the gamble is then

$$ER(\text{gamble}) = \sum_{k=1}^{\infty} 2^{k-1}(\tfrac{1}{2})^k = \sum_{k=1}^{\infty} (\tfrac{1}{2}).$$

But if a positive constant is summed an infinite number of times, the sum is infinite. Consequently, the expected payoff of the gamble is infinite. Because of the possibility of extremely high payoffs (even though they are extremely unlikely), the expected payoff is infinite (note that technically it would be possible to toss tails indefinitely and never toss heads). Despite this, it is doubtful that you would be willing to pay even, say, $20 for the gamble. Once again, the *ER* rule appears to be an unsatisfactory guide to action. Moreover, unlike the previous example, this result holds even if you have the opportunity to participate in the gamble repeatedly. This particular gambling situation is famous in probability theory, dating back to the eighteenth century, and it is known as the St. Petersburg paradox.

For a final illustration of a situation in which *ER* may not be satisfactory, consider a choice between two bets. In the first bet, you win $1 million if a coin comes up heads and you win $1 million if the coin comes up tails. In the second bet, you win $10 million if the coin comes up heads but you win nothing if the coin comes up tails. The expected payoffs of the two bets are

$$ER(\text{bet 1}) = \tfrac{1}{2}(\$1 \text{ million}) + \tfrac{1}{2}(\$1 \text{ million}) = \$1 \text{ million}$$

and

$$ER(\text{bet 2}) = \tfrac{1}{2}(\$10 \text{ million}) + \tfrac{1}{2}(\$0) = \$5 \text{ million}.$$

Bet 2 has a much larger expected payoff than bet 1. Which bet would you choose? In spite of the great difference in expected payoffs, most persons, if given this opportunity on a one-shot basis, would probably choose bet 1. This is because \$1 million is a very large sum of money, clearly enough to allow a person to lead a quite comfortable life (even at today's prices!). If this amount could be invested at 5 percent interest, you would receive \$20,000 per year without touching the principal. It would be even better, of course, to have \$10 million, but most people feel that \$10 million is not so preferable to \$1 million that it is worth a risk of winding up with nothing. Informally, this amounts to saying that to most persons, \$10 million is not worth 10 times as much as \$1 million. If it were, then bet 2 would be strongly preferred to bet 1.

A few comments are in order regarding this example. The amounts involved are much larger than they would be in most decision-making situations. Unless the decision at hand is a very important decision involving a large corporation, it would be expected that the amounts involved would be much smaller. The large amounts were chosen in order to stress the point, but other examples with smaller amounts could be found that give essentially the same results. An everyday example is the purchase of an insurance policy that has a negative expected payoff or an expected payoff less than that of a savings account in a bank. Also, the bets are to be offered strictly as a one-shot affair, as was pointed out. If the situation were to be repeated several times, bet 2 would surely be preferred to bet 1, for the chances of winding up with nothing would be greatly reduced. If the choice of bets is repeated just four times, the probability of winding up with nothing if bet 2 is chosen each time is just $(\frac{1}{2})^4$, or $\frac{1}{16}$. Of course, if the bets are to be repeated, then the decision maker must choose an overall strategy. He might, for instance, choose bet 1 on the first trial to assure himself of \$1 million and then choose bet 2 thereafter.

The examples illustrate the fact that the "value" of a dollar may differ from person to person and that for any specific person, the "value" of x dollars is not necessarily x times the "value" of a single dollar. As a result, the ER rule may not be satisfactory when the payoffs are expressed in terms of money. If it could somehow be possible to measure the true relative value to the decision maker of the various possible payoffs in a problem of decision making under uncertainty, expectations could be taken in terms of these "true" values. The theory of *utility* prescribes such a decision-making rule: the maximization of *expected utility*, or the EU criterion. In order to understand this criterion, it is first necessary to discuss briefly the concept of utility.

Essentially, the theory of utility makes it possible to measure the *relative* value to a decision maker of the payoffs, or consequences, in a decision problem. In the following discussion, the term "payoff" is used in a general sense to represent the "consequence" to the decision maker of taking a particular action and having a particular state of the world occur. This includes all aspects of the "consequence," monetary or otherwise, so you should be careful not to associate "payoffs" strictly with monetary rewards.

Formally, a utility function U can be interpreted in terms of a preference relationship. The two basic axioms of utility are as follows.

1. If payoff R_1 is preferred to payoff R_2, then $U(R_1) > U(R_2)$; if R_2 is preferred to R_1, then $U(R_2) > U(R_1)$; and if neither is preferred to the other, then $U(R_1) = U(R_2)$.
2. If you are indifferent between (a) receiving payoff R_1 for certain and (b) taking a bet or lottery in which you receive payoff R_2 with probability p and payoff R_3 with probability $(1 - p)$, then

$$U(R_1) = pU(R_2) + (1 - p)U(R_3).$$

It is important to note that a utility function is not unique, even for a specific individual. If, for a particular person, a function U satisfies the above axioms, then the function $W = c + dU$ also satisfies the axioms, where c and d are constants with d greater than 0. For Axiom 1, suppose that $U(R_1) > U(R_2)$. Then

$$W(R_1) = c + dU(R_1)$$

and

$$W(R_2) = c + dU(R_2).$$

It is obvious that $W(R_1) > W(R_2)$, since $d > 0$. Similarly, if $U(R_1) = pU(R_2) + (1 - p)U(R_3)$ from Axiom 2, then

$$W(R_1) = pW(R_2) + (1 - p)W(R_3).$$

To show this, observe that

$$\begin{aligned} pW(R_2) &+ (1 - p)W(R_3) \\ &= p[c + dU(R_2)] + (1 - p)[c + dU(R_3)] \\ &= pc + p\, dU(R_2) + (1 - p)c + (1 - p)\, dU(R_3) \\ &= c[p + (1 - p)] + d[pU(R_2) + (1 - p)U(R_3)] \\ &= c + d[pU(R_2) + (1 - p)U(R_3)]. \end{aligned}$$

But $pU(R_2) + (1 - p)U(R_3) = U(R_1)$, and therefore the right-hand side of the equation becomes $c + dU(R_1)$, which is equal to $W(R_1)$. This completes the proof that if U satisfies the axioms, then so does $W = c + dU$, where d must be positive. In words, it is said that a utility function is only unique up to a positive linear transformation.

It should be emphasized that the development of utility presented in this section and in the next three sections is only a brief, rough development of the most important points of the theory of utility, although it will suffice for the purpose of this book. In the next section, a practical problem of interest is considered: the assessment of a person's utilities.

Once utilities are assessed, they can be used in a decision-making problem, and the associated decision-making criterion is the maximization of expected utility.

5.6 THE ASSESSMENT OF UTILITY FUNCTIONS

Using the axioms of utility, it is possible for an individual to assess a *utility function*. Suppose that there are a number of possible payoffs in a decision-making problem facing you. Once again, it must be stressed that the term "payoff" is used in a general sense and is by no means restricted to monetary payoffs. In general, a payoff includes some aspects that are qualitative rather than quantitative.

First of all, it is necessary to determine what you consider to be the most preferable and the least preferable payoffs. If the problem is stated formally in such a manner that there could be an infinite number of possible payoffs (for instance, if the state of the world is represented by a continuous random variable), then it is theoretically possible for the set of payoffs to be unbounded, in which case there may be no "most preferable" or "least preferable" payoff. In practice, however, it can easily be assumed that the set of payoffs is bounded. Call the most preferable payoff R^* and the least preferable payoff R_*. Since a utility function is unique only up to a positive linear transformation, you can arbitrarily assign any values to $U(R^*)$ and $U(R_*)$, provided that $U(R^*) > U(R_*)$. Suppose that you let $U(R^*) = 1$ and $U(R_*) = 0$. The choice of 1 and 0 is arbitrary; you could just as easily choose 3 and -5, or 21.8 and 10.2, or any such pair of values, but 1 and 0 simplify the calculations somewhat, as you shall see.

Now consider any payoff R. From the choice of R^* and R_* and from the first axiom of utility presented in the preceding section, it must be true that

$$U(R_*) \leq U(R) \leq U(R^*),$$

or

$$0 \leq U(R) \leq 1.$$

To determine the value of $U(R)$ more precisely, consider the following choice of lotteries, or bets, where p is a canonical probability, as discussed in Section 2.4.

Lottery I: Receive R for certain.

Lottery II: Receive R^* with probability p and receive R_* with probability $(1 - p)$.

How should you choose between these lotteries? According to the *EU* criterion (essentially, according to the second axiom of utility presented in the preceding section), the lottery with the higher expected utility should be selected. The expected utilities are

$$EU(\text{Lottery I}) = U(R)$$

and

$$\begin{aligned} EU(\text{Lottery II}) &= pU(R^*) + (1 - p)U(R_*) \\ &= p(1) + (1 - p)(0) = p. \end{aligned}$$

Thus, if $U(R) > p$, Lottery I should be chosen; if $U(R) < p$, Lottery II should be chosen; and if $U(R) = p$, you should be indifferent between the two lotteries.

This relationship between $U(R)$ and p can be exploited to determine the utility of R. If you can determine a probability p that makes you indifferent between the two lotteries, then the utility of R must be equal to this value, p. In this manner, you can determine the utility of any consequence, or payoff, once the most and least preferable payoffs, R^* and R_*, have been determined. Because of the choice of $U(R^*) = 1$ and $U(R_*) = 0$, the utility for any other payoff can be interpreted in terms of an indifference probability.

In the umbrella example, the three possible payoffs, expressed in terms of money, are \$0, −\$1, and −\$60. Clearly, R^* is \$0 and R_* is −\$60. To find $U(-\$1)$, suppose that $U(R^*) = 1$ and $U(R_*) = 0$, and consider these lotteries:

Lottery I: Receive −\$1 for certain.

Lottery II: Receive \$0 with probability p and receive −\$60 with probability $(1 - p)$.

What value of p makes you indifferent between Lotteries I and II? Suppose it is $p = .97$. Then $U(-\$1) = .97$, and the payoff table for the umbrella example can be expressed in terms of utilities.

		STATE OF THE WORLD	
		Rain	*No rain*
ACTION	*Carry umbrella*	.97	.97
	Do not carry umbrella	0	1

If $P(\text{rain}) = 0.20$ and $P(\text{no rain}) = 0.80$, then

$$EU(\text{carry umbrella}) = .2(.97) + .8(.97) = .97$$

and

$$EU(\text{do not carry umbrella}) = .2(0) + .8(1) = .80.$$

The EU rule tells you to carry your umbrella.

For another example, consider the oil-drilling example presented in Section 2.5, with the following payoff table (in terms of dollars).

		STATE OF THE WORLD	
		Oil	*No oil*
ACTION	*Drill*	100,000	−30,000
	Do not drill	20,000	10,000

Here, $R^* = \$100{,}000$ and $R_* = -\$30{,}000$. To find the utility of \$20,000, the following lotteries are used:

Lottery I: Receive \$20,000 for certain.

Lottery II: Receive \$100,000 with probability p and receive −\$30,000 with probability $(1 - p)$.

The decision maker, considering his current financial position, might decide that Lottery I is clearly preferable if $p = .30$ and that Lottery II is clearly preferable if $p = .60$. After giving it some thought, he chooses $p = .45$ as his indifference point.

Next, to find the utility of \$10,000, the decision maker looks at these lotteries:

Lottery I: Receive \$10,000 for certain.

Lottery II: Receive \$100,000 with probability p and receive −\$30,000 with probability $(1 - p)$.

Suppose that his indifference point for these lotteries is $p = .35$. In terms of utilities, the payoff table is as follows.

		STATE OF THE WORLD	
		Oil	*No oil*
ACTION	*Drill*	1	0
	Do not drill	.45	.35

If $P(\text{oil}) = .3$, the expected utilities of the two actions are

$$EU(\text{drill}) = .3(1) + .7(0) = .30$$

and

$$EU(\text{do not drill}) = .3(.45) + .7(.35) = .38.$$

According to the EU criterion, the decision maker should not drill.

In the assessment of utility functions, as in the assessment of probabilities, an element of vagueness is involved. For a particular choice of lotteries, you may feel that you are roughly indifferent if p is between .3 and .4, but you find it difficult to come up with a single indifference point. By careful consideration of the lotteries, it should be possible to pin it down to a satisfactory degree of accuracy. Unless the decision-making problem is highly sensitive to slight changes in the utility function, it should not be too difficult to come up with a set of utilities that is a reasonable approximation of your preferences.

In any event, it should be noted that a person's utility function will not necessarily remain the same over time. As his asset position and other factors in his life change, his preferences change, thereby modifying his utility function in some manner. For instance, if someone gave you \$10,000, your preferences for various consequences might change significantly. Furthermore, it is important to note that different individuals (or different corporations) may have different utility functions, just as they may have different prior probability distributions. This is demonstrated by the fact that two persons with identical information (and hence identical probabilities) may make two different decisions in the same situation. One person may accept a certain bet while another rejects it. In any decision-making problem, then, the utility function of interest is the utility function of the person (or firm) that will receive the payoff. In deciding whether or not to take an umbrella with you, you must consider *your own* utilities, not those of anyone else.

Note that it is not at all necessary for the consequences, or payoffs, to be stated in terms of money. In the umbrella example, instead of working from the payoff table in terms of dollars, you could have considered the various consequences (including considerations such as getting wet, ruining your sport coat, suffering the inconvenience of carrying an umbrella) and assessed utilities directly for these consequences. This eliminates the intermediate step of determining *cash equivalents* for the consequences. Thus, it is possible to take into consideration both monetary and nonmonetary factors in determining the utility of a particular consequence. As indicated in the preceding section, nonmonetary factors such as social approval and environmental concern may be relevant to a person in a decision-making situation. In business decisions, such factors can be quite important, and it is useful to be able to consider them formally in decision theory. For instance, a corporation might be concerned about prestige, business ethics, and the satisfaction of employees and customers.

For a simple nonmonetary example, suppose that you must decide whether to go to a movie, to go on a boat ride, or to go to a football game. You are uncertain about the weather, and you feel that $P(\text{clear weather}) = .7$, $P(\text{light showers}) = .2$, and $P(\text{heavy rain}) = .1$. Thus, there are nine possible consequences, corresponding to the nine possible pairs of an action and a state of nature. You decide that the most preferable consequence would correspond to taking the boat ride in clear weather and that the least preferable consequence would correspond to taking the boat ride in a heavy rainstorm. Thus, $U(\text{boat ride, clear weather}) = 1$ and $U(\text{boat ride, heavy rain}) = 0$. To assess $U(\text{boat ride, light showers})$, consider the following lotteries.

Lottery I: Take a boat ride and encounter light showers.

Lottery II: Take a boat ride and encounter clear weather with probability p and heavy rain with probability $(1 - p)$.

Suppose that you feel that if $p = .9$, you prefer Lottery II, whereas if $p = .1$, you prefer Lottery I. After some consideration, you decide that $p = .6$ makes you roughly indifferent between the two lotteries, so that $U(\text{boat ride, light showers}) = .6$. Next, you decide that if you go to the movie, the weather is irrelevant, so that $U(\text{movie, clear weather}) = U(\text{movie, light showers}) = U(\text{movie, heavy rain})$. Furthermore, you are indifferent between the following two lotteries.

Lottery I: Go to a movie.

Lottery II: Take a boat ride and encounter clear weather with probability .8 and heavy rain with probability .2.

Thus, $U(\text{movie, clear weather}) = U(\text{movie, light showers}) = U(\text{movie,$

heavy rain) = .8. By considering similar lotteries, you assess U(football game, clear weather) = .9, U(football game, light showers) = .7, and U(football game, heavy rain) = .2.

The entire table of utilities for this problem is as follows.

		STATE OF THE WORLD		
		Clear weather	*Light showers*	*Heavy rain*
	Boat ride	1	.6	0
ACTION	*Movie*	.8	.8	.8
	Football game	.9	.7	.2

The expected utilities are

$$EU(\text{boat ride}) = .7(1) + .2(.6) + .1(0) = .82,$$
$$EU(\text{movie}) = .7(.8) + .2(.8) + .1(.8) = .80,$$
and
$$EU(\text{football game}) = .7(.9) + .2(.7) + .1(.2) = .79.$$

Using the EU criterion, you should go on the boat ride. This example illustrates the direct assessment of utilities without the consideration of factors such as money.

Although nonmonetary factors may be of importance, the relationship between money and utility is of some interest. In many situations, all of the consequences *can* be expressed in terms of money. This means that nonmonetary consequences can be expressed in terms of cash equivalents. In determining the relationship between utility and money over a given interval of monetary values for an individual, the technique introduced above can be applied. Suppose that you select the interval from $-\$100$ to $+\$100$ and that you let $U(-\$100) = 0$ and $U(+\$100) = 1$. Next, you take several values between $-\$100$ and $+\$100$ and determine their utilities. For instance, to determine the utility of \$0, consider the following two lotteries.

Lottery I: Receive \$0 for certain.

Lottery II: Receive \$100 with probability p and receive $-\$100$ with probability $(1 - p)$.

After considering several values of p, suppose that you decide that

$p = .70$ makes you indifferent between these two lotteries (that is, for $p < .70$, you prefer Lottery I; for $p > .70$, you prefer Lottery II). From the results of the previous section, this means that $U(\$0) = .70$. Even though $p = .5$ would make the expected payoffs of the two lotteries equal in terms of money, you feel that the probability of winning in Lottery II has to be .70 before you are indifferent between the sure thing in Lottery I and the risky situation in Lottery II. In a similar fashion, you assess $U(-\$50)$ to be .40 and $U(+\$50)$ to be .90. These values can be plotted on a graph, as in Figure 5.6.1, and a rough curve can be drawn through them. This curve is your *utility function for money*. The utility of any monetary payoff between $-\$100$ and $+\$100$ can be read directly from the curve.

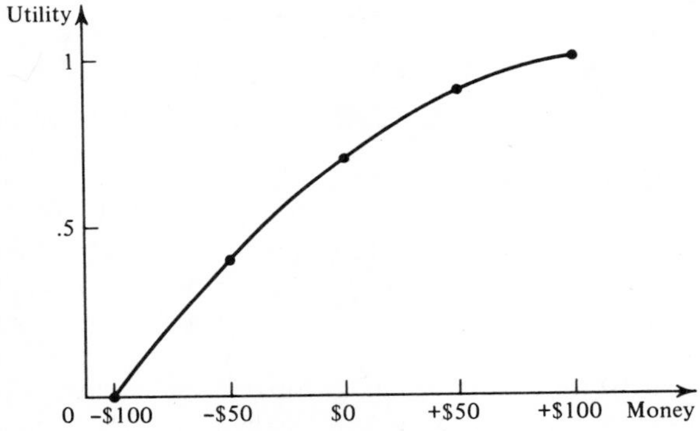

Figure 5.6.1

In the above procedure, the amount of money in Lottery I is fixed and you have to determine the probability p that would make you indifferent between the two lotteries. Alternatively, the probability p could be fixed and you could attempt to determine the amount to receive for certain in Lottery I that would make you indifferent between the lotteries. For example, suppose that $p = .50$, implying the following situation.

Lottery I: Receive $\$X$ for certain.

Lottery II: Receive $+\$100$ with probability $\frac{1}{2}$ and receive $-\$100$ with probability $\frac{1}{2}$.

Suppose that $X = -30$ makes you indifferent between the two lotteries. Then $U(-\$30) = .50$, and $-\$30$ can be thought of as a "cash equivalent" for the gamble in Lottery II. In this situation, \$30 is called a *risk premium;* it is the largest amount you are willing to pay to avoid the risk in Lottery II. Because of the risk in Lottery II, you are willing to pay a risk premium to avoid the bet even though the bet is perfectly fair in terms of money. A formal definition of the notion of a risk premium and a graphical interpretation of risk premiums are presented later in this section.

Several types, or classes, of utility functions can be distinguished, although there are utility functions not falling into any of the classes to be described. In Figures 5.6.2 through 5.6.4, three utility curves are presented. The curve in Figure 5.6.2 represents the utility function of a "*risk-avoider*," the curve in Figure 5.6.3 represents the utility function of a "*risk-taker*," and the curve in Figure 5.6.4 represents the utility function of a person who is neither a "risk-taker" nor a "risk-avoider." To see why these terms aptly describe the curves, consider the following bet: you win \$100 with probability $\frac{1}{2}$ and you lose \$100 with probability $\frac{1}{2}$. This can be thought of as a bet of \$100 on the toss of a fair coin. In terms of expected payoff, you should be indifferent about the bet since it has an expected payoff of zero. In terms of expected utility, however, the decision as to whether or not to take the bet depends on the shape of your utility function.

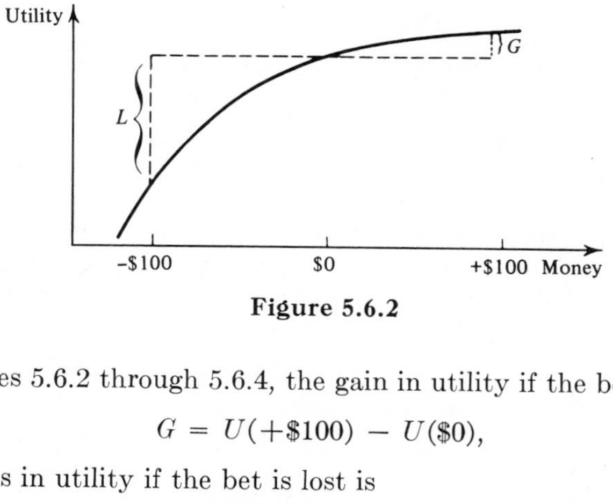

Figure 5.6.2

In Figures 5.6.2 through 5.6.4, the gain in utility if the bet is won is

$$G = U(+\$100) - U(\$0), \tag{5.6.1}$$

and the loss in utility if the bet is lost is

$$L = U(\$0) - U(-\$100). \tag{5.6.2}$$

The expected utility of the bet is

$$EU(\text{bet}) = \tfrac{1}{2}U(+\$100) + \tfrac{1}{2}U(-\$100).$$

The alternative action is not to bet, and the expected utility of this is just

$$EU(\text{not bet}) = U(\$0).$$

Under what circumstances would you take the bet? Using the EU rule, you would take the bet if

$$EU(\text{bet}) > EU(\text{not bet});$$

that is, if

$$EU(\text{bet}) - EU(\text{not bet}) > 0. \qquad (5.6.3)$$

But from the above equations,

$$EU(\text{bet}) - EU(\text{not bet}) = \tfrac{1}{2}U(+\$100) + \tfrac{1}{2}U(-\$100) - U(\$0).$$

The right-hand side of this equation can be written as

$$\tfrac{1}{2}U(+\$100) + \tfrac{1}{2}U(-\$100) - \tfrac{1}{2}U(\$0) - \tfrac{1}{2}U(\$0),$$

which is equal to

$$[\tfrac{1}{2}U(+\$100) - \tfrac{1}{2}U(\$0)] - [\tfrac{1}{2}U(\$0) - \tfrac{1}{2}U(-\$100)],$$

or, using Equations (5.6.1) and (5.6.2),

$$\tfrac{1}{2}G - \tfrac{1}{2}L = \tfrac{1}{2}(G - L).$$

From Equation (5.6.3) the decision rule is as follows:

Take the bet if $\frac{1}{2}(G - L) > 0$.
Do not take the bet if $\frac{1}{2}(G - L) < 0$.

Thus, in order to make the decision in this example, you need only look at the sign of $G - L$.

For the curve in Figure 5.6.2, G is smaller than L, so that $(G - L)$ is negative, and you should not take the bet. Since you will not take a bet with an expected monetary payoff of zero, you are called a *risk-avoider*. In fact, with this curve it is possible to find some bets with expected monetary payoffs *greater than zero* that you would consider unfavorable in terms of expected utility. Mathematically, the utility function graphed in Figure 5.6.2 is called a concave function.

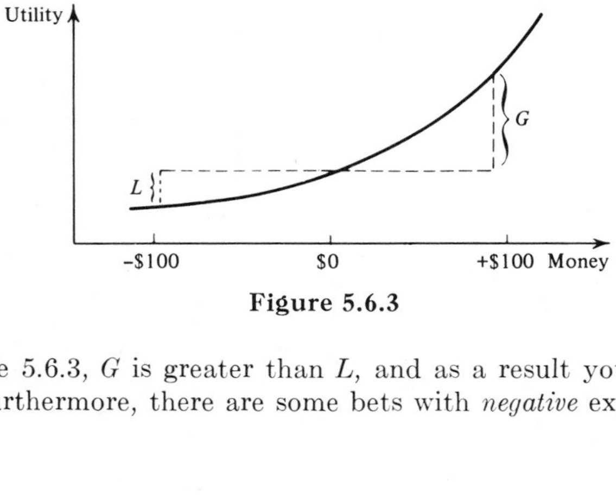

Figure 5.6.3

In Figure 5.6.3, G is greater than L, and as a result you *should* take the bet. Furthermore, there are some bets with *negative* expected monetary payoffs that you would consider to be favorable bets. As a result, this curve represents the utility function of a *risk-taker*. Mathematically, this type of function is called a convex function.

Finally, $G = L$ in Figure 5.6.4. In this case you are indifferent between taking the bet and not taking it, and thus you are neither a risk-taker nor a risk-avoider. For a person with a *linear* utility function (that is, the curve is a straight line), maximizing EU is equivalent to maximizing ER. To prove this, note that if the utility curve is linear, it can be written in the form

$$U = a + bR,$$

where R is expressed in terms of money and a and b are constants, with $b > 0$. The requirement that b be greater than 0 is necessary because the line must have a positive slope. By the first utility axiom of the preceding section, all utility functions are nondecreasing functions; that is, the utility of any given payoff must be at least as large as that of a lesser (less preferred) payoff.

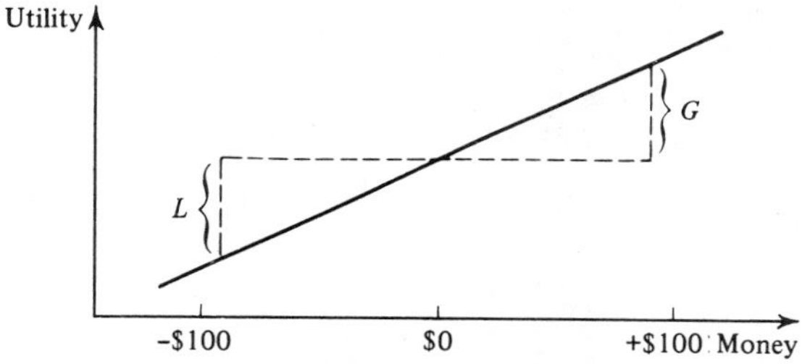

Figure 5.6.4

Since a and b are constants,

$$EU = a + bER,$$

and EU is at a maximum when ER is at a maximum. This is an important result: *if a person's utility function is linear with respect to money, then the ER criterion and the EU criterion are equivalent.* That is, the action that maximizes ER also maximizes EU. Of course, since the ER and EL criteria are equivalent, a linear utility function implies that the action that maximizes ER and EU also minimizes EL. The axiomatic development of utility theory provides strong justification for the EU criterion, as you shall see in the next section. To the extent that the assumption of a linear utility function for money is reasonable, then, strong justification is also provided for the ER and EL criteria.

In the above example, the probabilities for winning and losing are equal, and thus the decision depends only on the sign of $G - L$. If the probabilities are unequal, the difference in expected utilities of the two actions, and hence the decision, will depend on the relative magnitudes of G and L, not just on the sign of $G - L$. However, the results regarding the shapes of the curves are unchanged. Furthermore, even if the bet (or the decision-making problem) is a complex problem with numerous possible consequences, the results still hold provided that the expected payoff of the bet is \$0. You should keep in mind that a bet with an expected payoff of \$0 is a "fair" bet, regardless of the exact nature of the probabilities or the payoffs. For instance, a risk-avoider would not take a bet in which he could win \$10 with probability .1, win \$9 with probability .2, and lose \$4 with probability .7; the expected payoff is \$0, and risk-avoiders will not take a "fair" bet. In addition, be careful to note that not all decision-making problems are "fair bets." Thus, it is not true that a risk-avoider will decline all gambles. If a gamble is favorable enough (if the ER of the gamble is large enough), a risk-avoider will take the gamble. Similarly, a risk-taker will find some gambles (those with highly negative expected payoffs) unfavorable. Thus, refusing a gamble does not necessarily mean that the person is a risk-avoider, and taking a gamble does not necessarily imply a risk-taker. As you shall see later in the section, the sign of an individual's risk premium for a gamble reveals something about the individual's attitude toward risk.

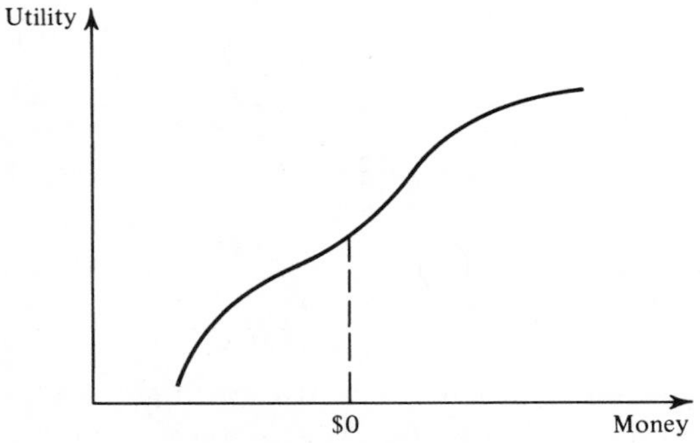

Figure 5.6.5

As you shall see in Section 5.8, mathematical functions are often used to approximate the relationship between money and utility. Of course, the curves discussed above are by no means the only possible forms for utility functions. Several other forms that are not described here have been proposed, and still other forms, many of which are not mathematically tractable (that is, they cannot be represented by a simple mathematical function), no doubt exist. For instance, some individuals appear to be risk-takers for some decisions (such as gambling) and risk-avoiders for other decisions (such as purchasing insurance). One possible form for such an individual's utility function is shown in Figure 5.6.5. It is an interesting exercise to select a range of values, such as $-\$100$ to $+\$100$, and to attempt to determine your own utility curve for money. Just as in the assessment of probabilities, the assessment of utilities is not as easy as it looks. It is often quite difficult to determine how you would act if given a choice between two particular bets.

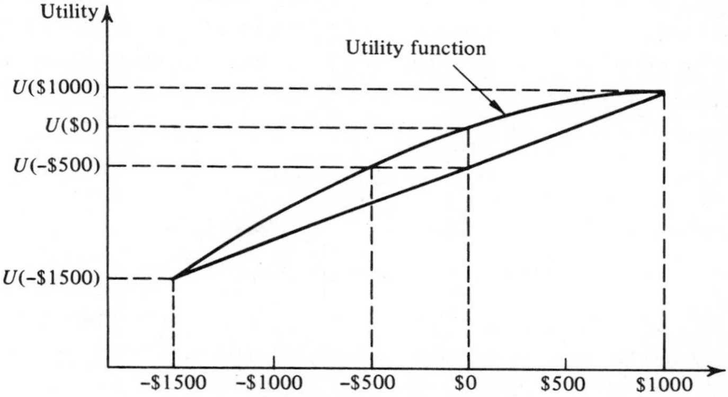

Figure 5.6.6

For a simple bet or lottery involving only two possible payoffs, it is easy to represent the decision graphically in terms of the decision maker's utility function. For instance, suppose that the decision maker is faced with a situation in which he will gain \$1000 with probability .6 and lose \$1500 with probability .4. This situation happens to result in an expected monetary payoff of 0, although this is not necessary for the graphical explanation of the choice of whether or not to take the "bet." The expected utility of the bet is

$$EU(\text{bet}) = .6U(\$1000) + .4U(-\$1500).$$

Graphically, this can be represented as a point on the straight line joining the points $[-\$1500, U(-\$1500)]$ and $[\$1000, U(\$1000)]$, shown in Figure 5.6.6. In particular, since the probability of obtaining \$1000 is .6, the exact point is the point on this straight line corresponding to

$$.6(\$1000) + .4(-\$1500) = \$0.$$

Note that since the straight line is everywhere below the utility function, $U(\$0)$ is higher than the point on the straight line corresponding to \$0. But this point on the line represents the expected utility of the bet, so that the expected utility of not betting, $U(\$0)$, is higher than $EU(\text{bet})$. Thus, the optimal act is not to bet. In fact, from Figure 5.6.6 it can be seen that the expected utility of the bet is equal to $U(-\$500)$. This means that if the decision maker is somehow forced to take the bet, he would be willing to pay up to \$500 to avoid having to take the bet. Thus, the cash equivalent of the bet to him is $-\$500$. The bet is a fair bet in terms of expected monetary payoff, but it is undesirable to the decision maker because of his risk-avoiding utility function. The amount that he is willing to pay to avoid the bet, \$500, is a *risk premium*. Paying \$500 to avoid the bet is similar to insuring against the risk in the bet, so a risk premium is analogous to an insurance premium. Formally, if the expected payoff under uncertainty is denoted by ER and the cash equivalent of the uncertain situation is denoted by CE, then the risk premium, denoted by RP, is

$$RP = ER - CE. \tag{5.6.1}$$

For the example, $ER = \$0$ and $CE = -\$500$; thus,

$$RP = \$0 - (-\$500) = \$500.$$

For another example, consider the utility function in Figure 5.6.7. For a bet in which the decision maker wins \$300 with probability .5 and loses \$200 with probability .5, the expected payoff is $ER(\text{bet}) = .5(\$300) + .5(-\$200) = \$50$. To find the utility of the bet, find the utility value corresponding to the point on the straight line above \$50. By reading to the left, this is equal to $U(\$10)$, so that the decision maker's cash equivalent for the bet is \$10. He would be willing to pay up to \$10 to be able to participate in the bet. This is still below the ER of \$50, since the utility function is that of a risk-avoider. The difference between the expected monetary payoff and the cash equivalent is the risk premium, which is \$40 in this example.

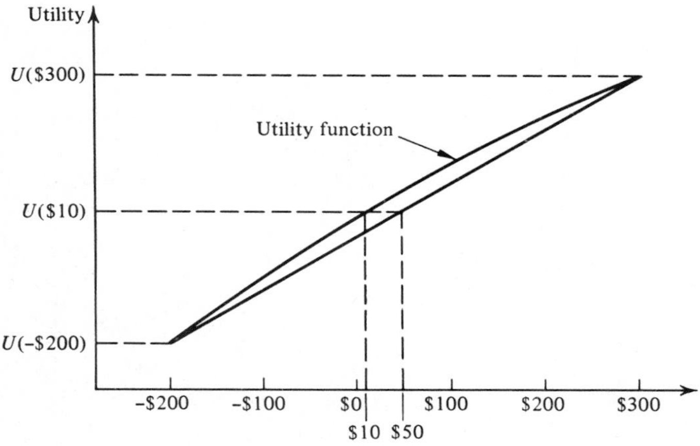

Figure 5.6.7

For an example of the reaction of a particular risk-taker to the same bet, see Figure 5.6.8. In this case, the utility of the bet is equal to $U(\$100)$, so \$100 is the cash equivalent for the bet. Although the expected monetary payoff is only \$50, this risk-taker is willing to pay up to \$100 to be able to participate in the bet. Here the risk premium is $-\$50$; it is negative because the decision maker, instead of being willing to pay a premium to *avoid* the risk in the bet, is willing to pay a premium (above and beyond the ER) in order to be able to participate in the bet.

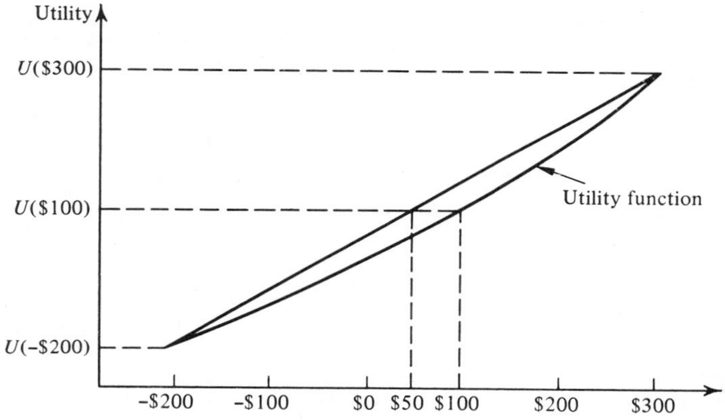

Figure 5.6.8

The concept of a risk premium provides another way to investigate the general shape of a person's utility function. For a risk-avoider, whose utility function is concave, as in Figure 5.6.2, the risk premium for a gamble (or for any situation in which the payoff is uncertain) is positive. For a risk-taker, whose utility function is convex, as in Figure 5.6.3, the risk premium is negative. For a person whose utility function is linear, as in Figure 5.6.4, the risk premium is always \$0, since the cash equivalent of a gamble is always equal to the expected monetary payoff (ER) of the gamble. For a person with a utility function such as that in Figure 5.6.5, the risk premium is positive for some gambles, negative for other gambles, and 0 for yet other gambles. For instance, for gambles involving large potential losses, the risk premium will generally be positive, whereas for gambles involving only small potential losses (and small or moderate potential gains), the risk premium is likely to be negative.

It should be noted that the horizontal axes in Figures 5.6.1 through 5.6.8 are expressed in terms of "changes in total monetary assets." In each case, \$0 represents "no change," $+\$100$ represents a gain of \$100, and so on. Sometimes it is more convenient to express utility in terms of total assets rather than changes in assets. For instance, if your assets currently total \$1000, then the values $-\$100$, \$0, and $+\$100$ on the horizontal axis in Figure 5.6.1 correspond to \$900, \$1000, and \$1100 in terms of total assets. A person's current asset position is an important factor in relation to his utility function; generally, as a person's asset position changes his utility function for changes in assets also changes. It makes no difference whether you express utility as a function of total assets or as a function of changes in total assets, as long as you are consistent in any given problem. In this book, unless explicitly stated otherwise, utility is always given as a function of changes in total assets, as in all of the previous examples.

It is also important to note once again that a person's utility function may change over time; it may be different tomorrow (or even an hour from now) than it is right now. This may be due in part to changes in the person's asset position, but it also may depend on the particular situation. For a trivial example, consider a person who has an opportunity to bet \$2 on a toss of a fair coin, winning \$2 if heads comes up and losing \$2 if tails comes up. If the person has only \$2 in his wallet today and if he needs that to buy his dinner, he would be unlikely to take the bet. On the other hand, if the same person has \$2 in his wallet the following day and would very much like to attend a concert that costs \$4, the bet may look very attractive. The differences in the two situations cause the utility function to change. In the first case, the utility of $-\$2$ is so low that EU(bet) is less than EU(no bet). In the second case, the utility of $+\$2$ is so high that EU(bet) is greater than EU(no bet).

5.7 UTILITY AND SUBJECTIVE PROBABILITY

The basis for the theory of utility, as well as for a joint theory of utility and subjective probability, is provided by certain *axioms of coherence*. These axioms, some of which were discussed very briefly in Chapter 2, provide a model of how a "rational man" makes decisions in the face of uncertainty. If a person satisfies the axioms of coherence, this implies the existence of a utility function that reflects his preferences for various payoffs (monetary and nonmonetary) and a subjective probability distribution that reflects his judgments about the uncertain quantities of interest in a decision-making problem. For purposes of decision theory, the most important thing to note is that the axioms of coherence imply that a person should make decisions in such a manner as to maximize expected utility. This provides a very powerful argument in favor of the EU criterion. If you are willing to accept the axioms of coherence, then you must accept the EU criterion, since it follows logically from the axioms. In order to investigate the EU criterion, then, it is useful to investigate the axioms of coherence. It should be pointed out that the set of axioms presented in this section is by no means unique; other sets of axioms yield the same results (that is, they imply the existence of utility and subjective probability and imply the optimality of the *EU* criterion). Generally, however, these other sets are derived from the set presented here by slightly modifying just one or two axioms or by replacing them with new axioms.

Axiom 1. Given any two payoffs R_1 and R_2 (these payoffs may involve monetary and/or nonmonetary factors; you can think of them as potential "consequences" in a decision-making problem), you can decide whether you prefer R_1 to R_2, you prefer R_2 to R_1, or you are indifferent between R_1 and R_2.

This axiom, which seems intuitively reasonable, states that a person can order any set of consequences in terms of preference. In the definition of subjective probability presented in Chapter 2 in terms of lotteries, it was assumed that for any two lotteries, Lottery A and Lottery B, an individual could decide whether he prefers Lottery A to Lottery B, prefers Lottery B to Lottery A, or is indifferent between the two lotteries. Admittedly, there may be a problem of vagueness at times (that is, at first glance you may not be sure which of two payoffs you prefer, especially if the payoffs are quite complex), but this difficulty should be eliminated with some careful thought. If after careful consideration you cannot decide which of two payoffs you prefer, then it should be safe to say that you are, for all practical purposes, indifferent between the two payoffs.

Axiom 2. If you prefer payoff R_1 to payoff R_2 and you prefer payoff R_2 to payoff R_3, then you must prefer R_1 to R_3.

Axiom 2 is the axiom of *transitivity of preferences*. For example, if you prefer wine instead of beer with your evening meal and if you prefer beer to water, then presumably you would prefer wine to water. If this axiom is violated, generally a re-examination of the payoffs in question will enable you to remove the violation. Transitivity can be extended to indifference: if you are indifferent between R_1 and R_2 and indifferent between R_2 and R_3, then you should be indifferent between R_1 and R_3. In general, transitivity is an important element in the theory of subjective probability and utility, and every set of axioms of coherence includes some type of transitivity axiom.

Before presenting the remaining axioms of coherence, it is convenient to simplify the terminology somewhat. Consider a bet in which you will receive payoff R_1 with probability p and you will receive payoff R_2 with probability $1 - p$. This will be called a *p-mixture* of payoffs R_1 and R_2. For example, suppose that you plan to go to a movie if it rains tomorrow and to go on a picnic if it does not rain. If you feel that the probability of rain is .20, then this is a p-mixture of a movie and a picnic, with $p = .20$. Note that the order in which the payoffs are written is important; while the above situation is a .20-mixture of a movie and a picnic, it is a .80-mixture of a picnic and a movie. The notion of a p-mixture is just a "shorthand" version of a simple lottery such as the lotteries used in Chapter 2 and in Section 5.6.

Axiom 3. If you prefer R_1 to R_2 and R_2 to R_3, then you can find some value of p such that a p-mixture of R_1 and R_3 is preferred to R_2; you can find some other value of p such that R_2 is preferred to a p-mixture of R_1 and R_3; and, finally, you can find yet another value of p such that you are indifferent between R_2 and a p-mixture of R_1 and R_3.

Note that Axiom 3 was used (implicitly) in the assessment of utility functions in Section 5.6. In that case, R_1 was the "most preferred" payoff, R_3 was the "least preferred" payoff, and R_2 was some other payoff. In order to find the utility of R_2, you need to find a value of p such that you are indifferent between R_2 and a p-mixture of the most-preferred and least-preferred payoffs. Axiom 3 states that such a value can be found. For instance, \$100 is preferred to \$50, which is preferred to \$0. Axiom 3 states that you can find a p such that a p-mixture of \$100 and \$0 is preferred to \$50 (try $p = .99$); that you can find a p such that the reverse is true (try $p = .01$); and that you can find a p such that you are indifferent between \$50 and a p-mixture of \$100 and \$0 (if your utility function is linear with respect to money, $p = .50$ would work; for most people, a larger p, such as .60, would be necessary for indifference).

Axiom 4. If R_1 is preferred to R_2 and if R_3 is some other payoff, then any p-mixture of R_1 and R_3 is preferred to the same p-mixture (that is, the mixture with the same value of p) of R_2 and R_3.

As an illustration of Axiom 4, consider a choice between a p-mixture of \$100 and \$0 and the same p-mixture of \$50 and \$0. Clearly the former is preferred. For example, if two bets provide you with the same payoff (\$0) if a coin comes up heads and if the bets pay \$100 and \$50, respectively, if the coin comes up tails, you would naturally prefer the bet with the \$100 payoff. In fact, the other bet would be an *inadmissible* action, since it provides the same payoff as the first bet if heads occurs and a smaller payoff if tails occurs. Axiom 4, then, provides the justification for the elimination of all inadmissible actions from consideration in a decision-making problem.

Axiom 5. If you are indifferent between R_1 and R_2, then they may be substituted for each other as payoffs in any decision-making problem.

Axiom 5 states simply that equally preferable payoffs may be substituted for each other in any decision-making problem. This axiom justifies the expression of nonmonetary consequences in terms of their cash equivalents, where each cash equivalent is determined such that you are indifferent between the nonmonetary payoff and the cash equivalent. In any given situation, of course, it may be difficult to determine a cash equivalent, just as it is difficult to assess a utility function.

Axiom 6. If R_1 is preferred to R_2, then a p-mixture of R_1 and R_2 is preferred to a q-mixture of R_1 and R_2 if and only if $p > q$.

As an illustration of Axiom 6, consider once again the situation in which you plan to go to a movie if it rains tomorrow and to go on a picnic if it does not rain. If you prefer the picnic to the movie, then clearly you will be happier if the probability of rain is .10 than if the probability of rain is .50. In a situation with just two payoffs, you always prefer a high probability of obtaining the "better" payoff to a low probability of obtaining that payoff. This axiom was used in the definition of subjective probability in terms of lotteries in Chapter 2, where it was assumed that one lottery is preferable to a second lottery with identical "prizes" if and only if the probability of the "better" (more preferred) prize is greater in the first lottery.

Note that Axiom 6 makes the first two parts of Axiom 3 superfluous. The third part of Axiom 3 states that you can find a p that makes you indifferent between R_2 and a p-mixture of R_1 and R_3. Axiom 6 then implies that a p-mixture of R_1 and R_3 is preferred to R_2 for any p larger than the indifference p (thus satisfying the first part of Axiom 3) and vice versa for any p smaller than the indifference p (thus satisfying the second part of Axiom 3).

These six axioms of coherence, then, appear to be intuitively reasonable, and most people would agree that a violation of any of the axioms in any decision-making situation would be considered unreasonable, or "irrational." This, of course, does not imply that the axioms are never violated in practice; violations may occur due to carelessness, complexity of the problem, vagueness, or various other reasons. Violations are particularly apt to occur when there are numerous payoffs and when any preferences among the payoffs are only slight. Because the assumptions seem reasonable, presumably you would be willing to modify your stated preferences and judgments to remove a violation if it were pointed out to you. The important point is that if you agree that the axioms are reasonable, then you should be willing to act in accordance with the axioms and you would not intentionally violate any of the axioms.

As noted at the beginning of this section, the set of axioms presented here is not unique. It is possible to consider slight modifications of some of the above axioms, and sometimes additional axioms are added, often seemingly trivial assumptions that are added to avoid certain mathematical problems. For instance, one such axiom states that there exist two payoffs R_1 and R_2 such that R_1 is preferred to R_2. The purpose of this axiom is to avoid the trivial, uninteresting situation in which the decision maker is indifferent among all of the potential payoffs.

The point of particular interest for the purposes of this book is as follows. (Because of lack of space and because it is not particularly instructive, the proof of this statement is not presented.) Given that a person obeys a set of axioms of coherence such as the set presented in this section, it can be shown that the person's preferences can be described by a utility function satisfying the two axioms of utility presented in Section 5.5; that the person's subjective judgments about uncertain quantities can be described by a probability distribution satisfying the three axioms of probability presented in Section 2.1; and, most important, that the person should make his decisions in such a manner as to maximize expected utility. In other words, *the axioms of coherence lead to a unified theory of utility and subjective probability, a theory that implies that the EU criterion should be used to make decisions under uncertainty.* This provides a strong justification for the *EU* criterion in particular and for the decision theory model presented in this book in general.

5.8 UTILITY AND DECISION MAKING

The criterion that has been adopted for making decisions under uncertainty in this book is the *EU* rule: select the action for which the expected utility is the highest. If a decision maker obeys the axioms of coherence presented in Section 5.7, then he *must* use the *EU* criterion in making decisions under uncertainty. If the utility of each consequence, or payoff, is known, this rule is simple to apply, and any problems are generally of a computational nature.

Often, however, the utilities of some consequences are not known. If there are numerous possible consequences, it may be a difficult task to determine their utilities. Even if there are only a few potential consequences, they may be extremely complex, involving numerous factors. For a simple example, suppose that you are looking for a new job and that by eliminating inadmissible jobs or by informally applying decision theory, you have narrowed your choice to three possible jobs. It is difficult for you to express your preferences among these three jobs because they involve different salaries, locations, opportunities for promotion, retirement benefits, and so on. This is a problem of *multidimensional* utility. One way to handle the multidimensional problem is to assess a multidimensional utility function. That is, instead of assessing a utility function for a single factor, such as money, and comparing the utilities of various amounts of money, the decision maker assesses a utility function for several factors, comparing various combinations of amounts of the different factors. For instance, you could directly compare consequences such as ($15,000 salary, Chicago, good opportunity for promotion), ($18,000 salary, San Francisco, little opportunity for promotion), ($18,000 salary, New York City, moderate opportunity for promotion), and so on. Thus, the utility of these consequences depends on three factors: salary, location, and opportunity for promotion. The choices among such multifactor consequences are very difficult to make. Furthermore, in a sense, considering such choices defeats one of the purposes of decision theory, which is to break a complex problem down into relatively simple components. Choosing among multifactor consequences is often equivalent to making a complex decision intuitively with no formal analysis.

An alternative approach to the multidimensional problem is to consider each of the factors separately. You might be able to determine your preferences, hence your utilities, for each dimension, or factor, independently of the others. For instance, you can determine a utility function for various salaries, you can assess utilities for the potential locations, and so on. The problem is the combination of these utilities across the different factors to come up with a single utility for each consequence that takes into consideration all factors. One possible solution to this problem is simply to add up the utilities for the different factors:

$$\begin{aligned} U(\text{job}) = {} & U(\text{salary}) + U(\text{location}) \\ & + U(\text{opportunities for promotion}) + U(\text{retirement benefits}). \end{aligned}$$

This is called an *additive* utility model. One difficulty is that this model seems to weigh the different factors equally. A variant of the model involves the assessment of weights reflecting the relative importance of the factors. For instance, you might have

$$\begin{aligned} U(\text{job}) = {} & 10U(\text{salary}) + 8U(\text{location}) \\ & + 8U(\text{opportunities for promotion}) + U(\text{retirement benefits}). \end{aligned}$$

This weighted additive model implies that you consider salary to be 10 times as important as retirement benefits, location 8 times as important as retirement benefits, and opportunities for promotion 8 times as important as retirement benefits.

This informal example illustrates the difficulties that may be encountered in multiple-factor situations. It is hard enough to assess utilities in terms of just a single factor, and the problem becomes much more complex in the multidimensional case. The additive and weighted additive models are two methods that have been proposed to simplify these problems. It should be mentioned, however, that such models represent informal approaches to the multidimensional problem that may be convenient and useful in some, but not all, situations. For example, if it is difficult for the decision maker to consider the factors independently, then the additive models may not apply.

If all of the consequences can be expressed in terms of a *single* factor, the task is somewhat easier. This factor may be, for instance, the number of patients cured if the decision-making problem involves a new drug. In many cases, however, the common factor is money. The notion of "cash equivalents" represents an attempt to express all of the relevant consequences in terms of money. As in Section 5.6, a few values can be selected, utilities determined for these values, and a rough utility curve drawn. The approximate utility of any monetary payoff can then be read from the curve. Even if there are many payoffs, this should not be difficult.

In the discussion of utility in this chapter, it has been assumed that a single person is making the decision and receiving the ultimate payoff. In many situations, particularly for important problems concerning large sums of money, there are several individuals involved. In business decision making, the difficulties are intensified. Who is to determine the utility function for a firm—the officers, the board of directors, the stockholders? Even if this question is resolved, the problem is that a single utility function for an entire group or firm represents a compromise; it will not accurately reflect the preferences of each individual involved in the decision-making problem.

In some situations, then, it may be difficult to assess a utility function to be used in a specific decision-making problem. When the utility function is not known, a convenient assumption is that the function is linear with respect to money. In that case, as shown in Section 5.6, the act with the highest expected utility is also the act with the highest expected monetary payoff. The *ER* criterion can thus be used to determine the optimal decision. Since the *EL* rule is equivalent to the *ER* rule, it too can be used.

How reasonable is the assumption that the utility function is linear with respect to money? Because of the problems mentioned above, this assumption seems fairly reasonable for a firm's utility function, except in cases involving very large sums of money. For an individual person's utility function, the applicability of the assumption varies from person to person. For most people, the assumption should not be too unrealistic unless the amounts of money involved are large. For small amounts, a straight line is probably a reasonably good approximation to the utility curve. For large amounts, it is likely to be a very bad approximation. Most individuals behave as risk-avoiders when large amounts are involved, and various simplifying models are available to represent the utility function of a risk-avoider. Two examples are *exponential* utility functions of the form

$$U = a - e^{-bR}$$

and *logarithmic* utility functions of the form

$$U = a \log (R + b),$$

where a and b are positive constants. Of course, what is a small amount to one person (say, a millionaire) may be a very large amount to another person (say, a student).

In any decision-making problem, the relevant question concerning the assumption of linear utility is this: would EU result in a different decision than ER? This is once again the question of the *sensitivity* of the ultimate decision to changes in the inputs. Previously, the sensitivity of the decision-making procedure to changes in the prior distribution was discussed. Sensitivity to changes in the utility function, including the use of models such as linear, exponential, or logarithmic utility functions, is also of interest.

In the examples involving money in the remainder of this book it is assumed, *unless stated otherwise,* that utility is linear with respect to money, so that ER and EL can be used as decision-making criteria. Alternatively, when payoffs or losses are discussed, you might think of them as being expressed in terms of utility. In some examples, payoff and loss tables are expressed in terms of utilities without worrying about the source of the utilities. These simplifications are useful in allowing you to concentrate on other important points. You should keep in mind that they are simplifications and that in any actual application of decision theory it is necessary to determine the utilities to be used as inputs to the decision-making problem and to investigate carefully the applicability of simplifying assumptions such as linear utility functions.

5.9 A FORMAL STATEMENT OF THE DECISION PROBLEM

Now that all of the inputs to a problem of decision under uncertainty have been discussed, the use of these inputs is considered and some new notation is introduced. In any decision-making problem, the decision maker starts out with a set of possible actions, or decisions. Call this set A, and denote a typical member of this set (that is, an action) by a. In addition, it is presumed that the uncertainty in the problem concerns the uncertain quantity (state of the world) $\tilde{\theta}$. The set of possible states of the world, or possible values of $\tilde{\theta}$, is labeled S. It is further assumed that the uncertainty about $\tilde{\theta}$ can be expressed in probabilistic terms, in the form of a probability distribution of $\tilde{\theta}$. This distribution is labeled $P(\theta)$ if $\tilde{\theta}$ is discrete and $f(\theta)$ if $\tilde{\theta}$ is continuous. Note that this distribution may be a posterior distribution $P''(\theta \mid y)$ or $f''(\theta \mid y)$ following an observed sample result y, or it may be a prior distribution $P'(\theta)$ or $f'(\theta)$ if there is no sample information. Unless it becomes necessary to do so (and it is when the value of information is discussed), you should not be concerned about whether the distribution is prior or posterior. There is also a set of consequences, with one consequence for each combination of an action and a state of the world: (a,θ). Finally, there is a utility function, denoted by U, giving the utility of each consequence. Since the consequence depends on the combination (a,θ), the utility function is written in the form $U(a,\theta)$.

Recapitulating, the following inputs to the decision-making problem are needed.

1. A, the set of actions, or acts.
2. S, the set of states of the world.
3. $P(\theta)$ or $f(\theta)$, the probability distribution of $\tilde{\theta}$, the state of the world.
4. $U(a,\theta)$, the utility function that associates a utility with each pair (a,θ), where a is an action and θ is a state of the world.

Given these inputs, one must make a decision, which amounts to selecting an act a from the set A. According to the development of utility and subjective probability presented in Section 5.7, the decision maker should select the act that is optimal under the EU rule. In other words, he should select the act that has the highest expected utility of all acts, or of all members of A. The expected utility of any act a can be calculated by using the definition of expectation, where the expectation is taken with regard to the distribution of $\tilde{\theta}$. The expected utility of act a is denoted by $EU(a)$. If $\tilde{\theta}$ is discrete, the expected utility of a is

$$EU(a) = \Sigma U(a,\theta)P(\theta), \tag{5.9.1}$$

where the summation is taken over all θ in S. If $\tilde{\theta}$ is continuous,

$$EU(a) = \int U(a,\theta)f(\theta)\, d\theta, \tag{5.9.2}$$

where the integration is over the set S. An act a^* is *optimal* with respect to the decision problem if

$$EU(a^*) \geq EU(a) \quad \text{for all acts } a \text{ in } A. \tag{5.9.3}$$

It should be mentioned that it is mathematically possible for such an optimal act not to exist, but that in real-world problems this will not be the case. Another point of interest is that if the "state of the world" in the decision-making problem is a future sample outcome, then the relevant probability distribution is a predictive distribution.

Note that the fourth input to the decision-making problem is a utility function that is simply a function of the pair (a,θ), where a is an action and θ is a state of the world. Assessing $U(a,\theta)$ directly in terms of pairs of the form (a,θ) avoids the problem of having to describe the "consequences" in terms of any factor or factors. Instead of comparing monetary amounts, then, the decision maker compares act-event pairs [that is, (a_1,θ_1) versus (a_2,θ_2), and so on]. This procedure was illustrated in the example presented in Section 5.6 concerning the choice among a boat ride, a movie, and a football game.

In some situations, of course, it is possible to express all of the consequences, or act-event pairs, in terms of a factor such as money. If the utility function U is linear with respect to money, then it is possible to work with the payoff function $R(a,\theta)$ or with the loss function $L(a,\theta)$, where $R(a,\theta)$ and $L(a,\theta)$ are expressed in terms of money. If $\tilde{\theta}$ is discrete, the expected payoff and expected loss of act a are given by

$$ER(a) = \Sigma R(a,\theta)P(\theta) \tag{5.9.4}$$

and

$$EL(a) = \Sigma L(a,\theta)P(\theta). \tag{5.9.5}$$

If $\tilde{\theta}$ is continuous,

$$ER(a) = \int R(a,\theta)f(\theta)\, d\theta \tag{5.9.6}$$

and

$$EL(a) = \int L(a,\theta)f(\theta)\, d\theta. \tag{5.9.7}$$

If U is linear with respect to money, the maximization of EU is equivalent to the maximization of ER or to the minimization of EL. In this case, an act a^* is optimal with respect to the decision problem if

$$ER(a^*) \geq ER(a) \quad \text{for all acts } a \text{ in } A, \tag{5.9.8}$$

or, equivalently, if

$$EL(a^*) \leq EL(a) \quad \text{for all acts } a \text{ in } A. \tag{5.9.9}$$

You should not be bothered by the overly formal nature of this section. The purpose is to take the parts that have been developed individually in Chapters 3 through 5 (primarily the prior or posterior distribution and the utility function) and to see how they fit together. In the process, the notation is also coordinated. There is nothing in this section that is really new, although a decision problem in which the state of the world is a continuous random variable has not yet been attacked (except for a brief example discussed in Sections 4.4 and 5.4).

This section provides a formal statement of the decision-making model presented in this book. This model is quite general and should be applicable to many situations, but there are still extensions that would make it even more general. For the purposes of this book, it is desirable to keep things simple so that the important points of interest are clear to the reader rather than being obscured by the complexity of the model. However, certain extensions are quite easy to make. For instance, the model assumes that the relevant consequences, or payoffs, are known with certainty; yet there are clearly many situations in which this is not true. It would be an easy matter to extend the model to include uncertain payoffs. As noted in Section 5.2, uncertain payoffs can be considered formally by adding another set of branches to the decision tree. In general, the development of a model for any decision-making problem involves many questions regarding the extension of the decision tree (the addition of branches to the tree) or the "pruning" of the tree (the deletion of branches from the tree). At any rate, in many situations involving uncertain payoffs, the *expected* payoff can be used as a *certainty equivalent*, where the expectation is taken with respect to the probability distribution for the payoffs [*not* the probability distribution for the state of the world, as in Equations (5.9.4) and (5.9.6)]. It is important to note that when the utility function for money is not linear, this cannot be done unless the payoffs are first converted into utilities.

In any real-world decision-making problem, you have considerable latitude in the way you set up the problem in terms of the model presented in this book. The first thing you need to do is to specify A, the set of potential actions, and S, the set of potential states of the world. For instance, perhaps you can think of, say, 50 possible actions; but you might immediately reject 45 of the actions as being clearly inferior to the remaining 5 actions. In this case, you are intuitively (rather than formally) deciding that the 45 rejected actions are inadmissible. Similarly, in trying to determine S, you may feel that there are, say, at least 10 uncertain quantities, or random variables, that could have some bearing on the eventual payoff. To keep the model reasonably simple, you might include only the two or three variables that you consider to be most relevant. Even after the sets of actions and states of the world are formally specified, it may be necessary to consider some simplifications in determining the other inputs to the problem. For instance, for a particular combination of an action and a state of the world, it may be impractical to consider all possible ramifications of the payoff. For a firm, this might include the effect on the price of the firm's stock, the reaction of competitors, and many other such considerations that are difficult to evaluate.

The preceding paragraph illustrates the problem of constructing a mathematical model of a real-world situation. In constructing such a model, the decision maker wants the model to be as realistic as possible, yet he wants it to be simple enough so that it is not too difficult to handle. If the model is too complex, it may be impossible to determine the optimal solution formally. The decision maker must attempt to incorporate the salient features of the real-world problem into his mathematical model. Of course, the decision as to what is salient is an informal one. In general, the construction of a mathematical model is an informal, subjective procedure (although if the model is used for repeated decisions, past results may suggest modifications), and some assumptions and approximations are necessary. Once the model is constructed and the necessary inputs to the model are determined, the solution of the problem is a formal, mathematical procedure. In this section, a framework has been presented for constructing mathematical models of decision-making problems. Within this framework, the choice of inputs (set of actions, set of states of the world, probability distributions, utility functions) is an informal procedure. Once the inputs are determined, the model can be formally solved by using the techniques presented in this book or extensions of these techniques.

5.10 DECISION MAKING UNDER UNCERTAINTY: AN EXAMPLE

To illustrate the application of the framework presented in this chapter for decision making under uncertainty, consider an oil-drilling venture. Assume that the decision maker owns drilling rights at a particular location and that he must decide how best to take advantage of these rights. To begin, he must determine A, the set of actions available to him. After giving this matter some thought, he arrives at five potential actions: (1) drilling with 100 percent interest; (2) finding a partner and drilling with 50 percent interest; (3) "farming out" the drilling rights for a $\frac{1}{8}$ override [this means that he will not have to share in the costs of drilling, but he will receive $\frac{1}{8}$ of the revenue, or net profit (net with respect to all costs except drilling costs), from any oil]; (4) "farming out" the drilling rights for \$10,000 and a $\frac{1}{16}$ override; and (5) not drilling. Obviously many other actions might be considered here; several partners could be found instead of just one, and numerous "farming out" arrangements could be devised. Perhaps the decision maker has chosen these five because he is sure that these particular options are available. That is, someone has already offered to come in as a 50-percent-partner, and someone else has proposed the two "farming out" deals. The decision maker might not be sure about the availability of other deals that have not yet been proposed, or he may feel that these five actions provide him with a reasonable range of alternatives and that it is not worthwhile to seek out additional alternatives. Sometimes the search for additional actions can be quite costly and time-consuming.

The next input to the problem involves S, the set of states of the world. In this oil-drilling venture, the states of the world correspond to the random variable $\tilde{\theta}$, which is defined as the amount of oil (in barrels) recovered from the well. This variable could be taken to be continuous, but in this case the decision maker decides to consider only five possible values: 0 (dry hole), 50,000 barrels, 100,000 barrels, 500,000 barrels, and 1,000,000 barrels. Based on his knowledge of the geological structure of the site, he considers intermediate values as somewhat unlikely, so he is willing to dismiss them from consideration. Even if he feels that intermediate values of $\tilde{\theta}$ are reasonably likely, he may use only five values as an approximation in the formal model because he wants to keep the model small and manageable.

In order to represent his uncertainty about $\tilde{\theta}$ in quantitative terms, the decision maker assesses a probability distribution for $\tilde{\theta}$. In doing so, he takes into consideration information about the success of other oil wells in the general vicinity or in similar situations in other locations and information about the geological structure at the site in question. If he feels somewhat unknowledgeable about the chances for the various values of $\tilde{\theta}$, he might hire a geologist to look at the site and to assess a probability distribution for $\tilde{\theta}$.

Suppose that the decision maker realizes from past experience that the odds are in favor of a dry hole. In particular, he feels indifferent between the following two lotteries.

Lottery A: Receive \$0 with probability .7 and \$100 with probability .3.

Lottery B: Receive \$0 if there is no oil (a dry hole) and \$100 if there is oil.

This implies that $P(\text{dry hole}) = P(\tilde{\theta} = 0) = .7$. Of the remaining four values of $\tilde{\theta}$, the decision maker judges $\tilde{\theta} = 100{,}000$ to be three times as likely as any of the other three values, which he considers to be approximately equally likely. This implies that

$$P(\tilde{\theta} = 50{,}000) = P(\tilde{\theta} = 500{,}000) = P(\tilde{\theta} = 1{,}000{,}000)$$

and

$$P(\tilde{\theta} = 100{,}000) = 3P(\tilde{\theta} = 50{,}000).$$

Since $P(\tilde{\theta} = 0) = .7$, it follows that

$$P(\tilde{\theta} = 50{,}000) + P(\tilde{\theta} = 100{,}000) + P(\tilde{\theta} = 500{,}000) + P(\tilde{\theta} = 1{,}000{,}000) = .3,$$

or

$$P(\tilde{\theta} = 50{,}000) + 3P(\tilde{\theta} = 50{,}000) + P(\tilde{\theta} = 50{,}000) + P(\tilde{\theta} = 50{,}000) = .3,$$

or

$$6P(\tilde{\theta} = 50{,}000) = .3.$$

Thus, $P(\tilde{\theta} = 50{,}000) = .05$, and since $\tilde{\theta} = 50{,}000$, $\tilde{\theta} = 500{,}000$, and $\tilde{\theta} = 1{,}000{,}000$ are equally likely, each has probability .05. The final value, $\tilde{\theta} = 100{,}000$, is three times as likely as any of these values, so $P(\tilde{\theta} = 100{,}000) = .15$.

Thus, the decision maker's probability distribution of $\tilde{\theta}$ is as follows:

$$P(\tilde{\theta} = 0) = .70,$$
$$P(\tilde{\theta} = 50{,}000) = .05,$$
$$P(\tilde{\theta} = 100{,}000) = .15,$$
$$P(\tilde{\theta} = 500{,}000) = .05,$$

and

$$P(\tilde{\theta} = 1{,}000{,}000) = .05.$$

The mean and the standard deviation of $\tilde{\theta}$ are 92,500 and 236,000, respectively.

Now the set of potential *actions*, the set of *states of the world* (values of $\tilde{\theta}$), and the *probability distribution* representing the decision maker's uncertainty about $\tilde{\theta}$ have been determined. The next step is to consider the *payoffs* associated with each particular combination of an action and a state of the world. With regard to the cost of drilling, the decision maker has a firm estimate of \$68,000. He is not absolutely certain about the value of any oil that is recovered, but he is willing to assume that the net profit from the oil will be approximately \$0.96 per barrel. With this information, he can construct a payoff table for the oil-drilling venture. If he drills with 100 percent interest, his payoff in dollars will be $.96\tilde{\theta} - 68{,}000$ (the total revenue from the oil, less the cost of drilling). If he takes a partner and drills with 50 percent interest, the payoff will be split into two halves, so that his payoff will be $.48\tilde{\theta} - 34{,}000$. If he farms out the drilling rights for a $\frac{1}{8}$ override, he will incur no drilling costs and will receive $\frac{1}{8}$ of the revenue from the oil, which will be $.96\tilde{\theta}/8 = .12\tilde{\theta}$. If he farms out the drilling rights for \$10,000 and a $\frac{1}{16}$ override, his payoff will consist of \$10,000 plus $\frac{1}{16}$ of the revenue, or $.06\tilde{\theta} + 10{,}000$. Finally, if he does not drill, his payoff will be 0. By calculating the values of the payoffs for each of the five values of $\tilde{\theta}$, the decision maker arrives at a payoff table for his problem.

ACTION	STATE OF THE WORLD (VALUE OF $\tilde{\theta}$): *0*	*50,000*	*100,000*	*500,000*	*1,000,000*
Drill with 100% interest	−68,000	−20,000	28,000	412,000	892,000
Drill with 50% interest	−34,000	−10,000	14,000	206,000	446,000
Farm out, $\frac{1}{8}$ override	0	6,000	12,000	60,000	120,000
Farm out, $10,000 and $\frac{1}{16}$ override	10,000	13,000	16,000	40,000	70,000
Do not drill	0	0	0	0	0

Alternatively, this could be represented in terms of a tree diagram, as in Figure 5.10.1.

From the payoff table, it is clear that "do not drill" is dominated by "farm out, \$10,000 and $\frac{1}{16}$ override" (in fact, it is also dominated by "farm out, $\frac{1}{8}$ override"). In "farm out, \$10,000 and $\frac{1}{16}$ override," the smallest payoff the decision maker can possibly get is \$10,000, whereas if he does not drill, the payoff is \$0. Therefore, "do not drill" is an inadmissible action, and the bottom line of the payoff table (and correspondingly, the lowest main branch of the decision tree) can be eliminated from further consideration. Note that it is unlikely that the payoff corresponding to "do not drill" is actually \$0, since the decision maker might be able to wait and drill at a later date. Of course, actions such as "do nothing now, then drill in a year" could be included in the set of actions. For the purposes of this example, assume that the drilling rights expire shortly. Therefore, it is not possible to postpone drilling, and "do not drill" is an inadmissible action.

For the remaining four actions, the expected payoffs are

$$\begin{aligned} ER(\text{drill with 100\% interest}) &= (-\$68{,}000)(.70) \\ &\quad + (-\$20{,}000)(.05) \\ &\quad + (\$28{,}000)(.15) \\ &\quad + (\$412{,}000)(.05) \\ &\quad + (\$892{,}000)(.05) = \$20{,}800, \\ ER(\text{drill with 50\% interest}) &= \$10{,}400, \\ ER(\text{farm out, } \tfrac{1}{8} \text{ override}) &= \$11{,}100, \\ \text{and } ER(\text{farm out, \$10,000 and } \tfrac{1}{16} \text{ override}) &= \$15{,}500. \end{aligned}$$

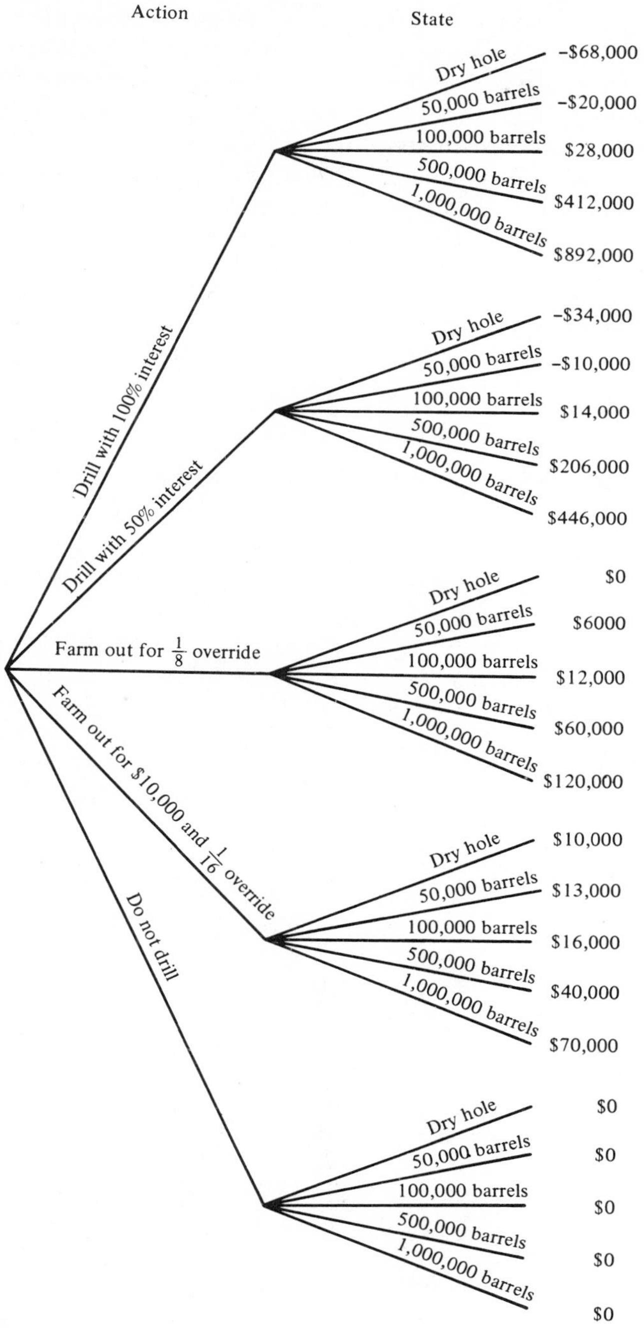
Action
State
Drill with 100% interest
Dry hole -$68,000
50,000 barrels -$20,000
100,000 barrels $28,000
500,000 barrels $412,000
1,000,000 barrels $892,000
Drill with 50% interest
Dry hole -$34,000
50,000 barrels -$10,000
100,000 barrels $14,000
500,000 barrels $206,000
1,000,000 barrels $446,000
Farm out for 1/8 override
Dry hole $0
50,000 barrels $6000
100,000 barrels $12,000
500,000 barrels $60,000
1,000,000 barrels $120,000
Farm out for $10,000 and 1/16 override
Dry hole $10,000
50,000 barrels $13,000
100,000 barrels $16,000
500,000 barrels $40,000
1,000,000 barrels $70,000
Do not drill
Dry hole $0
50,000 barrels $0
100,000 barrels $0
500,000 barrels $0
1,000,000 barrels $0

Figure 5.10.1

If the decision maker's utility function for money is linear, so that maximization of expected payoff is equivalent to maximization of expected utility, then the best action is to drill with 100 percent interest.

Under normal conditions, the decision maker feels that his utility function for money is approximately linear. However, he has suffered some unexpected losses recently, and his cash position is dangerously low. As a result, his current utility function for money is quite nonlinear and is shaped like the utility function of a risk-avoider. In order to graph his utility function, he considers various lotteries of the nature discussed in Section 5.6. First, he lets $U(\$900{,}000) = 1$ and $U(-\$100{,}000) = 0$, since all of the payoffs in his oil-drilling venture lie between these two values. Next, he considers the following two lotteries.

Lottery I: Receive \$0 for certain.

Lottery II: Receive \$900,000 with probability p and $-\$100{,}000$ with probability $(1 - p)$.

At first glance, he decides that he would surely prefer Lottery I if $p = .25$ and Lottery II if $p = .40$. After some serious thought, he chooses $p = .32$ as his indifference probability; if $p = .32$, he is indifferent between the two lotteries. It is obvious from this choice that the decision maker is a risk-avoider, for the expected monetary payoff for Lottery II with $p = .32$ is \$220,000; he is indifferent between \$0 for certain and a risky gamble with $ER = \$220{,}000$.

Next, the decision maker considers a second set of lotteries.

Lottery I: Receive \$100,000 for certain.

Lottery II: Receive \$900,000 with probability p and $-\$100{,}000$ with probability $(1 - p)$.

He judges his indifference probability for these two lotteries to be $p = .50$. Thus, $U(\$0) = .32$ and $U(\$100{,}000) = .50$. Similarly, he assesses utilities for several other amounts, plots these utilities as points on a graph, and attempts to draw a smooth curve that is a good fit to the assessed points. This is illustrated in Figure 5.10.2. After considering several additional gambles in order to check the "fit" of the utility function to his preferences, the decision maker decides that the smooth curve in Figure 5.10.2 is a very good approximation to his utility function and that he is willing to use this curve as his utility function in the oil-drilling venture. Using this utility function, he can express his payoff table in terms of utilities.

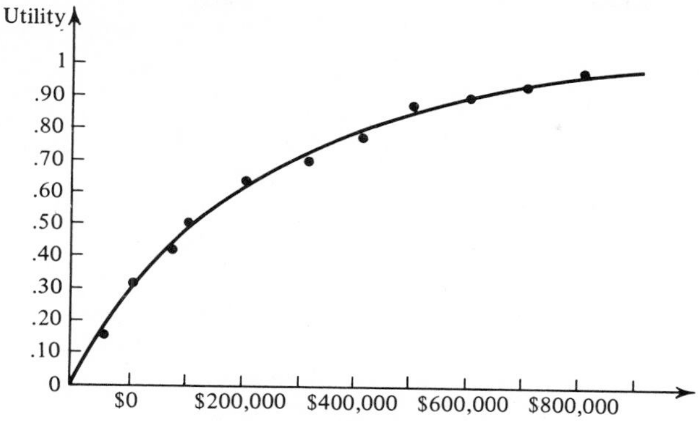

Figure 5.10.2

For instance,

$$U(\text{drill with 100 percent interest, dry hole}) = U(-\$68{,}000) = .11.$$

The entire table of utilities is as follows.

ACTION	STATE OF THE WORLD (VALUE OF $\tilde{\theta}$) 0	50,000	100,000	500,000	1,000,000
Drill with 100% interest	.11	.27	.38	.81	1.00
Drill with 50% interest	.23	.30	.35	.63	.84
Farm out, $\frac{1}{8}$ override	.32	.34	.35	.44	.53
Farm out, \$10,000 and $\frac{1}{16}$ override	.34	.35	.36	.40	.45

The expected utilities of the four actions to the decision maker are

$$\begin{aligned} EU(\text{drill with 100\% interest}) &= (.11)(.70) + (.27)(.05) \\ &\quad + (.38)(.15) + (.81)(.05) \\ &\quad + 1(.05) = .2380, \\ EU(\text{drill with 50\% interest}) &= .3020, \\ EU(\text{farm out, } \tfrac{1}{8} \text{ override}) &= .3420, \\ \text{and } EU(\text{farm out, \$10,000 and } \tfrac{1}{16} \text{ override}) &= .3520. \end{aligned}$$

Thus, the action with the largest expected utility is to farm out the drilling rights for \$10,000 and a $\frac{1}{16}$ override. Because of the risk-avoiding nature of the decision maker's utility function, the action with the largest expected payoff (drill with 100 percent interest) has the smallest expected utility. It just happens that this is the riskiest action in this oil-drilling venture. Note that "drill with 50 percent interest" has a much higher expected utility than "drill with 100 percent interest," although the expected monetary payoff of the former is only half as large as that of the latter. This illustrates the concept of "risk-sharing"; by finding a partner to take a 50 percent interest, the decision maker is reducing his risk. The same principle is illustrated by insurance companies when they "split up" a large insurance policy among several companies to spread the risk. An insurance company would much rather have a $\frac{1}{10}$ share in each of 10 large policies than have a 100 percent share in one of the policies.

By comparing the expected utilities of the four actions with the utility function in Figure 5.10.2, it is possible to determine cash equivalents for each action. For instance, a utility of .2380 corresponds roughly to −\$30,000, so the cash equivalent of "drill with 100 percent interest" is −\$30,000. The cash equivalents of the other three actions are approximately −\$10,000, \$10,000, and \$15,000, respectively. Thus, the best action (that is, the action with the largest expected utility) has a cash equivalent of \$15,000, so the decision maker should be willing to sell the drilling rights for \$15,000.

Of course, the above results depend on the inputs to the formal decision-making model. For example, if the decision maker could not find anyone who would offer him \$10,000 in cash and a $\frac{1}{16}$ override for the drilling rights, then this action would have to be eliminated from consideration. Perhaps someone might offer him \$5000 in cash and a $\frac{1}{16}$ override; he could then include this action in the model, determine the appropriate payoffs and utilities, and compute the expected utility. Another possible action would be created if a person offers to come in as a partner and take a 75 percent interest, leaving the decision maker with only a 25 percent interest. Any change in the set of actions, A, could change the result determined by the model.

Similarly, any change in the set of states of the world, S, could change matters. For instance, the decision maker might consider $\tilde{\theta}$ to be continuous rather than discrete. One possible distribution for $\tilde{\theta}$ is an exponential distribution, which is defined as follows:

$$f(\theta) = \begin{cases} .00001e^{-\theta/100{,}000} & \text{for } \theta \geq 0, \\ 0 & \text{for } \theta < 0. \end{cases}$$

The mean of this distribution is 100,000, and the standard deviation is also 100,000.

Even if the set of actions, S, is not changed, the probability distribution for $\tilde{\theta}$ could change. By looking at the table of utilities for the oil-drilling venture, you can see that if the probabilities for $\tilde{\theta} = 0$, $\tilde{\theta} = 50{,}000$, or $\tilde{\theta} = 100{,}000$ are increased, this will lend even more support to "farm out for \$10,000 and a $\frac{1}{16}$ override." Increases in the probabilities for $\tilde{\theta} = 500{,}000$ and $\tilde{\theta} = 1{,}000{,}000$, however, might make one of the other three actions optimal under the expected utility criterion. Suppose that the decision maker feels that the highest that either of these two probabilities could possibly be is .10. In order to see how sensitive the oil-drilling venture is to changes in the probabilities, then, he considers the following probability distribution:

$$\begin{aligned} P(\tilde{\theta} = 0) &= .60, \\ P(\tilde{\theta} = 50{,}000) &= .10, \\ P(\tilde{\theta} = 100{,}000) &= .10, \\ P(\tilde{\theta} = 500{,}000) &= .10, \\ \text{and} \qquad P(\tilde{\theta} = 1{,}000{,}000) &= .10. \end{aligned}$$

Using this distribution, the expected utilities are

$$\begin{aligned} EU(\text{drill with } 100\% \text{ interest}) &= .3120, \\ EU(\text{drill with } 50\% \text{ interest}) &\doteq .3500, \\ EU(\text{farm out, } \tfrac{1}{8} \text{ override}) &= .3580, \\ \text{and} \qquad EU(\text{farm out, \$10,000 and } \tfrac{1}{16} \text{ override}) &= .3600. \end{aligned}$$

The optimal action is still to farm out the drilling rights for \$10,000 and a $\frac{1}{16}$ override. Since the decision maker chose this probability distribution as an extreme distribution that would favor the other actions as much as possible (while still remaining somewhat in line with his judgments), it appears that the oil-drilling venture is fairly insensitive with respect to changes in the probability distribution.

Another input to the oil-drilling venture is the set of payoffs, which was determined on the basis of the decision maker's judgment that the costs of drilling are \$68,000 and that the revenue per barrel of oil is \$0.96. The decision maker is virtually certain that the \$68,000 figure is correct, but he is somewhat uncertain about the revenue per barrel of oil. Formally, he could represent this uncertainty in terms of probabilities and introduce a new probability distribution into the model. To investigate the sensitivity of the problem to changes in the revenue per barrel of oil, he decides to consider two values: a high value, \$1.12, which he is virtually certain will *not* be exceeded, and a low value, \$0.80, which he is virtually certain *will* be exceeded.

If the revenue per barrel of oil is \$1.12, the payoff table and the utility table are as follows.

	STATE OF THE WORLD (VALUE OF $\tilde{\theta}$)				
ACTION	*0*	*50,000*	*100,000*	*500,000*	*1,000,000*
Drill with 100% interest	−68,000	−12,000	44,000	492,000	1,052,000
Drill with 50% interest	−34,000	−6,000	22,000	246,000	526,000
Farm out, $\frac{1}{8}$ override	0	7,000	14,000	70,000	140,000
Farm out, \$10,000 and $\frac{1}{16}$ override	10,000	13,500	17,000	45,000	80,000

	STATE OF THE WORLD (VALUE OF $\tilde{\theta}$)				
ACTION	*0*	*50,000*	*100,000*	*500,000*	*1,000,000*
Drill with 100% interest	.11	.29	.41	.86	1.02
Drilt with 50% interest	.23	.31	.37	.68	.88
Farm out, $\frac{1}{8}$ override	.32	.34	.35	.45	.55
Farm out, \$10,000 and $\frac{1}{16}$ override	.34	.35	.36	.41	.47

The expected utilities for the four actions are .2470, .3100, .3435, and .3535, respectively, so that farming out the drilling rights for \$10,000 and a $\frac{1}{16}$ override is still the optimal action.

If the revenue per barrel of oil is $0.80, the payoff table and the utility table are as follows.

ACTION	STATE OF THE WORLD (VALUE OF $\tilde{\theta}$) 0	50,000	100,000	500,000	1,000,000
Drill with 100% interest	−68,000	−28,000	12,000	332,000	732,000
Drill with 50% interest	−34,000	−14,000	6,000	166,000	366,000
Farm out, $\frac{1}{8}$ override	0	5,000	10,000	50,000	100,000
Farm out, $10,000 and $\frac{1}{16}$ override	10,000	12,500	15,000	35,000	60,000

ACTION	STATE OF THE WORLD (VALUE OF $\tilde{\theta}$) 0	50,000	100,000	500,000	1,000,000
Drill with 100% interest	.11	.25	.35	.76	.97
Drill with 50% interest	.23	.29	.33	.59	.78
Farm out, $\frac{1}{8}$ override	.32	.33	.34	.42	.49
Farm out, $10,000 and $\frac{1}{16}$ override	.34	.35	.36	.39	.43

The expected utilities for the four actions are .2285, .2925, .3370, and .3505, respectively. Once again, the last action has the highest expected utility.

Thus, for all three values considered for the revenue per barrel of oil (a "high" value, an "intermediate" value, and a "low" value), the same action turns out to be optimal. The oil-drilling venture appears to be quite insensitive to changes in the revenue per barrel of oil. If the problem turned out to be fairly sensitive to changes in the revenue per barrel of oil, the decision maker might describe his uncertainty about this revenue in terms of a probability distribution. This additional source of uncertainty could be represented by a third set of branches on the tree diagram presented in Figure 5.10.1. The new set of branches would correspond to the possible values of $\tilde{\phi}$, the revenue per barrel of oil. In this manner, the uncertainty about $\tilde{\phi}$ could be included formally in the model.

Possible changes in the set of actions A and in the set of states of the world S have been informally discussed, and possible variations in the probability distribution of $\tilde{\theta}$ and in the payoffs have been formally investigated in terms of a sensitivity analysis. Similarly, variations in the decision maker's utility function could be investigated. Clearly, a risk-taker would always drill with 100 percent interest. Sensitivity analysis, whether conducted informally or formally, is an important part of decision making under uncertainty. Informally, it might suggest new actions; in the oil-drilling venture, the results indicate that maintaining only a part interest in the venture or farming the venture out is better for the decision maker than keeping a 100 percent interest. This result might cause the decision maker to work harder at seeking out partners or "farming out" deals. This could lead to possible new actions to be included in the model. Given the decision maker's aversion to risk, perhaps he should try to sell the drilling rights for a fixed amount of money. From the cash equivalents determined earlier in the section, any selling price over $15,000 would be preferable to "farm out, $10,000 and $\frac{1}{16}$ override."

Formally, a sensitivity analysis may pinpoint the crucial elements of the decision-making problem. In the oil-drilling venture, the problem appears to be quite insensitive to changes in the probability distribution and the profit per barrel of oil. It is also possible to vary the probabilities and the profit per barrel of oil simultaneously in order to investigate the sensitivity of the problem to various combinations of changes. In addition, sensitivity to variations in the other inputs might be of interest. For example, it looks as though the problem might be quite sensitive to changes in the utility function.

This oil-drilling example illustrates the application of the framework presented in this chapter for decision making under uncertainty. The importance of the determination of the inputs to the general framework has been stressed; this is largely a question of making the model for any particular application as realistic as possible without making it too complex to handle. Once all of the inputs are determined, the process of finding the optimal decision is merely a computational task.

5.11 REFERENCES AND SUGGESTIONS FOR FURTHER READING

Some general references to decision theory at a level roughly comparable to this chapter are Aitchison (1970*a*, Chapters 4, 6, and 7); Fishburn (1964); Hadley (1967, Chapters 2–4); LaValle (1970, Chapters 15–17); Lindley (1971, Chapters 1, 4, 5, and 9); Morris (1968, Chapters 4 and 5); Pratt, Raiffa, and Schlaifer (1965, Chapters 1–4, 6, 19); Raiffa (1968, Chapters 0–4, 8, 9); Roberts (1966*b*, Chapters 1 and 10); Schlaifer (1959, Chapter 7; 1969, Chapters 1–5); and Schmitt (1969, Chapter 8). Other general references of interest include Borch and Mossin (1968); Bross (1953); Edwards and Tversky (1967); Fishburn (1966*a*, 1966*b*); Forester (1968); Good (1952, 1962); Halter and Dean (1971); Hammond (1967); Hayes (1969*b*); Hirschleifer (1961); Howard (1965*a*, 1966*a*, 1966*b*, 1967, 1968, 1971); Machol (1960); Machol and Gray (1962); Magee (1964*a*); Matheson (1969); North (1968); Schrenk (1969); Smallwood (1968); Swalm (1966); Thrall, Coombs, and Davis (1954); Tribus (1969); and Woods (1966).

With regard to utility theory, some key historical references are Bernoulli (1738), Ramsey (1931), and von Neumann and Morgenstern (1947). The history of utility theory is discussed in Arrow (1951*a*), Edwards (1954), and Savage (1954), and several important references are reprinted in Page (1968). A recent review article on utility with an extensive bibliography is Fishburn (1968). Some other references emphasizing the foundations of utility are Anscombe and Aumann (1963); Arrow (1951*b*, 1958, 1971); Becker, De Groot, and Marschak (1963); Chipman (1960); Debreu (1959); De Groot (1970, Chapter 7); Ellsberg (1954); Fishburn (1966, 1970); Friedman and Savage (1948, 1952); Luce and Suppes (1965); Pratt, Raiffa, and Schlaifer (1964); Markowitz (1952); Marschak (1950); Samuelson (1952); Suppes (1956, 1961, 1969); and Suppes and Winet (1955). Most of these references deal at least in part with the axiomatic development of utility and/or subjective probability; for instance, the Pratt, Raiffa and Schlaifer article presents a different axiomatization from that given in Section 5.7. The axioms of utility and subjective probability are controversial; for instance, see Becker and Brownson (1964); Ellsberg (1961); Fellner (1961); Raiffa (1961); Roberts (1963*b*); Smith (1969); and Toda and Shuford (1965) (these references relate to Exercise 31, Chapter 2). Another example is the "Allais paradox," which is discussed in Allais (1953), Borch (1968*a*), and Morrison (1967); this situation is presented in Exercise 30 of this chapter.

Utility theory has also been investigated extensively from an experimental viewpoint, and four extensive review articles and bibliographies in this area are Edwards (1954, 1961, 1969) and Becker and McClintock (1967). Some experimentation has involved the axioms of utility; for instance, see MacCrimmon (1968), Marschak (1964), and Tversky (1969). Other experimentation has dealt primarily with the assessment of utility functions; some references are Becker, De Groot, and Marschak (1964); Davidson, Suppes, and Siegel (1957); De Groot (1963); Dolbear (1963); Mosteller and Nogee (1951); Nogee and Lieberman (1960); Preston and Baratta (1948); Royden, Suppes, and Walsh (1959); Stimson (1969); and Suppes and Walsh (1959).

Some other aspects of utility theory include risk aversion, which is discussed in Arrow (1965, 1971), Menezes and Hanson (1970), Meyer and Pratt (1968), Pratt (1964), and Zeckhauser and Keeler (1970); and cash equivalents, which are discussed in LaValle (1968). References concerning multidimensional utility are Fishburn (1967), Hausner (1954), Keeney (1968*a*, 1968*b*, 1969), and Raiffa (1969); and the time element in utility is discussed in Jamison (1969). Many of the preceding references include discussions of "shapes" of utility functions and the use of particular mathematical models to represent utility functions; for examples of specific references of this type, see Alderfer and Bierman (1970); Borch (1963, 1966); Friedman and Savage (1948, 1952); Hakansson (1970*a*); and Markowitz (1952). Finally, various aspects of group decision making are presented in Arrow (1951*b*), Kogan and Wallach (1964), and Wilson (1968*a*, 1968*b*).

In addition to the references given so far in this section, decision theory has been studied from viewpoints other than the statistical decision theory model presented in Section 5.9. Some decision-theoretic frameworks do not involve the use of subjective probabilities; historical references at the advanced level are Blackwell and Girshick (1954) and Wald (1950) [also, see Savage (1951)], and a more recent book in the same vein is Ferguson (1967). References at an elementary to intermediate level include Chernoff and Moses (1959) and Weiss (1961). A closely related topic is game theory, which is developed in Luce and Raiffa (1957) and Williams (1954) [also, see Harsanyi (1967, 1968*a*, 1968*b*)]. Another closely related area that includes models for decision making under certainty as well as probabilistic models is called operations research. For general discussions of operations research, see Hillier and Lieberman (1967), Horowitz (1972), and Wagner (1969). An example of a technique that comes under the heading of operations research is linear programming, which is developed in Dantzig (1963), Hadley (1962), and Spivey and Thrall (1970).

EXERCISES

1. Explain the difference between decision making under certainty and decision making under uncertainty.

2. You are given the following payoff table.

		STATE OF THE WORLD				
		A	*B*	*C*	*D*	*E*
ACTION	*1*	−50	80	20	100	0
	2	30	40	70	20	50
	3	10	30	−30	10	40
	4	−10	−50	−70	−20	200

(a) Are any of the actions inadmissible? If so, eliminate them from further consideration.

(b) Find the loss table corresponding to the above payoff table.

3. Consider the following *loss* table.

		STATE OF THE WORLD		
		I	*II*	*III*
ACTION	*1*	0	3	6
	2	1	1	0
	3	4	0	1

Given this loss table, complete the corresponding *payoff* table.

		STATE OF THE WORLD *I*	*II*	*III*
ACTION	*1*	12	7	9
	2			
	3			

Are any of the actions inadmissible?

4. The owner of a clothing store must decide how many men's shirts to order for the new season. For a particular type of shirt, he must order in quantities of 100 shirts. If he orders 100 shirts, his cost is \$10 per shirt; if he orders 200 shirts, his cost is \$9 per shirt; and if he orders 300 or more shirts, his cost is \$8.50 per shirt. His selling price for the shirt is \$12, but if any are left unsold at the end of the season, they will be sold in his famous "half-price, end-of-the-season sale." For the sake of simplicity, he is willing to assume that the demand for this type of shirt will be either 100, 150, 200, 250, or 300 shirts. Of course, he cannot sell more shirts than he stocks. He is also willing to assume that he will suffer no loss of goodwill among his customers if he understocks and the customers cannot buy all the shirts they want. Furthermore, he must place his order today for the entire season; he cannot wait to see how the demand is running for this type of shirt.
   (a) Given these details of the clothing-store owner's problem, list the set of actions available to him (including both admissible and inadmissible actions).
   (b) Construct a payoff table and eliminate any inadmissible actions.
   (c) Represent the problem in terms of a tree diagram.
   (d) If the owner decides that there is a loss of goodwill that is roughly equivalent to \$0.50 for each person who wants to buy the shirt but cannot because it is out of stock, make the appropriate changes in the payoff table. (Assume that the goodwill cost is \$0.50 *per shirt* if a customer wants to buy more than one shirt after the stock is exhausted.)

5. In Exercise 4(d), attempt to construct a loss table for the owner's problem *without* referring to the payoff table determined in that exercise. That is, using the concept of opportunity loss, try to express the consequences to the clothing-store owner in terms of losses. Then use the payoff table constructed in Exercise 4(d) to determine a loss table and compare the two loss tables, reconciling any differences.

6. A particular product is both manufactured and marketed by two different firms. The total demand for the product is virtually fixed, so neither firm

has advertised in the past. However, the owner of Firm A is considering an advertising campaign to woo customers away from Firm B. The ad campaign he has in mind will cost \$200,000. He is uncertain about the number of customers that will switch to his firm as a result of the advertising, but he is willing to assume that he will gain either 10 percent, 20 percent, or 30 percent of the market. For each 10 percent gain in market share, the firm's profits will increase by \$150,000. Construct the payoff table for this problem and find the corresponding loss table.

7. In Exercise 6, the owner of Firm A is worried that if he proceeds with the ad campaign, Firm B will do likewise, in which case the market shares of the two firms will remain constant. How does this affect the set of possible states of the world, S? Construct a modified payoff table to allow for this change.

8. For the payoff and loss tables in Exercise 2, find the actions that are optimal under the following decision-making criteria:
(a) maximin, (b) maximax, (c) minimax loss.

9. For the payoff and loss tables in Exercises 4(d) and 5, find the actions that are optimal under the maximin, maximax, and minimax loss rules.

10. For the payoff table in Exercise 6, find the actions that are optimal under the maximin and maximax rules. Comment on the implications of these particular rules in this example.

11. What is the primary disadvantage of decision-making rules such as maximin, maximax, and minimax loss?

12. One nonprobabilistic decision-making criterion not discussed in the text involves the consideration of a weighted average of the highest and lowest payoffs for each action. The weights, which must sum to 1, can be thought of as an optimism-pessimism index. The action with the highest weighted average of the highest and lowest payoffs is the action chosen by this criterion. Comment on this decision-making criterion and use it for the payoff table in Exercise 2, with the highest payoff in each row receiving a weight of .4 and the lowest payoff receiving a weight of .6.

13. Use the decision-making criterion discussed in Exercise 12 for the following payoff table, with the highest payoff in each row receiving a weight of .8 and the lowest payoff receiving a weight of .2.

		STATE OF THE WORLD	
		I	*II*
ACTION	*1*	10	4
	2	7	9

For this payoff table, the *ER* criterion would also involve a weighted average of the two payoffs in each row. Compare the criterion from Exercise 12 with the *ER* criterion.

14. In Exercise 2, find the expected payoff and the expected loss of each action and find the action that has the largest expected payoff and the smallest expected loss, given that the probabilities of the various states of the world are $P(A) = .10$, $P(B) = .20$, $P(C) = .25$, $P(D) = .10$, and $P(E) = .35$.

15. In Exercise 3, if $P(\text{I}) = .25$, $P(\text{II}) = .45$, and $P(\text{III}) = .30$, find the expected payoff and the expected loss of each of the three actions. Using *ER* (or *EL*) as a decision-making criterion, which action is optimal?

16. In Exercise 4(d), suppose that the owner decides that in comparing possible demands of 200 and 100, the demand is twice as likely to be 200 as it is to be 100. Similarly, 200 is three times as likely as 300, one and one-half times as likely as 150, and one and one-half times as likely as 250. Find the expected payoff of each action. How many shirts should the owner of the store order if he uses the *ER* criterion?

17. In Exercise 6, the owner of Firm A feels that if the ad campaign is initiated, the events "gain 20 percent of the market" and "gain 30 percent of the market" are equally likely and each of these events is three times as likely as the event "gain 10 percent of the market." Using the *ER* criterion, what should he do?

18. In Exercise 17, suppose that the probabilities given are conditional on the rival firm not advertising. If Firm B also advertises, then the owner of Firm A is certain (for all practical purposes) that there will be no change in the market share of either firm. He thinks that the chances are 2 in 3 that Firm B will advertise if Firm A does. What should the owner of Firm A do?

19. In Exercises 15, 16, and 17, determine the relationship between the expected payoffs of the various actions and the corresponding expected losses.

20. Explain how game-theory problems (that is, decision-making problems in which there is some "opponent") can be analyzed using the same techniques that are used in decision making under uncertainty. Instead of "states of the world" one must consider "actions of the opponent." Might this make it more difficult to determine the probabilities necessary to calculate expected payoffs and losses? Explain.

21. Consider a situation in which you and a friend both must choose a number from the two numbers 1 and 2. You must make your choices without communicating with each other. The relevant payoffs are as follows.

    If you choose 1 and he chooses 1, you both win \$10.
    If you choose 1 and he chooses 2, he wins \$15 and you win \$0.
    If you choose 2 and he chooses 1, you win \$15 and he wins \$0.
    If you choose 2 and he chooses 2, you both win \$5.

(a) How would you go about assigning probabilities to his two actions?
(b) On the basis of the probabilities assigned in (a), what is your optimal action?
(c) Could you have determined that this was your optimal action without using any probabilities? Why?
(d) Would you expect your action to be different if you could get together with your friend and make a bargain with him before the game is played? Explain.

22. A special type of decision-making problem frequently encountered in meteorology is called the "cost-loss" decision problem. The states of the world are "adverse weather" and "no adverse weather," and the actions are "protect against adverse weather" and "do not protect against adverse weather." C represents the cost of protecting against adverse weather, while L represents the cost that is incurred if you fail to protect against adverse weather and it turns out that the adverse weather occurs. (L is usually referred to as a "loss," but it is not a loss in an opportunity-loss sense.)
(a) Construct a payoff table and a decision tree for this decision-making problem.
(b) For what values of P(adverse weather) should you protect against adverse weather?
(c) Given the result in (b), is it necessary to know the absolute *magnitudes* of C and L?

23. A probabilistic decision-making criterion not discussed in the text is the maximum-likelihood criterion, which states that for each action, you find the most likely payoff, and then you choose the action with the largest most likely payoff. Use this criterion for the decision-making problems in Exercises 15, 16, and 17. Do you think that it is a good decision-making criterion? Explain.

24. In the automobile-salesman example presented in Section 3.4, the owner is particularly concerned with $\tilde{r}$, the number of cars the salesman will sell in the *next* six months (after he sells 10 cars in his first 24 days on the job). Hiring the salesman for that six-month period will cost the owner \$5000 in fixed salary (salary that will be paid to the salesman regardless of how many cars he sells), and the net profit to the owner is \$100 for each car that is sold. Assuming that the six-month period contains 130 working days for the salesman, should the owner hire him for this period?

25. In Exercise 24, suppose that the salesman is offered a job for the next six months and that he is given a choice of three salary plans: (1) a salary of \$5000 plus a commission of \$90 for each car that he sells, (2) a salary of \$8500 with no commission, or (3) a commission of \$200 for each car that he sells, with no fixed salary. At the time he is offered this choice, the salesman feels that the probability that he is a "great" salesman is .7, the probability that he is a "good" salesman is .29, and the probability that he is a "poor" salesman is .01. Which salary plan should he choose?

26. Suppose that a production manager is concerned about a particular run of five items from a certain manufacturing process. His prior distribution for $\tilde{p}$, the proportion of defective items produced by this process, is $P(\tilde{p} = .05) = .10$, $P(\tilde{p} = .10) = .20$, $P(\tilde{p} = .15) = .30$, $P(\tilde{p} = .20) = .30$, and $P(\tilde{p} = .25) = .10$. Furthermore, he has the option, for a cost of \$100, of adjusting the process and guaranteeing that none of the five items will be defective. The net profit per item is \$500, and it costs \$80 to repair a defective item.
    (a) Determine a predictive distribution for $\tilde{r}$, the number of defectives in the run of five items.
    (b) Set up a payoff table and a decision tree for this problem.
    (c) Should the production manager adjust the process?

27. What feature of Exercises 24, 25, and 26 distinguishes them from most of the decision-making problems considered in this chapter?

28. Why is the maximization of expected monetary value (that is, ER with the payoffs expressed in terms of money) not always a reasonable criterion for decision making? What problems does this create for the decision maker?

29. You must choose between three acts, where the payoff matrix is as follows (in terms of dollars).

		STATE OF THE WORLD		
		A	*B*	*C*
	1	100	70	20
ACTION	*2*	10	50	120
	3	50	80	30

What is the optimal act according to the expected utility criterion if $P(A) = .3$, $P(B) = .3$, $P(C) = .4$, and the utility function in the relevant range is (assuming M represents dollars):
    (a) $U(M) = 50 + 2M$,
    (b) $U(M) = 50 + 2M^2$,
    (c) $U(M) = M$,
    (d) $U(M) = M^2 + 5M + 6$.

30. Suppose that you are offered a choice between bets A and B.

Bet A: You win \$1,000,000 with certainty (that is, with probability 1).

Bet B: You win \$5,000,000 with probability .10.
You win \$1,000,000 with probability .89.
You win \$0 with probability .01.

Which bet would you choose? Similarly, choose between bets C and D.

Bet C: You win \$1,000,000 with probability .11.
You win \$0 with probability .89.

Bet D: You win \$5,000,000 with probability .10.
You win \$0 with probability .90.

Prove that if you chose bet A, then you should have chosen bet C, and if you chose bet B, then you should have chosen bet D. If you selected A and D or B and C, explain your choices. In light of the proof, would you change your choices?

31. Suppose that you are contemplating drilling an oil well, with the following payoff table (in terms of thousands of dollars).

		STATE OF THE WORLD	
		Oil	*No oil*
ACTION	*Drill*	100	−40
	Do not drill	0	0

If after consulting a geologist you decide that $P(\text{oil}) = .30$, find the optimal action according to the
(a) maximin criterion,
(b) maximax criterion,
(c) minimax loss criterion,
(d) *ER* criterion,
(e) *EL* criterion,
(f) *EU* criterion, where $U(0) = .40$, $U(100) = +1$, and $U(-40) = 0$.
Explain the differences among the results in parts (a) through (f).

32. Comment on the following statement: "Some individuals appear to be risk-takers for some decisions (such as gambling) and risk-avoiders for other decisions (such as purchasing insurance)." Can you explain why this phenomenon occurs? Can you draw a utility function that would explain it?

33. Attempt to determine your own utility function for money in the range from −\$500 to +\$500. If you were given actual decision-making situations, would you act in accordance with this utility function?

34. If someone gave you \$100,000 with "no strings attached," how would this affect your attitude toward risk? Assuming a gift of \$100,000, attempt to determine your utility function for money in the range from −\$500 to

+\$500. Comment on any differences between this utility function and the one assessed in Exercise 33.

35. Suppose that a person's utility function for *total* assets (*not* changes in assets) is

$$U(A) = 200A - A^2 \quad \text{for } 0 \leq A \leq 100,$$

where A represents total assets in thousands of dollars.

(a) Graph this utility function. How would you classify this person with regard to his attitude toward risk?

(b) If the person's total assets are currently \$10,000, should he take a bet in which he will win \$10,000 with probability .6 and lose \$10,000 with probability .4?

(c) If the person's total assets are currently \$90,000, should he take the bet given in (b)?

(d) Compare your answers to (b) and (c). Does the person's betting behavior seem reasonable to you? How could you intuitively explain such behavior?

36. For each of the following utility functions for changes in assets, graph the function and comment on the attitude toward risk that is implied by the function (M represents dollars). All of the functions are defined for $-1000 < M < 1000$.

(a) $U(M) = (M + 1000)^2$.
(b) $U(M) = -(1000 - M)^2$.
(c) $U(M) = 1000M + 2000$.
(d) $U(M) = \log (M + 1000)$.
(e) $U(M) = M^3$.
(f) $U(M) = 1 - e^{-M/100}$.

37. For each of the utility functions in Exercise 36, find out if the decision maker should take a bet in which he will win \$100 with probability p and lose \$50 with probability $(1 - p)$,

(a) if $p = \frac{1}{2}$,
(b) if $p = \frac{1}{3}$,
(c) if $p = \frac{1}{4}$.

38. Two persons, A and B, make the following bet: A wins \$40 if it rains tomorrow and B wins \$10 if it does not rain tomorrow.

(a) If they both agree that the probability of rain tomorrow is .10, what can you say about their utility functions?

(b) If they both agree that the probability of rain tomorrow is .30, what can you say about their utility functions?

(c) Given no information about their probabilities, is it possible that their utility functions could be identical? Explain.

(d) If they both agree that the probability of rain tomorrow is .20, is it possible that their utility functions could be identical? Explain.

39. Suppose that an investor is considering two investments. With each investment, he will double his money with probability .4 and lose half of his

money with probability .6. He has \$1000 to invest, and he can either put \$500 in each investment, put the entire \$1000 in one of the two investments, put \$500 in one investment and not invest the remaining \$500, or not invest at all. Assume that the outcomes of the two investments are independent and that the investor's utilities for *changes* in assets are $U(\$1000) = 1$, $U(\$500) = .85$, $U(\$250) = .75$, $U(\$0) = .5$, $U(-\$250) = .25$, and $U(-\$500) = 0$. What should the investor do? What commonly encountered financial concept does this illustrate?

40. In Exercise 16, suppose that $U(\$2000) = 1$, $U(-\$1000) = 0$, and the owner of the clothing store is indifferent between the following lotteries.

Lottery A: Receive \$1000 for certain.

Lottery B: Receive \$2000 with probability .87 and receive −\$1000 with probability .13.

What is $U(\$1000)$? By considering lotteries such as these, the owner assesses $U(-\$500) = .42$, $U(\$0) = .62$, $U(\$500) = .77$, and $U(\$1500) = .95$. Plot these points on a graph and attempt to fit a smooth curve to them. Using this curve, find the action that maximizes the owner's expected utility.

41. In Exercise 18, suppose that the owner of Firm A assesses $U(\$500{,}000) = 1$, $U(\$400{,}000) = .70$, $U(\$300{,}000) = .43$, $U(\$200{,}000) = .28$, $U(\$100{,}000) = .18$, $U(\$0) = .12$, $U(-\$100{,}000) = .06$, and $U(-\$200{,}000) = 0$. Graph these points, fit a smooth utility function to them, and use this utility function to find EU(ad campaign) and EU(no ad campaign).

42. For the utility function represented in Figure 5.6.5, the point (to the right of \$0) at which the utility function shifts from that of a risk-taker to that of a risk-avoider is sometimes called a "level of aspiration." Explain this terminology.

43. For each of the utility functions in Exercise 36, find the risk premiums for the following gambles.
    (a) You win \$100 with probability .5 and you lose \$100 with probability .5.
    (b) You win \$100 with probability .4 and you lose \$50 with probability .6.
    (c) You win \$70 with probability .3 and you lose \$30 with probability .7.
    (d) You win \$200 with probability .5 and you win \$50 with probability .5.

44. If your utility function for money is $U(M) = 40{,}000 - (200 - M)^2$ for $-200 \leq M \leq 200$, show the risk premium graphically for each of the gambles in Exercise 43.

45. One counterargument to the transitivity of indifference goes something like this. You are probably indifferent between a cup of black coffee and a cup of coffee with one grain of sugar. Similarly, you are indifferent between a cup of coffee with one grain of sugar and one with two grains of sugar. Therefore, if indifference is transitive, you should be indifferent between a

cup of black coffee and one with two grains of sugar. By adding a grain of sugar at a time, you can arrive at the conclusion that you should be indifferent between a cup of black coffee and one with a million grains of sugar. This seems to be an unreasonable conclusion; does this mean that transitivity of indifference is an unrealistic assumption? Discuss the issues raised by this example, both with regard to this specific assumption and with regard to the "axioms of coherence" in general.

46. How can the consideration of utility functions and subjective probabilities and the maximization of expected utility be justified formally in problems of decision making under uncertainty?

47. What is the multidimensional utility problem? Illustrate this problem with respect to a high-school senior trying to decide which college to attend.

48. How might a firm develop a utility function that would take into account profits *and* other factors such as pollution control? For a related problem, how might a governmental agency develop a utility function to decide among various potential courses of action?

49. Discuss the role of mathematical functions in providing convenient models for utility functions. Compare the linear, the exponential, and the logarithmic models presented in Section 5.8.

50. One complication that has been ignored in this chapter is the problem of how to handle cash flows over time. An investment may result in a sequence of payoffs rather than in a single payoff at a particular point in time. One approach to this problem is to choose an appropriate discount rate and to discount all future cash flows so that they can be expressed in present dollars rather than in future dollars. This approach obviously is a simplification, ignoring such factors as tax considerations, an individual's time preferences for payoffs, and so on. Discuss this problem and its relation to utility theory.

51. In Exercise 17, suppose that the owner of Firm A is uncertain about some of the elements in his payoff table. In particular, the ad campaign will cost either \$150,000, \$200,000, or \$250,000, with $P(\$150{,}000) = .2$, $P(\$200{,}000) = .4$, and $P(\$250{,}000) = .4$. In addition, the increase in profits for each 10 percent gain in market share will be either \$100,000, \$140,000, or \$180,000, with $P(\$100{,}000) = .3$, $P(\$140{,}000) = .5$, and $P(\$180{,}000) = .2$.
   (a) Express the problem in terms of a decision tree, including branches for the uncertainties about the payoffs.
   (b) What should the owner of Firm A do?

52. Suppose that one person (A) claims that $U(\$100) = .8$ and that $U(\$50) = .3$, whereas a second person (B) claims that $U(\$100) = .8$ and that $U(\$50) = .6$. What, if anything, can you say about the relative preferences of A and B? In general, is it meaningful to make interpersonal comparisons of utility functions? Explain.

53. Suppose that you could obtain an interest-free \$1000 loan with the restriction that it has to be invested in stocks, bonds, or savings accounts (or a combination of these types of investments) and that the loan has to be repaid in exactly one year. How could you set up this problem in terms of the framework presented in Section 5.9? First of all, how would you reduce the number of potential actions to a manageable number; second, how would you define S, the set of states of the world; third, how would you determine the possible payoffs or losses; fourth, how could you introduce a utility function; fifth, how could you quantify your judgments to assess probabilities for the various states of nature?

54. The problem in Exercise 53 is quite complex and must be simplified (for example, by eliminating actions) if it is to be expressed in terms of the formal decision-theoretic model. For an even more ill-structured problem, consider the high-school senior who must decide among numerous colleges, jobs, military service, and so on. Take a hypothetical high-school senior and try to express his problem in terms of the formal decision-theoretic framework. Which inputs might be most difficult to determine in this situation?

55. In the oil-drilling venture presented in Section 5.10, suppose that the decision maker finds three other persons who would be willing to take shares in the venture, thereby leaving the decision maker with a 25 percent share. Compute the appropriate payoffs and determine the expected utility of this new action, and compare this expected utility with the EU's computed in Section 5.10 for the other actions.

56. In Exercise 16, suppose that one of the employees in the clothing store assesses the following probabilities for the demand for shirts: $P(50) = .05$, $P(100) = .15$, $P(150) = .25$, $P(200) = .35$, $P(250) = .10$, $P(300) = .05$, and $P(350) = .05$. Note that two additional values for demand, 50 and 350, are being considered, so the payoff table must be expanded. Using the expanded payoff table and the employee's probabilities, which action has the largest expected payoff? Comment on the implications of this result with regard to the sensitivity of the shirt-ordering problem to changes in the set of states of the world, S, or to changes in the probabilities.

57. Discuss the importance of sensitivity analysis in problems of decision making under uncertainty. When might the results of a sensitivity analysis greatly simplify the decision maker's problem? When might the results of a sensitivity analysis make the decision maker's problem extremely difficult?

58. In the oil-drilling example presented in Section 5.10, the EU of the optimal action is approximately equal to $U(\$15{,}000)$. Thus, if the decision maker contemplates selling the drilling rights, the minimum selling price should be \$15,000. If the decision maker does not own the drilling rights, is it necessarily true that his maximum buying price for the rights should be \$15,000? In general, are the buying and selling prices determined from a given individual's utility function for, say, a lottery, always equal? Explain.

6

THE VALUE OF INFORMATION

6.1 TERMINAL DECISIONS AND PREPOSTERIOR DECISIONS

It was assumed in the previous chapter that a decision maker is faced with a problem of decision making under uncertainty and that he must make a decision on the basis of his current state of information. Such a decision is called a *terminal decision.* If the decision maker's current state of information is represented by a prior or posterior distribution, then his terminal decision will be based on this distribution (and, of course, on a payoff function or a loss function, which may be expressed in terms of money if the decision maker's utility function for money is linear, but which should be expressed in terms of utility otherwise).

Suppose that the decision maker has the option of obtaining more sample information before he makes his terminal decision. Such sample information might be valuable in reducing his uncertainty about the state of the world. For example, in a medical situation, it may be possible to run further diagnostic tests before making a decision regarding surgery; in oil drilling, the decision maker might run geological tests before deciding whether or not to drill; in an investment decision, the investor might be able to obtain more information about the investments under consideration (or about some related factor, such as the chances that taxes will be raised); in a marketing decision, it may be possible to obtain more information about consumers' intentions by conducting a market survey or to gain more information about the future actions of competitors by carefully observing their current actions (building new plants, initiating advertising campaigns, and so on). In these and many other examples, additional sample information could be quite useful to the decision maker.

Of course, there is a cost involved in sampling, so the decision maker must decide if the additional sample information is expected to be useful enough to justify its cost. This type of decision is called a *preposterior decision* because it involves the *potential* posterior distributions following the *proposed* sample. Note that this sample has not been observed yet; it is just being contemplated. The value of the sample to the decision maker usually depends on the observed result. Before taking the sample, he does not know what this result will be, although he *can* make some probabilistic statements about possible results. Probability distributions for future sample outcomes, which are called predictive distributions, were discussed in Sections 3.7 and 4.11.

The term "preposterior decision" can refer to such things as sample design, which can become quite complicated. Only decisions concerning sample size are considered here. It is assumed that the decision maker is contemplating a sample of some fixed sample design and wants to know the optimal sample size. Preposterior analysis can be used to determine this optimal sample size. It is possible for the optimal sample size to be zero, in which case the decision maker should not obtain any more sample information; he should make his terminal decision on the basis of his current state of information. Of course, preposterior analysis can be carried out repeatedly; after each sample, preposterior analysis can be used to determine an optimal sample size for the next sample. This process continues until the optimal sample size is zero, at which time the decision maker should stop sampling and should make his terminal decision. Alternatively, the decision maker could pause after each trial and decide whether to continue sampling or to stop sampling. This decision involves the consideration of an entire stream of potential sample results, together with the possibility of stopping to make a terminal decision after any trial. This procedure is called *sequential analysis* or *sequential decision making* because of the sequential nature of the procedure.

Before the problem of determining optimal sample size is attacked, a related, but somewhat easier, problem is investigated. Suppose that the decision maker has the opportunity to purchase "perfect" information (that is, perfect knowledge of the state of the world), so that his problem of decision under uncertainty could be changed into the easier problem of decision under certainty. How much should he be willing to pay for this perfect information?

6.2 THE VALUE OF PERFECT INFORMATION

The term "information" has been used in discussing the state of uncertainty facing the decision maker. Generally, additional information may reduce his uncertainty about the state of the world. In the extreme case, if he is able to get "perfect" information, then the problem of decision making under uncertainty becomes a problem of decision making under certainty. Decision making is simpler under certainty, as was pointed out in Section 5.1.

In terms of the notation presented in Section 5.9, suppose that the decision maker knows for certain that the value of the state of the world is $\tilde{\theta} = \theta$. Being certain that $\tilde{\theta} = \theta$ implies that

$$P(\tilde{\theta}) = \begin{cases} 1 & \text{if } \tilde{\theta} = \theta, \\ 0 & \text{if } \tilde{\theta} \neq \theta. \end{cases}$$

Taking expectations with regard to this distribution,

$$EU(a) = U(a,\theta),$$

since all of the terms in the summation $\Sigma U(a,\tilde{\theta})P(\tilde{\theta})$ are zero except the term $U(a,\theta)P(\theta)$, which is equal to $U(a,\theta)$. Thus, if it is known for certain that $\tilde{\theta} = \theta$, the optimal act a_θ is the act such that

$$U(a_\theta,\theta) \geq U(a,\theta) \quad \text{for all acts } a \text{ in } A. \tag{6.2.1}$$

If the utility function is linear with respect to money, then Equation (6.2.1) can be expressed in terms of payoffs or losses:

$$R(a_\theta,\theta) \geq R(a,\theta) \quad \text{for all acts } a \text{ in } A, \tag{6.2.2}$$
$$L(a_\theta,\theta) \leq L(a,\theta) \quad \text{for all acts } a \text{ in } A. \tag{6.2.3}$$

If the payoffs or losses in a particular problem are expressed in the form of a payoff or loss table, then perfect information enables the decision maker to neglect all of the columns of the table except one. In the umbrella example in Section 5.2, the payoff table was as follows.

		STATE OF THE WORLD *Rain*	*No rain*
ACTION	*Carry umbrella*	−1	−1
	Do not carry umbrella	−60	0

If you know for certain that it will rain, you will look only at the first column, and your decision should be to carry the umbrella. If you know for certain that it will not rain, then only the second column is relevant, in which case you should not carry the umbrella. Letting $\tilde{\theta} = 1$ if it rains and $\tilde{\theta} = 0$ if it does not rain, then a_θ = "carry the umbrella" if $\theta = 1$

and a_θ = "do not carry the umbrella" if $\theta = 0$. Clearly, the optimal act under perfect information, a_θ, depends on the particular value of θ.

In the advertising example in Section 5.2, the payoff table was as follows.

		STATE OF THE WORLD	
		Competitor introduces new product	*Competitor does not introduce new product*
	No advertising	100,000	700,000
ACTION	*Minor ad campaign*	300,000	600,000
	Major ad campaign	400,000	500,000

Letting $\tilde{\theta} = 1$ if the new product is introduced and $\tilde{\theta} = 0$ if the new product is not introduced, then a_θ = "major ad campaign" if $\theta = 1$ and a_θ = "no advertising" if $\theta = 0$.

For an example using a loss table, consider the stock-purchasing example (Section 5.3), with the following loss table.

		STATE OF THE WORLD			
		I	II	III	IV
	Buy A	0	0	600	0
ACTION	*Buy B*	200	400	200	0
	Buy C	400	1000	0	0

If you know that state I will occur, then you will just consider the first column. The action "buy A" has the smallest loss in this column, so this is the optimal action. Similarly, the optimal action under perfect knowledge indicating that state II (state III) will occur is to buy A (buy C). If you are sure that state IV will occur, then it makes no difference which action you select; the losses are the same. Letting $\tilde{\theta} = 1, 2, 3$, and 4 for states I, II, III, and IV, respectively, $a_\theta =$ "buy A" if $\theta = 1$ or $\theta = 2$, $a_\theta =$ "buy C" if $\theta = 3$, and any of the three actions may be taken as a_θ if $\theta = 4$.

Note from the above example that because of the way "losses" were defined in Section 5.2 (in terms of "opportunity loss"), the optimal act under certainty will always have a loss equal to zero. In every column of a loss table there is at least one zero, and there are never any negative losses. Therefore, if you have perfect information, your opportunity loss is zero. The perfect information must have some value to you, then, because under uncertainty (which can be thought of as "imperfect" information), the optimal act has some expected loss that is generally different from zero (although it is smaller than the expected loss of any other action).

It would be useful to find out exactly how much perfect information is worth to the decision maker. Under perfect information the loss is zero, as noted above. If a^* is optimal under the decision maker's current state of information, then

$$EL(a^*) \leq EL(a) \quad \text{for all actions } a \text{ in } A.$$

If the decision maker acts on the basis of his current information, he will choose act a^*, and his expected loss will be equal to $EL(a^*)$. Suppose that he contemplates purchasing perfect information. Under perfect information, his loss will be zero, so the *expected value of perfect information* (EVPI) must be equal to

$$\text{EVPI} = EL(a^*) - 0 = EL(a^*). \tag{6.2.4}$$

In words, the expected value of perfect information to the decision maker is equal to the expected loss of the action that is optimal under his current state of information. If the decision maker's utility function is not linear with respect to money, then it is assumed that the losses (and hence EVPI) are given in terms of utility.

In the stock-purchasing example, suppose that the probabilities of the four states of the world are .3, .2, .1, and .4. The expected losses of the three actions are then

$$EL(\text{buy A}) = 0(.3) + 0(.2) + 600(.1) + 0(.4) = 60,$$
$$EL(\text{buy B}) = 200(.3) + 400(.2) + 200(.1) + 0(.4) = 160,$$
and
$$EL(\text{buy C}) = 400(.3) + 1000(.2) + 0(.1) + 0(.4) = 320.$$

The optimal action, buy A, has an expected loss of \$60, so this is the EVPI. The decision maker should be willing to pay up to \$60 to obtain perfect information, since the information is worth \$60 to him.

The loss function $L(a,\theta)$ is used above to determine EVPI. It is also possible to think of EVPI in terms of the payoff function. If a^* is optimal under the decision maker's current state of information, then

$$ER(a^*) \geq ER(a) \quad \text{for all actions } a \text{ in } A.$$

Furthermore, suppose that the action a_θ is optimal under the perfect information that $\tilde{\theta} = \theta$. Of what value is this perfect information to the decision maker? If he takes the action a^* and if the state of the world $\tilde{\theta} = \theta$ occurs, his payoff is $R(a^*,\theta)$. With knowledge of the true value of $\tilde{\theta}$, he takes the action a_θ, so his payoff is $R(a_\theta,\theta)$. The value of the information is simply the difference between the two payoffs:

$$\text{VPI}(\theta) = R(a_\theta,\theta) - R(a^*,\theta). \tag{6.2.5}$$

This term, the *value of perfect information* that $\tilde{\theta} = \theta$, is of course dependent upon the distribution $P(\theta)$ [or $f(\theta)$] in the sense that it depends upon the optimal act under this distribution, which is a^*. Furthermore, $\text{VPI}(\theta)$ depends on the particular value of θ.

In the umbrella example, suppose that under your distribution of $\tilde{\theta}$, $P(\text{rain}) = .2$, so that the optimal act is to carry the umbrella. Then, if you obtain perfect knowledge that it will rain, the optimal act is still to carry the umbrella, so

$$\text{VPI(rain)} = R(\text{carry umbrella, rain}) - R(\text{carry umbrella, rain}) = 0.$$

If you obtain perfect knowledge that it will *not* rain, your optimal action is to leave the umbrella at home, and

$$\begin{aligned}\text{VPI(no rain)} &= R(\text{no umbrella, no rain}) - R(\text{umbrella, no rain}) \\ &= 0 - (-1) = 1.\end{aligned}$$

The value of perfect information clearly depends in this case on whether the perfect information is that there will be rain or whether it is that there will not be rain. How can the decision maker tell how much perfect information is worth to him? He cannot tell until he actually receives the perfect information. But then how can he determine how much he would be willing to pay to receive perfect information? The answer is simply that he should take the expectation of VPI with respect to his distribution of $\tilde{\theta}$. In the umbrella example, $P(\text{rain}) = .20$ and $P(\text{no rain}) = .80$. Thus, the probability is .20 that it will rain and the VPI will be 0, and the probability is .80 that it will not rain and the VPI will be 1. The *expected* value of perfect information is therefore

$$\begin{aligned}\text{EVPI} &= \text{VPI(rain)}P(\text{rain}) + \text{VPI(no rain)}P(\text{no rain}) \\ &= 0(.2) + 1(.8) = 0.80.\end{aligned}$$

You should be willing to pay up to \$0.80 to find out for sure whether or not it will rain.

In this manner, the *expected value of perfect information*, or *EVPI*, can be computed:

$$\text{EVPI} = \Sigma\ \text{VPI}(\theta)P(\theta) \tag{6.2.6}$$

or

$$\text{EVPI} = \int \text{VPI}(\theta)f(\theta)d\theta, \tag{6.2.7}$$

depending on whether $\tilde{\theta}$ is discrete or continuous. Notice that EVPI can never be negative because both VPI(θ) and $P(\theta)$ [or $f(\theta)$] can never be negative. By definition, since a_θ is the optimal act if $\tilde{\theta} = \theta$,

$$R(a_\theta,\theta) \geq R(a^*,\theta).$$

Therefore, from Equation (6.2.5), VPI is always greater than or equal to zero. Incidentally, Equation (6.2.4) is equivalent to Equations (6.2.6) and (6.2.7), since VPI(θ) is simply $L(a^*,\theta)$ [compare Equations (5.2.1) and (6.2.5)].

In the advertising example, suppose that the odds are 3 to 2 in favor of the new product being introduced by the competitor. Thus, P(new product is introduced) $= 3/(3 + 2) = .6$ and P(new product is not introduced) $= .4$. The expected payoffs are

$$\begin{aligned} ER(\text{no advertising}) &= \$340{,}000, \\ ER(\text{minor ad campaign}) &= \$420{,}000, \\ \text{and} \qquad ER(\text{major ad campaign}) &= \$440{,}000. \end{aligned}$$

Thus, a^* = "major ad campaign." The VPI's are

$$\text{VPI(new product is introduced)} = \$400{,}000 - \$400{,}000 = \$0$$

and

$$\text{VPI(new product is not introduced)} = \$700{,}000 - \$500{,}000 = \$200{,}000.$$

Thus,

$$\text{EVPI} = \$0(.6) + \$200{,}000(.4) = \$80{,}000.$$

The decision maker should be willing to pay up to $80,000 for perfect information about the competitor's decision regarding the new product.

A final method for calculating EVPI is as follows. Under the decision maker's current state of uncertainty, his optimal action is a^* and his expected payoff is $ER(a^*)$. With perfect information, his expected payoff (as calculated before he obtains the perfect information) is

$$\text{ERPI} = \sum_\theta R(a_\theta,\theta)P(\theta). \tag{6.2.8}$$

Note that $R(a_\theta,\theta)$ is simply the largest payoff in the column corresponding to the state of the world θ in the payoff table. Thus, the "expected payoff under perfect information," as given by Equation (6.2.8), is equal to T, as given by Equation (5.4.1). But from Equation (5.4.2), then,

$$EL(a) = \text{ERPI} - ER(a)$$

for any action a. Therefore, since EVPI $= EL(a^*)$,

$$\text{EVPI} = \text{ERPI} - ER(a^*); \tag{6.2.9}$$

the expected value of perfect information equals the difference between the expected payoff under perfect information and the expected payoff under the decision maker's current state of uncertainty.

For the umbrella example,

$$\begin{aligned} \text{ERPI} &= R(\text{carry umbrella, rain})P(\text{rain}) \\ &\qquad + R(\text{do not carry umbrella, no rain})\ P(\text{no rain}) \\ &= (-1)(.2) + (0)(.8) = -.2 \\ \text{and EVPI} &= \text{ERPI} - ER(\text{carry umbrella}) = -.2 - (-1) = .80. \end{aligned}$$

For the advertising example,

$$\begin{aligned} \text{ERPI} &= R(\text{major ad campaign, new product is introduced})P(\text{new product is introduced}) + R(\text{no advertising, new product is not introduced})P(\text{new product is not introduced}) \\ &= \$400{,}000(.6) + \$700{,}000(.4) = \$520{,}000 \\ \text{and EVPI} &= \text{ERPI} - ER(\text{major ad campaign}) \\ &= \$520{,}000 - \$440{,}000 = \$80{,}000. \end{aligned}$$

For another illustration of the calculation of the expected value of perfect information, consider the quality-control example from Section 3.3 once again. There are four states of the world, corresponding to four values of $\tilde{p}$, the proportion of defectives produced by the process: .01, .05, .10, and .25. The production manager has three actions from which to choose: he can have a mechanic make a major adjustment in the production process, he can have a minor adjustment made, or he can leave the process as is. The payoff table is as follows.

		STATE OF THE WORLD (VALUE OF $\tilde{p}$)			
		.01	*.05*	*.10*	*.25*
	Major adjustment	380	380	380*	380*
ACTION	*Minor adjustment*	455	375	375	375
	No adjustment	480*	400*	300	0

The largest entry in each column is starred to show the optimal action for that particular value of $\tilde{p}$.

The production manager's prior probabilities are

$$P(\tilde{p} = .01) = .60,$$
$$P(\tilde{p} = .05) = .30,$$
$$P(\tilde{p} = .10) = .08,$$
and
$$P(\tilde{p} = .25) = .02.$$

Under this prior distribution the expected payoffs of the three actions are

$$ER(\text{major adjustment}) = 380,$$
$$ER(\text{minor adjustment}) = 423,$$
and
$$ER(\text{no adjustment}) = 432.$$

The optimal act is thus "no adjustment," and the value of perfect information can be calculated for each possible value of $\tilde{p}$:

$$\text{VPI}(\tilde{p} = .01) = 480 - 480 = 0,$$
$$\text{VPI}(\tilde{p} = .05) = 400 - 400 = 0,$$
$$\text{VPI}(\tilde{p} = .10) = 380 - 300 = 80,$$
and
$$\text{VPI}(\tilde{p} = .25) = 380 - 0 = 380.$$

The expected value of perfect information is

$$\text{EVPI} = \Sigma\, \text{VPI}(p)P(p) = 0(.6) + 0(.3) + 80(.08) + 380(.02) = 14.$$

The loss table corresponding to the above payoff table is as follows.

		STATE OF THE WORLD (VALUE OF $\tilde{p}$)			
		.01	*.05*	*.10*	*.25*
	Major adjustment	100	20	0	0
ACTION	*Minor adjustment*	25	25	5	5
	No adjustment	0	0	80	380

Since the third action, no adjustment, is optimal under the prior distribution,

$$\text{EVPI} = EL(\text{no adjustment}) = 0(.6) + 0(.3) + 80(.08) + 380(.02) = 14,$$

which is equal to the result obtained from the payoff table.

The production manager takes a sample of size five, observing one defective item. By using Bayes' theorem, the following posterior probabilities can be calculated (see Section 3.3 for the calculations):

$$P(\tilde{p} = .01) = .232,$$
$$P(\tilde{p} = .05) = .492,$$
$$P(\tilde{p} = .10) = .212,$$
and
$$P(\tilde{p} = .25) = .064.$$

The expected losses are now

$$EL(\text{major adjustment}) = 33.04,$$
$$EL(\text{minor adjustment}) = 19.48,$$
and
$$EL(\text{no adjustment}) = 41.28.$$

The optimal action is to make a minor adjustment, and the expected value of perfect information is

$$\text{EVPI} = EL(\text{minor adjustment}) = 19.48.$$

As a result of seeing the sample, the decision maker is now willing to pay up to \$19.48 for perfect information, whereas before the sample he was only willing to pay \$14. How can this be? The sample information makes the higher values of $\tilde{p}$ appear more likely and the lower values of $\tilde{p}$ appear less likely. In this example, the posterior distribution has a larger variance than the prior distribution. The variance of the prior distribution is .0011, and the variance of the posterior distribution is .0033. This demonstrates that additional sample information will not always reduce the expected value of perfect information, although it often *will* reduce the EVPI, especially if the sample size is large.

Suppose that the production manager is concerned about this first sample result and that he takes another sample of size five, in which he observes two defective items. The new posterior distribution is

$$P(\tilde{p} = .01) = .005,$$
$$P(\tilde{p} = .05) = .245,$$
$$P(\tilde{p} = .10) = .359,$$
and
$$P(\tilde{p} = .25) = .391,$$

and the expected losses under this distribution are

$$EL(\text{major adjustment}) = 5.4,$$
$$EL(\text{minor adjustment}) = 10.0,$$
and
$$EL(\text{no adjustment}) = 177.3.$$

The optimal act following the second sample is to make a major adjustment, and the expected value of perfect information is now only \$5.40.

As new sample information is obtained, therefore, the expected value of perfect information may increase or decrease. At any point in time, if the decision maker can obtain perfect information at a cost that is smaller than his EVPI, then he should take advantage of this opportunity. In real-world decision-making situations, however, the decision maker seldom has the opportunity to obtain perfect information. No one knows for sure what the price of a stock will be one year from now, or, for that matter, what it will be one week from now. Similarly, it is doubtful that the production manager in the example would be able to find out *exactly* the true value of $\tilde{p}$. He might be able to obtain more sample information, however, and the value of sample information is now considered.

6.3 THE VALUE OF SAMPLE INFORMATION

From the expected value of perfect information (EVPI), the decision maker knows how much perfect information is worth to him. In other words, he knows how much he should be willing to pay for perfect information. Unfortunately, perfect information is seldom available, and the decision maker must take a sample if he wants more information. A sample generally provides some information, but not perfect information, about the uncertain quantity of interest. Since sampling involves some cost, it would be helpful if the decision maker could determine the worth to him of sample information so that he could decide whether or not to take a sample. In other words, he would like to determine the *expected value of sample information.* Since sample information can never be any better than perfect information, the expected value of sample information (EVSI) will always be less than or equal to the expected value of perfect information. Furthermore, as the sample size increases, EVSI approaches EVPI. Thus, although it is generally not possible to purchase perfect information, EVPI is still useful as an upper bound for the EVSI. In this section, the determination of the exact value of EVSI is discussed.

Suppose that the decision maker assesses a prior distribution $P'(\theta)$ or $f'(\theta)$ and that the optimal action under the prior distribution is a':

$$E'U(a') \geq E'U(a) \quad \text{for all acts } a \text{ in } A, \tag{6.3.1}$$

where the prime in $E'U$ indicates that the expectation is taken with regard to the prior distribution. If the decision maker's utility function for money is linear, this is equivalent to

$$E'R(a') \geq E'R(a) \quad \text{for all acts } a \text{ in } A \tag{6.3.2}$$

or

$$E'L(a') \leq E'L(a) \quad \text{for all acts } a \text{ in } A. \tag{6.3.3}$$

He then contemplates taking a sample. What would be the value to him of taking a sample and observing the sample result y? He could use this sample result to revise his prior distribution and arrive at a posterior distribution, $P''(\theta \mid y)$ or $f''(\theta \mid y)$. Denote the optimal act under this posterior distribution by a'', implying that

$$E''U(a'' \mid y) \geq E''U(a \mid y) \quad \text{for all acts } a \text{ in } A, \tag{6.3.4}$$

which, if utility is linear in money, reduces to

$$E''R(a'' \mid y) \geq E''R(a \mid y) \quad \text{for all } a \text{ in } A \tag{6.3.5}$$

or

$$E''L(a'' \mid y) \leq E''L(a \mid y) \quad \text{for all } a \text{ in } A. \tag{6.3.6}$$

The double primes in $E''U$, $E''R$, and $E''L$ indicate that the expectations are taken with regard to the posterior distribution.

Under the posterior distribution the optimal act is a'', whereas under the prior distribution it is a'. The value of this particular sample information y to the decision maker is simply

$$\text{VSI}(y) = E''U(a'' \mid y) - E''U(a' \mid y). \tag{6.3.7}$$

In terms of payoffs or losses (which are applicable if utility is linear in money),

$$\text{VSI}(y) = E''R(a'' \mid y) - E''R(a' \mid y) \tag{6.3.8}$$

or

$$\text{VSI}(y) = E''L(a' \mid y) - E''L(a'' \mid y). \tag{6.3.9}$$

In words, this is the posterior expected payoff of the now-optimal act a'' minus the posterior expected payoff of the previously-optimal act a'. From Section 5.4, $ER(a_1) - ER(a_2) = EL(a_2) - EL(a_1)$ for any two actions a_1 and a_2, so Equations (6.3.8) and (6.3.9) are equivalent. It should be emphasized that VSI depends on the particular sample result y.

For example, consider the following experiment. Suppose that there are two decks of cards. Deck A consists of 26 red cards and 26 green cards; deck B consists of 39 red cards and 13 green cards. A deck is chosen randomly from the two, and you are asked to guess which deck it is, subject to the following payoff table.

		STATE OF THE WORLD: *Deck A*	*Deck B*
ACTION (YOUR GUESS)	*Deck A*	4	2
	Deck B	1	6

Since the only information you have is that the deck has been chosen randomly, your prior probabilities should be

$$P'(\text{deck A}) = P'(\text{deck B}) = \tfrac{1}{2}.$$

Your expected payoffs are thus

$$E'R(\text{guess deck A}) = 4(\tfrac{1}{2}) + 2(\tfrac{1}{2}) = 3$$

and

$$E'R(\text{guess deck B}) = 1(\tfrac{1}{2}) + 6(\tfrac{1}{2}) = 3.5,$$

so that the optimal act is to guess deck B (that is, $a' =$ "guess deck B"). The expected loss of this action is $\frac{3}{2}$, so if the payoffs are expressed in terms of dollars, you should be willing to pay up to \$1.50 for perfect information.

Suppose that you can purchase sample information in the form of cards drawn randomly (with replacement) from the deck of cards that has been chosen. Each card observed (that is, each trial) will cost you \$0.20. Should you choose deck B immediately, or should you pay \$0.20 to see a single trial? The two possible sample outcomes are "red" and "green," and for each of these outcomes the posterior probabilities can be computed.

Sample outcome	*State of the world*	*Prior probability*	*Likelihood*	*Prior probability × likelihood*	*Posterior probability*
Red card	Deck A	.5	.50	.250	.40
	Deck B	.5	.75	.375	.60
				.625	
Green card	Deck A	.5	.50	.250	.67
	Deck B	.5	.25	.125	.33
				.375	

If the sample results in a red card, the posterior probabilities are .40 and .60, and the expected payoffs are

$$E''R(\text{guess deck A} \mid \text{red}) = 4(.4) + 2(.6) = 2.8$$

and

$$E''R(\text{guess deck B} \mid \text{red}) = 1(.4) + 6(.6) = 4.0.$$

The optimal act is still to guess deck B (that is, $a'' =$ "guess deck B"), from Equation (6.3.8), so the value of sample information if you see a red card is

$$\text{VSI(red card)} = E''R(\text{guess B} \mid \text{red}) - E''R(\text{guess B} \mid \text{red}) = 0.$$

If the sample outcome is a green card, the posterior probabilities are .67 and .33, and the expected payoffs are

$$E''R(\text{guess deck A} \mid \text{green}) = 4(.67) + 2(.33) = 3.33$$

and

$$E''R(\text{guess deck B} \mid \text{green}) = 1(.67) + 6(.33) = 2.67.$$

The optimal act in this case is to guess deck A (a'' = "guess deck A"), whereas before the sample guessing B was optimal, so from Equation (6.3.8),

$$\begin{aligned} \text{VSI(green card)} &= E''R(\text{guess A} \mid \text{green}) - E''R(\text{guess B} \mid \text{green}) \\ &= 3.33 - 2.67 = .66. \end{aligned}$$

The *value of sample information* (*VSI*) depends on the specific sample result y, just as the value of perfect information (VPI) depends on the specific value θ. You must decide whether or not to purchase sample information in advance of the actual sample, so that the values for VSI, taken alone, are of little help. What you need is the *expected value of sample information,* or *EVSI*. In this example, the EVSI is equal to

$$\text{EVSI} = \text{VSI(red card)}P(\text{red card}) + \text{VSI(green card)}P(\text{green card}).$$

But what are the probabilities P(red card) and P(green card)? They are predictive probabilities, and they can be calculated as follows (see Section 3.7):

$$\begin{aligned} P(\text{red}) &= P(\text{red} \mid \text{deck A})P(\text{deck A}) + P(\text{red} \mid \text{deck B})P(\text{deck B}) \\ &= .50(.5) + .75(.5) = .625 \end{aligned}$$

and

$$\begin{aligned} P(\text{green}) &= P(\text{green} \mid \text{deck A})P(\text{deck A}) + P(\text{green} \mid \text{deck B})P(\text{deck B}) \\ &= 50(.5) + .25(.5) = .375. \end{aligned}$$

Actually, these probabilities had already been calculated when the posterior probabilities were determined. Each is equal to a sum in the denominator of Bayes' theorem, or the sum of the column headed "prior probability × likelihood" in the appropriate part of the table used to illustrate the calculation of posterior probabilities for this example. Note that these sums are .625 and .375, respectively. Recall, from Section 3.7, that a predictive probability is the sum of a number of products of prior probabilities and likelihoods.

You can now determine the expected value of sample information corresponding to a potential sample of size one:

$$\begin{aligned} \text{EVSI} &= \text{VSI(red card)}P(\text{red card}) + \text{VSI(green card)}P(\text{green card}) \\ &= 0(.625) + .66(.375) \\ &= .25. \end{aligned}$$

The cost of taking the sample is \$0.20, and thus your expected net gain from the sample is \$0.25 − \$0.20 = \$0.05. Since this is greater than zero, you should take the sample. In general, the *expected net gain of sampling* (*ENGS*) is defined as the difference between the *expected value of sample information* (*EVSI*) and the *cost of sampling* (*CS*):

$$\text{ENGS} = \text{EVSI} - \text{CS}. \tag{6.3.10}$$

If the ENGS of a proposed sample is greater than zero, then the sample should be taken.

An alternative procedure for calculating EVSI involves taking the expected value (as calculated before the sample is observed) of the posterior expected payoff (assuming that optimal actions are always chosen) and subtracting from it the prior expected payoff $E'R(a')$. For the example, the optimal act under the prior distribution, a', is to guess deck B, and

$$E'R(a') = E'R(\text{guess deck B}) = 3.5.$$

If the sample results in a red card, the optimal act is still to guess deck B, and the posterior expected payoff is

$$E''R(a'' \mid \text{red card}) = E''R(\text{guess deck B} \mid \text{red card}) = 4.0.$$

On the other hand, if the sample results in a green card, the optimal act is to guess deck A, and the posterior expected payoff is

$$E''R(a'' \mid \text{green card}) = E''R(\text{guess deck A} \mid \text{green card}) = 3.33.$$

Since $P(\text{red card}) = .625$ and $P(\text{green card}) = .375$, the expected value of the posterior expected payoff of the act that is optimal under the posterior distribution, $E''R(a'')$ (as calculated before the sample is actually observed), is

$$\begin{aligned} E''R(a'') &= E''R(a'' \mid \text{red card})P(\text{red card}) \\ &\quad + E''R(a'' \mid \text{green card})P(\text{green card}) \\ &= 4.0(.625) + 3.33(.375) = 2.5 + 1.25 \\ &= 3.75. \end{aligned}$$

Furthermore, the expected value of the posterior expected payoff of the act that is optimal under the prior distribution, $E''R(a')$ (as calculated before the sample is actually observed), is

$$\begin{aligned} E''R(a') &= E''R(\text{guess deck B} \mid \text{red card})P(\text{red card}) \\ &\quad + E''R(\text{guess deck B} \mid \text{green card})P(\text{green card}) \\ &= 4.0(.625) + 2.67(.375) = 2.5 + 1.0 \\ &= 3.5. \end{aligned}$$

Note that $E''R(a') = E'R(a')$. In general, if $E'R(a')$ has been determined, it is unnecessary to calculate $E''R(a')$, since they are equal. In other words, *before seeing the sample*, the *expected* posterior ER of an action is equal to the prior ER of the same action. *After* seeing the sample, of course, the ER will usually change.

Given a particular sample result, *conditional* expectations such as $E''R(\text{guess deck A} \mid \text{green card})$ are of interest; *before* the sample result is observed, *unconditional* expectations such as $E''R(\text{guess deck B})$ are of interest. The EVSI is then

$$\text{EVSI} = E''R(a'') - E''R(a'). \qquad (6.3.11)$$

For the example, EVSI $= 3.75 - 3.5 = .25$. Note that this is the same answer arrived at by the first method presented for calculating EVSI. The first method can be thought of as an "incremental" approach in which the value of each possible sample is considered separately by determining how it would change your decision. Instead of calculating a VSI for each possible sample, the second method determines an "overall" posterior expected payoff of a'' and subtracts from it the "overall" posterior expected payoff of a'. The two methods will, of course, always yield identical results.

The process of investigating the value of sample information by using Equation (6.3.11) can be represented conveniently on a tree diagram. Such a tree diagram is presented for the above example in Figure 6.3.1. The first fork is an action fork corresponding to the preposterior decision, sample or do not sample. If the decision is not to sample (the lower branch), then the terminal decision is made, and the action fork at the end of the "do not sample" branch represents the terminal decision, guess deck A or guess deck B. Based on the prior probabilities (since no sample is taken), the expected payoffs for the two actions are \$3.00 and \$3.50, respectively, as calculated above. The optimal action is therefore to guess deck B, and the expected payoff, $E'R(a')$, is equal to \$3.50. Thus, if the decision maker does not sample, the terminal decision-making situation has an expected payoff to him of \$3.50, which is given at the end of the "do not sample" branch in Figure 6.3.1.

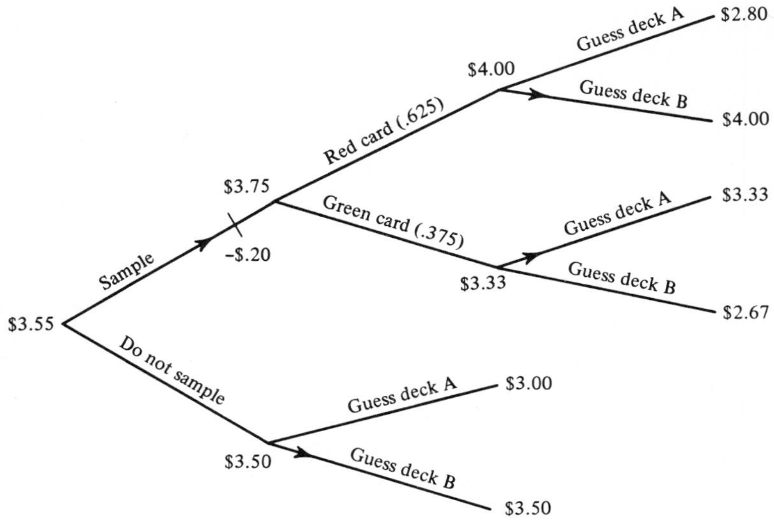

Figure 6.3.1

If the sample is taken, then the upper part of the decision tree is of interest. The "sample" branch is followed by a chance fork that represents the two possible sample results, a red card and a green card. At the end of each of the branches corresponding to a sample outcome, there is an action fork representing the terminal decision, guess deck A or guess deck B. The expected payoffs for these terminal actions are now posterior expected payoffs because the decision forks follow the branches representing sample information. As calculated above, the posterior expected payoffs are \$2.80 and \$4.00 if the sample yields a red card and \$3.33 and \$2.67 if the sample yields a green card. Thus, following a red card, the terminal decision-making situation has an expected payoff of $E''R(a'' \mid \text{red card}) = \$4.00$, since the decision maker will guess deck B. Following a green card, the terminal decision-making situation has an expected payoff of $E''R(a'' \mid \text{green card}) = \$3.33$, since the decision maker will guess deck A. Moving backward on the diagram, the overall posterior expected payoff, as computed before the sample is observed, can be determined by looking at the chance fork corresponding to the sample. The predictive probabilities are .625 and .375 for a red card and a green card, respectively, and the overall (unconditional) posterior expected payoff $E''R(a'')$ is \$3.75. Thus, before seeing the sample, the overall expected payoff is \$3.75 if the sample is taken and \$3.50 if the sample is not taken, as can be seen by looking at the ends of the "sample" and the "do not sample" branches on the decision tree. The expected value of the sample information, then, is $\$3.75 - \$3.50 = \$0.25$. The vertical slash on the "sample" branch represents the cost incurred by sampling; in evaluating the "sample" branch, it is necessary to subtract the cost of sampling from the overall expected payoff. This yields a *net* expected payoff of \$3.55 for the "sample" branch, as compared with \$3.50 for the "do not sample" branch, so the decision should be to sample.

The purpose of analyzing the decision tree in Figure 6.3.1 in such detail is to familiarize you with the use of decision trees in the investigation of the value of information. This problem is simple enough that it is easy to solve without using a tree diagram. Tree diagrams are particularly useful in complex problems, however, especially in problems involving sequences of actions, as you shall see in Section 6.5. The solution of a decision-making problem that is expressed in the form of a decision tree is sometimes called *backward induction*. In order to make the first decision (the decision regarding sampling), start at the right-hand side of the tree and work backward to determine the expected payoff at each fork in the diagram. The process of backward induction is also referred to as "averaging out and folding back." Starting at the right-hand side of the tree, simply "average out" (take expectations) at chance forks and "fold back" (choose the action with the largest overall expected payoff) at action forks. The arrows on the tree diagram in Figure 6.3.1 indicate the optimal actions at action forks.

Equation (6.3.10) gives the expected net gain of sampling for a particular sample. This may differ for different sample designs or different sample sizes. Suppose that the sample design is fixed and that it is necessary to determine how large a sample to take. One way to determine sample size is to consider the relationship between the sample size and the accuracy of estimation; if a desired degree of accuracy is specified, it is possible to find out how large a sample is necessary to attain this degree of accuracy. This is an informal procedure for determining sample size (although not as informal as the more common "oh, an n of about 20 should be sufficient"). Using decision theory, it is possible to determine an optimal sample size.

If the sample design is fixed and if the statistician wishes to determine n, the sample size, he can compute the expected net gain of sampling for various values of n. Equation (6.3.10) could be written in the form

$$\text{ENGS}(n) = \text{EVSI}(n) - \text{CS}(n). \tag{6.3.12}$$

The optimal sample size is then the value of n for which ENGS is maximized. Denoting this optimal sample size by n^*,

$$\text{ENGS}(n^*) \geq \text{ENGS}(n) \text{ for } n = 0, 1, 2, 3, \ldots. \tag{6.3.13}$$

It is entirely possible for n^* to be 0, in which case the statistician should make his terminal decision without further sampling.

A commonly encountered type of ENGS curve is shown in Figure 6.3.2. The ENGS curve represents the difference between the EVSI curve and the CS curve. In this graph, the expected value of sample information rises rather quickly and then levels off. The leveling off reflects the fact that as the sample size gets larger and larger, the additional value of each trial becomes smaller and smaller, and EVSI approaches EVPI. The cost of sampling, on the other hand, is shown as a linear function of n. This would be reasonable if each trial, or item sampled, costs a given amount and if there are no "economies of scale" associated with large values of n. Because the cost of sampling curve (CS) is continually increasing and the EVSI curve levels off, the ENGS curve rises until $n = n^*$ and then declines, eventually becoming negative. The curve presented in Figure 6.3.1 is by no means the only type of ENGS curve, although it is encountered quite often. In some cases the ENGS curve is never above the horizontal axis, and thus the statistician should take no sample. In yet other cases the curve may be below the axis for small n but may eventually rise above the axis, in which case there is an optimal n^* different from 0.

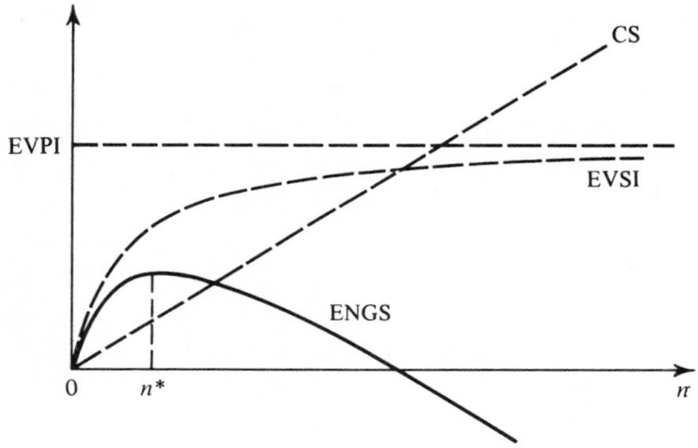

Figure 6.3.2

A problem with EVSI (and hence ENGS) is that it generally requires much more computational time and effort than, say, EVPI. In the example in the previous section, the computations for $n = 1$, while elementary, were somewhat time-consuming. Furthermore, the difficulty increases as n increases, for the number of possible sample results increases and each possible result must be considered. For $n = 10$ in the example, it would be necessary to calculate posterior probabilities and the value of sample information for the sample results (10 red, 0 green), (9 red, 1 green), (8 red, 2 green), and so on up to (0 red, 10 green). If there are more than two states of the world or more than two actions, the computational problem is intensified. In general, it is necessary to program the problem for a computer in order to find the optimal sample size. Even for reasonably complex problems, the computational burden associated with the determination of the expected value of sample information can be handled without difficulty on a computer. Thus, in the following example, you should not be bothered by the computational aspects of the problem; in practice, preposterior analysis often requires the use of a computer.

After the optimal sample size n^* is determined and the sample is taken (provided that $n^* > 0$), the decision maker can use Bayes' theorem to revise his probability distribution on the basis of the sample results. He then has two options once again: (1) to make a terminal decision on the basis of his newly determined posterior distribution, or (2) to take another sample. The second option involves the calculation of another ENGS(n) curve and the determination of another optimal sample size, n^{**}. Formally, the decision maker could continue this sequential process of taking a sample and conducting preposterior analysis until he reaches the point at which the optimal sample size is 0. He should then make a terminal decision. Incidentally, it is important to remember that sample information has no "value" by itself. The "value" in EVSI refers to the expected increase in expected payoff or decrease in expected loss *with respect to the terminal decision.*

In Section 6.2, the quality-control example from Section 3.3 was used to demonstrate the calculation of the expected value of perfect information. Suppose that the statistician is unable to purchase perfect information but that he *can* purchase sample information from the process. Starting from his original prior distribution (disregard the samples considered in Section 6.2), he contemplates three potential sample sizes: one, three, and five. In an actual situation he might also contemplate other values of n, but for the sake of simplicity, assume that he considers only these three values. The payoff table and the prior probabilities are given in Section 6.2, and the cost of sampling is \$1 per trial. That is, for \$1 the statistician can observe an item selected randomly from the items produced by the process, and he can find out if it is defective.

For the sample of size one, the two possible sample results are no defectives and one defective. The posterior probabilities for each of these two situations are calculated as shown in the table on page 316. For each value p of $\tilde{p}$, the likelihood is simply p if the sample produces a defective and $(1 - p)$ if the sample does not produce a defective. Note that the predictive probability that the item is defective is .0340 and the predictive probability that it is not defective is .9660.

From the payoff table in Section 6.2, it is possible to calculate the expected payoff for each action and each sample result. If a defective

Sample outcome (number of defectives in one trial)	*State of the world (value of $\tilde{p}$)*	*Prior probability*	*Likelihood*	*Prior probability × likelihood*	*Posterior probability*
One defective	.01	.60	.01	.0060	.177
	.05	.30	.05	.0150	.441
	.10	.08	.10	.0080	.235
	.25	.02	.25	.0050	.147
				.0340	
No defectives	.01	.60	.99	.5940	.615
	.05	.30	.95	.2850	.295
	.10	.08	.90	.0720	.075
	.25	.02	.75	.0150	.015
				.9660	

item is observed, the posterior probabilities are .177, .441, .235, and .147, and the expected payoffs are

$$E''R(\text{major adjustment} \mid \text{defective}) = 380.00,$$
$$E''R(\text{minor adjustment} \mid \text{defective}) = 389.12,$$
and
$$E''R(\text{no adjustment} \mid \text{defective}) = 331.76.$$

The optimal decision is to make a minor adjustment, and the expected payoff is 389.12. To avoid difficulties that might be caused by rounding errors, all expected payoffs in this example are computed from exact probabilities rather than from probabilities rounded to three decimal places (these rounded values are given in the tables representing the application of Bayes' theorem). For instance, in calculating the expected payoffs following the observance of a defective item, the posterior probabilities used are 6/34, 15/34, 8/34, and 5/34 rather than the rounded figures .177, .441, .235, and .147.

If a nondefective item is observed, the posterior probabilities are .615, .295, .075, and .015, and the expected payoffs are

$$E''R(\text{major adjustment} \mid \text{nondefective}) = 380.00,$$
$$E''R(\text{minor adjustment} \mid \text{nondefective}) = 424.19,$$
and
$$E''R(\text{no adjustment} \mid \text{nondefective}) = 435.53.$$

The optimal decision is to make no adjustment, and the expected payoff is 435.53.

Under the prior distribution, the optimal decision is to make no adjustment. Therefore, the value of sample information for each possible sample result can be computed, using Equation (6.3.8), as follows.

$$\begin{aligned}\text{VSI(defective)} &= E''R(\text{minor adjustment} \mid \text{defective}) \\ &\qquad - E''R(\text{no adjustment} \mid \text{defective}) \\ &= 389.12 - 331.76 = 57.36,\end{aligned}$$

and

$$\begin{aligned}\text{VSI(nondefective)} &= E''R(\text{no adjustment} \mid \text{nondefective}) \\ &\qquad - E''R(\text{no adjustment} \mid \text{nondefective}) \\ &= 435.53 - 435.53 = 0.\end{aligned}$$

The EVSI for a sample of size one is thus

$$\begin{aligned}\text{EVSI}(n = 1) &= \text{VSI(defective)}P(\text{defective}) \\ &\qquad + \text{VSI(nondefective)}P(\text{nondefective}) \\ &= 57.36(.0340) + 0(.9660) \\ &= 1.95.\end{aligned}$$

Alternatively, EVSI$(n = 1)$ could be computed by using Equation (6.3.11):

$$\begin{aligned}\text{EVSI}(n = 1) &= E''R(a'') - E''R(a') \\ &= [E''R(a'' \mid \text{defective})P(\text{defective}) \\ &\qquad + E''R(a'' \mid \text{nondefective})P(\text{nondefective})] - E'R(a') \\ &= [E''R(\text{minor adjustment} \mid \text{defective})P(\text{defective}) \\ &\qquad + E''R(\text{no adjustment} \mid \text{nondefective})P(\text{nondefective})] \\ &\qquad - E'R(\text{no adjustment}) \\ &= [389.12(.034) + 435.53(.966)] - 432.00 \\ &= 433.95 - 432.00 \\ &= 1.95.\end{aligned}$$

The statistician should be willing to pay up to \$1.95 for a sample of one item. But the cost of a sample of size one is \$1, so it is advantageous to him to take the sample, since

$$\text{ENGS}(n = 1) = \text{EVSI}(n = 1) - \text{CS}(n = 1) = 1.95 - 1.00 = 0.95.$$

For the contemplated sample of size three there are four possible sample outcomes: zero, one, two, or three defectives. The likelihoods can be calculated from the binomial distribution with $n = 3$ and $\tilde{p}$ equal to .01, .05, .10, or .25, as the case may be.

Sample outcome (number of defectives in three trials)	*State of the world (value of $\tilde{p}$)*	*Prior probability*	*Likelihood*	*Prior probability × likelihood*	*Posterior probability*
No defectives	.01	.60	.9703	.582180	.643
	.05	.30	.8574	.257220	.284
	.10	.08	.7290	.058320	.064
	.25	.02	.4219	.008438	.009
				.906158	
One defective	.01	.60	.0294	.017640	.205
	.05	.30	.1354	.040620	.471
	.10	.08	.2430	.019440	.226
	.25	.02	.4219	.008438	.098
				.086138	
Two defectives	.01	.60	.0003	.000180	.025
	.05	.30	.0071	.002130	.292
	.10	.08	.0270	.002160	.297
	.25	.02	.1406	.002812	.386
				.007282	
Three defectives	.01	.60	.0000	.000000	.000
	.05	.30	.0001	.000030	.071
	.10	.08	.0010	.000080	.190
	.25	.02	.0156	.000312	.739
				.000422	

For each action and each sample result, an expected payoff can be calculated from the relevant posterior probabilities. These expected payoffs are given in tabular form as follows, and the expected payoff corresponding to the best action for each sample result is starred.

		SAMPLE OUTCOME NUMBER OF DEFECTIVES IN SAMPLE OF SIZE THREE			
		0	*1*	*2*	*3*
ACTION	*Major adjustment*	380.00	380.00	380.00*	380.00*
	Minor adjustment	426.40	391.38*	376.98	375.00
	No adjustment	441.24*	354.63	217.85	85.31

To calculate the value of sample information, note once again that the act "no adjustment" was optimal under the prior distribution.

$$\begin{aligned}
\text{VSI(0 defectives)} &= E''R(\text{no adjustment} \mid \text{0 defectives}) \\
&\quad - E''R(\text{no adjustment} \mid \text{0 defectives}) = 0. \\
\text{VSI(1 defective)} &= E''R(\text{minor adjustment} \mid \text{1 defective}) \\
&\quad - E''R(\text{no adjustment} \mid \text{1 defective}) \\
&= 391.38 - 354.63 \\
&= 36.75. \\
\text{VSI(2 defectives)} &= E''R(\text{major adjustment} \mid \text{2 defectives}) \\
&\quad - E''R(\text{no adjustment} \mid \text{2 defectives}) \\
&= 380.00 - 217.85 \\
&= 162.15. \\
\text{VSI(3 defectives)} &= E''R(\text{major adjustment} \mid \text{3 defectives}) \\
&\quad - E''R(\text{no adjustment} \mid \text{3 defectives}) \\
&= 380.00 - 85.31 \\
&= 294.69.
\end{aligned}$$

The predictive probabilities of the various sample outcomes are

$$\begin{aligned}
P(\text{0 defectives}) &= .906158, \\
P(\text{1 defective}) &= .086138, \\
P(\text{2 defectives}) &= .007282,
\end{aligned}$$

and

$$P(\text{3 defectives}) = .000422,$$

so that the expected value of sample information for a sample of size three is

$$\begin{aligned}
\text{EVSI}(n = 3) &= 0(.906158) + 36.75(.086138) \\
&\quad + 162.15(.007282) + 294.69(.000422) \\
&= 4.47.
\end{aligned}$$

Therefore,

$$\text{ENGS}(n = 3) = 4.47 - 3(1) = 1.47.$$

For the contemplated sample of size five, there are six possible sample outcomes: zero, one, two, three, four, or five defectives. The likelihoods can be calculated from the binomial distribution with $n = 5$ and $\tilde{p}$ equal to .01, .05, .10, or .25, as the case may be.

Sample outcome (number of defectives in five trials)	*State of the world (value of $\tilde{p}$)*	*Prior probability*	*Likelihood*	*Prior probability × likelihood*	*Posterior probability*
No defectives	.01	.60	.9510	.570600	.668
	.05	.30	.7738	.232140	.272
	.10	.08	.5905	.047240	.055
	.25	.02	.2373	.004746	.005
				.854726	
One defective	.01	.60	.0480	.028800	.232
	.05	.30	.2036	.061080	.492
	.10	.08	.3280	.026240	.212
	.25	.02	.3955	.007910	.064
				.124030	
Two defectives	.01	.60	.0010	.000600	.033
	.05	.30	.0214	.006420	.354
	.10	.08	.0729	.005832	.322
	.25	.02	.2637	.005274	.291
				.018126	
Three defectives	.01	.60	.0000	.000000	.000
	.05	.30	.0011	.000330	.121
	.10	.08	.0081	.000648	.237
	.25	.02	.0879	.001758	.642
				.002736	
Four defectives	.01	.60	.0000	.000000	.000
	.05	.30	.0000	.000000	.000
	.10	.08	.0004	.000032	.099
	.25	.02	.0146	.000292	.901
				.000324	
Five defectives	.01	.60	.0000	.000000	.000
	.05	.30	.0000	.000000	.000
	.10	.08	.0000	.000000	.000
	.25	.02	.0010	.000020	1.000
				.000020	

It should be noted that the values of 0 occurring in the last three columns are not *exactly* 0. The numbers are very small, and when rounded off to the number of decimal places given in the table, they become 0. For all practical purposes, they can be considered equal to 0.

For each action and each sample result, an expected payoff can be calculated from the relevant posterior probabilities. These expected payoffs are presented in the following table.

		SAMPLE OUTCOME NUMBER OF DEFECTIVES IN SAMPLE OF SIZE FIVE					
		0	*1*	*2*	*3*	*4*	*5*
ACTION	*Major adjustment*	380.00	380.00	380.00*	380.00*	380.00*	380.00*
	Minor adjustment	428.41	393.57*	377.65	375.00	375.00	375.00
	No adjustment	445.66*	371.91	254.09	119.30	29.63	0.00

From this table and the predictive probabilities of the various sample outcomes,

$$P(0 \text{ defectives}) = .854726,$$
$$P(1 \text{ defective}) = .124030,$$
$$P(2 \text{ defectives}) = .018126,$$
$$P(3 \text{ defectives}) = .002736,$$
$$P(4 \text{ defectives}) = .000324,$$

and

$$P(5 \text{ defectives}) = .000020,$$

which do not sum exactly to 1 because of rounding error in the determination of the likelihoods from the binomial distribution, the expected value of sample information can be calculated. First,

$$\begin{aligned} \text{VSI}(0 \text{ defectives}) &= E''R(\text{no adjustment} \mid 0 \text{ defectives}) \\ &\quad - E''R(\text{no adjustment} \mid 0 \text{ defectives}) = 0, \\ \text{VSI}(1 \text{ defective}) &= E''R(\text{minor adjustment} \mid 1 \text{ defective}) \\ &\quad - E''R(\text{no adjustment} \mid 1 \text{ defective}) \\ &= 393.57 - 371.91 = 21.66, \end{aligned}$$

$$\begin{aligned}
\text{VSI(2 defectives)} &= E''R(\text{major adjustment} \mid \text{2 defectives}) \\
&\quad - E''R(\text{no adjustment} \mid \text{2 defectives}) \\
&= 380.00 - 254.09 = 125.91, \\
\text{VSI(3 defectives)} &= E''R(\text{major adjustment} \mid \text{3 defectives}) \\
&\quad - E''R(\text{no adjustment} \mid \text{3 defectives}) \\
&= 380.00 - 119.30 = 260.70, \\
\text{VSI(4 defectives)} &= E''R(\text{major adjustment} \mid \text{4 defectives}) \\
&\quad - E''R(\text{no adjustment} \mid \text{4 defectives}) \\
&= 380.00 - 29.63 = 350.37, \\
\text{and} \quad \text{VSI(5 defectives)} &= E''R(\text{major adjustment} \mid \text{5 defectives}) \\
&\quad - E''R(\text{no adjustment} \mid \text{5 defectives}) \\
&= 380.00 - 0.00 = 380.00.
\end{aligned}$$

The EVSI is then 5.80. The cost of a sample of size five is \$5, so

$$\text{ENGS}(n = 5) = \text{EVSI}(n = 5) - \text{CS}(n = 5) = 5.80 - 5.00 = 0.80.$$

Recall that

$$\text{ENGS}(n = 1) = \text{EVSI}(n = 1) - \text{CS}(n = 1) = 1.95 - 1.00 = 0.95$$

and

$$\text{ENGS}(n = 3) = \text{EVSI}(n = 3) - \text{CS}(n = 3) = 4.47 - 3.00 = 1.47.$$

Of the three samples considered, then, the sample of size three has the largest ENGS. Of course, it may be that some sample size other than one, three, or five has yet a larger ENGS. To carry the problem any further conveniently would require the use of a computer, in which case it would be necessary to take into account the cost of using the computer. The problem of determining optimal sample size can become quite complicated from a computational standpoint, as this example illustrates. As a result, decisions regarding sample size are often made in an informal manner, although for important decisions involving large payoffs it clearly would be worthwhile to apply preposterior analysis formally, hopefully on a high-speed computer.

A tree diagram for the quality-control example is presented in Figure 6.3.3. On the left-hand side of the tree diagram are the potential preposterior decisions, which in this case are $n = 0$ (do not sample), $n = 1$, $n = 3$, and $n = 5$. Of course, other values of n could be considered, in which case there would be more "branches" on the tree diagram. If the

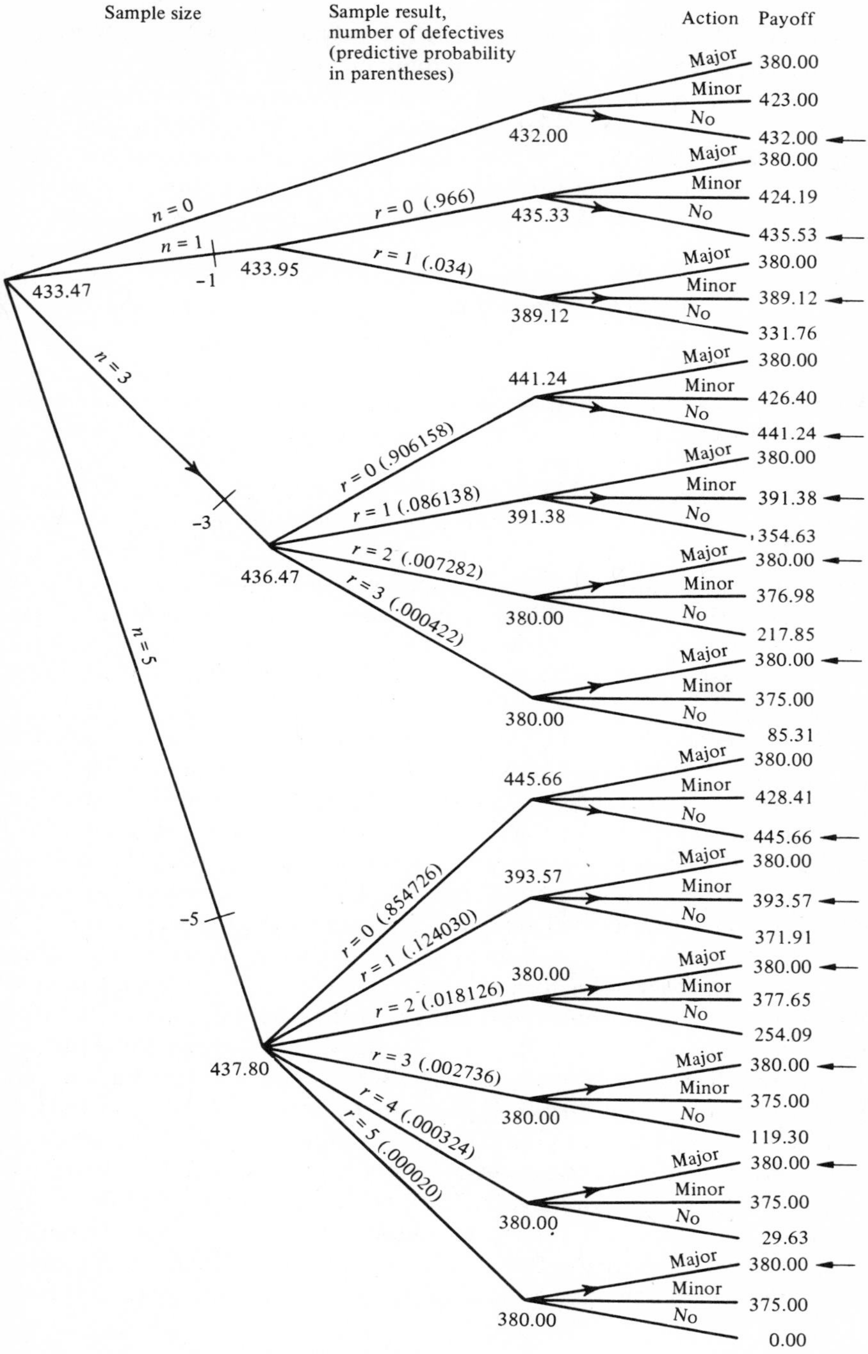

Sample size
Sample result, number of defectives (predictive probability in parentheses)
Action
Payoff
n = 0
n = 1
-1
n = 3
-3
n = 5
-5
433.47
433.95
436.47
437.80
432.00
r = 0 (.966)
435.33
r = 1 (.034)
389.12
r = 0 (.906158)
441.24
r = 1 (.086138)
391.38
r = 2 (.007282)
380.00
r = 3 (.000422)
380.00
r = 0 (.854726)
445.66
r = 1 (.124030)
393.57
r = 2 (.018126)
380.00
r = 3 (.002736)
380.00
r = 4 (.000324)
380.00
r = 5 (.000020)
380.00
Major
Minor
No
380.00
423.00
432.00
380.00
424.19
435.53
380.00
389.12
331.76
380.00
426.40
441.24
380.00
391.38
354.63
380.00
376.98
217.85
380.00
375.00
85.31
380.00
428.41
445.66
380.00
393.57
371.91
380.00
377.65
254.09
380.00
375.00
119.30
380.00
375.00
29.63
380.00
375.00
0.00

Figure 6.3.3

preposterior decision is not to sample, the terminal decision is made on the basis of the current information by finding the action with the highest expected payoff. If the preposterior decision is to sample, then the sample result is observed and used to determine a posterior distribution; the terminal decision is based on expected payoffs calculated from this posterior distribution. Using the backward induction procedure, the values of $E''R(a'')$ are 433.95, 436.47, and 437.80. Thus, using Equation (6.3.11),

$$\text{EVSI}(n = 1) = 433.95 - 432.00 = 1.95,$$
$$\text{EVSI}(n = 3) = 436.47 - 432.00 = 4.47,$$

and

$$\text{EVSI}(n = 5) = 437.80 - 432.00 = 5.80.$$

After the costs of sampling are subtracted, the optimal action, as noted above, is to take a sample of size three, and the net expected payoff is 433.47.

It should be stressed that the "value" of the sample information is to be interpreted with respect to the terminal decision. In the framework presented in Section 5.9 for decision making under uncertainty, the decision of primary interest is the terminal decision. Sample information is valuable to the extent that it may be of some use to the decision maker in terms of his terminal decision. Thus, if all potential sample results lead to the same decision, then the expected value of sample information is 0.

In this section the value of sample information has been discussed for a single sample. It is also possible to conduct a new preposterior analysis after each sample to determine the optimal sample size for the next sample. The sampling ends and a terminal decision is made when the result of such a preposterior analysis is an optimal sample size of 0. In terms of the tree diagram in Figure 6.3.3, more branches could be added to allow for repeated sampling. For instance, suppose that $n = 3$ and $r = 1$; following these branches, one arrives at a fork representing the possible actions. This fork implicitly assumes that no second sample will be taken (that is, that $n = 0$ for the second sample); additional forks could be added for possible sample sizes for the second sample. Alternatively, the decision maker could revise his probability distribution and decide whether or not to continue sampling after *each* trial, so that it would be unnecessary to consider the question of optimal sample size. After observing a trial, the decision maker is at a decision fork with two possible actions, "continue sampling" or "stop sampling." The choice between these actions involves the consideration of an entire stream of potential future sample results, together with the possibility of stopping after any trial and making a terminal decision. This sequential procedure is discussed in Section 6.5, and you shall see that a tree diagram for sequential analysis can quickly become quite complex!

6.4 PREPOSTERIOR ANALYSIS: AN EXAMPLE

For an example of the application of preposterior analysis and its relation to terminal analysis, consider the oil-drilling venture presented in Section 5.10. The decision maker assumed this model: the set of actions, A, consists of

a_1 = "Drill with 100% interest,"
a_2 = "Drill with 50% interest,"
a_3 = "Farm out drilling rights for a $\frac{1}{8}$ override,"
and a_4 = "Farm out drilling rights for \$10,000 and a $\frac{1}{16}$ override"

(a fifth action, "do not drill," was considered but found to be inadmissible); the set of states of the world, which are values of $\tilde{\theta}$, the amount of oil (in barrels) recovered from the well, consists of $\tilde{\theta} = 0$, $\tilde{\theta} = 50{,}000$, $\tilde{\theta} = 100{,}000$, $\tilde{\theta} = 500{,}000$, and $\tilde{\theta} = 1{,}000{,}000$. The probabilities for the five possible values of $\tilde{\theta}$ are .70, .05, .15, .05, and .05, respectively. The payoff table (in terms of dollars) is as follows.

ACTION	STATE OF THE WORLD (VALUE OF $\tilde{\theta}$) 0	50,000	100,000	500,000	1,000,000
Drill with 100% interest	−68,000	−20,000	28,000	412,000	892,000
Drill with 50% interest	−34,000	−10,000	14,000	206,000	446,000
Farm out, $\frac{1}{8}$ override	0	6,000	12,000	60,000	120,000
Farm out, \$10,000 and $\frac{1}{16}$ override	10,000	13,000	16,000	40,000	70,000

The decision maker is a risk-avoider, and his utility function is graphed in Figure 5.10.2. In terms of utility, the payoff table in the oil-drilling venture is as follows.

ACTION	STATE OF THE WORLD (VALUE OF $\tilde{\theta}$) 0	50,000	100,000	500,000	1,000,000
Drill with 100% interest	.11	.27	.38	.81	1.00
Drill with 50% interest	.23	.30	.35	.63	.84
Farm out, $\frac{1}{8}$ override	.32	.34	.35	.44	.53
Farm out, \$10,000 and $\frac{1}{16}$ override	.34	.35	.36	.40	.45

The expected utilities of the four actions to the decision maker are .2380, .3020, .3420, and .3520, respectively. Thus, in order to maximize his expected utility, the decision maker should farm out the drilling rights for \$10,000 and a $\frac{1}{16}$ override. The cash equivalent for this action is approximately \$15,000.

Before the decision maker makes his terminal decision, suppose that he is offered the opportunity to purchase perfect information concerning $\tilde{\theta}$ at a cost of \$30,000. The procedure for determining $\tilde{\theta}$ is reasonably complicated, and hence the perfect information costs \$30,000. (Note that the cost of drilling is \$68,000, so that for another \$38,000 the decision maker could get perfect information *and* have a well drilled.) If he decides to buy perfect information, his payoff table in dollars is as follows.

ACTION	STATE OF THE WORLD (VALUE OF $\tilde{\theta}$) 0	50,000	100,000	500,000	1,000,000
Drill with 100% interest	−98,000	−50,000	−2,000	382,000	862,000
Drill with 50% interest	−64,000	−40,000	−16,000	176,000	416,000
Farm out, $\frac{1}{8}$ override	−30,000	−24,000	−18,000	30,000	90,000
Farm out, \$10,000 and $\frac{1}{16}$ override	−20,000	−17,000	−14,000	10,000	40,000

This payoff table is found by subtracting \$30,000 from each payoff in the previous payoff table. (For the second action, note that since the decision maker has not yet decided whether to find a partner, he must bear the full cost of the information himself.)

From Figure 5.10.2, the utility table corresponding to this payoff table is approximately as follows.

ACTION	STATE OF THE WORLD (VALUE OF $\tilde{\theta}$) 0	50,000	100,000	500,000	1,000,000
Drill with 100% interest	0	.17	.32	.79	.99
Drill with 50% interest	.12	.21	.28	.60	.82
Farm out, $\frac{1}{8}$ override	.24	.26	.28	.39	.48
Farm out, \$10,000 and $\frac{1}{16}$ override	.27	.28	.29	.34	.40

Under perfect information, the decision maker will choose the action with the highest utility. If $\tilde{\theta} = 0$, this action is a_4, and the utility is .27; if $\tilde{\theta} = 50{,}000$, the optimal action is a_4, and the utility is .28; for the other values of $\tilde{\theta}$, the optimal action is a_1, and the utilities are .32, .79, and .99, respectively. Thus, *before* he knows the true value of $\tilde{\theta}$, the overall expected utility to the assessor under the condition of perfect information is

$$EU(\text{perfect information}) = .70(.27) + .05(.28) + .15(.32) + .05(.79) + .05(.99) = .3400.$$

The cash equivalent corresponding to this expected utility is, from Figure 5.10.2, approximately \$10,000. But under imperfect information (that is, under the decision maker's current state of uncertainty), the expected utility of the optimal action is

$$EU(\text{farm out, \$10,000 and } \tfrac{1}{16} \text{ override}) = .70(.34) + .05(.35) + .15(.36) + .05(.40) + .05(.45) = .3520,$$

which corresponds to a cash equivalent of approximately \$15,000. Hence, the decision maker is better off under his current condition of uncertainty than he is if he pays \$30,000 for perfect information about $\tilde{\theta}$. The EVPI is less than \$30,000, and therefore he should not buy perfect information.

Another possibility is available to the decision maker, however; he can purchase seismic information concerning the site in question. This seismic information is not perfect information (it will not enable him to find the value of $\tilde{\theta}$ for certain); instead, it can be thought of as sample information. The seismic information, which costs $10,000, is used to determine the geological structure at the site. Either there is no structure, an open structure, or a closed structure. No structure tends to indicate the absence of oil, whereas a closed structure tends to indicate the presence of oil; an open structure is intermediate between no structure and a closed structure.

In order to formally consider the purchase of sample information in the form of seismic information, the decision maker must assess conditional probabilities for the various sample outcomes, given different values of $\tilde{\theta}$. This situation is slightly different from the examples considered in Section 6.3, for there is no statistical model such as the Bernoulli model, the Poisson model, or the normal model available to help the decision maker determine his conditional probabilities, or likelihoods. However, the decision maker has considerable experience with oil-drilling ventures, so he feels that he can assess these conditional probabilities directly without the aid of a statistical model. He judges that the conditional probability distribution of the sample outcomes given that $\tilde{\theta} = 0$ (a dry hole) consists of the probabilities

$$P(\text{no structure} \mid \tilde{\theta} = 0) = .60,$$
$$P(\text{open structure} \mid \tilde{\theta} = 0) = .30,$$
and
$$P(\text{closed structure} \mid \tilde{\theta} = 0) = .10.$$

This means that given a dry hole, "no structure" is six times as likely as "closed structure" and twice as likely as "open structure." In assessing conditional probabilities, it may be useful for the decision maker to consider conditional bets or lotteries, as discussed in Section 2.6. By using such devices, the decision maker assesses the other conditional probabilities that are required for his preposterior analysis:

$$P(\text{no structure} \mid \tilde{\theta} = 50{,}000) = .50,$$
$$P(\text{open structure} \mid \tilde{\theta} = 50{,}000) = .30,$$
and
$$P(\text{closed structure} \mid \tilde{\theta} = 50{,}000) = .20;$$

$$P(\text{no structure} \mid \tilde{\theta} = 100{,}000) = .30,$$
$$P(\text{open structure} \mid \tilde{\theta} = 100{,}000) = .30,$$
and
$$P(\text{closed structure} \mid \tilde{\theta} = 100{,}000) = .40;$$

$$P(\text{no structure} \mid \tilde{\theta} = 500{,}000) = .20,$$
$$P(\text{open structure} \mid \tilde{\theta} = 500{,}000) = .20,$$
and
$$P(\text{closed structure} \mid \tilde{\theta} = 500{,}000) = .60;$$

$$P(\text{no structure} \mid \tilde{\theta} = 1{,}000{,}000) = .10,$$
$$P(\text{open structure} \mid \tilde{\theta} = 1{,}000{,}000) = .20,$$
and
$$P(\text{closed structure} \mid \tilde{\theta} = 1{,}000{,}000) = .70.$$

With these conditional probabilities as likelihoods, Bayes' theorem can be used to revise the distribution of $\tilde{\theta}$. For the purposes of preposterior analysis, it is necessary to apply Bayes' theorem three times, once for each potential sample result.

Sample result	$\tilde{\theta}$	*Prior probability*	*Likelihood*	*Prior probability* × *likelihood*	*Posterior probability*
	0	.70	.6	.420	.83
	50,000	.05	.5	.025	.05
No structure	100,000	.15	.3	.045	.09
	500,000	.05	.2	.010	.02
	1,000,000	.05	.1	.005	.01
				.505	1.00
	0	.70	.3	.210	.72
	50,000	.05	.3	.015	.05
Open structure	100,000	.15	.3	.045	.15
	500,000	.05	.2	.010	.04
	1,000,000	.05	.2	.010	.04
				.290	1.00
	0	.70	.1	.070	.34
	50,000	.05	.2	.010	.05
Closed structure	100,000	.15	.4	.060	.29
	500,000	.05	.6	.030	.15
	1,000,000	.05	.7	.035	.17
				.205	1.00

In this example, the prior probabilities $P(\theta)$ and the conditional probabilities $P(y \mid \theta)$ have been assessed and have been used to determine the posterior probabilities $P(\theta \mid y)$ and the predictive probabilities $P(y)$. An alternative procedure that is useful when the decision maker's information bears directly on the posterior and predictive probabilities is simply to assess $P(\theta \mid y)$ and $P(y)$ directly, thus bypassing the formal use of Bayes' theorem to revise probabilities. For instance, a geologist might find it easier to assess the probabilities of different amounts of oil given a particular structure [$P(\theta \mid y)$] than to assess probabilities for different structures given a certain amount of oil [$P(y \mid \theta)$]. The model is flexible in that the desired probabilities generally can be determined in more than one way.

Since the cost of the seismic information to the decision maker is $10,000, the original payoff table for the oil-drilling venture should be modified by subtracting $10,000 from each payoff. The modified payoff table in dollars is as follows.

		STATE OF THE WORLD (VALUE OF $\tilde{\theta}$)				
		0	*50,000*	*100,000*	*500,000*	*1,000,000*
ACTION	*Drill with 100% interest*	−78,000	−30,000	18,000	402,000	882,000
	Drill with 50% interest	−44,000	−20,000	4,000	196,000	436,000
	Farm out, $\frac{1}{8}$ override	−10,000	−4,000	2,000	50,000	110,000
	Farm out, $10,000 and $\frac{1}{16}$ override	0	3,000	6,000	30,000	60,000

From Figure 5.10.2, utilities for these payoffs can be found, resulting in the following approximate payoff table in terms of utility.

		STATE OF THE WORLD (VALUE OF $\tilde{\theta}$)				
		0	*50,000*	*100,000*	*500,000*	*1,000,000*
ACTION	*Drill with 100% interest*	.08	.24	.36	.81	1.00
	Drill with 50% interest	.19	.27	.33	.62	.83
	Farm out, $\frac{1}{8}$ override	.30	.32	.33	.42	.51
	Farm out, $10,000 and $\frac{1}{16}$ override	.32	.33	.34	.39	.43

For each combination of a sample result and an action, an expected utility can be found by using the appropriate posterior probabilities and the table of utilities on page 330. For instance,

$$EU(\text{drill with } 100\% \text{ interest} \mid \text{no structure})$$
$$= .83(.08) + .05(.24) + .09(.36) + .02(.81) + .01(1.00)$$
$$= .1370.$$

These expected utilities are given in the following table.

ACTION	SAMPLE RESULT *No structure*	 *Open structure*	 *Closed structure*
Drill with 100% interest	.1370	.1960	.4351
Drill with 50% interest	.2216	.2578	.4079
Farm out, $\frac{1}{8}$ override	.3082	.3187	.3634
Farm out, \$10,000 and $\frac{1}{16}$ override	.3248	.3307	.3555

After observing the seismic information, the decision maker will choose the action with the highest expected utility. If the information indicates no structure, this action is a_4, with expected utility .3248; if the information indicates an open structure, the optimal action is a_4, with expected utility .3307; finally, if the information indicates a closed structure, the optimal action is a_1, with expected utility .4351.

Before he purchases the seismic information, the decision maker is uncertain about the sample result, and he can express his uncertainty in terms of the predictive probabilities of the three possible outcomes. These predictive probabilities, which are the appropriate sums of the "prior probability × likelihood" columns in the application of Bayes' theorem (see Section 3.7), are

$$P(\text{no structure}) = .505,$$
$$P(\text{open structure}) = .290,$$

and

$$P(\text{closed structure}) = .205.$$

Thus, *before* he obtains the seismic information, the overall expected utility to the decision maker given that he purchases this information is

$$EU(\text{seismic information}) = .505(.3248) + .290(.3307) + .205(.4351)$$
$$= .349.$$

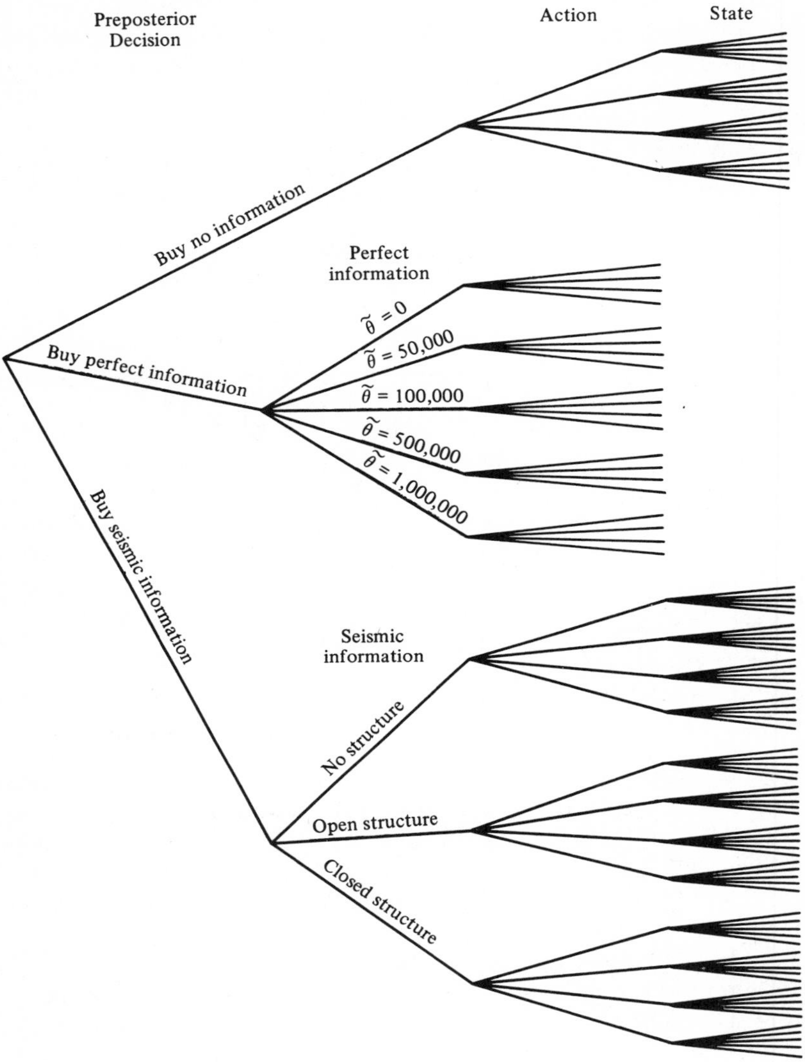

Figure 6.4.1

The certainty equivalent corresponding to this expected utility is, from Figure 5.10.2, approximately \$13,000. But under the decision maker's current state of uncertainty (without seeing any seismic information), the expected utility of the optimal action is

$$EU(\text{farm out, \$10,000 and } \tfrac{1}{16} \text{ override}) = .3520,$$

which corresponds to a cash equivalent of approximately $15,000. Hence, the decision maker is better off under his current condition of uncertainty than he is if he pays $10,000 for seismic information. The EVSI, then, must be less than $10,000.

In this section, the oil-drilling venture introduced in Section 5.10 has been extended to allow for consideration of perfect information and sample information. Given the costs of perfect information and sample information, however, it turns out that the decision maker should make his terminal decision on the basis of his current information rather than buy any information. The preposterior analysis in this section is schematically represented on a tree diagram in Figure 6.4.1. The choice of preposterior decisions is to buy perfect information, sample (seismic) information, or no information. If no information is purchased, the terminal decision is made by comparing the expected utilities of the four actions. If perfect information is purchased, then the terminal decision is made by comparing the utilities of the four actions in the column of the utility matrix corresponding to the true value of $\tilde{\theta}$. If seismic information is purchased, then the posterior distribution is determined following the observation of the sample result, and the terminal decision is made by comparing the expected utilities of the four actions with respect to this posterior distribution.

6.5 SEQUENTIAL ANALYSIS

The discussion in Sections 6.3 and 6.4 involved the expected value of sample information for a particular *single-stage* sampling plan, as opposed to a *sequential* sampling plan. As noted in those sections, it is possible for the decision maker to revise his probabilities after each trial (or each new bit of information) and to decide whether to continue sampling or whether to stop and make a terminal decision. After observing a trial, the decision maker is at a decision fork with two possible actions, "continue sampling" and "stop sampling." The choice between these two actions involves the consideration of an entire stream of potential future sample results, together with the possibility of stopping after any trial and making a terminal decision. This procedure, called *sequential analysis*, is discussed in this section.

In sequential analysis it is convenient to represent a decision-making problem in terms of a decision tree. The problem can then be solved by using the process of backward induction, or averaging out and folding back, as discussed in Section 6.3. This process enables the decision maker to determine his optimal action (continue sampling or stop sampling) at any point in the sampling procedure.

As an illustration of sequential analysis, consider the quality-control example presented in Section 3.3 and discussed in terms of the value of information in Sections 6.2 and 6.3. Suppose that the decision maker considers a sequential sampling plan instead of a single-stage sampling plan (the latter type of sampling plan was considered for the quality-control example in Section 6.3). In particular, to simplify the calculations, suppose that the maximum number of trials is three; perhaps the decision maker has time for at most three trials before he must make his terminal decision, or perhaps he has subjectively decided that he will consider no more than three trials. His sequential decision-making problem is presented in terms of a tree diagram in Figure 6.5.1. The first fork is an action fork, "sample" or "do not sample." If the "do not sample" branch is chosen, then the terminal decision is made on the basis of the prior information by comparing the prior expected payoffs of each of the three actions. The optimal action is to make no adjustment, and the expected payoff is 432.

If the "sample" branch is chosen, the analysis becomes more complicated. The chance fork at the end of the "sample" branch represents the two possible sample results at the first trial; the item is defective (D) or it is not defective (G, representing a good item). At the end of each of these branches (D or G), there is another action fork, corresponding to the decision to continue sampling or to stop sampling and make a terminal decision. Essentially, each of these action forks is identical to the first action fork at the left-hand side of the tree diagram; the choice is to sample or not to sample. The rest of the decision tree in Figure 6.5.1 is similar to the portion of the tree that has already been described. After a sample result is observed, the decision maker is faced with a choice of continuing to sample (in which case he encounters a chance fork with the possible sample results, followed by another choice to continue sampling or not) or terminating the sampling and making a terminal decision.

Note that if no upper limit is placed on the total sample size, the tree continues to grow indefinitely. In this example it is assumed that the total sample size cannot exceed three, so that a terminal decision must be made after a third item is observed from the process (of course, it is also possible to stop sampling earlier).

Once the framework of the tree is completed, the next step is to determine the expected payoffs at all of the "terminating" branches and the probabilities at all of the chance forks. If no sample is taken, the expected payoffs of the three actions (major adjustment, minor adjustment, and no adjustment) are 380, 423, and 432, respectively. These payoffs have been computed in previous discussions of the quality-control example, and they may be found, for instance, in Figure 6.3.3. Similarly, if sampling is terminated after one trial, the expected payoffs, which are calculated

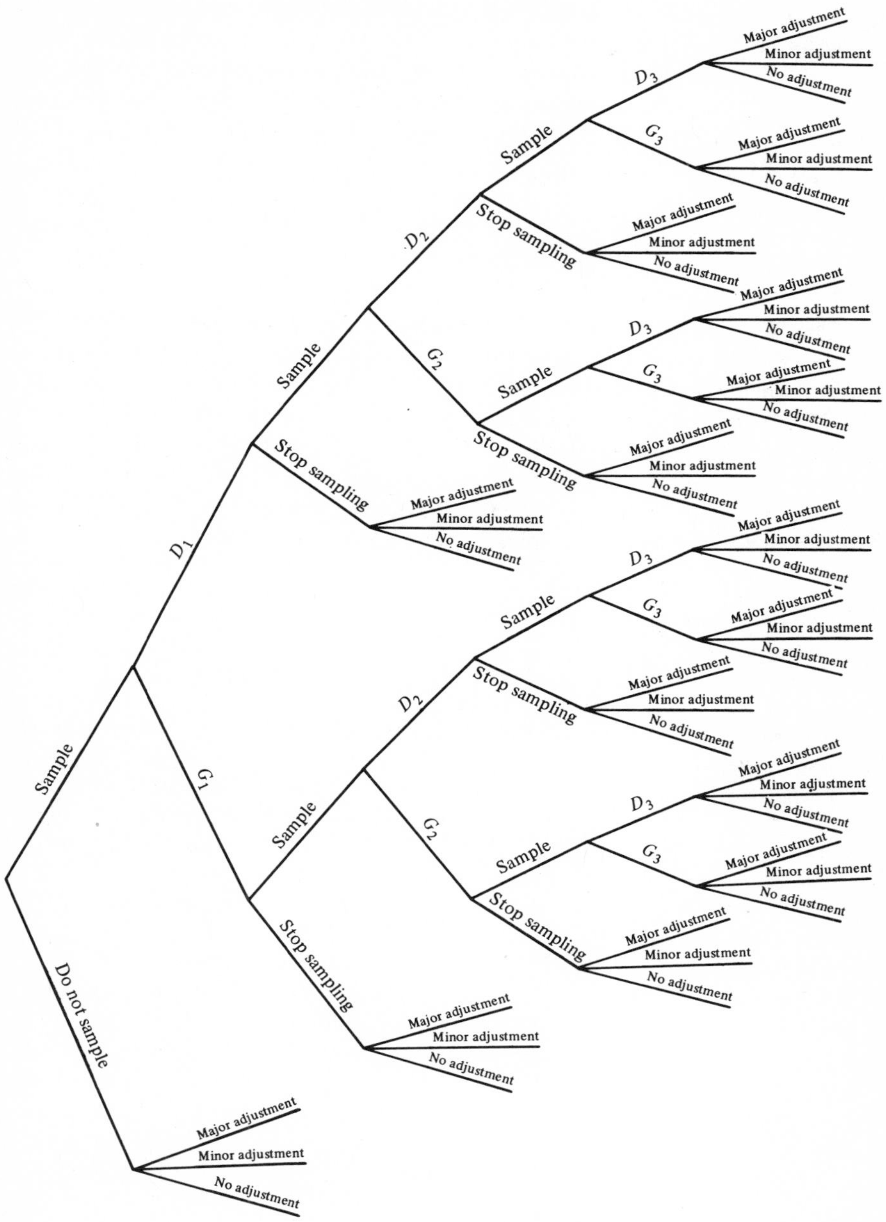
Sample
Do not sample
D_1
G_1
Sample
Stop sampling
D_2
G_2
D_3
G_3
Major adjustment
Minor adjustment
No adjustment

Figure 6.5.1

with respect to the posterior distribution following the one trial, are given in Figure 6.3.3. If the single trial yields a defective item, the expected payoffs are

$$E''R(\text{major adjustment} \mid D) = 380.00,$$
$$E''R(\text{minor adjustment} \mid D) = 389.12,$$

and

$$E''R(\text{no adjustment} \mid D) = 331.76;$$

if the trial yields a good item, the expected payoffs are

$$E''R(\text{major adjustment} \mid G) = 380.00,$$
$$E''R(\text{minor adjustment} \mid G) = 424.19,$$

and

$$E''R(\text{no adjustment} \mid G) = 435.53.$$

The next sets of terminating branches occur if sampling is terminated after two items are observed. Only samples of one, three, and five items were considered in Section 6.3, so the posterior distributions following two trials were not determined there. The possible sample results are two defective items, one defective item and one good item, and two good items. The likelihoods can be calculated from the binomial distribution with $n = 2$ and $\tilde{p}$ equal to .01, .05, .10, and .25. Applying Bayes' theorem in tabular form, the posterior distributions are as follows.

Sample outcome (number of defectives in two trials)	*State of the world (value of $\tilde{p}$)*	*Prior probability*	*Likelihood*	*Prior probability × likelihood*	*Posterior probability*
No defectives	.01	.60	.9801	.588060	.629
	.05	.30	.9025	.270750	.290
	.10	.08	.8100	.064800	.069
	.25	.02	.5625	.011250	.012
				.934860	
One defective	.01	.60	.0198	.011880	.191
	.05	.30	.0950	.028500	.458
	.10	.08	.1800	.014400	.231
	.25	.02	.3750	.007500	.120
				.062280	
Two defectives	.01	.60	.0001	.000060	.021
	.05	.30	.0025	.000750	.262
	.10	.08	.0100	.000800	.280
	.25	.02	.0625	.001250	.437
				.002860	

For each action and each sample result, an expected payoff can be calculated from the relevant posterior distribution. (As in Section 6.3, the posterior probabilities used to calculate the expected payoffs are the exact posterior probabilities rather than those given above, which are rounded to three decimal places.) These expected payoffs are given in tabular form, and they are also entered in the decision tree in Figure 6.5.2.

		SAMPLE OUTCOME (NUMBER OF DEFECTIVES IN SAMPLE OF SIZE TWO)		
		0	*1*	*2*
	Major adjustment	380.00	380.00	380.00*
ACTION	*Minor adjustment*	425.32	390.26*	376.68
	No adjustment	438.58*	343.97	198.88

The final set of expected payoffs is entered at the far right-hand side of the tree diagram. These expected payoffs, calculated after three items are observed, may be found in Figure 6.3.3, so their calculation is not discussed here. It might be noted that some expected payoffs appear more than once in Figure 6.5.2. The expected payoffs following the observance of one defective item in three trials are 380.00, 391.38, and 354.63; these values are found three times in Figure 6.5.2, since the decision tree itemizes the sample results trial-by-trial and one defective item in three trials can occur in three ways: (D, G, G), (G, D, G), and (G, G, D).

The other inputs needed for the tree diagram are the probabilities at the various chance forks. Since the chance forks involve sample outcomes, these probabilities are predictive probabilities. For instance, at the first chance fork (reading from the left), the probabilities of interest are the predictive probabilities for the first sample outcome; these probabilities are $P(\text{defective}) = .034$ and $P(\text{good}) = .966$. The next chance forks involve the second trial, and thus the predictive probabilities are conditional upon the result at the first trial. In other words, they are based on the posterior distribution following the first trial (which is, of course, the prior distribution with respect to the second trial). If the first trial yields a

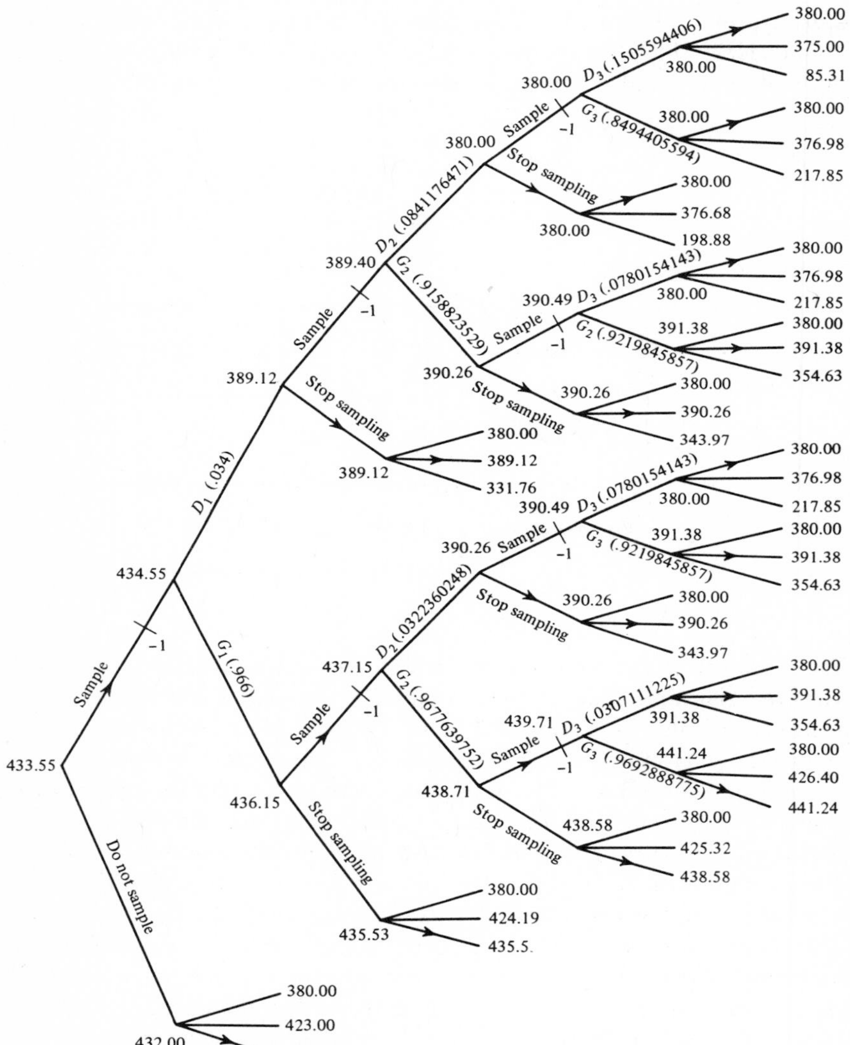
433.55
Sample
-1
Do not sample
434.55
D1 (.034)
G1 (.966)
389.12
Sample
-1
Stop sampling
389.40
D2 (.0841176471)
G2 (.9158823529)
380.00
Sample
-1
Stop sampling
380.00
D3 (.1505594406)
G3 (.8494405594)
380.00
380.00
375.00
85.31
380.00
380.00
376.98
217.85
380.00
380.00
376.68
198.88
390.26
Sample
-1
Stop sampling
390.49
D3 (.0780154143)
G2 (.9219845857)
380.00
380.00
376.98
217.85
391.38
380.00
391.38
354.63
390.26
380.00
390.26
343.97
389.12
380.00
389.12
331.76
436.15
Sample
-1
Stop sampling
437.15
D2 (.0322360248)
G2 (.9677639752)
390.26
Sample
-1
Stop sampling
390.49
D3 (.0780154143)
G3 (.9219845857)
380.00
380.00
376.98
217.85
391.38
380.00
391.38
354.63
390.26
380.00
390.26
343.97
438.71
Sample
-1
Stop sampling
439.71
D3 (.0307111225)
G3 (.9692888775)
391.38
380.00
391.38
354.63
441.24
380.00
426.40
441.24
438.58
380.00
425.32
438.58
435.53
380.00
424.19
435.5.
432.00
380.00
423.00
432.00

Figure 6.5.2

defective item, the predictive probabilities for the second trial are

$$P(D_2 \mid D_1) = \sum_p P(D_2 \mid p, D_1) P(p \mid D_1) \tag{6.5.1}$$

and

$$P(G_2 \mid D_1) = \sum_p P(G_2 \mid p, D_1) P(p \mid D_1), \tag{6.5.2}$$

where D_1 refers to a defective item on the first trial, D_2 refers to a defective item on the second trial, and so on. If the first trial yields a good item, the predictive probabilities for the second trial are

$$P(D_2 \mid G_1) = \sum_p P(D_2 \mid p, G_1) P(p \mid G_1) \tag{6.5.3}$$

and

$$P(G_2 \mid G_1) = \sum_p P(G_2 \mid p, G_1) P(p \mid G_1). \tag{6.5.4}$$

From Section 6.3, the posterior probabilities following one defective item in one trial are, to three decimal places,

$$\begin{aligned} P(\tilde{p} = .01 \mid D_1) &= .177, \\ P(\tilde{p} = .05 \mid D_1) &= .441, \\ P(\tilde{p} = .10 \mid D_1) &= .235, \end{aligned}$$

and

$$P(\tilde{p} = .25 \mid D_1) = .147.$$

Under the assumption that the production process behaves like a Bernoulli process, the likelihoods for the second trial do not depend on the result of the first trial. Thus,

$$P(D_2 \mid p, D_1) = P(D_2 \mid p, G_1) = P(D_2 \mid p)$$

and

$$P(G_2 \mid p, D_1) = P(G_2 \mid p, G_1) = P(G_2 \mid p).$$

The likelihoods are

$$\begin{aligned} P(D_2 \mid \tilde{p} = .01) &= .01, \\ P(D_2 \mid \tilde{p} = .05) &= .05, \\ P(D_2 \mid \tilde{p} = .10) &= .10, \\ P(D_2 \mid \tilde{p} = .25) &= .25, \\ P(G_2 \mid \tilde{p} = .01) &= .99, \\ P(G_2 \mid \tilde{p} = .05) &= .95, \\ P(G_2 \mid \tilde{p} = .10) &= .90, \end{aligned}$$

and

$$P(G_2 \mid \tilde{p} = .25) = .75.$$

Thus, from Equations (6.5.1) and (6.5.2), the predictive probabilities for the second trial following a defective item on the first trial are approximately

$$P(D_2 \mid D_1) = .01(.177) + .05(.441) + .10(.235) + .25(.147)$$

and

$$P(G_2 \mid D_1) = .99(.177) + .95(.441) + .90(.235) + .75(.147).$$

Calculated precisely (using exact posterior probabilities $P(p \mid D_1)$ instead of probabilities rounded to three decimal places), these predictive probabilities, to 10 decimal places, are $P(D_2 \mid D_1) = .0841176471$ and $P(G_2 \mid D_1) = .9158823529$. Similarly, the predictive probabilities following a good item on the first trial are $P(D_2 \mid G_1) = .0322360248$ and $P(G_2 \mid G_1) = .9677639752$. Note that before the first trial, the predictive probabilities are $P(D_1) = .034$ and $P(G_1) = .966$. If the first trial results in a defective item, the predictive probability of a defective item increases from .034 to approximately .084; if the first trial results in a good item, the predictive probability of a defective item decreases slightly from .034 to about .032.

The final set of chance forks involves the third trial, and the predictive probabilities of interest here are conditional upon the results of the first two trials. For instance,

$$P(D_3 \mid D_1,D_2) = \sum_p P(D_3 \mid p)P(p \mid D_1,D_2), \tag{6.5.5}$$

where $P(p \mid D_1,D_2)$ is a posterior probability following two defective items on the first two trials and $P(D_3 \mid p)$ is a likelihood. The predictive probabilities for a defective item on the third trial are

$$P(D_3 \mid D_1,D_2) = .1505594406,$$

$$P(D_3 \mid G_1,D_2) = P(D_3 \mid D_1,G_2) = .0780154143,$$

and

$$P(D_3 \mid G_1,G_2) = .0307111225.$$

The predictive probabilities for a good item on the third trial can be obtained by subtracting from one the corresponding predictive probabilities for a defective item on the third trial.

Now that all of the inputs to the sequential problem have been determined, the tree diagram for the problem is reproduced in Figure 6.5.2 with these inputs included. The process of backward induction is carried out, resulting in the expected payoffs shown at the various forks in the diagram. After all payoffs (including the cost of sampling) are considered, the net expected payoff of the sequential sampling plan is 433.55, as compared with an expected payoff of 432.00 for making a terminal decision with no sampling. Thus, the expected net gain of sampling from the sequential sampling plan is

$$\text{ENGS} = 433.55 - 432.00 = 1.55.$$

Recall that among the single-stage sampling plans considered in Section 6.3, the optimal sample size is three, and the ENGS is 1.47. Thus, the sequential sampling plan with a maximum total sample size of three has a larger ENGS than the single-stage sampling plan with a fixed sample size of three. This might be expected, since the sequential sampling plan offers more flexibility. From Figure 6.5.2, the sequential plan leads to the following optimal strategy:

Sample one item. If it is defective, stop sampling and make a minor adjustment; if it is good, sample a second item. If the second item is defective, stop sampling and make a minor adjustment; if it is good, sample a third item. If the third item is defective, make a minor adjustment; if it is good, make no adjustment.

In general, if all other things (such as the cost of sampling) are equal, a sequential sampling plan with a maximum total sample size of n will have a greater ENGS than a single-stage sampling plan with a fixed sample size of n. In practice, however, the costs of sampling are often different for the two plans. In addition to a cost per unit sampled ($1 per unit in the quality-control example), there may be a fixed cost incurred with each sample, where the fixed cost is the same regardless of the sample size. For a single-stage sample of n trials, the fixed cost is incurred once. For a sequential sample in which each trial is considered a separate sample, the fixed cost may be incurred up to n times. In sampling from a production process, for example, it may be necessary to stop the process in order to take a sample at a particular point in the process, and a fixed cost of stopping the process and starting it again may be incurred. If a sequential sample necessitates stopping the process separately for each individual trial, the total cost of sampling will clearly be greater than with a single-stage sample. Of course, the benefits of the sequential plan, as measured by the expected net gain of sampling, may also be considerably greater than the benefits of the single-stage plan. The point is simply that if there is no fixed cost associated with sampling, so that the cost of sampling is strictly a variable cost per unit sampled, then sequential plans yield higher values of ENGS than do single-stage plans; if there is a fixed cost, then the sequential plan may be better or worse than the single-stage plan, depending on the exact situation.

For a second example of sequential analysis, consider an extension of the oil-drilling example in the previous section. In particular, suppose that if seismic information is obtained and it indicates "open structure" or "closed structure," then it is possible to obtain further information by conducting another geological test. This second test costs \$5000, and the results can be classified into two categories, "favorable" (to the presence of oil) and "unfavorable." Note that if seismic information is not purchased or if seismic information is purchased and indicates "no structure," then the second test cannot be conducted. Recall from Section 6.4 that on the basis of the expected net gain from sampling, seismic information alone is not worth purchasing at the stated cost of \$10,000. The combination of seismic information with the possibility of conducting the second geological test might be worth purchasing, however. This is a sequential sampling situation, since the decision maker first has the option of purchasing seismic information and then has the option, provided that he purchases the seismic information and that it indicates "open structure" or "closed structure," of "purchasing" the second test. A decision tree for this example is presented in Figure 6.5.3.

In order to consider the second test in a formal decision-theoretic sense, the decision maker must assess conditional probabilities for the two possible sample outcomes, "favorable" and "unfavorable." This situation is slightly different from the quality-control example, for there is no statistical model such as the Bernoulli model, the Poisson model, or the normal model available to help the decision maker determine his conditional probabilities, or likelihoods. Furthermore, under the assumption of a Bernoulli process, the likelihoods in the quality-control example remain the same from trial to trial because of the stationarity assumption of the Bernoulli process. In the oil-drilling example, the likelihoods for the second test would probably depend to a great extent on the result of the seismic test; that is, the two tests are conditionally dependent (see Section 3.6).

On the basis of considerable experience with similar situations and a brief consultation with a geologist, the decision maker assesses the conditional probabilities required for the model. He feels that

$$P(\text{favorable} \mid \text{open structure}, \tilde{\theta} = 0) = .1$$

and

$$P(\text{unfavorable} \mid \text{open structure}, \tilde{\theta} = 0) = .9.$$

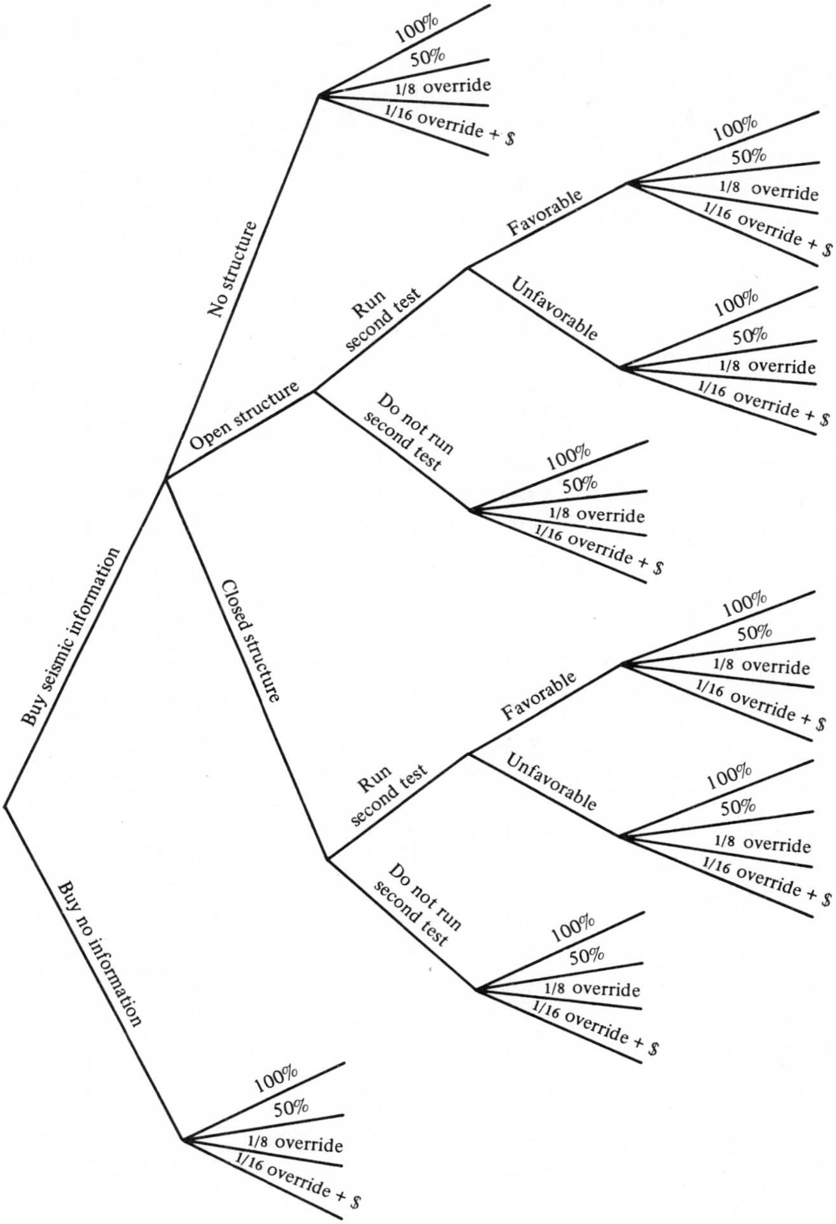

Buy seismic information
No structure
100%
50%
1/8 override
1/16 override + $
Open structure
Run second test
Favorable
100%
50%
1/8 override
1/16 override + $
Unfavorable
100%
50%
1/8 override
1/16 override + $
Do not run second test
100%
50%
1/8 override
1/16 override + $
Closed structure
Run second test
Favorable
100%
50%
1/8 override
1/16 override + $
Unfavorable
100%
50%
1/8 override
1/16 override + $
Do not run second test
100%
50%
1/8 override
1/16 override + $
Buy no information
100%
50%
1/8 override
1/16 override + $

Figure 6.5.3

This means that given a seismic report of "open structure" and given a dry hole, the second test is nine times as likely to produce an unfavorable report as it is to produce a favorable report. In assessing probabilities such as these, the decision maker might find it useful to consider conditional bets or lotteries, as discussed in Section 2.6. By using such devices, he assesses the other conditional probabilities of interest concerning the second test:

$$P(\text{favorable} \mid \text{open structure}, \tilde{\theta} = 50{,}000) = .2,$$
$$P(\text{unfavorable} \mid \text{open structure}, \tilde{\theta} = 50{,}000) = .8,$$

$$P(\text{favorable} \mid \text{open structure}, \tilde{\theta} = 100{,}000) = .2,$$
$$P(\text{unfavorable} \mid \text{open structure}, \tilde{\theta} = 100{,}000) = .8,$$

$$P(\text{favorable} \mid \text{open structure}, \tilde{\theta} = 500{,}000) = .3,$$
$$P(\text{unfavorable} \mid \text{open structure}, \tilde{\theta} = 500{,}000) = .7,$$

$$P(\text{favorable} \mid \text{open structure}, \tilde{\theta} = 1{,}000{,}000) = .3,$$
$$P(\text{unfavorable} \mid \text{open structure}, \tilde{\theta} = 1{,}000{,}000) = .7,$$

$$P(\text{favorable} \mid \text{closed structure}, \tilde{\theta} = 0) = .1,$$
$$P(\text{unfavorable} \mid \text{closed structure}, \tilde{\theta} = 0) = .9,$$

$$P(\text{favorable} \mid \text{closed structure}, \tilde{\theta} = 50{,}000) = .2,$$
$$P(\text{unfavorable} \mid \text{closed structure}, \tilde{\theta} = 50{,}000) = .8,$$

$$P(\text{favorable} \mid \text{closed structure}, \tilde{\theta} = 100{,}000) = .6,$$
$$P(\text{unfavorable} \mid \text{closed structure}, \tilde{\theta} = 100{,}000) = .4,$$

$$P(\text{favorable} \mid \text{closed structure}, \tilde{\theta} = 500{,}000) = .8,$$
$$P(\text{unfavorable} \mid \text{closed structure}, \tilde{\theta} = 500{,}000) = .2,$$

$$P(\text{favorable} \mid \text{closed structure}, \tilde{\theta} = 1{,}000{,}000) = .9,$$

and $$P(\text{unfavorable} \mid \text{closed structure}, \tilde{\theta} = 1{,}000{,}000) = .1.$$

Using these conditional probabilities as likelihoods, Bayes' theorem can be used to revise the distribution of $\tilde{\theta}$. The prior probabilities (prior to the second test) for this application of Bayes' theorem are simply the posterior probabilities following the seismic information. If the seismic information indicates "no structure," the second test cannot be conducted, so it is not necessary to consider that possibility with regard to the second test.

Seismic information	*Second test result*	$\tilde{\theta}$	*Prior probability*	*Likelihood*	*Prior probability × likelihood*	*Posterior probability*
Open structure	Favorable	0	.72	.1	.072	.53
		50,000	.05	.2	.010	.07
		100,000	.15	.2	.030	.22
		500,000	.04	.3	.012	.09
		1,000,000	.04	.3	.012	.09
					.136	
	Unfavorable	0	.72	.9	.648	.75
		50,000	.05	.8	.040	.05
		100,000	.15	.8	.120	.14
		500,000	.04	.7	.028	.03
		1,000,000	.04	.7	.028	.03
					.864	
Closed structure	Favorable	0	.34	.1	.034	.07
		50,000	.05	.2	.010	.02
		100,000	.29	.6	.174	.35
		500,000	.15	.8	.120	.25
		1,000,000	.17	.9	.153	.31
					.491	
	Unfavorable	0	.34	.9	.306	.60
		50,000	.05	.8	.040	.08
		100,000	.29	.4	.116	.23
		500,000	.15	.2	.030	.06
		1,000,000	.17	.1	.017	.03
					.509	

This table gives not only the posterior probabilities following the second test but also the predictive probabilities for the second test, which are

$$P(\text{favorable} \mid \text{open structure}) = .136,$$
$$P(\text{unfavorable} \mid \text{open structure}) = .864,$$
$$P(\text{favorable} \mid \text{closed structure}) = .491,$$

and

$$P(\text{unfavorable} \mid \text{closed structure}) = .509.$$

Since the cost of the second test to the decision maker is $5000, the original payoff table for the oil-drilling venture, presented in Section 5.10, must be modified by subtracting $15,000 from each payoff ($10,000 for the seismic information, $5000 for the second test). The modified payoff table in dollars is as follows.

ACTION	STATE OF THE WORLD (VALUE OF $\tilde{\theta}$): 0	50,000	100,000	500,000	1,000,000
Drill with 100% interest	−83,000	−35,000	13,000	397,000	875,000
Drill with 50% interest	−49,000	−25,000	−1,000	191,000	431,000
Farm out, $\frac{1}{8}$ override	−15,000	−9,000	−3,000	45,000	105,000
Farm out, $10,000 and $\frac{1}{16}$ override	−5,000	−2,000	1,000	25,000	55,000

From Figure 5.10.2, utilities for these payoffs can be found, resulting in the following approximate payoff table in terms of utility.

ACTION	STATE OF THE WORLD (VALUE OF $\tilde{\theta}$): 0	50,000	100,000	500,000	1,000,000
Drill with 100% interest	.07	.22	.34	.81	1.00
Drill with 50% interest	.17	.25	.32	.62	.83
Farm out, $\frac{1}{8}$ override	.28	.30	.31	.41	.51
Farm out, $10,000 and $\frac{1}{16}$ override	.31	.31	.32	.37	.43

For each combination of seismic information and a result from the second test, an expected utility can be found for each action by using the appropriate posterior probabilities and the above table of utilities. For instance,

$$EU(\text{drill with 100\% interest} \mid \text{open structure, favorable})$$
$$= .53(.07) + .07(.22) + .22(.34) + .09(.81) + .09(1.00) = .2902.$$

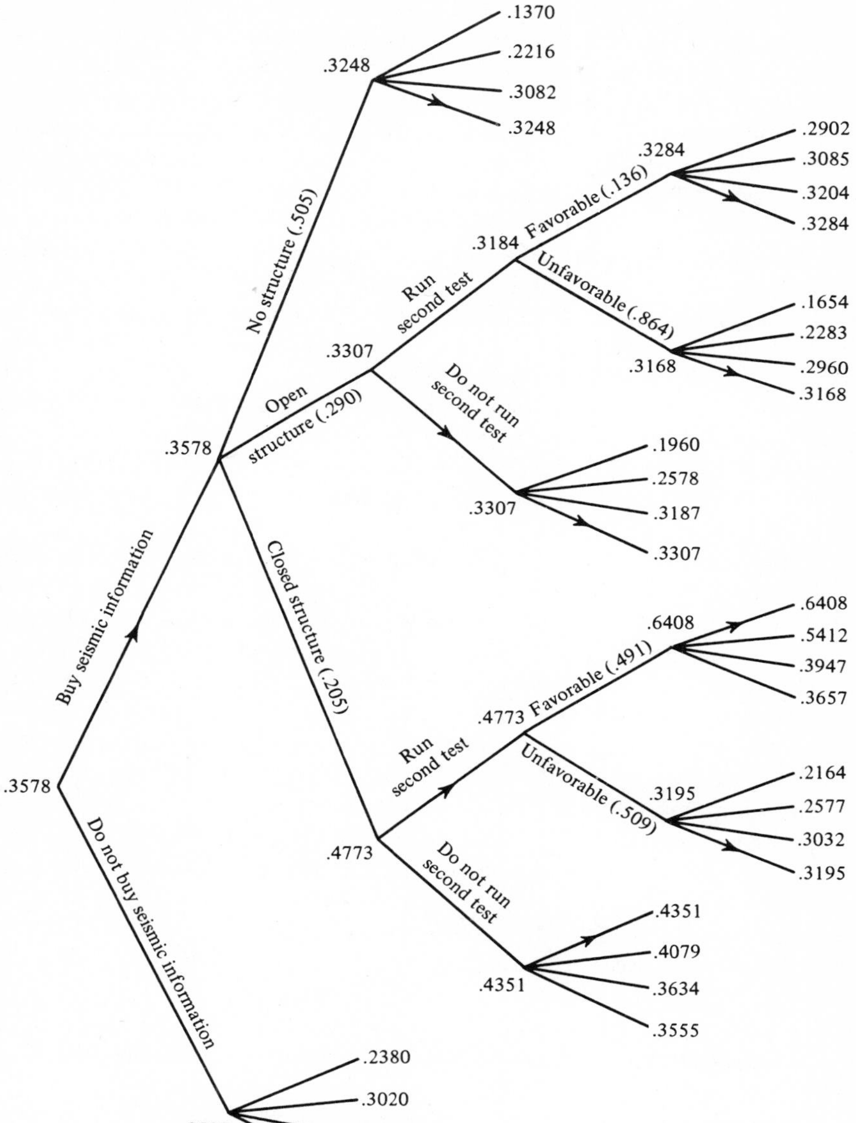
.3578
Buy seismic information
Do not buy seismic information
.3578
No structure (.505)
Open structure (.290)
Closed structure (.205)
.3248
.1370
.2216
.3082
.3248
.3307
Run second test
Do not run second test
.3184
Favorable (.136)
Unfavorable (.864)
.3284
.2902
.3085
.3204
.3284
.3168
.1654
.2283
.2960
.3168
.3307
.1960
.2578
.3187
.3307
.4773
.4773
.6408
Favorable (.491)
Unfavorable (.509)
.6408
.5412
.3947
.3657
.3195
.2164
.2577
.3032
.3195
.4351
.4351
.4079
.3634
.3555
.3520
.2380
.3020
.3420
.3520

Figure 6.5.4

The expected utilities are given at the appropriate "terminal branches" of the decision tree in Figure 6.5.4. It should be mentioned that some of the numbers given in Figure 6.5.4 (those not involving the second test) were determined in Section 6.4.

The process of backward induction results in the completed tree in Figure 6.5.4. The overall expected utility of purchasing the seismic information with the possibility of purchasing the second test as well is .3578, which corresponds to a cash equivalent of approximately $19,000. If no information is purchased, the optimal action is to farm out the oil-drilling venture for $10,000 and a $\frac{1}{16}$ override, and the expected utility is .3520, which corresponds to a certainty equivalent of approximately $15,000. Thus, the decision maker should purchase the seismic information. If it indicates "no structure," he should farm out for $10,000 and a $\frac{1}{16}$ override; if it indicates "open structure," he should farm out for $10,000 and a $\frac{1}{16}$ override (that is, he should not conduct the second test); if it indicates "closed structure," he should conduct the second test. If the result of the second test is favorable, he should drill with 100 percent interest; if it is unfavorable, he should farm out for $10,000 and a $\frac{1}{16}$ override.

The quality-control and oil-drilling examples illustrate sequential decision making, which is sometimes called multistage decision making or, because of the time element involved in the sequential procedure, dynamic decision making. Many real-world decision-making problems involve sequences of decisions to be made over time and frequent opportunities to purchase new information, where the action to be taken at any point may depend on previous actions or events and on possible future actions and events. As noted at the beginning of the section, it is convenient to represent a sequential problem in terms of a decision tree, in which case the problem can be solved by using the process of backward induction, or averaging out and folding back. The result of this process is an overall strategy for the decision maker that tells him what action to take (in order to maximize expected utility) at any action fork in the decision tree.

6.6 LINEAR PAYOFF FUNCTIONS: THE TWO-ACTION PROBLEM

In examples concerning the value of information, only the situation in which there are a finite number of states of the world and a finite number of actions has been considered. Under these conditions, it is possible to set up a payoff or loss table and to use this table in the calculation of EVPI and EVSI. In many decision-making problems, however, the state of the world is represented by a continuous random variable. The quality-control example seems more realistic if it is assumed that $\tilde{p}$ is continuous and can take on any value between 0 and 1 instead of being limited to .01, .05, .10, and .25. In the stock-purchasing example, the four given states of the world are by no means the only possible states; it would be more reasonable to assume that the price of a stock one year from now is a continuous random variable. In the oil-drilling venture, a continuous distribution for $\tilde{\theta}$ would seem more realistic than a discrete distribution with only five possible values. In numerous such decision problems, the state of the world can be thought of as continuous [although this is an idealization—the state of the world may actually take on a very large (but finite) number of values].

If the state of the world is continuous, it is not possible to represent the payoffs in a table, because the table would have an (uncountably) infinite number of columns. In general, an attempt is made to specify the payoffs in functional form. In this section it is assumed that the functional form is *linear in terms of the state of the world.* In other words, if $\tilde{\theta}$ is the state of the world, then the payoff function can be expressed in the form

$$R(a,\theta) = r + s\theta, \tag{6.6.1}$$

where r and s are constants. It is assumed in the following discussion that the decision maker has a utility function that is linear in R, so that maximizing ER is equivalent to maximizing EU (alternatively, you can think of the payoff function as already being expressed in terms of utility).

Payoff functions of the form (6.6.1) greatly simplify the decision-making problem because expected payoffs can be written as

$$ER(a) = E(r + s\tilde{\theta}) = r + sE(\tilde{\theta}). \tag{6.6.2}$$

This result implies that in order to make a terminal decision, all the knowledge that is needed about the distribution of $\tilde{\theta}$ is the mean, $E(\tilde{\theta})$. Thus, it is necessary to know only $E(\tilde{\theta})$, not the entire distribution. In making a decision, the decision maker can act as though the mean is equal to the true value of $\tilde{\theta}$ with certainty; hence, the mean of the distribution is called a *certainty equivalent* in this situation. Certainty equivalents simplify the determination of one of the inputs to the decision problem, for it was pointed out in Chapter 4 that the assessment of entire distributions may be a difficult and time-consuming task, especially in the continuous case. The importance of such *prior* certainty equivalents is somewhat limited, for the assessed distribution is often combined with sample information to form a posterior distribution. Then the *posterior* mean is a certainty equivalent. In order to determine the posterior mean, however, it is generally necessary to assess the entire prior distribution before Bayes' theorem can be applied to determine the posterior distribution.

For an example of a problem with linear payoff functions, consider the quality-control example presented in Section 4.4. In that example the uncertain quantity of interest (that is, the state of the world) was $\tilde{p}$, the proportion of defective items produced by a certain production process. The production manager had to decide whether or not to adjust the process, and the payoff functions were

$$R(\text{no adjustment}, p) = 500 - 2000p$$

and

$$R(\text{adjustment}, p) = 400 - 1000p.$$

These linear payoff functions are illustrated in Figure 6.6.1. The production manager's prior distribution for $\tilde{p}$ was a beta distribution with $r' = 1$ and $n' = 20$, so that the mean of the distribution, $E'(\tilde{p}) = r'/n' = .05$, is a certainty equivalent in this example.

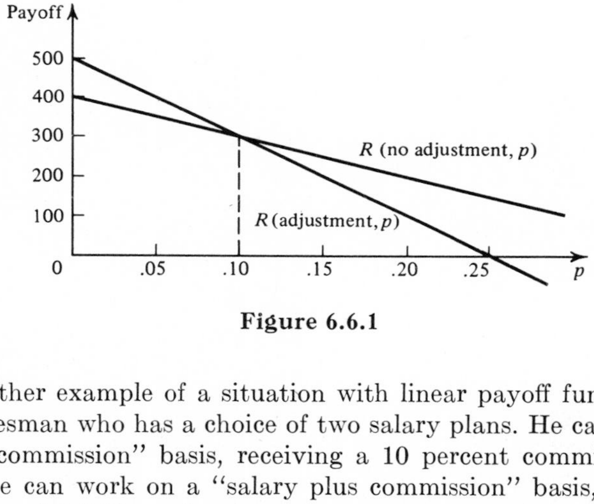

Figure 6.6.1

For another example of a situation with linear payoff functions, consider a salesman who has a choice of two salary plans. He can work on a "straight commission" basis, receiving a 10 percent commission on all sales, or he can work on a "salary plus commission" basis, receiving a fixed salary of \$5000 yearly plus a five percent commission on all sales. If the uncertain quantity $\tilde{\theta}$ denotes the salesman's total sales in the coming year (it is assumed that he can switch plans at the end of any year if he desires, so it is not necessary to consider potential sales in future years), then

$$R(\text{straight commission}, \theta) = .1\theta$$

and

$$R(\text{salary plus commission}, \theta) = 5000 + .05\theta.$$

These payoff functions are illustrated in Figure 6.6.2. If the salesman assesses a probability distribution for $\tilde{\theta}$, then the mean of the distribution (that is, his expected total sales in the coming year) is a certainty equivalent for this problem.

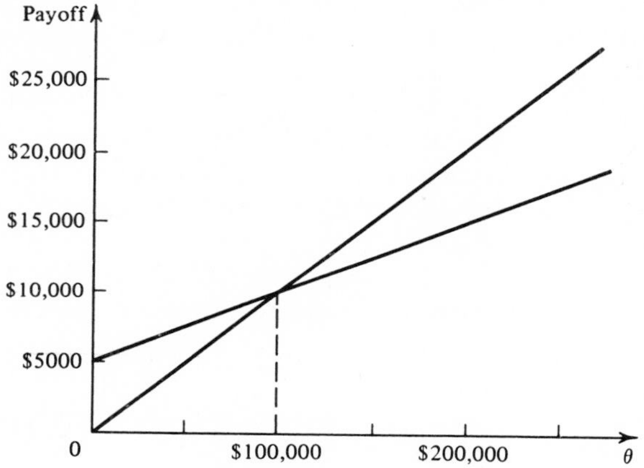

Figure 6.6.2

There are many other situations in which the payoff functions for a given decision-making problem are linear. For instance, the payoff functions in the oil-drilling example presented in Sections 5.10 and 6.4 are linear:

$$R(\text{farm out for \$10,000 and } \tfrac{1}{16} \text{ override}, \theta) = .06\theta + 10{,}000,$$
$$R(\text{farm out for } \tfrac{1}{8} \text{ override}, \theta) = .12\theta,$$
$$R(\text{drill with 50\% interest}, \theta) = .48\theta - 34{,}000,$$

and

$$R(\text{drill with 100\% interest}, \theta) = .96\theta - 68{,}000,$$

where $\tilde{\theta}$ represents the number of barrels of oil that can be recovered if a well is drilled. In producing and selling a product, a firm's profits often can be expressed in the form

$$R(a,\theta) = -r + s\theta,$$

where r represents fixed costs, s represents net profit per unit sold (price per unit minus variable profit per unit), and $\tilde{\theta}$ is the number of units sold. The point is simply that linear payoff functions are widely applicable in real-world decision-making problems.

Suppose that there are just two possible actions, a_1 and a_2, and that their payoff functions are

$$R(a_1,\theta) = r_1 + s_1\theta$$

and (6.6.3)

$$R(a_2,\theta) = r_2 + s_2\theta,$$

where $s_2 > s_1$. Under what conditions will a_1 be optimal under a prior distribution $f(\theta)$, and under what conditions will a_2 be optimal? The first act, a_1, will be optimal if

$$ER(a_1) > ER(a_2). \tag{6.6.4}$$

But all that is needed from $f(\theta)$ in order to compute the expected payoffs is the mean, $E(\tilde{\theta})$, so Equation (6.6.4) is equivalent to

$$r_1 + s_1E(\tilde{\theta}) > r_2 + s_2E(\tilde{\theta}).$$

Subtracting r_2 and $s_1E(\tilde{\theta})$ from both sides, we get

$$r_1 - r_2 > E(\tilde{\theta})(s_2 - s_1).$$

Since $s_2 > s_1$, the term $(s_2 - s_1)$ is greater than 0, so dividing both sides by this term will not affect the inequality:

$$\frac{r_1 - r_2}{s_2 - s_1} > E(\tilde{\theta}). \tag{6.6.5}$$

Therefore, if Equation (6.6.5) is satisfied, a_1 is the optimal action; if the inequality is reversed, a_2 is the optimal action. For this decision-making problem, θ_b is called the *breakeven value* of $\tilde{\theta}$:

$$\theta_b = \frac{r_1 - r_2}{s_2 - s_1}. \tag{6.6.6}$$

If the expected value of $\tilde{\theta}$ is lower than θ_b, then a_1 is optimal; if it is greater than θ_b, then a_2 is optimal. If $E(\tilde{\theta}) = \theta_b$, the expected payoffs of a_1 and a_2 are equal.

If the decision maker is basing his terminal decision on a prior distribution with mean $E'(\tilde{\theta})$, then a_1 is optimal if

$$E'(\tilde{\theta}) < \theta_b,$$

and a_2 is optimal if

$$E'(\tilde{\theta}) > \theta_b.$$

If the decision maker is basing his terminal decision on a posterior distribution with mean $E''(\tilde{\theta})$, then the above inequalities hold, with $E''(\tilde{\theta})$ substituted for $E'(\tilde{\theta})$. Similarly, suppose that the decision maker knows for certain that $\tilde{\theta} = \theta$. In this case, a_1 is optimal if $\theta < \theta_b$ and a_2 is optimal if $\theta > \theta_b$. Thus, it is easy to determine which act is optimal under a prior distribution, under a posterior distribution, or under certainty.

A graph should serve to illustrate the above points. In Figure 6.6.3, both payoff functions are linear, and hence they are represented on the graph by straight lines. The condition $s_2 > s_1$ means that the line corresponding to $R(a_2,\theta)$ has a greater slope than the line corresponding to $R(a_1,\theta)$. This is clearly true in Figure 6.6.3, for the former line has a positive slope and the latter line has a negative slope. The breakeven value, θ_b, is the value of $\tilde{\theta}$ at which the two lines intersect. To the left of this point, the line representing $R(a_1,\theta)$ is higher, so that a_1 yields a larger payoff than a_2, and to the right of the point the line representing $R(a_2,\theta)$ is higher, so that a_2 yields a higher payoff than a_1.

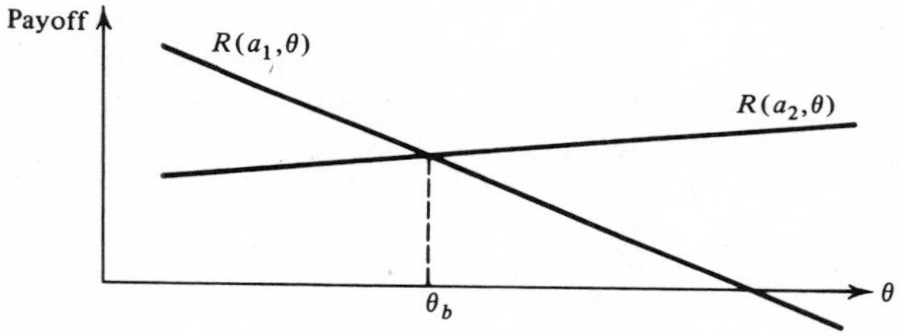

Figure 6.6.3

In the quality-control example illustrated in Figure 6.6.1, $-1000 > -2000$, so a_1 = "no adjustment" and a_2 = "adjustment." Thus, the breakeven value of $\tilde{p}$ is

$$p_b = \frac{r_1 - r_2}{s_2 - s_1} = \frac{500 - 400}{-1000 - (-2000)} = \frac{100}{1000} = .10,$$

as shown in Figure 6.6.1. Since $f(p)$ is a beta distribution with $r' = 1$ and $n' = 20$, $E'(\tilde{p}) = r'/n' = .05$, which is less than the breakeven value. Hence, a_1, no adjustment, is the optimal decision.

In the example involving two salary plans, $.1 > .05$, so a_1 = "salary plus commission" and a_2 = "straight commission." The breakeven value of $\tilde{\theta}$ is

$$\theta_b = \frac{r_1 - r_2}{s_2 - s_1} = \frac{5000 - 0}{.1 - .05} = \frac{5000}{.05} = 100{,}000,$$

as illustrated in Figure 6.6.2. If the salesman's probability distribution for $\tilde{\theta}$, total sales in the coming year, has a mean of $E(\tilde{\theta}) = \$150{,}000$, then a_2, straight commission, is his best choice in terms of maximizing ER.

For a numerical example involving the use of Bayes' theorem, suppose that you are interested in the mean $\tilde{\mu}$ of a normal data-generating process with known variance 100. The prior distribution of $\tilde{\mu}$ is a normal distribution with mean 100 and variance 25. There are two possible actions, and their payoff functions are linear:

$$R(a_1,\mu) = -45 + .5\mu$$

and

$$R(a_2,\mu) = -94 + \mu.$$

The breakeven value of $\tilde{\mu}$ is

$$\mu_b = \frac{-45 - (-94)}{1 - .50} = \frac{49}{.5} = 98.$$

Since $E'(\mu) = 100$ and $\mu_b = 98$, the optimal act under the prior distribution is a_2.

Consider a sample of size 10 from the above process, with sample mean $m = 95$. Using Bayes' theorem, the posterior distribution is normal with parameters m'' and σ''^2, where

$$\frac{1}{\sigma''^2} = \frac{1}{\sigma'^2} + \frac{n}{\sigma^2} = \frac{1}{25} + \frac{10}{100} = .04 + .10 = .14,$$

$$\sigma''^2 = \frac{1}{.14} = 7.14,$$

and $$m'' = \frac{(1/\sigma'^2)m' + (n/\sigma^2)m}{(1/\sigma'^2) + (n/\sigma^2)} = \frac{.04(100) + .10(95)}{.04 + .10} = 96.43.$$

Therefore, the posterior mean is less than the breakeven value, and the optimal act under the posterior distribution is a_1. If you were somehow able to obtain perfect information that $\tilde{\mu} = \mu = 97.5$, then the optimal act under certainty would be a_1. This example is illustrated in Figure 6.6.4.

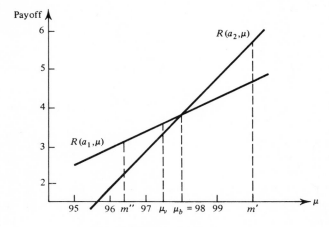

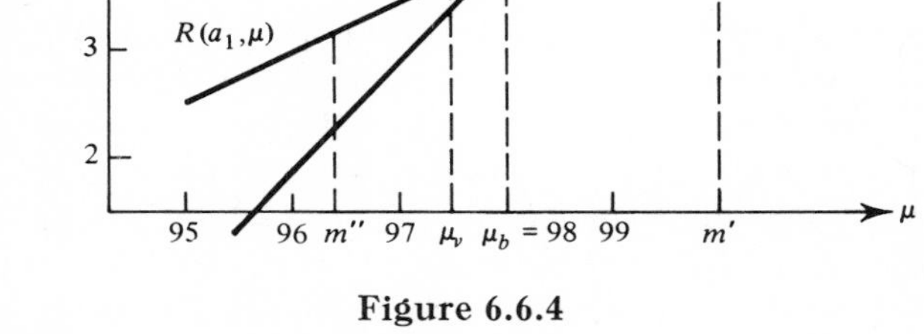

Figure 6.6.4

You are probably interested in how this example could be interpreted in terms of a real-world problem. The two actions could involve a certain investment (for example, in a stock or in an oil-drilling venture). Action a_1 might correspond to a half interest in the investment. The half interest costs \$45, and the payoff is $\$\mu/2$. Action a_2 corresponds to a full interest, which costs \$94 and pays $\$\mu$. The sample is taken from other previous investments that are judged to be virtually identical to the current investment, and the payoff from each of these investments is assumed to be normally distributed with mean $\tilde{\mu}$ and known standard deviation \$10.

In summary, then, linear payoff functions are quite useful in two respects. First, payoff functions in real-world decision problems are often linear or very nearly linear. Second, when there are just two competing actions, the decision criterion of maximization of expected payoff reduces to the simple process of comparing two numbers: the breakeven value, θ_b, and the mean of the decision maker's probability distribution, $E(\tilde{\theta})$.

6.7 LOSS FUNCTIONS AND LINEAR PAYOFF FUNCTIONS

If the payoff functions for both acts in a two-action problem are linear with respect to the state of the world, then the corresponding loss functions are linear also. In the example illustrated in Figure 6.6.4, the optimal action under the prior distribution is a_2. If this action is taken, what does the loss function look like? The loss function, like the payoff function, is a function of $\tilde{\mu}$. If the true value of $\tilde{\mu}$ is greater than the breakeven value, 98, then the act chosen under the prior distribution is in fact the better of the two acts, and the loss is 0 (as always, losses are defined in terms of the "opportunity loss" concept). On the other hand, suppose that the true value of $\tilde{\mu}$ is less than 98. Then a_1 gives the higher payoff, and the loss suffered by choosing a_2 is the difference in payoffs,

$$R(a_1,\mu) - R(a_2,\mu).$$

From the payoff functions,

$$R(a_1,\mu) = -45 + .5\mu$$

and

$$R(a_2,\mu) = -94 + \mu,$$

the loss function (given that act a_2 is chosen) can be written as

$$L(a_2,\mu) = \begin{cases} (-45 + .5\mu) - (-94 + \mu) = 49 - .5\mu & \text{if } \mu < 98, \\ 0 & \text{if } \mu \geq 98. \end{cases}$$

This loss function is illustrated in Figure 6.7.1.

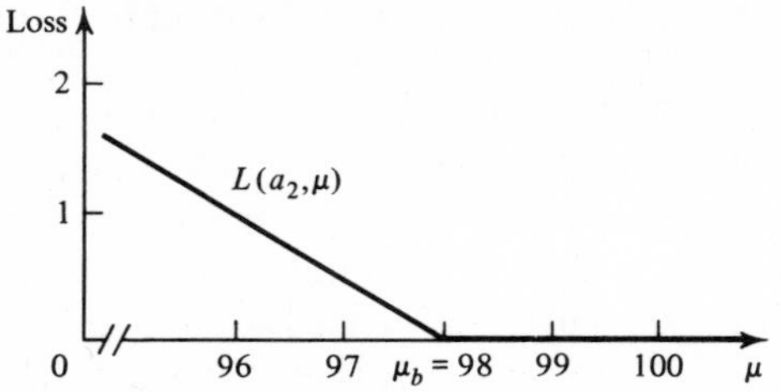

Figure 6.7.1

If act a_1 is chosen (as it would be under the posterior distribution given in the example), then the loss depends once again on $\tilde{\mu}$. If $\tilde{\mu}$ is less than 98, the best act was chosen, and the loss is 0. If $\tilde{\mu}$ is greater than the breakeven value, the loss is the difference

$$R(a_2,\mu) - R(a_1,\mu).$$

The loss function for act a_1 can thus be written

$$L(a_1,\mu) = \begin{cases} 0 & \text{if } \mu \leq 98, \\ (-94 + \mu) - (-45 + .5\mu) = -49 + .5\mu & \text{if } \mu > 98. \end{cases}$$

The loss function is shown in Figure 6.7.2.

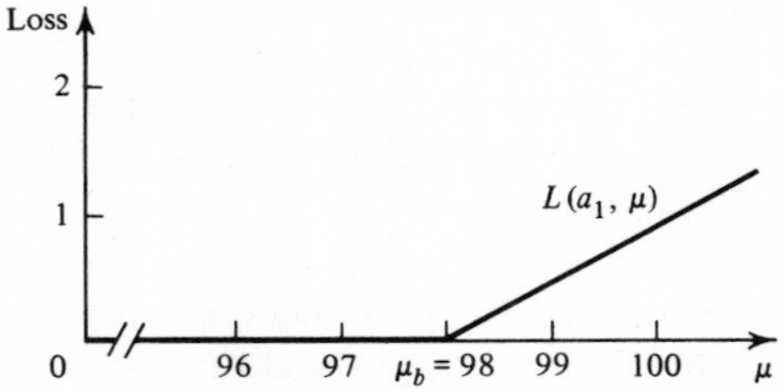

Figure 6.7.2

Consider a specific value or two. If a_2 is chosen and the true value of $\tilde{\mu}$ is 95, then the loss incurred is

$$L(a_2,95) = 49 - .5(95) = 49 - 47.5 = 1.5.$$

If a_1 is chosen and the true value of $\tilde{\mu}$ is 95, then the loss is 0. If a_1 is chosen and the true value of $\tilde{\mu}$ is 110, however, then the loss is

$$L(a_1,110) = -49 + .5(110) = -49 + 55 = 6.$$

In Figures 6.7.1 and 6.7.2 the loss function is equal to 0 on one side of the breakeven value and is a straight line on the other side of the breakeven value. In general, suppose that you have two acts with payoff functions of the form given by Equation (6.6.3):

$$R(a_1,\theta) = r_1 + s_1\theta$$

and

$$R(a_2,\theta) = r_2 + s_2\theta.$$

These payoff functions are shown in Figure 6.7.3. Consider the payoff function that represents the best you can do for any value of $\tilde{\theta}$; this is the V-shaped function with the dots on it in the graph (such a function is called piecewise linear). Now suppose that a_1 is the act that is chosen. The loss suffered at any particular value of $\tilde{\theta}$ is simply the vertical distance between the line representing $R(a_1,\theta)$ and the piecewise linear function that represents the best you can do. But this distance is 0 when $\tilde{\theta} < \theta_b$, and it is the difference

$$R(a_2,\theta) - R(a_1,\theta)$$

when $\tilde{\theta} > \theta_b$. The loss function is thus of the form

$$L(a_1,\theta) = \begin{cases} 0 & \text{if } \theta \leq \theta_b, \\ (r_2 + s_2\theta) - (r_1 + s_1\theta) = (r_2 - r_1) + (s_2 - s_1)\theta & \text{if } \theta > \theta_b. \end{cases} \tag{6.7.1}$$

The graph of this loss function will look like Figure 6.7.2. Similarly, you can see that

$$L(a_2,\theta) = \begin{cases} (r_1 + s_1\theta) - (r_2 + s_2\theta) = (r_1 - r_2) + (s_1 - s_2)\theta & \text{if } \theta < \theta_b, \\ 0 & \text{if } \theta \geq \theta_b. \end{cases} \tag{6.7.2}$$

The graph of this loss function will look like Figure 6.7.1.

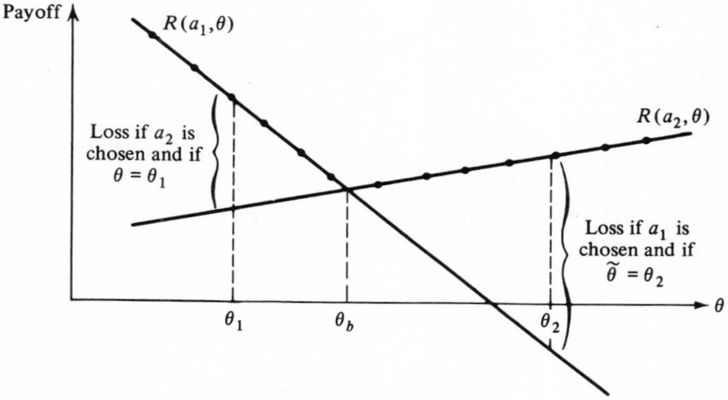

Figure 6.7.3

In the quality-control example in the previous section, $r_1 = 500$, $r_2 = 400$, $s_1 = -2000$, $s_2 = -1000$, and $p_b = .10$, so that

$$L(a_1,p) = \begin{cases} 0 & \text{if } p \leq .10, \\ -100 + 1000p & \text{if } p > .10, \end{cases}$$

and

$$L(a_2,p) = \begin{cases} 100 - 1000p & \text{if } p < .10, \\ 0 & \text{if } p \geq .10. \end{cases}$$

These loss functions are graphed in Figures 6.7.4 and 6.7.5. In the example from the previous section involving two salary plans, $r_1 = 5000$, $r_2 = 0$, $s_1 = .05$, $s_2 = .10$, and $\theta_b = 100{,}000$, so that

$$L(a_1,\theta) = \begin{cases} 0 & \text{if } \theta \leq 100{,}000, \\ -5000 + .05\theta & \text{if } \theta > 100{,}000, \end{cases}$$

and

$$L(a_2,\theta) = \begin{cases} 5000 - .05\theta & \text{if } \theta < 100{,}000, \\ 0 & \text{if } \theta \geq 100{,}000. \end{cases}$$

These loss functions are graphed in Figures 6.7.6 and 6.7.7.

Why are these loss functions of interest? The decision maker may want to compute EVPI, which is equal to the expected loss of the act that is optimal under the distribution of $\tilde{\theta}$. For instance, assume for the moment that a_1 is optimal and that the distribution is represented by $f(\theta)$. Then the VPI is given by $L(a_1,\theta)$, and the EVPI is equal to

$$EL(a_1) = \int_{-\infty}^{\infty} L(a_1,\theta)f(\theta)\,d\theta.$$

Using Equation (6.7.1),

$$EL(a_1) = \int_{-\infty}^{\theta_b} 0f(\theta)\,d\theta + \int_{\theta_b}^{\infty} [(r_2 - r_1) + (s_2 - s_1)\theta]f(\theta)\,d\theta,$$

or

$$EL(a_1) = \int_{\theta_b}^{\infty} [(r_2 - r_1) + (s_2 - s_1)\theta]f(\theta)\,d\theta. \tag{6.7.3}$$

Similarly, if a_2 is optimal under $f(\theta)$, then the EVPI is equal to

$$EL(a_2) = \int_{-\infty}^{\infty} L(a_2,\theta)f(\theta)\,d\theta,$$

$$= \int_{-\infty}^{\theta_b} [(r_1 - r_2) + (s_1 - s_2)\theta]f(\theta)\,d\theta + \int_{\theta_b}^{\infty} 0f(\theta)\,d\theta,$$

or

$$= \int_{-\infty}^{\theta_b} [(r_1 - r_2) + (s_1 - s_2)\theta]f(\theta)\,d\theta. \tag{6.7.4}$$

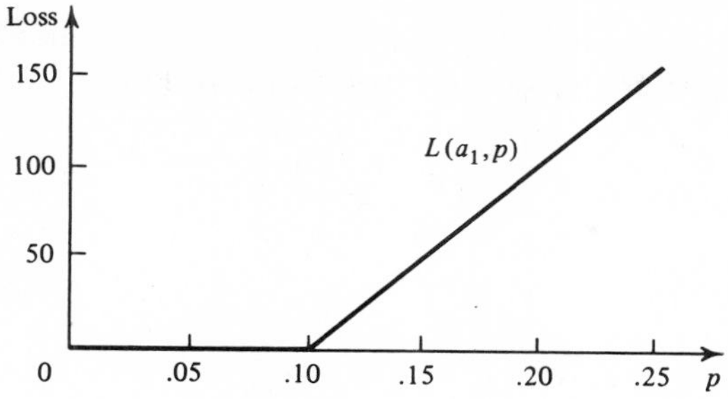

Figure 6.7.4

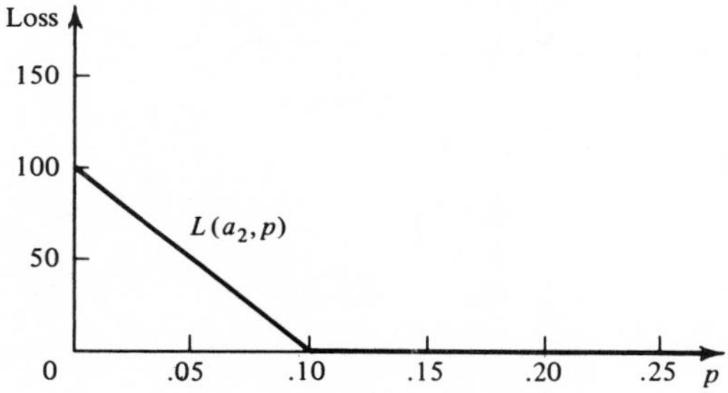

Figure 6.7.5

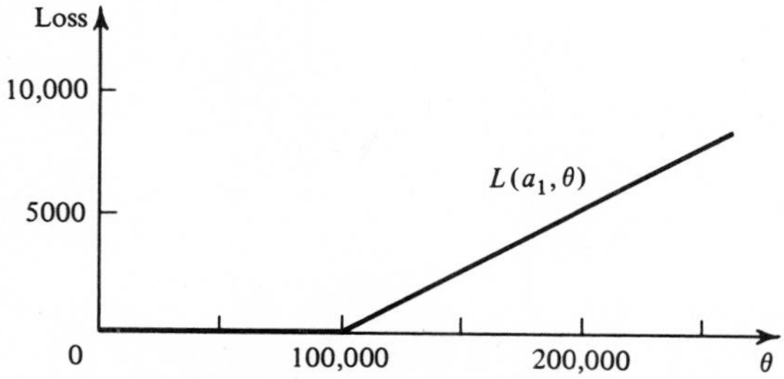

Figure 6.7.6

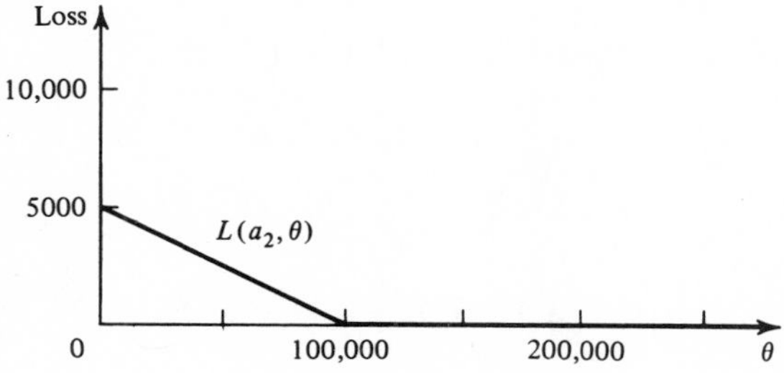

Figure 6.7.7

From Figures 6.7.1 and 6.7.2, it is clear that the loss functions are related to the breakeven point, θ_b. In particular, for $\theta > \theta_b$,

$$L(a_1,\theta) = (r_2 - r_1) + (s_2 - s_1)\theta = (s_2 - s_1)\left[\frac{(r_2 - r_1)}{(s_2 - s_1)}\right] + (s_2 - s_1)\theta.$$

But from Equation (6.5.6), $\theta_b = (r_1 - r_2)/(s_2 - s_1)$. Therefore, for $\theta > \theta_b$,

$$L(a_1,\theta) = (s_2 - s_1)(-\theta_b) + (s_2 - s_1)\theta = (s_2 - s_1)(\theta - \theta_b).$$

This agrees with the graph in Figure 6.7.2, which implies that to the right of θ_b, the loss function is a linear function of $(\theta - \theta_b)$, the distance from θ to θ_b. The entire loss function is simply

$$L(a_1,\theta) = \begin{cases} 0 & \text{if } \theta \leq \theta_b, \\ (s_2 - s_1)(\theta - \theta_b) & \text{if } \theta \geq \theta_b. \end{cases} \tag{6.7.5}$$

The loss function corresponding to the second action can be written in the form

$$L(a_2,\theta) = \begin{cases} (s_2 - s_1)(\theta_b - \theta) & \text{if } \theta \leq \theta_b, \\ 0 & \text{if } \theta \geq \theta_b. \end{cases} \tag{6.7.6}$$

For the first example in this section, $s_1 = .5$, $s_2 = 1$, and the break-even value of the uncertain quantity $\tilde{\mu}$ is $\mu_b = 98$. Thus,

$$L(a_1,\mu) = \begin{cases} 0 & \text{if } \mu \leq \mu_b, \\ .5(\mu - \mu_b) & \text{if } \mu \geq \mu_b, \end{cases}$$

and

$$L(a_2,\mu) = \begin{cases} .5(\mu_b - \mu) & \text{if } \mu \leq \mu_b, \\ 0 & \text{if } \mu \geq \mu_b. \end{cases}$$

For the quality-control example,

$$L(a_1,p) = \begin{cases} 0 & \text{if } p \leq .10, \\ 1000(p - p_b) & \text{if } p \geq .10, \end{cases}$$

and

$$L(a_2,p) = \begin{cases} 1000(p_b - p) & \text{if } p \leq .10, \\ 0 & \text{if } p \geq .10. \end{cases}$$

When $L(a_1,\theta)$ and $L(a_2,\theta)$ are expressed in the forms of Equations (6.7.5) and (6.7.6), the expected losses are

$$EL(a_1) = (s_2 - s_1) \int_{\theta_b}^{\infty} (\theta - \theta_b) f(\theta)\, d\theta \tag{6.7.7}$$

and

$$EL(a_2) = (s_2 - s_1) \int_{-\infty}^{\theta_b} (\theta_b - \theta) f(\theta)\, d\theta. \tag{6.7.8}$$

The EVPI is equal to the smaller of the two expected losses. The expressions given by Equations (6.7.7) and (6.7.8) [or by Equations (6.7.3) and (6.7.4)], as well as similar expressions for EVSI, may be difficult to calculate because the integration in each case is over only part of the real line, representing the part for which the loss function is nonzero. Such expressions are called *partial expectations* in general, and the integrals in Equations (6.7.7) and (6.7.8) are called right-hand and left-hand *linear loss integrals,* respectively. In the next two sections methods are presented for evaluating these integrals when $f(\theta)$ is a normal distribution or a beta distribution.

6.8 LINEAR LOSS FUNCTIONS AND THE NORMAL DISTRIBUTION

Under the assumption that $f(\theta)$ is a normal distribution, it is possible to compute the values of Equations (6.7.7) and (6.7.8) with the aid of a table of the "unit normal linear loss integral," which is defined as

$$L_N(D) = \int_D^{\infty} (\theta - D) f_N(\theta)\, d\theta, \tag{6.8.1}$$

where $f_N(\theta)$ is the standard normal density function. In addition, this table, which is presented at the end of the book, facilitates the computation of the expected value of sample information when the loss functions are linear and the distribution is normal. The equations by which EVPI and EVSI can be computed from the table are stated without proof to avoid mathematical details.

Assume that a decision maker is interested in the mean, $\tilde{\mu}$, of a normal process with known variance σ^2. He has a prior distribution $f'(\mu)$ and a two-action problem with linear loss functions (linear in terms of $\tilde{\mu}$) of the forms (6.7.1) and (6.7.2). The prior distribution is a normal distribution with mean m' and variance σ'^2. The EVPI is then

$$\text{EVPI} = |s_1 - s_2|\sigma' L_N(D), \tag{6.8.2}$$

where

$$D = \left| \frac{\mu_b - m'}{\sigma'} \right| \tag{6.8.3}$$

and the value $L_N(D)$ can be read from the table of the unit normal linear loss integral.

First, note that in Equations (6.7.5) and (6.7.6), the slope of the nonzero portion of the loss function is $(s_2 - s_1)$. Since the actions are labeled so that $s_2 > s_1$, $(s_2 - s_1) = |s_2 - s_1|$, and that is represented by the first factor in Equation (6.8.2). As this slope increases (in absolute value), the EVPI increases, which is entirely reasonable, since a greater slope implies greater losses for wrong decisions (in terms of the payoff functions, $|s_1 - s_2|$ is the absolute difference between the slopes of the two payoff functions, as illustrated in Figure 6.7.3). But as the losses due to wrong decisions increase, then it is more important that the right decision be made, so that perfect information is worth more to the decision maker. The second factor in the equation for EVPI is the prior standard deviation, which also affects the third factor, $L_N(D)$, through D. The smaller the prior standard deviation, the more current information the decision maker has about $\tilde{\mu}$. But the more information he has, the less perfect information is worth to him. Thus, the second factor makes sense intuitively. Finally, consider the factor $L_N(D)$. As D increases (note that it will always be positive, since it is an absolute value), $L_N(D)$ decreases, as can be seen from the table of $L_N(D)$. What does D represent? D is a "standardized" variable that represents the number of prior standard deviations σ' that separate the prior mean, m', and the breakeven value, μ_b. The larger D is, the further m' is from μ_b in terms of standardized units, and the more certain the decision maker is that the optimal decision under the prior distribution is the correct decision. As D increases, the probability that $\tilde{\mu}$ is on the opposite side of the breakeven value from m' decreases, implying that the sample is less likely to change the optimal decision. In this case, EVPI should decrease, and you can see from the table of values of $L_N(D)$ that $L_N(D)$ decreases as D increases, implying that EVPI also decreases.

Consider the example from the previous section, in which

$$L(a_1,\mu) = \begin{cases} 0 & \text{if } \mu \leq 98, \\ -49 + .5\mu & \text{if } \mu \geq 98, \end{cases}$$

and

$$L(a_2,\mu) = \begin{cases} 49 - .5\mu & \text{if } \mu \leq 98, \\ 0 & \text{if } \mu \geq 98. \end{cases}$$

The breakeven value is $\mu_b = 98$. The decision maker's prior distribution is a normal distribution with mean 100 and variance 25, and the known variance of the process (or population) is 100. Since $m' > \mu_b$, a_2 is the optimal action under the prior distribution, and $L(a_2,\mu)$ is the relevant loss function. The prior distribution of $\tilde{\mu}$ and the loss function $L(a_2,\mu)$ are illustrated in Figure 6.8.1. The slope of the positive portion of $L(a_2,\mu)$ is $-.5$, so $|s_1 - s_2|$, the absolute value of this slope, is equal to .5. The prior standard deviation, σ', is $\sqrt{25}$, or 5, and

$$D = \left|\frac{\mu_b - m'}{\sigma'}\right| = \left|\frac{98 - 100}{5}\right| = |-.4| = .4.$$

The expected value of perfect information, from Equation (6.8.2), is

$$\text{EVPI} = .5(5)L_N(.4).$$

From the table of $L_N(D)$, $L_N(.4) = .2304$, so that

$$\text{EVPI} = .5(5)(.2304) = .5760.$$

If the losses are in terms of dollars, then the decision maker should be willing to pay up to \$0.57 for perfect information. If the losses are in terms of thousands of dollars (and they might be if this is, say, a major investment decision), he should be willing to pay up to \$576 for perfect information.

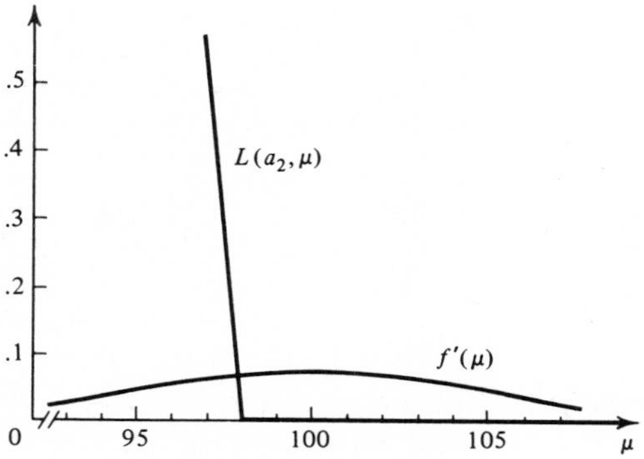

Figure 6.8.1

Perhaps the decision maker is contemplating a sample of size n and would like to know the expected value of such a sample. The formula for EVSI, which is very similar to Equation (6.8.2), is

$$\text{EVSI} = |s_1 - s_2|\sigma^* L_N(D^*), \tag{6.8.4}$$

where

$$\sigma^{*2} = \frac{n\sigma'^4}{\sigma^2 + n\sigma'^2} \tag{6.8.5}$$

and

$$D^* = \left| \frac{\mu_b - m'}{\sigma^*} \right|. \qquad (6.8.6)$$

The term denoted by σ^{*2} is essentially the reduction in variance due to the sample, or the prior variance minus the posterior variance. From Equation (4.5.8),

$$\frac{1}{\sigma''^2} = \frac{1}{\sigma'^2} + \frac{n}{\sigma^2} = \frac{\sigma^2 + n\sigma'^2}{\sigma'^2\sigma^2}.$$

Hence,

$$\sigma''^2 = \frac{\sigma'^2\sigma^2}{\sigma^2 + n\sigma'^2},$$

and the reduction in variance is then

$$\begin{aligned} \sigma'^2 - \sigma''^2 &= \sigma'^2 - \frac{\sigma'^2\sigma^2}{\sigma^2 + n\sigma'^2} \\ &= \frac{\sigma'^2\sigma^2 + n\sigma'^4 - \sigma'^2\sigma^2}{\sigma^2 + n\sigma'^2} \\ &= \frac{n\sigma'^4}{\sigma^2 + n\sigma'^2}, \end{aligned}$$

which is identical with the expression for σ^{*2} given in Equation (6.8.5). The term σ^* is therefore the square root of the reduction in variance due to the sample of size n.

Note that the EVSI, just like the EVPI, becomes smaller as the difference between μ_b and m' becomes larger. As this difference increases, it becomes less likely that the sample information will change the decision. But the "value" of the sample to the decision maker depends solely on how the sample affects his terminal decision. If the sample does not change his decision, the information does not affect his eventual payoff (although it may make him more certain that he is taking the best action). Thus, in preposterior analysis, if it is unlikely that a proposed sample will change the choice of a terminal decision, the EVSI will be small. Remember that in the decision-making framework considered in this book, sample information has no "value" by itself; its value relates to how it affects the terminal decision. For instance, if you are faced with an investment decision, information concerning the possibility of a tax increase might be of some interest to you. However, if you would choose the same investment regardless of whether or not taxes are raised, then information concerning a possible tax increase is of no value to you in terms of your investment decision.

Returning to the example, suppose that the decision maker contemplates a sample of size 10. The reduction in variance would be

$$\sigma^{*2} = \frac{n\sigma'^4}{\sigma^2 + n\sigma'^2} = \frac{10(5)^4}{100 + 10(25)} = \frac{6250}{350} = 17.9,$$

and

$$\sigma^* = \sqrt{17.9} = 4.23.$$

The value of D^* is

$$D^* = \left| \frac{\mu_b - m'}{\sigma^*} \right| = \left| \frac{98 - 100}{4.23} \right| = |-.47| = .47,$$

and from Equation (6.7.4),

$$\text{EVSI} = .5(4.23)L_N(.47) = .5(4.23)(.2072) = .438.$$

This is the value to the decision maker of a sample of size 10.

In a similar manner, the value of a sample of any given sample size could be computed. Some values are given in the following table.

Sample size (n)	σ^*	D^*	$EVSI(n)$
1	2.24	.89	.115
2	2.89	.69	.210
3	3.28	.61	.272
4	3.54	.56	.319
5	3.75	.53	.354
6	3.88	.51	.378
7	3.99	.50	.396
8	4.09	.49	.411
9	4.16	.48	.424
10	4.23	.47	.438
11	4.29	.47	.444

Notice that as n gets larger, EVSI(n) also gets larger, but the rate of increase of EVSI(n) becomes smaller. In terms of a graph, the EVSI curve is beginning to "level off." It is worthwhile to note that as n gets larger, the sample information is closer to what has been termed "perfect information," and the EVSI gets closer to the EVPI, which in this case has been shown to be .576. For instance, if $n = 100$, EVSI $= .556$. Obviously EVSI can never exceed EVPI, because sample information can never be more valuable than perfect information.

In general, as n increases, the posterior variance becomes smaller, implying that the reduction in variance, σ^{*2}, becomes larger. As σ^{*2} becomes larger, D^* becomes smaller, since σ^* is in the denominator of D^*. Finally, as D^* becomes smaller, $L_N(D^*)$ becomes larger. The EVSI consists of three factors: $|s_1 - s_2|$, σ^*, and $L_N(D^*)$. As n gets larger, the first term remains constant and the other two terms increase, as the above discussion indicates. Thus, as n gets larger, the EVSI gets larger. By comparing Equations (6.8.2) and (6.8.4) as the sample size increases, you can see that the EVSI approaches the EVPI. The posterior variance approaches 0 as n increases, so that $\sigma^{*2} = \sigma'^2 - \sigma''^2$ approaches σ'^2. But the only difference between Equations (6.8.2) and (6.8.4) is that σ^* is used in the formula for EVSI in place of σ', which is used in the formula for EVPI. Since σ^* approaches σ' as n increases, EVSI must approach EVPI as n increases.

In determining optimal sample size, it is necessary to consider the cost of sampling, CS(n), as well as EVSI(n). Suppose that

$$\text{CS}(n) = .03n.$$

That is, the cost of sampling is a constant .03 per trial. Then, for each value of n, we can determine ENGS(n) = EVSI(n) − CS(n).

Sample size (n)	*EVSI(n)*	*CS(n)*	*ENGS(n)*
1	.115	.030	.085
2	.210	.060	.150
3	.272	.090	.182
4	.319	.120	.199
5	.354	.150	.204
6	.378	.180	.198
7	.396	.210	.186
8	.411	.240	.171
9	.424	.270	.154
10	.438	.300	.138
11	.444	.330	.114

The optimal sample size is $n = 5$. Note, however, that the ENGS is reasonably insensitive to slight deviations from this optimal sample size. A sample size as small as two or as large as nine results in approximately a 25 percent reduction in ENGS. As n increases, the ENGS increases until the EVSI starts to level off, while the cost of sampling continues to increase at a constant rate. In Figure 6.8.2, values in the preceding table are graphed, and three curves, representing EVSI, CS, and ENGS, are drawn through the three sets of points.

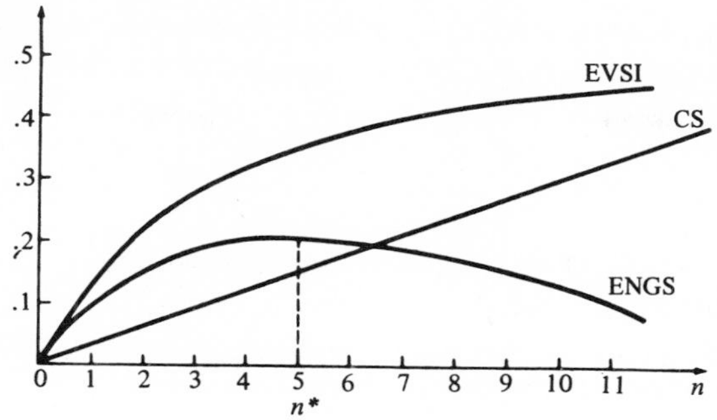

Figure 6.8.2

Of course, preposterior analysis could be carried out repeatedly. After taking the sample of size five (since $n = 5$ is the optimal sample size), the decision maker can revise his prior distribution and use the resulting posterior distribution to conduct another preposterior analysis. As pointed out in Section 6.1, this sequential procedure may continue until a preposterior analysis leads to an optimal sample size of 0, at which time the decision maker should stop sampling and make his terminal decision.

6.9 LINEAR LOSS FUNCTIONS AND THE BETA DISTRIBUTION

In the previous section the determination of the value of perfect information and sample information was discussed under the assumption that the loss functions were linear and that the distribution of $\tilde{\theta}$ was normal. If $f(\theta)$ is not a normal distribution, the general idea is the same, but different linear loss integrals must be considered. Recall that L_N is the unit *normal* linear loss integral. If $f(\theta)$ is a normal distribution or if it can be closely approximated by a normal distribution, then the table of values of L_N can be used in determining EVPI and EVSI. Otherwise, it is necessary to look at different linear loss integrals, such as the loss integrals associated with a beta distribution.

If the uncertain quantity of interest is the parameter $\tilde{p}$ of a Bernoulli process, then the family of conjugate distributions is the family of beta distributions, as discussed in Section 4.3. Consider a decision-making problem in which there are two actions and the payoff (and hence loss) functions for both actions are linear:

$$R(a_1,p) = r_1 + s_1p$$

and

$$R(a_2,p) = r_2 + s_2p,$$

where $s_2 > s_1$. These payoff functions are of the form of Equation (6.6.3). The breakeven value of $\tilde{p}$, p_b, is determined in the usual manner:

$$p_b = \frac{r_1 - r_2}{s_2 - s_1}.$$

Given a prior distribution, the optimal decision can be found by comparing the prior mean and the breakeven value. If $E'(\tilde{p}) < p_b$, then a_1 is optimal; if $E'(\tilde{p}) > p_b$, then a_2 is optimal. If the prior distribution is a beta distribution with parameters r' and n', the expected value of perfect information is given by

$$\text{EVPI} = \begin{cases} |s_1 - s_2|[E'(\tilde{p})P_\beta(\tilde{p} > p_b \mid r' + 1, n' + 1) \\ \qquad - p_b P_\beta(\tilde{p} > p_b \mid r',n')] \quad \text{if } E'(\tilde{p}) \leq p_b, \\ \\ |s_1 - s_2|[p_b P_\beta(\tilde{p} < p_b \mid r',n') \\ \quad - E'(\tilde{p})P_\beta(\tilde{p} < p_b \mid r' + 1, n' + 1)] \quad \text{if } E'(\tilde{p}) \geq p_b. \end{cases} \tag{6.9.1}$$

This formula for EVPI is stated without proof in order to avoid mathematical details. It should be noted, however, that the terms enclosed in square brackets in Equation (6.9.1) are linear loss integrals for the beta distribution.

For example, consider the quality-control example from Section 4.4, in which the production manager had to decide whether or not to make an adjustment in a certain production process. The uncertain quantity of interest was $\tilde{p}$, the proportion of defective items produced by the process, and the production manager's prior distribution for $\tilde{p}$ was a beta distribution with $r' = 1$ and $n' = 20$. The payoff functions were

$$R(a_1,p) = 500 - 2000p$$

and

$$R(a_2,p) = 400 - 1000p,$$

where action a_1 stands for "no adjustment" and action a_2 stands for "adjustment." In this example,

$$p_b = \frac{500 - 400}{-1000 - (-2000)} = .10,$$

and the mean of the prior distribution is

$$E'(\tilde{p}) = \frac{r'}{n'} = \frac{1}{20} = .05.$$

Thus, $E'(\tilde{p}) < p_b$, and a_1 is optimal. The EVPI, from Equation (6.9.1), is then

$$\begin{aligned}\text{EVPI} &= |s_1 - s_2|[E'(\tilde{p})P_\beta(\tilde{p} > p_b \mid r' + 1, n' + 1) - p_b P_\beta(\tilde{p} > p_b \mid r',n')] \\ &= (1000)[.05P_\beta(\tilde{p} > .10 \mid 2,21) - .10P_\beta(\tilde{p} > .10 \mid 1,20)].\end{aligned}$$

Cumulative beta probabilities can be expressed conveniently in terms of the binomial distribution, as indicated in Section 4.4 [Equations (4.3.4) and (4.3.5)]:

$$P_\beta(\tilde{p} < p \mid r,n) = \begin{cases} P_{\text{bin}}(\tilde{r} \geq r \mid n - 1,p) & \text{if } p \leq .5, \\ P_{\text{bin}}(\tilde{r} < n - r \mid n - 1, 1 - p) & \text{if } p \geq 5. \end{cases}$$

For the production example, this yields

$$\begin{aligned}P_\beta(\tilde{p} > .10 \mid 2,21) &= 1 - P_\beta(\tilde{p} < .10 \mid 2,21) \\ &= 1 - P_{\text{bin}}(\tilde{r} \geq 2 \mid 20,.10) \\ &= 1 - .6083 = .3917\end{aligned}$$

and

$$\begin{aligned}P_\beta(\tilde{p} > .10 \mid 1,20) &= 1 - P_\beta(\tilde{p} < .10 \mid 1,20) \\ &= 1 - P_{\text{bin}}(\tilde{r} \geq 1 \mid 19,.10) \\ &= 1 - .8649 = .1351,\end{aligned}$$

so that

$$\begin{aligned}\text{EVPI} &= (1000)[.05(.3917) - .10(.1351)] \\ &= 6.075.\end{aligned}$$

Thus, the production manager should be willing to pay up to \$6.07 to know $\tilde{p}$ for certain.

The formula for EVSI in the case of a beta-distributed random variable $\tilde{p}$ is very similar to the formula for EVPI given in Equation (6.9.1):

$$\text{EVSI}(n) = \begin{cases} |s_1 - s_2|[E'(\tilde{p})P_{\beta\text{bin}}(\tilde{r} > p_b n'' - r' \mid r' + 1, n' + 1, n) \\ \quad - p_b P_{\beta\text{bin}}(\tilde{r} > p_b n'' - r' \mid r',n',n)] \quad \text{if } E'(\tilde{p}) \leq p_b, \\ \\ |s_1 - s_2|[p_b P_{\beta\text{bin}}(\tilde{r} \leq p_b n'' - r' \mid r',n',n) \\ \quad - E'(\tilde{p})P_{\beta\text{bin}}(\tilde{r} \leq p_b n'' - r' \mid r' + 1, n' + 1, n)] \\ \qquad\qquad \text{if } E'(\tilde{p}) \geq p_b. \end{cases} \tag{6.9.2}$$

The main difference between the formulas for EVSI and EVPI is that the probabilities involve the beta-binomial distribution in the calculation of EVSI and the beta distribution in the calculation of EVPI. As noted in Section 4.11, the beta-binomial distribution is the predictive distribution of $\tilde{r}$ when the data-generating process follows a Bernoulli model and the prior distribution is a beta distribution. Since beta-binomial probabilities may be difficult to compute unless n is small (although approximations are available and the problem can be handled easily on a computer), no examples of the calculation of EVSI from Equation (6.9.2) are presented here. As you should expect, EVPI from Equation (6.9.1) is an upper bound for EVSI; as n increases, the beta-binomial probabilities in the formula for EVSI approach the beta probabilities in the formula for EVPI, and the two formulas are identical in all other respects.

This section and the previous section have involved linear payoff and loss functions in situations in which the variable of interest has a normal distribution or a beta distribution. The analysis is similar for other distributions; essentially, the only thing that changes in the determination of EVPI or EVSI is the exact form of the linear loss integral, which may be tedious to compute for some distributions. Since the general approach is similar to the approach presented in this section and in the previous section, loss integrals other than the normal and beta linear loss integrals are not discussed here.

6.10 THE GENERAL FINITE-ACTION PROBLEM

The past four sections have involved *two-action problems* in which the payoff functions (and hence the loss functions) of both of the actions are linear. In some situations there may be more than two possible actions; in the quality-control and salesman examples in Section 6.6, another type of adjustment of the production process might be available and there could be several different "salary plus commission" plans. As long as the number of possible actions is finite, these situations can be referred to as *finite-action problems*. Under linear payoff functions, the *finite-action problem* is similar to its special case, the two-action problem. The formulas for the loss functions, EVPI, and EVSI are fairly complex, however, so just a brief nonmathematical discussion is presented and the important points are illustrated with graphs.

Formally, suppose that there are k possible actions, so that the set of actions, A, can be represented by $a_1, a_2, \ldots, a_k$. The payoff function of the action a_i is given by

$$R(a_i,\theta) = r_i + s_i\theta \quad \text{for } i = 1, 2, \ldots, k. \tag{6.10.1}$$

If $k = 2$, this reduces to a two-action problem. Figure 6.10.1 illustrates the payoff functions in a four-action problem. Notice that there are *three* "breakeven values," which are labeled θ_{b_1}, θ_{b_2}, and θ_{b_3}. Under perfect information about $\tilde{\theta}$, the following decision rule maximizes ER:

$$\begin{aligned}
&\text{Choose } a_1 \quad \text{if } \tilde{\theta} \leq \theta_{b_1},\\
&\text{Choose } a_2 \quad \text{if } \theta_{b_1} \leq \tilde{\theta} \leq \theta_{b_2},\\
&\text{Choose } a_3 \quad \text{if } \theta_{b_2} \leq \tilde{\theta} \leq \theta_{b_3},\\
\text{and} \quad &\text{Choose } a_4 \quad \text{if } \tilde{\theta} \geq \theta_{b_3}.
\end{aligned} \tag{6.10.2}$$

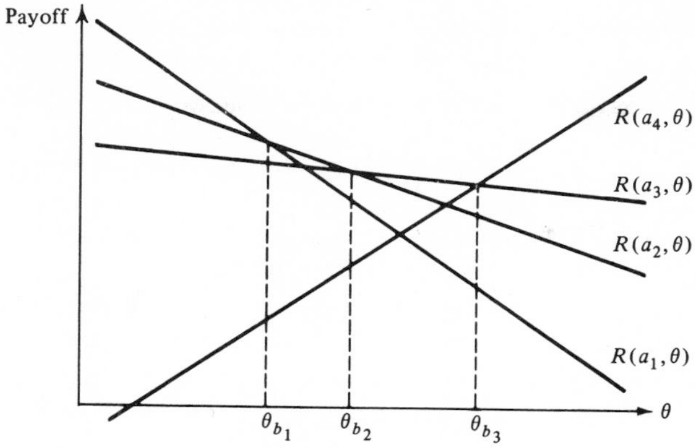

Figure 6.10.1

Under uncertainty, the decision rule is as above with $E(\tilde{\theta})$ substituted for $\tilde{\theta}$. The reason for this is that the payoff functions are linear, and therefore the mean of the decision maker's distribution is a certainty equivalent. In general, for a k-action problem, there will be $(k - 1)$ breakeven values. These values will divide the real line into k intervals, and each of the k actions will be optimal over exactly one of the intervals.

This framework can be illustrated by an extension of the quality-control example presented in Section 4.4. Suppose that the production manager can leave the process as is (make no adjustment), he can make a "minor" adjustment that costs \$40 and reduces $\tilde{p}$ by a factor of $\frac{1}{4}$ (that is, it reduces $\tilde{p}$ to $\frac{3}{4}$ of its original value), he can make a "moderate" adjustment that costs \$100 and halves $\tilde{p}$, or he can make a "major" adjustment that costs \$180 and reduces $\tilde{p}$ to only $\frac{1}{4}$ of its original value. The decision concerns a run of 1000 items, the net profit per item (exclusive of the costs of adjustment and replacing defectives) is \$0.50, and the cost of replacing a defective item is \$2.00. Given this information, the payoff functions for the four actions are

$$\begin{aligned} R(\text{no adjustment}, p) &= 500 - 2000p, \\ R(\text{minor adjustment}, p) &= 460 - 1500p, \\ R(\text{moderate adjustment}, p) &= 400 - 1000p, \\ \text{and} \qquad R(\text{major adjustment}, p) &= 320 - 500p. \end{aligned}$$

These payoff functions are graphed in Figure 6.10.2. The breakeven values are $p_{b_1} = .08$, $p_{b_2} = .12$, and $p_{b_3} = .16$. The production manager should make no adjustment if $E(\tilde{p}) \leq .08$, a minor adjustment if $.08 \leq E(\tilde{p}) \leq .12$, a moderate adjustment if $.12 \leq E(\tilde{p}) \leq .16$, and a major adjustment if $E(\tilde{p}) \geq .16$. If the prior distribution for $\tilde{p}$ is a beta distribution with parameters $r' = 1$ and $n' = 20$, then $E(\tilde{p}) = r'/n' = 1/20 = .05$, and the production manager should leave the process as is.

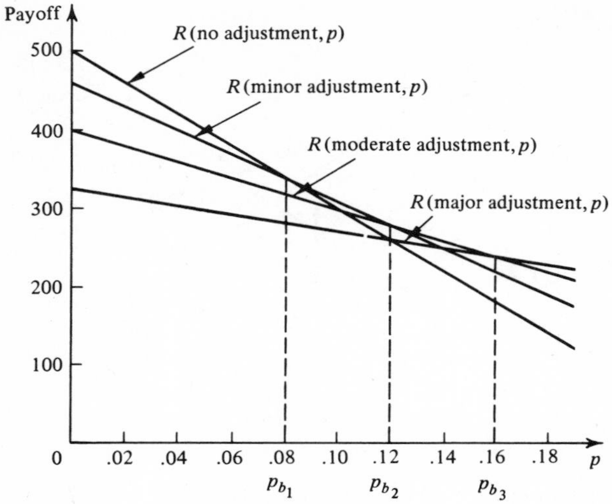

Figure 6.10.2

The loss function in a k-action problem is a piecewise linear function. That is, it is made up of a series of k line segments with different slopes. In the four-action problem, suppose that a_2 is the optimal act under a prior distribution. From Figure 6.10.1,

$$L(a_2,\theta) = \begin{cases} R(a_1,\theta) - R(a_2,\theta) & \text{if } \theta \leq \theta_{b_1}, \\ 0 & \text{if } \theta_{b_1} \leq \theta \leq \theta_{b_2}, \\ R(a_3,\theta) - R(a_2,\theta) & \text{if } \theta_{b_2} \leq \theta \leq \theta_{b_3}, \\ R(a_4,\theta) - R(a_2,\theta) & \text{if } \theta \geq \theta_{b_3}. \end{cases} \tag{6.10.3}$$

At any point, the loss is equal to the difference between the highest payoff for that value of $\tilde{\theta}$ and the payoff obtained with action a_2. Figure 6.10.3 demonstrates the shape of $L(a_2,\theta)$. For the quality-control example, the optimal action is to make no adjustment, and the loss function is

$$L(\text{no adjustment, } p) = \begin{cases} 0 & \text{if } p \leq .08, \\ -40 + 500p & \text{if } .08 \leq p \leq .12, \\ -100 + 1000p & \text{if } .12 \leq p \leq .16, \\ -180 + 1500p & \text{if } p \geq .16. \end{cases}$$

This loss function is graphed in Figure 6.10.4.

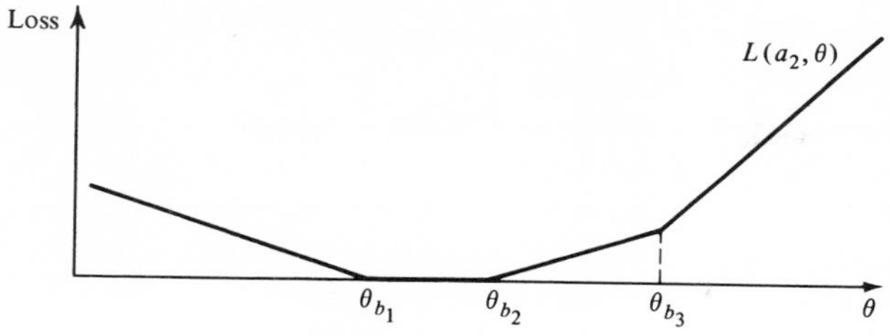

Figure 6.10.3

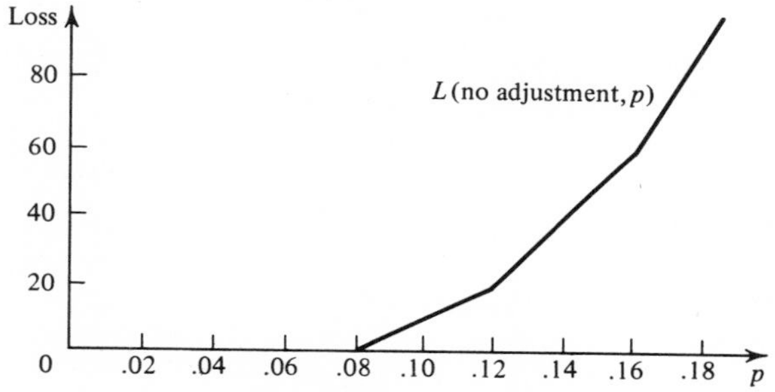

Figure 6.10.4

To evaluate the expected loss (and hence the EVPI) in the case of linear payoffs and more than two actions, it is necessary to consider a series of terms involving linear loss integrals. Since the general idea is similar to the discussion in the previous sections, little would be gained by presenting these reasonably complicated formulas. The point of this section is simply to note that the two-action problem with linear losses can be generalized to a finite-action problem with linear losses and to point out the similarity between the two types of problems. It might also be mentioned that *infinite-action problems* with linear loss functions are considered in the next chapter under the discussion of point estimation as a decision-making problem.

Sections 6.6 through 6.10 have dealt with problems in which the payoff functions and loss functions are linear in terms of the uncertain quantity of interest in a decision-making problem. The framework involving linear payoff and loss functions is quite valuable because such functions are encountered in quite a few real-world problems. However, it should be emphasized that there are many other problems in which, for one reason or another, there are *nonlinear* payoff and loss functions. For instance, consider an "all-or-nothing" form of payoff function, with a payoff of zero if the uncertain quantity $\tilde{\theta}$ is below some given value and a fixed payoff, say \$100, if $\tilde{\theta}$ is greater than or equal to that value. A generalization of this payoff function would be a "step function" payoff, with the payoff increasing in "steps" much like a discrete cumulative distribution function instead of increasing linearly with $\tilde{\theta}$. In yet other situations, the payoff might increase exponentially or logarithmically rather than linearly.

Even if the payoff function *is* linear in terms of money, the decision maker might have a nonlinear utility function for money, as discussed in Section 5.6 (for instance, he may be a strong risk-avoider). If the utility function for money is nonlinear and if the payoff function in a decision-making problem is linear in terms of money, the payoff function will generally be nonlinear when it is expressed in terms of utility. This result occurs in the oil-drilling example presented in Section 5.10. The payoff functions for the four actions in that example are linear in terms of money, as noted in Section 6.6, but they are nonlinear in terms of utility, as illustrated in Figure 6.10.5. The possibility of nonlinear utility functions provides perhaps the strongest argument for the applicability of nonlinear payoff and loss functions.

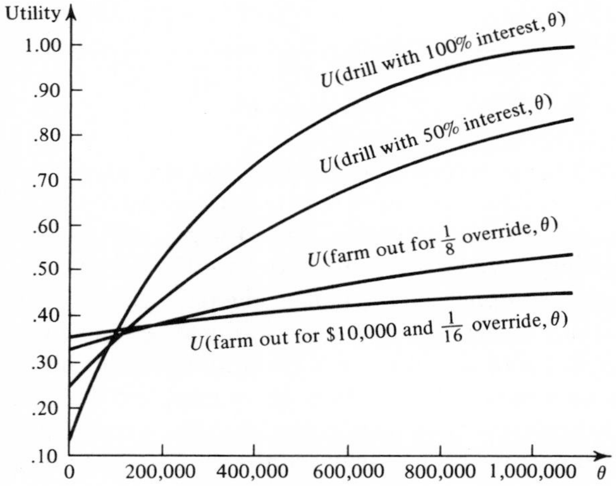

Figure 6.10.5

The point of this brief discussion concerning nonlinear payoff and loss functions is not to minimize the value of linear payoff and loss functions. Instead, it is to point out that there are many decision-making problems for which the payoff or loss functions are nonlinear. In general, such nonlinearity makes the problem more difficult to solve. For instance, when the payoff functions are not linear, the mean of the decision maker's probability distribution usually is no longer a certainty equivalent, and the rule for finding the optimal decision may be much more complex. In calculations involving the value of information, linear loss integrals may be replaced by more complicated expressions. This implies that it is convenient if the payoff and the loss functions in a decision-making problem are linear or close enough to being linear that they can be well approximated by linear functions, since the analysis is generally easier in the linear case. This raises the question of the sensitivity of the decision problem to deviations from linear payoff and loss functions. If the mean of the decision maker's distribution is not near a breakeven point, the problem should be reasonably insensitive; if the mean is very close to a breakeven point, then the problem might be quite sensitive.

There is a large variety of potential nonlinear payoff and loss functions, which when taken together have wide applicability. No *single* nonlinear form appears to have nearly as much applicability as linear payoff and loss functions, however, and the general approach (but not the equations) is the same in the nonlinear case, so no specific nonlinear payoff and loss functions are presented in the context of the value of information (a quadratic function is used in the next chapter as a loss function in an estimation problem). Nevertheless, you should bear in mind that nonlinear payoff and loss functions are frequently encountered in real-world problems.

6.11 REFERENCES AND SUGGESTIONS FOR FURTHER READING

Some general references at a level roughly comparable to this chapter are Aitchison (1970*a*, Chapter 7); Fishburn (1964, Chapter 12); Hadley (1967, Chapters 3, 8, and 11); Lindley (1971, Chapters 7 and 8); Morris (1968, Chapters 5 and 11); Pratt, Raiffa, and Schlaifer (1965, Chapters 12 and 14); Raiffa (1968, Chapter 7); Schlaifer (1959, Chapters 22, 33–38; 1969, Chapter 14); and Schmitt (1969, Chapter 8). A slightly more advanced reference is Raiffa and Schlaifer (1961, Chapters 4 and 5).

With regard to sequential analysis, the key historical reference is Wald (1947), and a recent reference that is also at an advanced level is De Groot (1970, Chapters 12–14). Other references, including both theory and applications, are Kaufman (1963*a*); Lindley (1961*a*); Lindley and Barnett (1965); McCall (1965); De Groot (1968); and Wetherill (1961, 1966). A related topic is dynamic programming, which is a method for solving decision-making problems (under certainty or uncertainty) involving a series of decisions over time; some references in this general area are Bellman (1957), Bellman and Dreyfus (1962), Hadley (1964), Howard (1960, 1965*b*), and Nemhauser (1966). Also related are investigations of stochastic systems; see Aoki (1967); Bather (1963); Sawaragi, Sunahara, and Nakamizo (1967); Sheridan (1970); White (1967); and Ying (1967, 1969) (these should be considered as intermediate to advanced level references).

In this chapter only simple random sampling with fixed sample size and sequential sampling were considered. More complex sampling plans, such as stratified sampling plans, have been extensively studied from the Bayesian viewpoint. Some references in the area of sampling theory are Campling (1968); De Groot and Starr (1969); Draper and Guttman (1968); Ericson (1965, 1967*a*, 1967*b*, 1968, 1969*b*, 1969*c*); Kaufman (1968); Pfanzagl (1963); Schleifer (1969); Solomon and Zacks (1970); and Wetherill and Campling (1966). A related problem is the design of experiments; Draper and Hunter (1967), for example, involves a Bayesian approach to experimental design. Complex sampling plans and experimental designs generally involve the consideration of several variables, and most of these references are at an intermediate to advanced level.

In addition to theoretical developments and applications of the methods discussed in this chapter, some experimental work has been conducted. For examples of psychological experiments involving information purchasing and sequential decision making, see Green, Halbert, and Minas (1964); Hershman and Levine (1970); Rapoport (1964, 1967, 1968); Rapoport and Tversky (1970); Tversky and Edwards (1966); and Wendt (1969).

It is possible to investigate the sensitivity of a decision-making problem to variations in any of the inputs, including the prior probabilities, the likelihoods, the sample size, the sample design, and the payoffs or utilities. Some references regarding sensitivity analysis are Antelman (1965); Fishburn, Murphy, and Isaacs (1968); Isaacs (1963); Pierce and Folks (1969); and Rappaport (1967).

EXERCISES

1. What is the difference between a terminal decision and a preposterior decision? Are they at all related?

2. In Exercises 14 and 15 in Chapter 5, find the EVPI.

3. In Exercise 16, Chapter 5, how much should the store owner be willing to pay for perfect information about the demand for the type of shirts he is about to order?

4. In Exercise 18, Chapter 5, what is the value of perfect information that the rival firm, Firm B, will advertise if Firm A advertises? What is the value of perfect information that Firm B will not advertise and the increase in Firm A's market share will be 10 percent? What is the value of perfect information that Firm B will not advertise and the increase in Firm A's market share will be 20 percent? Finally, what is the value of perfect information that Firm B will not advertise and the increase in Firm A's market share will be 30 percent? From these VPI's, calculate the EVPI.

5. In Exercise 22, Chapter 5, give a general expression for the expected value of perfect information regarding the weather and find the EVPI if
   (a) P(adverse weather) $= .4$, $C = 3.5$, and $L = 10$,
   (b) P(adverse weather) $= .3$, $C = 3.5$, and $L = 10$,
   (c) P(adverse weather) $= .4$, $C = 10$, and $L = 10$,
   (d) P(adverse weather) $= .3$, $C = 2$, and $L = 8$.

6. In Exercise 5, if P(adverse weather) $= .3$ and $C/L = \frac{1}{2}$, how much should you be willing to pay for perfect information concerning the absolute magnitudes of C and L? Explain.

7. In Exercise 24, Chapter 5, how much should the owner be willing to pay for perfect information concerning $\tilde{r}$? If he cannot obtain perfect information concerning $\tilde{r}$, how much should he be willing to pay for perfect information concerning $\tilde{\lambda}$?

8. In decision-making problems for which the uncertain quantity of primary interest can be viewed as a future sample outcome $\tilde{y}$, the relevant distribution of interest to the decision maker is the predictive distribution of $\tilde{y}$.
   (a) Given a prior distribution $f(\theta)$ and a likelihood function $f(y|\theta)$, how would you find the expected value of perfect information about $\tilde{y}$?
   (b) Given a prior distribution $f(\theta)$ and a likelihood function $f(y|\theta)$, how would you find the expected value of perfect information about $\tilde{\theta}$?
   (c) Explain the difference between your answers to (a) and (b).

9. In Exercise 26, Chapter 5, the uncertain quantity of interest is $\tilde{r}$, the number of defective items in a sample of size five from a production process.
   (a) How much should the production manager be willing to pay for perfect information about $\tilde{r}$?
   (b) How much should he be willing to pay for perfect information about $\tilde{p}$?
   (c) If he obtains perfect information that $\tilde{p} = .15$, how much should he be willing to pay for perfect information about $\tilde{r}$?
   (d) If he obtains perfect information that $\tilde{r} = 1$, how much should he be willing to pay for perfect information about $\tilde{p}$?

10. In Exercise 8, is one type of perfect information (that is, perfect information about $\tilde{y}$ or perfect information about $\tilde{\theta}$) more "valuable" to the decision maker than the other? Explain your answer and illustrate with respect to Exercise 9.

11. In Exercise 25, Chapter 5, find the expected value of perfect information (a) about $\tilde{r}$, (b) about $\tilde{\lambda}$.

12. In Exercise 51, Chapter 5, find the expected value of perfect information about (a) the cost of the ad campaign, (b) the increase in profits for each 10 percent gain in market share, and (c) both the cost of the ad campaign *and* the increase in profits for each 10 percent gain in market share. Does the answer to part (c) equal the sum of the answers to parts (a) and (b)? Explain.

13. In Exercise 31, Chapter 5, assuming that your utility function for money is linear, find the value of perfect information that there is oil and the value of perfect information that there is no oil. From these results, how much would you be willing to pay for a geological test that will tell you for certain whether or not there is oil?

14. In Exercise 4, show that the EVPI is equal to the expected loss of the action that is optimal under the decision maker's probability distribution.

15. A store must decide whether or not to stock a new item. The decision depends on the reaction of consumers to the item, and the payoff table (in dollars) is as follows.

		PROPORTION OF CONSUMERS PURCHASING THE ITEM				
		.10	*.20*	*.30*	*.40*	*.50*
	Stock 100	−10	−2	12	22	40
DECISION	*Stock 50*	−4	6	12	16	16
	Do not stock	0	0	0	0	0

If $P(.10) = .2$, $P(.20) = .3$, $P(.30) = .3$, $P(.40) = .1$, and $P(.50) = .1$, what decision should be made? If perfect information is available, find VPI for each of the five possible states of the world and compute EVPI.

16. In Exercise 15, suppose that sample information is available in the form of a random sample of consumers. For a sample of size *one,*
    (a) find the posterior distribution if the one person sampled will purchase the item, and find the value of this sample information;
    (b) find the posterior distribution if the one person sampled will *not* purchase the item, and find the value of this sample information;
    (c) find the expected value of sample information.

17. In Exercise 16, suppose that you also want to consider other sample sizes.
    (a) Find EVSI for a sample of size 2.
    (b) Find EVSI for a sample of size 5.
    (c) Find EVSI for a sample of size 10.
    (d) If the cost of sampling is \$0.50 per unit sampled, find the expected net gain of sampling (ENGS) for samples of sizes 1, 2, 5, and 10.

18. Consider a bookbag filled with 100 poker chips. You know that either 70 of the chips are red and the remainder blue, or 70 are blue and the remainder

red. You must guess whether the bookbag has 70R–30B or 70B–30R. If you guess correctly, you win \$5. If you guess incorrectly, you lose \$3. Your prior probability that the bookbag contains 70R–30B is .40.

(a) If you had to make your guess on the basis of the prior information, what would you guess?

(b) If you could purchase perfect information, what is the most that you should be willing to pay for it?

(c) If you could purchase sample information in the form of one draw from the bookbag, how much should you be willing to pay for it?

(d) If you could purchase sample information in the form of five draws (with replacement) from the bookbag, how much should you be willing to pay for it?

19. Do Exercise 18 with the following payoff table (in dollars).

		STATE OF THE WORLD	
		70R–30B	*70B–30R*
YOUR GUESS	*70R–30B*	6	−2
	70B–30R	−6	10

20. In the automobile-salesman example discussed in Section 3.4, suppose that the owner of the dealership must decide whether or not to hire a new salesman. The payoff table (in terms of dollars) is as follows.

		STATE OF THE WORLD		
		Great salesman	*Good salesman*	*Poor salesman*
ACTION	*Hire*	20,000	5,000	−10,000
	Do not hire	0	0	0

The prior probabilities for the three states of the world are $P(\text{great}) = .10$, $P(\text{good}) = .50$, and $P(\text{poor}) = .40$. The process of selling cars is assumed to behave according to a Poisson process with $\tilde{\lambda} = \frac{1}{2}$ per day for a great salesman, $\tilde{\lambda} = \frac{1}{4}$ per day for a good salesman, and $\tilde{\lambda} = \frac{1}{8}$ per day for a poor salesman.

(a) Find VPI(great salesman), VPI(good salesman), and VPI(poor salesman).

(b) Find the expected value of perfect information.

(c) Suppose that the owner of the dealership can purchase sample information at the rate of \$10 per day by hiring the salesman on a temporary basis. He must sample in units of four days, however, for a salesman's work week consists of four days. He considers hiring the salesman for one week (four days). Find EVSI and ENGS for this proposed sample.

21. In Exercise 20, suppose that the owner hires the salesman for one week and that he sells two cars. At the end of the week, the owner has three choices: hire the salesman permanently, fire the salesman, or hire the salesman temporarily for another week (at the extra cost of \$10 per day). What should he do?

22. In Exercise 18, Chapter 5, suppose that the owner of Firm A can obtain information about Firm B's reaction to advertising by Firm A. In particular, he can ask a contact he has in Firm B. However, the information from the contact cannot be regarded as perfect information. If Firm B would really advertise if Firm A does, the chances are 4-in-5 that the contact would report this correctly. However, if Firm B would not advertise, the chances are only 2-in-3 that the contact would report this correctly.
(a) If the contact reports that Firm B will advertise if Firm A does, what is the value of this sample information?
(b) If the contact reports that Firm B will not advertise even if Firm A does, what is the value of this sample information?
(c) What is the expected value of sample information?

23. In the oil-drilling example in Section 6.4, what is the expected value of the seismic information if the decision maker's utility function is linear with respect to money?

24. A firm is considering the marketing of a new product. For convenience, suppose that the events of interest are simply θ_1 = "new product is a success" and θ_2 = "new product is a failure." The prior probabilities are $P(\theta_1) = .3$ and $P(\theta_2) = .7$. If the product is marketed and is a failure, the firm suffers a loss of \$300,000. If the product is not marketed and it would be a success, the firm suffers an opportunity loss of \$500,000. The firm is considering two separate surveys, A and B, and the results from each survey can be classified as favorable, neutral, and unfavorable. The conditional probabilities for survey A are

$$P(\text{favorable} \mid \theta_1) = .6,\ P(\text{neutral} \mid \theta_1) = .3,\ P(\text{unfavorable} \mid \theta_1) = .1,$$
$$P(\text{favorable} \mid \theta_2) = .1,\ P(\text{neutral} \mid \theta_2) = .2, \text{ and } P(\text{unfavorable} \mid \theta_2) = .7.$$

The conditional probabilities for survey B are

$$P(\text{favorable} \mid \theta_1) = .8,\ P(\text{neutral} \mid \theta_1) = .1,\ P(\text{unfavorable} \mid \theta_1) = .1,$$
$$P(\text{favorable} \mid \theta_2) = .1,\ P(\text{neutral} \mid \theta_2) = .4, \text{ and } P(\text{unfavorable} \mid \theta_2) = .5.$$

Survey A costs \$20,000 and survey B costs \$30,000.

(a) Find the expected value of perfect information about $\tilde{\theta}$.
(b) Find the expected net gain from survey A.

(c) Find the expected net gain from survey B.
(d) Suppose that the firm has a choice. It can use no survey, survey A, or survey B, but not both surveys. What is the optimal course of action?

25. In Exercise 24, suppose that the firm also has the option of using *both* surveys. Furthermore, since both surveys would be conducted by the same marketing research firm, the total cost of the two surveys is only \$40,000, provided that a decision is made in advance to use both surveys (that is, the firm cannot use one survey and then decide whether or not to use the other). Given any one of the three events θ_1, θ_2, and θ_3, the results of survey B are considered to be independent of the results of survey A. What should the firm do?

26. In Exercise 25, suppose that the firm has the additional option of using survey A and, after seeing the results of survey A, deciding whether or not to use survey B. The reverse procedure, using survey B first and then considering survey A, is not possible. Of course, if the sequential plan is used and both surveys are taken, the total cost of the surveys is \$50,000 rather than \$40,000. Find the expected net gain of the sequential plan and compare this with the expected net gains of the sampling plans considered in Exercises 24 and 25.

27. In Exercise 17, consider a sequential sampling plan with a maximum total sample size of two and analyze the problem as follows.
(a) Represent the situation in terms of a tree diagram.
(b) Using backward induction, find the ENGS for the sequential plan.
(c) Compare the sequential plan with a single-stage plan having $n = 2$.

28. Repeat Exercise 27 for sequential samples with maximum total sample sizes of three, four, and five. Discuss your results.

29. In Exercise 18, assuming that the cost of sampling is 25 cents per draw, find the ENGS for
(a) a sequential sampling plan with a maximum total sample size of two,
(b) a sequential sampling plan with a maximum total sample size of four,
(c) a single-stage sampling plan with a sample size of two,
(d) a single-stage sampling plan with a sample size of four.

30. Repeat Exercise 29 under the assumption that the cost of sampling is 15 cents per draw plus a fixed cost of 50 cents for each separate sample. Comment on any differences between the results of Exercise 29 and this exercise.

31. Repeat Exercise 29 under the assumption that the sampling from the bookbag is done *without* replacement. Comment on any differences between the results of Exercise 29 and this exercise. In particular, for each sampling plan, is there any difference between the ENGS for sampling with replacement and the ENGS for sampling without replacement? Explain.

32. The preceding exercises illustrate the idea of sequential analysis, or sequential decision making. In a sequential procedure, how does the decision maker know when to stop sampling and make his terminal decision? Is it possible for a sequential sampling plan to have a positive ENGS even though no sampling plan with a fixed sample size has a positive ENGS in the same situation? Explain.

33. Comment on the statement, "Linear payoff and loss functions have wide applicability in real-world decision-making problems." Give some examples of problems in which the payoff and loss functions are linear or approximately linear.

34. Suppose that the payoff functions (in dollars) of two actions are

$$R(a_1,\mu) = 70 - .5\mu$$

and

$$R(a_2,\mu) = 50 + .5\mu,$$

where $\tilde{\mu}$ is the mean of a normal process with variance 1200.

(a) Find the breakeven value, μ_b.
(b) If the prior distribution is a normal distribution with mean $m' = 25$ and variance $\sigma'^2 = 400$, which action should you choose?
(c) What is the value of perfect information that $\tilde{\mu} = 15$?
(d) What is the value of perfect information that $\tilde{\mu} = 21$?
(e) What is the expected value of perfect information?
(f) Graph the payoff functions and the associated loss functions.
(g) What would the expected value of perfect information be if the prior mean $m' = 30$? Explain the difference between this answer and the answer to (e).

35. The payoff in a certain decision-making problem depends on $\tilde{p}$, the parameter of a Bernoulli process. The payoff functions are

$$R(a_1,p) = 50p$$

and

$$R(a_2,p) = -10 + 100p.$$

(a) What is the breakeven value of $\tilde{p}$?
(b) Graph the payoff functions and the associated loss functions.
(c) If the prior distribution is a beta distribution with $r' = 4$ and $n' = 24$, which action should be chosen?
(d) If a sample of size 15 is taken, with 4 successes, find the posterior distribution and the optimal action under this distribution.

36. In Exercise 34, find the EVSI and ENGS for samples of size 1, 2, 3, 4, 5, 6, 7, 8, 9, and 10, assuming that the cost of sampling is \$0.15 per trial. On a graph, draw curves representing EVSI, ENGS, and CS.

37. In Exercise 35(c), find the expected value of perfect information about $\tilde{p}$.

38. A firm is contemplating the purchase of 500 typewriter ribbons. One supplier, supplier A, offers the ribbons at \$1.50 each, guarantees each ribbon, and will replace all defective ribbons free. A second supplier, supplier B, offers the ribbons at \$1.40 each with no guarantee. However, supplier B will replace defective ribbons with good ribbons for \$1.00 per ribbon. Suppose that the proportion of defective ribbons produced by supplier B is denoted by $\tilde{p}$, and suppose that the prior distribution for $\tilde{p}$ is a beta distribution with parameters $r' = 2$ and $n' = 50$.
   (a) What should the firm do on the basis of the prior distribution?
   (b) How much is it worth to the firm to know the proportion of defective ribbons for certain?
   (c) Suppose that supplier B can provide a randomly chosen sample of 10 ribbons. What is the expected value of this sample?
   (d) In the sample of 10 ribbons, 1 is defective. Find the posterior distribution of $\tilde{p}$ and use this distribution to determine which supplier the firm should deal with.

39. Suppose that an investor has a choice of four investments (assume that he must choose one, and only one, of the four—he cannot divide his money among them):
   A: a savings account in a bank;
   B: a stock that moves counter to the general stock market;
   C: a growth stock that moves at a faster pace than the market;
   D: a second growth stock that is similar to C but not quite as speculative.
   The investor has decided that his payoff depends on his choice of investment and on the *next year's change* in the Dow-Jones Industrials (DJI), which we shall call $\tilde{\theta}$. (Ignore the fact that he should be interested in the change in price of the investments themselves—for example, by assuming perfect correlations between the DJI and the investments.) His payoff functions are as follows:

$$R(A,\theta) = .05,$$
$$R(B,\theta) = .05 - \theta,$$
$$R(C,\theta) = -.15 + 2\theta,$$

and

$$R(D,\theta) = -.05 + \theta.$$

   (a) Draw the graphs of these payoff functions and determine a decision rule for choice of an investment under certainty about $\tilde{\theta}$.
   (b) Suppose that the investor's prior distribution is normal with mean .08 and standard deviation .05. Which investment should he choose?
   (c) Show graphically and algebraically the loss function. If the investor obtains perfect information and finds that $\tilde{\theta} = .02$, what is the VPI?
   (d) Determine and graph the loss functions for the four actions A, B, C, and D.

40. In the oil-drilling venture example presented in Section 5.10, suppose that $\tilde{\theta}$ is assumed to be continuous. Given the information in Section 5.10, find the payoff functions for the four actions and graph these functions. Furthermore, assume that the mean of the decision maker's prior distribution for $\tilde{\theta}$

is 125,000. If the decision maker's utility function is taken to be linear with respect to money, find the optimal action under the prior distribution. Why is it necessary to know only the mean of the prior distribution in order to find the optimal action?

41. What are certainty equivalents and how can they help simplify problems of decision making under uncertainty?

42. Give some examples of decision-making situations in which the payoff or loss functions might be nonlinear.

43. Discuss the importance of sensitivity analysis in relation to both terminal and preposterior decisions.

44. In the example presented in Section 6.8, find the optimal sample size
    (a) if $CS(n) = .10 + .02n$,
    (b) if $CS(n) = .20 + .01n$.

45. The cost of sampling is often taken to be a fixed multiple of the sample size, so that the sample costs, say, $5 per item sampled. The cost functions in Exercise 44, however, include both a cost per item *and* a fixed cost. Under what circumstances might such cost functions arise, and what are the implications for sequential sampling?

46. In Exercise 35, assume that the payoff functions are given in terms of dollars and that the decision maker's utility function for money is of the form

$$U(M) = (M + 100)^2 \quad \text{for } -100 \leq M \leq 100,$$

where M represents dollars. Graph the payoff functions and the associated loss functions in terms of utility, determine which of the two actions has the higher expected utility under the prior distribution, and determine which of the two actions has the higher expected utility under the posterior distribution. [*Hint:* The "laws of expectation" presented in Section 3.1 are useful here.]

47. In Exercise 35, suppose that a third action is considered, with payoff function

$$R(a_3,p) = -20 + 140p.$$

    (a) Graph the three payoff functions and determine a decision rule for choosing the optimal action under uncertainty about $\tilde{p}$.
    (b) Given the posterior distribution from Exercise 35, what action should be chosen?
    (c) If the decision maker's utility function for money is as given in Exercise 46, graph the three payoff functions in terms of utility and determine the optimal action under the posterior distribution.

48. In Exercise 34, suppose that the decision maker's utility function for money is of the form

$$U(M) = M^3,$$

where M represents dollars. Graph the payoff functions in terms of utility and determine the optimal action under the prior distribution given in Exercise 34. [*Hint:* For a normal distribution with mean μ and variance σ^2, the third moment about the origin is $\mu(\mu^2 + 3\sigma^2)$.]

49. The payoff functions for three actions are

$$R(a_1,p) = -16 + 100p,$$
$$R(a_2,p) = 1000(p - .20)^2,$$
and
$$R(a_3,p) = 600(p - .15)^2,$$

where $\tilde{p}$ is the parameter of a Bernoulli process. If the prior distribution of $\tilde{p}$ is a beta distribution with $r' = 10$ and $n' = 50$, find the expected payoffs of the three actions.

7

INFERENCE AND DECISION

The objective of inferential statistics is to make inferences about some uncertain quantity. In the Bayesian approach to statistics, the application of Bayes' theorem is an inferential procedure and the resulting posterior distribution is an inferential statement about the uncertain quantity of interest. Thus, the material in Chapters 3 and 4 is classified as *Bayesian inference.* In decision theory, as opposed to inference, one must take some *action* rather than just make some inference about an uncertain quantity. As a result, one additional input is required in decision theory: a loss function, or payoff function. The material in Chapters 5 and 6 comes under the heading of *Bayesian decision theory.*

According to a "non-Bayesian" school of thought, inferences about an uncertain quantity should be based solely on sample information. In other words, prior information should not be included in the formal statistical analysis. This school of thought is usually referred to as the "classical," or "sampling theory," approach to statistics. Most classical statisticians follow the frequency interpretation of probability, which was discussed in Chapter 2, and they do not admit subjective probabilities into their formal statistical analysis. Frequentists claim that a population parameter such as $\tilde{\theta}$, $\tilde{p}$, $\tilde{\mu}$, or $\tilde{\lambda}$ has a certain value, which may be unknown, and that it is senseless to talk of the probability that the parameter equals some number; either it does or it does not. The subjectivist, on the other hand, argues that the frequentist is ignoring the uncertainty about the parameter and that because of this uncertainty, the parameter is an uncertain quantity, or random variable, and probability statements can be made about it.

In Section 2.2, it was noted that the frequency and subjective interpretations are not the only interpretations of probability, although they are the most commonly encountered interpretations. Some followers of the frequency school of thought do interpret some probability statements concerning population parameters (notably, interval estimation statements) as probability statements about an uncertain quantity, calling such probabilities "fiducial" probabilities. Another interpretation is called the "necessary" view of probability; essentially, this approach is based on symmetry arguments. Interpretations other than the frequency and subjective interpretations are not discussed in this book, although references to discussions of such interpretations are given in Sections 2.8 and 7.10.

In classical statistics, all inferences are based on the sample information (although it is too broad and highly simplified, this is taken as the "working definition" of classical statistics for the purpose of this chapter). The Bayesian statistician also includes any information that may be present prior to taking a sample. This information is expressed in the form of a prior distribution, which is combined with a likelihood function via Bayes' theorem to form a posterior distribution on which inferences and decisions are based. The Bayesian argues that all available information about a parameter, whether it is of an "objective" or a "subjective" nature, should be utilized formally in making inferences or decisions. Because of their implications for decision theory, Bayesian methods have been used widely in business and economics. In these areas, decisions often must be made in situations in which there is little or no sample information but a great deal of information of a more subjective nature. Of course, Bayesian methods are applicable in all problems of statistical inference or decision making, regardless of the area of application. Since *all* probabilities are ultimately subjective, as pointed out in Chapter 2, the refusal to include nonsample information appears to be somewhat arbitrary, and the Bayesian approach provides a better framework for problems of inference and decision. (It should be noted, however, that the terms "Bayesian" and "subjective" are not synonymous; it is possible to develop "Bayesian" procedures without using subjective probability.)

Nevertheless, much work has been done in the area of classical statistics, and it is useful to compare this work with the Bayesian approach. As this chapter indicates, Bayesian procedures may in some cases be considered to be *extensions* of classical procedures, including additional inputs such as a prior distribution and a payoff or loss function. It should be mentioned that in some situations a classical statistician may use prior information, although he uses it in an *informal* manner. For example, he may use prior information to help determine the form of his model, to suggest hypotheses of interest, and so on. As was emphasized in Chapters 3 and 4, the choice of a statistical model to represent a real-world data-generating process is ultimately a subjective choice, involving a trade-off between the realism and the complexity of the model. Thus, the likelihoods used to represent sample information in probabilistic form are ultimately subjective. The Bayesian statistician, like the classical statistician, uses prior information in an informal manner to define the sample space, to decide on a model for the data-generating process, and so on. However, the Bayesian statistician also *formally* includes prior information in his analysis in the form of a prior distribution.

In classical statistics, inferential procedures are generally divided into two categories: estimation and hypothesis testing. Since this book is concerned with the Bayesian approach to statistical inference and decision, a full development of classical techniques is not presented. References to such developments are given in Section 7.10. In this chapter, Bayesian extensions of classical estimation and hypothesis testing procedures are presented. You should keep in mind that there are many inferential and decision-making situations that do not fall into either of the two categories considered in this chapter. As noted previously, the use of Bayes' theorem to revise a probability distribution is an inferential procedure, and the entire posterior distribution constitutes an inferential statement; for many (perhaps most) purposes, the entire distribution is much more informative and useful than any summarizations in the form of estimates or tests of hypotheses.

7.1 DIFFUSE PRIOR DISTRIBUTIONS AND CLASSICAL STATISTICS

When a diffuse prior distribution is used in a Bayesian analysis, the posterior distribution is virtually identical to the likelihood function, as pointed out in Section 4.10. Thus, any inferences and/or decisions based on the posterior distribution will in reality depend almost solely on the sample information, as summarized by the likelihood function. But a classical statistician bases inferences and decisions solely on the sample information. Under a diffuse prior state, then, Bayesian and classical statistical procedures are essentially based on the same set of information. If relevant prior information is available, the Bayesian's posterior distribution will reflect both this information and that of the sample, and the Bayesian results are likely to be quite different from the corresponding classical results. In the special case of a diffuse prior distribution, however, classical and Bayesian results are quite similar, being based on essentially the same information.

As you shall see, under a diffuse prior distribution Bayesian techniques often result in *numerical* results identical to classical results. However, there is still an important difference in *interpretation*. The Bayesian thinks of $\tilde{\theta}$ as a random variable, or uncertain quantity, and is willing to make probability statements concerning $\tilde{\theta}$. The prior and posterior distributions are probability distributions of $\tilde{\theta}$. The classical statistician, on the other hand, claims that $\tilde{\theta}$ is a fixed parameter, not a random variable, and that it makes no sense to talk of the probability of values of $\tilde{\theta}$ occurring. Thus, he would write θ, not $\tilde{\theta}$, since the tilde indicates a random variable; to avoid confusion, the notation $\tilde{\theta}$ is maintained in this book whether the discussion involves the classical or the Bayesian approach. You should keep in mind, however, that the classical statistician does *not* consider $\tilde{\theta}$ to be a random variable.

The classical statistician's inferences are based on $f(y \mid \theta)$, which may be interpreted as a sampling distribution or as a likelihood function. Some classical statisticians consider $f(y \mid \theta)$ as a function of $\tilde{y}$, with θ fixed; this is the *sampling distribution* of $\tilde{y}$, given $\tilde{\theta} = \theta$. Other classical statisticians consider $f(y \mid \theta)$ as a function of $\tilde{\theta}$, with y fixed ($\tilde{y} = y$ is simply the observed sample result); this is the *likelihood function* of $\tilde{y} = y$, given different values of $\tilde{\theta}$. The likelihood interpretation is consistent with the Bayesian's use of $f(y \mid \theta)$, whereas the sampling distribution interpretation is not. *After* the sample is observed, the Bayesian (and the classical "likelihoodist") is concerned about making inferences about $\tilde{\theta}$ based on the *observed* sample result, $\tilde{y} = y$. He is *not* concerned with any other values of $\tilde{y}$ that did *not* occur. Thus, he is not concerned with the sampling distribution of $\tilde{y}$, given $\tilde{\theta} = \theta$, since this distribution involves the various possible values of $\tilde{y}$ (*before* observing the sample, of course, he might be interested in this distribution).

For instance, in sampling from a Bernoulli process with fixed sample size n, the sample statistic of interest is $\tilde{r}$, the number of "successes" in the n trials, and the sampling distribution of $\tilde{r}$, given the Bernoulli parameter $\tilde{p} = p$, is the binomial distribution, as discussed in Section 3.3:

$$P(\tilde{r} = r \mid n, p) = \binom{n}{r} p^r(1 - p)^{n-r}.$$

After the sample is observed, with $\tilde{r} = r$, the Bayesian (and the likelihoodist) considers binomial probabilities $P(\tilde{r} = r \mid n, p)$ for fixed r and different values of p. This is illustrated in Section 3.3. Note that *after* the sample, the Bayesian is *not* concerned with the sampling distribution of $\tilde{r}$ for a fixed $\tilde{p} = p$, since $\tilde{r}$ has already been observed. Similarly, in sampling from a Poisson process for a fixed amount of time t, the Bayesian (and the likelihoodist) considers Poisson probabilities $P(\tilde{r} = r \mid \lambda t)$ for fixed r (the observed value of $\tilde{r}$) and different values of λ, as illustrated in Section 3.4. Of course, the Bayesian and the classical likelihoodist, although they use the sample information in the same manner, differ in that the Bayesian admits a prior distribution into the formal analysis, whereas the classical likelihoodist does not. These differences in interpretation are explored at greater length in succeeding sections.

The use of the likelihood function to represent sample information is based on the *likelihood principle,* which states essentially that the likelihood function contains all of the information from the sample that is relevant for inferential and decision-making purposes. The concept of a *sufficient statistic,* discussed in Section 4.2, is related to the likelihood principle. Recall that if the posterior distribution of $\tilde{\theta}$ given y is identical to the posterior distribution of $\tilde{\theta}$ given all of the details of the sample, then y is a sufficient statistic. In terms of the likelihood principle, y is a sufficient statistic if knowledge of y enables the statistician to specify the likelihood function. For example, under the assumption of a Bernoulli process with fixed sample size n, the likelihood function can be taken to be any positive multiple of $p^r(1 - p)^{n-r}$. Therefore, knowledge of r and n enables the statistician to specify the likelihood function, and $y = (r,n)$ is a sufficient statistic.

Because diffuse distributions, as discussed in Section 4.10, are valuable to the Bayesian for several reasons, they deserve some discussion at this point. Their first and most obvious use occurs when the statistician has no prior information or very little information relative to the information contained in the sample. Diffuse conjugate distributions such as those presented in Section 4.10 provide a convenient way to express a diffuse prior state of information. The application of Bayes' theorem is simple, and the posterior distribution is based almost entirely on the sample information.

A second, equally important use of diffuse prior distributions relates to scientific reporting. When reporting the results of a statistical analysis, the statistician is faced with this question: What should be reported? One possible answer is to report the posterior distribution and any resulting decisions. The entire posterior distribution itself is much more informative than any summarizations in the form of point or interval estimates. Furthermore, the posterior distribution is informative in the sense that it reflects not only the sample results, but also the prior judgments of the statistician. What if someone else wants to investigate the results of the statistical analysis? He might want to assess his own prior distribution (which could be quite different from the statistician's distribution) and then use Bayes' theorem to revise the distribution. But unless the statistician reports the sample results separately (that is, other than as a part of the posterior distribution), this person will be unable to carry out the analysis using his own prior distribution.

One possible solution to the problem is for the statistician to carry out a separate analysis using a diffuse prior distribution. By removing the effect of the statistician's own prior judgments, this analysis will enable anyone to include his own prior judgments and to proceed accordingly. Also, it will allow the reader to see approximately what results would have been obtained using classical techniques. Alternatively, the statistician could report the likelihood function, for by the likelihood principle, this constitutes the entire evidence of the sample. As noted in Section 4.10, under a diffuse prior distribution the posterior distribution is approximately proportional to the likelihood function. Bear in mind that the statistician should base his *own* inferences and decisions on his *own* posterior distribution. For the benefit of others, it is useful for him to present the results that follow from the use of a diffuse prior distribution. Such results may be of some interest to the statistician himself, since he can then investigate the influence of his prior distribution on the results. It should be noted that there may be some difficulties involved in the choice of exactly what is taken as a representation of diffuseness, as indicated in Section 4.10. Unless there is very little sample information, however, these difficulties should cause no concern in practice.

This discussion of diffuse prior distributions with respect to scientific reporting holds only if there is substantial agreement on the form of the likelihood function. That is, there must be substantial agreement on the use of a particular statistical model to represent the data-generating process (in some situations, of course, no well-known statistical model is appropriate). The choice of such a model is ultimately subjective, as emphasized in Section 3.6, so the likelihoods determined from the model are ultimately subjective in nature. To the extent that different people agree about the choice of a statistical model, then, they also agree about the likelihoods, whereas they are much less likely to agree about prior probabilities. Thus, the reporting of results based on a diffuse prior distribution may be of some general interest.

7.2 THE POSTERIOR DISTRIBUTION AND ESTIMATION

One type of inferential procedure is that of *point estimation*, in which a single point, or single value, is used to estimate an unknown parameter. The classical statistician, of course, bases point estimates entirely on the sample information. Thus, a sample statistic is used to estimate an unknown parameter; for instance, the sample mean m might be used to estimate the mean $\tilde{\mu}$ of a particular process (or population), and the sample proportion r/n might be used to estimate the parameter $\tilde{p}$ of a Bernoulli process. A number of properties have been put forth by classical statisticians as "good" properties for estimators to possess. Examples of such properties are sufficiency, unbiasedness, and efficiency. The concept of sufficiency has been covered in Sections 4.2 and 7.1, and the classical definition of a sufficient statistic is equivalent to the Bayesian definition. Thus, sufficiency is not discussed further at this point, although it should be noted that it is the most important of the three properties listed above.

A sample statistic $\tilde{y}$ is said to be an *unbiased* estimator of an unknown parameter $\tilde{\theta}$ if $E(\tilde{y} \mid \tilde{\theta} = \theta) = \theta$ for all values θ of $\tilde{\theta}$. For example, $E(\tilde{m} \mid \tilde{\mu} = \mu) = \mu$, so the sample mean is an unbiased estimator of the process (or population) mean $\tilde{\mu}$; for sampling from a Bernoulli process with fixed sample size n (binomial sampling), $E(\tilde{r}/n \mid \tilde{p} = p) = p$, so the sample proportion is an unbiased estimator of the parameter $\tilde{p}$. The property of unbiasedness is interpreted by the classical statistician as follows: In repeated samples from the process or population of interest, the average value of $\tilde{y}$ will be equal to the true value of $\tilde{\theta}$. This interpretation is in terms of relative frequencies over repeated samples, as discussed in Section 2.2. Even aside from the limitations of this interpretation (see Sections 2.3 through 2.5), the property of unbiasedness is no guarantee of a "good" estimator. For instance, if a sample of 100 items is taken from some process, the value of the first item is an unbiased estimator of the process mean $\tilde{\mu}$, although it ignores the information contained in the other 99 items in the sample (it clearly is not a sufficient statistic). Note that the property of unbiasedness is based on the sampling distribution interpretation of $f(y \mid \theta)$ [$\tilde{\theta} = \theta$ fixed, $\tilde{y}$ variable] rather than on the likelihood interpretation [$\tilde{y} = y$ fixed, $\tilde{\theta}$ variable]. Hence it is inconsistent with the likelihood principle. It somehow does not seem right that any estimator with an agreeable-sounding property like unbiasedness could be a poor estimator, but that is often the case.

Another property of estimators is called *efficiency*. Essentially, one estimator $\tilde{y}_1$ is said to be a more efficient estimator of $\tilde{\theta}$ than a second estimator $\tilde{y}_2$ if $\text{MSE}(\tilde{y}_1 \mid \tilde{\theta} = \theta)$ is smaller than $\text{MSE}(\tilde{y}_2 \mid \tilde{\theta} = \theta)$, where $\text{MSE}(\tilde{y} \mid \tilde{\theta} = \theta)$ is the *mean-square error* of an estimator:

$$\text{MSE}(\tilde{y} \mid \tilde{\theta} = \theta) = E(\tilde{y} - \theta)^2. \tag{7.2.1}$$

If the *bias* of an estimator y is defined as

$$B(\tilde{y} \mid \tilde{\theta} = \theta) = E(\tilde{y} \mid \tilde{\theta} = \theta) - \theta, \tag{7.2.2}$$

then the mean-square error of $\tilde{y}$ can be written as the sum of the variance of $\tilde{y}$ and the square of the bias:

$$\text{MSE}(\tilde{y} \mid \tilde{\theta} = \theta) = V(\tilde{y} \mid \tilde{\theta} = \theta) + [B(\tilde{y} \mid \tilde{\theta} = \theta)]^2. \tag{7.2.3}$$

Thus, if $\tilde{y}_1$ and $\tilde{y}_2$ are unbiased estimators of $\tilde{\theta}$, so that

$$B(\tilde{y}_1 \mid \tilde{\theta} = \theta) = B(\tilde{y}_2 \mid \tilde{\theta} = \theta) = 0,$$

then $\tilde{y}_1$ is a more efficient estimator of $\tilde{\theta}$ than $\tilde{y}_2$ if $V(\tilde{y}_1 \mid \theta) < V(y_2 \mid \theta)$.

That is, an estimator with a small variance is preferred to an estimator with a larger variance. For instance, in a sample of size n from a normal population or data-generating process, the sample mean $\tilde{m}$ has variance $V(\tilde{m} \mid \mu,\sigma^2) = \sigma^2/n$, which is smaller than the variance of any other estimator of $\tilde{\mu}$. In this situation, one might consider the sample median as an estimator of $\tilde{\mu}$; the variance of the sample median, however, is approximately $V(\widetilde{md} \mid \mu,\sigma^2) = 1.57\sigma^2/n$, which is 1.57 times as large as the variance of the sample mean. Since both the sample mean and the sample median are unbiased estimators of $\tilde{\mu}$ in this situation, $\text{MSE}(\tilde{m} \mid \mu,\sigma^2) = V(\tilde{m} \mid \mu,\sigma^2)$ and $\text{MSE}(\widetilde{md} \mid \mu,\sigma^2) = V(\widetilde{md} \mid \mu,\sigma^2)$. The property of efficiency implies that a more precise estimator is better than a less precise estimator, which seems intuitively reasonable. Efficiency is, in general, a much more useful property than unbiasedness. There are other properties of point estimators that are considered by classical statisticians to be "good" properties. These are not discussed here; for more complete discussions of classical point estimation, references are given at the end of the chapter.

The Bayesian statistician is generally more concerned with problems of decision making than with problems of inference (the relation of decision theory to point estimation is discussed in Sections 7.3 and 7.4). With respect to inference, he considers the entire posterior distribution (or any probability determined from this distribution) as an inferential statement, and he may not be interested in a single point estimate. If such an estimate is desired, it will of course be based on the posterior distribution rather than just on the sample information. Since properties such as unbiasedness and efficiency involve only the sample information, they are not of much interest to the Bayesian, although in a rough sense, they do have Bayesian "counterparts." For instance, a Bayesian counterpart of efficiency might be the desire to have a small posterior variance. Some potential estimators of $\tilde{\theta}$ based on the posterior distribution $P''(\theta \mid y)$ or $f''(\theta \mid y)$ are the posterior mean, the posterior median, the posterior mode, and so on. For a normal data-generating process with known variance σ^2, the posterior distribution is normal with mean m'' and variance σ''^2, assuming a conjugate prior distribution. Since the distribution is unimodal and symmetric, the posterior mean, m'', is equal to the posterior median and to the posterior mode. Thus, m'' should be a "good" estimator of $\tilde{\mu}$. For a Bernoulli process, the posterior distribution is a beta distribution with parameters r'' and n'', assuming a conjugate prior distribution. This distribution is *not* symmetric in general; it is symmetric only when $r''/n'' = \frac{1}{2}$. The mean of the distribution is r''/n'' and, provided that $r'' > 1$ and $n'' - r'' > 1$, the mode is $(r'' - 1)/(n'' - 2)$.

If the prior distribution is diffuse, it is interesting to see how estimates based on the posterior distribution relate to classical estimates. The usual classical estimator of $\tilde{\mu}$ is simply the sample mean in the case of a normal data-generating process. But, from Section 4.10, the posterior mean m'' is equal to the sample mean m if a diffuse normal prior distribution with $n' = 0$ is assumed. Similarly, in sampling from a Bernoulli process, the usual classical estimator of $\tilde{p}$ is r/n under binomial sampling (fixed sample size). Under a diffuse beta prior distribution with $r' = n' = 0$, the posterior mean r''/n'' is equal to r/n. Of course, relationships such as these do not always hold. In Pascal sampling from a Bernoulli process (that is, sampling with a fixed number of "successes" rather than a fixed sample size), one classical estimator of $\tilde{p}$ is $(r - 1)/(n - 1)$, which has no obvious Bayesian counterpart. As indicated in Section 4.3, the posterior distribution following a sample from a Bernoulli process is the same under binomial sampling and Pascal sampling. Thus, any estimator based on the posterior distribution would be the same under the two sampling procedures.

One important method for determining estimators that classical statisticians consider to be "good" estimators is the *method of maximum likelihood.* This method finds the value of $\tilde{\theta}$ that maximizes the likelihood function. Because the estimator is based on the likelihood function, the method of maximum likelihood is consistent with the likelihood principle. Graphically, in terms of $\tilde{\theta}$, the maximum-likelihood estimator is the value corresponding to the highest point on the graph of $f(y \mid \theta)$ when $f(y \mid \theta)$ is plotted as a function of θ. But under a diffuse prior distribution, the posterior distribution is approximately proportional to the likelihood function, so the maximum-likelihood estimator in this case should roughly correspond to the highest point, the mode, of the posterior distribution. For sampling from a normal process with known variance and a normal prior distribution for $\tilde{\mu}$, the posterior mode is m'', which is equal to the maximum-likelihood estimator, m, under a diffuse prior distribution with $n' = 0$. For sampling from a Bernoulli process with a beta prior distribution for $\tilde{p}$, the posterior mode is $(r'' - 1)/(n'' - 2)$ [provided that $r'' > 1$ and $n'' - r'' > 1$], which is equal to $(r - 1)/(n - 2)$ under a diffuse prior distribution with $r' = n' = 0$ and is equal to the maximum-likelihood estimator, r/n, under a different diffuse prior distribution with $r' = 1$ and $n' = 2$ (see Section 4.10).

Even though the maximum-likelihood estimate and the mode of the posterior distribution are roughly equal under a diffuse prior distribution, their interpretation is not the same. The mode of the posterior distribution is "the most likely value of $\tilde{\theta}$." The maximum-likelihood estimator is "the value of $\tilde{\theta}$ that makes the *observed sample results* appear most likely." The difference is subtle, but it is nevertheless a difference. The Bayesian makes a probability statement about $\tilde{\theta}$; the classical statistician makes a probability statement concerning $\tilde{y}$, the sample results.

The above discussion concerns *point estimation.* A second form of estimation is *interval estimation,* in which the estimate is an interval of values rather than a single value, or a single point. From the posterior distribution, the probability of any interval of values of the uncertain quantity $\tilde{\theta}$ can be determined. To find an interval containing 90 percent of the probability, one could consider the interval from the .05 fractile to the .95 fractile, the interval from the .01 fractile to the .91 fractile, and so on. The conventional choice of an interval containing γ (Greek gamma), or 100γ percent, of the probability is the interval from the $(1 - \gamma)/2$ fractile of the distribution to the $(1 + \gamma)/2$ fractile. Thus, an equal amount of probability [exactly $(1 - \gamma)/2$] is "cut off" in each tail of the distribution, as indicated in Figure 7.2.1. For instance, an interval estimate for $\tilde{\mu}$ with $\gamma = .95$ in the case of a normal posterior distribution with mean m'' and variance σ''^2 has limits

$$m'' - 1.96\sigma'' \qquad \text{and} \qquad m'' + 1.96\sigma''.$$

In terms of m'' and n'', where $n'' = \sigma^2/\sigma''^2$ (see Section 4.5), the limits of the interval are

$$m'' - \frac{1.96\sigma}{\sqrt{n''}} \qquad \text{and} \qquad m'' + \frac{1.96\sigma}{\sqrt{n''}}.$$

This is called a 95 percent *credible interval* for $\tilde{\mu}$. By comparison, the corresponding classical interval estimate has limits

$$m - \frac{1.96\sigma}{\sqrt{n}} \qquad \text{and} \qquad m + \frac{1.96\sigma}{\sqrt{n}}.$$

This is called a 95 percent *confidence interval* for $\tilde{\mu}$.

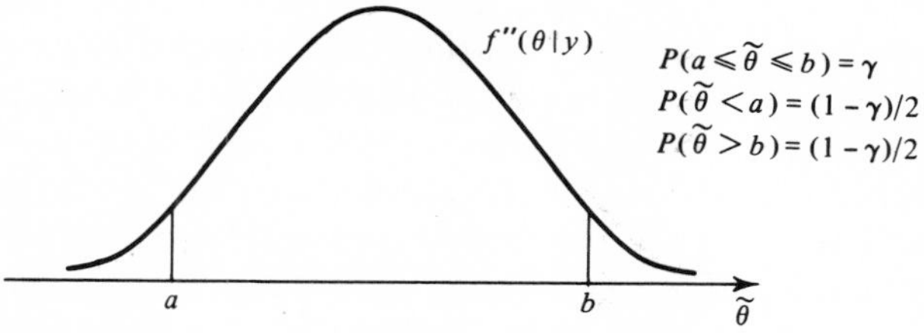

Figure 7.2.1

Under a diffuse normal prior distribution for $\tilde{\mu}$, $m'' = m$ and $n'' = n$, and therefore the Bayesian limits are identical to the classical limits.

Once again, the interpretation is different. The Bayesian would say, "the probability is .95 that $\tilde{\mu}$ lies within the specified interval," and the classical statistician would say, "in the long run, 95 percent of all such intervals will contain the true value of $\tilde{\mu}$." The classical statement is based on long-run frequency considerations, and the Bayesian statement concerns not the long run, but the specific case at hand. To differentiate between the two statements, the classical interval estimate is called a "confidence interval" and the Bayesian interval is called a "credible interval." In general, the classical interpretation of an interval estimate in terms of long-run frequency considerations is extremely counterintuitive; users of such intervals tend to interpret them in the subjective sense as probability statements about a random variable $\tilde{\mu}$, despite the classical statistician's emphasis on the frequency interpretation. Also, the classical approach requires that γ, the probability associated with the interval estimate, be chosen in advance of sampling. The Bayesian may wish to look at intervals for several different values of γ (not necessarily chosen in advance). For instance, an interval estimate for $\tilde{\mu}$ with $\gamma = .80$ in the case of a normal posterior distribution with mean m'' and variance σ''^2 has limits

$$m'' - 1.28\sigma'' \qquad \text{and} \qquad m'' + 1.28\sigma''.$$

Here, -1.28 is the $(1 - \gamma)/2$ fractile of the standard normal distribution and $+1.28$ is the $(1 + \gamma)/2$ fractile. Alternatively, the Bayesian may be interested in the posterior probability of a certain interval, in which case he starts with the interval and calculates γ instead of starting with γ and calculating the limits of an interval.

It has been emphasized that under certain conditions (a diffuse prior distribution), Bayesian and classical procedures *may* produce similar results (this is not always the case), although the interpretations are different. In actual applications, particularly in decision-making applications, the assumption of a diffuse prior distribution is seldom reasonable. In most business decision-making situations, for example, some prior information is known. For an investment decision, some prior information should be available concerning the past performance of investments similar to those being evaluated. This information may be in the form of recorded statistics or it may be the subjective information of an investment broker; usually, both types of information are available. In medicine, a surgeon should have some judgments regarding the probability that a patient will survive a particular operation. Even if this type of operation has never been performed before, the surgeon is familiar with the medical factors involved and with past results of related surgical techniques. In marketing, a marketing manager should have some idea about the proportion of consumers that will purchase a new product. Even in scientific research, some prior information is generally available to the scientist. This prior information may come from some scientific theories or it may come from previous experimental work. The list of examples is endless, and the point is simply this: diffuse prior distributions are quite useful, primarily for reporting purposes, but they are often quite unrealistic in actual inferential or decision-making situations.

If the prior distribution is not diffuse, point estimates and interval estimates based on the posterior distribution will probably differ from similar estimates based on the likelihood function. In the production process example of Section 3.3, the maximum-likelihood estimate of $\tilde{p}$ is $r/n = 3/10 = .30$, and the posterior mean turns out to be .146. In the modification of the same example, presented in Section 4.4, the maximum-likelihood estimate is .30, and the posterior mean is $r''/n'' = 4/30 = .133$. In the example concerning the weekly sales of a store (Section 4.6), the maximum-likelihood estimate, the sample mean, equals 1500. The mean of the posterior distribution is 1387.5. A 95 percent *credible* interval based on the posterior distribution is

$$(m'' - 1.96\sigma'', m'' + 1.96\sigma''),$$

or

$$(1327.5,1447.5).$$

A 95 percent *confidence* interval based on the sample information alone is

$$\left(m - \frac{1.96\sigma}{\sqrt{n}}, m + \frac{1.96\sigma}{\sqrt{n}}\right),$$

or

$$(1424,1576).$$

These examples demonstrate that the inclusion of prior information can have an important effect on point and interval estimates. Next, estimation is looked at in yet another light: estimation as a decision-making procedure. In decision making, it is necessary to include not only prior information, but also information concerning the potential "losses" due to errors in estimation.

7.3 DECISION THEORY AND POINT ESTIMATION

As noted in the previous section, in the classical approach to point estimation, which is not covered in any detail in this book, estimates are based solely on the sample information. Since estimates should reflect all of the available information, according to the Bayesian, they should be based on the posterior distribution, which incorporates both sample information and information available prior to sampling. Both of these approaches are informal procedures in the sense that there is no single "best" estimator. For the classical statistician, the choice of an estimator may be a matter of deciding whether unbiasedness is more important than efficiency, and so on. For example, $\sum_{i=1}^{n} (\tilde{x}_i - \tilde{m})^2/n$ is a maximum-likelihood estimator of the variance of a normal population, but it is biased. The corresponding unbiased estimator, $\sum_{i=1}^{n} (\tilde{x}_i - \tilde{m})^2/(n - 1)$, on the other hand, is not a maximum-likelihood estimator of σ^2. Which should be used to estimate σ^2? The choice between the two must be made subjectively. Similarly, the Bayesian judgmentally chooses among such estimators as the posterior mean, the posterior median, and the posterior mode. If all that is wanted by the statistician is a "rough-and-ready" estimate or a general idea of the value of the parameter of interest, such an informal approach should prove satisfactory. If the estimate is to be used in a situation where errors in estimation may lead to serious consequences, a more formal approach is desirable.

In the Bayesian decision-theoretic framework, the choice of an estimate can be thought of as a problem of decision making under uncertainty. Here the actions correspond to the possible estimates. If $\tilde{\theta}$ is being estimated and if S is the set of possible values of $\tilde{\theta}$, then the set of actions A is identical to the set S. If a payoff function $R(a,\theta)$ or a loss function $L(a,\theta)$ can be determined that gives the payoff or loss associated with an incorrect estimate a, then the problem of point estimation can be treated in accordance with the general decision framework presented in Section 5.9.

As example should help to clarify the decision-theoretic approach to point estimation. As usual, the example is simplified to emphasize the points of interest. Consider an automobile dealer who sells a line of luxury automobiles. He has sold his entire inventory of cars, and therefore he must order more from the manufacturer. Furthermore, next year's model will be introduced shortly and this will be his last chance to order the current model. How many automobiles should he order? This is essentially a point estimation problem, for he must estimate the demand for the cars—the number of customers that will come in and want to purchase a car. The economic factors are quite simple; the dealer will make a profit of \$500 for every car he sells before the new model comes out. Once the new model comes out, any of the current models that are still in inventory will have to be sold at a large discount, and he will suffer a loss of \$300 for each car which must be sold at this discount.

The automobile dealer's loss function can be found from the facts given above. Remember that an action, a, corresponds to the number of cars that he orders from the manufacturer; a state of the world, or value of the uncertain quantity $\tilde{\theta}$, corresponds to the number of cars that are demanded by his customers. Both a and $\tilde{\theta}$ can take on the values 0, 1, 2, . . . (all nonnegative integers). If $a = \theta$, then the number ordered is exactly equal to the number demanded, and $L(a,\theta) = 0$ because the losses are "opportunity losses," or "regrets." Given any value θ of $\tilde{\theta}$, the best possible action is $a = \theta$. If $a > \theta$, then the number of cars ordered is greater than the number demanded. In this case there is a loss associated with the "extra" cars. For each car not sold before the new models are introduced, the loss is \$300. Thus, if $a > \theta$, the loss function is $L(a,\theta) = 300(a - \theta)$. Finally, suppose that $a < \theta$. In this situation more cars are demanded than the dealer has available, and for each car demanded but not available he suffers an opportunity loss of \$500, the "lost profit." Therefore, if $a < \theta$, the loss function is $L(a,\theta) = 500(\theta - a)$. The entire loss function is thus

$$L(a,\theta) = \begin{cases} 500(\theta - a) & \text{if } a < \theta, \\ 0 & \text{if } a = \theta, \\ 300(a - \theta) & \text{if } a > \theta. \end{cases} \tag{7.3.1}$$

In order to make a decision, the dealer first must specify the other input to the problem: the probability distribution for demand. Because it is very close to the introduction of next year's model and because the cars are quite expensive, the dealer is certain that demand will be either 0, 1, 2, or 3. His subjective probability distribution consists of the probabilities

$$P(\tilde{\theta} = 0) = .1,$$
$$P(\tilde{\theta} = 1) = .2,$$
$$P(\tilde{\theta} = 2) = .3,$$

and

$$P(\tilde{\theta} = 3) = .4.$$

Since there is only a finite number of possible values of $\tilde{\theta}$, there will be only a finite number of possible values of a (the dealer cannot order a negative number of cars, and he would not be foolish enough to order more than three as long as he thinks that demand cannot possibly be greater than three). The loss function in Equation (7.3.1) can be used to find the following loss table.

ACTION (NUMBER OF CARS ORDERED)	STATE OF THE WORLD (NUMBER OF CARS DEMANDED) 0	1	2	3
0	0	500	1000	1500
1	300	0	500	1000
2	600	300	0	500
3	900	600	300	0

For any act a, the expected loss is

$$EL(a) = \sum_{\theta=0}^{3} L(a,\theta)P(\theta). \tag{7.3.2}$$

The expected losses for the four acts are

$$\begin{aligned} &EL(a = 0) = 0(.1) + 500(.2) + 1000(.3) + 1500(.4) = 1000,\\ &EL(a = 1) = 300(.1) + 0(.2) + 500(.3) + 1000(.4) = 580,\\ &EL(a = 2) = 600(.1) + 300(.2) + 0(.3) + 500(.4) = 320,\\ \text{and}\ &EL(a = 3) = 900(.1) + 600(.2) + 300(.3) + 0(.4) = 300. \end{aligned}$$

The optimal act is thus $a = 3$; the dealer should order three cars from the manufacturer. In a decision-theoretic sense, then, his "point estimate" of demand is three. This point estimate depends both on the loss function and on the probability distribution. If the last two probabilities are switched around, so that $P(\tilde{\theta} = 2) = .4$ and $P(\tilde{\theta} = 3) = .3$, the expected losses are 950, 530, 270, and 330. In this case the optimal point estimate, or number of cars to order, is two. Incidentally, it can be shown formally that any action other than 0, 1, 2, or 3 is inadmissible in this case. Consider $a = 4$. The values in the loss table for the row $a = 4$ would be 1200, 900, 600, and 300. But in all four columns, these numbers are higher than the corresponding numbers for $a = 3$. Therefore, the action $a = 4$ is dominated by $a = 3$; hence, using the definition of inadmissible actions that was presented in Section 5.2, $a = 4$ is inadmissible. The same is true for any action $a > 3$.

This example illustrates the decision-theoretic approach to point estimation. Whenever it is possible to determine a loss function $L(a,\theta)$ or a payoff function $R(a,\theta)$ in a point estimation situation, where a is the estimate and θ is the true value of the variable of interest, this procedure is applicable. It is then possible to find an optimal, or "best" (in the sense of minimizing EL or maximizing ER) estimate. In the next section, a quick way to determine this optimal estimate under certain assumptions regarding the loss function is given. Of course, not all point estimation problems involve losses. If a "rough-and-ready" estimate is all that is wanted, the Bayesian inferential approach in Section 7.2, using the posterior distribution but no loss function, should prove satisfactory. If losses are involved and if the potential losses are important, then formal decision theory should be applied to the problem.

7.4 POINT ESTIMATION: LINEAR AND QUADRATIC LOSS FUNCTIONS

The example in the last section illustrates the idea of point estimation as decision making under uncertainty. It is clear, however, that such a problem could be quite difficult to solve if the number of possible values of $\tilde{\theta}$ (and hence the number of possible estimates) is large. Instead of automobiles, suppose that the dealer sells radios and that the demand for radios could be as low as 0 or as high as 100. It would appear to be more trouble than it is worth to calculate the expected losses for all 101 possible actions. Fortunately, if certain assumptions can be made about the loss functions, it is not necessary to calculate all of these EL's. If the expected loss can be expressed in a certain form, it is possible to use the calculus to find the point at which EL is smallest. This involves taking the derivative of EL with respect to a, setting it equal to 0, and solving the resulting equation. Because EL is expressed as an integral or as a sum, where the limits of integration or limits of summation are related to a, the differentiation of EL is not as elementary as might be thought. As a result, some general statements about the optimal act are presented without proof.

First of all, suppose that the loss function is of the following form:

$$L(a,\theta) = \begin{cases} k_u(\theta - a) & \text{if } a < \theta, \\ 0 & \text{if } a = \theta, \\ k_o(a - \theta) & \text{if } a > \theta, \end{cases} \tag{7.4.1}$$

where k_u and k_o are positive constants. Such a loss function is called a *linear loss function with respect to a point estimation problem.* This should not be confused with the concept of a utility function that is linear with respect to money or a linear payoff or loss function in a two-action problem. This particular loss function is linear with respect to the difference between the estimate a and the true state of the world θ (that is, it is linear with respect to the error of estimation).

Notice that if $a < \theta$, the loss is $k_u(\theta - a)$. In this case a is smaller than θ, so $\tilde{\theta}$ has been underestimated. The *cost per unit of underestimation* is k_u. Similarly, if $a > \theta$, $\tilde{\theta}$ has been overestimated, and the *cost per unit of overestimation* is k_o. Note that a could take on any real value, so this problem is sometimes called an infinite-action problem. It can be shown mathematically that for the loss function given by Equation (7.4.1), the optimal point estimate is simply a $k_u/(k_u + k_o)$ fractile of the decision maker's distribution of $\tilde{\theta}$. This is a value a such that

$$P(\tilde{\theta} \leq a) \geq \frac{k_u}{k_u + k_o} \quad \text{and} \quad P(\tilde{\theta} \geq a) \geq 1 - \frac{k_u}{k_u + k_o}. \tag{7.4.2}$$

If $\tilde{\theta}$ is continuous, this reduces to

$$P(\tilde{\theta} \leq a) = \frac{k_u}{k_u + k_o}, \tag{7.4.3}$$

as discussed in Sections 4.3 and 4.8. This greatly simplifies the problem from a computational standpoint. Instead of calculating a number of expected losses, it is only necessary, given k_u, k_o, and the distribution of $\tilde{\theta}$, to find a single fractile of the distribution.

In the example presented in the previous section, the loss function is linear with respect to the difference between a and θ. The values of k_u and k_o are 500 and 300, and thus

$$\frac{k_u}{k_u + k_o} = \frac{500}{500 + 300} = \frac{5}{8}.$$

To find the $\frac{5}{8}$ fractile of the dealer's probability distribution $P(\theta)$, consider the corresponding CDF, presented in Figure 7.4.1. A horizontal line passing through $\frac{5}{8}$ on the vertical axis intersects the CDF at the "step" at $\tilde{\theta} = 3$. Thus, the $\frac{5}{8}$ fractile is 3 [note that this value satisfies Equation (7.4.2), since $P(\tilde{\theta} \leq 3) = 1 > \frac{5}{8}$ and $P(\tilde{\theta} \geq 3) = .4 > \frac{3}{8}$], so the optimal value of a is 3, just as determined in the last section by calculating the expected losses for all of the possible actions. In the discrete case, it is possible for a horizontal line to intersect the CDF at a "step," as in the example, or over an entire line segment. For instance, suppose that you want to find the .6 fractile of the distribution of $\tilde{\theta}$. A horizontal line passing through .6 on the vertical axis intersects the CDF in a line segment for which $2 \leq \tilde{\theta} < 3$ and also intersects the "step" at $\tilde{\theta} = 3$ (see Figure 7.4.2). Technically, all real numbers between 2 and 3 (including the end points, 2 and 3) satisfy the definition of a .6 fractile, as given by Equation (7.4.2). To determine a single number to use as a .6 fractile, one convention is to take the midpoint of the line segment, which is 2.5. In the decision-making problem, it is impossible to order 2.5 cars. If $k_u/(k_u + k_o) = .6$, $EL(a = 2) = EL(a = 3)$, so it makes no difference whether 2 or 3 cars are ordered.

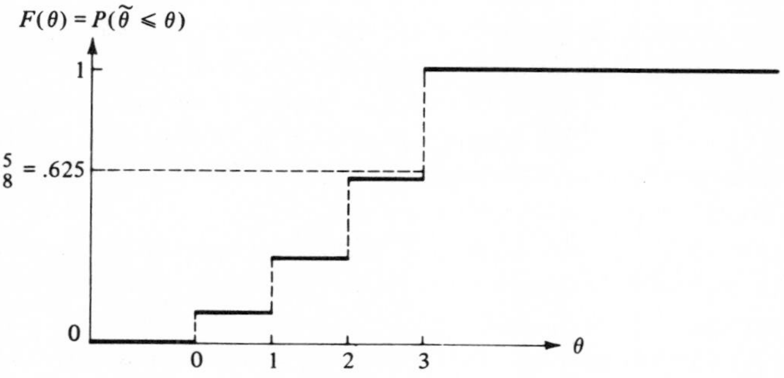

Figure 7.4.1

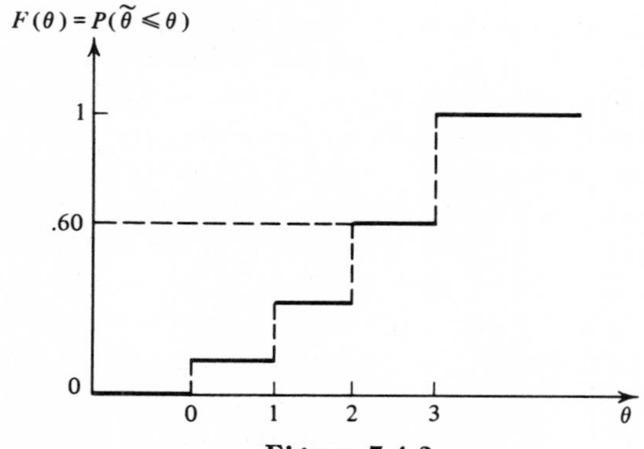

Figure 7.4.2

The general result stated above is applicable whether the variable $\tilde{\theta}$ is discrete or continuous. Suppose that a grocer has to decide how many oranges to order for a particular week. The oranges cost the grocer \$0.60 per dozen, and he sells them for \$0.80 per dozen. Any oranges that remain unsold at the end of the week will spoil and will therefore be worthless. If the grocer orders more oranges than are demanded, he suffers a loss of \$0.60 per dozen, his cost, for all unsold oranges. If he orders fewer than are demanded, he suffers an opportunity loss of \$0.20 per dozen, reflecting lost profits, for all oranges demanded over and above the amount stocked. His loss function is of the form of Equation (7.4.1), with $k_u = .20$ and $k_o = .60$.

Based on past experience, the grocer feels that the weekly demand for oranges is approximately normally distributed with a mean of 80 dozen and a standard deviation of 25 dozen. Perhaps the grocer has kept extensive records and, upon investigating his data carefully, he finds that the sample mean is 80 dozen, the sample variance is 625, and the normal model appears to provide a reasonable "fit" to the data. How many dozen oranges should the grocer order? He needs to find the $k_u/(k_u + k_o) = .2/(.2 + .6) = \frac{1}{4}$ fractile of the distribution of $\tilde{\theta}$. From tables of the normal distribution, the $\frac{1}{4}$ fractile of the standard normal distribution is approximately $-.675$. This implies that the $\frac{1}{4}$ fractile of a normal distribution with mean m' and standard deviation σ' is simply .675 standard deviations to the left of the mean, or $m' - .675\sigma'$. The $\frac{1}{4}$ fractile of the grocer's distribution is then $80 - .675(25)$, or approximately 63 dozen oranges. To minimize expected loss, the grocer should order 63 dozen oranges.

In these two examples, the linear loss function seems to be realistic. In many cases, especially in problems such as these involving inventories, the loss function is linear or almost linear. The result given by Equations (7.4.2) and (7.4.3) should therefore have wide applicability; is it intuitively reasonable? If $k_u = k_o$, the optimal estimate is the .50 fractile, or the median, of the decision maker's distribution. This is reasonable; if the costs of overestimation and underestimation are equal, the optimal act should be somewhere in the "middle" of the distribution. Why specifically the median rather than the mean? Because if the median is chosen, the probability of overestimating is equal to the probability of underestimating. If $k_u > k_o$, then $k_u/(k_u + k_o)$ is greater than .50. The estimate is higher than the median. In this case, the cost of underestimation is greater than the cost of overestimation, so the decision maker "hedges" his estimate upward to avoid the greater cost of underestimation. In the automobile-dealer example, the cost of ordering too many cars is only \$300 per extra car, as compared with the \$500 per car opportunity loss if too few cars are ordered. The dealer would rather overestimate slightly than underestimate, so his optimal estimate is three cars. In the limiting case, as k_o becomes very small in relation to k_u, the decision maker will move further and further toward the right-hand tail of the distribution in an attempt to avoid suffering the high cost of underestimation. If $k_u < k_o$, just the opposite is true. In the grocer example, the cost of not having enough oranges on hand is just the lost profit of \$0.20 per dozen, while any extra oranges that spoil will cost him \$0.60

per dozen. To avoid the higher cost of overestimation, he orders only 63 dozen oranges, although the mean of his distribution is 80 dozen. For highly perishable goods, k_o is often greater than k_u. For nonperishables, k_o is usually lower than k_u, since the only cost of overestimation is the cost of storing the unsold inventory. At any rate, the result in Equations (7.4.2) and (7.4.3) seems to be intuitively reasonable; if $k_o < k_u$, the estimate will be to the right of the median, and if $k_o > k_u$, the estimate will be to the left of the median. In each case, the greater the *relative* difference between k_u and k_o (as represented by the ratio of the larger of the two to the smaller), the further from the median the estimate will be.

In decision-making problems of the nature considered in this section, the decision maker might be interested in the value of perfect information. From Equation (6.2.4), the expected value of perfect information is equal to the expected loss of the action that is optimal under the decision maker's current state of information. In the example in Section 7.2 involving the automobile dealer, the optimal action is to order three cars, and the expected loss of this action is \$300. Thus, EVPI = 300; it should be worth \$300 to the dealer to know exactly how many cars will be demanded. For the example involving the grocer, it is more difficult to compute expected losses because $\tilde{\theta}$ is taken to be a continuous random variable. If $\tilde{\theta}$ is normally distributed with mean m' and variance σ'^2, and if the loss function is given by Equation (7.4.1), then

$$\text{EVPI} = (k_u + k_o)f(z \mid 0,1)\sigma', \tag{7.4.4}$$

where a^* is the optimal action given by Equation (7.4.3) and $z = (a^* - m')/\sigma'$ is the corresponding standardized random variable. Note that whereas the optimal action depends only on the *relative* magnitudes of k_u and k_o [via the ratio $k_u/(k_u + k_o)$], the EVPI depends on the *absolute* magnitudes [via the sum $(k_u + k_o)$]. In the grocer example, $k_u = .2$, $k_o = .6$, $z = -.675$, and $\sigma' = 25$. From the table of the standard normal density function, $f(-.675 \mid 0,1) = .3176$, so that

$$\text{EVPI} = (.2 + .6)(.3176)(25) = 6.352.$$

Thus, the grocer should be willing to pay up to \$6.35 to know for certain what the demand for oranges will be during the next week.

A linear loss function of the form of Equation (7.4.1) is often quite realistic, as has been shown. There may be situations in which it is not applicable, however. Suppose that the costs of underestimation and overestimation are equal (let $k_o = k_u = k$) and that an error in estimation of two units is *more* than twice as costly as an error of just one unit. In particular, assume that an error of two units is four times as costly as an error of just one unit and that in general, the loss associated with an error is proportional to the *square* of the error. The resulting loss function, which is called a *quadratic loss function* or a *squared-error loss function,* can be written as follows:

$$L(a,\theta) = k(a - \theta)^2. \tag{7.4.5}$$

For this loss function, the optimal point estimate a is the mean of the decision maker's distribution of $\tilde{\theta}$. That is, the optimal estimate is

$$a = E(\tilde{\theta}). \tag{7.4.6}$$

It is possible to prove the above result without using calculus. It is necessary to find a so as to minimize

$$EL(a) = E[k(a - \tilde{\theta})^2] = kE(a - \tilde{\theta})^2. \tag{7.4.7}$$

Clearly, since k is a positive constant, if a can be found so as to minimize $E(a - \tilde{\theta})^2$, this a will also minimize $EL(a)$. Adding and subtracting $E(\tilde{\theta})$ within the parentheses yields.

$$\begin{aligned} E(a - \tilde{\theta})^2 &= E[a - E(\tilde{\theta}) + E(\tilde{\theta}) - \tilde{\theta}]^2 \\ &= E[(a - E(\tilde{\theta})) + (E(\tilde{\theta}) - \tilde{\theta})]^2 \\ &= E[(a - E(\tilde{\theta}))^2 + 2(a - E(\tilde{\theta}))(E(\tilde{\theta}) - \tilde{\theta}) + (E(\tilde{\theta}) - \tilde{\theta})^2] \\ &= E(a - E(\tilde{\theta}))^2 + 2E[(a - E(\tilde{\theta}))(E(\tilde{\theta}) - \tilde{\theta})] + E(E(\tilde{\theta}) - \tilde{\theta})^2. \end{aligned}$$

The expectation is being taken with respect to the distribution of $\tilde{\theta}$, so a and $E(\tilde{\theta})$ are constants. Thus, using the rules for dealing with expectations,

$$E(a - \tilde{\theta})^2 = (a - E(\tilde{\theta}))^2 + 2(a - E(\tilde{\theta}))E[E(\tilde{\theta}) - \tilde{\theta}] + E(E(\tilde{\theta}) - \tilde{\theta})^2.$$

But the middle term is 0, since

$$E[E(\tilde{\theta}) - \tilde{\theta}] = E(\tilde{\theta}) - E(\tilde{\theta}) = 0.$$

The last term does not involve a, so it is irrelevant to the choice of an optimal a. This leaves the first term, which is

$$(a - E(\tilde{\theta}))^2.$$

Since this is a squared term, it must be either positive or zero. But it is zero only when $a = E(\tilde{\theta})$, so this is the value of a that minimizes EL.

The basic result of this proof is that the expected squared deviation (that is, the mean-square error) is smallest when calculated from the mean. This is the reason that $E(\tilde{\theta})$ is the optimal estimate for a quadratic, or squared-error, loss function, and it is also one of the reasons that the mean is considered by statisticians to be a "good" estimator. Estimation with a quadratic loss function is a rough Bayesian counterpart to the classical notion of efficient estimators, which was discussed briefly in Section 7.2. In taking the posterior mean as an estimator of the population mean, the Bayesian statistician is essentially acting as though he had a quadratic loss function, although it should be noted that there are also many other loss functions for which the mean is an optimal estimator. If the loss function is difficult to define exactly but if it appears to be similar to the quadratic loss function in Equation (7.4.5), then the statistician is justified, in a decision-theoretic sense, in using the mean as his estimator. Also, if the loss function appears to be similar to the linear loss function in Equation (7.4.1) with $k_u = k_o$, then there is justification for using the median as the estimator.

In a point estimation problem with a quadratic loss function of the form of Equation (7.4.5), the optimal point estimate is the mean of the distribution of $\tilde{\theta}$. It is only necessary to know the mean in order to make a decision, and the decision maker can act as though the mean is equal to the true value of $\tilde{\theta}$ with certainty. Recall, from Section 6.5, that the mean is called a *certainty equivalent* in a situation like this. Similarly, if the loss function is linear of the form of Equation (7.4.1), the $k_u/(k_u + k_o)$ fractile of the distribution is a certainty equivalent. It should also be noted that a useful property of decision-theoretic estimates is that they are *invariant*. For instance, if the optimal estimate of $\tilde{\theta}$ in terms of minimizing expected loss is 25, then the optimal estimate of $\tilde{\theta}^2$ is $(25)^2 = 625$. The optimal estimate is invariant with respect to the transformation from $\tilde{\theta}$ to $\tilde{\theta}^2$. Maximum-likelihood estimates are also invariant, but classical estimates are not, in general, invariant.

Suppose that in the quality-control example presented in Section 4.4, the production manager wants to estimate $\tilde{p}$, the proportion of defective items. His prior distribution is a beta distribution with $r' = 1$ and $n' = 20$. Suppose that he feels that his loss function is linear with $k_u = 9k_o$. That is, the cost of underestimation, which is associated with the error of leaving the process alone when it should be adjusted, is judged to be nine times the cost of overestimation, which is associated with the error of adjusting the process needlessly. The optimal estimate is thus the $k_u/(k_u + k_o) = 9k_o/(9k_o + k_o) = .9$ fractile of the distribution of $\tilde{p}$. From the table of fractiles of beta distributions that is presented in the back of the book, the .9 fractile of the beta distribution with $r' = 1$ and $n' = 20$ is approximately .12. If, on the other hand, his loss function is quadratic, then the optimal estimate is the mean of the distribution, $E'(\tilde{p}) = \frac{1}{20} = .05$. In the first case, the .9 fractile of the prior distribution is a certainty equivalent; in the second case, the mean of the prior distribution is a certainty equivalent.

When the loss function is quadratic, the determination of EVPI is much easier than in the linear case. Since the mean is the optimal estimate, the expected loss of this optimal estimate is, from Equation (7.4.7), simply

$$EL[E(\tilde{\theta})] = E[k(E(\tilde{\theta}) - \tilde{\theta})^2] = kV(\tilde{\theta}). \tag{7.4.8}$$

If the distribution of $\tilde{\theta}$ is a normal distribution with mean m' and variance σ'^2, then,

$$\text{EVPI} = k\sigma'^2. \tag{7.4.9}$$

If the distribution of $\tilde{\theta}$ is a beta distribution with parameters r' and n', then $V(\tilde{\theta}) = r'(n' - r')/n'^2(n' + 1)$, so that

$$\text{EVPI} = \frac{kr'(n' - r')}{n'^2(n' + 1)}. \tag{7.4.10}$$

The expected value of sample information also turns out to be relatively easy to compute. For a sample of size n from a normal process with known variance σ^2, where the prior distribution of $\tilde{\mu}$ is normal with mean m' and variance $\sigma'^2 = \sigma^2/n'$,

$$\text{EVSI}(n) = k\left(\frac{n}{n + n'}\right)\sigma'^2. \tag{7.4.11}$$

For a sample of size n (binomial sampling) from a Bernoulli process when the prior distribution is beta with parameters r' and n',

$$\text{EVSI}(n) = k\,\frac{\left(\frac{n}{n + n'}\right)r'(n' - r')}{n'^2(n' + 1)}. \tag{7.4.12}$$

Note in both cases that as n increases, EVSI(n) approaches EVPI. In the quality-control example with a quadratic loss function and $k = 1000$,

$$\begin{aligned}\text{EVPI} &= \frac{kr'(n' - r')}{n'^2(n' + 1)}\\ &= \frac{1000(1)(19)}{(20)^2(21)} = 2.26,\end{aligned}$$

whereas

$$\begin{aligned}\text{EVSI}(n) &= k\left(\frac{n}{n + n'}\right)\frac{r'(n' - r')}{n'^2(n' + 1)}\\ &= \left(\frac{n}{n + n'}\right)\text{EVPI},\end{aligned}$$

so that

$$\begin{aligned}\text{EVSI}(n = 1) &= (1/21)(2.26) = .11,\\ \text{EVSI}(n = 10) &= (10/30)(2.26) = .75,\\ \text{EVSI}(n = 20) &= (20/40)(2.26) = 1.13,\\ \text{EVSI}(n = 60) &= (60/80)(2.26) = 1.70,\\ \text{EVSI}(n = 100) &= (100/120)(2.26) = 1.88,\end{aligned}$$

and so on. As n increases, $n/(n + n')$ approaches 1 and EVSI(n) approaches EVPI.

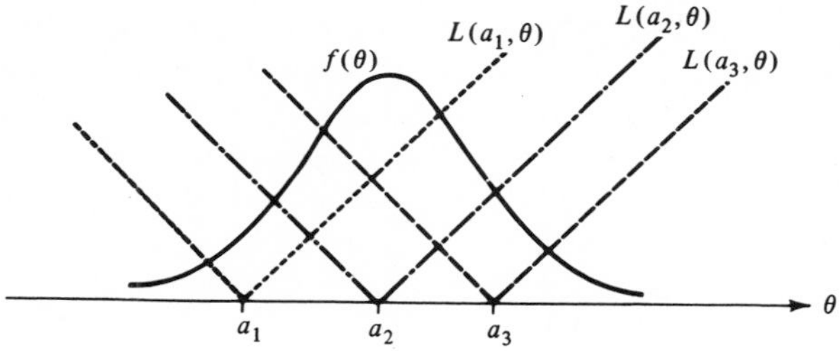

Figure 7.4.3

Linear and quadratic loss functions have been presented in algebraic form, and it might be useful to see how they can be represented on a graph. In Figure 7.4.3, a linear loss function with $k_u = k_o = 1$ is presented for three possible acts: a_1, a_2, and a_3. The density function $f(\theta)$ is also included in the graph. When the expected loss of an action is computed, the loss function is "weighted" by the density function. In the graph, a_2 appears to have a lower expected loss than a_1 or a_3. This is because the loss function of a_2 is low when the density function is high, and vice versa. For both a_1 and a_3, the loss function is quite high at the same time that the density function is high, so the expected loss would be greater than is the case with a_2. A similar result occurs in Figure 7.4.4 for quadratic loss functions. Once again, a_2 would appear to have the smallest expected loss. Of course, a_1, a_2, and a_3 are by no means the only possible estimates. Since $\tilde{\theta}$ is continuous, there is an infinite number of potential estimates. In the linear situation, the optimal estimate is the median; in the quadratic situation, the optimal estimate is the mean. Notice that if $f(\theta)$ is a symmetric distribution, the mean and the median are equal. If $k_o = k_u$, then, the difference between the linear and quadratic loss functions is only important when the distribution is skewed.

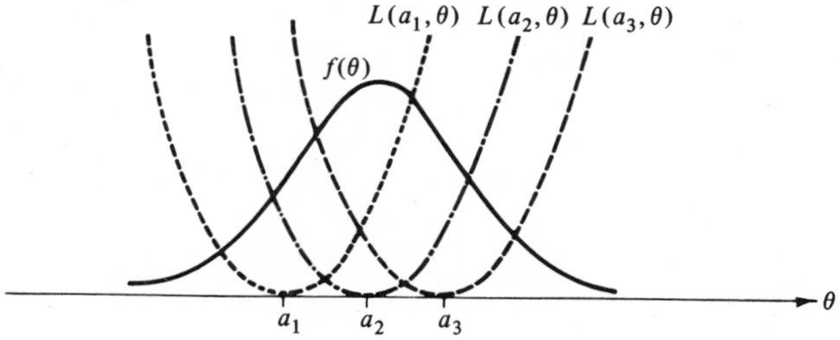

Figure 7.4.4

In Figure 7.4.5, loss functions with $k_o = 4$ and $k_u = 1$ are presented for three actions, a_1, a_2, and a_3. In this situation, because the cost of overestimation is four times the cost of underestimation, it appears that a_1 is the best of the three estimators. For a_2 and a_3, the line corresponding to the cost of overestimation (which is the line going to the left, since to the left of a we have $\theta < a$) rises so quickly that the loss function is quite high at the same time that the density function is reasonably high. Of course, there are more than three possible estimators; in this example, the $k_u/(k_u + k_o) = .20$ fractile is the best estimator.

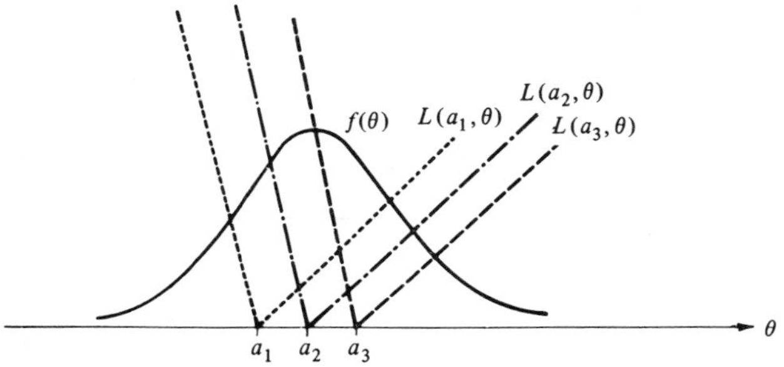

Figure 7.4.5

A final point of interest with regard to the decision-theoretic approach to point estimation is that the framework presented in this section can also be used for "point predictions" of future sample outcomes. For instance, if you want to predict the sample mean $\tilde{m}$ for a sample of size n subject to a loss function $L(a,m)$, then the relevant distribution is the predictive distribution of $\tilde{m}$. If the loss function is of the form of Equation (7.4.1), the optimal prediction is the $k_u/(k_u + k_o)$ fractile of the predictive distribution; if the loss function is of the form of Equation (7.4.5), the optimal prediction is the mean of the predictive distribution. The example in Section 7.3 could be reformulated as a predictive problem with an underlying Poisson process generating sales. The parameter of the Poisson process is $\tilde{\lambda}$ (the mean number of cars sold per unit of time), whereas the uncertain quantity of interest to the decision maker is $\tilde{r}$ (the actual demand for cars in the period until the new models are introduced). Similarly, the grocer example could be expressed in terms of an underlying normal process with parameter $\tilde{\mu}$ [mean sales of oranges (in dozens) per day], in which case the problem at hand concerns $\tilde{m}$ [the sample mean sales for a sample of size seven (the seven days in the coming week)]. For discussions regarding predictive distributions, see Sections 3.7 and 4.11.

7.5 PRIOR AND POSTERIOR ODDS RATIOS

Instead of estimating a certain parameter, a statistician may wish to test a particular *hypothesis* concerning that parameter against some particular alternative hypothesis. For instance, if the data-generating process is assumed to follow a Bernoulli model, the statistician might be interested in the hypotheses H_1: $\tilde{p} = .2$ and H_2: $\tilde{p} = .3$. In investigating such hypotheses, it is convenient to consider the *odds ratio* of one hypothesis, H_1, to the other hypothesis, H_2. In this section, the concept of odds ratios is discussed and Bayes' theorem is presented in terms of odds ratios; this relates to the discussion of probability versus betting odds presented in Section 2.4. In the following sections, such odds ratios are related to hypothesis testing, both from an inferential viewpoint and from a decision-theoretic viewpoint.

Suppose that a politician is interested in the proportion of votes that he will receive in a forthcoming election. Call this proportion $\tilde{p}$. The politician's advisors think that there are two possible values for $\tilde{p}$, .40 and .60. Furthermore, the prior probabilities of these two values are

$$P'(.40) = .25 \qquad \text{and} \qquad P'(.60) = .75.$$

A sample of 40 voters is taken, and 18 of them claim that they will vote for this politician. For the purposes of this example, possible differences between stated voting intentions and actual voting behavior are ignored. A table can be set up for the computation of the posterior probabilities (the likelihoods are based on the assumption that the process is a Bernoulli process).

Value of $\tilde{p}$	*Prior probability*	*Likelihood*	*Prior probability* × *likelihood*	*Posterior probability*
.40	.25	.10255	.02564	.628
.60	.75	.02026	.01519	.372
	1.00		.04083	1.000

Bayes' theorem can be written in terms of odds instead of probabilities. For this example, let A correspond to the event that $\tilde{p} = .40$, and let B correspond to the event that $\tilde{p} = .60$. Then the *prior odds ratio of A to B*, denoted by Ω', is defined as the ratio of the prior probabilities,

$$\Omega' = \frac{P'(A)}{P'(B)} = \frac{.25}{.75} = \frac{1}{3}. \tag{7.5.1}$$

If this ratio is greater than 1, then A is more likely than B; if it is less than 1, B is more likely than A; and if it is equal to 1, A and B are equally likely. The same holds for the *posterior odds ratio of A to B*, denoted by Ω'':

$$\Omega'' = \frac{P''(A \mid y)}{P''(B \mid y)} = \frac{.628}{.372} = 1.69, \tag{7.5.2}$$

where y represents the sample results (18 "successes" in 40 trials). But from Bayes' theorem, the posterior probability is proportional to the product of the prior probability and the likelihood. Thus, a *ratio* of posterior probabilities should be equal to the product of a *ratio* of prior probabilities and a *ratio* of likelihoods. The *likelihood ratio*, denoted by LR, is of the form

$$LR = \frac{f(y \mid A)}{f(y \mid B)} = \frac{.10255}{.02026} = 5.06. \tag{7.5.3}$$

From Equation (3.2.3), Bayes' theorem for discrete probability models can be written in the form

$$P(\theta_j \mid y) = \frac{P(y \mid \theta_j)P(\theta_j)}{\sum_{i=1}^{J} P(y \mid \theta_i)P(\theta_i)}.$$

For any two values θ_a and θ_b of $\tilde{\theta}$, the posterior odds ratio of θ_a to θ_b is

$$\Omega'' = \frac{P(\theta_a \mid y)}{P(\theta_b \mid y)} = \frac{P(y \mid \theta_a)P(\theta_a) \Big/ \sum_{i=1}^{J} P(y \mid \theta_i)P(\theta_i)}{P(y \mid \theta_b)P(\theta_b) \Big/ \sum_{i=1}^{J} P(y \mid \theta_i)P(\theta_i)}$$

$$= \frac{P(y \mid \theta_a)P(\theta_a)}{P(y \mid \theta_b)P(\theta_b)}.$$

But

$$\Omega' = \frac{P(\theta_a)}{P(\theta_b)}$$

and

$$LR = \frac{P(y \mid \theta_a)}{P(y \mid \theta_b)}.$$

Thus, Bayes' theorem can be written in terms of odds:

$$\Omega'' = \Omega' \times (LR). \tag{7.5.4}$$

In words, the *posterior odds ratio* of any two events or values of an uncertain quantity is equal to the product of the *prior odds ratio* and the *likelihood ratio.* Note that this is true for *any* two values of an uncertain quantity for which Ω' and LR are defined; they need not be the only possible values, as in the above numerical example.

In the election example,

$$\Omega'' = \Omega' \times (LR) = (\tfrac{1}{3})(5.06) = 1.69,$$

which agrees with the result determined directly from the posterior probabilities in Equation (7.5.2). Whereas the prior odds ratio favors B, the sample information is such that the posterior odds ratio favors A, as is indicated by Ω''.

For another example, suppose that the occurrence of accidents along a particular stretch of highway is assumed to follow roughly a Poisson process, and you feel that the intensity of accidents, $\tilde{\lambda}$, is either two per week or three per week. Your prior judgments are such that you feel that the odds in favor of $\tilde{\lambda} = 3$ are 2 to 1. Thus, the prior odds ratio is

$$\Omega' = \frac{P'(\tilde{\lambda} = 2)}{P'(\tilde{\lambda} = 3)} = \frac{1}{2}.$$

You then obtain sample information in the form of the accident records for the highway for a six-week period. In this six-week period, 17 accidents occurred, so the likelihood ratio is

$$LR = \frac{P[\tilde{r} = 17 \mid \lambda t = 2(6) = 12]}{P[\tilde{r} = 17 \mid \lambda t = 3(6) = 18]} = \frac{.0383}{.0936} = .4092.$$

Thus, the posterior odds ratio is

$$\Omega'' = \Omega' \times (LR) = (.5)(.4092) = .2046.$$

This implies that the odds in favor of $\tilde{\lambda} = 2$ are .2046 to 1, so the odds in favor of $\tilde{\lambda} = 3$ are 1 to .2046, or 4.9 to 1. The sample information increased the odds in favor of $\tilde{\lambda} = 3$ from 2 to 1 to 4.9 to 1.

For a third example, consider the weight of items produced by a certain manufacturing process. Suppose that this weight is normally distributed with mean $\tilde{\mu}$ and variance $\sigma^2 = 4$; furthermore, it is assumed that the mean weight $\tilde{\mu}$ is either 9 or 10 (perhaps it is 9 if the process is "in control," but it is 10 if the process is "out of control"). The prior probabilities are $P(\tilde{\mu} = 9) = .85$ and $P(\tilde{\mu} = 10) = .15$, for the process tends to be out of control about 15 percent of the time. Thus, the prior odds ratio is

$$\Omega' = \frac{P'(\tilde{\mu} = 9)}{P'(\tilde{\mu} = 10)} = \frac{.85}{.15} = 5.67.$$

A sample of size 25 is taken from the process, and the mean weight of the 25 items is 9.2. The likelihood ratio is a ratio of two density values:

$$LR = \frac{f'(\tilde{m} = 9.2 \mid \tilde{\mu} = 9, \sigma^2/n = .16)}{f'(\tilde{m} = 9.2 \mid \tilde{\mu} = 10, \sigma^2/n = .16)}.$$

Using Equation (4.5.5),

$$LR = \frac{f\left(\tilde{z} = \dfrac{9.2 - 9}{.4}\right) \Big/ .4}{f\left(\tilde{z} = \dfrac{9.2 - 10}{.4}\right) \Big/ .4}$$

$$= \frac{f(\tilde{z} = .5)/.4}{f(\tilde{z} = -2)/.4} = \frac{.3521/.4}{.0540/.4} = 6.52.$$

Thus,

$$\Omega'' = \Omega' \times (LR) = 5.67(6.52) = 36.97.$$

The odds are strongly in favor of the process being in control.

What implications does Equation (7.5.4) hold for hypothesis testing? In the election example, the two possible values of $\tilde{p}$ can be thought of as two hypotheses:

$$H_1\colon \tilde{p} = .40$$

and

$$H_2\colon \tilde{p} = .60.$$

If the politician is interested in which hypothesis is more likely, he need only look at the posterior odds ratio, Ω''. In the example, $\Omega'' = 1.69$, implying that H_1 is more likely than H_2. The odds in favor of H_1 are 1.7 to 1. H_1 is more likely than H_2 if the posterior odds ratio of H_1 to H_2 is greater than 1, and H_2 is more likely than H_1 if this odds ratio is less than 1. In the accident example, if the hypotheses are

$$H_1\colon \tilde{\lambda} = 2$$

and

$$H_2\colon \tilde{\lambda} = 3,$$

then H_2 is more likely because $\Omega'' = .2046$. In the weight example, if the hypotheses are

$$H_1\colon \tilde{\mu} = 9$$

and

$$H_2\colon \tilde{\mu} = 10,$$

then H_1 is more likely because $\Omega'' = 36.97$.

These examples admittedly are somewhat artificial, for it is assumed in each case that there are only two possible values of the random variable of interest. In the election example, it would be more realistic for the politician to assume that $\tilde{p}$ is continuous and to assess a continuous posterior distribution for $\tilde{p}$. If he is primarily interested in whether $\tilde{p}$ is greater than or less than $\frac{1}{2}$, he can determine $P(\tilde{p} > \frac{1}{2})$ and $P(\tilde{p} < \frac{1}{2})$ from the posterior distribution and then compute the odds ratio $P(\tilde{p} > \frac{1}{2})/P(\tilde{p} < \frac{1}{2})$. In the accident example, $\tilde{\lambda}$ might be thought of as continuous, in which case the odds ratio of interest might be

$$P(\tilde{\lambda} > 2)/P(\tilde{\lambda} < 2).$$

For instance, if $\tilde{\lambda} > 2$ the speed limit might be reduced by the state, while if $\tilde{\lambda} < 2$, the speed limit might not be changed. In the weight example, $\tilde{\mu}$ could be thought of as continuous. For instance, if the posterior distribution of $\tilde{\mu}$ turns out to be a normal distribution with mean $m'' = 9.2$ and variance $\sigma''^2 = .0225$, then the production manager might be interested in

$$\frac{P(\tilde{\mu} > 9.4)}{P(\tilde{\mu} < 9.4)} = \frac{P\left(\tilde{z} > \dfrac{9.4 - 9.2}{.15}\right)}{P\left(\tilde{z} < \dfrac{9.4 - 9.2}{.15}\right)} = \frac{P(\tilde{z} > 1.33)}{P(\tilde{z} < 1.33)} = \frac{.092}{.908} = .101.$$

Thus, it is possible to determine posterior odds ratios when the hypotheses of interest are composite hypotheses. (A *composite* hypothesis, such as $\tilde{\mu} > 9.4$, involves more than one value of the random variable of interest; a *simple* hypothesis, such as $\tilde{\mu} = 9$, involves just a single value of the random variable.) However, when the hypotheses are composite it is generally necessary to compute the posterior distribution in the usual manner and then to determine the odds ratio from this distribution [as opposed to applying Equation (7.5.4) directly]. For a composite hypothesis such as H_1: $\tilde{\mu} > 9.4$, there is no single value of $\tilde{\mu}$ that can be used to compute the likelihood $f(y \mid H_1)$. Therefore, it is difficult to determine the likelihood ratio. Consequently, the Bayesian approach to composite hypotheses may be slightly different from that presented in this section, as you shall see in Section 7.7. The general idea is the same, however.

It should be noted that the posterior odds ratio is the *product* of the prior odds ratio and the likelihood ratio. If logarithms are taken, the logarithm of the posterior odds ratio is equal to the *sum* of the logarithm of the prior odds ratio and the logarithm of the likelihood ratio:

$$\log \Omega'' = \log \Omega' + \log LR. \qquad (7.5.5)$$

Sometimes it is useful to work with "log odds" and "log likelihood ratios" instead of odds and likelihood ratios.

The concepts of odds ratios and likelihood ratios are quite important in hypothesis testing, as you shall continue to see in the next few sections. The standard classical hypothesis testing procedures generally are related to the likelihood ratio; this relationship is examined in the next section. Odds ratios are useful to the Bayesian in informally comparing composite hypotheses as well as in comparing simple hypotheses, as you shall see in Section 7.7. Finally, in Section 7.8, hypothesis testing is discussed as a decision-making procedure, in which case a decision is made by comparing an odds ratio and a ratio of losses.

7.6 THE LIKELIHOOD RATIO AND CLASSICAL HYPOTHESIS TESTING

In the election example in the previous section, a classical statistician would have based his inferences solely on the sample information. Given the hypotheses

$$H_1\colon \tilde{p} = .40$$

and

$$H_2\colon \tilde{p} = .60,$$

the likelihood ratio, LR, summarizes all of the sample information relevant to these two hypotheses. This ratio is of the form

$$LR = \frac{f(y \mid \tilde{p} = .40)}{f(y \mid \tilde{p} = .60)} = \frac{P(\tilde{r} = r \mid n, \tilde{p} = .40)}{P(\tilde{r} = r \mid n, \tilde{p} = .60)}. \tag{7.6.1}$$

If $LR > 1$, the observed sample result appears more likely given H_1 than given H_2; if $LR < 1$, the reverse is true. Thus, the larger the likelihood ratio, the more likely it appears that H_1 is the true hypothesis; the smaller the likelihood ratio, the more likely it appears that H_2 is the true hypothesis. The likelihood ratio can be used, then, to investigate the relative merits of the two hypotheses. In order to decide whether to "accept" the hypothesis H_1 or to "reject" H_1 in favor of the alternative hypothesis H_2, the classical statistician can determine a *rejection region* in terms of LR. This rejection region, which is a region of values of LR for which H_1 would be rejected in favor of H_2, is of the form

$$LR \leq c, \tag{7.6.2}$$

where c is some constant. The region is of this form because low values of LR tend to favor H_2, whereas high values tend to favor H_1. If c is known and LR is known, the statistician just compares the two to determine whether to accept or to reject H_1 (that is, whether to accept H_1 or to accept H_2).

For the election example, what would a rejection region look like in terms of r, the number of successes in a sample of size n? Looking at the two hypotheses, it is obvious that a rejection region would be of the form

$$r \geq d, \tag{7.6.3}$$

where d is some constant, since large values of r clearly favor H_2. Could this be equivalent to the rejection region given by Equation (7.6.2)? For sampling from a Bernoulli process with fixed n, the binomial distribution is applicable, and the likelihood ratio is

$$LR = \frac{f(y \mid \tilde{p} = .40)}{f(y \mid \tilde{p} = .60)} = \frac{\binom{n}{r}(.40)^r(.60)^{n-r}}{\binom{n}{r}(.60)^r(.40)^{n-r}}.$$

Canceling and combining terms,

$$LR = \frac{(.40)^{2r-n}}{(.60)^{2r-n}} = \left(\frac{.40}{.60}\right)^{2r-n} = \left(\frac{2}{3}\right)^{2r-n}.$$

For a fixed n, what happens to LR as r is varied? As r gets larger, LR gets smaller [because the number being raised to the $(2r - n)$th power is $\frac{2}{3}$, which is less than 1]. Conversely, as r gets smaller, LR gets larger. If you are not convinced that this is so, try some specific values of r (for fixed n) and calculate LR. Now, according to Equation (7.6.2), H_1 will be rejected when LR is small. But when LR is small, r must be large, so the rejection regions of the forms (7.6.2) and (7.6.3) are equivalent. In order for LR to be smaller than some constant c, r must be greater than some other constant d.

In the accident example in the previous section, the hypotheses of interest are

$$H_1: \tilde{\lambda} = 2$$

and

$$H_2: \tilde{\lambda} = 3.$$

Suppose that a sample of t weeks is observed. It is clear from looking at H_1 and H_2 that if r denotes the number of accidents in the t weeks, high values of r tend to favor H_2 and small values of r tend to favor H_1. Thus, a rejection region should be of the form $r \geq d$, where d is some constant. Is this consistent with Equation (7.6.2)? The likelihood ratio in this example is a ratio of Poisson probabilities:

$$LR = \frac{P(\tilde{r} = r \mid \tilde{\lambda}t = 2t)}{P(\tilde{r} = r \mid \tilde{\lambda}t = 3t)} = \frac{e^{-2t}(2t)^r/r!}{e^{-3t}(3t)^r/r!}$$

$$= e^{3t-2t}\left(\frac{2t}{3t}\right)^r = e^t\left(\frac{2}{3}\right)^r.$$

As r gets larger, LR gets smaller, so the rejection region of the form $r \geq d$ is equivalent to the rejection region given by Equation (7.6.2).

As these examples illustrate, the rejection region in classical hypothesis testing generally can be expressed in the form $LR \leq c$, where c is some constant. For a *specific* rejection region in any given situation, c must be specified. The classical statistician considers the following table of possible errors in hypothesis testing:

		STATE OF THE WORLD	
		H_1 *true*	H_1 *false*
DECISION	*Accept* H_1	No error	Type II error
	Reject H_1	Type I error	No error

The rejection region is determined by considering the relative seriousness of the two types of error; usually, the hypotheses are set up so that a type I error is more serious than a type II error. Thus, the statistician wants to keep the probability of a type I error quite low, so he specifies that this probability must be no larger than some small number α (Greek alpha). The constant c is then determined so that

$$P(\text{type I error}) = P(\text{reject } H_1 \mid H_1 \text{ true}) \leq \alpha.$$

Since the rejection region is of the form $LR \leq c$, c is then the largest value such that

$$P(LR \leq c \mid H_1 \text{ true}) \leq \alpha. \tag{7.6.4}$$

In the election example, suppose that the sample size is 10. For each possible sample result, a likelihood ratio can be determined from the table of binomial probabilities.

r	$LR = \dfrac{P(r \mid n = 10, \tilde{p} = .40)}{P(r \mid n = 10, \tilde{p} = .60)}$	$P(LR \mid H_1) = P(LR \mid \tilde{p} = .40)$
0	.0060/.0001 = 60.00	.0060
1	.0403/.0016 = 25.19	.0403
2	.1209/.0106 = 11.41	.1209
3	.2150/.0425 = 5.06	.2150
4	.2508/.1115 = 2.25	.2508
5	.2007/.2007 = 1.00	.2007
6	.1115/.2508 = .44	.1115
7	.0425/.2150 = .20	.0425
8	.0106/.1209 = .09	.0106
9	.0016/.0403 = .04	.0016
10	.0001/.0060 = .02	.0001

First, note that as r becomes larger, LR becomes smaller. The rejection region in terms of r will be of the form $r \geq d$. Next, suppose that the statistician decides that $\alpha = .10$. Note that $P(\tilde{r} \geq 6 \mid \tilde{p} = .40) = .1663$, whereas $P(\tilde{r} \geq 7 \mid \tilde{p} = .40) = .0548$. Thus, converting from r to LR, $P(LR \leq .44 \mid \tilde{p} = .40) = .1663$ and $P(LR \leq .20 \mid \tilde{p} = .40) = .0548$. Of the values of LR given in the table, $c = .20$ is the largest value satisfying Equation (7.6.4); the next larger value is .44, and this value does not satisfy the equation, since $P(LR \leq .44 \mid H_1 \text{ true}) = .1663$, which is larger than $\alpha = .10$. Therefore, the rejection region is $LR \leq .20$, which is equivalent to $r \geq 7$, so $d = 7$. It should be noted that because r (and hence LR) is discrete in this example, intermediate values of c such as .25, .34, and so on (any value such that $.20 \leq c < .44$) will give equivalent results. These values of c correspond to intermediate values of d $(6 < d \leq 7)$.

If the statistician *carefully* considers the relative seriousness of the two types of error in a hypothesis testing problem in order to determine a value for α, the above approach is a reasonable classical procedure. Too often, however, α is arbitrarily chosen to be some very small number without careful consideration of the relative seriousness of the two types of error and of the probability of committing a type II error (the probability of committing a type II error is denoted as β). This represents a weakness in the *application* of the classical framework, not a weakness in the framework itself. The likelihood ratio approach allows the classical statistician to investigate the relative merits of two hypotheses in an inferential sense. If he wishes to extend this to a decision-theoretic context, considering the losses associated with the two types of error and actually making a decision by choosing one of the two hypotheses, then he should formally consider losses, as you shall see in Section 7.8.

In general, the classical "likelihood ratio" approach to hypothesis testing suggests that for two hypotheses H_1 and H_2, H_2 should be rejected if

$$LR = \frac{f(y \mid H_1)}{f(y \mid H_2)} \leq c,$$

where c is some constant. As long as H_1 and H_2 are simple hypotheses, LR is uniquely determined. If either H_1 or H_2 is composite, however, LR is not unique, because the likelihoods are not unique. If the hypothesis H_1 in the above example is of the form

$$H_1\colon \tilde{p} \leq .40,$$

how can the likelihood in the numerator of LR be uniquely determined? One method is to take the largest of the potential likelihoods; that is, to find the value of $\tilde{p}$ that maximizes $f(y \mid p)$, subject to the constraint that $\tilde{p}$ must be no greater than .40. This is similar to the principle of maximum likelihood, which is used in point estimation (see Section 7.2). You need not be concerned about the rationale behind this method, although it is worthwhile to reiterate at this point that all of the usual tests used by classical statisticians are based on generalizations of this likelihood ratio approach. For more complete discussions of classical hypothesis testing procedures, references are presented in Section 7.10.

7.7 THE POSTERIOR DISTRIBUTION AND HYPOTHESIS TESTING

As indicated in Section 7.5, the Bayesian approach to hypothesis testing can be presented in terms of odds ratios, with the posterior odds ratio equalling the product of the prior odds ratio and the likelihood ratio. The relation of the Bayesian approach to the classical likelihood ratio concept should be clear; instead of looking just at LR, the Bayesian considers $\Omega'' = \Omega' \times LR$. This means that the numerical results will be the same only if the prior odds ratio, Ω', is equal to 1. This might be thought of as a diffuse prior state of information, since neither hypothesis is favored. If the prior odds ratio differs from 1, the posterior odds ratio differs from the likelihood ratio, and hence Bayesian and classical inferences will not necessarily be the same numerically (in terms of interpretation, they will *never* be the same).

For simple hypotheses such as $\tilde{p} = .40$, the use of odds ratios and their revision by using likelihood ratios is reasonably straightforward, as you have seen. In most applications, however, at least one of the hypotheses is composite (for example, $\tilde{p} < .40$). If this is the case, then the posterior distribution should be determined by using the methods in Chapters 3 and 4 rather than by using odds ratios and likelihood ratios. Once the posterior distribution has been determined, probabilities of the hypotheses of interest can be found and the posterior odds ratio can be formed from these probabilities.

To begin the discussion of the Bayesian approach to composite hypotheses, suppose that $\tilde{\mu}$ is the mean of a normal data-generating process and assume that the hypotheses of interest concerning $\tilde{\mu}$ are "one-tailed," or "directional." That is, the hypotheses are either of the form

$$H_1\colon \tilde{\mu} \geq \mu_0$$

and

$$H_2\colon \tilde{\mu} < \mu_0$$

or of the form

$$H_1\colon \tilde{\mu} \leq \mu_0$$

and

$$H_2\colon \tilde{\mu} > \mu_0,$$

where μ_0 is some given value of $\tilde{\mu}$. Since the Bayesian is willing to make probability statements about $\tilde{\mu}$, he can make direct probability statements concerning the two hypotheses. If he has a posterior distribution for $\tilde{\mu}$, he can use this distribution to calculate probabilities such as

$$P''(H_1 \text{ is true}) = P''(\tilde{\mu} \leq \mu_0)$$

and

$$P''(H_2 \text{ is true}) = P''(\tilde{\mu} > \mu_0). \tag{7.7.1}$$

In the example involving weekly sales that was presented in Section 4.6, suppose that the retailer is interested in the hypotheses

$$H_1\colon \tilde{\mu} \leq 1300$$

and

$$H_2\colon \tilde{\mu} > 1300,$$

where $\tilde{\mu}$ is the mean of the distribution of weekly sales. His posterior distribution is a normal distribution with mean 1387.5 and standard deviation 30.6. From this distribution, standardizing and using the cumulative normal table yields

$$\begin{aligned} P''(H_1) = P''(\tilde{\mu} \leq 1300) &= P\left(\frac{\tilde{\mu} - m''}{\sigma''} \leq \frac{1300 - 1387.5}{30.6}\right) \\ &= P(\tilde{z} \leq -2.86) = .0021 \end{aligned}$$

and

$$P''(H_2) = P''(\tilde{\mu} > 1300) = P(\tilde{z} > -2.86) = .9979.$$

Under the retailer's posterior distribution, the hypothesis H_1 appears extremely unlikely, and the posterior odds ratio of H_1 to H_2 is $\Omega'' = P''(H_1)/P''(H_2) = .0021/.9979 = .0021$.

For another example, suppose that you are sampling from a normally distributed population and that the following hypotheses are of interest:

$$H_1\colon \tilde{\mu} \leq 100$$

and

$$H_2\colon \tilde{\mu} > 100.$$

It is assumed that the variance of the population is known to be equal to 400, so that the standard deviation is 20. A sample of size four is taken, with $m = 106$. Under a diffuse normal prior distribution (Section 4.10), the posterior distribution is a normal distribution with $m'' = m = 106$ and $\sigma'' = \sigma/\sqrt{n''} = \sigma/\sqrt{n} = 20/2 = 10$. Then, from the normal tables,

$$P''(H_1) = P''(\tilde{\mu} \leq 100) = P\left(\tilde{z} \leq \frac{100 - 106}{10}\right) = P(\tilde{z} \leq -.60) = .274$$

and

$$P''(H_2) = P(\tilde{z} > -.60) = .726.$$

The posterior probability that H_1 is true is equal to .274, and the posterior odds ratio is

$$\Omega'' = \frac{P''(H_1)}{P''(H_2)} = \frac{.274}{.726} = .377.$$

How would a classical statistician handle this type of problem? He would ask the following question: How likely or unlikely does the hypothesis appear on the basis of the sample information? To answer this question, he would compute the *significance level* of the sample results. Essentially, the significance level calculated from a particular sample measures the chance of obtaining a sample result as "unusual" as or more "unusual" than the one actually observed, given that the hypothesis H_1 is true. In this context, "unusual" means "extreme" and depends both on the hypothesis of interest, H_1, and on the alternate hypothesis, H_2. If $\tilde{\mu} = 100$, how "unusual" is the sample result of $m = 106$ in the preceding example? The standardized value corresponding to $m = 106$ is

$$z = \frac{m - 100}{20/\sqrt{4}} = \frac{106 - 100}{10} = 0.60.$$

Since the alternative hypothesis H_2 is one-tailed to the right, the significance level is equal to $P(\tilde{z} \geq .60)$, which, from the table of the cumulative standard normal distribution, is equal to .274. (If the hypotheses are reversed, with H_1: $\tilde{\mu} \geq 100$ and $H_2 = \tilde{\mu} < 100$, then the alternative hypothesis H_2 is one-tailed to the left, and the significance level is equal to $P(\tilde{z} \leq .60) = .726$.) The smaller the significance level, the more unlikely the sample result is given that H_1 is true (the more "unusual" the sample result is if H_1 is true), and thus the more support is given H_2 as opposed to H_1.

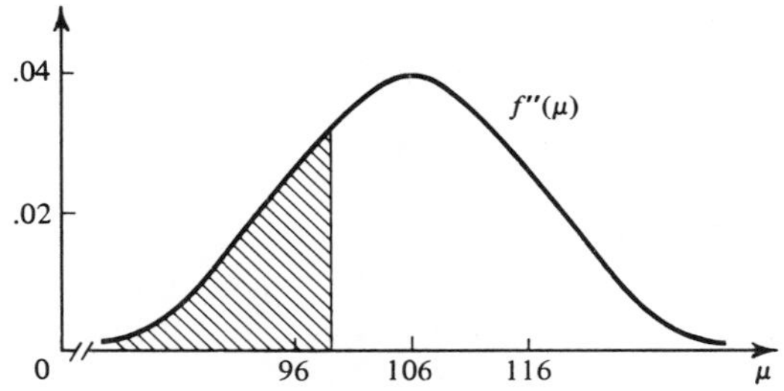

Figure 7.7.1

Note that in this example, the significance level is equal to $P''(H_1)$, which was computed above under the assumption of a diffuse prior distribution. Although these results are numerically equal, the interpretation is not. The difference is demonstrated in Figures 7.7.1 and 7.7.2. The curve centered about $\tilde{\mu} = 106$ (Figure 7.7.1) is the Bayesian's posterior distribution (assuming a diffuse prior distribution), and the shaded area under this curve represents the posterior probability that H_1 is true. The density function centered about $\tilde{m} = 100$ (Figure 7.7.2) is the sampling distribution of $\tilde{m}$, given that $\tilde{\mu} = 100$. The cross-hatched area represents the significance level, or the "unusualness" of the sample result given the sampling distribution specified by H_1 (a smaller significance level implies a more "unusual" sample result).

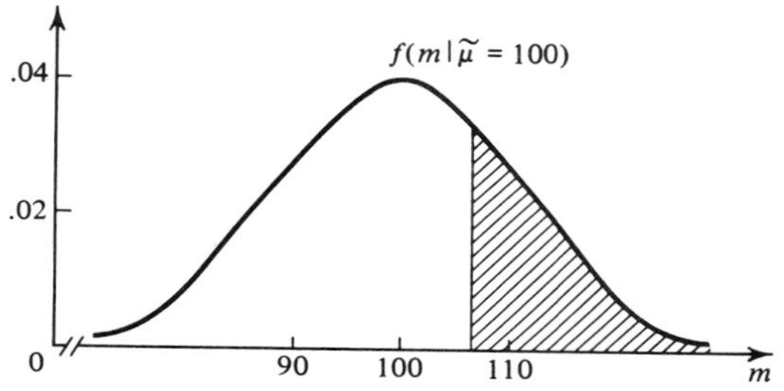

Figure 7.7.2

If the prior distribution is diffuse, then, the results obtained from the classical significance level approach in a one-tailed test can be compared to the results obtained using the posterior distribution. The results may be very similar numerically, although the interpretation is different. This statement holds for hypotheses concerning parameters other than $\tilde{\mu}$, although the example given above only involves $\tilde{\mu}$, the mean of a normal population with known variance. If, as is usually the case, the prior distribution is not diffuse, the two methods generally will produce different results. It is important to emphasize that the preceding comparison between Bayesian and classical procedures applies only to "one-tailed" tests. Next the problems that arise for "two-tailed" tests are discussed.

A "two-tailed" test is one in which the hypotheses are not "directional"; instead, one hypothesis generally is a simple hypothesis, such as $\tilde{\mu} = \mu_0$ (it may also be an interval of finite length), and the other hypothesis involves both directions ($\tilde{\mu} \neq \mu_0$). For example, suppose that the retailer had been interested in a test of the hypotheses

$$H_1\colon \tilde{\mu} = 1300$$

and

$$H_2\colon \tilde{\mu} \neq 1300.$$

The classical approach to this type of problem is similar to the classical approach to "one-tailed" tests. A significance level is determined by computing a one-tailed significance level and doubling it to allow for the other tail of the distribution. In the above example, $m = 106$, and the corresponding standardized value is $z = .60$. The one-tailed level of significance is .274, and the corresponding two-tailed level of significance is $2(.274) = .548$, as illustrated in Figure 7.7.3.

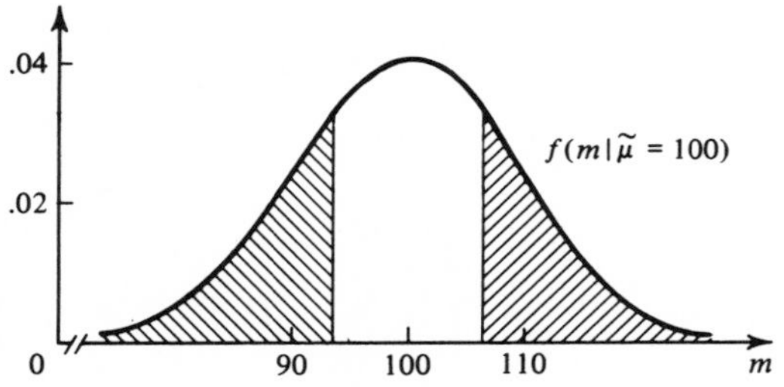

Figure 7.7.3

Unlike the one-tailed situation, there is no obvious Bayesian interpretation of the classical two-tailed test. From the posterior distribution,

$$P''(H_1) = P''(\tilde{\mu} = 1300) = P(\tilde{z} = -2.86) = 0$$

and

$$P''(H_2) = P''(\tilde{\mu} \neq 1300) = P(\tilde{z} \neq -2.86) = 1.$$

At first glance, it appears that H_1 is essentially impossible. The posterior distribution must not agree at all with the hypothesis H_1. But suppose that $m'' = 1300$ (certainly no other value of m'' would make H_1 appear more likely). Then

$$P''(H_1) = P''(\tilde{\mu} = 1300) = P(\tilde{z} = 0) = 0$$

and

$$P''(H_2) = P''(\tilde{\mu} \neq 1300) = P(\tilde{z} \neq 0) = 1.$$

H_1 is still an essentially "impossible" event! This is certainly a strange result.

The explanation for the above result is quite simple. The hypothesis H_1 consists of just a single point. The posterior distribution of $\tilde{\mu}$ is continuous, and thus the probability of any single value of $\tilde{\mu}$ is equal to 0. This does not reflect a weakness in the Bayesian approach; instead, it reflects the absurdity of the hypothesis H_1. Under the classical two-tailed testing situation there usually is a difference between the *statistical* hypotheses and the corresponding *real-world* hypotheses, and exact hypotheses such as $\tilde{\mu} = 1300$ are unrealistic and should be modified. What the retailer probably means by H_1 is "the mean is close to the value 1300." He surely cannot literally mean that "the mean is *exactly* equal to the value 1300." Under the classical theory of hypothesis testing, it is convenient to specify the hypothesis as

$$H_1: \tilde{\mu} = 1300,$$

even though this statement, taken literally, is quite unrealistic. If the posterior distribution is strictly continuous, it makes no sense for the Bayesian to consider such a hypothesis.

One alternative would be to place a mass of probability at $\tilde{\mu} = 1300$ when assessing the prior distribution. The distribution would then be a "mixture" of distributions, the discrete distribution consisting of a mass $P'(\tilde{\mu} = 1300)$ at 1300 and the continuous distribution consisting of a normal distribution with mean m' and variance σ'^2 representing the remaining $1 - P'(\tilde{\mu} = 1300)$. The resulting posterior distribution would also be a "mixture" of distributions, with a mass $P''(\tilde{\mu} = 1300)$ at 1300 and a continuous distribution representing the remaining $1 - P''(\tilde{\mu} = 1300)$. If this approach is used, an interesting paradox arises when it is compared with the classical two-tailed testing procedure. This result, which is called *Lindley's paradox*, can be stated for this example as follows. It is possible to find a sample result that simultaneously satisfies the following two conditions.

1. The classical level of significance is equal to an arbitrarily small number, implying that the sample result favors H_2.
2. The posterior mass $P''(\tilde{\mu} = 1300)$ following the sample result is arbitrarily close to 1, implying that the sample result favors H_1.

In general, the paradox occurs when the sample size is large. As the sample size increases, the sampling distribution of the sample mean $\tilde{m}$ has smaller variance, since $V(\tilde{m}) = \sigma^2/n$. But from Figure 7.7.3, the level of significance corresponds to area in the tails of this sampling distribution. For the sales example, $\sigma^2 = 90{,}000$, so that $V(\tilde{m}) = 90{,}000/n$. As the variance of the distribution decreases, the distribution is less "spread out," so that to maintain a fixed level of significance, the observed value of $\tilde{m}$ must be closer to 1300. For a sample of size 1, $\tilde{m} = 601$ (or $\tilde{m} = 1999$) yields a significance level of .01; for a sample of size 25, on the other hand, $\tilde{m} = 1160$ (or $\tilde{m} = 1440$) yields a significance level of .01; for a sample of size 100, $\tilde{m} = 1230$ (or $\tilde{m} = 1370$) yields a significance level of .01; for a sample of size 10,000, $\tilde{m} = 1293$ (or $\tilde{m} = 1307$) yields a significance level of .01; and so on. The same phenomenon occurs for any given level of significance. As the sample size increases, the classical test becomes sensitive to smaller and smaller deviations of $\tilde{m}$ from 1300. At the same time, as $\tilde{m}$ gets closer and closer to 1300, the posterior mass $P''(\tilde{\mu} = 1300)$ increases.

The implications are very simple: by taking a large enough sample, a classical statistician can be virtually assured of getting a very low significance level, which will generally lead him to reject the hypothesis H_1. This will be true even if the actual value of $\tilde{\mu}$ is so close to the value hypothesized in H_1 as to be almost identical with it (for example, identical to three decimal places). This appears to be a severe weakness in the classical procedure, for it means that the classical statistician can reject virtually any exact hypothesis H_1 in a two-tailed hypothesis testing situation. Although some classical statisticians carefully take the sample size into consideration when drawing inferences, many users of the two-tailed testing procedure are unaware of or ignore the problem. Such two-tailed tests are not used by Bayesians, nor do they have any obvious Bayesian interpretations. If the Bayesian places a mass at the value of $\tilde{\mu}$ hypothesized in H_1, as indicated above, then sample means very close to this value will support H_1, as you might expect. Using the classical approach, such values might *not* support H_1, especially if the sample size is large.

The above Bayesian approach is not applicable unless the statistician feels that there should be a prior mass exactly at the point specified in the hypothesis H_1. In the sales example, the retailer is highly unlikely to feel that the mean sales could be *exactly* 1300. Why, then, should the retailer's distribution be altered to fit the hypothesis? It should not, unless such a distribution truly reflects his information about $\tilde{\mu}$ (in some circumstances, such a distribution may be reasonable). A better solution from the Bayesian point of view would be to modify the statistical hypotheses H_1 and H_2 so that they correspond with the real-world hypotheses. The retailer's hypotheses might be expressed as follows:

$$H_1\colon 1250 \leq \tilde{\mu} \leq 1350$$

and

$$H_2\colon \tilde{\mu} < 1250 \text{ or } \tilde{\mu} > 1350.$$

Now, from the retailer's posterior distribution,

$$\begin{aligned} P''(H_1) &= P''(1250 \leq \tilde{\mu} \leq 1350) \\ &= P\left(\frac{1250 - 1387.5}{30.6} \leq \frac{\tilde{\mu} - m''}{\sigma''} \leq \frac{1350 - 1387.5}{30.6}\right) \\ &= P(-4.49 \leq \tilde{z} \leq -1.23) = .1093, \end{aligned}$$

and

$$\begin{aligned} P''(H_2) &= P''(\tilde{\mu} < 1250) + P''(\tilde{\mu} > 1350) = 1 - P''(1250 \leq \tilde{\mu} \leq 1350) \\ &= 1 - .1093 = .8907. \end{aligned}$$

Therefore, $\Omega'' = P''(H_1)/P''(H_2) = .1093/.8907 = .1227$.

By modifying the hypotheses and making them more realistic, the statistician can make inferences from the posterior distribution when one of the hypotheses is two-tailed. In general, if he is interested in a parameter $\tilde{\theta}$ and specifies the hypotheses

$$H_1\colon \theta_0 \leq \tilde{\theta} \leq \theta_1$$

and

$$H_2\colon \tilde{\theta} < \theta_0 \text{ or } \tilde{\theta} > \theta_1, \tag{7.7.2}$$

where θ_0 and θ_1 are constants, then the posterior distribution $f''(\theta \mid y)$ can be used to determine

$$P''(H_1) = P''(\theta_0 \leq \tilde{\theta} \leq \theta_1)$$

and

$$P''(H_2) = P''(\tilde{\theta} < \theta_0) + P''(\tilde{\theta} > \theta_1) = 1 - P''(H_1).$$

This is illustrated in Figure 7.7.4. As long as θ_0 and θ_1 are not equal, these probabilities should prove meaningful to the statistician. If they are equal, then $P''(H_1) = 0$ regardless of the observed sample outcome if the posterior distribution is continuous.

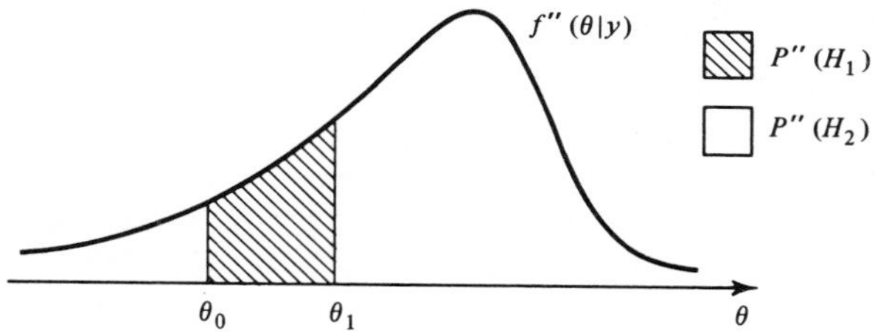

Figure 7.7.4

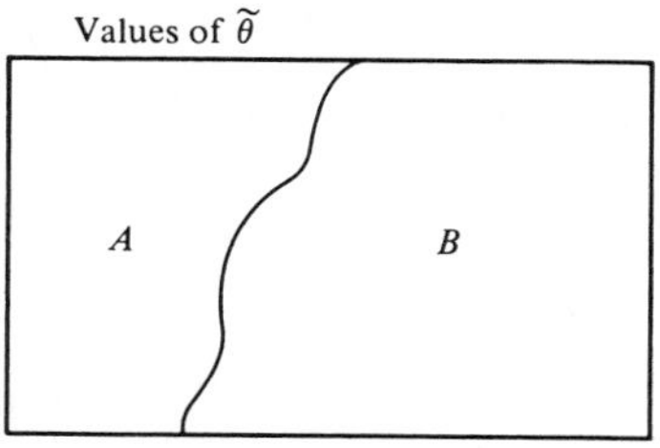

Figure 7.7.5

The hypothesis H_1 need not consist of just a single interval; it could easily consist of a series of intervals. If the entire set of possible values of $\tilde{\theta}$ is partitioned into two mutually exclusive and exhaustive sets A and B (that is, each possible value of $\tilde{\theta}$ is included in one set or the other, but no value is in both sets), as illustrated in a Venn diagram in Figure 7.7.5, then the hypotheses can be written in the form

$$H_1\colon \tilde{\theta} \in A$$

and (7.7.3)

$$H_2\colon \tilde{\theta} \in B.$$

Here "$\in$" stands for "is an element of the set." The posterior probabilities of interest are the probability that $\tilde{\theta}$ is an element of the set A and the probability that $\tilde{\theta}$ is an element of the set B:

$$P''(H_1) = P''(\tilde{\theta} \in A)$$

and (7.7.4)

$$P''(H_2) = P''(\tilde{\theta} \in B) = 1 - P''(H_1).$$

No restrictions are placed on A and B other than that they must be mutually exclusive and exhaustive. In the modified two-tailed test, $A = \{\theta \mid \theta_0 \leq \theta \leq \theta_1\}$ (this stands for: "The set A consists of all values θ such that θ is between θ_0 and θ_1, inclusive") and $B = \{\theta \mid \theta < \theta_0 \text{ or } \theta > \theta_1\}$ ("The set B consists of all values θ such that either θ is less than θ_0 or θ is greater than θ_1"). In the example of such a test that is presented above, $A = \{\mu \mid 1250 \leq \mu \leq 1350\}$ and $B = \{\mu \mid \mu < 1250 \text{ or } \mu > 1350)$. The restriction that A and B must be exhaustive is not crucial to the analysis and can be eliminated. Of course, this would mean that the sum of the probabilities $P''(H_1)$ and $P''(H_2)$ would not necessarily equal 1.

It should be pointed out that Equations (7.7.3) and (7.7.4) are applicable whether $\tilde{\theta}$ is discrete or continuous. In the above examples, it has been convenient to use the continuous case. However, the discrete case can also be handled by this framework. Also, in the discrete case it may be reasonable for one or both of the sets A and B to consist of just a single value of $\tilde{\theta}$. Consider the following hypotheses:

$$H_1\colon \tilde{\theta} = 10$$

and

$$H_2\colon \tilde{\theta} \neq 10.$$

If $\tilde{\theta}$ is continuous, then $P(H_1) = P(\tilde{\theta} = 10) = 0$, and it makes no sense to compare these two hypotheses from a Bayesian viewpoint. If $\tilde{\theta}$ is discrete, however, then $P(H_1)$ may be greater than 0, in which case it *does* make sense to make inferences about the two hypotheses based on a prior or posterior distribution of $\tilde{\theta}$.

For a discrete example, consider the production-process example presented in Section 3.3 and suppose that the production manager is interested in the hypotheses

$$H_1\colon \tilde{p} = .05$$

and

$$H_2\colon \tilde{p} \neq .05.$$

From the prior distribution, $P'(H_1) = .30$ and $P'(H_2) = .70$. After a sample of five items yields one defective item, $P''(H_1) = .492$ and $P''(H_2) = .508$ (for the calculation of the posterior distribution, see Section 3.3). Thus, the posterior odds ratio is $\Omega'' = P''(H_1)/P''(H_2) = .9685$.

The framework provided by Equations (7.7.3) and (7.7.4) is quite general and is much more flexible than the classical approach. Classical procedures could be developed to handle most tests of this nature, but they would not be as easy to apply or as intuitively appealing as the Bayesian approach. Given the posterior distribution, the Bayesian need only calculate certain posterior probabilities, which is not too difficult, especially if the posterior distribution is a member of a conjugate family (or, for that matter, of *any* family for which tables are readily available).

There is no reason why the statistician should be limited to just two hypotheses. He may have several hypotheses in mind and may want to know which of the hypotheses appears to be the most likely to be true. Extending the results of the previous section, consider the set of hypotheses $\{H_1, H_2, \ldots, H_k\}$, where H_i is of the form

$$H_i\colon \tilde{\theta} \in A_i, \quad i = 1, 2, \ldots, K. \tag{7.7.5}$$

The statistician can compute the posterior probabilities of the K hypotheses:

$$P''(H_i) = P''(\tilde{\theta} \in A_i), \quad i = 1, 2, \ldots, K. \tag{7.7.6}$$

The K sets $A_1, A_2, \ldots, A_K$ do not have to be mutually exclusive and exhaustive. If they are, of course, the sum of the probabilities, $\sum_{i=1}^{K} P''(H_i)$, is equal to 1. If the statistician just wants to know which hypothesis is most likely to be true, he can compare the $P''(H_i)$ and find the hypothesis with the largest posterior probability. For example, the hypotheses might correspond to competing economic theories, in which case the statistician might be interested in knowing which theory seems to be the most likely according to the posterior distribution.

The interest in problems involving two or more hypotheses is primarily decision-oriented. A test of a hypothesis H_1 versus an alternative hypothesis H_2 can be thought of as a decision-making procedure in which there are two actions available to the statistician: accept H_1 or reject H_1 in favor of H_2 (assuming that the statistician must make a decision and cannot suspend judgment). In making a decision, the statistician should consider the probabilities and losses associated with the two possible errors. If there are K hypotheses, hypothesis testing can be considered as a K-action problem, where the statistician must choose one of the K hypotheses. Again, his decision should be based on both the posterior probabilities associated with the hypotheses, which have been discussed, and the losses due to erroneous decisions, which are discussed in the next section.

7.8 DECISION THEORY AND HYPOTHESIS TESTING

In general, hypothesis testing can be thought of as a decision-making problem in which the two decisions are (1) to accept H_1 and (2) to reject H_1 in favor of H_2. Thus, the losses associated with the two types of error should be assessed and the rejection region determined so as to minimize expected loss. Suppose that there are two hypotheses of interest,

$$H_1\colon \tilde{\theta} \in A$$

and (7.8.1)

$$H_2\colon \tilde{\theta} \in B.$$

Furthermore, assume that A and B are mutually exclusive and exhaustive. From the decision maker's distribution of $\tilde{\theta}$, it is possible to compute

$$P(H_1) = P(\tilde{\theta} \in A)$$

and (7.8.2)

$$P(H_2) = P(\tilde{\theta} \in B).$$

Since A and B are mutually exclusive and exhaustive, $P(H_1) + P(H_2) = 1$. This development is identical to the development of Section 7.7.

The other input to the decision problem is a loss function. In hypothesis testing, there are two types of error, as indicated in Section 7.6. A *type I error* is committed if H_1 is rejected when in fact it is true, and a *type II error* is committed if H_1 is accepted when in fact it is false. Denote the loss associated with a type I error by $L(\text{I})$, and denote the loss associated with a type II error by $L(\text{II})$. The loss function can be represented by the following loss table, since there are just two actions and two relevant states of the world.

		STATE OF THE WORLD	
		H_1 true ($\theta \in A$)	*H_1 false* ($\theta \in B$)
ACTION	*Accept H_1*	0	$L(\text{II})$
	Reject H_1	$L(\text{I})$	0

Calculating expected losses in the usual manner,

$$EL(\text{accept } H_1) = 0P(H_1) + L(\text{II})P(H_2) = L(\text{II})P(H_2) \quad (7.8.3)$$

and

$$EL(\text{reject } H_1) = L(\text{I})P(H_1) + 0P(H_2) = L(\text{I})P(H_1). \quad (7.8.4)$$

According to the EL criterion for making decisions, H_1 will be rejected if

$$EL(\text{reject } H_1) < EL(\text{accept } H_1). \quad (7.8.5)$$

Using Equations (7.8.3) and (7.8.4), this is equivalent to

$$L(\text{I})P(H_1) < L(\text{II})P(H_2).$$

Dividing both sides by $L(\text{I})$ and by $P(H_2)$ yields

$$\frac{P(H_1)}{P(H_2)} < \frac{L(\text{II})}{L(\text{I})}. \quad (7.8.6)$$

If Equation (7.8.6) is satisfied, the optimal decision is to reject H_1. If the inequality is reversed, the optimal decision is to accept H_1; and if the two sides of the equation are equal, the two expected losses are equal, so neither action is better than the other.

The left-hand side of Equation (7.8.6) is an odds ratio, and the right-hand side is a ratio of losses, or a *loss ratio*. In order to make a decision in a hypothesis-testing situation, it is necessary only to compare an odds ratio with a loss ratio. In fact, suppose that in Equation (7.8.6), both sides of the equation are divided by the loss ratio $L(\text{II})/L(\text{I})$. The result is

$$\frac{P(H_1)}{P(H_2)} \times \frac{L(\text{I})}{L(\text{II})} < 1. \quad (7.8.7)$$

Denoting the odds ratio by Ω and the loss ratio by $L_{\text{I/II}}$, the decision rule can be written as follows:

$$\text{reject } H_1 \text{ if } \Omega \times L_{\text{I/II}} < 1,$$

and (7.8.8)

$$\text{accept } \text{H}_1 \text{ if } \Omega \times L_{\text{I/II}} > 1.$$

Does Equation (7.8.8) seem reasonable? For a given loss ratio $L_{\text{I/II}}$, you are more likely to reject H_1 when Ω gets smaller, that is, when $P(H_1)$ gets smaller relative to $P(H_2)$. This makes sense; as the probability of H_1 gets smaller, H_1 appears to be less likely to be true, so you should be more likely to reject H_1. For a given odds ratio Ω, you are more likely to reject H_1 when $L(\text{I})$ gets smaller relative to $L(\text{II})$. But $L(\text{I})$ is the loss associated with the error of rejecting H_1 when it is in fact true. As this loss gets smaller relative to $L(\text{II})$, you should be more likely to reject H_1 for the same reason that you are more likely to underestimate in a point estimation problem as the cost of underestimation gets smaller relative to the cost of overestimation.

The framework developed in this section obviously is related to the Bayesian approach to hypothesis testing that has been discussed in the preceding sections. From Equation (7.5.4), the posterior odds ratio is equal to the product of the prior odds ratio and the likelihood ratio:

$$\Omega'' = \Omega' \times (LR). \tag{7.8.9}$$

If it is not feasible to assess the losses $L(\text{I})$ and $L(\text{II})$, the posterior odds ratio will tell you which hypothesis is more likely but will not enable you to make a decision. If you informally decide to reject H_1 if H_2 is more likely (that is, if $\Omega'' < 1$) and to accept H_1 if H_1 itself is more likely (that is, if $\Omega'' > 1$), then your decision rule would resemble Equation (7.8.8) without the loss ratio. This amounts to assuming that $L(\text{I}) = L(\text{II})$, so that $L_{\text{I/II}} = 1$. The inclusion of the loss ratio, $L_{\text{I/II}}$, represents an *extension* of the informal decision rule to a more formal rule that satisfies the EL criterion.

How does this relate to classical hypothesis testing? It was pointed out in Section 7.6 that classical tests of hypotheses are based on the likelihood ratio, LR. In Equation (7.8.9), if the prior odds ratio is equal to 1 (that is, if the two hypotheses are equally likely under the prior information), then the posterior odds ratio is equal to the likelihood ratio:

$$\Omega'' = \Omega' \times (LR) = LR.$$

In this case, the decision rule given by Equation (7.8.8) becomes

$$\text{reject } H_1 \text{ if } LR \times L_{\text{I/II}} < 1$$

and

$$\text{accept } H_1 \text{ if } LR \times L_{\text{I/II}} > 1. \tag{7.8.10}$$

Dividing both sides by $L_{\text{I/II}}$, the rejection region becomes

$$LR < \frac{1}{L_{\text{I/II}}} = \frac{L(\text{II})}{L(\text{I})}. \tag{7.8.11}$$

Now, from Equation (7.6.2), the rejection region for a classical test is of the form

$$LR \leq c.$$

Thus, the classical "likelihood ratio" test is essentially equivalent to the test given by Equation (7.8.11), with $c = L(\text{II})/L(\text{I})$. The difference is that the classical statistician usually does not formally assess $L(\text{I})$ and $L(\text{II})$. Instead, he typically uses a convention that implies that $L(\text{I})$ is more serious than, and therefore higher than, $L(\text{II})$. He chooses his rejection region so that the probability of a type I error is guaranteed to be small (this probability is usually taken to be .01, .05, or .10). In this manner, he has no direct control over the probability of a type II error, although he has indirect control (for instance, if the probability of a type I error is held constant, then increasing the sample size will reduce the probability of a type II error). This is a highly informal approach to the hypothesis-testing problem.

The *formal* assessment of losses and the use of the *EL* criterion is an extension of the informal analysis. It is not even necessary to assess the two losses $L(\text{I})$ and $L(\text{II})$ separately, since all that is needed is their ratio. Sometimes it is convenient to think in terms of loss ratios. For example, it is difficult to quantify the losses due to (1) convicting an innocent man or (2) setting a guilty man free. It may, however, be possible to assess the loss *ratio* of (1) to (2). Some persons might assess this ratio as 100; others might think it to be 10; others might choose 1, saying that the two losses are equal; and still others might select .1. The point is simply that it is often much easier to assess a loss *ratio* than it is to assess directly the two losses comprising the loss ratio. This point relates to the consideration of odds ratios in the assessment of prior probabilities, as discussed in Section 3.5.

For a simple (and brief) example of hypothesis testing as a decision-making procedure, consider a manufacturer who is concerned about the potential demand for a new product he is thinking of producing. For the sake of simplicity, assume that he is interested in the following two hypotheses, where $\tilde{p}$ is the proportion of consumers who will purchase the new product:

$$H_1\colon \tilde{p} = .20$$

and

$$H_2\colon \tilde{p} = .30.$$

If the manufacturer accepts H_1, then he will not produce the new product; if he rejects H_1 in favor of H_2, then the product will be produced.

The manufacturer's prior probabilities are

$$P'(H_1) = P'(\tilde{p} = .20) = .25$$

and

$$P'(H_2) = P'(\tilde{p} = .30) = .75.$$

He takes a sample of 20 consumers and finds that only 2 will buy the new product. Thus, assuming that the Bernoulli model is applicable, the likelihood ratio is (from the binomial distribution):

$$LR = \frac{P(\tilde{r} = 2 \mid n = 20, \tilde{p} = .20)}{P(\tilde{r} = 2 \mid n = 20, \tilde{p} = .30)} = \frac{.1369}{.0278} = 4.92.$$

Multiplying this by the prior odds ratio of $.25/.75 = 1/3$,

$$\Omega'' = \Omega' \times (LR) = \tfrac{1}{3}(4.92) = 1.64.$$

If the manufacturer decides just to accept the most likely hypothesis, then he would accept H_1, since the posterior odds ratio is greater than 1. Suppose that he takes potential losses into consideration. If he accepts H_1 (and therefore does not produce the new product), he might suffer a large opportunity loss in the form of lost profits if H_2 is true. This is $L(\text{II})$. On the other hand, if he rejects H_1 (and produces the new product), the new product might result in losses to the firm if the demand is not high enough (that is, if H_1 is true). This is $L(\text{I})$. After some serious thought, the manufacturer decides that $L(\text{II})$ is twice as large as $L(\text{I})$. This means that

$$L_{\text{I/II}} = \frac{L(\text{I})}{L(\text{II})} = \frac{1}{2}.$$

Multiplying the posterior odds ratio by the loss ratio,

$$\Omega'' \times L_{\text{I/II}} = (1.64)(\tfrac{1}{2}) = 0.82.$$

Since this is less than 1, the optimal decision is to reject H_1 and to produce the new product. Even though the posterior odds ratio favors H_1, the losses are such that H_2 is the "better" hypothesis from a decision-theoretic standpoint.

This example purposely was made simple in order to demonstrate clearly the idea of hypothesis testing as a decision-making procedure. In particular, both H_1 and H_2 were taken to be simple hypotheses. It should be mentioned that although the general procedure is still valid when the hypotheses are *not* both simple, there are certain subtleties involved in some situations that are not considered here.

7.9 THE DIFFERENT APPROACHES TO STATISTICAL PROBLEMS

In this section an attempt is made to recapitulate very briefly the relationship among classical inference, Bayesian inference, and decision theory. First of all, classical inference prescribes inferential techniques that are based on sample information alone. If prior information is considered, it is only considered in an informal manner. Inferential methods in Bayesian statistics are based on the posterior distribution, which is a combination of the prior distribution, representing the prior state of information, and the likelihood function, representing the information of a sample. Thus, the prior distribution is an input to Bayesian methods but not to classical methods. Because they utilize all available information,

Bayesian methods sometimes lead to quite different (and more plausible) results than classical methods, and they are particularly valuable in decision-making problems.

The objective of inferential statistics is to make inferences about some uncertain quantity. These inferences may be in the form of entire probability distributions, point estimates, interval estimates, and so on. In decision theory, as opposed to inferential statistics, one is interested in taking some *action* rather than merely in making some inference about an uncertain quantity. As a result, one additional input is required in decision theory: a loss function, or a payoff function (where the losses or payoffs may be given in terms of utility or they may be just in terms of money or some other units). The inclusion of the loss function (or payoff function) enables the decision maker to "optimize" in the sense of minimizing expected loss or maximizing expected payoff.

For important decisions, the formal decision-theoretic approach is most useful because it takes into account all available information that bears on the decision, both information concerning the state of the world (the value of the uncertain quantity of interest) and information concerning the consequences of potential decisions. If the decision is not too important, an informal analysis may be preferred, since applying decision theory may be quite time-consuming. It is in this sense that the Bayesian inferential techniques may be useful, since they are somewhat easier to apply, requiring no loss function and specifying no single "optimizing" criterion. They may be useful in other situations as well, even situations in which the decision is important, if the statistician is unable to determine some of the inputs to the problem. In complex decision-making situations, it may not be clear what the relevant losses or payoffs are. If a big decision problem consists of a combination of several smaller problems, it may be difficult to see exactly how the "smaller" decisions relate to the major decision. In a way, this involves model-building, as discussed in Section 5.9; how can you state the problem in decision-theoretic terms and still have it be as realistic as possible? Decision theory provides a useful framework for making decisions under uncertainty. It is up to the decision maker to attempt to express real-world problems in this framework without sacrificing realism. For instance, if the decision maker makes assumptions that are unrealistic, then the optimal solution to the decision theory problem may not be the same as the best solution to the real-world problem. If this is the case, an informal analysis may prove to be just as useful as a formal decision-theoretic approach. Of course, as noted above, the formal approach is particularly valuable for important problems, in which case it is surely worthwhile to develop realistic decision models.

7.10 REFERENCES AND SUGGESTIONS FOR FURTHER READING

Some general references at a level roughly comparable to this chapter are Aitchison (1970*a*, Chapter 8); Hadley (1967, Chapter 9); LaValle (1970, Chapters 18–22); Pratt, Raiffa, and Schlaifer (1965, Appendix 3); Raiffa (1968, Chapter 10); Roberts (1966*b*, Chapters 11A, 11B, and 12); and Schlaifer (1959, Chapters 39–42). References dealing primarily with the likelihood principle include Barnard (1967*b*); Birnbaum (1962*b*, 1968, 1970); Durbin (1970); Fraser (1963); Hartigan (1967); Savage (1970*a*); and Sprott and Kalbfleisch (1965). A point related to the likelihood principle concerns stopping rules, which were mentioned briefly in Section 4.4 and which are discussed in Raiffa and Schlaifer (1961, Chapter 2) and Roberts (1967). Hildreth (1963) discusses the question of scientific reporting. Other general references regarding the foundations of statistics include Anscombe (1963, 1964); Barnard (1967*a*); Barnard, Jenkins, and Winsten (1962); Birnbaum (1962*a*, 1969); Bross (1961); Cornfield (1966*b*, 1967, 1969); de Finetti (1961, 1969); Dempster (1963, 1968); Godambe and Sprott (1971); Hacking (1965); Hartley (1963); Jeffrey (1965); Jeffreys (1961); Kyburg (1966); Levi (1967); Lindley (1953, 1961*b*); Plackett (1966); Popper (1959); Roberts (1966*a*); Salmon (1966); Savage (1954, 1961, 1970*c*); Savage, *et al.* (1962); Smith (1961, 1965); Stone (1969); Tukey (1960); and Wolfowitz (1962).

Specific comparisons of Bayesian and classical inferential procedures are found in Bartholomew (1965), Chambers (1970), and Pratt (1965). References with particular emphasis on Bayesian estimation are Aitchison (1964); Aitchison and Dunsmore (1968); Easterling and Weeks (1970); Guttman (1970); Hayes (1969*c*); Lee, Judge, and Zellner (1968); Raiffa and Schlaifer (1961, Chapter 6); and Thatcher (1964). A discussion of difficulties encountered in the chi-square test, as reported in Berkson (1938), is related to Lindley's paradox [see Lindley (1957) and Edwards (1965)]. For other discussions of Bayesian estimation, Bayesian hypothesis testing, and related topics, see Box and Draper (1965); Cornfield (1966*a*); Dickey (1971); Dickey and Lientz (1970); Dunsmore (1966, 1968, 1969); Edwards, Lindman, and Savage (1963); Good (1967); and Jeffreys (1961). Two interesting applications are found in Good (1969) and Mosteller and Wallace (1964); the latter relates to Exercise 31 of this chapter.

For elementary to intermediate level discussions with emphasis on classical inference, see Clelland *et al.* (1966); Freund (1962); Harnett (1970); Hays (1963); Hodges and Lehmann (1970); Hoel (1962); Lindgren (1968); Lippman (1971); Mood and Graybill (1963); and Wallis and Roberts (1956). An extensive discussion of classical methods is found in Kendall and Stuart (1958, 1961, 1966), and three classical books at an advanced level are Cramér (1946), Lehmann (1959), and Wilks (1962). Three statisticians closely associated with the development of different branches of the classical school are R. A. Fisher, J. Neyman, and E. S. Pearson; for historical references, see Bartlett (1965); Fisher (1958, 1959, 1960); Neyman (1952); and Neyman and Pearson (1928, 1933).

As noted in the introduction to this chapter, the terms "classical" and "Bayesian," as applied to approaches to inferential and decision-making problems, are too broad and highly simplified. The methods discussed in this chapter are by no means exhaustive. For example, an alternative set of procedures, known as the "empirical Bayes" approach, is discussed in Maritz (1970), Neyman (1962), and Robbins (1955, 1964); and an approach called "data analysis" is exposited by Tukey (1962). In a less specific vein, some thoughts on "the future of statistics" (including the role of computers) are presented in Kendall (1966, 1968) and Watts (1968).

EXERCISES

1. Comment on the statement, "If the classical statistician uses prior information at all, he uses it in an informal manner, whereas the Bayesian formally incorporates it into his inferential and decision-making framework."

2. Discuss the relationship between classical statistical procedures and Bayesian procedures under a diffuse prior distribution, both with regard to numerical results and with regard to the interpretation of results.

3. The conditional probability $P(y \mid \theta)$ or the conditional density $f(y \mid \theta)$ can be interpreted as a sampling distribution or as a likelihood function; explain the difference between these two interpretations.

4. Discuss the importance of the likelihood principle and the concept of sufficient statistics with regard to problems of statistical inference and decision.

5. In Exercise 23, Chapter 3, find a point estimate for $\tilde{p}$, the proportion of consumers who will purchase the product, based (a) on the prior distribution alone, (b) on the sample information alone, and (c) on the posterior distribution.

6. Do Exercise 5 under the assumption that the prior distribution of $\tilde{p}$ is a beta distribution with $r' = 4$ and $n' = 10$.

7. In Exercise 33, Chapter 3, find a point estimate for $\tilde{\lambda}$, the intensity of occurrence of accidents along a particular stretch of highway,
   (a) based on the prior distribution alone,
   (b) based on the sample information alone,
   (c) based on the posterior distribution.

8. In Exercise 42, Chapter 3, find a point estimate for $\tilde{\theta}$
   (a) given that $\tilde{\phi} = 1$,
   (b) given that $\tilde{\phi} = 0$,
   (c) given only that $P(\tilde{\phi} = 1) = .3$.

9. Suppose that a sample from a given population with known variance σ^2 consists of four independent trials and that the sample information on the ith trial is represented by the random variable $\tilde{x}_i$, $i = 1, 2, 3, 4$. Consider the following estimators of the population mean, $\tilde{\mu}$:

$$\tilde{y}_1 = \tilde{x}_1, \qquad \tilde{y}_2 = (\tilde{x}_1 + \tilde{x}_2 + \tilde{x}_3 + \tilde{x}_4)/4,$$
$$\tilde{y}_3 = (4\tilde{x}_1 + 3\tilde{x}_2 + 2\tilde{x}_3 + \tilde{x}_4)/10, \qquad \text{and } \tilde{y}_4 = (\tilde{x}_1 + \tilde{x}_2 + \tilde{x}_3 + \tilde{x}_4)/3.$$

   (a) Find $E(\tilde{y}_i \mid \tilde{\mu} = \mu)$ for $i = 1, 2, 3, 4$.
   (b) Find $V(\tilde{y}_i \mid \tilde{\mu} = \mu)$ for $i = 1, 2, 3, 4$.
   (c) Find $B(\tilde{y}_i \mid \tilde{\mu} = \mu)$ for $i = 1, 2, 3, 4$.
   (d) Find $\text{MSE}(\tilde{y}_i \mid \tilde{\mu} = \mu)$ for $i = 1, 2, 3, 4$.
   (e) Comment on the merits of $\tilde{y}_1$, $\tilde{y}_2$, $\tilde{y}_3$, and $\tilde{y}_4$ as classical estimators with respect to unbiasedness and efficiency.

10. In Exercise 28, Chapter 4, find
   (a) a point estimate for $\tilde{\mu}$ based on the prior distribution alone,
   (b) a point estimate for $\tilde{\mu}$ based on the sample information alone,
   (c) a point estimate for $\tilde{\mu}$ based on the posterior distribution,
   (d) a 90 percent credible interval for $\tilde{\mu}$ based on the prior distribution alone,
   (e) a 90 percent confidence interval for $\tilde{\mu}$ based on the sample information alone,
   (f) a 90 percent credible interval for $\tilde{\mu}$ based on the posterior distribution.
   Comment on any differences in your answers to (a) through (c) and to (d) through (f).

11. In Exercise 33, Chapter 4, find
   (a) an 86.6 percent confidence interval for $\tilde{\mu}$,
   (b) a point estimate for $\tilde{\mu}$ based on the posterior distribution,
   (c) a 38.3 percent credible interval for $\tilde{\mu}$.

12. In Exercise 6, find
   (a) a 90 percent credible interval for $\tilde{p}$ based on the prior distribution,
   (b) a 90 percent credible interval for $\tilde{p}$ based on the posterior distribution,
   (c) a 98 percent credible interval for $\tilde{p}$ based on the prior distribution,
   (d) a 50 percent credible interval for $\tilde{p}$ based on the prior distribution.

13. *Carefully* distinguish between a classical confidence interval and a Bayesian credible interval and also between the classical and Bayesian approaches to point estimation.

14. Suppose that a contractor must decide whether or not to build any speculative houses (houses for which he would have to find a buyer), and if so, how many. The houses that this contractor builds are sold for a price of

\$30,000, and they cost him \$26,000 to build. Since the contractor cannot afford to have too much cash tied up at once, any houses that remain unsold three months after they are completed will have to be sold to a realtor for \$25,000. The contractor's prior distribution for $\tilde{\theta}$, the number of houses that will be sold within three months of completion, is:

θ	$P(\tilde{\theta} = \theta)$
0	.05
1	.10
2	.10
3	.20
4	.25
5	.20
6	.10

If the contractor's utility function is linear with respect to money, how many houses should he build? How much should he be willing to pay to find out for certain how many houses will be sold within three months?

15. A hot-dog vendor at a football game must decide in advance how many hot dogs to order. He makes a profit of \$0.10 on each hot dog that is sold, and he suffers a \$0.20 loss on hot dogs that are unsold. If his distribution of the number of hot dogs that will be demanded at the football game is a normal distribution with mean 10,000 and standard deviation 2000, how many hot dogs should he order? How much is it worth to the vendor to know in advance exactly how many hot dogs will be demanded?

16. A sales manager is asked to forecast the total sales of his division for a forthcoming period of time. He feels that his loss function is linear as a function of the difference between his estimate and the true value, but he also feels that an error of overestimation is three times as serious as an error of underestimation (given that the magnitudes of the errors are equal). He feels this way because his superiors will criticize him if the division does worse than he predicts, but they will be happy if the division does better than predicted and will be less likely to be concerned about an error in predicting. If his actual judgments can be represented by a normal distribution with mean 50,000 and standard deviation 10,000, what value should he report as his forecast of sales?

17. Suppose that you want to estimate $\tilde{p}$, the proportion of the market for a particular product that will be obtained by a new brand of the product. Your prior distribution for $\tilde{p}$ is a diffuse beta distribution with $r' = 0$ and $n' = 0$, and you are willing to assume that the process behaves like a Bernoulli process. You take a random sample of 200 purchasers of the product and find that 24 of them buy the new brand. If the loss function for the estimation problem is linear and if the per unit cost of overestimation is three times the per unit cost of underestimation, find the optimal estimate of $\tilde{p}$.

18. Prove that if a decision maker's loss function in a point estimation problem is given by Equation (7.4.1), then a $k_u/(k_u + k_o)$ fractile of the decision maker's distribution of $\tilde{\theta}$ is an optimal point estimate.

19. An economist is asked to predict a future value of a particular economic indicator, and he thinks that his loss function is linear with $k_u = 4k_o$. If his distribution is a uniform distribution on the interval from 650 to 680, what should his estimate be?

20. In Exercises 14, 15, 16, and 19, what are the optimal estimates if the loss functions are quadratic instead of linear [that is, if the loss functions are of the form (7.4.5)]? In each case, if $k = 1$, what is the expected value of perfect information?

21. In Exercise 17, what is the optimal estimate if the loss function is quadratic instead of linear? Find the EVPI and, assuming that a second sample of purchasers of the product is being considered, find the EVSI for samples of size 50, 100, 200, and 500.

22. Comment on the following statement: "In taking the sample mean as an estimator of the population mean, the classical statistician is acting essentially as though he had a quadratic loss function."

23. If a person faces a point estimation problem with a linear loss function with $k_u = 4$ and $k_o = 3$, does he need to assess an entire probability distribution or can he determine a certainty equivalent? Explain.

24. If a statistician wishes to estimate an uncertain quantity $\tilde{\theta}$ subject to a loss function that is linear with $k_o = 2k_u$ and if his distribution of $\tilde{\theta}$ is an exponential distribution,

$$f(\theta \mid \lambda) = \begin{cases} \lambda e^{-\lambda\theta} & \text{if } \theta > 0, \\ 0 & \text{elsewhere,} \end{cases}$$

with $\lambda = 4$, what is his optimal estimate of $\tilde{\theta}$?

25. If the loss function in a point estimation problem is of the form

$$L(a,\theta) = \begin{cases} 0 & \text{if } |a - \theta| < k, \\ 1 & \text{otherwise,} \end{cases}$$

where k is some very small positive number, what is the optimal estimate? Can you think of any realistic situations in which the loss function might be of this form?

26. In Exercise 6, determine an estimate of $\tilde{p}$ from the posterior distribution if the loss function is

$$L(a,\theta) = \begin{cases} 19(a - \theta) & \text{if } \theta \le a, \\ (\theta - a) & \text{if } \theta \ge a. \end{cases}$$

How does this differ from the estimate obtained from the *prior* distribution using the same loss function?

27. In Exercise 10, the production manager must make an estimate of the mean weight of items turned out by the process in question. His loss function is linear with $k_o = k_u$. What should his estimate be?

28. In Exercise 10, suppose that instead of estimating $\tilde{\mu}$, the production manager wants to predict $\tilde{m}$, the sample mean from a sample of size 10. He is making this prediction after having seen the first sample, and his loss function for the prediction problem is linear with $k_u = 6$ and $k_o = 2$. What is his optimal prediction of $\tilde{m}$? If the sample size is 100 rather than 10, what is the optimal prediction of $\tilde{m}$?

29. In Exercise 27, suppose that the production manager wants an interval estimate for $\tilde{\mu}$, given the following loss function: the interval must be of length .5, and the loss is 1 if the interval includes $\tilde{\mu}$ and 0 if it does not include $\tilde{\mu}$. In general, what type of interval is optimal for this "all-or-nothing" type of loss function?

30. Suppose that a marketing manager is interested in $\tilde{p}$, the proportion of consumers that will buy a particular new product. He considers the following two hypotheses:

$$H_1\colon \tilde{p} = .10$$

and

$$H_2\colon \tilde{p} = .20.$$

His prior probabilities are $P(\tilde{p} = .10) = .85$ and $P(\tilde{p} = .20) = .15$, and a random sample of eight consumers results in three consumers who state that they will buy the product if it is marketed.

(a) What is the prior odds ratio?
(b) What is the likelihood ratio?
(c) What is the posterior odds ratio?
(d) The manager decides that the posterior probability of H_1 must be no larger than .40 to make it worthwhile to market the product. Should the product be marketed?
(e) On the basis of the sample information alone, should the product be marketed? If the decision is made on this basis, what is implied about the prior distribution?

31. Suppose that you are uncertain about which of two authors, A or B, wrote a particular essay, and you feel that you would be indifferent between the following lotteries.

Lottery I: You win \$10 if A wrote the essay, \$0 otherwise.
Lottery II: You win \$10 with probability .3, \$0 with probability .7.

You know from past analyses of the writings of A and B that A uses a certain key word with an average frequency of four times in 100 words, and B uses this word with an average frequency of twice in 100 words (assume that the process is stationary and independent). In the particular essay of inter-

est, there are 500 words and the key word appears eight times. You are interested in finding out which author wrote the essay.
(a) What statistical model would you use to represent the data-generating process?
(b) Express the "real-world" hypotheses in terms of the statistical model chosen in (a).
(c) Find the prior odds ratio.
(d) Find the likelihood ratio.
(e) Find the posterior odds ratio.
(f) What would you conclude about the authorship of the essay?

32. Comment on the statement, "Hypothesis testing can sometimes be thought of as a problem of classification or as a problem of discrimination."

33. In Exercise 10, assume that the prior distribution is discrete with $P(\tilde{\mu} = 109) = .2$ and $P(\tilde{\mu} = 110) = .8$. For H_1: $\tilde{\mu} = 109$ and H_2: $\tilde{\mu} = 110$, find
(a) the prior odds ratio,
(b) the likelihood ratio,
(c) the posterior odds ratio.

34. In Exercise 30, it would be more realistic to consider hypotheses such as the following:

$$H_1: \tilde{p} \leq .15$$

and

$$H_2: \tilde{p} > .15.$$

If the prior distribution is a beta distribution with $r' = 2$ and $n' = 18$, find the posterior distribution and use this posterior distribution to find the posterior odds ratio of H_1 to H_2.

35. In Exercise 10, suppose that the production manager is interested in the hypotheses H_1: $\tilde{\mu} \leq 110$ and H_2: $\tilde{\mu} > 110$. From the prior distribution, find the prior odds ratio; also, from the posterior distribution, find the posterior odds ratio. If the losses involved in a decision-making problem involving the production process are such that H_1 should be accepted only if the posterior odds ratio is greater than three, what decision should be made?

36. A statistician is interested in the mean $\tilde{\mu}$ of a normal population, and his prior distribution for $\tilde{\mu}$ is normal with mean 800 and variance 12. The variance of the population is known to be 72. How large a sample is needed to guarantee that the variance of the posterior distribution will be no larger than 1? How large a sample is needed to guarantee that the variance of the posterior distribution will be no larger than 0.1?

37. In testing a hypothesis concerning the mean of a normal process with known variance against a one-tailed alternative, discuss the relationship between the classical significance level and the posterior probability $P''(H_1)$
(a) if the prior distribution is diffuse,
(b) if the prior distribution is not diffuse.

38. Comment on the following statement: "A hypothesis such as $\tilde{\mu} = 100$ is not realistic and should be modified somewhat; otherwise, the Bayesian approach to hypothesis testing may not be applicable."

39. Suppose that you are sampling from a normal data-generating process with unknown mean $\tilde{\mu}$ and known variance $\sigma^2 = 25$. You are interested in the hypotheses $H_1: \tilde{\mu} = 50$ versus $H_2: \tilde{\mu} \neq 50$. Find the classical two-tailed level of significance if
(a) $n = 1$ and $m = 51$,
(b) $n = 25$ and $m = 51$,
(c) $n = 100$ and $m = 51$,
(d) $n = 10{,}000$ and $m = 51$,
(e) $n = 1$ and $m = 50.1$,
(f) $n = 25$ and $m = 50.1$,
(g) $n = 100$ and $m = 50.1$,
(h) $n = 10{,}000$ and $m = 50.1$.
Comment on these results, especially with respect to Lindley's paradox.

40. An automobile manufacturer claims that the average mileage per gallon of gas for a particular model is normally distributed with $m' = 20$ and $\sigma' = 4$, provided that the car is driven on a level road at a constant speed of 30 miles per hour. A rival manufacturer decides to use this as a prior distribution and to obtain additional information by conducting an experiment. In the experiment, it is assumed that the variance of mileage is $\sigma^2 = 96$ and that the mileage is normally distributed. The experiment is conducted on 10 randomly chosen cars, with the sample mean (the average mileage in the sample) equalling 18. Find the posterior distribution for $\tilde{\mu}$, the average mileage per gallon of gas. On the basis of this posterior distribution, what is the probability that $\tilde{\mu}$ is greater than or equal to 20? What is the probability that $\tilde{\mu}$ is between 19 and 21?

41. In Exercise 35, suppose that the manager is interested in the hypotheses $H_1: \tilde{\mu} = 110$ and $H_2: \tilde{\mu} \neq 110$. From the posterior distribution, what is the posterior odds ratio of H_1 to H_2? What is the posterior odds ratio if the hypotheses are $H_1: 109 < \tilde{\mu} < 111$ and $H_2: \tilde{\mu} \leq 109$ or $\tilde{\mu} \geq 111$?

42. Suppose that a statistician is interested in $H_1: \tilde{\mu} = 50$ and $H_2: \tilde{\mu} \neq 50$. His prior distribution consists of a mass of probability of 0.25 at $\tilde{\mu} = 50$, with the remaining .75 of probability distributed uniformly over the interval from $\tilde{\mu} = 40$ to $\tilde{\mu} = 60$. Find the prior probability that
(a) $45 < \tilde{\mu} < 55$,
(b) $47 \leq \tilde{\mu} \leq 50$,
(c) H_1 is exactly true,
(d) the hypothesis $49 < \tilde{\mu} < 51$ is true.

43. Suppose that $\tilde{x}$ is normally distributed with mean $\tilde{\mu}$ and variance $\sigma^2 = 400$. The prior distribution of $\tilde{\mu}$ is normally distributed with mean $m' = -60$

and variance $\sigma'^2 = 40$. Furthermore, a sample of size 20 is taken, with sample mean $m = -69$.

(a) Find the posterior distribution of $\tilde{\mu}$.
(b) Find a 63 percent credible interval for $\tilde{\mu}$.
(c) Find $P(H_1)$ from the posterior distribution, where H_1 is the hypothesis that $\tilde{\mu}$ is greater than or equal to -70.

44. In Exercise 43, suppose that three hypotheses are under consideration:

$$H_1: \tilde{\mu} \leq -70,$$
$$H_2: -70 < \tilde{\mu} < -60,$$
and
$$H_3: \tilde{\mu} \geq -60.$$

(a) Find the posterior probabilities of these three hypotheses.
(b) If the decision problem at hand involves three actions (corresponding to the three hypotheses) and if the hypothesis to be selected is the one that is most probable, which hypothesis would be selected under the posterior distribution?
(c) In part (b), which hypothesis would be selected on the basis of the *prior* distribution?

45. Suppose that a statistician is interested in the difference in the means of two normal populations, $\tilde{\mu}_1$ and $\tilde{\mu}_2$. Let $\tilde{\mu} = \tilde{\mu}_1 - \tilde{\mu}_2$. Each population has variance 100, and the two populations are independent. A sample of size 25 is taken from the first population, with sample mean $m_1 = 80$, and a sample of size 25 is taken from the second population, with sample mean $m_2 = 60$.

(a) Suppose that the prior distribution of $\tilde{\mu} = \tilde{\mu}_1 - \tilde{\mu}_2$ is normally distributed with mean $m' = 10$ and variance $\sigma'^2 = 50$. Find $P'(H_1)$ and $P'(H_2)$, where the hypotheses are $H_1: \tilde{\mu} \leq 0$ and $H_2: \tilde{\mu} > 0$.
(b) Find the posterior distribution of $\tilde{\mu} = \tilde{\mu}_1 - \tilde{\mu}_2$.
(c) From the posterior distribution, find the posterior odds ratio of H_1 to H_2.

46. Exercise 45 suggests a Bayesian approach to inferences regarding the difference between two means under the conditions that the two populations of interest are normally distributed and the variances are known. Using the notation of Exercise 45, determine a general formula for finding the posterior distribution of the difference between two means if the prior distribution has mean m' and variance σ'^2.

47. In Exercise 30, the marketing manager decides that the loss that will be suffered if the company markets the product and $\tilde{p}$ is in fact only .10 is three times as great as the loss that will be suffered if the company fails to market the product and $\tilde{p}$ is actually .20. Should the product be marketed?

48. In Exercise 31, the loss due to claiming incorrectly that A authored the essay is considered to be three times as great as the loss due to claiming incorrectly that B authored the essay. If you must make a claim concerning the authorship, which one should you claim as the author?

49. In Exercise 33, if $L_{\text{I}} = 7$ and $L_{\text{II}} = 5$, should you accept H_1 or H_2?

50. A statistician is interested in testing the hypotheses

$$H_1 \colon \tilde{\mu} \geq 120$$

and

$$H_2 \colon \tilde{\mu} < 120,$$

where $\tilde{\mu}$ is the mean of a normally distributed population with variance 144. The prior distribution for $\tilde{\mu}$ is a normal distribution with mean $m' = 115$ and variance $\sigma'^2 = 36$. A sample of size eight is taken, with sample mean $m = 121$.

(a) Find the prior odds ratio of H_1 to H_2.

(b) Find the posterior odds ratio of H_1 to H_2.

(c) If $L_{\mathrm{I}} = 4$ and $L_{\mathrm{II}} = 6$, which hypothesis should be accepted according to the posterior distribution?

51. For a sample of size one from a normal population with known variance 25, show that the classical test of

$$H_1 \colon \tilde{\mu} = 50$$

versus

$$H_2 \colon \tilde{\mu} = 60$$

is a likelihood ratio test. That is, show that the rejection region can be expressed in the form $LR \leq c$, where LR is the likelihood ratio and c is some constant. [*Hint:* Consider the equation $LR \leq c$ and attempt to manipulate it algebraically to get the result $x \geq k$, where x is the sample outcome and k is some constant, since you know that the rejection region must be of this form.]

52. Consider the hypotheses $H_1 \colon \tilde{\mu} = 10$ and $H_2 \colon \tilde{\mu} = 12$, where $\tilde{\mu}$ is the mean of a normally distributed population with variance 1. The prior distribution is diffuse (take a normal distribution with $n' = 0$), and a sample of size 10 is to be taken.

(a) If $L_{\mathrm{I}} = 100$ and $L_{\mathrm{II}} = 50$, find the region of rejection (the sample results for which you would reject H_1 in favor of H_2), using the decision-theoretic approach.

(b) From the region of rejection determined in (a), find P(type I error) and P(type II error).

(c) Do (a) and (b) if $L_{\mathrm{I}} = 50$ and $L_{\mathrm{II}} = 50$.

53. In classical hypothesis testing, the rejection region is often determined simply by choosing an arbitrarily small value, such as .05, and requiring that P(type I error) equal this value. What advantages or disadvantages does this procedure have in comparison with the decision-theoretic approach, as illustrated in Exercise 52?

54. Comment on the statement, "The entire posterior distribution constitutes an inferential statement, and for many (perhaps most) purposes, the entire distribution is much more informative and useful than any summarizations in the form of estimates or tests of hypotheses."

55. Carefully distinguish among classical inferential statistics, Bayesian inferential statistics, and decision theory.

ADDITIONAL REFERENCES AND SUGGESTIONS FOR FURTHER READING

Most of the examples and exercises in this book are small-scale problems rather than large cases. Some extensive cases regarding decision making under uncertainty are presented in Schlaifer (1968). The oil-drilling example in Sections 5.10, 6.4, and 6.5 is based on Grayson (1960), which includes a much more extensive discussion of the application of decision theory to oil-drilling decisions; also, see Kaufman (1963b). A general discussion of current applications of decision theory in business is presented in Brown (1970), and numerous applications and potential applications of the theory of statistical inference and decision to problems in a wide variety of areas have been reported. In marketing, for instance, see Brown (1969); Buzzell, Cox, and Brown (1969, Chapters 9–11); Cook (1968); Frank and Green (1967); Green (1962, 1963); Green and Frank (1966); and Roberts (1963a). In meteorology, see Epstein (1962), Murphy (1966, 1969), and Nelson and Winter (1964). For developments of economic theory under uncertainty, see Baron (1970); Cyert and De Groot (1970); Hakansson (1971b); Horowitz (1970); Mills (1959); Sandmo (1970, 1971); and Tisdell (1968). Other references regarding statistical inference and decision in economics and finance include Bierman and Hausman (1970); Borch (1968b); Bracken and Soland (1969); Fama (1970); Farrar (1962); Hakansson (1970b, 1971a); Hanoch and Levy (1970); Hausman and White (1968); Hayes (1969a); Hertz (1964, 1968); Hespos and Strassmann (1965); Kalymon (1971); Levy and Hanoch (1970); Magee (1964a, 1964b); Mao and Särndal (1966); Markowitz (1959); Sharpe (1970); Spetzler (1968); and Wilson (1969). Applications in medical decision making include Aitchison (1970b); Betaque and Gorry (1971); Ginsberg (1970); Ginsberg and Offensend (1968); Gustafson *et al.* (1969); Lusted (1968);

and Savage (1970b). Some bidding models are discussed in Christenson (1965), Friedman (1956), LaValle (1967), Mercer and Russell (1969), Rothkopf (1969), and Wilson (1967); and inventory models are discussed from a Bayesian viewpoint in Hayes (1969c) and Murray and Silver (1966). Bayesian inference and decision in accounting are discussed in Hakansson (1969), Sorenson (1969), and Tracy (1969); a reference from the field of education is Pritsker (1965); applications in insurance include Borch (1967), Moore (1966), and Seal (1969); and four references regarding reliability and quality control are Brender (1968a, 1968b) and Soland (1968, 1969).

The above references provide a reasonable and (hopefully) representative cross section of applications of statistical inference and decision in various fields. They also serve to demonstrate the wide applicability of and the increasing interest in the types of models discussed in this book. As suggested by the recency of many of the references cited in this book, interest in the Bayesian approach has greatly increased in the past decade. Although some pathbreaking work was done earlier (see the historical references cited in several of the chapters), recent developments have been quite extensive. Furthermore, since there is a time lag between the development and the actual application of theory and methodology, most of the applications-oriented references are very recent. Indeed, the Bayesian approach to statistical inference and decision is being applied more and more frequently. At the current time, a considerable amount of work of both a theoretical nature and an applied nature is being conducted in the area of statistical inference and decision.

TABLES

Table 1. Poisson Probabilities

This table gives values of the Poisson probability mass function,

$$P(\tilde{r} = r \mid t,\lambda) = e^{-\lambda t}(\lambda t)^r/r!,$$

for

$$\lambda t = .1(.1)10(1)20$$

and suitable values of r.

Examples:

$$P(\tilde{r} = 2 \mid t = 5, \lambda = 1.5) = .0156$$

and

$$P(\tilde{r} = 14 \mid t = 2, \lambda = 6) = .0905.$$

Table 1

	λt									
r	**0.1**	**0.2**	**0.3**	**0.4**	**0.5**	**0.6**	**0.7**	**0.8**	**0.9**	**1.0**
0	.9048	.8187	.7408	.6703	.6065	.5488	.4966	.4493	.4066	.3679
1	.0905	.1637	.2222	.2681	.3033	.3293	.3476	.3595	.3659	.3679
2	.0045	.0164	.0333	.0536	.0758	.0988	.1217	.1438	.1647	.1839
3	.0002	.0011	.0033	.0072	.0126	.0198	.0284	.0383	.0494	.0613
4	.0000	.0001	.0002	.0007	.0016	.0030	.0050	.0077	.0111	.0153
5	.0000	.0000	.0000	.0001	.0002	.0004	.0007	.0012	.0020	.0031
6	.0000	.0000	.0000	.0000	.0000	.0000	.0001	.0002	.0003	.0005
7	.0000	.0000	.0000	.0000	.0000	.0000	.0000	.0000	.0000	.0001

	λt									
r	**1.1**	**1.2**	**1.3**	**1.4**	**1.5**	**1.6**	**1.7**	**1.8**	**1.9**	**2.0**
0	.3329	.3012	.2725	.2466	.2231	.2019	.1827	.1653	.1496	.1353
1	.3662	.3614	.3543	.3452	.3347	.3230	.3106	.2975	.2842	.2707
2	.2014	.2169	.2303	.2417	.2510	.2584	.2640	.2678	.2700	.2707
3	.0738	.0867	.0998	.1128	.1255	.1378	.1496	.1607	.1710	.1804
4	.0203	.0260	.0324	.0395	.0471	.0551	.0636	.0723	.0812	.0902
5	.0045	.0062	.0084	.0111	.0141	.0176	.0216	.0260	.0309	.0361
6	.0008	.0012	.0018	.0026	.0035	.0047	.0061	.0078	.0098	.0120
7	.0001	.0002	.0003	.0005	.0008	.0011	.0015	.0020	.0027	.0034
8	.0000	.0000	.0001	.0001	.0001	.0002	.0003	.0005	.0006	.0009
9	.0000	.0000	.0000	.0000	.0000	.0000	.0001	.0001	.0001	.0002

	λt									
r	**2.1**	**2.2**	**2.3**	**2.4**	**2.5**	**2.6**	**2.7**	**2.8**	**2.9**	**3.0**
0	.1225	.1108	.1003	.0907	.0821	.0743	.0672	.0608	.0550	.0498
1	.2572	.2438	.2306	.2177	.2052	.1931	.1815	.1703	.1596	.1494
2	.2700	.2681	.2652	.2613	.2565	.2510	.2450	.2384	.2314	.2240
3	.1890	.1966	.2033	.2090	.2138	.2176	.2205	.2225	.2237	.2240
4	.0992	.1082	.1169	.1254	.1336	.1414	.1488	.1557	.1622	.1680
5	.0417	.0476	.0538	.0602	.0668	.0735	.0804	.0872	.0940	.1008
6	.0146	.0174	.0206	.0241	.0278	.0319	.0362	.0407	.0455	.0540
7	.0044	.0055	.0068	.0083	.0099	.0118	.0139	.0163	.0188	.0216
8	.0011	.0015	.0019	.0025	.0031	.0038	.0047	.0057	.0068	.0081
9	.0003	.0004	.0005	.0007	.0009	.0011	.0014	.0018	.0022	.0027
10	.0001	.0001	.0001	.0002	.0002	.0003	.0004	.0005	.0006	.0008
11	.0000	.0000	.0000	.0000	.0000	.0001	.0001	.0001	.0002	.0002
12	.0000	.0000	.0000	.0000	.0000	.0000	.0000	.0000	.0000	.0001

	λt									
r	**3.1**	**3.2**	**3.3**	**3.4**	**3.5**	**3.6**	**3.7**	**3.8**	**3.9**	**4.0**
0	.0450	.0408	.0369	.0344	.0302	.0273	.0247	.0224	.0202	.0183
1	.1397	.1304	.1217	.1135	.1057	.0984	.0915	.0850	.0789	.0733
2	.2165	.2087	.2008	.1929	.1850	.1771	.1692	.1615	.1539	.1465
3	.2237	.2226	.2209	.2186	.2158	.2125	.2087	.2046	.2001	.1954
4	.1734	.1781	.1823	.1858	.1888	.1912	.1931	.1944	.1951	.1954

Table 1 (continued)

r	3.1	3.2	3.3	3.4	3.5	3.6	3.7	3.8	3.9	4.0
5	.1075	.1140	.1203	.1264	.1322	.1377	.1429	.1477	.1522	.1563
6	.0555	.0608	.0662	.0716	.0771	.0826	.0881	.0936	.0989	.1042
7	.0246	.0278	.0312	.0348	.0385	.0425	.0466	.0508	.0551	.0595
8	.0095	.0111	.0129	.0148	.0169	.0191	.0215	.0241	.0269	.0298
9	.0033	.0040	.0047	.0056	.0066	.0076	.0089	.0102	.0116	.0132
10	.0010	.0013	.0016	.0019	.0023	.0028	.0033	.0039	.0045	.0053
11	.0003	.0004	.0005	.0006	.0007	.0009	.0011	.0013	.0016	.0019
12	.0001	.0001	.0001	.0002	.0002	.0003	.0003	.0004	.0005	.0006
13	.0000	.0000	.0000	.0000	.0001	.0001	.0001	.0001	.0002	.0002
14	.0000	.0000	.0000	.0000	.0000	.0000	.0000	.0000	.0000	.0001

λt

r	4.1	4.2	4.3	4.4	4.5	4.6	4.7	4.8	4.9	5.0
0	.0166	.0150	.0136	.0123	.0111	.0101	.0091	.0082	.0074	.0067
1	.0679	.0630	.0583	.0540	.0500	.0462	.0427	.0395	.0365	.0337
2	.1393	.1323	.1254	.1188	.1125	.1063	.1005	.0948	.0894	.0842
3	.1904	.1852	.1798	.1743	.1687	.1631	.1574	.1517	.1460	.1404
4	.1951	.1944	.1933	.1917	.1898	.1875	.1849	.1820	.1789	.1755
5	.1600	.1633	.1662	.1687	.1708	.1725	.1738	.1747	.1753	.1755
6	.1093	.1143	.1191	.1237	.1281	.1323	.1362	.1398	.1432	.1462
7	.0640	.0686	.0732	.0778	.0824	.0869	.0914	.0959	.1002	.1044
8	.0328	.0360	.0393	.0428	.0463	.0500	.0537	.0575	.0614	.0653
9	.0150	.0168	.0188	.0209	.0232	.0255	.0280	.0307	.0334	.0363
10	.0061	.0071	.0081	.0092	.0104	.0118	.0132	.0147	.0164	.0181
11	.0023	.0027	.0032	.0037	.0043	.0049	.0056	.0064	.0073	.0082
12	.0008	.0009	.0011	.0014	.0016	.0019	.0022	.0026	.0030	.0034
13	.0002	.0003	.0004	.0005	.0006	.0007	.0008	.0009	.0011	.0013
14	.0001	.0001	.0001	.0001	.0002	.0002	.0003	.0003	.0004	.0005
15	.0000	.0000	.0000	.0000	.0001	.0001	.0001	.0001	.0001	.0002

λt

r	5.1	5.2	5.3	5.4	5.5	5.6	5.7	5.8	5.9	6.0
0	.0061	.0055	.0050	.0045	.0041	.0037	.0033	.0030	.0027	.0025
1	.0311	.0287	.0265	.0244	.0225	.0207	.0191	.0176	.0162	.0149
2	.0793	.0746	.0701	.0659	.0618	.0580	.0544	.0509	.0477	.0446
3	.1348	.1293	.1239	.1185	.1133	.1082	.1033	.0985	.0938	.0892
4	.1719	.1681	.1641	.1600	.1558	.1515	.1472	.1428	.1383	.1339
5	.1753	.1748	.1740	.1728	.1714	.1697	.1678	.1656	.1632	.1606
6	.1490	.1515	.1537	.1555	.1571	.1584	.1594	.1601	.1605	.1606
7	.1086	.1125	.1163	.1200	.1234	.1267	.1298	.1326	.1353	.1377
8	.0692	.0731	.0771	.0810	.0849	.0887	.0925	.0962	.0998	.1033
9	.0392	.0423	.0454	.0486	.0519	.0552	.0586	.0620	.0654	.0688

Table 1 (continued)

r	λt = 5.1	5.2	5.3	5.4	5.5	5.6	5.7	5.8	5.9	6.0
10	.0200	.0220	.0241	.0262	.0285	.0309	.0334	.0359	.0386	.0413
11	.0093	.0104	.0116	.0129	.0143	.0157	.0173	.0190	.0207	.0225
12	.0039	.0045	.0051	.0058	.0065	.0073	.0082	.0092	.0102	.0113
13	.0015	.0018	.0021	.0024	.0028	.0032	.0036	.0041	.0046	.0052
14	.0006	.0007	.0008	.0009	.0011	.0013	.0015	.0017	.0019	.0022
15	.0002	.0002	.0003	.0003	.0004	.0005	.0006	.0007	.0008	.0009
16	.0001	.0001	.0001	.0001	.0001	.0002	.0002	.0002	.0003	.0003
17	.0000	.0000	.0000	.0000	.0000	.0001	.0001	.0001	.0001	.0001

r	λt = 6.1	6.2	6.3	6.4	6.5	6.6	6.7	6.8	6.9	7.0
0	.0022	.0020	.0018	.0017	.0015	.0014	.0012	.0011	.0010	.0009
1	.0137	.0126	.0116	.0106	.0098	.0090	.0082	.0076	.0070	.0064
2	.0417	.0390	.0364	.0340	.0318	.0296	.0276	.0258	.0240	.0223
3	.0848	.0806	.0765	.0726	.0688	.0652	.0617	.0584	.0552	.0521
4	.1294	.1249	.1205	.1162	.1118	.1076	.1034	.0992	.0952	.0912
5	.1579	.1549	.1519	.1487	.1454	.1420	.1385	.1349	.1314	.1277
6	.1605	.1601	.1595	.1586	.1575	.1562	.1546	.1529	.1511	.1490
7	.1399	.1418	.1435	.1450	.1462	.1472	.1480	.1486	.1489	.1490
8	.1066	.1099	.1130	.1160	.1188	.1215	.1240	.1263	.1284	.1304
9	.0723	.0757	.0791	.0825	.0858	.0891	.0923	.0954	.0985	.1014
10	.0441	.0469	.0498	.0528	.0558	.0588	.0618	.0649	.0679	.0710
11	.0245	.0265	.0285	.0307	.0330	.0353	.0377	.0401	.0426	.0452
12	.0124	.0137	.0150	.0164	.0179	.0194	.0210	.0227	.0245	.0264
13	.0058	.0065	.0073	.0081	.0089	.0098	.0108	.0119	.0130	.0142
14	.0025	.0029	.0033	.0037	.0041	.0046	.0052	.0058	.0064	.0071
15	.0010	.0012	.0014	.0016	.0018	.0020	.0023	.0026	.0029	.0033
16	.0004	.0005	.0005	.0006	.0007	.0008	.0010	.0011	.0013	.0014
17	.0001	.0002	.0002	.0002	.0003	.0003	.0004	.0004	.0005	.0006
18	.0000	.0001	.0001	.0001	.0001	.0001	.0001	.0002	.0002	.0002
19	.0000	.0000	.0000	.0000	.0000	.0000	.0000	.0001	.0001	.0001

r	λt = 7.1	7.2	7.3	7.4	7.5	7.6	7.7	7.8	7.9	8.0
0	.0008	.0007	.0007	.0006	.0006	.0005	.0005	.0004	.0004	.0003
1	.0059	.0054	.0049	.0045	.0041	.0038	.0035	.0032	.0029	.0027
2	.0208	.0194	.0180	.0167	.0156	.0145	.0134	.0125	.0116	.0107
3	.0492	.0464	.0438	.0413	.0389	.0366	.0345	.0324	.0305	.0286
4	.0874	.0836	.0799	.0764	.0729	.0696	.0663	.0632	.0602	.0573
5	.1241	.1204	.1167	.1130	.1094	.1057	.1021	.0986	.0951	.0916
6	.1468	.1445	.1420	.1394	.1367	.1339	.1311	.1282	.1252	.1221
7	.1489	.1486	.1481	.1474	.1465	.1454	.1442	.1428	.1413	.1396
8	.1321	.1337	.1351	.1363	.1373	.1382	.1388	.1392	.1395	.1396
9	.1042	.1070	.1096	.1121	.1144	.1167	.1187	.1207	.1224	.1241
10	.0740	.0770	.0800	.0829	.0858	.0887	.0914	.0941	.0967	.0993
11	.0478	.0504	.0531	.0558	.0585	.0613	.0640	.0667	.0695	.0722

Table 1 (continued)

r	7.1	7.2	7.3	7.4	7.5	7.6	7.7	7.8	7.9	8.0
12	.0283	.0303	.0323	.0344	.0366	.0388	.0411	.0434	.0457	.0481
13	.0154	.0168	.0181	.0196	.0211	.0227	.0243	.0260	.0278	.0296
14	.0078	.0086	.0095	.0104	.0113	.0123	.0134	.0145	.0157	.0169
15	.0037	.0041	.0046	.0051	.0057	.0062	.0069	.0075	.0083	.0090
16	.0016	.0019	.0021	.0024	.0026	.0030	.0033	.0037	.0041	.0045
17	.0007	.0008	.0009	.0010	.0012	.0013	.0015	.0017	.0019	.0021
18	.0003	.0003	.0004	.0004	.0005	.0006	.0006	.0007	.0008	.0009
19	.0001	.0001	.0001	.0002	.0002	.0002	.0003	.0003	.0003	.0004
20	.0000	.0000	.0001	.0001	.0001	.0000	.0001	.0001	.0001	.0002
21	.0000	.0000	.0000	.0000	.0000	.0000	.0000	.0000	.0001	.0001

λt

r	8.1	8.2	8.3	8.4	8.5	8.6	8.7	8.8	8.9	9.0
0	.0003	.0003	.0002	.0002	.0002	.0002	.0002	.0002	.0001	.0001
1	.0025	.0023	.0021	.0019	.0017	.0016	.0014	.0013	.0012	.0011
2	.0100	.0092	.0086	.0079	.0074	.0068	.0063	.0058	.0054	.0050
3	.0269	.0252	.0237	.0222	.0208	.0195	.0183	.0171	.0160	.0150
4	.0544	.0517	.0491	.0466	.0443	.0420	.0398	.0377	.0357	.0337
5	.0882	.0849	.0816	.0784	.0752	.0722	.0692	.0663	.0635	.0607
6	.1191	.1160	.1128	.1097	.1066	.1034	.1003	.0972	.0941	.0911
7	.1378	.1358	.1338	.1317	.1294	.1271	.1247	.1222	.1197	.1171
8	.1395	.1392	.1388	.1382	.1375	.1366	.1356	.1344	.1332	.1318
9	.1256	.1269	.1280	.1290	.1299	.1306	.1311	.1315	.1317	.1318
10	.1017	.1040	.1063	.1084	.1104	.1123	.1140	.1157	.1172	.1186
11	.0749	.0776	.0802	.0828	.0853	.0878	.0902	.0925	.0948	.0970
12	.0505	.0530	.0555	.0579	.0604	.0629	.0654	.0679	.0703	.0728
13	.0315	.0334	.0354	.0374	.0395	.0416	.0438	.0459	.0481	.0504
14	.0182	.0196	.0210	.0225	.0240	.0256	.0272	.0289	.0306	.0324
15	.0098	.0107	.0116	.0126	.0136	.0147	.0158	.0169	.0182	.0194
16	.0050	.0055	.0060	.0066	.0072	.0079	.0086	.0093	.0101	.0109
17	.0024	.0026	.0029	.0033	.0036	.0040	.0044	.0048	.0053	.0058
18	.0011	.0012	.0014	.0015	.0017	.0019	.0021	.0024	.0026	.0029
19	.0005	.0005	.0006	.0007	.0008	.0009	.0010	.0011	.0012	.0014
20	.0002	.0002	.0002	.0003	.0003	.0004	.0004	.0005	.0005	.0006
21	.0001	.0001	.0001	.0001	.0001	.0002	.0002	.0002	.0002	.0003
22	.0000	.0000	.0000	.0000	.0001	.0001	.0001	.0001	.0001	.0001

λt

r	9.1	9.2	9.3	9.4	9.5	9.6	9.7	9.8	9.9	10
0	.0001	.0001	.0001	.0001	.0001	.0001	.0001	.0001	.0001	.0000
1	.0010	.0009	.0009	.0008	.0007	.0007	.0006	.0005	.0005	.0005
2	.0046	.0043	.0040	.0037	.0034	.0031	.0029	.0027	.0025	.0023
3	.0140	.0131	.0123	.0115	.0107	.0100	.0093	.0087	.0081	.0076
4	.0319	.0302	.0285	.0269	.0254	.0240	.0226	.0213	.0201	.0189

Table 1 (continued)

r	λt 9.1	9.2	9.3	9.4	9.5	9.6	9.7	9.8	9.9	10
5	.0581	.0555	.0530	.0506	.0483	.0460	.0439	.0418	.0398	.0378
6	.0881	.0851	.0822	.0793	.0764	.0736	.0709	.0682	.0656	.0631
7	.1145	.1118	.1091	.1064	.1037	.1010	.0982	.0955	.0928	.0901
8	.1302	.1286	.1269	.1251	.1232	.1212	.1191	.1170	.1148	.1126
9	.1317	.1315	.1311	.1306	.1300	.1293	.1284	.1274	.1263	.1251
10	.1198	.1210	.1219	.1228	.1235	.1241	.1245	.1249	.1250	.1251
11	.0991	.1012	.1031	.1049	.1067	.1083	.1098	.1112	.1125	.1137
12	.0752	.0776	.0799	.0822	.0844	.0866	.0888	.0908	.0928	.0948
13	.0526	.0549	.0572	.0594	.0617	.0640	.0662	.0685	.0707	.0729
14	.0342	.0361	.0380	.0399	.0419	.0439	.0459	.0479	.0500	.0521
15	.0208	.0221	.0235	.0250	.0265	.0281	.0297	.0313	.0330	.0347
16	.0118	.0127	.0137	.0147	.0157	.0168	.0180	.0192	.0204	.0217
17	.0063	.0069	.0075	.0081	.0088	.0095	.0103	.0111	.0119	.0128
18	.0032	.0035	.0039	.0042	.0046	.0051	.0055	.0060	.0065	.0071
19	.0015	.0017	.0019	.0021	.0023	.0026	.0028	.0031	.0034	.0037
20	.0007	.0008	.0009	.0010	.0011	.0012	.0014	.0015	.0017	.0019
21	.0003	.0003	.0004	.0004	.0005	.0006	.0006	.0007	.0008	.0009
22	.0001	.0001	.0002	.0002	.0002	.0002	.0003	.0003	.0004	.0004
23	.0000	.0001	.0001	.0001	.0001	.0001	.0001	.0001	.0002	.0002
24	.0000	.0000	.0000	.0000	.0000	.0000	.0000	.0001	.0001	.0001

r	λt 11	12	13	14	15	16	17	18	19	20
0	.0000	.0000	.0000	.0000	.0000	.0000	.0000	.0000	.0000	.0000
1	.0002	.0001	.0000	.0000	.0000	.0000	.0000	.0000	.0000	.0000
2	.0010	.0004	.0002	.0001	.0000	.0000	.0000	.0000	.0000	.0000
3	.0037	.0018	.0008	.0004	.0002	.0001	.0000	.0000	.0000	.0000
4	.0102	.0053	.0027	.0013	.0006	.0003	.0001	.0001	.0000	.0000
5	.0224	.0127	.0070	.0037	.0019	.0010	.0005	.0002	.0001	.0001
6	.0411	.0255	.0152	.0087	.0048	.0026	.0014	.0007	.0004	.0002
7	.0646	.0437	.0281	.0174	.0104	.0060	.0034	.0018	.0010	.0005
8	.0888	.0655	.0457	.0304	.0194	.0120	.0072	.0042	.0024	.0013
9	.1085	.0874	.0661	.0473	.0324	.0213	.0135	.0083	.0050	.0029
10	.1194	.1048	.0859	.0663	.0486	.0341	.0230	.0150	.0095	.0058
11	.1194	.1144	.1015	.0844	.0663	.0496	.0355	.0245	.0164	.0106
12	.1094	.1144	.1099	.0984	.0829	.0661	.0504	.0368	.0259	.0176
13	.0926	.1056	.1099	.1060	.0956	.0814	.0658	.0509	.0378	.0271
14	.0728	.0905	.1021	.1060	.1024	.0930	.0800	.0655	.0541	.0387
15	.0534	.0724	.0885	.0989	.1024	.0992	.0906	.0786	.0650	.0516
16	.0367	.0543	.0719	.0866	.0960	.0992	.0963	.0884	.0772	.0646
17	.0237	.0383	.0550	.0713	.0847	.0934	.0963	.0936	.0863	.0760
18	.0145	.0256	.0397	.0554	.0706	.0830	.0909	.0936	.0911	.0844
19	.0084	.0161	.0272	.0409	.0557	.0699	.0814	.0887	.0911	.0888
20	.0046	.0097	.0177	.0286	.0418	.0559	.0692	.0798	.0866	.0888
21	.0024	.0055	.0109	.0191	.0299	.0426	.0560	.0684	.0783	.0846
22	.0012	.0030	.0065	.0121	.0204	.0310	.0433	.0560	.0676	.0769
23	.0006	.0016	.0037	.0074	.0133	.0216	.0320	.0438	.0559	.0669
24	.0003	.0008	.0020	.0043	.0083	.0144	.0226	.0328	.0442	.0557

Table 1 (continued)

r	λt 11	12	13	14	15	16	17	18	19	20
25	.0001	.0004	.0010	.0024	.0050	.0092	.0154	.0237	.0336	.0446
26	.0000	.0002	.0005	.0013	.0029	.0057	.0101	.0164	.0246	.0343
27	.0000	.0001	.0002	.0007	.0016	.0034	.0063	.0109	.0173	.0254
28	.0000	.0000	.0001	.0003	.0009	.0019	.0038	.0070	.0117	.0181
29	.0000	.0000	.0001	.0002	.0004	.0011	.0023	.0044	.0077	.0125
30	.0000	.0000	.0000	.0001	.0002	.0006	.0013	.0026	.0049	.0083
31	.0000	.0000	.0000	.0000	.0001	.0003	.0007	.0015	.0030	.0054
32	.0000	.0000	.0000	.0000	.0001	.0001	.0004	.0009	.0018	.0034
33	.0000	.0000	.0000	.0000	.0000	.0001	.0002	.0005	.0010	.0020
34	.0000	.0000	.0000	.0000	.0000	.0000	.0001	.0002	.0006	.0012
35	.0000	.0000	.0000	.0000	.0000	.0000	.0000	.0001	.0003	.0007
36	.0000	.0000	.0000	.0000	.0000	.0000	.0000	.0001	.0002	.0004
37	.0000	.0000	.0000	.0000	.0000	.0000	.0000	.0000	.0001	.0002
38	.0000	.0000	.0000	.0000	.0000	.0000	.0000	.0000	.0000	.0001
39	.0000	.0000	.0000	.0000	.0000	.0000	.0000	.0000	.0000	.0001

Table 2. Binomial Probabilities

This table gives values of the binomial probability mass function,

$$P(\tilde{r} = r \mid n,p) = \binom{n}{r} p^r(1-p)^{n-r},$$

for

$$n = 1(1)20,\ 50,\ 100,$$
$$r = 0(1)n,$$

and

$$p = .01(.01).99.$$

The values of r at the *left* of any section of the tables are to be used in conjunction with the values of p at the *top* of that section; the values of r at the *right* of any section are to be used in conjunction with the values of p at the *bottom* of that section.

Examples:

$$P(\tilde{r} = 3 \mid n = 8,\ p = .25) = .2076$$

and

$$P(\tilde{r} = 2 \mid n = 5,\ p = .62) = .2109.$$

Table 2

n = 1

r	*p* .01	.02	.03	.04	.05	.06	.07	.08	.09	.10	
0	.9900	.9800	.9700	.9600	.9500	.9400	.9300	.9200	.9100	.9000	1
1	.0100	.0200	.0300	.0400	.0500	.0600	.0700	.0800	.0900	.1000	0
	.99	.98	.97	.96	.95	.94	.93	.92	.91	.90 *p*	*r*

r	*p* .11	.12	.13	.14	.15	.16	.17	.18	.19	.20	
0	.8900	.8800	.8700	.8600	.8500	.8400	.8300	.8200	.8100	.8000	1
1	.1100	.1200	.1300	.1400	.1500	.1600	.1700	.1800	.1900	.2000	0
	.89	.88	.87	.86	.85	.84	.83	.82	.81	.80 *p*	*r*

r	*p* .21	.22	.23	.24	.25	.26	.27	.28	.29	.30	
0	.7900	.7800	.7700	.7600	.7500	.7400	.7300	.7200	.7100	.7000	1
1	.2100	.2200	.2300	.2400	.2500	.2600	.2700	.2800	.2900	.3000	0
	.79	.78	.77	.76	.75	.74	.73	.72	.71	.70 *p*	*r*

r	*p* .31	.32	.33	.34	.35	.36	.37	.38	.39	.40	
0	.6900	.6800	.6700	.6600	.6500	.6400	.6300	.6200	.6100	.6000	1
1	.3100	.3200	.3300	.3400	.3500	.3600	.3700	.3800	.3900	.4000	0
	.69	.68	.67	.66	.65	.64	.63	.62	.61	.60 *p*	*r*

r	*p* .41	.42	.43	.44	.45	.46	.47	.48	.49	.50	
0	.5900	.5800	.5700	.5600	.5500	.5400	.5300	.5200	.5100	.5000	1
1	.4100	.4200	.4300	.4400	.4500	.4600	.4700	.4800	.4900	.5000	0
	.59	.58	.57	.56	.55	.54	.53	.52	.51	.50 *p*	*r*

n = 2

r	*p* .01	.02	.03	.04	.05	.06	.07	.08	.09	.10	
0	.9801	.9604	.9409	.9216	.9025	.8836	.8649	.8464	.8281	.8100	2
1	.0198	.0392	.0582	.0768	.0950	.1128	.1302	.1472	.1638	.1800	1
2	.0001	.0004	.0009	.0016	.0025	.0036	.0049	.0064	.0081	.0100	0
	.99	.98	.97	.96	.95	.94	.93	.92	.91	.90 *p*	*r*

r	*p* .11	.12	.13	.14	.15	.16	.17	.18	.19	.20	
0	.7921	.7744	.7569	.7396	.7225	.7056	.6889	.6724	.6561	.6400	2
1	.1958	.2112	.2262	.2408	.2550	.2688	.2822	.2952	.3078	.3200	1
2	.0121	.0144	.0169	.0196	.0225	.0256	.0289	.0324	.0361	.0400	0
	.89	.88	.87	.86	.85	.84	.83	.82	.81	.80 *p*	*r*

r	*p* .21	.22	.23	.24	.25	.26	.27	.28	.29	.30	
0	.6241	.6084	.5929	.5776	.5625	.5476	.5329	.5184	.5041	.4900	2
1	.3318	.3432	.3542	.3648	.3750	.3848	.3942	.4032	.4118	.4200	1
2	.0441	.0484	.0529	.0576	.0625	.0676	.0729	.0784	.0841	.0900	0
	.79	.78	.77	.76	.75	.74	.73	.72	.71	.70 *p*	*r*

r	*p* .31	.32	.33	.34	.35	.36	.37	.38	.39	.40	
0	.4761	.4624	.4489	.4356	.4225	.4096	.3969	.3844	.3721	.3600	2
1	.4278	.4352	.4422	.4488	.4550	.4608	.4662	.4712	.4758	.4800	1
2	.0961	.1024	.1089	.1156	.1225	.1296	.1369	.1444	.1521	.1600	0
	.69	.68	.67	.66	.65	.64	.63	.62	.61	.60 *p*	*r*

r	*p* .41	.42	.43	.44	.45	.46	.47	.48	.49	.50	
0	.3481	.3364	.3249	.3136	.3025	.2916	.2809	.2704	.2601	.2500	2
1	.4838	.4872	.4902	.4928	.4950	.4968	.4982	.4992	.4998	.5000	1
2	.1681	.1764	.1849	.1936	.2025	.2116	.2209	.2304	.2401	.2500	0
	.59	.58	.57	.56	.55	.54	.53	.52	.51	.50 *p*	*r*

r	*p* .21	.22	.23	.24	.25	.26	.27	.28	.29	.30	
0	.3895	.3702	.3515	.3336	.3164	.2999	.2840	.2687	.2541	.2401	4
1	.4142	.4176	.4200	.4214	.4219	.4214	.4201	.4180	.4152	.4116	3
2	.1651	.1767	.1882	.1996	.2109	.2221	.2331	.2439	.2544	.2646	2
3	.0293	.0332	.0375	.0420	.0469	.0520	.0575	.0632	.0693	.0756	1
4	.0019	.0023	.0028	.0033	.0039	.0046	.0053	.0061	.0071	.0081	0
	.79	.78	.77	.76	.75	.74	.73	.72	.71	.70 *p*	*r*

r	*p* .31	.32	.33	.34	.35	.36	.37	.38	.39	.40	
0	.2267	.2138	.2015	.1897	.1785	.1678	.1575	.1478	.1385	.1296	4
1	.4074	.4025	.3970	.3910	.3845	.3775	.3701	.3623	.3541	.3456	3
2	.2745	.2841	.2933	.3021	.3105	.3185	.3260	.3330	.3396	.3456	2
3	.0822	.0891	.0963	.1038	.1115	.1194	.1276	.1361	.1447	.1536	1
4	.0092	.0105	.0119	.0134	.0150	.0168	.0187	.0209	.0231	.0256	0
	.69	.68	.67	.66	.65	.64	.63	.62	.61	.60 *p*	*r*

r	*p* .41	.42	.43	.44	.45	.46	.47	.48	.49	.50	
0	.1212	.1132	.1056	.0983	.0915	.0850	.0789	.0731	.0677	.0625	4
1	.3368	.3278	.3185	.3091	.2995	.2897	.2799	.2700	.2600	.2500	3
2	.3511	.3560	.3604	.3643	.3675	.3702	.3723	.3738	.3747	.3750	2
3	.1627	.1719	.1813	.1908	.2005	.2102	.2201	.2300	.2400	.2500	1
4	.0283	.0311	.0342	.0375	.0410	.0448	.0488	.0531	.0576	.0625	0
	.59	.58	.57	.56	.55	.54	.53	.52	.51	.50 *p*	*r*

n = 5

r	*p* .01	.02	.03	.04	.05	.06	.07	.08	.09	.10	
0	.9510	.9039	.8587	.8154	.7738	.7339	.6957	.6591	.6240	.5905	5
1	.0480	.0922	.1328	.1699	.2036	.2342	.2618	.2866	.3086	.3280	4
2	.0010	.0038	.0082	.0142	.0214	.0299	.0394	.0498	.0610	.0729	3
3	.0000	.0001	.0003	.0006	.0011	.0019	.0030	.0043	.0060	.0081	2
4	.0000	.0000	.0000	.0000	.0000	.0001	.0001	.0002	.0003	.0004	1
	.99	.98	.97	.96	.95	.94	.93	.92	.91	.90 *p*	*r*

r	*p* .11	.12	.13	.14	.15	.16	.17	.18	.19	.20	
0	.5584	.5277	.4984	.4704	.4437	.4182	.3939	.3707	.3487	.3277	5
1	.3451	.3598	.3724	.3829	.3915	.3983	.4034	.4069	.4089	.4096	4
2	.0853	.0981	.1113	.1247	.1382	.1517	.1652	.1786	.1919	.2048	3
3	.0105	.0134	.0166	.0203	.0244	.0289	.0338	.0392	.0450	.0512	2
4	.0007	.0009	.0012	.0017	.0022	.0028	.0035	.0043	.0053	.0064	1
5	.0000	.0000	.0000	.0001	.0001	.0001	.0001	.0002	.0002	.0003	0
	.89	.88	.87	.86	.85	.84	.83	.82	.81	.80 *p*	*r*

r	*p* .21	.22	.23	.24	.25	.26	.27	.28	.29	.30	
0	.3077	.2887	.2707	.2536	.2373	.2219	.2073	.1935	.1804	.1681	5
1	.4090	.4072	.4043	.4003	.3995	.3898	.3834	.3762	.3685	.3601	4
2	.2174	.2297	.2415	.2529	.2637	.2739	.2836	.2926	.3010	.3087	3
3	.0578	.0648	.0721	.0798	.0879	.0962	.1049	.1138	.1229	.1323	2
4	.0077	.0091	.0108	.0126	.0146	.0169	.0194	.0221	.0251	.0283	1
5	.0004	.0005	.0006	.0008	.0010	.0012	.0014	.0017	.0021	.0024	0
	.79	.78	.77	.76	.75	.74	.73	.72	.71	.70 *p*	*r*

n = 3

r	p .01	.02	.03	.04	.05	.06	.07	.08	.09	.10	
0	.9703	.9412	.9127	.8847	.8574	.8306	.8044	.7787	.7536	.7290	3
1	.0294	.0576	.0847	.1106	.1354	.1590	.1816	.2031	.2236	.2430	2
2	.0003	.0012	.0026	.0046	.0071	.0102	.0137	.0177	.0221	.0270	1
3	.0000	.0000	.00000	.0001	.0001	.0002	.0003	.0005	.0007	.0010	0
	.99	.98	.97	.96	.95	.94	.93	.92	.91	.90 p	r

r	p .11	.12	.13	.14	.15	.16	.17	.18	.19	.20	
0	.7050	.6815	.6585	.6361	.6141	.5927	.5718	.5514	.5314	.5120	3
1	.2614	.2788	.2952	.3106	.3251	.3387	.3513	.3631	.3740	.3840	2
2	.0323	.0380	.0441	.0506	.0574	.0645	.0720	.0797	.0877	.0960	1
3	.0013	.0017	.0022	.0027	.0034	.0041	.0049	.0058	.0069	.0080	0
	.89	.88	.87	.86	85	.84	.83	.82	.81	.80 p	r

r	p .21	.22	.23	.24	.25	.26	.27	.28	.29	.30	
0	.4930	.4746	.4565	.4390	.4219	.4052	.3890	.3732	.3579	.3430	3
1	.3932	.4015	.4091	.4159	.4219	.4271	.4316	.4355	.4386	.4410	2
2	.1045	.1133	.1222	.1313	.1406	.1501	.1597	.1693	.1791	.1890	1
3	.0093	.0106	.0122	.0138	.0156	.0176	.0197	.0220	.0244	.0270	0
	.79	.78	.77	.76	.75	.74	.73	.72	.71	.70 p	r

r	p .31	.32	.33	.34	.35	.36	.37	.38	.39	.40	
0	.3285	.3144	.3008	.2875	.2746	.2621	.2500	.2383	.2270	.2160	3
1	.4428	.4439	.4444	.4443	.4436	.4424	.4406	.4382	.4354	.4320	2
2	.1989	.2089	.2189	.2289	.2389	.2488	.2587	.2686	.2783	.2880	1
3	.0298	.0328	.0359	.0393	.0429	.0467	.0507	.0549	.0593	.0640	0
	.69	.68	.67	.66	.65	.64	.63	.62	.61	.60 p	r

r	p .41	.42	.43	.44	.45	.46	.47	.48	.49	.50	
0	.2054	.1951	.1852	.1756	.1664	.1575	.1489	.1406	.1327	.1250	3
1	.4282	.4239	.4191	.4140	.4084	.4024	.3961	.3894	.3823	.3750	2
2	.2975	.3069	.3162	.3252	.3341	.3428	.3512	.3594	.3674	.3750	1
3	.0689	.0741	.0795	.0852	.0911	.0973	.1038	.1106	.1176	.1250	0
	.59	.58	.57	.56	.55	.54	.53	.52	.51	.50 p	r

n = 4

r	p .01	.02	.03	.04	.05	.06	.07	.08	.09	.10	
0	.9606	.9224	.8853	.8493	.8145	.7807	.7481	.7164	.6857	.6561	4
1	.0388	.0753	.1095	.1416	.1715	.1993	.2252	.2492	.2713	.2916	3
2	.0006	.0023	.0051	.0088	.0135	.0191	.0254	.0325	.0402	.0486	2
3	.0000	.0000	.0001	.0002	.0005	.0008	.0013	.0019	.0027	.0036	1
4	.0000	.0000	.0000	.0000	.0000	.0000	.0000	.0000	.0001	.0001	0
	.99	.98	.97	.96	.95	.94	.93	.92	.91	.90 p	r

r	p .11	.12	.13	.14	.15	.16	.17	.18	.19	.20	
0	.6274	.5997	.5729	.5470	.5220	.4979	.4746	.4521	.4305	.4096	4
1	.3102	.3271	.3424	.3562	.3685	.3793	.3888	.3970	.4039	.4096	3
2	.0575	.0669	.0767	.0870	.0975	.1084	.1195	.1307	.1421	.1536	2
3	.0047	.0061	.0076	.0094	.0115	.0138	.0163	.0191	.0222	.0256	1
4	.0001	.0002	.0003	.0004	.0005	.0007	.0008	.0010	.0013	.0016	0
	.89	.88	.87	.86	.85	.84	.83	.82	.81	.80 p	r

Note: Read down entire column.

r	p .31	.32	.33	.34	.35	.36	.37	.38	.39	.40	
0	.1564	.1454	.1350	.1252	.1160	.1074	.0992	.0916	.0845	.0778	5
1	.3513	.3421	.3325	.3226	.3124	.3020	.2914	.2808	.2700	.2592	4
2	.3157	.3220	.3275	.3323	.3364	.3397	.3423	.3441	.3452	.3456	3
3	.1418	.1515	.1613	.1712	.1811	.1911	.2010	.2109	.2207	.2304	2
4	.0319	.0357	.0397	.0441	.0488	.0537	.0590	.0646	.0706	.0768	1
5	.0029	.0034	.0039	.0045	.0053	.0060	.0069	.0079	.0090	.0102	0
	.69	.68	.67	.66	.65	.64	.63	.62	.61	.60 p	r

r	p .41	.42	.43	.44	.45	.46	.47	.48	.49	.50	
0	.0715	.0656	.0602	.0551	.0503	.0459	.0418	.0380	.0345	.0313	5
1	.2484	.2376	.2270	.2164	.2059	.1956	.1854	.1755	.1657	.1562	4
2	.3452	.3442	.3424	.3400	.3369	.3332	.3289	.3240	.3185	.3125	3
3	.2399	.2492	.2583	.2671	.2757	.2838	.2916	.2990	.3060	.3125	2
4	.0834	.0902	.0974	.1049	.1128	.1209	.1293	.1380	.1470	.1562	1
5	.0116	.0131	.0147	.0165	.0185	.0206	.0229	.0255	.0282	.0312	0
	.59	.58	.57	.56	.55	.54	.53	.52	.51	.50 p	r

n = 6

r	p .01	.02	.03	.04	.05	.06	.07	.08	.09	.10	
0	.9415	.8858	.8330	.7828	.7351	.6899	.6470	.6064	.5679	.5314	6
1	.0571	.1085	.1546	.1957	.2321	.2642	.2922	.3164	.3370	.3543	5
2	.0014	.0055	.0120	.0204	.0305	.0422	.0550	.0688	.0833	.0984	4
3	.0000	.0002	.0005	.0011	.0021	.0036	.0055	.0080	.0110	.0146	3
4	.0000	.0000	.0000	.0000	.0001	.0002	.0003	.0005	.0008	.0012	2
5	.0000	.0000	.0000	.0000	.0000	.0000	.0000	.0000	.0000	.0001	1
	.99	.98	.97	.96	.95	.94	.93	.92	.91	.90 p	r

r	p .11	.12	.13	.14	.15	.16	.17	.18	.19	.20	
0	.4970	.4644	.4336	.4046	.3771	.3513	.3269	.3040	.2824	.2621	6
1	.3685	.3800	.3888	.3952	.3993	.4015	.4018	.4004	.3975	.3932	5
2	.1139	.1295	.1452	.1608	.1762	.1912	.2057	.2197	.2331	.2458	4
3	.0188	.0236	.0289	.0349	.0415	.0486	.0562	.0643	.0729	.0819	3
4	.0017	.0024	.0032	.0043	.0055	.0069	.0086	.0106	.0128	.0154	2
5	.0001	.0001	.0002	.0003	.0004	.0005	.0007	.0009	.0012	.0015	1
6	.0000	.0000	.0000	.0000	.0000	.0000	.0000	.0000	.0000	.0001	0
	.89	.88	.87	.86	.85	.84	.83	.82	.81	80 p	r

r	p .21	.22	.23	.24	.25	.26	.27	.28	.29	.30	
0	.2431	.2252	.2084	.1927	.1780	.1642	.1513	.1393	.1281	.1176	6
1	.3877	.3811	.3735	.3651	.3560	.3462	.3358	.3251	.3139	.3025	5
2	.2577	.2687	.2789	.2882	.2966	.3041	.3105	.3160	.3206	.3241	4
3	.0913	.1011	.1111	.1214	.1318	.1424	.1531	.1639	.1746	.1852	3
4	.0182	.0214	.0249	.0287	.0330	.0375	.0425	.0478	.0535	.0595	2
5	.0019	.0024	.0030	.0036	.0044	.0053	.0063	.0074	.0087	.0102	1
6	.0001	.0001	.0001	.0002	.0002	.0003	.0004	.0005	.0006	.0007	0
	.79	.78	.77	.76	.75	.74	.73	.72	.71	.70 p	r

r	p .31	.32	.33	.34	.35	.36	.37	.38	.39	.40	
0	.1079	.0989	.0905	.0827	.0754	.0687	.0625	.0568	.0515	.0467	6
1	.2909	.2792	.2673	.2555	.2437	.2319	.2203	.2089	.1976	.1866	5
2	.3267	.3284	.3292	.3290	.3280	.3261	.3235	.3201	.3159	.3110	4
3	.1957	.2061	.2162	.2260	.2355	.2446	.2533	.2616	.2693	.2765	3
4	.0660	.0727	.0799	.0873	.0951	.1032	.1116	.1202	.1291	.1382	2

Table 2 (continued)

r											
5	.0119	.0137	.0157	.0180	.0205	.0232	.0262	.0295	.0330	.0369	1
6	.0009	.0011	.0013	.0015	.0018	.0022	.0026	.0030	.0035	.0041	0
	.69	.68	.67	.66	.65	.64	.63	.62	.61	.60 p	r

r \ p	.41	.42	.43	.44	.45	.46	.47	.48	.49	.50	
0	.0422	.0381	.0343	.0308	.0277	.0248	.0222	.0198	.0176	.0156	6
1	.1759	.1654	.1552	.1454	.1359	.1267	.1179	.1095	.1014	.0937	5
2	.3055	.2994	.2928	.2856	.2780	.2699	.2615	.2527	.2436	.2344	4
3	.2831	.2891	.2945	.2992	.3032	.3065	.3091	.3110	.3121	.3125	3
4	.1475	.1570	.1666	.1763	.1861	.1958	.2056	.2153	.2249	.2344	2
5	.0410	.0455	.0503	.0554	.0609	.0667	.0729	.0795	.0864	.0937	1
6	.0048	.0055	.0063	.0073	.0083	.0095	.0108	.0122	.0138	.0156	0
	.59	.58	.57	.56	.55	.54	.53	.52	.51	.50 p	r

$n = 7$

r \ p	.01	.02	.03	.04	.05	.06	.07	.08	.09	.10	
0	.9321	.8681	.8080	.7514	.6983	.6485	.6017	.5578	.5168	.4783	7
1	.0659	.1240	.1749	.2192	.2573	.2897	.3170	.3396	.3578	.3720	6
2	.0020	.0076	.0162	.0274	.0406	.0555	.0716	.0886	.1061	.1240	5
3	.0000	.0003	.0008	.0019	.0036	.0059	.0090	.0128	.0175	.0230	4
4	.0000	.0000	.0000	.0001	.0002	.0004	.0007	.0011	.0017	.0026	3
5	.0000	.0000	.0000	.0000	.0000	.0000	.0000	.0001	.0001	.0002	2
	.99	.98	.97	.96	.95	.94	.93	.92	.91	.90 p	r

r \ p	.11	.12	.13	.14	.15	.16	.17	.18	.19	.20	
0	.4423	.4087	.3773	.3479	.3206	.2951	.2714	.2493	.2288	.2097	7
1	.3827	.3901	.3946	.3965	.3960	.3935	.3891	.3830	.3756	.3670	6
2	.1419	.1596	.1769	.1936	.2097	.2248	.2391	.2523	.2643	.2753	5
3	.0292	.0363	.0441	.0525	.0617	.0714	.0816	.0923	.1033	.1147	4
4	.0036	.0049	.0066	.0086	.0109	.0136	.0167	.0203	.0242	.0287	3
5	.0003	.0004	.0006	.0008	.0012	.0016	.0021	.0027	.0034	.0043	2
6	.0000	.0000	.0000	.0000	.0001	.0001	.0001	.0002	.0003	.0004	1
	.89	.88	.87	.86	.85	.84	.83	.82	.81	.80 p	r

r \ p	.21	.22	.23	.24	.25	.26	.27	.28	.29	.30	
0	.1920	.1757	.1605	.1465	.1335	.1215	.1105	.1003	.0910	.0824	7
1	.3573	.3468	.3356	.3237	.3115	.2989	.2860	.2731	.2600	.2471	6
2	.2850	.2935	.3007	.3067	.3115	.3150	.3174	.3186	.3186	.3177	5
3	.1263	.1379	.1497	.1614	.1730	.1845	.1956	.2065	.2169	.2269	4
4	.0336	.0389	.0447	.0510	.0577	.0648	.0724	.0803	.0886	.0972	3
5	.0054	.0066	.0080	.0097	.0115	.0137	.0161	.0187	.0217	.0250	2
6	.0005	.0006	.0008	.0010	.0013	.0016	.0020	.0024	.0030	.0036	1
7	.0000	.0000	.0000	.0000	.0001	.0001	.0001	.0001	.0002	.0002	0
	.79	.78	.77	.76	.75	.74	.73	.72	.71	.70 p	r

r \ p	.31	.32	.33	.34	.35	.36	.37	.38	.39	.40	
0	.0745	.0672	.0606	.0546	.0490	.0440	.0394	.0352	.0314	.0280	7
1	.2342	.2215	.2090	.1967	.1848	.1732	.1619	.1511	.1407	.1306	6
2	.3156	.3127	.3088	.3040	.2985	.2922	.2853	.2778	.2698	.2613	5
3	.2363	.2452	.2535	.2610	.2679	.2740	.2793	.2838	.2875	.2903	4
4	.1062	.1154	.1248	.1345	.1442	.1541	.1640	.1739	.1838	.1935	3

r											
5	.0527	.0591	.0659	.0732	.0808	.0888	.0971	.1058	.1147	.1239	3
6	.0118	.0139	.0162	.0188	.0217	.0250	.0285	.0324	.0367	.0413	2
7	.0015	.0019	.0023	.0028	.0033	.0040	.0048	.0057	.0067	.0079	1
8	.0001	.0001	.0001	.0002	.0002	.0003	.0004	.0004	.0005	.0007	0
	.69	.68	.67	.66	.65	.64	.63	.62	.61	.60 p	r

r \ p	.41	.42	.43	.44	.45	.46	.47	.48	.49	.50	
0	.0147	.0128	.0111	.0097	.0084	.0072	.0062	.0053	.0046	.0039	8
1	.0816	.0742	.0672	.0608	.0548	.0493	.0442	.0395	.0352	.0312	7
2	.1985	.1880	.1776	.1672	.1569	.1469	.1371	.1275	.1183	.1094	6
3	.2759	.2723	.2679	.2627	.2568	.2503	.2431	.2355	.2273	.2187	5
4	.2397	.2465	.2526	.2580	.2627	.2665	.2695	.2717	.2730	.2734	4
5	.1332	.1428	.1525	.1622	.1719	.1816	.1912	.2006	.2098	.2187	3
6	.0463	.0517	.0575	.0637	.0703	.0774	.0848	.0926	.1008	.1094	2
7	.0092	.0107	.0124	.0143	.0164	.0188	.0215	.0244	.0277	.0312	1
8	.0008	.0010	.0012	.0014	.0017	.0020	.0024	.0028	.0033	.0039	0
	.59	.58	.57	.56	.55	.54	.53	.52	.51	.50 p	r

$n = 9$

r \ p	.01	.02	.03	.04	.05	.06	.07	.08	.09	.10	
0	.9135	.8337	.7602	.6925	.6302	.5730	.5204	.4722	.4279	.3874	9
1	.0830	.1531	.2116	.2597	.2985	.3292	.3525	.3695	.3809	.3874	8
2	.0034	.0125	.0262	.0433	.0629	.0840	.1061	.1285	.1507	.1722	7
3	.0001	.0006	.0019	.0042	.0077	.0125	.0186	.0261	.0348	.0446	6
4	.0000	.0000	.0001	.0003	.0006	.0012	.0021	.0034	.0052	.0074	5
5	.0000	.0000	.0000	.0000	.0000	.0001	.0002	.0003	.0005	.0008	4
6	.0000	.0000	.0000	.0000	.0000	.0000	.0000	.0000	.0000	.0001	3
	.99	.98	.97	.96	.95	.94	.93	.92	.91	.90 p	r

r \ p	.11	.12	.13	.14	.15	.16	.17	.18	.19	.20	
0	.3504	.3165	.2885	.2573	.2316	.2082	.1869	.1676	.1501	.1342	9
1	.3897	.3884	.3840	.3770	.3679	.3569	.3446	.3312	.3169	.3020	8
2	.1927	.2119	.2295	.2455	.2597	.2720	.2823	.2908	.2973	.3020	7
3	.0556	.0674	.0800	.0933	.1069	.1209	.1349	.1489	.1627	.1762	6
4	.0103	.0138	.0179	.0228	.0283	.0345	.0415	.0490	.0573	.0661	5
5	.0013	.0019	.0027	.0037	.0050	.0066	.0085	.0108	.0134	.0165	4
6	.0001	.0002	.0003	.0004	.0006	.0008	.0012	.0016	.0021	.0028	3
7	.0000	.0000	.0000	.0000	.0000	.0001	.0001	.0001	.0002	.0003	2
	.89	.88	.87	.86	.85	.84	.83	.82	.81	.80 p	r

r \ p	.21	.22	.23	.24	.25	.26	.27	.28	.29	.30	
0	.1199	.1069	.0952	.0846	.0751	.0665	.0589	.0520	.0458	.0404	9
1	.2867	.2713	.2558	.2404	.2253	.2104	.1960	.1820	.1685	.1556	8
2	.3049	.3061	.3056	.3037	.3003	.2957	.2899	.2831	.2754	.2668	7
3	.1891	.2014	.2130	.2238	.2336	.2424	.2502	.2569	.2624	.2668	6
4	.0754	.0852	.0954	.1060	.1168	.1278	.1388	.1499	.1608	.1715	5
5	.0200	.0240	.0285	.0335	.0389	.0449	.0513	.0583	.0657	.0735	4
6	.0036	.0045	.0057	.0070	.0087	.0105	.0127	.0151	.0179	.0210	3
7	.0004	.0005	.0007	.0010	.0012	.0016	.0020	.0025	.0031	.0039	2
8	.0000	.0000	.0001	.0001	.0001	.0001	.0002	.0002	.0003	.0004	1
	.79	.78	.77	.76	.75	.74	.73	.72	.71	.70 p	r

r	p	.31	.32	.33	.34	.35	.36	.37	.38	.39	.40		
5		.0286	.0326	.0369	.0416	.0466	.0520	.0578	.0640	.0705	.0774		2
6		.0043	.0051	.0061	.0071	.0084	.0098	.0113	.0131	.0150	.0172		1
7		.0003	.0003	.0004	.0005	.0006	.0008	.0009	.0011	.0014	.0016		0
		.69	.68	.67	.66	.65	.64	.63	.62	.61	.60	p	r

r	p	.41	.42	.43	.44	.45	.46	.47	.48	.49	.50		
0		.0249	.0221	.0195	.0173	.0152	.0134	.0117	.0103	.0090	.0078		7
1		.1211	.1119	.1032	.0950	.0872	.0798	.0729	.0664	.0604	.0547		6
2		.2524	.2431	.2336	.2239	.2140	.2040	.1940	.1840	.1760	.1641		5
3		.2923	.2934	.2937	.2932	.2918	.2897	.2867	.2830	.2786	.2734		4
4		.2031	.2125	.2216	.2304	.2388	.2468	.2543	.2612	.2676	.2734		3
5		.0847	.0923	.1003	.1086	.1172	.1261	.1353	.1447	.1543	.1641		2
6		.0196	.0223	.0252	.0284	.0320	.0358	.0400	.0445	.0494	.0547		1
7		.0019	.0023	.0027	.0032	.0037	.0044	.0051	.0059	.0068	.0078		0
		.59	.58	.57	.56	.55	.54	.53	.52	.51	.50	p	r

$n = 8$

r	p	.01	.02	.03	.04	.05	.06	.07	.08	.09	.10		
0		.9227	.8508	.7837	.7214	.6634	.6096	.5596	.5132	.4703	.4305		8
1		.0746	.1389	.1939	.2405	.2793	.3113	.3370	.3570	.3721	.3826		7
2		.0026	.0099	.0210	.0351	.0515	.0695	.0888	.1087	.1288	.1488		6
3		.0001	.0004	.0013	.0029	.0054	.0089	.0134	.0189	.0255	.0331		5
4		.0000	.0000	.0001	.0002	.0004	.0007	.0013	.0021	.0031	.0046		4
5		.0000	.0000	.0000	.0000	.0000	.0000	.0001	.0001	.0002	.0004		3
		.99	.98	.97	.96	.95	.94	.93	.92	.91	.90	p	r

r	p	.11	.12	.13	.14	.15	.16	.17	.18	.19	.20		
0		.3937	.3596	.3282	.2992	.2725	.2479	.2252	.2044	.1853	.1678		8
1		.3892	.3923	.3923	.3897	.3847	.3777	.3691	.3590	.3477	.3355		7
2		.1684	.1872	.2052	.2220	.2376	.2518	.2646	.2758	.2855	.2936		6
3		.0416	.0511	.0613	.0723	.0839	.0959	.1084	.1211	.1339	.1468		5
4		.0064	.0087	.0115	.0147	.0185	.0228	.0277	.0332	.0393	.0459		4
5		.0006	.0009	.0014	.0019	.0026	.0035	.0045	.0058	.0074	.0092		3
6		.0000	.0001	.0001	.0002	.0002	.0003	.0005	.0006	.0009	.0011		2
7		.0000	.0000	.0000	.0000	.0000	.0000	.0000	.0000	.0001	.0001		1
		.89	.88	.87	.86	.85	.84	.83	.82	.81	.80	p	r

r	p	.21	.22	.23	.24	.25	.26	.27	.28	.29	.30		
0		.1517	.1370	.1236	.1113	.1001	.0899	.0806	.0722	.0646	.0576		8
1		.3226	.3092	.2953	.2812	.2670	.2527	.2386	.2247	.2110	.1977		7
2		.3002	.3052	.3087	.3108	.3115	.3108	.3089	.3058	.3017	.2965		6
3		.1596	.1722	.1844	.1963	.2076	.2184	.2285	.2379	.2464	.2541		5
4		.0530	.0607	.0689	.0775	.0865	.0959	.1056	.1156	.1258	.1361		4
5		.0113	.0137	.0165	.0196	.0231	.0270	.0313	.0360	.0411	.0467		3
6		.0015	.0019	.0025	.0031	.0038	.0047	.0058	.0070	.0084	.0100		2
7		.0001	.0002	.0002	.0003	.0004	.0005	.0006	.0008	.0010	.0012		1
8		.0000	.0000	.0000	.0000	.0000	.0000	.0000	.0000	.0001	.0001		0
		.79	.78	.77	.76	.75	.74	.73	.72	.71	.70	p	r

r	p	.31	.32	.33	.34	.35	.36	.37	.38	.39	.40		
0		.0514	.0457	.0406	.0360	.0319	.0281	.0248	.0218	.0192	.0168		8
1		.1847	.1721	.1600	.1484	.1373	.1267	.1166	.1071	.0981	.0896		7
2		.2904	.2835	.2758	.2675	.2587	.2494	.2397	.2297	.2194	.2090		6
3		.2609	.2668	.2717	.2756	.2786	.2805	.2815	.2815	.2806	.2787		5
4		.1465	.1569	.1673	.1775	.1875	.1973	.2067	.2157	.2242	.2322		4

Note: Read down entire column.

r	p	.31	.32	.33	.34	.35	.36	.37	.38	.39	.40		
0		.0355	.0311	.0272	.0238	.0207	.0180	.0156	.0135	.0117	.0101		9
1		.1433	.1317	.1206	.1102	.1004	.0912	.0826	.0747	.0673	.0605		8
2		.2576	.2478	.2376	.2270	.2162	.2052	.1941	.1831	.1721	.1612		7
3		.2701	.2721	.2731	.2729	.2716	.2693	.2660	.2618	.2567	.2508		6
4		.1820	.1921	.2017	.2109	.2194	.2272	.2344	.2407	.2462	.2508		5
5		.0818	.0904	.0994	.1086	.1181	.1278	.1376	.1475	.1574	.1672		4
6		.0245	.0284	.0326	.0373	.0424	.0479	.0539	.0603	.0671	.0743		3
7		.0047	.0057	.0069	.0082	.0098	.0116	.0136	.0158	.0184	.0212		2
8		.0005	.0007	.0008	.0011	.0013	.0016	.0020	.0024	.0029	.0035		1
9		.0000	.0000	.0000	.0001	.0001	.0001	.0001	.0002	.0002	.0003		0
		.69	.68	.67	.66	.65	.64	.63	.62	.61	.60	p	r

r	p	.41	.42	.43	.44	.45	.46	.47	.48	.49	.50		
0		.0087	.0074	.0064	.0054	.0046	.0039	.0033	.0028	.0023	.0020		9
1		.0542	.0484	.0431	.0383	.0339	.0299	.0263	.0231	.0202	.0176		8
2		.1506	.1402	.1301	.1204	.1110	.1020	.0934	.0853	.0776	.0703		7
3		.2442	.2369	.2291	.2207	.2119	.2027	.1933	.1837	.1739	.1641		6
4		.2545	.2573	.2592	.2601	.2600	.2590	.2571	.2543	.2506	.2461		5
5		.1769	.1863	.1955	.2044	.2128	.2207	.2280	.2347	.2408	.2461		4
6		.0819	.0900	.0983	.1070	.1160	.1253	.1348	.1445	.1542	.1641		3
7		.0244	.0279	.0318	.0360	.0407	.0458	.0512	.0571	.0635	.0703		2
8		.0042	.0051	.0060	.0071	.0083	.0097	.0114	.0132	.0153	.0176		1
9		.0003	.0004	.0005	.0006	.0008	.0009	.0011	.0014	.0016	.0020		0
		.59	.58	.57	.56	.55	.54	.53	.52	.51	.50	p	r

$n = 10$

r	p	.01	.02	.03	.04	.05	.06	.07	.08	.09	.10		
0		.9044	.8171	.7374	.6648	.5987	.5386	.4840	.4344	.3894	.3487		10
1		.0914	.1667	.2281	.2770	.3151	.3438	.3643	.3777	.3851	.3874		9
2		.0042	.0153	.0317	.0519	.0746	.0988	.1234	.1478	.1714	.1937		8
3		.0001	.0008	.0026	.0058	.0105	.0168	.0248	.0343	.0452	.0574		7
4		.0000	.0000	.0001	.0004	.0010	.0019	.0033	.0052	.0078	.0112		6
5		.0000	.0000	.0000	.0000	.0001	.0001	.0003	.0005	.0009	.0015		5
6		.0000	.0000	.0000	.0000	.0000	.0000	.0000	.0000	.0001	.0001		4
		.99	.98	.97	.96	.95	.94	.93	.92	.91	.90	p	r

r	p	.11	.12	.13	.14	.15	.16	.17	.18	.19	.20		
0		.3118	.2785	.2484	.2213	.1969	.1749	.1552	.1374	.1216	.1074		10
1		.3854	.3798	.3712	.3603	.3474	.3331	.3178	.3017	.2852	.2684		9
2		.2143	.2330	.2496	.2639	.2759	.2856	.2929	.2980	.3010	.3020		8
3		.0706	.0847	.0995	.1146	.1298	.1450	.1600	.1745	.1883	.2013		7
4		.0153	.0202	.0260	.0326	.0401	.0483	.0573	.0670	.0773	.0881		6
5		.0023	.0033	.0047	.0064	.0085	.0011	.0141	.0177	.0218	.0264		5
6		.0002	.0004	.0006	.0009	.0012	.0018	.0024	.0032	.0043	.0055		4
7		.0000	.0000	.0000	.0001	.0001	.0002	.0003	.0004	.0006	.0008		3
8		.0000	.0000	.0000	.0000	.0000	.0000	.0000	.0000	.0001	.0001		2
		.89	.88	.87	.86	.85	.84	.83	.82	.81	.80	p	r

r	p	.21	.22	.23	.24	.25	.26	.27	.28	.29	.30		
0		.0947	.0834	.0733	.0643	.0563	.0492	.0430	.0374	.0326	.0282		10
1		.2517	.2351	.2188	.2030	.1877	.1730	.1590	.1456	.1330	.1211		9
2		.3011	.2984	.2942	.2885	.2816	.2735	.2646	.2548	.2444	.2335		8
3		.2134	.2244	.2343	.2429	.2503	.2563	.2609	.2642	.2662	.2668		7
4		.0993	.1108	.1225	.1343	.1460	.1576	.1689	.1798	.1903	.2001		6

Table 2 (continued)

r											
5	.0317	.0375	.0439	.0509	.0584	.0664	.0750	.0839	.0933	.1029	5
6	.0070	.0088	.0109	.0134	.0162	.0195	.0231	.0272	.0317	.0368	4
7	.0011	.0014	.0019	.0024	.0031	.0039	.0049	.0060	.0074	.0090	3
8	.0001	.0002	.0002	.0003	.0004	.0005	.0007	.0009	.0011	.0014	2
9	.0000	.0000	.0000	.0000	.0000	.0000	.0001	.0001	.0001	.0001	1
	.79	.78	.77	.76	.75	.74	.73	.72	.71	.70 p	r

r p	.31	.32	.33	.34	.35	.36	.37	.38	.39	.40	
0	.0245	.0211	.0182	.0157	.0135	.0115	.0098	.0084	.0071	.0060	10
1	.1099	.0995	.0898	.0808	.0725	.0649	.0578	.0514	.0456	.0403	9
2	.2222	.2107	.1990	.1873	.1757	.1642	.1529	.1419	.1312	.1209	8
3	.2662	.2644	.2614	.2573	.2522	.2462	.2394	.2319	.2237	.2150	7
4	.2093	.2177	.2253	.2320	.2377	.2424	.2461	.2487	.2503	.2508	6
5	.1128	.1229	.1332	.1434	.1536	.1636	.1734	.1829	.1920	.2007	5
6	.0422	.0482	.0547	.0616	.0689	.0767	.0849	.0934	.1023	.1115	4
7	.0108	.0130	.0154	.0181	.0212	.0247	.0285	.0327	.0374	.0425	3
8	.0018	.0023	.0028	.0035	.0043	.0052	.0063	.0075	.0090	.0106	2
9	.0002	.0002	.0003	.0004	.0005	.0006	.0008	.0010	.0013	.0016	1
10	.0000	.0000	.0000	.0000	.0000	.0000	.0000	.0001	.0001	.0001	0
	.69	.68	.67	.66	.65	.64	.63	.62	.61	.60 p	r

r p	.41	.42	.43	.44	.45	.46	.47	.48	.49	.50	
0	.0051	.0043	.0036	.0030	.0025	.0021	.0017	.0014	.0012	.0010	10
1	.0355	.0312	.0273	.0238	.0207	.0180	.0155	.0133	.0114	.0098	9
2	.1111	.1017	.0927	.0843	.0763	.0688	.0619	.0554	.0494	.0439	8
3	.2058	.1963	.1865	.1765	.1665	.1564	.1464	.1364	.1267	.1172	7
4	.2503	.2488	.2462	.2427	.2384	.2331	.2271	.2204	.2130	.2051	6
5	.2087	.2162	.2229	.2289	.2340	.2383	.2417	.2441	.2456	.2461	5
6	.1209	.1304	.1401	.1499	.1596	.1692	.1786	.1878	.1966	.2051	4
7	.0480	.0540	.0604	.0673	.0746	.0824	.0905	.0991	.1080	.1172	3
8	.0125	.0147	.0171	.0198	.0229	.0263	.0301	.0343	.0389	.0439	2
9	.0019	.0024	.0029	.0035	.0042	.0050	.0059	.0070	.0083	.0098	1
10	.0001	.0002	.0002	.0003	.0003	.0004	.0005	.0006	.0008	.0010	0
	.59	.58	.57	.56	.55	.54	.53	.52	.51	.50 p	r

$n = 11$

r p	.01	.02	.03	.04	.05	.06	.07	.08	.09	.10	
0	.8953	.8007	.7153	.6382	.5688	.5063	.4501	.3996	.3544	.3138	11
1	.0995	.1798	.2433	.2925	.3293	.3555	.3727	.3823	.3855	.3835	10
2	.0050	.0183	.0376	.0609	.0867	.1135	.1403	.1662	.1906	.2131	9
3	.0002	.0011	.0035	.0076	.0137	.0217	.0317	.0434	.0566	.0710	8
4	.0000	.0000	.0002	.0006	.0014	.0028	.0048	.0075	.0112	.0158	7
5	.0000	.0000	.0000	.0000	.0001	.0002	.0005	.0009	.0015	.0025	6
6	.0000	.0000	.0000	.0000	.0000	.0000	.0000	.0001	.0002	.0003	5
	.99	.98	.97	.96	.95	.94	.93	.92	.91	.90 p	r

r p	.11	.12	.13	.14	.15	.16	.17	.18	.19	.20	
0	.2775	.2451	.2161	.1903	.1673	.1469	.1288	.1127	.0985	.0859	11
1	.3773	.3676	.3552	.3408	.3248	.3078	.2901	.2721	.2541	.2362	10
2	.2332	.2507	.2654	.2774	.2866	.2932	.2971	.2987	.2980	.2953	9
3	.0865	.1025	.1190	.1355	.1517	.1675	.1826	.1967	.2097	.2215	8
4	.0214	.0280	.0356	.0441	.0536	.0638	.0748	.0864	.0984	.1107	7

$n = 12$

r p	.01	.02	.03	.04	.05	.06	.07	.08	.09	.10	
0	.8864	.7847	.6938	.6127	.5404	.4759	.4186	.3677	.3225	.2824	12
1	.1074	.1922	.2575	.3064	.3413	.3645	.3781	.3837	.3827	.3766	11
2	.0060	.0216	.0438	.0702	.0988	.1280	.1565	.1835	.2082	.2301	10
3	.0002	.0015	.0045	.0098	.0173	.0272	.0393	.0532	.0686	.0852	9
4	.0000	.0001	.0003	.0009	.0021	.0039	.0067	.0104	.0153	.0213	8
5	.0000	.0000	.0000	.0001	.0002	.0004	.0008	.0014	.0024	.0038	7
6	.0000	.0000	.0000	.0000	.0000	.0000	.0001	.0001	.0003	.0005	6
	.99	.98	.97	.96	.95	.94	.93	.92	.91	.90 p	r

r p	.11	.12	.13	.14	.15	.16	.17	.18	.19	.20	
0	.2470	.2157	.1880	.1637	.1422	.1234	.1069	.0924	.0798	.0687	12
1	.3663	.3529	.3372	.3197	.3012	.2821	.2627	.2434	.2245	.2062	11
2	.2490	.2647	.2771	.2863	.2924	.2955	.2960	.2939	.2897	.2835	10
3	.1026	.1203	.1380	.1553	.1720	.1876	.2021	.2151	.2265	.2362	9
4	.0285	.0369	.0464	.0569	.0683	.0804	.0931	.1062	.1195	.1329	8
5	.0056	.0081	.0111	.0148	.0193	.0245	.0305	.0373	.0449	.0532	7
6	.0008	.0013	.0019	.0028	.0040	.0054	.0073	.0096	.0123	.0155	6
7	.0001	.0001	.0002	.0004	.0006	.0009	.0013	.0018	.0025	.0033	5
8	.0000	.0000	.0000	.0000	.0001	.0001	.0002	.0002	.0004	.0005	4
9	.0000	.0000	.0000	.0000	.0000	.0000	.0000	.0000	.0000	.0001	3
	.89	.88	.87	.86	.85	.84	.83	.82	.81	.80 p	r

r p	.21	.22	.23	.24	.25	.26	.27	.28	.29	.30	
0	.0591	.0507	.0434	.0371	.0317	.0270	.0229	.0194	.0164	.0138	12
1	.1885	.1717	.1557	.1407	.1267	.1137	.1016	.0906	.0804	.0712	11
2	.2756	.2663	.2558	.2444	.2323	.2197	.2068	.1937	.1807	.1678	10
3	.2442	.2503	.2547	.2573	.2581	.2573	.2549	.2511	.2460	.2397	9
4	.1460	.1589	.1712	.1828	.1936	.2034	.2122	.2197	.2261	.2311	8
5	.0621	.0717	.0818	.0924	.1032	.1143	.1255	.1367	.1477	.1585	7
6	.0193	.0236	.0285	.0340	.0401	.0469	.0542	.0620	.0704	.0792	6
7	.0044	.0057	.0073	.0092	.0115	.0141	.0172	.0207	.0246	.0291	5
8	.0007	.0010	.0014	.0018	.0024	.0031	.0040	.0050	.0063	.0078	4
9	.0001	.0001	.0002	.0003	.0004	.0005	.0007	.0009	.0011	.0015	3
10	.0000	.00000	.0000	.0000	.0000	.0001	.0001	.0001	.0001	.0002	2
	.79	.78	.77	.76	.75	.74	.73	.72	.71	.70 p	r

r p	.31	.32	.33	.34	.35	.36	.37	.38	.39	.40	
0	.0116	.0098	.0082	.0068	.0057	.0047	.0039	.0032	.0027	.0022	12
1	.0628	.0552	.0484	.0422	.0368	.0319	.0276	.0237	.0204	.0174	11
2	.1552	.1429	.1310	.1197	.1088	.0986	.0890	.0800	.0716	.0639	10
3	.2324	.2241	.2151	.2055	.1954	.1849	.1742	.1634	.1526	.1419	9
4	.2349	.2373	.2384	.2382	.2367	.2340	.2302	.2254	.2195	.2128	8
5	.1688	.1787	.1879	.1963	.2039	.2106	.2163	.2210	.2246	.2270	7
6	.0885	.0981	.1079	.1180	.1281	.1382	.1482	.1580	.1675	.1766	6
7	.0341	.0396	.0456	.0521	.0591	.0666	.0746	.0830	.0918	.1009	5
8	.0096	.0116	.0140	.0168	.0199	.0234	.0274	.0318	.0367	.0420	4
9	.0019	.0024	.0031	.0038	.0048	.0059	.0071	.0087	.0104	.0125	3
10	.0003	.0003	.0005	.0006	.0008	.0010	.0013	.0016	.0020	.0025	2
11	.0000	.0000	.0000	.0001	.0001	.0001	.0001	.0002	.0002	.0003	1
	.69	.68	.67	.66	.65	.64	.63	.62	.61	.60 p	r

r											r
5	.0037	.0053	.0074	.0101	.0132	.0170	.0214	.0265	.0323	.0388	6
6	.0005	.0007	.0011	.0016	.0023	.0032	.0044	.0058	.0076	.0097	5
7	.0000	.0001	.0001	.0002	.0003	.0004	.0006	.0009	.0013	.0017	4
8	.0000	.0000	.0000	.0000	.0000	.0000	.0001	.0001	.0001	.0002	3
	.89	.88	.87	.86	.85	.84	.83	.82	.81	.80 p	r

r p	.21	.22	.23	.24	.25	.26	.27	.28	.29	.30	
0	.0748	.0650	.0564	.0489	.0422	.0364	.0314	.0270	.0231	.0198	11
1	.2187	.2017	.1854	.1697	.1549	.1408	.1276	.1153	.1038	.0932	10
2	.2907	.2845	.2768	.2680	.2581	.2474	.2360	.2242	.2121	.1998	9
3	.2318	.2407	.2481	.2539	.2581	.2608	.2619	.2616	.2599	.2568	8
4	.1232	.1358	.1482	.1603	.1721	.1832	.1937	.2035	.2123	.2201	7
5	.0459	.0536	.0620	.0709	.0803	.0901	.1003	.1108	.1214	.1321	6
6	.0122	.0151	.0185	.0224	.0268	.0317	.0371	.0431	.0496	.0566	5
7	.0023	.0030	.0039	.0050	.0064	.0079	.0098	.0120	.0145	.0173	4
8	.0003	.0004	.0006	.0008	.0011	.0014	.0018	.0023	.0030	.0037	3
9	.0000	.0000	.0001	.0001	.0001	.0002	.0002	.0003	.0004	.0005	2
	.79	.78	.77	.76	.75	.74	.73	.72	.71	.70 p	r

r p	.31	.32	.33	.34	.35	.36	.37	.38	.39	.40	
0	.0169	.0144	.0122	.0104	.0088	.0074	.0062	.0052	.0044	.0036	11
1	.0834	.0744	.0662	.0587	.0518	.0457	.0401	.0351	.0306	.0266	10
2	.1874	.1751	.1630	.1511	.1395	.1284	.1177	.1075	.0978	.0887	9
3	.2526	.2472	.2408	.2335	.2254	.2167	.2074	.1977	.1876	.1774	8
4	.2269	.2326	.2372	.2406	.2428	.2438	.2436	.2423	.2399	.2365	7
5	.1427	.1533	.1636	.1735	.1830	.1920	.2003	.2079	.2148	.2207	6
6	.0641	.0721	.0806	.0894	.0985	.1080	.1176	.1274	.1373	.1471	5
7	.0206	.0242	.0283	.0329	.0379	.0434	.0494	.0558	.0627	.0701	4
8	.0046	.0057	.0070	.0085	.0102	.0122	.0145	.0171	.0200	.0234	3
9	.0007	.0009	.0011	.0015	.0018	.0023	.0028	.0035	.0043	.0052	2
10	.0001	.0001	.0001	.0001	.0002	.0003	.0003	.0004	.0005	.0007	1
	.69	.68	.67	.66	.65	.64	.63	.62	.61	.60 p	r

r p	.41	.42	.43	.44	.45	.46	.47	.48	.49	.50	
0	.0030	.0025	.0021	.0017	.0014	.0011	.0009	.0008	.0006	.0005	11
1	.0231	.0199	.0171	.0147	.0125	.0107	.0090	.0076	.0065	.0064	10
2	.0801	.0721	.0646	.0577	.0513	.0454	.0401	.0352	.0308	.0269	9
3	.1670	.1566	.1462	.1359	.1259	.1161	.1067	.0976	.0888	.0806	8
4	.2321	.2267	.2206	.2136	.2060	.1978	.1892	.1801	.1707	.1611	7
5	.2258	.2299	.2329	.2350	.2360	.2359	.2348	.2327	.2296	.2256	6
6	.1569	.1664	.1757	.1846	.1931	.2010	.2083	.2148	.2206	.2256	5
7	.0779	.0861	.0947	.1036	.1128	.1223	.1319	.1416	.1514	.1611	4
8	.0271	.0312	.0357	.0407	.0462	.0521	.0585	.0654	.0727	.0806	3
9	.0063	.0075	.0090	.0107	.0126	.0148	.0173	.0201	.0233	.0269	2
10	.0009	.0011	.0014	.0017	.0021	.0025	.0031	.0037	.0045	.0054	1
11	.0001	.0001	.0001	.0001	.0002	.0002	.0002	.0003	.0004	.0005	0
	.59	.58	.57	.56	.55	.54	.53	.52	.51	.50 p	r

r p	.41	.42	.43	.44	.45	.46	.47	.48	.49	.50	
0	.0018	.0014	.0012	.0010	.0008	.0006	.0005	.0004	.0003	.0002	12
1	.0148	.0126	.0106	.0090	.0075	.0063	.0052	.0043	.0036	.0029	11
2	.0567	.0502	.0442	.0388	.0339	.0294	.0255	.0220	.0189	.0161	10
3	.1314	.1211	.1111	.1015	.0923	.0836	.0754	.0676	.0604	.0537	9
4	.2054	.1973	.1886	.1794	.1700	.1602	.1504	.1405	.1306	.1208	8
5	.2284	.2285	.2276	.2256	2225	.2184	.2134	.2075	.2008	.1934	7
6	.1851	.1931	.2003	.2068	.2124	.2171	.2208	.2234	.2250	.2256	6
7	.1103	.1198	.1295	.1393	.1489	.1585	.1678	.1768	.1853	.1934	5
8	.0479	.0542	.0611	.0684	.0762	.0844	.0930	.1020	.1113	.1208	4
9	.0148	.0175	.0205	.0239	.0277	.0319	.0367	.0418	.0475	.0537	3
10	.0031	.0038	.0046	.0056	.0068	.0082	.0098	.0116	.0137	.0161	2
11	.0004	.0005	.0006	.0008	.0010	.0013	.0016	.0019	.0024	.0029	1
12	.0000	.0000	.0000	.0001	.0001	.0001	.0001	.0001	.0002	.0002	0
	.59	.58	.57	.56	.55	.54	.53	.52	.51	.50 p	r

n = 13

r p	.01	.02	.03	.04	.05	.06	.07	.08	.09	.10	
0	.8775	.7690	.6730	.5882	.5133	.4474	.3893	.3383	.2935	.2542	13
1	.1152	.2040	.2706	.3186	.3512	.3712	.3809	.3824	.3773	.3672	12
2	.0070	.0250	.0502	.0797	.1109	.1422	.1720	.1995	.2239	.2448	11
3	.0003	.0019	.0057	.0122	.0214	.0333	.0475	.0636	.0812	.0997	10
4	.0000	.0001	.0004	.0013	.0028	.0053	.0089	.0138	.0201	.0277	9
5	.0000	.0000	.0000	.0001	.0003	.0006	.0012	.0022	.0036	.0055	8
6	.0000	.0000	.0000	.0000	.0000	.0001	.0001	.0003	.0005	.0008	7
7	.0000	.0000	.0000	.0000	.0000	.0000	.0000	.0000	.0000	.0001	6
	.99	.98	.97	.96	.95	.94	.93	.92	.91	.90 p	r

r p	.11	.12	.13	.14	.15	.16	.17	.18	.19	.20	
0	.2198	.1898	.1636	.1408	.1209	.1037	.0887	.0758	.0646	.0550	13
1	.3532	.3364	.3178	.2979	.2774	.2567	.2362	.2163	.1970	.1787	12
2	.2619	.2753	.2849	.2910	.2937	.2934	.2903	.2848	.2773	.2680	11
3	.1187	.1376	.1561	.1737	.1900	.2049	.2180	.2293	.2385	.2457	10
4	.0367	.0469	.0583	.0707	.0838	.0976	.1116	.1258	.1399	.1535	9
5	.0082	.0115	.0157	.0207	.0266	.0335	.0412	.0497	.0591	.0691	8
6	.0013	.0021	.0031	.0045	.0063	.0085	.0112	.0145	.0185	.0230	7
7	.0002	.0003	.0005	.0007	.0011	.0016	.0023	.0032	.0043	.0058	6
8	.0000	.0000	.0001	.0001	.0001	.0002	.0004	.0005	.0008	.0011	5
9	.0000	.0000	.0000	.0000	.0000	.0000	.0000	.0001	.0001	.0001	4
	.89	.88	.87	.86	.85	.84	.83	.82	.81	.80 p	r

r p	.21	.22	.23	.24	.25	.26	.27	.28	.29	.30	
0	.0467	.0396	.0334	.0282	.0238	.0200	.0167	.0140	.0117	.0097	13
1	.1613	.1450	.1299	.1159	.1029	.0911	.0804	.0706	.0619	.0540	12
2	.2573	.2455	.2328	.2195	.2059	.1921	.1784	.1648	.1516	.1388	11
3	.2508	.2539	.2550	.2542	.2517	.2475	.2419	.2351	.2271	.2181	10
4	.1667	.1790	.1904	.2007	.2097	.2174	.2237	.2285	.2319	.2337	9
5	.0797	.0909	.1024	.1141	.1258	.1375	.1489	.1600	.1705	.1803	8
6	.0283	.0342	.0408	.0480	.0559	.0644	.0734	.0829	.0928	.1030	7
7	.0075	.0096	.0122	.0152	.0186	.0226	.0272	.0323	.0379	.0442	6
8	.0015	.0020	.0027	.0036	.0047	.0060	.0075	.0094	.0116	.0142	5
9	.0002	.0003	.0005	.0006	.0009	.0012	.0015	.0020	.0026	.0034	4

Note: Read down entire column.

Table 2 (continued)

10	.0000	.0000	.0001	.0001	.0001	.0002	.0002	.0003	.0004	.0006	3
11	.0000	.0000	.0000	.0000	.0000	.0000	.0000	.0000	.0000	.0001	2
	.79	.78	.77	.76	.75	.74	.73	.72	.71	.70	*p* / *r*

r / *p*	.31	.32	.33	.34	.35	.36	.37	.38	.39	.40	
0	.0080	.0066	.0055	.0045	.0037	.0030	.0025	.0020	.0016	.0013	13
1	.0469	.0407	.0351	.0302	.0259	.0221	.0188	.0159	.0135	.0113	12
2	.1265	.1148	.1037	.0933	.0836	.0746	.0663	.0586	.0516	.0453	11
3	.2084	.1981	.1874	.1763	.1651	.1538	.1427	.1317	.1210	.1107	10
4	.2341	.2331	.2307	.2270	.2222	.2163	.2095	.2018	.1934	.1845	9
5	.1893	.1974	.2045	.2105	.2154	.2190	.2215	.2227	.2226	.2214	8
6	.1134	.1239	.1343	.1446	.1546	.1643	.1734	.1820	.1898	.1968	7
7	.0509	.0583	.0662	.0745	.0833	.0924	.1019	.1115	.1213	.1312	6
8	.0172	.0206	.0244	.0288	.0336	.0390	.0449	.0513	.0582	.0656	5
9	.0043	.0054	.0067	.0082	.0101	.0122	.0146	.0175	.0207	.0243	4
10	.0008	.0010	.0013	.0017	.0022	.0027	.0034	.0043	.0053	.0065	3
11	.0001	.0001	.0002	.0002	.0003	.0004	.0006	.0007	.0009	.0012	2
12	.0000	.0000	.0000	.0000	.0000	.0000	.0001	.0001	.0001	.0001	1
	.69	.68	.67	.66	.65	.64	.63	.62	.61	.60	*p* / *r*

r / *p*	.41	.42	.43	.44	.45	.46	.47	.48	.49	.50	
0	.0010	.0008	.0007	.0005	.0004	.0003	.0003	.0002	.0002	.0001	13
1	.0095	.0079	.0066	.0054	.0045	.0037	.0030	.0024	.0020	.0016	12
2	.0395	.0344	.0298	.0256	.0220	.0188	.0160	.0135	.0114	.0095	11
3	.1007	.0913	.0823	.0739	.0660	.0587	.0519	.0457	.0401	.0349	10
4	.1750	.1653	.1553	.1451	.1350	.1250	.1151	.1055	.0962	.0873	9
5	.2189	.2154	.2108	.2053	.1989	.1917	.1838	.1753	.1664	.1571	8
6	.2029	.2080	.2121	.2151	.2169	.2177	.2173	.2158	.2131	.2095	7
7	.1410	.1506	.1600	.1690	.1775	.1854	.1927	.1992	.2048	.2095	6
8	.0735	.0818	.0905	.0996	.1089	.1185	.1282	.1379	.1476	.1571	5
9	.0284	.0329	.0379	.0435	.0495	.0561	.0631	.0707	.0788	.0873	4
10	.0079	.0095	.0114	.0137	.0162	.0191	.0224	.0261	.0303	.0349	3
11	.0015	.0019	.0024	.0029	.0036	.0044	.0054	.0066	.0079	.0095	2
12	.0002	.0002	.0003	.0004	.0005	.0006	.0008	.0010	.0013	.0016	1
13	.0000	.0000	.0000	.0000	.0000	.0000	.0001	.0001	.0001	.0001	0
	.59	.58	.57	.56	.55	.54	.53	.52	.51	.50	*p* / *r*

$n = 14$

r / *p*	.01	.02	.03	.04	.05	.06	.07	.08	.09	.10	
0	.8687	.7536	.6528	.5647	.4877	.4205	.3620	.3112	.2670	.2288	14
1	.1229	.2153	.2827	.3294	.3593	.3758	.3815	.3788	.3698	.3559	13
2	.0081	.0286	.0568	.0892	.1229	.1559	.1867	.2141	.2377	.2570	12
3	.0003	.0023	.0070	.0149	.0259	.0398	.0562	.0745	.0940	.1142	11
4	.0000	.0001	.0006	.0017	.0037	.0070	.0116	.0178	.0256	.0349	10
5	.0000	.0000	.0000	.0001	.0004	.0009	.0018	.0031	.0051	.0078	9
6	.0000	.0000	.0000	.0000	.0000	.0001	.0002	.0004	.0008	.0013	8
7	.0000	.0000	.0000	.0000	.0000	.0000	.0000	.0000	.0001	.0002	7
	.99	.98	.97	.96	.95	.94	.93	.92	.91	.90	*p* / *r*

5	.2009	.1943	.1869	.1788	.1701	.1610	.1515	.1418	.1320	.1222	9
6	.2094	.2111	.2115	.2108	.2088	.2057	.2015	.1963	.1902	.1833	8
7	.1663	.1747	.1824	.1892	.1952	.2003	.2043	.2071	.2089	.2095	7
8	.1011	.1107	.1204	.1301	.1398	.1493	.1585	.1673	.1756	.1833	6
9	.0469	.0534	.0605	.0682	.0762	.0848	.0937	.1030	.1125	.1222	5
10	.0163	.0193	.0228	.0268	.0312	.0361	.0415	.0475	.0540	.0611	4
11	.0041	.0051	.0063	.0076	.0093	.0112	.0134	.0160	.0189	.0222	3
12	.0007	.0009	.0012	.0015	.0019	.0024	.0030	.0037	.0045	.0056	2
13	.0001	.0001	.0001	.0002	.0002	.0003	.0004	.0005	.0007	.0009	1
14	.0000	.0000	.0000	.0000	.0000	.0000	.0000	.0000	.0000	.0001	0
	.59	.58	.57	.56	.55	.54	.53	.52	.51	.50	*p* / *r*

$n = 15$

r / *p*	.01	.02	.03	.04	.05	.06	.07	.08	.09	.10	
0	.8601	.7386	.6333	.5421	.4633	.3953	.3367	.2863	.2430	.2059	15
1	.1303	.2261	.2938	.3388	.3658	.3785	.3801	.3734	.3605	.3432	14
2	.0092	.0323	.0636	.0988	.1348	.1691	.2003	.2273	.2496	.2669	13
3	.0004	.0029	.0085	.0178	.0307	.0468	.0653	.0857	.1070	.1285	12
4	.0000	.0002	.0008	.0022	.0049	.0090	.0148	.0223	.0317	.0428	11
5	.0000	.0000	.0001	.0002	.0006	.0013	.0024	.0043	.0069	.0105	10
6	.0000	.0000	.0000	.0000	.0000	.0001	.0003	.0006	.0011	.0019	9
7	.0000	.0000	.0000	.0000	.0000	.0000	.0000	.0001	.0001	.0003	8
	.99	.98	.97	.96	.95	.94	.93	.92	.91	.90	*p* / *r*

r / *p*	.11	.12	.13	.14	.15	.16	.17	.18	.19	.20	
0	.1741	.1470	.1238	.1041	.0874	.0731	.0611	.0510	.0424	.0352	15
1	.3228	.3006	.2775	.2542	.2312	.2090	.1878	.1678	.1492	.1319	14
2	.2793	.2870	.2903	.2897	.2856	.2787	.2692	.2578	.2449	.2309	13
3	.1496	.1696	.1880	.2044	.2184	.2300	.2389	.2452	.2489	.2501	12
4	.0555	.0694	.0843	.0998	.1156	.1314	.1468	.1615	.1752	.1876	11
5	.0151	.0208	.0277	.0357	.0449	.0551	.0662	.0780	.0904	.1032	10
6	.0031	.0047	.0069	.0097	.0132	.0175	.0226	.0285	.0353	.0430	9
7	.0005	.0008	.0013	.0020	.0030	.0043	.0059	.0081	.0107	.0138	8
8	.0001	.0001	.0002	.0003	.0005	.0008	.0012	.0018	.0025	.0035	7
9	.0000	.0000	.0000	.0000	.0001	.0001	.0002	.0003	.0005	.0007	6
10	.0000	.0000	.0000	.0000	.0000	.0000	.0000	.0000	.0001	.0001	5
	.89	.88	.87	.86	.85	.84	.83	.82	.81	.80	*p* / *r*

r / *p*	.21	.22	.23	.24	.25	.26	.27	.28	.29	.30	
0	.0291	.0241	.0198	.0163	.0134	.0109	.0089	.0072	.0059	.0047	15
1	.1162	.1018	.0889	.0772	.0668	.0576	.0494	.0423	.0360	.0305	14
2	.2162	.2010	.1858	.1707	.1559	.1416	.1280	.1150	.1029	.0916	13
3	.2490	.2457	.2405	.2336	.2252	.2156	.2051	.1939	.1821	.1700	12
4	.1986	.2079	.2155	.2213	.2252	.2273	.2276	.2262	.2231	.2186	11
5	.1161	.1290	.1416	.1537	.1651	.1757	.1852	.1935	.2005	.2061	10
6	.0514	.0606	.0705	.0809	.0917	.1029	.1142	.1254	.1365	.1472	9
7	.0176	.0220	.0271	.0329	.0393	.0465	.0543	.0627	.0717	.0811	8
8	.0047	.0062	.0081	.0104	.0131	.0163	.0201	.0244	.0293	.0348	7
9	.0010	.0014	.0019	.0025	.0034	.0045	.0058	.0074	.0093	.0116	6

r p	.11	.12	.13	.14	.15	.16	.17	.18	.19	.20	
0	.1956	.1670	.1423	.1211	.1028	.0871	.0736	.0621	.0523	.0440	14
1	.3385	.3188	.2977	.2759	.2539	.2322	.2112	.1910	.1719	.1539	13
2	.2720	.2826	.2892	.2919	.2912	.2875	.2811	.2725	.2620	.2501	12
3	.1345	.1542	.1728	.1901	.2056	.2190	.2303	.2393	.2459	.2501	11
4	.0457	.0578	.0710	.0851	.0998	.1147	.1297	.1444	.1586	.1720	10
5	.0113	.0158	.0212	.0277	.0352	.0437	.0531	.0634	.0744	.0860	9
6	.0021	.0032	.0048	.0068	.0093	.0125	.0163	.0209	.0262	.0322	8
7	.0003	.0005	.0008	.0013	.0019	.0027	.0038	.0052	.0070	.0092	7
8	.0000	.0001	.0001	.0002	.0003	.0005	.0007	.0010	.0014	.0020	6
9	.0000	.0000	.0000	.0000	.0000	.0001	.0001	.0001	.0002	.0003	5
	.89	.88	.87	.86	.85	.84	.83	.82	.81	.80 p	r

r p	.21	.22	.23	.24	.25	.26	.27	.28	.29	.30	
0	.0369	.0309	.0258	.0214	.0178	.0148	.0122	.0101	.0083	.0068	14
1	.1372	.1218	.1077	.0948	.0832	.0726	.0632	.0548	.0473	.0407	13
2	.2371	.2234	.2091	.1946	.1802	.1659	.1519	.1385	.1256	.1134	12
3	.2521	.2520	.2499	.2459	.2402	.2331	.2248	.2154	.2052	.1943	11
4	.1843	.1955	.2052	.2135	.2202	.2252	.2286	.2304	.2305	.2290	10
5	.0980	.1103	.1226	.1348	.1468	.1583	.1691	.1792	.1883	.1963	9
6	.0391	.0466	.0549	.0639	.0734	.0834	.0938	.1045	.1153	.1262	8
7	.0119	.0150	.0188	.0231	.0280	.0335	.0397	.0464	.0538	.0618	7
8	.0028	.0037	.0049	.0064	.0082	.0103	.0128	.0158	.0192	.0232	6
9	.0005	.0007	.0010	.0013	.0018	.0024	.0032	.0041	.0052	.0066	5
10	.0001	.0001	.0001	.0002	.0003	.0004	.0006	.0008	.0011	.0014	4
11	.0000	.0000	.0000	.0000	.0000	.0001	.0001	.0001	.0002	.0002	3
	.79	.78	.77	.76	.75	.74	.73	.72	.71	.70 p	r

r p	.31	.32	.33	.34	.35	.36	.37	.38	.39	.40	
0	.0055	.0045	.0037	.0030	.0024	.0019	.0016	.0012	.0010	.0008	14
1	.0349	.0298	.0253	.0215	.0181	.0152	.0128	.0106	.0088	.0073	13
2	.1018	.0911	.0811	.0719	.0634	.0557	.0487	.0424	.0367	.0317	12
3	.1830	.1715	.1598	.1481	.1366	.1253	.1144	.1039	.0940	.0845	11
4	.2261	.2219	.2164	.2098	.2022	.1938	.1848	.1752	.1652	.1549	10
5	.2032	.2088	.2132	.2161	.2178	.2181	.2170	.2147	.2112	.2066	9
6	.1369	.1474	.1575	.1670	.1759	.1840	.1912	.1974	.2026	.2066	8
7	.0703	.0793	.0886	.0983	.1082	.1183	.1283	.1383	.1480	.1574	7
8	.0276	.0326	.0382	.0443	.0510	.0582	.0659	.0742	.0828	.0918	6
9	.0083	.0102	.0125	.0152	.0183	.0218	.0258	.0303	.0353	.0408	5
10	.0019	.0024	.0031	.0039	.0049	.0061	.0076	.0093	.0113	.0136	4
11	.0003	.0004	.0006	.0007	.0010	.0013	.0016	.0021	.0026	.0033	3
12	.0000	.0000	.0001	.0001	.0001	.0002	.0002	.0003	.0004	.0005	2
13	.0000	.0000	.0000	.0000	.0000	.0000	.0000	.0000	.0000	.0001	1
	.69	.68	.67	.66	.65	.64	.63	.62	.61	.60 p	r

r p	.41	.42	.43	.44	.45	.46	.47	.48	.49	.50	
0	.0006	.0005	.0004	.0003	.0002	.0002	.0001	.0001	.0001	.0001	14
1	.0060	.0049	.0040	.0033	.0027	.0021	.0017	.0014	.0011	.0009	13
2	.0272	.0233	.0198	.0168	.0141	.0118	.0099	.0082	.0068	.0056	12
3	.0757	.0674	.0597	.0527	.0462	.0403	.0350	.0303	.0260	.0222	11
4	.1446	.1342	.1239	.1138	.1040	.0945	.0854	.0768	.0687	.0611	10

Note: Read down entire column.

r											
10	.0002	.0002	.0003	.0005	.0007	.0009	.0013	.0017	.0023	.0030	5
11	.0000	.0000	.0000	.0001	.0001	.0002	.0002	.0003	.0004	.0006	4
12	.0000	.0000	.0000	.0000	.0000	.0000	.0000	.0000	.0001	.0001	3
	.79	.78	.77	.76	.75	.74	.73	.72	.71	.70 p	r

r p	.31	.32	.33	.34	.35	.36	.37	.38	.39	.40	
0	.0038	.0031	.0025	.0020	.0016	.0012	.0010	.0008	.0006	.0005	15
1	.0258	.0217	.0182	.0152	.0126	.0104	.0086	.0071	.0058	.0047	14
2	.0811	.0715	.0627	.0547	.0476	.0411	.0354	.0303	.0259	.0219	13
3	.1579	.1457	.1338	.1222	.1110	.1002	.0901	.0805	.0716	.0634	12
4	.2128	.2057	.1977	.1888	.1792	.1692	.1587	.1481	.1374	.1268	11
5	.2103	.2130	.2142	.2140	.2123	.2093	.2051	.1997	.1933	.1859	10
6	.1575	.1671	.1759	.1837	.1906	.1963	.2008	.2040	.2059	.2066	9
7	.0910	.1011	.1114	.1217	.1319	.1419	.1516	.1608	.1693	.1771	8
8	.0409	.0476	.0549	.0627	.0710	.0798	.0890	.0985	.1082	.1181	7
9	.0143	.0174	.0210	.0251	.0298	.0349	.0407	.0470	.0538	.0612	6
10	.0038	.0049	.0062	.0078	.0096	.0118	.0143	.0173	.0206	.0245	5
11	.0008	.0011	.0014	.0018	.0024	.0030	.0038	.0048	.0060	.0074	4
12	.0001	.0002	.0002	.0003	.0004	.0006	.0007	.0010	.0013	.0016	3
13	.0000	.0000	.0000	.0000	.0001	.0001	.0001	.0001	.0002	.0003	2
	.69	.68	.67	.66	.65	.64	.63	.62	.61	.60 p	r

r p	.41	.42	.43	.44	.45	.46	.47	.48	.49	.50	
0	.0004	.0003	.0002	.0002	.0001	.0001	.0001	.0001	.0000	.0000	15
1	.0038	.0031	.0025	.0020	.0016	.0012	.0010	.0008	.0006	.0005	14
2	.0185	.0156	.0130	.0108	.0090	.0074	.0060	.0049	.0040	.0032	13
3	.0558	.0489	.0426	.0369	.0318	.0272	.0232	.0197	.0166	.0139	12
4	.1163	.1061	.0963	.0869	.0780	.0696	.0617	.0545	.0478	.0417	11
5	.1778	.1691	.1598	.1502	.1404	.1304	.1204	.1106	.1010	.0916	10
6	.2060	.2041	.2010	.1967	.1914	.1851	.1780	.1702	.1617	.1527	9
7	.1840	.1900	.1949	.1987	.2013	.2028	.2030	.2020	.1997	.1964	8
8	.1279	.1376	.1470	.1561	.1647	.1727	.1800	.1864	.1919	.1964	7
9	.0691	.0775	.0863	.0954	.1048	.1144	.1241	.1338	.1434	.1527	6
10	.0288	.0337	.0390	.0450	.0515	.0585	.0661	.0741	.0827	.0916	5
11	.0091	.0111	.0134	.0161	.0191	.0226	.0266	.0311	.0361	.0417	4
12	.0021	.0027	.0034	.0042	.0052	.0064	.0079	.0096	.0116	.0139	3
13	.0003	.0004	.0006	.0008	.0010	.0013	.0016	.0020	.0026	.0032	2
14	.0000	.0000	.0001	.0001	.0001	.0002	.0002	.0003	.0004	.0005	1
	.59	.58	.57	.56	.55	.54	.53	.52	.51	.50 p	r

$n = 16$

r p	.01	.02	.03	.04	.05	.06	.07	.08	.09	.10	
0	.8515	.7238	.6143	.5204	.4401	.3716	.3131	.2634	.2211	.1853	16
1	.1376	.2363	.3040	.3469	.3706	.3795	.3771	.3665	.3499	.3294	15
2	.0104	.0362	.0705	.1084	.1463	.1817	.2129	.2390	.2596	.2745	14
3	.0005	.0034	.0102	.0211	.0359	.0541	.0748	.0970	.1198	.1423	13
4	.0000	.0002	.0010	.0029	.0061	.0112	.0183	.0274	.0385	.0514	12
5	.0000	.0000	.0001	.0003	.0008	.0017	.0033	.0057	.0091	.0137	11
6	.0000	.0000	.0000	.0000	.0001	.0002	.0005	.0009	.0017	.0028	10
7	.0000	.0000	.0000	.0000	.0000	.0000	.0000	.0001	.0002	.0004	9
8	.0000	.0000	.0000	.0000	.0000	.0000	.0000	.0000	.0000	.0001	8
	.99	.98	.97	.96	.95	.94	.93	.92	.91	.90 p	r

Table 2 (continued)

r	p .11	.12	.13	.14	.15	.16	.17	.18	.19	.20	
0	.1550	.1293	.1077	.0895	.0743	.0614	.0507	.0418	.0343	.0281	16
1	.3065	.2822	.2575	.2332	.2097	.1873	.1662	.1468	.1289	.1126	15
2	.2841	.2886	.2886	.2847	.2775	.2675	.2554	.2416	.2267	.2111	14
3	.1638	.1837	.2013	.2163	.2285	.2378	.2441	.2475	.2482	.2463	13
4	.0658	.0814	.0977	.1144	.1311	.1472	.1625	.1766	.1892	.2001	12
5	.0195	.0266	.0351	.0447	.0555	.0673	.0799	.0930	.1065	.1201	11
6	.0044	.0067	.0096	.0133	.0180	.0235	.0300	.0374	.0458	.0550	10
7	.0008	.0013	.0020	.0031	.0045	.0064	.0088	.0117	.0153	.0197	9
8	.0001	.0002	.0003	.0006	.0009	.0014	.0020	.0029	.0041	.0055	8
9	.0000	.0000	.0000	.0001	.0001	.0002	.0004	.0006	.0008	.0012	7
10	.0000	.0000	.0000	.0000	.0000	.0000	.0001	.0001	.0001	.0002	6
	.89	.88	.87	.86	.85	.84	.83	.82	.81	.80 p	r

r	p .21	.22	.23	.24	.25	.26	.27	.28	.29	.30	
0	.0230	.0188	.0153	.0124	.0100	.0081	.0065	.0052	.0042	.0033	16
1	.0979	.0847	.0730	.0626	.0535	.0455	.0385	.0325	.0273	.0228	15
2	.1952	.1792	.1635	.1482	.1336	.1198	.1068	.0947	.0835	.0732	14
3	.2421	.2359	.2279	.2185	.2079	.1964	.1843	.1718	.1591	.1465	13
4	.2092	.2162	.2212	.2242	.2252	.2243	.2215	.2171	.2112	.2040	12
5	.1334	.1464	.1586	.1699	.1802	.1891	.1966	.2026	.2071	.2099	11
6	.0650	.0757	.0869	.0984	.1101	.1218	.1333	.1445	.1551	.1649	10
7	.0247	.0305	.0371	.0444	.0524	.0611	.0704	.0803	.0905	.1010	9
8	.0074	.0097	.0125	.0158	.0197	.0242	.0293	.0351	.0416	.0487	8
9	.0017	.0024	.0033	.0044	.0058	.0075	.0096	.0121	.0151	.0185	7
10	.0003	.0005	.0007	.0010	.0014	.0019	.0025	.0033	.0043	.0056	6
11	.0000	.0001	.0001	.0002	.0002	.0004	.0005	.0007	.0010	.0013	5
12	.0000	.0000	.0000	.0000	.0000	.0001	.0001	.0001	.0002	.0002	4
	.79	.78	.77	.76	.75	.74	.73	.72	.71	.70 p	r

r	p .31	.32	.33	.34	.35	.36	.37	.38	.39	.40	
0	.0026	.0021	.0016	.0013	.0010	.0008	.0006	.0005	.0004	.0003	16
1	.0190	.0157	.0130	.0107	.0087	.0071	.0058	.0047	.0038	.0030	15
2	.0639	.0555	.0480	.0413	.0353	.0301	.0255	.0215	.0180	.0150	14
3	.1341	.1220	.1103	.0992	.0888	.0790	.0699	.0615	.0538	.0468	13
4	.1958	.1865	.1766	.1662	.1553	.1444	.1333	.1224	.1118	.1014	12
5	.2111	.2107	.2088	.2054	.2008	.1949	.1879	.1801	.1715	.1623	11
6	.1739	.1818	.1885	.1940	.1982	.2010	.2024	.2024	.2010	.1983	10
7	.1116	.1222	.1326	.1428	.1524	.1615	.1698	.1772	.1836	.1889	9
8	.0564	.0647	.0735	.0827	.0923	.1022	.1122	.1222	.1320	.1417	8
9	.0225	.0271	.0322	.0379	.0442	.0511	.0586	.0666	.0750	.0840	7
10	.0071	.0089	.0111	.0137	.0167	.0201	.0241	.0286	.0336	.0392	6
11	.0017	.0023	.0030	.0038	.0049	.0062	.0077	.0095	.0117	.0142	5
12	.0003	.0004	.0006	.0008	.0011	.0014	.0019	.0024	.0031	.0040	4
13	.0000	.0001	.0001	.0001	.0002	.0003	.0003	.0005	.0006	.0008	3
14	.0000	.0000	.0000	.0000	.0000	.0000	.0000	.0001	.0001	.0001	2
	.69	.68	.67	.66	.65	.64	.63	.62	.61	.60 p	r

r	p .41	.42	.43	.44	.45	.46	.47	.48	.49	.50	
0	.0002	.0002	.0001	.0001	.0001	.0001	.0000	.0000	.0000	.0000	16
1	.0024	.0019	.0015	.0012	.0009	.0007	.0005	.0004	.0003	.0002	15
2	.0125	.0103	.0085	.0069	.0056	.0046	.0037	.0029	.0023	.0018	14
3	.0405	.0349	.0299	.0254	.0215	.0181	.0151	.0126	.0104	.0085	13
4	.0915	.0821	.0732	.0649	.0572	.0501	.0436	.0378	.0325	.0278	12

10	.0006	.0009	.0013	.0018	.0025	.0033	.0044	.0058	.0074	.0095	7
11	.0001	.0002	.0002	.0004	.0005	.0007	.0010	.0014	.0019	.0026	6
12	.0000	.0000	.0000	.0001	.0001	.0001	.0002	.0003	.0004	.0006	5
13	.0000	.0000	.0000	.0000	.0000	.0000	.0000	.0000	.0001	.0001	4
	.79	.78	.77	.76	.75	.74	.73	.72	.71	.70 p	r

r	p .31	.32	.33	.34	.35	.36	.37	.38	.39	.40	
0	.0018	.0014	.0011	.0009	.0007	.0005	.0004	.0003	.0002	.0002	17
1	.0139	.0114	.0093	.0075	.0060	.0048	.0039	.0031	.0024	.0019	16
2	.0500	.0428	.0364	.0309	.0260	.0218	.0182	.0151	.0125	.0102	15
3	.1123	.1007	.0898	.0795	.0701	.0614	.0534	.0463	.0398	.0341	14
4	.1766	.1659	.1547	.1434	.1320	.1208	.1099	.0993	.0892	.0796	13
5	.2063	.2030	.1982	.1921	.1849	.1767	.1677	.1582	.1482	.1379	12
6	.1854	.1910	.1952	.1979	.1991	.1988	.1970	.1939	.1895	.1839	11
7	.1309	.1413	.1511	.1602	.1685	.1757	.1818	.1868	.1904	.1927	10
8	.0735	.0831	.0930	.1032	.1134	.1235	.1335	.1431	.1521	.1606	9
9	.0330	.0391	.0458	.0531	.0611	.0695	.0784	.0877	.0973	.1070	8
10	.0119	.0147	.0181	.0219	.0263	.0313	.0368	.0430	.0498	.0571	7
11	.0034	.0044	.0057	.0072	.0090	.0112	.0138	.0168	.0202	.0242	6
12	.0008	.0010	.0014	.0018	.0024	.0031	.0040	.0051	.0065	.0081	5
13	.0001	.0002	.0003	.0004	.0005	.0007	.0009	.0012	.0016	.0021	4
14	.0000	.0000	.0000	.0001	.0001	.0001	.0002	.0002	.0003	.0004	3
15	.0000	.0000	.0000	.0000	.0000	.0000	.0000	.0000	.0000	.0001	2
	.69	.68	.67	.66	.65	.64	.63	.62	.61	.60 p	r

r	p .41	.42	.43	.44	.45	.46	.47	.48	.49	.50	
0	.0001	.0001	.0001	.0001	.0000	.0000	.0000	.0000	.0000	.0000	17
1	.0015	.0012	.0009	.0007	.0005	.0004	.0003	.0002	.0002	.0001	16
2	.0084	.0068	.0055	.0044	.0035	.0028	.0022	.0017	.0013	.0010	15
3	.0290	.0246	.0207	.0173	.0144	.0119	.0097	.0079	.0064	.0052	14
4	.0706	.0622	.0546	.0475	.0411	.0354	.0302	.0257	.0217	.0182	13
5	.1276	.1172	.1070	.0971	.0875	.0784	.0697	.0616	.0541	.0472	12
6	.1773	.1697	.1614	.1525	.1432	.1335	.1237	.1138	.1040	.0944	11
7	.1936	.1932	.1914	.1883	.1841	.1787	.1723	.1650	.1570	.1484	10
8	.1682	.1748	.1805	.1850	.1883	.1903	.1910	.1904	.1886	.1855	9
9	.1169	.1266	.1361	.1453	.1540	.1621	.1694	.1758	.1812	.1855	8
10	.0650	.0733	.0822	.0914	.1008	.1105	.1202	.1298	.1393	.1484	7
11	.0287	.0338	.0394	.0457	.0525	.0599	.0678	.0763	.0851	.0944	6
12	.0100	.0122	.0149	.0179	.0215	.0255	.0301	.0352	.0409	.0472	5
13	.0027	.0034	.0043	.0054	.0068	.0084	.0103	.0125	.0151	.0182	4
14	.0005	.0007	.0009	.0012	.0016	.0020	.0026	.0033	.0041	.0052	3
15	.0001	.0001	.0001	.0002	.0003	.0003	.0005	.0006	.0008	.0010	2
16	.0000	.0000	.0000	.0000	.0000	.0000	.0001	.0001	.0001	.0001	1
	.59	.58	.57	.56	.55	.54	.53	.52	.51	.50 p	r

$n = 18$

r	p .01	.02	.03	.04	.05	.06	.07	.08	.09	.10	
0	.8345	.6951	.5780	.4796	.3972	.3283	.2708	.2229	.1831	.1501	18
1	.1517	.2554	.3217	.3597	.3763	.3772	.3669	.3489	.3260	.3002	17
2	.0130	.0443	.0846	.1274	.1683	.2047	.2348	.2579	.2741	.2835	16
3	.0007	.0048	.0140	.0283	.0473	.0697	.0942	.1196	.1446	.1680	15
4	.0000	.0004	.0016	.0044	.0093	.0167	.0266	.0390	.0536	.0700	14

r											
5	.1526	.1426	.1325	.1224	.1123	.1024	.0929	.0837	.0749	.0667	11
6	.1944	.1894	.1833	.1762	.1684	.1600	.1510	.1416	.1319	.1222	10
7	.1930	.1959	.1975	.1978	.1969	.1947	.1912	.1867	.1811	.1746	9
8	.1509	.1596	.1676	.1749	.1812	.1865	.1908	.1939	.1958	.1964	8
9	.0932	.1027	.1124	.1221	.1318	.1413	.1504	.1591	.1672	.1746	7
10	.0453	.0521	.0594	.0672	.0755	.0842	.0934	.1028	.1124	.1222	6
11	.0172	.0206	.0244	.0288	.0337	.0391	.0452	.0518	.0589	.0667	5
12	.0050	.0062	.0077	.0094	.0115	.0139	.0167	.0199	.0236	.0278	4
13	.0011	.0014	.0018	.0023	.0029	.0036	.0046	.0057	.0070	.0085	3
14	.0002	.0002	.0003	.0004	.0005	.0007	.0009	.0011	.0014	.0018	2
15	.0000	.0000	.0000	.0000	.0001	.0001	.0001	.0001	.0002	.0002	1
	.59	.58	.57	.56	.55	.54	.53	.52	.51	.50 p	r

$n = 17$

r p	.01	.02	.03	.04	.05	.06	.07	.08	.09	.10	
0	.8429	.7093	.5958	.4996	.4181	.3493	.2912	.2423	.2012	.1668	17
1	.1447	.2461	.3133	.3539	.3741	.3790	.3726	.3582	.3383	.3150	16
2	.0117	.0402	.0775	.1180	.1575	.1935	.2244	.2492	.2677	.2800	15
3	.0006	.0041	.0120	.0246	.0415	.0618	.0844	.1083	.1324	.1556	14
4	.0000	.0003	.0013	.0036	.0076	.0138	.0222	.0330	.0458	.0605	13
5	.0000	.0000	.0001	.0004	.0010	.0023	.0044	.0075	.0118	.0175	12
6	.0000	.0000	.0000	.0000	.0001	.0003	.0007	.0013	.0023	.0039	11
7	.0000	.0000	.0000	.0000	.0000	.0000	.0001	.0002	.0004	.0007	10
8	.0000	.0000	.0000	.0000	.0000	.0000	.0000	.0000	.0000	.0001	9
	.99	.98	.97	.96	.95	.94	.93	.92	.91	.90 p	r

r p	.11	.12	.13	.14	.15	.16	.17	.18	.19	.20	
0	.1379	.1138	.0937	.0770	.0631	.0516	.0421	.0343	.0278	.0225	17
1	.2898	.2638	.2381	.2131	.1893	.1671	.1446	.1279	.1109	.0957	16
2	.2865	.2878	.2846	.2775	.2673	.2547	.2402	.2245	.2081	.1914	15
3	.1771	.1963	.2126	.2259	.2359	.2425	.2460	.2464	.2441	.2393	14
4	.0766	.0937	.1112	.1287	.1457	.1617	.1764	.1893	.2004	.2093	13
5	.0246	.0332	.0432	.0545	.0668	.0801	.0939	.1081	.1222	.1361	12
6	.0061	.0091	.0129	.0177	.0236	.0305	.0385	.0474	.0573	.0680	11
7	.0012	.0019	.0030	.0045	.0065	.0091	.0124	.0164	.0211	.0267	10
8	.0002	.0003	.0006	.0009	.0014	.0022	.0032	.0045	.0062	.0084	9
9	.0000	.0000	.0001	.0002	.0003	.0004	.0006	.0010	.0015	.0021	8
10	.0000	.0000	.0000	.0000	.0000	.0001	.0001	.0002	.0003	.0004	7
11	.0000	.0000	.0000	.0000	.0000	.0000	.0000	.0000	.0000	.0001	6
	.89	.88	.87	.86	.85	.84	.83	.82	.81	.80 p	r

r p	.21	.22	.23	.24	.25	.26	.27	.28	.29	.30	
0	.0182	.0146	.0118	.0094	.0075	.0060	.0047	.0038	.0030	.0023	17
1	.0822	.0702	.0597	.0505	.0426	.0357	.0299	.0248	.0206	.0169	16
2	.1747	.1584	.1427	.1277	.1136	.1005	.0883	.0772	.0672	.0581	15
3	.2322	.2234	.2131	.2016	.1893	.1765	.1634	.1502	.1372	.1245	14
4	.2161	.2205	.2228	.2228	.2209	.2170	.2115	.2044	.1961	.1868	13
5	.1493	.1617	.1730	.1830	.1914	.1982	.2033	.2067	.2083	.2081	12
6	.0794	.0912	.1034	.1156	.1276	.1393	.1504	.1608	.1701	.1784	11
7	.0332	.0404	.0485	.0573	.0668	.0769	.0874	.0982	.1092	.1201	10
8	.0110	.0143	.0181	.0226	.0279	.0338	.0404	.0478	.0558	.0644	9
9	.0029	.0040	.0054	.0071	.0093	.0119	.0150	.0186	.0228	.0276	8

r											
5	.0000	.0000	.0001	.0005	.0014	.0030	.0056	.0095	.0148	.0218	13
6	.0000	.0000	.0000	.0000	.0002	.0004	.0009	.0018	.0032	.0052	12
7	.0000	.0000	.0000	.0000	.0000	.0000	.0001	.0003	.0005	.0010	11
8	.0000	.0000	.0000	.0000	.0000	.0000	.0000	.0000	.0001	.0002	10
	.99	.98	.97	.96	.95	.94	.93	.92	.91	.90 p	r

r p	.11	.12	.13	.14	.15	.16	.17	.18	.19	.20	
0	.1227	.1002	.0815	.0662	.0536	.0434	.0349	.0281	.0225	.0180	18
1	.2731	.2458	.2193	.1940	.1704	.1486	.1288	.1110	.0951	.0811	17
2	.2869	.2850	.2785	.2685	.2556	.2407	.2243	.2071	.1897	.1723	16
3	.1891	.2072	.2220	.2331	.2406	.2445	.2450	.2425	.2373	.2297	15
4	.0877	.1060	.1244	.1423	.1592	.1746	.1882	.1996	.2087	.2153	14
5	.0303	.0405	.0520	.0669	.0787	.0931	.1079	.1227	.1371	.1507	13
6	.0081	.0120	.0168	.0229	.0301	.0384	.0479	.0584	.0697	.0816	12
7	.0017	.0028	.0043	.0064	.0091	.0126	.0168	.0220	.0280	.0350	11
8	.0003	.0005	.0009	.0014	.0022	.0033	.0047	.0066	.0090	.0120	10
9	.0000	.0001	.0001	.0003	.0004	.0007	.0011	.0016	.0024	.0033	9
10	.0000	.0000	.0000	.0000	.0001	.0001	.0002	.0003	.0005	.0008	8
11	.0000	.0000	.0000	.0000	.0000	.0000	.0000	.0001	.0001	.0001	7
	.89	.88	.87	.86	.85	.84	.83	.82	.81	.80 p	r

r p	.21	.22	.23	.24	.25	.26	.27	.28	.29	.30	
0	.0144	.0114	.0091	.0072	.0056	.0044	.0035	.0027	.0021	.0016	18
1	.0687	.0580	.0487	.0407	.0338	.0280	.0231	.0189	.0155	.0126	17
2	.1553	.1390	.1236	.1092	.0958	.0836	.0725	.0626	.0537	.0458	16
3	.2202	.2091	.1969	.1839	.1704	.1567	.1431	.1298	.1169	.1046	15
4	.2195	.2212	.2205	.2177	.2130	.2065	.1985	.1892	.1790	.1681	14
5	.1634	.1747	.1845	.1925	.1988	.2031	.2055	.2061	.2048	.2017	13
6	.0941	.1067	.1194	.1317	.1436	.1546	.1647	.1736	.1812	.1873	12
7	.0429	.0516	.0611	.0713	.0820	.0931	.1044	.1157	.1269	.1376	11
8	.0157	.0200	.0251	.0310	.0376	.0450	.0531	.0619	.0713	.0811	10
9	.0046	.0063	.0083	.0109	.0139	.0176	.0218	.0267	.0323	.0386	9
10	.0011	.0016	.0022	.0031	.0042	.0056	.0073	.0094	.0119	.0149	8
11	.0002	.0003	.0005	.0007	.0010	.0014	.0020	.0026	.0035	.0046	7
12	.0000	.0001	.0001	.0001	.0002	.0003	.0004	.0006	.0008	.0012	6
13	.0000	.0000	.0000	.0000	.0000	.0000	.0001	.0001	.0002	.0002	5
	.79	.78	.77	.76	.75	.74	.73	.72	.71	.70 p	r

r p	.31	.32	.33	.34	.35	.36	.37	.38	.39	.40	
0	.0013	.0010	.0007	.0006	.0004	.0003	.0002	.0002	.0001	.0001	18
1	.0102	.0082	.0066	.0052	.0042	.0033	.0026	.0020	.0016	.0012	17
2	.0388	.0327	.0275	.0229	.0190	.0157	.0129	.0105	.0086	.0069	16
3	.0930	.0822	.0722	.0630	.0547	.0471	.0604	.0344	.0292	.0246	15
4	.1567	.1450	.1333	.1217	.1104	.0994	.0890	.0791	.0699	.0614	14
5	.1971	.1911	.1838	.1755	.1664	.1566	.1463	.1358	.1252	.1146	13
6	.1919	.1948	.1962	.1959	.1941	.1908	.1862	.1803	.1734	.1655	12
7	.1478	.1572	.1656	.1730	.1792	.1840	.1875	.1895	.1900	.1892	11
8	.0913	.1017	.1122	.1226	.1327	.1423	.1514	.1597	.1671	.1734	10
9	.0456	.0532	.0614	.0701	.0794	.0890	.0988	.1087	.1187	.1284	9
10	.0184	.0225	.0272	.0325	.0385	.0450	.0522	.0600	.0683	.0771	8
11	.0060	.0077	.0097	.0122	.0151	.0184	.0223	.0267	.0318	.0374	7
12	.0016	.0021	.0028	.0037	.0047	.0060	.0076	.0096	.0118	.0145	6
13	.0003	.0005	.0006	.0009	.0012	.0016	.0021	.0027	.0035	.0045	5
14	.0001	.0001	.0001	.0002	.0002	.0003	.0004	.0006	.0008	.0011	4

Note: Read down entire column.

Table 2 (continued)

r	p										
15	.0000	.0000	.0000	.0000	.0000	.0000	.0001	.0001	.0001	.0002	3
	.69	.68	.67	.66	.65	.64	.63	.62	.61	.60 p	r

r p	.41	.42	.43	.44	.45	.46	.47	.48	.49	.50	
0	.0001	.0001	.0000	.0000	.0000	.0000	.0000	.0000	.0000	.0000	18
1	.0009	.0007	.0005	.0004	.0003	.0002	.0002	.0001	.0001	.0001	17
2	.0055	.0044	.0035	.0028	.0022	.0017	.0013	.0010	.0008	.0006	16
3	.0206	.0171	.0141	.0116	.0095	.0077	.0062	.0050	.0039	.0031	15
4	.0536	.0464	.0400	.0342	.0291	.0246	.0206	.0172	.0142	.0117	14
5	.1042	.0941	.0844	.0753	.0666	.0586	.0512	.0444	.0382	.0327	13
6	.1569	.1477	.1380	.1281	.1181	.1081	.0983	.0887	.0796	.0708	12
7	.1869	.1833	.1785	.1726	.1657	.1579	.1494	.1404	.1310	.1214	11
8	.1786	.1825	.1852	.1864	.1864	.1850	.1822	.1782	.1731	.1669	10
9	.1379	.1469	.1552	.1628	.1694	.1751	.1795	.1828	.1848	.1855	9
10	.0862	.0957	.1054	.1151	.1248	.1342	.1433	.1519	.1598	.1669	8
11	.0436	.0504	.0578	.0658	.0742	.0831	.0924	.1020	.1117	.1214	7
12	.0177	.0213	.0254	.0301	.0354	.0413	.0478	.0549	.0626	.0708	6
13	.0057	.0071	.0089	.0109	.0134	.0162	.0196	.0234	.0278	.0327	5
14	.0014	.0018	.0024	.0031	.0039	.0049	.0062	.0077	.0095	.0117	4
15	.0003	.0004	.0005	.0006	.0009	.0011	.0015	.0019	.0024	.0031	3
16	.0000	.0000	.0001	.0001	.0001	.0002	.0002	.0003	.0004	.0006	2
17	.0000	.0000	.0000	.0000	.0000	.0000	.0000	.0000	.0000	.0001	1
	.59	.58	.57	.56	.55	.54	.53	.52	.51	.50 p	r

$n = 19$

r p	.01	.02	.03	.04	.05	.06	.07	.08	.09	.10	
0	.8262	.6812	.5606	.4604	.3774	.3086	.2519	.2051	.1666	.1351	19
1	.1586	.2642	.3294	.3645	.3774	.3743	.3602	.3389	.3131	.2852	18
2	.0144	.0485	.0917	.1367	.1787	.2150	.2440	.2652	.2787	.2852	17
3	.0008	.0056	.0161	.0323	.0533	.0778	.1041	.1307	.1562	.1796	16
4	.0000	.0005	.0020	.0054	.0112	.0199	.0313	.0455	.0618	.0798	15
5	.0000	.0000	.0002	.0007	.0018	.0038	.0071	.0119	.0183	.0266	14
6	.0000	.0000	.0000	.0001	.0002	.0006	.0012	.0024	.0042	.0069	13
7	.0000	.0000	.0000	.0000	.0000	.0001	.0002	.0004	.0008	.0014	12
8	.0000	.0000	.0000	.0000	.0000	.0000	.0000	.0001	.0001	.0002	11
	.99	.98	.97	.96	.95	.94	.93	.92	.91	.90 p	r

r p	.11	.12	.13	.14	.15	.16	.17	.18	.19	.20	
0	.1092	.0881	.0709	.0569	.0456	.0364	.0290	.0230	.0182	.0144	19
1	.2565	.2284	.2014	.1761	.1529	.1318	.1129	.0961	.0813	.0685	18
2	.2854	.2803	.2708	.2581	.2428	.2259	.2081	.1898	.1717	.1540	17
3	.1999	.2166	.2293	.2381	.2428	.2439	.2415	.2361	.2282	.2182	16
4	.0988	.1181	.1371	.1550	.1714	.1858	.1979	.2073	.2141	.2182	15
5	.0366	.0483	.0614	.0757	.0907	.1062	.1216	.1365	.1507	.1636	14
6	.0106	.0154	.0214	.0288	.0374	.0472	.0581	.0699	.0825	.0955	13
7	.0024	.0039	.0059	.0087	.0122	.0167	.0221	.0285	.0359	.0443	12
8	.0004	.0008	.0013	.0021	.0032	.0048	.0068	.0094	.0126	.0166	11
9	.0001	.0001	.0002	.0004	.0007	.0011	.0017	.0025	.0036	.0051	10
10	.0000	.0000	.0000	.0001	.0001	.0002	.0003	.0006	.0009	.0013	9
11	.0000	.0000	.0000	.0000	.0000	.0000	.0001	.0001	.0002	.0003	8
	.89	.88	.87	.86	.85	.84	.83	.82	.81	.80 p	r

r											
11	.0611	.0694	.0783	.0875	.0970	.1066	.1163	.1259	.1352	.1442	8
12	.0283	.0335	.0394	.0458	.0529	.0606	.0688	.0775	.0866	.0961	7
13	.0106	.0131	.0160	.0194	.0233	.0278	.0328	.0385	.0448	.0518	6
14	.0032	.0041	.0052	.0065	.0082	.0101	.0125	.0152	.0185	.0222	5
15	.0007	.0010	.0013	.0017	.0022	.0029	.0037	.0047	.0059	.0074	4
16	.0001	.0002	.0002	.0003	.0005	.0006	.0008	.0011	.0014	.0018	3
17	.0000	.0000	.0000	.0000	.0001	.0001	.0001	.0002	.0002	.0003	2
	.59	.58	.57	.56	.55	.54	.53	.52	.51	.50 p	r

$n = 20$

r p	.01	.02	.03	.04	.05	.06	.07	.08	.09	.10	
0	.8179	.6676	.5438	.4420	.3585	.2901	.2342	.1887	.1516	.1216	20
1	.1652	.2725	.3364	.3683	.3774	.3703	.3526	.3282	.3000	.2702	19
2	.0159	.0528	.0988	.1458	.1887	.2246	.2521	.2711	.2818	.2852	18
3	.0010	.0065	.0183	.0364	.0596	.0860	.1139	.1414	.1672	.1901	17
4	.0000	.0006	.0024	.0065	.0133	.0233	.0364	.0523	.0703	.0898	16
5	.0000	.0000	.0002	.0009	.0022	.0048	.0088	.0145	.0222	.0319	15
6	.0000	.0000	.0000	.0001	.0003	.0008	.0017	.0032	.0055	.0089	14
7	.0000	.0000	.0000	.0000	.0000	.0001	.0002	.0005	.0011	.0020	13
8	.0000	.0000	.0000	.0000	.0000	.0000	.0000	.0001	.0002	.0004	12
9	.0000	.0000	.0000	.0000	.0000	.0000	.0000	.0000	.0000	.0001	11
	99	98	97	96	95	94	93	92	91	90 p	r

r p	.11	.12	.13	.14	.15	.16	.17	.18	.19	.20	
0	.0972	.0776	.0617	.0490	.0388	.0306	.0241	.0189	.0148	.0115	20
1	.2403	.2115	.1844	.1595	.1368	.1165	.0986	.0829	.0693	.0576	19
2	.2822	.2740	.2618	.2466	.2293	.2109	.1919	.1730	.1545	.1369	18
3	.2093	.2242	.2347	.2409	.2428	.2410	.2358	.2278	.2175	.2054	17
4	.1099	.1299	.1491	.1666	.1821	.1951	.2053	.2125	.2168	.2182	16
5	.0435	.0567	.0713	.0868	.1028	.1189	.1345	.1493	.1627	.1746	15
6	.0134	.0193	.0266	.0353	.0454	.0566	.0689	.0819	.0954	.1091	14
7	.0033	.0053	.0080	.0115	.0160	.0216	.0282	.0360	.0448	.0545	13
8	.0007	.0012	.0019	.0030	.0046	.0067	.0094	.0128	.0171	.0222	12
9	.0001	.0002	.0004	.0007	.0011	.0017	.0026	.0038	.0053	.0074	11
10	.0000	.0000	.0001	.0001	.0002	.0004	.0006	.0009	.0014	.0020	10
11	.0000	.0000	.0000	.0000	.0000	.0001	.0001	.0002	.0003	.0005	9
12	.0000	.0000	.0000	.0000	.0000	.0000	.0000	.0000	.0001	.0001	8
	.89	.88	.87	.86	.85	.84	.83	.82	.81	.80 p	r

r p	.21	.22	.23	.24	.25	.26	.27	.28	.29	.30	
0	.0090	.0069	.0054	.0041	.0032	.0024	.0018	.0014	.0011	.0008	20
1	.0477	.0392	.0321	.0261	.0211	.0170	.0137	.0109	.0087	.0068	19
2	.1204	.1050	.0910	.0783	.0669	.0569	.0480	.0403	.0336	.0278	18
3	.1920	.1777	.1631	.1484	.1339	.1199	.1065	.0940	.0823	.0716	17
4	.2169	.2131	.2070	.1991	.1897	.1790	.1675	.1553	.1429	.1304	16
5	.1845	.1923	.1979	.2012	.2023	.2013	.1982	.1933	.1868	.1789	15
6	.1226	.1356	.1478	.1589	.1686	.1768	.1833	.1879	.1907	.1916	14
7	.0652	.0765	.0883	.1003	.1124	.1242	.1356	.1462	.1558	.1643	13
8	.0282	.0351	.0429	.0515	.0609	.0709	.0815	.0924	.1034	.1144	12
9	.0100	.0132	.0171	.0217	.0271	.0332	.0402	.0479	.0563	.0654	11

r	p .21	.22	.23	.24	.25	.26	.27	.28	.29	.30	
0	.0113	.0089	.0070	.0054	.0042	.0033	.0025	.0019	.0015	.0011	19
1	.0573	.0477	.0396	.0326	.0268	.0219	.0178	.0144	.0116	.0093	18
2	.1371	.1212	.1064	.0927	.0803	.0692	.0592	.0503	.0426	.0358	17
3	.2065	.1937	.1800	.1659	.1517	.1377	.1240	.1109	.0985	.0869	16
4	.2196	.2185	.2151	.2096	.2023	.1935	.1835	.1726	.1610	.1491	15
5	.1751	.1849	.1928	.1986	.2023	.2040	.2036	.2013	.1973	.1916	14
6	.1086	.1217	.1343	.1463	.1574	.1672	.1757	.1827	.1880	.1916	13
7	.0536	.0637	.0745	.0858	.0974	.1091	.1207	.1320	.1426	.1525	12
8	.0214	.0270	.0334	.0406	.0487	.0575	.0670	.0770	.0874	.0981	11
9	.0069	.0093	.0122	.0157	.0198	.0247	.0303	.0366	.0436	.0514	10
10	.0018	.0026	.0036	.0050	.0066	.0087	.0112	.0142	.0178	.0220	9
11	.0004	.0006	.0009	.0013	.0018	.0025	.0034	.0045	.0060	.0077	8
12	.0001	.0001	.0002	.0003	.0004	.0006	.0008	.0012	.0016	.0022	7
13	.0000	.0000	.0000	.0000	.0001	.0001	.0002	.0002	.0004	.0005	6
14	.0000	.0000	.0000	.0000	.0000	.0000	.0000	.0000	.0001	.0001	5
	.79	.78	.77	.76	.75	.74	.73	.72	.71	.70 p	r

r	p .31	.32	.33	.34	.35	.36	.37	.38	.39	.40	
0	.0009	.0007	.0005	.0004	.0003	.0002	.0002	.0001	.0001	.0001	19
1	.0074	.0059	.0046	.0036	.0029	.0022	.0017	.0013	.0010	.0008	18
2	.0299	.0249	.0206	.0169	.0138	.0112	.0091	.0073	.0058	.0046	17
3	.0762	.0664	.0574	.0494	.0422	.0358	.0302	.0253	.0211	.0175	16
4	.1370	.1249	.1131	.1017	.0909	.0806	.0710	.0621	.0540	.0467	15
5	.1846	.1764	.1672	.1572	.1468	.1360	.1251	.1143	.1036	.0933	14
6	.1935	.1936	.1921	.1890	.1844	.1785	.1714	.1634	.1546	.1451	13
7	.1615	.1692	.1757	.1808	.1844	.1865	.1870	.1860	.1835	.1797	12
8	.1088	.1195	.1298	.1397	.1489	.1573	.1647	.1710	.1760	.1797	11
9	.0597	.0687	.0782	.0880	.0980	.1082	.1182	.1281	.1375	.1464	10
10	.0268	.0323	.0385	.0453	.0528	.0608	.0694	.0785	.0879	.0976	9
11	.0099	.0124	.0155	.0191	.0233	.0280	.0334	.0394	.0460	.0532	8
12	.0030	.0039	.0051	.0066	.0083	.0105	.0131	.0161	.0196	.0237	7
13	.0007	.0010	.0014	.0018	.0024	.0032	.0041	.0053	.0067	.0085	6
14	.0001	.0002	.0003	.0004	.0006	.0008	.0010	.0014	.0018	.0024	5
15	.0000	.0000	.0000	.0001	.0001	.0001	.0002	.0003	.0004	.0005	4
16	.0000	.0000	.0000	.0000	.0000	.0000	.0000	.0000	.0001	.0001	3
	.69	.68	.67	.66	.65	.64	.63	.62	.61	.60 p	r

r	p .41	.42	.43	.44	.45	.46	.47	.48	.49	.50	
1	.0006	.0004	.0003	.0002	.0002	.0001	.0001	.0001	.0001	.0000	18
2	.0037	.0029	.0022	.0017	.0013	.0010	.0008	.0006	.0004	.0003	17
3	.0144	.0118	.0096	.0077	.0062	.0049	.0039	.0031	.0024	.0018	16
4	.0400	.0341	.0289	.0243	.0203	.0168	.0138	.0113	.0092	.0074	15
5	.0834	.0741	.0653	.0572	.0497	.0429	.0368	.0313	.0265	.0222	14
6	.1353	.1252	.1150	.1049	.0949	.0853	.0761	.0674	.0593	.0518	13
7	.1746	.1683	.1611	.1530	.1443	.1350	.1254	.1156	.1058	.0961	12
8	.1820	.1829	.1823	.1803	.1771	.1725	.1668	.1601	.1525	.1442	11
9	.1546	.1618	.1681	.1732	.1771	.1796	.1808	.1806	.1791	.1762	10
10	.1074	.1172	.1268	.1361	.1449	.1530	.1603	.1667	.1721	.1762	9

Note: Read down entire column.

r											
10	.0029	.0041	.0056	.0075	.0099	.0128	.0163	.0205	.0253	.0308	10
11	.0007	.0010	.0015	.0022	.0030	.0041	.0055	.0072	.0094	.0120	9
12	.0001	.0002	.0003	.0005	.0008	.0011	.0015	.0021	.0029	.0039	8
13	.0000	.0000	.0001	.0001	.0002	.0002	.0003	.0005	.0007	.0010	7
14	.0000	.0000	.0000	.0000	.0000	.0000	.0001	.0001	.0001	.0002	6
	.79	.78	.77	.76	.75	.74	.73	.72	.71	.70 p	r

r	p .31	.32	.33	.34	.35	.36	.37	.38	.39	.40	
0	.0006	.0004	.0003	.0002	.0002	.0001	.0001	.0001	.0001	.0000	20
1	.0054	.0042	.0033	.0025	.0020	.0015	.0011	.0009	.0007	.0005	19
2	.0229	.0188	.0153	.0124	.0100	.0080	.0064	.0050	.0040	.0031	18
3	.0619	.0531	.0453	.0383	.0323	.0270	.0224	.0185	.0152	.0123	17
4	.1181	.1062	.0947	.0839	.0738	.0645	.0559	.0482	.0412	.0350	16
5	.1698	.1599	.1493	.1384	.1272	.1161	.1051	.0945	.0843	.0746	15
6	.1907	.1881	.1839	.1782	.1712	.1632	.1543	.1447	.1347	.1244	14
7	.1714	.1770	.1811	.1836	.1844	.1836	.1812	.1774	.1722	.1659	13
8	.1251	.1354	.1450	.1537	.1614	.1678	.1730	.1767	.1790	.1797	12
9	.0750	.0849	.0952	.1056	.1158	.1259	.1354	.1444	.1526	.1597	11
10	.0370	.0440	.0516	.0598	.0686	.0779	.0875	.0974	.1073	.1171	10
11	.0151	.0188	.0231	.0280	.0336	.0398	.0467	.0542	.0624	.0710	9
12	.0051	.0066	.0085	.0108	.0136	.0168	.0206	.0249	.0299	.0355	8
13	.0014	.0019	.0026	.0034	.0045	.0058	.0074	.0094	.0118	.0146	7
14	.0003	.0005	.0006	.0009	.0012	.0016	.0022	.0029	.0038	.0049	6
15	.0001	.0001	.0001	.0002	.0003	.0004	.0005	.0007	.0010	.0013	5
16	.0000	.0000	.0000	.0000	.0000	.0001	.0001	.0001	.0002	.0003	4
	.69	.68	.67	.66	.65	.64	.63	.62	.61	.60 p	r

r	p .41	.42	.43	.44	.45	.46	.47	.48	.49	.50	
1	.0004	.0003	.0002	.0001	.0001	.0001	.0001	.0000	.0000	.0000	19
2	.0024	.0018	.0014	.0011	.0008	.0006	.0005	.0003	.0002	.0002	18
3	.0100	.0080	.0064	.0051	.0040	.0031	.0024	.0019	.0014	.0011	17
4	.0295	.0247	.0206	.0170	.0139	.0113	.0092	.0074	.0059	.0046	16
5	.0656	.0573	.0496	.0427	.0365	.0309	.0260	.0217	.0180	.0148	15
6	.1140	.1037	.0936	.0839	.0746	.0658	.0577	.0501	.0432	.0370	14
7	.1585	.1502	.1413	.1318	.1221	.1122	.1023	.0925	.0830	.0739	13
8	.1790	.1768	.1732	.1683	.1623	.1553	.1474	.1388	.1296	.1201	12
9	.1658	.1707	.1742	.1763	.1771	.1763	.1742	.1708	.1661	.1602	11
10	.1268	.1359	.1446	.1524	.1593	.1652	.1700	.1734	.1755	.1762	10
11	.0801	.0895	.0991	.1089	.1185	.1280	.1370	.1455	.1533	.1602	9
12	.0417	.0486	.0561	.0642	.0727	.0818	.0911	.1007	.1105	.1201	8
13	.0178	.0217	.0260	.0310	.0366	.0429	.0497	.0572	.0653	.0739	7
14	.0062	.0078	.0098	.0122	.0150	.0183	.0221	.0264	.0314	.0370	6
15	.0017	.0023	.0030	.0038	.0049	.0062	.0078	.0098	.0121	.0148	5
16	.0004	.0005	.0007	.0009	.0013	.0017	.0022	.0028	.0036	.0046	4
17	.0001	.0001	.0001	.0002	.0002	.0003	.0005	.0006	.0008	.0011	3
18	.0000	.0000	.0000	.0000	.0000	.0000	.0001	.0001	.0001	.0002	2
	.59	.58	.57	.56	.55	.54	.53	.52	.51	.50 p	r

Table 2 (continued)

n = 50

r	*p* .01	.02	.03	.04	.05	.06	.07	.08	.09	.10	
0	.6050	.3642	.2181	.1299	.0769	.0453	.0266	.0155	.0090	.0052	50
1	.3056	.3716	.3372	.2706	.2025	.1447	.0999	.0672	.0443	.0286	49
2	.0756	.1858	.2555	.2762	.2611	.2262	.1843	.1433	.1073	.0779	48
3	.0122	.0607	.1264	.1842	.2199	.2311	.2219	.1993	.1698	.1386	47
4	.0015	.0145	.0459	.0902	.1360	.1733	.1963	.2037	.1973	.1809	46
5	.0001	.0027	.0131	.0346	.0658	.1018	.1359	.1629	.1795	.1849	45
6	.0000	.0004	.0030	.0108	.0260	.0487	.0767	.1063	.1332	.1541	44
7	.0000	.0001	.0006	.0028	.0086	.0195	.0363	.0581	.0828	.1076	43
8	.0000	.0000	.0001	.0006	.0024	.0067	.0147	.0271	.0440	.0643	42
9	.0000	.0000	.0000	.0001	.0006	.0020	.0052	.0110	.0203	.0333	41
10	.0000	.0000	.0000	.0000	.0001	.0005	.0016	.0039	.0082	.0152	40
11	.0000	.0000	.0000	.0000	.0000	.0001	.0004	.0012	.0030	.0061	39
12	.0000	.0000	.0000	.0000	.0000	.0000	.0001	.0004	.0010	.0022	38
13	.0000	.0000	.0000	.0000	.0000	.0000	.0000	.0001	.0003	.0007	37
14	.0000	.0000	.0000	.0000	.0000	.0000	.0000	.0000	.0001	.0002	36
15	.0000	.0000	.0000	.0000	.0000	.0000	.0000	.0000	.0000	.0001	35
	.99	.98	.97	.96	.95	.94	.93	.92	.91	.90 *p*	*r*

r	*p* .11	.12	.13	.14	.15	.16	.17	.18	.19	.20	
0	.0029	.0017	.0009	.0005	.0003	.0002	.0001	.0000	.0000	.0000	50
1	.0182	.0114	.0071	.0043	.0026	.0016	.0009	.0005	.0003	.0002	49
2	.0552	.0382	.0259	.0172	.0113	.0073	.0046	.0029	.0018	.0011	48
3	.1091	.0833	.0619	.0449	.0319	.0222	.0151	.0102	.0067	.0044	47
4	.1584	.1334	.1086	.0858	.0661	.0496	.0364	.0262	.0185	.0128	46
5	.1801	.1674	.1493	.1286	.1072	.0869	.0687	.0530	.0400	.0295	45
6	.1670	.1712	.1674	.1570	.1419	.1242	.1055	.0872	.0703	.0554	44
7	.1297	.1467	.1572	.1606	.1575	.1487	.1358	.1203	.1037	.0870	43
8	.0862	.1075	.1262	.1406	.1493	.1523	.1495	.1420	.1307	.1169	42
9	.0497	.0684	.0880	.1068	.1230	.1353	.1429	.1454	.1431	.1364	41
10	.0252	.0383	.0539	.0713	.0890	.1057	.1200	.1309	.1376	.1398	40
11	.0113	.0190	.0293	.0422	.0571	.0732	.0894	.1045	.1174	.1271	39
12	.0045	.0084	.0142	.0223	.0328	.0453	.0595	.0745	.0895	.1033	38
13	.0016	.0034	.0062	.0106	.0169	.0252	.0356	.0478	.0613	.0755	37
14	.0005	.0012	.0025	.0046	.0079	.0127	.0193	.0277	.0380	.0499	36
15	.0002	.0004	.0009	.0018	.0033	.0058	.0095	.0146	.0214	.0299	35
16	.0000	.0001	.0003	.0006	.0013	.0024	.0042	.0070	.0110	.0164	34
17	.0000	.0000	.0001	.0002	.0005	.0009	.0017	.0031	.0052	.0082	33
18	.0000	.0000	.0000	.0001	.0001	.0003	.0007	.0012	.0022	.0037	32
19	.0000	.0000	.0000	.0000	.0000	.0001	.0002	.0005	.0009	.0016	31
20	.0000	.0000	.0000	.0000	.0000	.0000	.0001	.0002	.0003	.0006	30
21	.0000	.0000	.0000	.0000	.0000	.0000	.0000	.0000	.0001	.0002	29
22	.0000	.0000	.0000	.0000	.0000	.0000	.0000	.0000	.0000	.0001	28
	.89	.88	.87	.86	.85	.84	.83	.82	.81	.80 *p*	*r*

r	*p* 21	22	23	24	25	26	27	28	29	30	
1	.0001	.0001	.0000	.0000	.0000	.0000	.0000	.0000	.0000	.0000	49
2	.0007	.0004	.0002	.0001	.0001	.0000	.0000	.0000	.0000	.0000	48
3	.0028	.0018	.0011	.0007	.0004	.0002	.0001	.0001	.0000	.0000	47
4	.0088	.0059	.0039	.0025	.0016	.0010	.0006	.0004	.0002	.0001	46
5	.0214	.0152	.0106	.0073	.0049	.0033	.0021	.0014	.0009	.0006	45
29	.0000	.0001	.0002	.0003	.0005	.0008	.0012	.0019	.0029	.0043	21
30	.0000	.0000	.0001	.0001	.0002	.0003	.0005	.0008	.0013	.0020	20
31	.0000	.0000	.0000	.0000	.0001	.0001	.0002	.0003	.0005	.0009	19
32	.0000	.0000	.0000	.0000	.0000	.0000	.0001	.0001	.0002	.0003	18
33	.0000	.0000	.0000	.0000	.0000	.0000	.0000	.0000	.0001	.0001	17
	.69	.68	.67	.66	.65	.64	.63	.62	.61	.60 *p*	*r*

r	*p* .41	.42	.43	.44	.45	.46	.47	.48	.49	.50	
8	.0001	.0001	.0000	.0000	.0000	.0000	.0000	.0000	.0000	.0000	42
9	.0003	.0002	.0001	.0001	.0000	.0000	.0000	.0000	.0000	.0000	41
10	.0009	.0006	.0004	.0002	.0001	.0001	.0001	.0000	.0000	.0000	40
11	.0024	.0016	.0010	.0007	.0004	.0003	.0002	.0001	.0001	.0000	39
12	.0054	.0037	.0026	.0017	.0011	.0007	.0005	.0003	.0002	.0001	38
13	.0109	.0079	.0057	.0040	.0027	.0018	.0012	.0008	.0005	.0003	37
14	.0200	.0152	.0113	.0082	.0059	.0041	.0029	.0019	.0013	.0008	36
15	.0334	.0264	.0204	.0155	.0116	.0085	.0061	.0043	.0030	.0020	35
16	.0508	.0418	.0337	.0267	.0207	.0158	.0118	.0086	.0062	.0044	34
17	.0706	.0605	.0508	.0419	.0339	.0269	.0209	.0159	.0119	.0087	33
18	.0899	.0803	.0703	.0604	.0508	.0420	.0340	.0270	.0210	.0160	32
19	.1053	.0979	.0893	.0799	.0700	.0602	.0507	.0419	.0340	.0270	31
20	.1134	.1099	.1044	.0973	.0888	.0795	.0697	.0600	.0506	.0419	30
21	.1126	.1137	.1126	.1092	.1038	.0967	.0884	.0791	.0695	.0598	29
22	.1031	.1086	.1119	.1131	.1119	.1086	.1033	.0963	.0880	.0788	28
23	.0872	.0957	.1028	.1082	.1115	.1126	.1115	.1082	.1029	.0960	27
24	.0682	.0780	.0872	.0956	.1026	.1079	.1112	.1124	.1112	.1080	26
25	.0493	.0587	.0684	.0781	.0873	.0956	.1026	.1079	.1112	.1123	25
26	.0329	.0409	.0497	.0590	.0687	.0783	.0875	.0957	.1027	.1080	24
27	.0203	.0263	.0333	.0412	.0500	.0593	.0690	.0786	.0877	.0960	23
28	.0116	.0157	.0206	.0266	.0336	.0415	.0502	.0596	.0692	.0788	22
29	.0061	.0086	.0118	.0159	.0208	.0268	.0338	.0417	.0504	.0598	21
30	.0030	.0044	.0062	.0087	.0119	.0160	.0210	.0270	.0339	.0419	20
31	.0013	.0020	.0030	.0044	.0063	.0088	.0120	.0161	.0210	.0270	19
32	.0006	.0009	.0014	.0021	.0031	.0044	.0063	.0088	.0120	.0160	18
33	.0002	.0003	.0006	.0009	.0014	.0021	.0031	.0044	.0063	.0087	17
34	.0001	.0001	.0002	.0003	.0006	.0009	.0014	.0020	.0030	.0044	16
35	.0000	.0000	.0001	.0001	.0002	.0003	.0005	.0009	.0013	.0020	15
36	.0000	.0000	.0000	.0000	.0001	.0001	.0002	.0003	.0005	.0008	14
37	.0000	.0000	.0000	.0000	.0000	.0000	.0001	.0001	.0002	.0003	13
38	.0000.	.0000	.0000	.0000	.0000	.0000	.0000	.0000	.0001	.0001	12
	.59	.58	.57	.56	.55	.54	.53	.52	.51	.50 *p*	*r*

n = 100

r	*p* .01	.02	.03	.04	.05	.06	.07	.08	.09	.10	
0	.3660	.1326	.0476	.0169	.0059	.0021	.0007	.0002	.0001	.0000	100
1	.3697	.2707	.1471	.0703	.0312	.0131	.0053	.0021	.0008	.0003	99
2	.1849	.2734	.2252	.1450	.0812	.0414	.0198	.0090	.0039	.0016	98
3	.0610	.1823	.2275	.1973	.1396	.0864	.0486	.0254	.0125	.0059	97
4	.0149	.0902	.1706	.1994	.1781	.1338	.0888	.0536	.0301	.0159	96

6		.0427	.0322	.0238	.0173	.0123	.0087	.0060	.0040	.0027	.0018	44
7		.0713	.0571	.0447	.0344	.0259	.0191	.0139	.0099	.0069	.0048	43
8		.1019	.0865	.0718	.0583	.0463	.0361	.0276	.0207	.0152	.0110	42
9		.1263	.1139	.1001	.0859	.0721	.0592	.0476	.0375	.0290	.0220	41
10		.1377	.1317	.1226	.1113	.0985	.0852	.0721	.0598	.0485	.0386	40
11		.1331	.1351	.1332	.1278	.1194	.1089	.0970	.0845	.0721	.0602	39
12		.1150	.1238	.1293	.1311	.1294	.1244	.1166	.1068	.0957	.0838	38
13		.0894	.1021	.1129	.1210	.1261	.1277	.1261	.1215	.1142	.1050	37
14		.0628	.0761	.0891	.1010	.1110	.1186	.1233	.1248	.1233	.1189	36
15		.0400	.0515	.0639	.0766	.0888	.1000	.1094	.1165	.1209	.1223	35
16		.0233	.0318	.0417	.0529	.0648	.0769	.0885	.0991	.1080	.1147	34
17		.0124	.0179	.0249	.0334	.0432	.0540	.0655	.0771	.0882	.0983	33
18		.0060	.0093	.0137	.0193	.0264	.0348	.0444	.0550	.0661	.0772	32
19		.0027	.0044	.0069	.0103	.0148	.0206	.0277	.0360	.0454	.0558	31
20		.0011	.0019	.0032	.0050	.0077	.0112	.0159	.0217	.0288	.0370	30
21		.0004	.0008	.0014	.0023	.0036	.0056	.0084	.0121	.0168	.0227	29
22		.0001	.0003	.0005	.0009	.0016	.0026	.0041	.0062	.0090	.0128	28
23		.0000	.0001	.0002	.0004	.0006	.0011	.0018	.0029	.0045	.0067	27
24		.0000	.0000	.0001	.0001	.0002	.0004	.0008	.0013	.0021	.0032	26
25		.0000	.0000	.0000	.0000	.0001	.0002	.0003	.0005	.0009	.0014	25
26		.0000	.0000	.0000	.0000	.0000	.0001	.0001	.0002	.0003	.0006	24
27		.0000	.0000	.0000	.0000	.0000	.0000	.0000	.0001	.0001	.0002	23
28		.0000	.0000	.0000	.0000	.0000	.0000	.0000	.0000	.0000	.0001	22
		.79	.78	.77	.76	.75	.74	.73	.72	.71	.70 p	r

r	p	.31	.32	.33	.34	.35	.36	.37	.38	.39	.40	
4		.0001	.0000	.0000	.0000	.0000	.0000	.0000	.0000	.0000	.0000	46
5		.0003	.0002	.0001	.0001	.0000	.0000	.0000	.0000	.0000	.0000	45
6		.0011	.0007	.0005	.0003	.0002	.0001	.0001	.0000	.0000	.0000	44
7		.0032	.0022	.0014	.0009	.0006	.0004	.0002	.0001	.0001	.0000	43
8		.0078	.0055	.0037	.0025	.0017	.0011	.0007	.0004	.0003	.0002	42
9		.0164	.0120	.0086	.0061	.0042	.0029	.0019	.0013	.0008	.0005	41
10		.0301	.0231	.0174	.0128	.0093	.0066	.0046	.0032	.0022	.0014	40
11		.0493	.0395	.0311	.0240	.0182	.0136	.0099	.0071	.0050	.0035	39
12		.0719	.0604	.0498	.0402	.0319	.0248	.0189	.0142	.0105	.0076	38
13		.0944	.0831	.0717	.0606	.0502	.0408	.0325	.0255	.0195	.0147	37
14		.1121	.1034	.0933	.0825	.0714	.0607	.0505	.0412	.0330	.0260	36
15		.1209	.1168	.1103	.1020	.0923	.0819	.0712	.0606	.0507	.0415	35
16		.1188	.1202	.1189	.1149	.1088	.1008	.0914	.0813	.0709	.0606	34
17		.1068	.1132	.1171	.1184	.1171	.1133	.1074	.0997	.0906	.0808	33
18		.0880	.0976	.1057	.1118	.1156	.1169	.1156	.1120	.1062	.0987	32
19		.0666	.0774	.0877	.0970	.1048	.1107	.1144	.1156	.1144	.1109	31
20		.0463	.0564	.0670	.0775	.0875	.0965	.1041	.1098	.1134	.1146	30
21		.0297	.0379	.0471	.0570	.0673	.0776	.0874	.0962	.1035	.1091	29
22		.0176	.0235	.0306	.0387	.0478	.0575	.0676	.0777	.0873	.0959	28
23		.0096	.0135	.0183	.0243	.0313	.0394	.0484	.0580	.0679	.0778	27
24		.0049	.0071	.0102	.0141	.0190	.0249	.0319	.0400	.0489	.0584	26
25		.0023	.0035	.0052	.0075	.0106	.0146	.0195	.0255	.0325	.0405	25
26		.0010	.0016	.0025	.0037	.0055	.0079	.0110	.0150	.0200	.0259	24
27		.0004	.0007	.0011	.0017	.0026	.0039	.0058	.0082	.0113	.0154	23
28		.0001	.0003	.0004	.0007	.0012	.0018	.0028	.0041	.0060	.0084	22

5		.0029	.0353	.1013	.1595	.1800	.1639	.1283	.0895	.0571	.0339	95
6		.0005	.0114	.0496	.1052	.1500	.1657	.1529	.1233	.0895	.0596	94
7		.0001	.0031	.0206	.0589	.1060	.1420	.1545	.1440	.1188	.0889	93
8		.0000	.0007	.0074	.0285	.0649	.1054	.1352	.1455	.1366	.1148	92
9		.0000	.0002	.0023	.0121	.0349	.0687	.1040	.1293	.1381	.1304	91
10		.0000	.0000	.0007	.0046	.0167	.0399	.0712	.1024	.1243	.1319	90
11		.0000	.0000	.0002	.0016	.0072	.0209	.0439	.0728	.1006	.1199	89
12		.0000	.0000	.0000	.0005	.0028	.0099	.0245	.0470	.0738	.0988	88
13		.0000	.0000	.0000	.0001	.0010	.0043	.0125	.0276	.0494	.0743	87
14		.0000	.0000	.0000	.0000	.0003	.0017	.0058	.0149	.0304	.0513	86
15		.0000	.0000	.0000	.0000	.0001	.0006	.0025	.0074	.0172	.0327	85
16		.0000	.0000	.0000	.0000	.0000	.0002	.0010	.0034	.0090	.0193	84
17		.0000	.0000	.0000	.0000	.0000	.0001	.0004	.0015	.0044	.0106	83
18		.0000	.0000	.0000	.0000	.0000	.0000	.0001	.0006	.0020	.0054	82
19		.0000	.0000	.0000	.0000	.0000	.0000	.0000	.0002	.0009	.0026	81
20		.0000	.0000	.0000	.0000	.0000	.0000	.0000	.0001	.0003	.0012	80
21		.0000	.0000	.0000	.0000	.0000	.0000	.0000	.0000	.0001	.0005	79
22		.0000	.0000	.0000	.0000	.0000	.0000	.0000	.0000	.0000	.0002	78
23		.0000	.0000	.0000	.0000	.0000	.0000	.0000	.0000	.0000	.0001	77
		.99	.98	.97	.96	.95	.94	.93	.92	.91	.90 p	r

r	p	.11	.12	.13	.14	.15	.16	.17	.18	.19	.20	
1		.0001	.0000	.0000	.0000	.0000	.0000	.0000	.0000	.0000	.0000	99
2		.0007	.0003	.0001	.0000	.0000	.0000	.0000	.0000	.0000	.0000	98
3		.0027	.0012	.0005	.0002	.0001	.0000	.0000	.0000	.0000	.0000	97
4		.0080	.0038	.0018	.0008	.0003	.0001	.0001	.0000	.0000	.0000	96
5		.0189	.0100	.0050	.0024	.0011	.0005	.0002	.0001	.0000	.0000	95
6		.0369	.0215	.0119	.0063	.0031	.0015	.0007	.0003	.0001	.0001	94
7		.0613	.0394	.0238	.0137	.0075	.0039	.0020	.0009	.0004	.0002	93
8		.0881	.0625	.0414	.0259	.0153	.0086	.0047	.0024	.0012	.0006	92
9		.1112	.0871	.0632	.0430	.0276	.0168	.0098	.0054	.0029	.0015	91
10		.1251	.1080	.0860	.0637	.0444	.0292	.0182	.0108	.0062	.0034	90
11		.1265	.1205	.1051	.0849	.0640	.0454	.0305	.0194	.0118	.0069	89
12		.1160	.1219	.1165	.1025	.0838	.0642	.0463	.0316	.0206	.0128	88
13		.0970	.1125	.1179	.1130	.1001	.0827	.0642	.0470	.0327	.0216	87
14		.0745	.0954	.1094	.1143	.1098	.0979	.0817	.0641	.0476	.0335	86
15		.0528	.0745	.0938	.1067	.1111	.1070	.0960	.0807	.0640	.0481	85
16		.0347	.0540	.0744	.0922	.1041	.1082	.1044	.0941	.0798	.0638	84
17		.0212	.0364	.0549	.0742	.0908	.1019	.1057	.1021	.0924	.0789	83
18		.0121	.0229	.0379	.0557	.0739	.0895	.0998	.1033	.1000	.0909	82
19		.0064	.0135	.0244	.0391	.0563	.0736	.0882	.0979	.1012	.0981	81
20		.0032	.0074	.0148	.0258	.0402	.0567	.0732	.0870	.0962	.0993	80
21		.0015	.0039	.0084	.0160	.0270	.0412	.0571	.0728	.0859	.0966	79
22		.0007	.0019	.0045	.0094	.0171	.0282	.0420	.0574	.0724	.0849	78
23		.0003	.0009	.0023	.0052	.0103	.0182	.0292	.0427	.0576	.0720	77
24		.0001	.0004	.0011	.0027	.0058	.0111	.0192	.0301	.0433	.0577	76
25		.0000	.0002	.0005	.0013	.0031	.0064	.0119	.0201	.0309	.0439	75
26		.0000	.0001	.0002	.0006	.0016	.0035	.0071	.0127	.0209	.0316	74
27		.0000	.0000	.0001	.0003	.0008	.0018	.0040	.0076	.0134	.0217	73
28		.0000	.0000	.0000	.0001	.0004	.0009	.0021	.0044	.0082	.0141	72
29		.0000	.0000	.0000	.0000	.0002	.0004	.0011	.0024	.0048	.0088	71
30		.0000	.0000	.0000	.0000	.0001	.0002	.0005	.0012	.0027	.0052	70

Note: Read down entire column.

Table 2 (continued)

31		.0000	.0000	.0000	.0000	.0000	.0001	.0002	.0006	.0014	.0029	69
32		.0000	.0000	.0000	.0000	.0000	.0000	.0001	.0003	.0007	.0016	68
33		.0000	.0000	.0000	.0000	.0000	.0000	.0000	.0001	.0003	.0008	67
34		.0000	.0000	.0000	.0000	.0000	.0000	.0000	.0001	.0002	.0004	66
35		.0000	.0000	.0000	.0000	.0000	.0000	.0000	.0000	.0001	.0002	65
36		.0000	.0000	.0000	.0000	.0000	.0000	.0000	.0000	.0000	.0001	64
		.89	.88	.87	.86	.85	.84	.83	.82	.81	.80 *p*	*r*

r	*p*	.21	.22	.23	.24	.25	.26	.27	.28	.29	.30	
7		.0001	.0000	.0000	.0000	.0000	.0000	.0000	.0000	.0000	.0000	93
8		.0003	.0001	.0001	.0000	.0000	.0000	.0000	.0000	.0000	.0000	92
9		.0007	.0003	.0002	.0001	.0000	.0000	.0000	.0000	.0000	.0000	91
10		.0018	.0009	.0004	.0002	.0001	.0000	.0000	.0000	.0000	.0000	90
11		.0038	.0021	.0011	.0005	.0003	.0001	.0001	.0000	.0000	.0000	89
12		.0076	.0043	.0024	.0012	.0006	.0003	.0001	.0001	.0000	.0000	88
13		.0136	.0082	.0048	.0027	.0014	.0007	.0004	.0002	.0001	.0000	87
14		.0225	.0144	.0089	.0052	.0030	.0016	.0009	.0004	.0002	.0001	86
15		.0343	.0233	.0152	.0095	.0057	.0033	.0018	.0010	.0005	.0002	85
16		.0484	.0350	.0241	.0159	.0100	.0061	.0035	.0020	.0011	.0006	84
17		.0636	.0487	.0356	.0248	.0165	.0106	.0065	.0038	.0022	.0012	83
18		.0780	.0634	.0490	.0361	.0254	.0171	.0111	.0069	.0041	.0024	82
19		.0895	.0772	.0631	.0492	.0365	.0259	.0177	.0115	.0072	.0044	81
20		.0963	.0881	.0764	.0629	.0493	.0369	.0264	.0182	.0120	.0076	80
21		.0975	.0947	.0869	.0756	.0626	.0494	.0373	.0269	.0186	.0124	79
22		.0931	.0959	.0932	.0858	.0749	.0623	.0495	.0376	.0273	.0190	78
23		.0839	.0917	.0944	.0919	.0847	.0743	.0621	.0495	.0378	.0277	77
24		.0716	.0830	.0905	.0931	.0906	.0837	.0736	.0618	.0496	.0380	76
25		.0578	.0712	.0822	.0893	.0918	.0894	.0828	.0731	.0615	.0496	75
26		.0444	.0579	.0708	.0814	.0883	.0906	.0883	.0819	.0725	.0613	74
27		.0323	.0448	.0580	.0704	.0806	.0873	.0896	.0873	.0812	.0720	73
28		.0224	.0329	.0451	.0580	.0701	.0799	.0864	.0886	.0864	.0804	72
29		.0148	.0231	.0335	.0455	.0580	.0697	.0793	.0855	.0876	.0856	71
30		.0093	.0154	.0237	.0340	.0458	.0580	.0694	.0787	.0847	.0868	70
31		.0056	.0098	.0160	.0242	.0344	.0460	.0580	.0691	.0781	.0840	69
32		.0032	.0060	.0103	.0165	.0248	.0349	.0462	.0579	.0688	.0776	68
33		.0018	.0035	.0063	.0107	.0170	.0252	.0352	.0464	.0579	.0685	67
34		.0009	.0019	.0037	.0067	.0112	.0175	.0257	.0356	.0466	.0579	66
35		.0005	.0010	.0021	.0040	.0070	.0116	.0179	.0261	.0359	.0468	65
36		.0002	.0005	.0011	.0023	.0042	.0073	.0120	.0183	.0265	.0362	64
37		.0001	.0003	.0006	.0012	.0024	.0045	.0077	.0123	.0187	.0268	63
38		.0000	.0001	.0003	.0006	.0013	.0026	.0047	.0079	.0127	.0191	62
39		.0000	.0001	.0001	.0003	.0007	.0015	.0028	.0049	.0082	.0130	61
40		.0000	.0000	.0001	.0002	.0004	.0008	.0016	.0029	.0051	.0085	60
41		.0000	.0000	.0000	.0001	.0002	.0004	.0008	.0017	.0031	.0053	59
42		.0000	.0000	.0000	.0000	.0001	.0002	.0004	.0009	.0018	.0032	58
43		.0000	.0000	.0000	.0000	.0000	.0001	.0002	.0005	.0010	.0019	57
44		.0000	.0000	.0000	.0000	.0000	.0000	.0001	.0002	.0005	.0010	56
45		.0000	.0000	.0000	.0000	.0000	.0000	.0000	.0001	.0003	.0005	55
46		.0000	.0000	.0000	.0000	.0000	.0000	.0000	.0001	.0001	.0003	54

35		.0578	.0680	.0763	.0816	.0834	.0816	.0765	.0688	.0593	.0491	65
36		.0469	.0578	.0678	.0759	.0811	.0829	.0811	.0761	.0685	.0591	64
37		.0365	.0471	.0578	.0676	.0755	.0806	.0824	.0807	.0757	.0682	63
38		.0272	.0367	.0472	.0577	.0674	.0752	.0802	.0820	.0803	.0754	62
39		.0194	.0275	.0369	.0473	.0577	.0672	.0749	.0799	.0816	.0799	61
40		.0133	.0197	.0277	.0372	.0474	.0577	.0671	.0746	.0795	.0812	60
41		.0087	.0136	.0200	.0280	.0373	.0475	.0577	.0670	.0744	.0792	59
42		.0055	.0090	.0138	.0203	.0282	.0375	.0476	.0576	.0668	.0742	58
43		.0033	.0057	.0092	.0141	.0205	.0285	.0377	.0477	.0576	.0667	57
44		.0019	.0035	.0059	.0094	.0143	.0207	.0287	.0378	.0477	.0576	56
45		.0011	.0020	.0036	.0060	.0096	.0145	.0210	.0289	.0380	.0478	55
46		.0006	.0011	.0021	.0037	.0062	.0098	.0147	.0211	.0290	.0381	54
47		.0003	.0006	.0012	.0022	.0038	.0063	.0099	.0149	.0213	.0292	53
48		.0001	.0003	.0007	.0012	.0023	.0039	.0064	.0101	.0151	.0215	52
49		.0001	.0002	.0003	.0007	.0013	.0023	.0040	.0066	.0102	.0152	51
50		.0000	.0001	.0002	.0004	.0007	.0013	.0024	.0041	.0067	.0103	50
51		.0000	.0000	.0001	.0002	.0004	.0007	.0014	.0025	.0042	.0068	49
52		.0000	.0000	.0000	.0001	.0002	.0004	.0008	.0014	.0025	.0042	48
53		.0000	.0000	.0000	.0000	.0001	.0002	.0004	.0008	.0015	.0026	47
54		.0000	.0000	.0000	.0000	.0000	.0001	.0002	.0004	.0008	.0015	46
55		.0000	.0000	.0000	.0000	.0000	.0000	.0001	.0002	.0004	.0008	45
56		.0000	.0000	.0000	.0000	.0000	.0000	.0000	.0001	.0002	.0004	44
57		.0000	.0000	.0000	.0000	.0000	.0000	.0000	.0001	.0001	.0002	43
58		.0000	.0000	.0000	.0000	.0000	.0000	.0000	.0000	.0001	.0001	42
59		.0000	.0000	.0000	.0000	.0000	.0000	.0000	.0000	.0000	.0001	41
		.69	.68	.67	.66	.65	.64	.63	.62	.61	.60 *p*	*r*

r	*p*	.41	.42	.43	.44	.45	.46	.47	.48	.49	.50	
23		.0001	.0000	.0000	.0000	.0000	.0000	.0000	.0000	.0000	.0000	77
24		.0002	.0001	.0000	.0000	.0000	.0000	.0000	.0000	.0000	.0000	76
25		.0003	.0002	.0001	.0000	.0000	.0000	.0000	.0000	.0000	.0000	75
26		.0007	.0003	.0002	.0001	.0000	.0000	.0000	.0000	.0000	.0000	74
27		.0013	.0007	.0004	.0002	.0001	.0000	.0000	.0000	.0000	.0000	73
28		.0023	.0013	.0007	.0004	.0002	.0001	.0000	.0000	.0000	.0000	72
29		.0039	.0024	.0014	.0008	.0004	.0002	.0001	.0000	.0000	.0000	71
30		.0065	.0040	.0024	.0014	.0008	.0004	.0002	.0001	.0001	.0000	70
31		.0102	.0066	.0041	.0025	.0014	.0008	.0004	.0002	.0001	.0001	69
32		.0152	.0103	.0067	.0042	.0025	.0015	.0008	.0004	.0002	.0001	68
33		.0218	.0154	.0104	.0068	.0043	.0026	.0015	.0008	.0004	.0002	67
34		.0298	.0219	.0155	.0105	.0069	.0043	.0026	.0015	.0009	.0005	66
35		.0391	.0299	.0220	.0156	.0106	.0069	.0044	.0026	.0015	.0009	65
36		.0491	.0391	.0300	.0221	.0157	.0107	.0070	.0044	.0027	.0016	64
37		.0590	.0490	.0391	.0300	.0222	.0157	.0107	.0070	.0044	.0027	63
38		.0680	.0588	.0489	.0391	.0301	.0222	.0158	.0108	.0071	.0045	62
39		.0751	.0677	.0587	.0489	.0391	.0301	.0223	.0158	.0108	.0071	61
40		.0796	.0748	.0675	.0586	.0488	.0391	.0301	.0223	.0159	.0108	60
41		.0809	.0793	.0745	.0673	.0584	.0487	.0391	.0301	.0223	.0159	59
42		.0790	.0806	.0790	.0743	.0672	.0583	.0487	.0390	.0301	.0223	58
43		.0740	.0787	.0804	.0788	.0741	.0670	.0582	.0486	.0390	.0301	57
44		.0666	.0739	.0785	.0802	.0786	.0739	.0669	.0581	.0485	.0390	56
45		.0576	.0666	.0737	.0784	.0800	.0784	.0738	.0668	.0580	.0485	55
46		.0479	.0576	.0665	.0736	.0782	.0798	.0783	.0737	.0667	.0580	54
47		.0382	.0480	.0576	.0665	.0736	.0781	.0797	.0781	.0736	.0666	53

47	.0000	.0000	.0000	.0000	.0000	.0000	.0000	.0000	.0001	.0001	53
48	.0000	.0000	.0000	.0000	.0000	.0000	.0000	.0000	.0000	.0001	52
	.79	.78	.77	.76	.75	.74	.73	.72	.71	.70 *p*	*r*

r *p*	.31	.32	.33	.34	.35	.36	.37	.38	.39	.40	
15	.0001	.0001	.0000	.0000	.0000	.0000	.0000	.0000	.0000	.0000	85
16	.0003	.0001	.0001	.0000	.0000	.0000	.0000	.0000	.0000	.0000	84
17	.0006	.0003	.0002	.0001	.0000	.0000	.0000	.0000	.0000	.0000	83
18	.0013	.0007	.0004	.0002	.0001	.0000	.0000	.0000	.0000	.0000	82
19	.0025	.0014	.0008	.0004	.0002	.0001	.0000	.0000	.0000	.0000	81
20	.0046	.0027	.0015	.0008	.0004	.0002	.0001	.0001	.0000	.0000	80
21	.0079	.0049	.0029	.0016	.0009	.0005	.0002	.0001	.0001	.0000	79
22	.0127	.0082	.0051	.0030	.0017	.0010	.0005	.0003	.0001	.0001	78
23	.0194	.0131	.0085	.0053	.0032	.0018	.0010	.0006	.0003	.0001	77
24	.0280	.0198	.0134	.0088	.0055	.0033	.0019	.0011	.0006	.0003	76
25	.0382	.0283	.0201	.0137	.0090	.0057	.0035	.0020	.0012	.0006	75
26	.0496	.0384	.0286	.0204	.0140	.0092	.0059	.0036	.0021	.0012	74
27	.0610	.0495	.0386	.0288	.0207	.0143	.0095	.0060	.0037	.0022	73
28	.0715	.0608	.0495	.0387	.0290	.0209	.0145	.0097	.0062	.0038	72
29	.0797	.0710	.0605	.0495	.0388	.0292	.0211	.0147	.0098	.0063	71
30	.0848	.0791	.0706	.0603	.0494	.0389	.0294	.0213	.0149	.0100	70
31	.0860	.0840	.0785	.0702	.0601	.0494	.0389	.0295	.0215	.0151	69
32	.0833	.0853	.0833	.0779	.0698	.0599	.0493	.0390	.0296	.0217	68
33	.0771	.0827	.0846	.0827	.0774	.0694	.0597	.0493	.0390	.0297	67
34	.0683	.0767	.0821	.0840	.0821	.0769	.0691	.0595	.0492	.0391	66

48	.0293	.0383	.0480	.0577	.0665	.0735	.0781	.0797	.0781	.0735	52
49	.0216	.0295	.0384	.0481	.0577	.0664	.0735	.0780	.0796	.0780	51
50	.0153	.0218	.0296	.0385	.0482	.0577	.0665	.0735	.0780	.0796	50
51	.0104	.0155	.0219	.0297	.0386	.0482	.0578	.0665	.0735	.0780	49
52	.0068	.0105	.0156	.0220	.0298	.0387	.0483	.0578	.0665	.0735	48
53	.0043	.0069	.0106	.0156	.0221	.0299	.0388	.0483	.0579	.0666	47
54	.0026	.0044	.0070	.0107	.0157	.0221	.0299	.0388	.0484	.0580	46
55	.0015	.0026	.0044	.0070	.0108	.0158	.0222	.0300	.0389	.0485	45
56	.0008	.0015	.0027	.0044	.0071	.0108	.0158	.0222	.0300	.0390	44
57	.0005	.0009	.0016	.0027	.0045	.0071	.0108	.0158	.0223	.0301	43
58	.0002	.0005	.0009	.0016	.0027	.0045	.0071	.0108	.0159	.0223	42
59	.0001	.0002	.0005	.0009	.0016	.0027	.0045	.0071	.0109	.0159	41
60	.0001	.0001	.0002	.0005	.0009	.0016	.0027	.0045	.0071	.0108	40
61	.0000	.0001	.0001	.0002	.0005	.0009	.0016	.0027	.0045	.0071	39
62	.0000	.0000	.0001	.0001	.0002	.0005	.0009	.0016	.0027	.0045	38
63	.0000	.0000	.0000	.0001	.0001	.0002	.0005	.0009	.0016	.0027	37
64	.0000	.0000	.0000	.0000	.0001	.0001	.0002	.0005	.0009	.0016	36
65	.0000	.0000	.0000	.0000	.0000	.0001	.0001	.0002	.0005	.0009	35
66	.0000	.0000	.0000	.0000	.0000	.0000	.0001	.0001	.0002	.0005	34
67	.0000	.0000	.0000	.0000	.0000	.0000	.0000	.0001	.0001	.0002	33
68	.0000	.0000	.0000	.0000	.0000	.0000	.0000	.0000	.0001	.0001	32
69	.0000	.0000	.0000	.0000	.0000	.0000	.0000	.0000	.0000	.0001	31
	.59	.58	.57	.56	.55	.54	.53	.52	.51	.50 *p*	*r*

Note: Read down entire column.

Table 3. Cumulative Standard Normal Probabilities

This table gives values of the cumulative distribution function of the standard normal distribution,

$$F(z) = \int_{-\infty}^{z} \frac{1}{\sqrt{2\pi}} e^{-z^2/2}\, dz,$$

for

$$z = 0(.01)2.60(.10)3.00(.20)4.00(.50)5.50.$$

To find $F(z)$ for $z < 0$, take $F(z) = 1 - F(-z)$.

Examples:

$$F(1.12) = .8686431$$

and $$F(-.67) = 1 - F(.67) = 1 - .7485711 = .2514289.$$

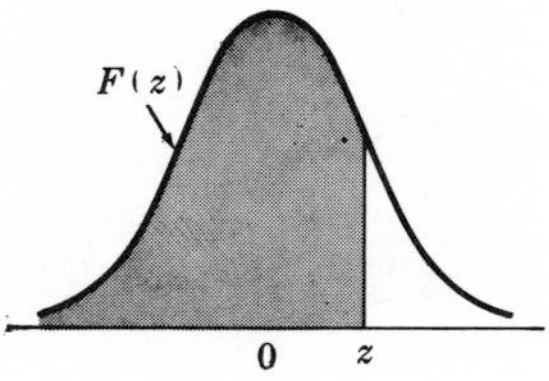

This table is condensed from Table 1 of *Biometrika Tables for Statisticians,* Vol. 1, by Pearson and Hartley, and is reproduced with the permission of E. S. Pearson and the trustees of *Biometrika.*

Table 3

z	F(z)	z	F(z)	z	F(z)	z	F(z)
.00	.5000000	.36	.6405764	.72	.7642375	1.08	.8599289
.01	.5039894	.37	.6443088	.73	.7673049	1.09	.8621434
.02	.5079783	.38	.6480273	.74	.7703500	1.10	.8643339
.03	.5119665	.39	.6517317	.75	.7733726	1.11	.8665005
.04	.5159534	.40	.6554217	.76	.7763727	1.12	.8686431
.05	.5199388	.41	.6590970	.77	.7793501	1.13	.8707619
.06	.5239222	.42	.6627573	.78	.7823046	1.14	.8728568
.07	.5279032	.43	.6664022	.79	.7852361	1.15	.8749281
.08	.5318814	.44	.6700314	.80	.7881446	1.16	.8769756
.09	.5358564	.45	.6736448	.81	.7910299	1.17	.8789995
.10	.5398278	.46	.6772419	.82	.7938919	1.18	.8809999
.11	.5437953	.47	.6808225	.83	.7967306	1.19	.8829768
.12	.5477584	.48	.6843863	.84	.7995458	1.20	.8849303
.13	.5517168	.49	.6879331	.85	.8023375	1.21	.8868606
.14	.5556700	.50	.6914625	.86	.8051055	1.22	.8887676
.15	.5596177	.51	.6949743	.87	.8078498	1.23	.8906514
.16	.5635595	.52	.6984682	.88	.8105703	1.24	.8925123
.17	.5674949	.53	.7019440	.89	.8132671	1.25	.8943502
.18	.5714237	.54	.7054015	.90	.8159399	1.26	.8961653
.19	.5753454	.55	.7088403	.91	.8185887	1.27	.8979577
.20	.5792597	.56	.7122603	.92	.8212136	1.28	.8997274
.21	.5831662	.57	.7156612	.93	.8238145	1.29	.9014747
.22	.5870604	.58	.7190427	.94	.8263912	1.30	.9031995
.23	.5909541	.59	.7224047	.95	.8289439	1.31	.9049021
.24	.5948349	.60	.7257469	.96	.8314724	1.32	.9065825
.25	.5987063	.61	.7290691	.97	.8339768	1.33	.9082409
.26	.6025681	.62	.7323711	.98	.8364569	1.34	.9098773
.27	.6064199	.63	.7356527	.99	.8389129	1.35	.9114920
.28	.6102612	.64	.7389137	1.00	.8413447	1.36	.9130850
.29	.6140919	.65	.7421539	1.01	.8437524	1.37	.9146565
.30	.6179114	.66	.7453731	1.02	.8461358	1.38	.9162067
.31	.6217195	.67	.7485711	1.03	.8484950	1.39	.9177356
.32	.6255158	.68	.7517478	1.04	.8508300	1.40	.9192433
.33	.6293000	.69	.7549029	1.05	.8531409	1.41	.9207302
.34	.6330717	.70	.7580363	1.06	.8554277	1.42	.9221962
.35	.6368307	.71	.7611479	1.07	.8576903	1.43	.9236415

Table 3 (continued)

z	$F(z)$	z	$F(z)$	z	$F(z)$	z	$F(z)$
1.44	.9250663	1.77	.9616364	2.10	.9821356	2.43	.9924506
1.45	.9264707	1.78	.9624620	2.11	.9825708	2.44	.9926564
1.47	.9278550	1.79	.9632730	2.12	.9829970	2.45	.9928572
1.47	.9292191	1.80	.9640697	2.13	.9834142	2.46	.9930531
1.48	.9305634	1.81	.9648521	2.14	.9838226	2.47	.9932443
1.49	.9318879	1.82	.9656205	2.15	.9842224	2.48	.9934309
1.50	.9331928	1.83	.9663750	2.16	.9846137	2.49	.9936128
1.51	.9344783	1.84	.9671159	2.17	.9849966	2.50	.9937903
1.52	.9357445	1.85	.9678432	2.18	.9853713	2.51	.9939634
1.53	.9369916	1.86	.9685572	2.19	.9857379	2.52	.9941323
1.54	.9382198	1.87	.9692581	2.20	.9860966	2.53	.9942969
1.55	.9394292	1.88	.9699460	2.21	.9864474	2.54	.9944574
1.56	.9406201	1.89	.9706210	2.22	.9867906	2.55	.9946139
1.57	.9417924	1.90	.9712834	2.23	.9871263	2.56	.9947664
1.58	.9429466	1.91	.9719334	2.24	.9874545	2.57	.9949151
1.59	.9440826	1.92	.9725711	2.25	.9877755	2.58	.9950600
1.60	.9452007	1.93	.9731966	2.26	.9880894	2.59	.9952012
1.61	.9463011	1.94	.9738102	2.27	.9883962	2.60	.9953388
1.62	.9473839	1.95	.9744119	2.28	.9886962	2.70	.9965330
1.63	.9484493	1.96	.9750021	2.29	.9889893	2.80	.9974449
1.64	.9494974	1.97	.9755808	2.30	.9892759	2.90	.9981342
1.65	.9505285	1.98	.9761482	2.31	.9895559	3.00	.9986501
1.66	.9515428	1.99	.9767045	2.32	.9898296	3.20	.9993129
1.67	.9525403	2.00	.9772499	2.33	.9900969	3.40	.9996631
1.68	.9535213	2.01	.9777844	2.34	.9903581	3.60	.9998409
1.69	.9544860	2.02	.9783083	2.35	.9906133	3.80	.9999277
1.70	.9554345	2.03	.9788217	2.36	.9908625	4.00	.9999683
1.71	.9563671	2.04	.9793248	2.37	.9911060	4.50	.9999966
1.72	.9572838	2.05	.9798178	2.38	.9913437	5.00	.9999997
1.73	.9581849	2.06	.9803007	2.39	.9915758	5.50	.9999999
1.74	.9590705	2.07	.9807738	2.40	.9918025		
1.75	.9599408	2.08	.9812372	2.41	.9920237		
1.76	.9607961	2.09	.9816911	2.42	.9922397		

Table 4. Standard Normal Density Function

This table gives values of the standard normal density function,

$$f(z) = \frac{1}{\sqrt{2\pi}} e^{-z^2/2},$$

for

$$z = 0(.01)4.29.$$

To find $f(z)$ for $z < 0$, take $f(z) = f(-z)$.

Examples:

$$f(2.16) = .0387$$

and

$$f(-1.57) = f(1.57) = .1163.$$

These Tables of "Gaussian Density Function" appear in *Analysis of Decisions under Uncertainty* by Robert Schlaifer, published by McGraw-Hill Book Company, Inc., 1969. They are reproduced here by specific permission of the copyright holder, The President and Fellows of Harvard College.

Table 4

z	.00	.01	.02	.03	.04	.05	.06	.07	.08	.09
0.0	.3989	.3989	.3989	.3988	.3986	.3984	.3982	.3980	.3977	.3973
0.1	.3970	.3965	.3961	.3956	.3951	.3945	.3939	.3932	.3925	.3918
0.2	.3910	.3902	.3894	.3885	.3876	.3867	.3857	.3847	.3836	.3825
0.3	.3814	.3802	.3790	.3778	.3765	.3752	.3739	.3725	.3712	.3697
0.4	.3683	.3668	.3653	.3637	.3621	.3605	.3589	.3572	.3555	.3538
0.5	.3521	.3503	.3485	.3467	.3448	.3429	.3410	.3391	.3372	.3352
0.6	.3332	.3312	.3292	.3271	.3251	.3230	.3209	.3187	.3166	.3144
0.7	.3123	.3101	.3079	.3056	.3034	.3011	.2989	.2966	.2943	.2920
0.8	.2897	.2874	.2850	.2827	.2803	.2780	.2756	.2732	.2709	.2685
0.9	.2661	.2637	.2613	.2589	.2565	.2541	.2516	.2492	.2468	.2444
1.0	.2420	.2396	.2371	.2347	.2323	.2299	.2275	.2251	.2227	.2203
1.1	.2179	.2155	.2131	.2107	.2083	.2059	.2036	.2012	.1989	.1965
1.2	.1942	.1919	.1895	.1872	.1849	.1826	.1804	.1781	.1758	.1736
1.3	.1714	.1691	.1669	.1647	.1626	.1604	.1582	.1561	.1539	.1518
1.4	.1497	.1476	.1456	.1435	.1415	.1394	.1374	.1354	.1334	.1315
1.5	.1295	.1276	.1257	.1238	.1219	.1200	.1182	.1163	.1145	.1127
1.6	.1109	.1092	.1074	.1057	.1040	.1023	.1006	.0989	.0973	.0957
1.7	.0940	.0925	.0909	.0893	.0878	.0863	.0848	.0833	.0818	.0804
1.8	.0790	.0775	.0761	.0748	.0734	.0721	.0707	.0694	.0681	.0669
1.9	.0656	.0644	.0632	.0620	.0608	.0596	.0584	.0573	.0562	.0551
2.0	.0540	.0529	.0519	.0508	.0498	.0488	.0478	.0468	.0459	.0449
2.1	.0440	.0431	.0422	.0413	.0404	.0396	.0387	.0379	.0371	.0363
2.2	.0355	.0347	.0339	.0332	.0325	.0317	.0310	.0303	.0297	.0290
2.3	.0283	.0277	.0270	.0264	.0258	.0252	.0246	.0241	.0235	.0229
2.4	.0224	.0219	.0213	.0208	.0203	.0198	.0194	.0189	.0184	.0180
2.5	.0175	.0171	.0167	.0163	.0158	.0154	.0151	.0147	.0143	.0139
2.6	.0136	.0132	.0129	.0126	.0122	.0119	.0116	.0113	.0110	.0107
2.7	.0104	.0101	.0099	.0096	.0093	.0091	.0088	.0086	.0084	.0081
2.8	.0079	.0077	.0075	.0073	.0071	.0069	.0067	.0065	.0063	.0061
2.9	.0060	.0058	.0056	.0055	.0053	.0051	.0050	.0048	.0047	.0046
3.0	.0044	.0043	.0042	.0040	.0039	.0038	.0037	.0036	.0035	.0034
3.1	.0033	.0032	.0031	.0030	.0029	.0028	.0027	.0026	.0025	.0025
3.2	.0024	.0023	.0022	.0022	.0021	.0020	.0020	.0019	.0018	.0018
3.3	.0017	.0017	.0016	.0016	.0015	.0015	.0014	.0014	.0013	.0013
3.4	.0012	.0012	.0012	.0011	.0011	.0010	.0010	.0010	.0009	.0009
3.5	.0009	.0008	.0008	.0008	.0008	.0007	.0007	.0007	.0007	.0006
3.6	.0006	.0006	.0006	.0005	.0005	.0005	.0005	.0005	.0005	.0004
3.7	.0004	.0004	.0004	.0004	.0004	.0004	.0003	.0003	.0003	.0003
3.8	.0003	.0003	.0003	.0003	.0003	.0002	.0002	.0002	.0002	.0002
3.9	.0002	.0002	.0002	.0002	.0002	.0002	.0002	.0002	.0001	.0001
4.0	.0001	.0001	.0001	.0001	.0001	.0001	.0001	.0001	.0001	.0001
4.1	.0001	.0001	.0001	.0001	.0001	.0001	.0001	.0001	.0001	.0001
4.2	.0001	.0001	.0001	.0001	.0000	.0000	.0000	.0000	.0000	.0000

Table 5. Fractiles of the Beta Distribution

This table gives fractiles of the beta distribution, which are values p_f such that

$$P(\tilde{p} \leq p_f \mid r,n) = \int_0^{p_f} \frac{(n-1)!}{(r-1)!(n-r-1)!} p^{r-1}(1-p)^{n-r-1}\,dp = f.$$

For all integral values of r and n and for

$$f = .001, .01, .05, .1, .25, .5, .75, .9, .95, .99, .999,$$

the table either gives p_f directly or tells how to approximate it. For each combination of r and n, there is a row in which either (1) the value of p_f can be read in the f column or (2) reference is made to a footnote telling how to find p_f or how to compute an approximation to p_f.

Examples:

Suppose that $f = .01$, $r = 27$, and $n = 150$; in the table for $r = 27$ and $n = 150$, read $p_{.01} = .1139$.

Suppose that $f = .01$, $r = 27$, and $n = 220$; in the table for $r = 27$ and $n = 203+$, read 16.40* in the .01 column; then follow the instructions in the footnote referenced by * and compute

$$p_{.01} = 16.40/(n-1) = 16.40/219 = .0749.$$

Suppose that $f = .01$, $r = 27$, and $n = 80$; in the table for $r = 27$ and $42 \leq n \leq 124$, read "see note ***"; following the instructions in the note, compute

$$m = r/n = 27/80 = .3375,$$
$$s = \sqrt{m(1-m)/(n+1)} = \sqrt{.3375(.6625)/81} = \sqrt{.002760} = .0525,$$
$$z_{.01} = -2.326 \text{ (from Table 3)},$$

and

$$p_{.01} = m + z_{.01}s = .3375 - 2.326(.0525) = .2154.$$

Suppose that $f = .01$, $r = 27$, and $n = 40$; in the table for $r = 27$ and $28 \leq n \leq 41$, read "see note **"; following the instructions in the note, compute $n^* = 40 - 27 = 13$, then read $p^*_{.99} = .5052$ in the table for $r = 13$ and $n = 40$ and compute

$$p_{.01} = 1 - .5052 = .4948.$$

Suppose that $f = .01$, $r = 27$, and $n = 28$; in the table for $r = 27$ and $28 \leq n \leq 41$, read "see note **"; following the instructions in the note, compute $n^* = 28 - 27 = 1$, then read 4.605* in the table for $f = .99$, $r = 1$ and $n = 20+$; following the instructions in *this* footnote compute $p^*_{.99} = 4.605/(n-1) = 4.605/27 = .1706$ and

$$p_{.01} = 1 - .1706 = .8294.$$

Table 5

r	n	.001	.01	.05	.1	.25	f .5	.75	.9	.95	.99	.999
1	2	.0010	.0100	.0500	.1000	.2500	.5000	.7500	.9000	.9500	.9900	.9990
	3	.0005	.0050	.0253	.0513	.1340	.2929	.5000	.6838	.7764	.9000	.9684
	4	.0003	.0033	.0170	.0345	.0914	.2063	.3700	.5358	.6316	.7846	.9000
	5	.0003	.0025	.0127	.0260	.0694	.1591	.2929	.4377	.5271	.6838	.8222
	6	.0002	.0020	.0102	.0209	.0559	.1294	.2421	.3690	.4507	.6019	.7488
	7	.0002	.0017	.0085	.0174	.0468	.1091	.2063	.3187	.3930	.5358	.6838
	8	.0001	.0014	.0073	.0149	.0403	.0943	.1797	.2803	.3482	.4821	.6272
	9	.0001	.0013	.0064	.0131	.0353	.0830	.1591	.2501	.3123	.4377	.5783
	10	.0001	.0011	.0057	.0116	.0315	.0741	.1428	.2257	.2831	.4005	.5358
	11	.0001	.0010	.0051	.0105	.0284	.0670	.1294	.2057	.2589	.3690	.4988
	12	.0001	.0009	.0047	.0095	.0258	.0611	.1184	.1889	.2384	.3421	.4663
	13	.0001	.0008	.0043	.0087	.0237	.0561	.1091	.1746	.2209	.3187	.4377
	14	.0001	.0008	.0039	.0081	.0219	.0519	.1011	.1623	.2058	.2983	.4122
	15	.0001	.0007	.0037	.0075	.0203	.0483	.0943	.1517	.1926	.2803	.3895
	16	.0001	.0007	.0034	.0070	.0190	.0452	.0883	.1423	.1810	.2644	.3690
	17	.0001	.0006	.0032	.0066	.0178	.0424	.0830	.1340	.1707	.2501	.3506
	18	.0001	.0006	.0030	.0062	.0168	.0400	.0783	.1267	.1616	.2373	.3339
	19	.0001	.0006	.0028	.0058	.0159	.0378	.0741	.1201	.1533	.2257	.3187
	20+	.0010*	.0101*	.0513*	.1054*	.2877*	.6931*	1.386*	2.303*	2.996*	4.605*	6.908*
2	3	SEE NOTE**										
	4	.0184	.0589	.1354	.1958	.3264	.5000	.6736	.8042	.8646	.9411	.9816
	5	.0130	.0420	.0976	.1426	.2430	.3857	.5437	.6795	.7514	.8591	.9360
	6	.0101	.0327	.0764	.1122	.1938	.3138	.4542	.5839	.6574	.7779	.8780
	7	.0083	.0268	.0628	.0926	.1612	.2644	.3895	.5103	.5818	.7057	.8186
	8	.0070	.0227	.0534	.0788	.1380	.2285	.3407	.4526	.5207	.6434	.7625
	9	.0060	.0197	.0464	.0686	.1206	.2011	.3027	.4062	.4707	.5899	.7113
	10	.0053	.0174	.0410	.0608	.1072	.1796	.2723	.3684	.4291	.5440	.6651
	11	.0048	.0155	.0368	.0545	.0964	.1623	.2474	.3368	.3942	.5044	.6237
	12	.0043	.0141	.0333	.0495	.0876	.1480	.2266	.3102	.3644	.4698	.5866
	13	.0039	.0128	.0305	.0452	.0803	.1360	.2091	.2875	.3387	.4395	.5534
	14	.0036	.0118	.0281	.0417	.0741	.1258	.1941	.2678	.3163	.4128	.5234
	15	.0034	.0110	.0260	.0387	.0688	.1170	.1810	.2507	.2967	.3891	.4963
	16	.0031	.0102	.0242	.0360	.0642	.1094	.1697	.2356	.2794	.3679	.4718
	17	.0029	.0095	.0227	.0337	.0602	.1027	.1596	.2222	.2640	.3488	.4495
	18	.0027	.0090	.0213	.0317	.0566	.0968	.1507	.2102	.2501	.3316	.4292
	19	.0026	.0085	.0201	.0299	.0535	.0915	.1427	.1995	.2377	.3160	.4105
	20	.0025	.0080	.0190	.0283	.0507	.0868	.1355	.1898	.2264	.3018	.3934
	21	.0023	.0076	.0181	.0269	.0481	.0825	.1290	.1810	.2161	.2888	.3776
	22	.0022	.0072	.0172	.0256	.0458	.0786	.1232	.1729	.2067	.2768	.3630
	23	.0021	.0069	.0164	.0244	.0437	.0751	.1178	.1656	.1981	.2658	.3494
	24	.0020	.0066	.0157	.0234	.0418	.0719	.1128	.1588	.1902	.2557	.3369
	25	.0019	.0063	.0150	.0224	.0401	.0690	.1083	.1526	.1829	.2462	.3252
	26	.0019	.0060	.0144	.0215	.0385	.0662	.1041	.1469	.1761	.2375	.3142
	27	.0018	.0058	.0138	.0206	.0370	.0637	.1002	.1415	.1698	.2293	.3040
	28	.0017	.0056	.0133	.0199	.0356	.0614	.0966	.1366	.1640	.2217	.2944
	29	.0016	.0054	.0128	.0192	.0344	.0592	.0933	.1319	.1585	.2146	.2854
	30+	.0454*	.1486*	.3554*	.5318*	.9613*	1.678*	2.693*	3.890*	4.744*	6.638*	9.233*
3	4-5	SEE NOTE**										
	6	.0476	.1056	.1893	.2466	.3594	.5000	.6406	.7534	.8107	.8944	.9524
	7	.0379	.0847	.1532	.2009	.2969	.4214	.5532	.6668	.7287	.8269	.9060
	8	.0316	.0708	.1288	.1696	.2531	.3641	.4861	.5962	.6587	.7637	.8562
	9	.0270	.0608	.1111	.1469	.2206	.3205	.4332	.5382	.5997	.7068	.8073
	10	.0237	.0533	.0977	.1295	.1955	.2862	.3905	.4901	.5496	.6563	.7612
	11	.0210	.0475	.0873	.1158	.1756	.2586	.3554	.4496	.5069	.6117	.7185
	12	.0189	.0428	.0788	.1048	.1593	.2358	.3261	.4152	.4701	.5723	.6793
	13	.0172	.0390	.0719	.0957	.1459	.2167	.3012	.3855	.4381	.5373	.6436
	14	.0158	.0358	.0660	.0880	.1345	.2004	.2798	.3598	.4101	.5062	.6110
	15	.0146	.0331	.0611	.0815	.1248	.1865	.2612	.3372	.3854	.4783	.5812
	16	.0135	.0307	.0568	.0759	.1163	.1743	.2450	.3173	.3634	.4532	.5539
	17	.0126	.0287	.0531	.0710	.1090	.1637	.2306	.2996	.3438	.4305	.5290
	18	.0119	.0269	.0499	.0667	.1025	.1542	.2178	.2837	.3262	.4099	.5060
	19	.0112	.0254	.0470	.0629	.0968	.1458	.2064	.2694	.3103	.3912	.4849
	20	.0105	.0240	.0445	.0595	.0916	.1383	.1961	.2565	.2958	.3741	.4654
	21	.0100	.0227	.0422	.0564	.0870	.1314	.1867	.2448	.2826	.3583	.4474
	22	.0095	.0216	.0401	.0537	.0828	.1253	.1783	.2340	.2706	.3439	.4306
	23	.0090	.0206	.0382	.0512	.0790	.1197	.1705	.2242	.2595	.3305	.4151
	24	.0086	.0196	.0365	.0489	.0756	.1146	.1634	.2152	.2492	.3181	.4006
	25	.0083	.0188	.0350	.0468	.0724	.1099	.1569	.2069	.2398	.3066	.3870

*To approximate p_f, divide the starred entry under f by $(n - 1)$.
**Compute $r^* = n - r$, find p^*_{1-f} for r^* and n, and compute $p_f = 1 - p^*_{1-f}$.

Table 5 (continued)

							f					
r	*n*	.001	.01	.05	.1	.25	.5	.75	.9	.95	.99	.999
3	26	.0079	.0180	.0335	.0449	.0695	.1055	.1509	.1991	.2310	.2959	.3743
	27	.0076	.0173	.0322	.0432	.0668	.1015	.1453	.1920	.2229	.2859	.3624
	28	.0073	.0166	.0310	.0415	.0643	.0978	.1401	.1853	.2153	.2766	.3512
	29	.0070	.0160	.0298	.0400	.0620	.0944	.1353	.1791	.2082	.2679	.3407
	30	.0068	.0155	.0288	.0386	.0599	.0911	.1308	.1733	.2016	.2596	.3308
	31	.0066	.0149	.0278	.0373	.0579	.0881	.1265	.1678	.1953	.2519	.3214
	32	.0063	.0144	.0269	.0361	.0560	.0853	.1226	.1627	.1895	.2446	.3126
	33	.0061	.0140	.0260	.0349	.0542	.0827	.1189	.1579	.1839	.2377	.3042
	34	.0059	.0135	.0252	.0339	.0526	.0802	.1153	.1533	.1787	.2312	.2963
	35	.0058	.0131	.0245	.0329	.0510	.0779	.1121	.1490	.1738	.2251	.2887
	36	.0056	.0127	.0238	.0319	.0496	.0757	.1090	.1450	.1692	.2192	.2815
	37	.0054	.0124	.0231	.0310	.0482	.0736	.1060	.1412	.1647	.2137	.2747
	38+	.1905*	.4360*	.8177*	1.102*	1.727*	2.674*	3.920*	5.322*	6.296*	8.406*	11.23*
4	5–7	SEE NOTE**										
	8	.0767	.1423	.2253	.2786	.3788	.5000	.6212	.7214	.7747	.8577	.9233
	9	.0648	.1210	.1929	.2397	.3291	.4402	.5555	.6554	.7108	.8018	.8804
	10	.0562	.1053	.1688	.2104	.2910	.3931	.5020	.5994	.6551	.7500	.8371
	11	.0496	.0932	.1500	.1876	.2609	.3551	.4577	.5517	.6066	.7029	.7954
	12	.0444	.0837	.1351	.1692	.2364	.3238	.4205	.5108	.5644	.6604	.7559
	13	.0402	.0759	.1229	.1542	.2162	.2976	.3888	.4753	.5273	.6222	.7192
	14	.0368	.0695	.1127	.1416	.1991	.2753	.3615	.4443	.4946	.5878	.6851
	15	.0338	.0640	.1040	.1309	.1846	.2561	.3377	.4170	.4657	.5567	.6537
	16	.0314	.0594	.0967	.1218	.1720	.2394	.3169	.3928	.4398	.5285	.6246
	17	.0292	.0554	.0903	.1138	.1611	.2247	.2985	.3712	.4166	.5029	.5977
	18	.0273	.0519	.0846	.1068	.1514	.2118	.2821	.3519	.3956	.4796	.5729
	19	.0257	.0488	.0797	.1006	.1429	.2002	.2674	.3344	.3767	.4583	.5499
	20	.0242	.0461	.0753	.0951	.1353	.1899	.2541	.3186	.3594	.4387	.5286
	21	.0229	.0436	.0714	.0902	.1284	.1805	.2421	.3042	.3437	.4207	.5087
	22	.0218	.0414	.0678	.0858	.1222	.1721	.2312	.2910	.3292	.4041	.4903
	23	.0207	.0394	.0646	.0817	.1166	.1644	.2212	.2789	.3159	.3887	.4730
	24	.0198	.0376	.0617	.0781	.1114	.1573	.2120	.2678	.3036	.3745	.4569
	25	.0189	.0360	.0590	.0747	.1068	.1509	.2036	.2575	.2923	.3612	.4419
	26	.0181	.0345	.0566	.0717	.1024	.1449	.1958	.2480	.2817	.3488	.4277
	27	.0174	.0331	.0543	.0688	.0985	.1394	.1886	.2392	.2719	.3372	.4144
	28	.0167	.0318	.0522	.0662	.0948	.1343	.1819	.2309	.2627	.3264	.4019
	29	.0161	.0306	.0503	.0638	.0914	.1296	.1756	.2232	.2542	.3162	.3901
	30	.0155	.0295	.0485	.0615	.0882	.1252	.1698	.2160	.2461	.3066	.3790
	31	.0149	.0285	.0469	.0594	.0852	.1210	.1644	.2093	.2386	.2976	.3685
	32	.0144	.0275	.0453	.0575	.0824	.1172	.1592	.2030	.2315	.2891	.3585
	33	.0140	.0267	.0438	.0556	.0798	.1136	.1544	.1970	.2248	.2811	.3490
	34	.0135	.0258	.0425	.0539	.0774	.1101	.1499	.1914	.2185	.2735	.3401
	35	.0131	.0250	.0412	.0523	.0751	.1069	.1456	.1860	.2125	.2663	.3315
	36	.0127	.0243	.0400	.0508	.0729	.1039	.1416	.1810	.2069	.2594	.3234
	37	.0124	.0236	.0389	.0493	.0709	.1011	.1378	.1762	.2015	.2530	.3157
	38	.0120	.0229	.0378	.0480	.0690	.0983	.1342	.1717	.1964	.2468	.3083
	39	.0117	.0223	.0368	.0467	.0671	.0958	.1307	.1674	.1916	.2409	.3013
	40	.0114	.0217	.0358	.0455	.0654	.0933	.1275	.1633	.1870	.2353	.2945
	41	.0111	.0212	.0349	.0443	.0638	.0910	.1244	.1594	.1826	.2299	.2881
	42	.0108	.0206	.0340	.0432	.0622	.0888	.1214	.1557	.1784	.2248	.2819
	43	.0105	.0201	.0332	.0422	.0607	.0867	.1186	.1522	.1744	.2199	.2760
	44	.0103	.1096	.0324	.0412	.0593	.0847	.1159	.1488	.1706	.2152	.2703
	45	.0100	.0192	.0317	.0402	.0579	.0828	.1134	.1456	.1669	.2107	.2649
	46+	.4286*	.8232*	1.266*	1.745*	2.535*	3.672*	5.109*	6.681*	7.754*	10.05*	13.06*
5	6–9	SEE NOTE**										
	10	.1025	.1710	.2514	.3010	.3920	.5000	.6080	.6990	.7486	.8290	.8975
	11	.0898	.1504	.2224	.2673	.3507	.4517	.5555	.6458	.6965	.7817	.8587
	12	.0799	.1344	.1996	.2405	.3173	.4119	.5111	.5995	.6502	.7378	.8206
	13	.0721	.1215	.1810	.2187	.2898	.3785	.4731	.5590	.6091	.6976	.7841
	14	.0656	.1108	.1657	.2005	.2668	.3502	.4403	.5234	.5726	.6609	.7497
	15	.0602	.1019	.1527	.1851	.2471	.3258	.4117	.4920	.5400	.6274	.7173
	16	.0556	.0944	.1417	.1720	.3201	.3045	.3865	.4640	.5108	.5969	.6871
	17	.0517	.0878	.1321	.1606	.2154	.2859	.3642	.4389	.4844	.5690	.6590
	18	.0483	.0822	.1238	.1506	.2024	.2694	.3444	.4164	.4605	.5434	.6328
	19	.0454	.0772	.1164	.1418	.1909	.2547	.3265	.3960	.4389	.5199	.6083
	20	.0427	.0728	.1099	.1339	.1806	.2415	.3105	.3775	.4191	.4983	.5856
	21	.0404	.0688	.1041	.1269	.1714	.2297	.2959	.3607	.4010	.4783	.5643
	22	.0383	.0653	.0988	.1206	.1631	.2189	.2826	.3452	.3844	.4598	.5444

***To approximate p_f, compute $m = r/n$ and $s = \sqrt{m(1-m)/(n+1)}$, find z_f in Table 3, and compute $p_f \doteq m + z_f s$.

Table 5 (continued)

r	n	.001	.01	.05	.1	.25	f .5	.75	.9	.95	.99	.999
5	23	.0364	.0621	.0941	.1149	.1556	.2091	.2705	.3310	.3691	.4426	.5258
	24	.0347	.0593	.0898	.1097	.1487	.2001	.2593	.3180	.3549	.4267	.5084
	25	.0331	.0566	.0859	.1050	.1424	.1919	.2490	.3059	.3418	.4118	.4920
	26	.0317	.0542	.0823	.1006	.1366	.1843	.2396	.2947	.3296	.3979	.4766
	27	.0304	.0520	.0790	.0966	.1313	.1774	.2308	.2842	.3182	.3849	.4621
	28	.0292	.0500	.0759	.0929	.1264	.1709	.2226	.2745	.3076	.3727	.4485
	29	.0281	.0481	.0731	.0895	.1218	.1648	.2150	.2655	.2977	.3613	.4356
	30	.0271	.0463	.0705	.0863	.1175	.1592	.2079	.2570	.2884	.3505	.4234
	31	.0261	.0447	.0681	.0834	.1136	.1540	.2012	.2490	.2796	.3403	.4118
	32	.0252	.0432	.0658	.0806	.1099	.1491	.1950	.2415	.2714	.3307	.4009
	33	.0244	.0418	.0637	.0780	.1064	.1444	.1891	.2344	.2636	.3216	.3905
	34	.0236	.0405	.0617	.0756	.1031	.1401	.1836	.2278	.2563	.3130	.3806
	35	.0229	.0392	.0598	.0733	.1001	.1360	.1784	.2215	.2493	.3049	.3712
	36	.0222	.0381	.0580	.0712	.0972	.1322	.1734	.2155	.2427	.2971	.3622
	37	.0215	.0370	.0564	.0691	.0944	.1285	.1688	.2099	.2365	.2898	.3537
	38	.0209	.0359	.0548	.0672	.0919	.1251	.1644	.2045	.2305	.2828	.3455
	39	.0203	.0349	.0533	.0654	.0894	.1218	.1602	.1994	.2249	.2761	.3377
	40	.0198	.0340	.0519	.0637	.0871	.1187	.1562	.1946	.2195	.2697	.3303
	41	.0193	.0331	.0506	.0621	.0849	.1158	.1524	.1900	.2144	.2636	.3231
	42	.0188	.0323	.0493	.0605	.0828	.1130	.1488	.1856	.2095	.2578	.3163
	43	.0183	.0315	.0481	.0591	.0808	.1103	.1453	.1814	.2048	.2522	.3097
	44	.0179	.0307	.0470	.0577	.0789	.1078	.1421	.1774	.2004	.2469	.3034
	45	.0175	.0300	.0459	.0563	.0771	.1054	.1389	.1735	.1961	.2418	.2974
	46	.0171	.0293	.0448	.0550	.0754	.1030	.1359	.1698	.1920	.2369	.2916
	47	.0167	.0287	.0438	.0538	.0738	.1008	.1330	.1663	.1880	.2321	.2860
	48	.0163	.0280	.0429	.0527	.0722	.0987	.1303	.1629	.1843	.2276	.2806
	49	.0160	.0274	.0420	.0515	.0707	.0966	.1276	.1597	.1806	.2232	.2754
	50	.0156	.0269	.0411	.0505	.0692	.0947	.1251	.1566	.1771	.2190	.2704
	51	.0153	.0263	.0402	.0494	.0678	.0928	.1226	.1535	.1738	.2150	.2656
	52	.0150	.0258	.0394	.0485	.0665	.0910	.1203	.1507	.1706	.2111	.2609
	53	.0147	.0253	.0387	.0475	.0652	.0892	.1180	.1479	.1674	.2073	.2564
	54+	.7394*	1.279*	1.970*	2.433*	3.369*	4.671*	6.274*	7.994*	9.154*	11.60*	14.79*
6	7–11	SEE NOTE**										
	12	.1249	.1940	.2712	.3177	.4016	.5000	.5984	.6823	.7288	.8060	.8751
	13	.1120	.1746	.2453	.2882	.3663	.4595	.5547	.6377	.6848	.7651	.8401
	14	.1016	.1588	.2240	.2637	.3368	.4251	.5167	.5982	.6452	.7271	.8062
	15	.0929	.1457	.2061	.2432	.3117	.3954	.4835	.5631	.6096	.6920	.7738
	16	.0857	.1346	.1909	.2256	.2902	.3697	.4543	.5317	.5774	.6597	.7432
	17	.0794	.1251	.1778	.2104	.2714	.3471	.4283	.5035	.5483	.6299	.7143
	18	.0741	.1168	.1664	.1972	.2549	.3270	.4051	.4781	.5219	.6025	.6872
	19	.0694	.1096	.1563	.1855	.2404	.3092	.3843	.4550	.4978	.5772	.6617
	20	.0653	.1032	.1475	.1751	.2274	.2932	.3655	.4340	.4758	.5538	.6378
	21	.0616	.0975	.1396	.1659	.2157	.2788	.3484	.4149	.4556	.5321	.6154
	22	.0584	.0925	.1324	.1575	.2052	.2657	.3329	.3973	.4370	.5120	.5944
	23	.0554	.0879	.1260	.1500	.1956	.2538	.3187	.3812	.4198	.4933	.5747
	24	.0528	.0838	.1202	.1432	.1870	.2430	.3056	.3663	.4039	.4758	.5561
	25	.0504	.0800	.1149	.1369	.1790	.2330	.2936	.3525	.3891	.4595	.5386
	26	.0482	.0765	.1101	.1312	.1717	.2238	.2824	.3397	.3754	.4443	.5222
	27	.0462	.0734	.1056	.1260	.1650	.2153	.2721	.3277	.3626	.4300	.5066
	28	.0443	.0705	.1015	.1211	.1588	.2074	.2625	.3166	.3506	.4166	.4919
	29	.0426	.0678	.0977	.1166	.1530	.2001	.2536	.3062	.3394	.4039	.4780
	30	.0410	.0653	.0942	.1125	.1476	.1933	.2452	.2965	.3289	.3920	.4649
	31	.0395	.0630	.0909	.1086	.1426	.1869	.2374	.2874	.3190	.3808	.4524
	32	.0382	.0609	.0878	.1050	.1380	.1809	.2300	.2788	.3096	.3702	.4406
	33	.0369	.0588	.0850	.1016	.1336	.1753	.2231	.2707	.3008	.3601	.4293
	34	.0357	.0570	.0823	.0984	.1295	.1701	.2166	.2630	.2925	.3506	.4186
	35	.0346	.0552	.0798	.0954	.1256	.1651	.2105	.2558	.2846	.3416	.4084
	36	.0335	.0535	.0774	.0926	.1220	.1605	.2047	.2490	.2772	.3330	.3987
	37	.0326	.0520	.0752	.0900	.1186	.1560	.1992	.2425	.2701	.3248	.3894
	38	.0316	.0505	.0731	.0875	.1153	.1519	.1940	.2363	.2634	.3170	.3805
	39	.0307	.0491	.0711	.0851	.1123	.1479	.1891	.2305	.2570	.3096	.3721
	40	.0299	.0478	.0692	.0829	.1093	.1441	.1844	.2249	.2509	.3025	.3639
	41	.0291	.0466	.0674	.0807	.1066	.1406	.1799	.2196	.2450	.2957	.3562
	42	.0284	.0454	.0657	.0787	.1040	.1372	.1757	.2145	.2395	.2893	.3487
	43	.0277	.0442	.0641	.0768	.1015	.1339	.1716	.2097	.2342	.2831	.3416
	44	.0270	.0432	.0626	.0750	.0991	.1308	.1677	.2051	.2291	.2771	.3347
	45	.0263	.0422	.0611	.0732	.0968	.1279	.1640	.2006	.2242	.2714	.3281

*To approximate p_f, divide the starred entry under f by $(n - 1)$.

**Compute $r^* = n - r$, find p^*_{1-f} for r^* and n, and compute $p_f = 1 - p^*_{1-f}$.

Table 5 (continued)

r	n	f = .001	.01	.05	.1	.25	.5	.75	.9	.95	.99	.999
6	46	.0257	.0412	.0597	.0716	.0946	.1251	.1605	.1964	.2195	.2659	.3217
	47	.0252	.0402	.0584	.0700	.0925	.1224	.1571	.1923	.2151	.2607	.3156
	48	.0246	.0394	.0571	.0684	.0906	.1198	.1538	.1884	.2108	.2556	.3097
	49	.0241	.0385	.0559	.0670	.0887	.1173	.1507	.1847	.2066	.2508	.3041
	50	.0235	.0377	.0547	.0656	.0868	.1149	.1477	.1811	.2027	.2461	.2986
	51	.0231	.0369	.0536	.0643	.0851	.1126	.1448	.1776	.1988	.2416	.2933
	52	.0226	.0362	.0525	.0630	.0834	.1105	.1421	.1743	.1952	.2372	.2882
	53	.0221	.0354	.0515	.0617	.0818	.1083	.1394	.1711	.1916	.2330	.2833
	54	.0213	.0341	.0495	.0594	.0787	.1044	.1344	.1650	.1849	.2250	.2739
	56	.0209	.0335	.0486	.0583	.0773	.1025	.1320	.1621	.1817	.2213	.2695
	57	.0205	.0328	.0477	.0573	.0759	.1006	.1297	.1593	.1786	.2176	.2652
	58	.0201	.0322	.0468	.0562	.0746	.0989	.1274	.1566	.1756	.2141	.2610
	59	.0198	.0317	.0460	.0552	.0733	.0972	.1253	.1541	.1728	.2106	.2569
	60	.0194	.0311	.0452	.0543	.0720	.0956	.1232	.1515	.1700	.2073	.2530
	61	.0191	.0306	.0445	.0534	.0708	.0940	.1212	.1491	.1673	.2041	.2492
	62+	1.107*	1.785*	2.613*	3.152*	4.219*	5.670*	7.423*	9.275*	10.51*	13.11*	16.45*
7	8–13	SEE NOTE**										
	14–15	SEE NOTE***										
	16	.1209	.1795	.2437	.2822	.3528	.4348	.5204	.5965	.6404	.7177	.7936
	17	.1119	.1665	.2267	.2629	.3289	.4082	.4909	.5654	.6090	.6866	.7645
	18	.1041	.1552	.2119	.2461	.3088	.3847	.4646	.5374	.5803	.6577	.7369
	19	.0974	.1454	.1990	.2314	.2910	.3637	.4409	.5118	.5540	.6309	.7108
	20	.0915	.1368	.1875	.2183	.2752	.3449	.4195	.4886	.5300	.6060	.6861
	21	.0862	.1292	.1773	.2067	.2610	.3280	.4000	.4673	.5078	.5829	.6628
	22	.0816	.1223	.1682	.1962	.2482	.3126	.3823	.4477	.4874	.5613	.6409
	23	.0774	.1162	.1599	.1867	.2366	.2986	.3660	.4297	.4685	.5412	.6202
	24	.0736	.1107	.1525	.1782	.2260	.2858	.3511	.4131	.4510	.5224	.6007
	25	.0702	.1056	.1457	.1703	.2164	.2741	.3373	.3976	.4347	.5048	.5823
	26	.0671	.1010	.1395	.1632	.2075	.2632	.3246	.3833	.4195	.4884	.5649
	27	.0642	.0968	.1338	.1566	.1994	.2532	.3128	.3700	.4054	.4729	.5485
	28	.0616	.0929	.1285	.1505	.1918	.2440	.3018	.3575	.3921	.4584	.5329
	29	.0592	.0894	.1237	.1449	.1848	.2354	.2915	.3459	.3797	.4447	.5181
	30	.0570	.0860	.1192	.1397	.1784	.2274	.2819	.3349	.3680	.4317	.5041
	31	.0549	.0830	.1150	.1348	.1723	.2199	.2730	.3247	.3570	.4195	.4908
	32	.0530	.0801	.1111	.1303	.1666	.2128	.2646	.3150	.3467	.4080	.4782
	33	.0512	.0774	.1074	.1261	.1613	.2063	.2566	.3059	.3369	.3970	.4662
	34	.0495	.0749	.1040	.1221	.1564	.2001	.2492	.2973	.3276	.3866	.4547
	35	.0480	.0726	.1008	.1184	.1517	.1942	.2422	.2892	.3189	.3767	.4438
	36	.0465	.0704	.0978	.1149	.1473	.1887	.2355	.2815	.3106	.3674	.4334
	37	.0451	.0683	.0950	.1116	.1431	.1836	.2292	.2742	.3027	.3584	.4234
	38	.0438	.0664	.0923	.1085	.1392	.1786	.2232	.2673	.2952	.3499	.4139
	39	.0426	.0645	.0898	.1056	.1355	.1740	.2176	.2607	.2880	.3418	.4048
	40	.0414	.0628	.0874	.1028	.1320	.1696	.2122	.2544	.2812	.3340	.3960
	41	.0403	.0611	.0851	.1001	.1286	.1654	.2071	.2485	.2747	.3266	.3877
	42	.0393	.0596	.0830	.0976	.1255	.1614	.2022	.2427	.2685	.3195	.3796
	43	.0383	.0581	.0809	.0952	.1224	.1575	.1975	.2373	.2626	.3127	.3719
	44	.0373	.0567	.0790	.0929	.1196	.1539	.1931	.2321	.2569	.3062	.3645
	45	.0365	.0553	.0771	.0908	.1168	.1504	.1888	.2271	.2515	.2999	.3574
	46	.0356	.0540	.0754	.0887	.1142	.1471	.1847	.2223	.2463	.2939	.3506
	47	.0348	.0528	.0737	.0867	.1117	.1439	.1808	.2177	.2413	.2882	.3440
	48	.0340	.0516	.0720	.0848	.1093	.1409	.1771	.2133	.2365	.2826	.3376
	49	.0333	.0505	.0705	.0830	.1070	.1380	.1735	.2091	.2319	.2773	.3315
	50	.0325	.0495	.0690	.0813	.1048	.1352	.1701	.2050	.2274	.2721	.3256
	51	.0319	.0484	.0676	.0796	.1026	.1325	.1668	.2011	.2232	.2671	.3198
	52	.0312	.0474	.0662	.0780	.1006	.1299	.1636	.1974	.2191	.2624	.3143
	53	.0306	.0465	.0649	.0765	.0987	.1274	.1605	.1937	.2151	.2577	.3090
	54	.0300	.0456	.0637	.0750	.0968	.1251	.1576	.1903	.2113	.2533	.3038
	55	.0294	.0447	.0625	.0736	.0950	.1227	.1547	.1869	.2076	.2490	.2989
	56	.0288	.0439	.0613	.0722	.0932	.1205	.1520	.1836	.2040	.2448	.2940
	57	.0283	.0431	.0602	.0709	.0915	.1184	.1493	.1805	.2006	.2408	.2894
	58	.0278	.0423	.0591	.0697	.0899	.1163	.1468	.1775	.1972	.2369	.2848
	59	.0273	.0415	.0580	.0684	.0884	.1143	.1443	.1745	.1940	.2331	.2804
	60	.0268	.0408	.0570	.0672	.0869	.1124	.1419	.1717	.1909	.2295	.2762
	61	.0263	.0401	.0561	.0662	.0854	.1105	.1396	.1689	.1879	.2259	.2720
	62	.0259	.0394	.0551	.0650	.0840	.1087	.1373	.1663	.1849	.2225	.2680
	63	.0255	.0388	.0542	.0639	.0826	.1070	.1352	.1637	.1821	.2191	.2641
	64	.0250	.0381	.0533	.0629	.0813	.1053	.1331	.1612	.1793	.2159	.2603
	65	.0246	.0375	.0525	.0619	.0800	.1037	.1310	.1588	.1767	.2128	.2567

***To approximate p_f, compute $m = r/n$ and $s = \sqrt{m(1-m)/(n+1)}$, find z_f in Table 3, and compute $p_f \doteq m + z_f s$.

Table 5 (continued)

r	n	.001	.01	.05	.1	.25	f .5	.75	.9	.95	.99	.999
7	66	.0242	.0369	.0516	.0609	.0788	.1021	.1291	.1564	.1741	.2097	.2531
	67	.0239	.0363	.0508	.0600	.0776	.1005	.1271	.1541	.1716	.2067	.2496
	68	.0235	.0358	.0501	.0591	.0764	.0991	.1253	.1519	.1691	.2039	.2462
	69+	1.520*	2.330*	3.285*	3.895*	5.083*	6.670*	8.558*	10.53*	11.84*	14.57*	18.06*
8	9–14	SEE NOTE**										
	15–18	SEE NOTE***										
	19	.1289	.1844	.2440	.2792	.3427	.4182	.4964	.5667	.6078	.6814	.7560
	20	.1209	.1733	.2297	.2633	.3239	.3966	.4725	.5413	.5819	.6553	.7309
	21	.1139	.1634	.2171	.2491	.3071	.3771	.4507	.5 80	.5580	.6309	.7070
	22	.1076	.1546	.2057	.2363	.2920	.3594	.4308	.4966	.5359	.6082	.6844
	23	.1020	.1468	.1956	.2248	.2783	.3433	.4126	.4768	.5155	.5868	.6630
	24	.0969	.1397	.1863	.2144	.2658	.3286	.3959	.4586	.4964	.5669	.6427
	25	.0923	.1332	.1780	.2049	.2544	.3151	.3804	.4416	.4787	.5481	.6235
	26	.0882	.1273	.1703	.1962	.2440	.3027	.3661	.4258	.5622	.5306	.6054
	27	.0844	.1220	.1633	.1883	.2344	.2912	.3528	.4111	.4468	.5140	.5881
	28	.0809	.1170	.1568	.1809	.2255	.2806	.3405	.3974	.4323	.4984	.5718
	29	.0777	.1125	.1509	.1741	.2172	.2707	.3290	.3845	.4187	.4837	.5562
	30	.0747	.1083	.1453	.1678	.2096	.2614	.3182	.3725	.4060	.4699	.5415
	31	.0720	.1044	.1402	.1620	.2024	.2528	.3081	.3611	.3939	.4567	.5274
	32	.0694	.1007	.1354	.1565	.1958	.2447	.2986	.3505	.3826	.4443	.5141
	33	.0671	.0973	.1309	.1514	.1895	.2372	.2897	.3404	.3719	.4325	.5013
	34	.0649	.0942	.1268	.1466	.1837	.2301	.2813	.3309	.3618	.4213	.4892
	35	.0628	.0912	.1228	.1421	.1782	.2234	.2734	.3219	.3522	.4107	.4776
	36	.0608	.0884	.1191	.1379	.1730	.2170	.2659	.3134	.3430	.4005	.4665
	37	.0590	.0858	.1157	.1339	.1681	.2111	.2588	.3053	.3344	.3909	.4559
	38	.0573	.0833	.1124	.1302	.1635	.2054	.2521	.2976	.3262	.3817	.4458
	39	.0557	.0810	.1093	.1266	.1591	.2001	.2457	.2903	.3183	.3729	.4361
	40	.0541	.0788	.1064	.1233	.1550	.1950	.2397	.2834	.3108	.3645	.4268
	41	.0527	.0767	.1036	.1201	.1510	.1901	.2339	.2767	.3037	.3565	.4179
	42	.0513	.0747	.1010	.1171	.1473	.1855	.2284	.2704	.2969	.3488	.4093
	43	.0500	.0729	.0985	.1142	.1437	.1812	.2231	.2643	.2904	.3414	.4011
	44	.0488	.0711	.0961	.1115	.1403	.1770	.2181	.2586	.2841	.3343	.3932
	45	.0476	.0694	.0938	.1089	.1371	.1730	.2133	.2530	.2781	.3276	.3856
	46	.0465	.0678	.0927	.1064	.1340	.1692	.2087	.2477	.2724	.3210	.3783
	47	.0454	.0662	.0896	.1040	.1311	.1655	.2043	.2426	.2669	.3148	.3712
	48	.0444	.0647	.0876	.1017	.1282	.1620	.2001	.2377	.2616	.3087	.3644
	49	.0434	.0633	.0857	.0995	.1255	.1587	.1961	.2330	.2565	.3029	.3578
	50	.0425	.0620	.0839	.0974	.1229	.1555	.1922	.2285	.2516	.2973	.3515
	51	.0416	.0607	.0822	.0954	.1205	.1524	.1884	.2242	.2469	.2920	.3454
	52	.0407	.0594	.0805	.0935	.1181	.1494	.1848	.2200	.2424	.2868	.3395
	53	.0399	.0582	.0789	.0917	.1158	.1465	.1814	.2160	.2380	.2817	.3338
	54	.0391	.0571	.0774	.0899	.1136	.1438	.1781	.2121	.2338	.2769	.3282
	55	.0383	.0560	.0759	.0882	.1114	.1411	.1748	.2084	.2297	.2722	.3229
	56	.0376	.0549	.0745	.0866	.1094	.1386	.1717	.2047	.2258	.2677	.3177
	57	.0369	.0539	.0731	.0850	.1074	.1361	.1688	.2012	.2220	.2633	.3127
	58	.0362	.0529	.0718	.0835	.1055	.1338	.1659	.1979	.2183	.2591	.3078
	59	.0356	.0520	.0705	.0820	.1037	.1315	.1631	.1946	.2148	.2549	.3031
	60	.0349	.0511	.0693	.0806	.1019	.1293	.1604	.1914	.2113	.2510	.2985
	61	.0343	.0502	.0681	.0792	.1002	.1271	.1578	.1884	.2080	.2471	.2941
	62	.0337	.0493	.0670	.0779	.0985	.1250	.1552	.1854	.2048	.2434	.2898
	63	.0332	.0485	.0659	.0766	.0969	.1230	.1528	.1826	.2016	.2397	.2856
	64	.0326	.0477	.0648	.0754	.0954	.1211	.1504	.1798	.1986	.2362	.2815
	65	.0321	.9469	.0637	.0742	.0939	.1192	.1481	.1771	.1956	.2328	.2776
	66	.0316	.0462	.0627	.0730	.0924	.1174	.1459	.1744	.1928	.2295	.2737
	67	.0311	.0455	.0618	.0719	.0910	.1156	.1437	.1719	.1900	.2262	.2700
	68	.0306	.0448	.0608	.0708	.0896	.1139	.1416	.1694	.1873	.2231	.2663
	69	.0301	.0441	.0599	.0697	.0883	.1122	.1396	.1670	.1847	.2200	.2628
	70	.0297	.0434	.0590	.0687	.0870	.1106	.1376	.1647	.1821	.2171	.2593
	71	.0292	.0428	.0582	.0677	.0858	.1090	.1357	.1624	.1796	.2142	.2560
	72	.0288	.0422	.0573	.0667	.0845	.1075	.1338	.1602	.1772	.2113	.2527
	73	.0284	.0416	.0565	.0658	.0834	.1060	.1320	.1581	.1749	.2086	.2495
	74	.0280	.0410	.0557	.0649	.0822	.1046	.1302	.1560	.1726	.2059	.2464
	75	.0276	.0404	.0549	.0640	.0811	.1032	.1285	.1539	.1703	.2033	.2433
	76+	1.971*	2.906*	3.981*	4.656*	5.956*	7.669*	9.684*	11.77*	13.15*	16.00*	19.63*
9	10–16	SEE NOTE**										
	17–21	SEE NOTE***										
	22	.1362	.1891	.2450	.2778	.3366	.4063	.4787	.5442	.5828	.6528	.7252
	23	.1289	.1793	.2327	.2642	.3207	.3881	.4585	.5227	.5609	.6304	.7033

*To approximate p_f, divide the starred entry under f by $(n - 1)$.
**Compute $r^* = n - r$, find p^*_{1-f} for r^* and n, and compute $p_f = 1 - p^*_{1-f}$.

Table 5 (continued)

							f					
r	*n*	.001	.01	.05	.1	.25	.5	.75	.9	.95	.99	.999
9	24	.1224	.1705	.2216	.2518	.3062	.3715	.4400	.5029	.5405	.6094	.6824
	25	.1166	.1625	.2116	.2406	.2931	.3562	.4229	.4845	.5214	.5896	.6626
	26	.1112	.1553	.2024	.2303	.2810	.3422	.4071	.4673	.5036	.5711	.6437
	27	.1064	.1467	.1940	.2209	.2698	.3292	.3924	.4513	.4870	.5535	.6258
	28	.1019	.1426	.1862	.2122	.2596	.3171	.3787	.4364	.4714	.5370	.6088
	29	.0978	.1370	.1791	.2042	.2500	.3059	.3660	.4224	.4567	.5214	.5926
	30	.0940	.1318	.1725	.1968	.2412	.2955	.3540	.4092	.4429	.5066	.5771
	31	.0905	.1270	.1663	.1899	.2330	.2858	.3428	.3968	.4299	.4927	.5624
	32	.0873	.1225	.1606	.1834	.2253	.2766	.3323	.3852	.4177	.4794	.5484
	33	.0843	.1184	.1553	.1774	.2181	.2681	.3224	.3742	.4061	.4668	.5350
	34	.0815	.1145	.1503	.1718	.2113	.2600	.3131	.3638	.3951	.4549	.5223
	35	.0788	.1109	.1456	.1665	.2050	.2525	.3043	.3540	.3847	.4435	.5101
	36	.0764	.1075	.1412	.1615	.1990	.2453	.2960	.3446	.3748	.4327	.4984
	37	.0741	.1043	.1371	.1569	.1933	.2386	.2882	.3358	.3654	.4223	.4872
	38	.0719	.1012	.1332	.1524	.1880	.2322	.2807	.3274	.3564	.4125	.4765
	39	.0698	.0984	.1295	.1483	.1830	.2261	.2736	.3194	.3479	.4031	.4663
	40	.0679	.0957	.1260	.1443	.1782	.2204	.2669	.3118	.3398	.3941	.4565
	41	.0661	.0931	.1227	.1406	.1737	.2149	.2604	.3045	.3320	.3854	.4470
	42	.0643	.0907	.1196	.1370	.1694	.2097	.2543	.2976	.3246	.3772	.4380
	43	.0627	.0884	.1166	.1337	.1653	.2048	.2485	.2909	.3175	.3693	.4292
	44	.0611	.0863	.1138	.1304	.1614	.2000	.2429	.2846	.3107	.3617	.4209
	45	.0596	.0842	.1111	.1274	.1576	.1955	.2376	.2785	.3042	.3544	.4128
	46	.0582	.0822	.1085	.1245	.1541	.1912	.2325	.2727	.2980	.3474	.4050
	47	.0568	.0803	.1060	.1217	.1507	.1871	.2276	.2671	.2920	.3407	.3975
	48	.0556	.0785	.1037	.1190	.1474	.1831	.2229	.2617	.2862	.3342	.3903
	49	.0543	.0768	.1015	.1164	.1443	.1794	.2184	.2566	.2807	.3279	.3833
	50	.0531	.0751	.0993	.1140	.1413	.1757	.2141	.2516	.2754	.3219	.3766
	51	.0520	.0736	.0972	.1116	.1385	.1722	.2099	.2469	.2702	.3161	.3701
	52	.0509	.0721	.0953	.1094	.1357	.1689	.2059	.2423	.2653	.3105	.3638
	53	.0499	.0706	.0934	.1072	.1331	.1656	.2021	.2379	.2605	.3051	.3577
	54	.0489	.0692	.0916	.1052	.1305	.1625	.1984	.2336	.2559	.2999	.3519
	55	.0479	.0679	.0898	.1032	.1281	.1595	.1948	.2295	.2513	.2948	.3462
	56	.0470	.0666	.0881	.1012	.1257	.1567	.1913	.2255	.2472	.2900	.3406
	57	.0461	.0653	.0865	.0994	.1235	.1539	.1880	.2217	.2430	.2852	.3353
	58	.0453	.0641	.0849	.0976	.1213	.1512	.1848	.2180	.2390	.2807	.3301
	59	.0445	.0630	.0834	.0959	.1191	.1486	.1817	.2144	.2351	.2762	.3251
	60	.0437	.0619	.0820	.0942	.1171	.1461	.1787	.2109	.2314	.2719	.3202
	61	.0429	.0608	.0805	.0926	.1151	.1437	.1758	.2075	.2277	.2678	.3155
	62	.0422	.0598	.0792	.0910	.1132	.1413	.1730	.2043	.2242	.2637	.3109
	63	.0415	.0588	.0779	.0895	.1114	.1391	.1703	.2011	.2208	.2598	.3064
	64	.0408	.0578	.0766	.0881	.1096	.1369	.1676	.1981	.2175	.2560	.3021
	65	.0401	.0569	.0754	.0867	.1079	.1347	.1651	.1951	.2143	.2523	.2979
	66	.0394	.0559	.0742	.0853	.1062	.1327	.1626	.1922	.2111	.2487	.2938
	67	.0388	.0551	.0730	.0840	.1046	.1307	.1602	.1894	.2081	.2452	.2898
	68	.0382	.0542	.0719	.0827	.1030	.1287	.1578	.1867	.2051	.2418	.2859
	69	.0376	.0534	.0708	.0815	.1015	.1269	.1556	.1841	.2023	.2385	.2821
	70	.0371	.0526	.0698	.0803	.1000	.1250	.1533	.1815	.1995	.2353	.2784
	71	.0365	.0518	.0687	.0791	.0985	.1233	.1512	.1790	.1968	.2322	.2748
	72	.0360	.0511	.0678	.0780	.0971	.1215	.1491	.1766	.1941	.2292	.2713
	73	.0355	.0503	.0668	.0769	.0958	.1198	.1471	.1742	.1916	.2262	.2679
	74	.0349	.0496	.0658	.0758	.0945	.1182	.1451	.1719	.1891	.2233	.2646
	75	.0345	.0489	.0649	.0747	.0932	.1166	.1432	.1697	.1866	.2205	.2613
	76	.0340	.0482	.0641	.0737	.0919	.1151	.1413	.1675	.1842	.2177	.2581
	77	.0335	.0476	.0632	.0727	.0907	.1136	.1395	.1654	.1819	.2150	.2550
	78	.0331	.0469	.0623	.0718	.0895	.1121	.1377	.1633	.1797	.2124	.2520
	79	.0326	.0463	.0615	.0708	.0883	.1107	.1360	.1613	.1775	.2099	.2491
	80	.0322	.0457	.0607	.0699	.0872	.1093	.1343	.1593	.1753	.2074	.2462
	81	.0318	.0451	.0600	.0690	.0861	.1079	.1326	.1574	.1732	.2049	.2434
	82	.0314	.0446	.0592	.0682	.0850	.1066	.1310	.1555	.1712	.2026	.2406
	83+	2.452*	3.507*	4.695*	5.432*	6.838*	8,669*	10.80*	12.99*	14.43*	17.40*	21.16*
10	11–17	SEE NOTE**										
	18–25	SEE NOTE***										
	26	.1361	.1848	.2356	.2653	.3184	.3816	.4476	.5080	.5439	.6100	.6802
	27	.1300	.1768	.2257	.2544	.3058	.3671	.4315	.4907	.5262	.5916	.6617
	28	.1245	.1695	.2166	.2443	.2941	.3537	.4165	.4746	.5095	.5742	.6441
	29	.1194	.1627	.2082	.2350	.2833	.3412	.4025	.4594	.4938	.5578	.6273
	30	.1148	.1565	.2005	.2264	.2732	.3296	.3895	.4452	.4790	.5422	.6113

***To approximate p_f, compute $m = r/n$ and $s = \sqrt{m(1-m)/(n+1)}$, find z_f *in* Table 3, and compute $p_f \doteq m + z_f s$.

Table 5 (continued)

							f					
r	*n*	.001	.01	.05	.1	.25	.5	.75	.9	.95	.99	.999
10	31	.1105	.1508	.1933	.2184	.2638	.3187	.3772	.4319	.4651	.5274	.5960
	32	.1065	.1454	.1866	.2110	.2551	.3086	.3657	.4193	.4519	.5134	.5814
	33	.1027	.1404	.1804	.2040	.2469	.2990	.3548	.4074	.4394	.5001	.5674
	34	.0993	.1358	.1746	.1975	.2393	.2900	.3446	.3961	.4276	.4874	.5541
	35	.0960	.1315	.1691	.1914	.2321	.2816	.3350	.3855	.4165	.4754	.5413
	36	.0930	.1274	.1640	.1857	.2253	.2736	.3258	.3754	.4058	.4639	.5291
	37	.0902	.1236	.1591	.1803	.2189	.2661	.3172	.3658	.3957	.4529	.5174
	38	.0875	.1199	.1546	.1752	.2128	.2590	.3090	.3567	.3861	.4424	.5062
	39	.0850	.1165	.1503	.1704	.2071	.2522	.3012	.3480	.3769	.4324	.4954
	40	.0826	.1133	.1462	.1658	.2017	.2458	.2938	.3397	.3682	.4228	.4851
	41	.0803	.1103	.1424	.1615	.1966	.2397	.2867	.3318	.3598	.4137	.4752
	42	.0782	.1074	.1387	.1574	.1917	.2339	.2800	.3243	.3518	.4049	.4657
	43	.0762	.1047	.1353	.1535	.1870	.2284	.2736	.3171	.3442	.3964	.4565
	44	.0743	.1021	.1320	.1498	.1826	.2231	.2675	.3102	.3368	.3883	.4477
	45	.0724	.0996	.1288	.1463	.1784	.2181	.2616	.3036	.3298	.3806	.4392
	46	.0707	.0973	.1258	.1429	.1744	.2133	.2560	.2973	.3231	.3731	.4310
	47	.0691	.0950	.1230	.1397	.1705	.2087	.2506	.2912	.3166	.3659	.4231
	48	.0675	.0929	.1202	.1366	.1668	.2043	.2455	.2854	.3104	.3590	.4154
	49	.0660	.0908	.1176	.1337	.1633	.2000	.2405	.2798	.3044	.3523	.4081
	50	.0645	.0889	.1151	.1309	.1599	.1960	.2358	.2744	.2986	.3459	.4010
	51	.0631	.0870	.1127	.1282	.1567	.1921	.2312	.2692	.2931	.3397	.3941
	52	.0618	.0852	.1104	.1256	.1535	.1883	.2268	.2642	.2877	.3337	.3875
	53	.0606	.0835	.1082	.1231	.1505	.1847	.2226	.2594	.2826	.3279	.3810
	54	.0593	.0818	.1061	.1207	.1477	.1813	.2185	.2548	.2776	.3224	.3748
	55	.0582	.0802	.1041	.1184	.1449	.1779	.2146	.2503	.2728	.3170	.3688
	56	.0571	.0787	.1021	.1162	.1422	.1747	.2108	.2460	.2682	.3117	.3629
	57	.0560	.0772	.1002	.1141	.1397	.1716	.2071	.2418	.2637	.3067	.3573
	58	.0549	.0758	.0984	.1120	.1372	.1686	.2036	.2378	.2594	.3018	.3518
	59	.0539	.0744	.0966	.1100	.1348	.1657	.2002	.2339	.2552	.2970	.3465
	60	.0530	.0731	.0949	.1081	.1325	.1630	.1969	.2301	.2511	.2925	.3413
	61	.0520	.0718	.0933	.1063	.1302	.1603	.1937	.2264	.2472	.2880	.3363
	62	.0511	.0706	.0917	.1045	.1281	.1576	.1906	.2229	.2433	.2837	.3314
	63	.0503	.0694	.0902	.1027	.1260	.1551	.1876	.2195	.2396	.2795	.3267
	64	.0494	.0683	.0887	.1011	.1240	.1527	.1847	.2161	.2361	.2754	.3221
	65	.0486	.0672	.0873	.0995	.1220	.1503	.1819	.2129	.2326	.2714	.3176
	66	.0478	.0661	.0859	.0979	.1201	.1480	.1791	.2098	.2292	.2676	.3133
	67	.0471	.0650	.0846	.0964	.1183	.1458	.1765	.2067	.2259	.2639	.3090
	68	.0463	.0640	.0833	.0949	.1165	.1436	.1739	.2038	.2227	.2602	.3049
	69	.0456	.0631	.0820	.0935	.1147	.1415	.1714	.2009	.2196	.2567	.3009
	70	.0449	.0621	.0808	.0921	.1131	.1394	.1690	.1981	.2166	.2532	.2970
	71	.0442	.0612	.0796	.0907	.1114	.1375	.1666	.1954	.2136	.2499	.2931
	72	.0436	.0603	.0785	.0894	.1098	.1355	.1643	.1927	.2108	.2466	.2894
	73	.0430	.0594	.0773	.0882	.1083	.1337	.1621	.1901	.2080	.2434	.2858
	74	.0423	.0586	.0762	.0869	.1068	.1318	.1599	.1876	.2053	.2403	.2823
	75	.0417	.0578	.0752	.0857	.1054	.1301	.1578	.1852	.2026	.2373	.2788
	76	.0412	.0570	.0742	.0846	.1039	.1283	.1557	.1828	.2001	.2343	.2754
	77	.0406	.0562	.0732	.0834	.1026	.1267	.1537	.1805	.1975	.2315	.2722
	78	.0401	.0554	.0722	.0823	.1012	.1250	.1518	.1782	.1951	.2287	.2689
	79	.0395	.0547	.0712	.0813	.0999	.1234	.1499	.1760	.1927	.2259	.2658
	80	.0390	.0540	.0703	.0802	.0986	.1219	.1480	.1739	.1904	.2232	.2627
	81	.0385	.0533	.0694	.0792	.0973	.1204	.1462	.1718	.1881	.2206	.2597
	82	.0380	.0526	.0685	.0782	.0962	.1189	.1444	.1697	.1859	.2181	.2568
	83	.0375	.0519	.0677	.0772	.0950	.1174	.1427	.1677	.1837	.2156	.2540
	84	.0370	.0513	.0668	.0763	.0938	.1160	.1410	.1658	.1816	.2131	.2512
	85	.0366	.0507	.0660	.0753	.0927	.1146	.1393	.1639	.1795	.2108	.2484
	86	.0361	.0501	.0652	.0744	.0916	.1133	.1377	.1620	.1775	.2084	.2457
	87	.0357	.0495	.0645	.0736	.0905	.1120	.1362	.1602	.1755	.2061	.2431
	88	.0353	.0489	.0637	.0727	.0895	.1107	.1346	.1584	.1736	.2039	.2405
	89	.0349	.0483	.0630	.0719	.0885	.1095	.1331	.1566	.1717	.2017	.2380
	90+	2.961*	4.130*	5.425*	6.221*	7.726*	9.669*	11.91*	14.21*	15.71*	18.78*	22.66*
11	12–19	SEE NOTE**										
	20–28	SEE NOTE***										
	29	.1425	.1896	.2383	.2665	.3169	.3765	.4388	.4958	.5300	.5930	.6606
	30	.1368	.1823	.2293	.2567	.3056	.3637	.4246	.4806	.5143	.5767	.6441
	31	.1316	.1755	.2211	.2476	.2951	.3517	.4112	.4663	.4994	.5612	.6282
	32	.1268	.1693	.2134	.2391	.2853	.3405	.3987	.4528	.4854	.5465	.6131
	33	.1223	.1634	.2062	.2312	.2761	.3299	.3869	.4400	.4721	.5324	.5986
	34	.1182	.1580	.1995	.2238	.2675	.3200	.3758	.4279	.4596	.5191	.5848
	35	.1143	.1529	.1932	.2168	.2594	.3107	.3653	.4165	.4476	.5064	.5715

*To approximate p_f, divide the starred entry under *f* by $(n - 1)$.
**Compute $r^* = n - r$, find p^*_{1-f} for r^* and n, and compute $p_f = 1 - p^*_{1-f}$.

Table 5 (continued)

r	n	.001	.01	.05	.1	.25	f .5	.75	.9	.95	.99	.999
11	36	.1106	.1481	.1873	.2103	.2518	.3019	.3554	.4056	.4363	.4943	.5588
	37	.1072	.1436	.1818	.2042	.2447	.2936	.3460	.3953	.4255	.4827	.5466
	38	.1040	.1394	.1765	.1984	.2379	.2857	.3371	.3855	.4152	.4716	.5349
	39	.1010	.1354	.1716	.1929	.2315	.2783	.3286	.3762	.4054	.4610	.5237
	40	.0981	.1317	.1669	.1877	.2254	.2712	.3205	.3673	.3960	.4509	.5129
	41	.0954	.1281	.1625	.1828	.2197	.2645	.3129	.3588	.3871	.4412	.5025
	42	.0929	.1248	.1583	.1782	.2142	.2581	.3055	.3507	.3785	.4319	.4925
	43	.0905	.1216	.1544	.1738	.2090	.2520	.2985	.3429	.3703	.4230	.4829
	44	.0882	.1185	.1506	.1695	.2040	.2462	.2919	.3355	.3625	.4144	.4737
	45	.0860	.1157	.1470	.1655	.1993	.2406	.2855	.3284	.3549	.4062	.4648
	46	.0839	.1129	.1436	.1617	.1948	.2353	.2794	.3216	.3477	.3982	.4562
	47	.0819	.1103	.1403	.1581	.1905	.2302	.2735	.3150	.3408	.3906	.4479
	48	.0801	.1078	.1372	.1546	.1864	.2254	.2679	.3087	.3341	.3833	.4399
	49	.0783	.1054	.1342	.1512	.1824	.2207	.2625	.3027	.3277	.3762	.4321
	50	.0765	.1031	.1313	.1480	.1786	.2162	.2573	.2969	.3215	.3694	.4247
	51	.0749	.1009	.1286	.1450	.1750	.2119	.2524	.2913	.3156	.3628	.4175
	52	.0733	.0988	.1259	.1420	.1715	.2078	.2476	.2859	.3099	.3564	.4105
	53	.0718	.0968	.1234	.1392	.1682	.2039	.2430	.2807	.3043	.3503	.4037
	54	.0703	.0949	.1210	.1365	.1649	.2000	.2385	.2757	.2990	.3444	.3972
	55	.0690	.0930	.1186	.1339	.1618	.1963	.2342	.2709	.2938	.3386	.3908
	56	.0676	.0913	.1164	.1314	.1589	.1928	.2301	.2662	.2889	.3331	.3847
	57	.0663	.0895	.1142	.1290	.1560	.1894	.2261	.2617	.2841	.3277	.3787
	58	.0651	.0879	.1122	.1266	.1532	.1861	.2222	.2574	.2794	.3225	.3729
	59	.0639	.0863	.1102	.1244	.1505	.1829	.2185	.2531	.2749	.3174	.3673
	60	.0627	.0848	.1082	.1222	.1479	.1798	.2149	.2491	.2705	.3126	.3619
	61	.0616	.0833	.1064	.1201	.1454	.1768	.2114	.2451	.2663	.3078	.3566
	62	.0606	.0818	.1045	.1181	.1430	.1739	.2081	.2413	.2622	.3032	.3515
	63	.0595	.0805	.1028	.1162	.1407	.1711	.2048	.2376	.2582	.2987	.3465
	64	.0585	.0791	.1011	.1143	.1384	.1684	.2016	.2340	.2544	.2944	.3416
	65	.0576	.0778	.0995	.1124	.1362	.1658	.1986	.2305	.2506	.2902	.3369
	66	.0566	.0766	.0979	.1107	.1341	.1633	.1956	.2271	.2470	.2861	.3323
	67	.0557	.0754	.0964	.1089	.1321	.1608	.1927	.2238	.2434	.2821	.3278
	68	.0548	.0742	.0949	.1073	.1301	.1584	.1899	.2206	.2400	.2782	.3235
	69	.0540	.0731	.0934	.1057	.1281	.1561	.1871	.2175	.2367	.2745	.3192
	70	.0532	.0720	.0921	.1041	.1263	.1539	.1845	.2145	.2334	.2708	.3151
	71	.0524	.0709	.0907	.1026	.1244	.1517	.1819	.2115	.2303	.2672	.3111
	72	.0516	.0698	.0894	.1011	.1227	.1496	.1794	.2087	.2272	.2637	.3071
	73	.0508	.0688	.0881	.0997	.1209	.1475	.1770	.2059	.2242	.2603	.3033
	74	.0501	.0679	.0869	.0983	.1193	.1455	.1746	.2032	.2213	.2570	.2996
	75	.0494	.0669	.0857	.0969	.1176	.1435	.1723	.2005	.2184	.2538	.2959
	76	.0487	.0660	.0845	.0956	.1161	.1416	.1701	.1980	.2157	.2507	.2924
	77	.0480	.0651	.0833	.0943	.1145	.1398	.1679	.1955	.2130	.2476	.2889
	78	.0474	.0642	.0822	.0930	.1130	.1380	.1657	.1930	.2103	.2446	.2855
	79	.0468	.0633	.0811	.0918	.1115	.1362	.1636	.1906	.2078	.2417	.2822
	80	.0461	.0625	.0801	.0906	.1101	.1345	.1616	.1883	.2052	.2388	.2789
	81	.0455	.0617	.0791	.0895	.1087	.1328	.1596	.1860	.2028	.2360	.2758
	82	.0450	.0609	.0781	.0884	.1074	.1312	.1577	.1838	.2004	.2333	.2727
	83	.0444	.0602	.0771	.0873	.1061	.1296	.1558	.1817	.1981	.2307	.2696
	84	.0438	.0594	.0761	.0862	.1048	.1280	.1540	.1795	.1958	.2281	.2667
	85	.0433	.0587	.0752	.0851	.1035	.1265	.1522	.1775	.1936	.2255	.2638
	86	.0427	.0580	.0743	.0841	.1023	.1250	.1504	.1755	.1914	.2230	.2609
	87	.0422	.0573	.0734	.0831	.1011	.1236	.1487	.1735	.1893	.2206	.2582
	88	.0417	.0566	.0726	.0821	.0999	.1222	.1470	.1716	.1872	.2182	.2555
	89	.0412	.0559	.0717	.0812	.0988	.1208	.1454	.1697	.1851	.2159	.2528
	90	.0407	.0553	.0709	.0803	.0976	.1194	.1438	.1678	.1831	.2136	.2502
	91	.0403	.0546	.0701	.0794	.0965	.1181	.1422	.1660	.1812	.2114	.2476
	92	.0398	.0540	.0693	.0785	.0955	.1168	.1407	.1642	.1793	.2092	.2451
	93	.0394	.0534	.0685	.0776	.0944	.1155	.1392	.1625	.1774	.2070	.2427
	94	.0389	.0528	.0678	.0768	.0934	.1143	.1377	.1608	.1756	.2049	.2403
	95	.0385	.0522	.0670	.0759	.0924	.1131	.1363	.1592	.1738	.2029	.2379
	96	.0381	.0517	.0663	.0751	.0914	.1119	.1348	.1575	.1720	.2008	.2356
	97+	3.491*	4.771*	6.169*	7.021*	8.620*	10.67*	13.02*	15.41*	16.96*	20.14*	24.13*
12	13–29	SEE NOTE**										
	21–32	SEE NOTE***										
	33	.1430	.1873	.2326	.2589	.3055	.3609	.4188	.4721	.5042	.5639	.6288
	34	.1380	.1810	.2250	.2505	.2960	.3500	.4068	.4592	.4909	.5499	.6145
	35	.1334	.1751	.2179	.2427	.2870	.3398	.3954	.4470	.4782	.5366	.6007

***To approximate p_f, compute $m = r/n$ and $s = \sqrt{m(1-m)/(n+1)}$, find z_f in Table 3, and compute $p_f \doteq m + z_f s$.

Table 5 (continued)

r	n	.001	.01	.05	.1	.25	f .5	.75	.9	.95	.99	.999
12	36	.1291	.1696	.2112	.2354	.2786	.3302	.3847	.4354	.4661	.5239	.5875
	37	.1251	.1644	.2049	.2285	.2707	.3211	.3746	.4244	.4547	.5117	.5749
	38	.1213	.1595	.1990	.2220	.2632	.3125	.3649	.4139	.4438	.5001	.5628
	39	.1178	.1550	.1934	.2158	.2561	.3044	.3558	.4039	.4333	.4890	.5511
	40	.1144	.1506	.1881	.2100	.2493	.2966	.3471	.3944	.4234	.4783	.5398
	41	.1113	.1466	.1831	.2045	.2430	.2893	.3388	.3853	.4139	.4681	.5290
	42	.1083	.1427	.1784	.1993	.2369	.2823	.3309	.3767	.4048	.4583	.5187
	43	.1054	.1390	.1739	.1943	.2311	.2756	.3233	.3684	.3961	.4489	.5086
	44	.1027	.1355	.1696	.1896	.2256	.2693	.3161	.3604	.3877	.4399	.4990
	45	.1002	.1322	.1655	.1851	.2204	.2632	.3092	.3258	.3797	.4312	.4897
	46	.0977	.1291	.1617	.1808	.2154	.2574	.3026	.3455	.3720	.4228	.4807
	47	.0954	.1260	.1580	.1767	.2106	.2518	.2963	.3385	.3646	.4148	.4721
	48	.0932	.1232	.1544	.1728	.2061	.2465	.2902	.3318	.3575	.4071	.4637
	49	.0911	.1204	.1510	.1691	.2017	.2414	.2843	.3253	.3507	.3996	.4556
	50	.0891	.1178	.1478	.1655	.1975	.2365	.2787	.3191	.3441	.3924	.4478
	51	.0872	.1153	.1447	.1620	.1935	.2318	.2734	.3131	.3378	.3854	.4402
	52	.0853	.1129	.1417	.1588	.1896	.2273	.2682	.3073	.3317	.3787	.4329
	53	.0836	.1106	.1389	.1556	.1859	.2230	.2632	.3018	.3258	.3722	.4259
	54	.0819	.1084	.1362	.1526	.1823	.2188	.2584	.2964	.3201	.3659	.4190
	55	.0802	.1062	.1335	.1496	.1789	.2147	.2538	.2912	.3146	.3599	.4124
	56	.0787	.1042	.1310	.1468	.1756	.2109	.2493	.2862	.3093	.3540	.4059
	57	.0772	.1022	.1286	.1441	.1724	.2071	.2450	.2814	.3041	.3483	.3997
	58	.0757	.1003	.1262	.1415	.1694	.2035	.2408	.2767	.2992	.3428	.3936
	59	.0743	.0985	.1239	.1390	.1664	.2000	.2368	.2722	.2943	.3375	.3877
	60	.0730	.0968	.1218	.1366	.1635	.1967	.2329	.2678	.2897	.3323	.3820
	61	.0717	.0951	.1197	.1342	.1608	.1934	.2291	.2636	.2852	.3273	.3765
	62	.0704	.0934	.1176	.1320	.1581	.1902	.2254	.2595	.2808	.3224	.3711
	63	.0692	.0918	.1157	.1298	.1555	.1872	.2219	.2555	.2765	.3177	.3658
	64	.0680	.0903	.1137	.1276	.1530	.1842	.2185	.2516	.2724	.3131	.3607
	65	.0669	.0888	.1119	.1256	.1506	.1814	.2151	.2479	.2684	.3086	.3558
	66	.0658	.0874	.1101	.1236	.1482	.1786	.2119	.2442	.2645	.3043	.3509
	67	.0648	.0860	.1084	.1217	.1460	.1759	.2088	.2407	.2608	.3001	.3462
	68	.0637	.0847	.1067	.1198	.1438	.1733	.2058	.2373	.2571	.2959	.3417
	69	.0628	.0834	.1051	.1180	.1416	.1708	.2028	.2339	.2535	.2919	.3372
	70	.0618	.0821	.1035	.1163	.1395	.1683	.1999	.2307	.2501	.2881	.3329
	71	.0609	.0809	.1020	.1146	.1375	.1659	.1971	.2275	.2467	.2843	.3286
	72	.0600	.0797	.1005	.1129	.1356	.1636	.1944	.2245	.2434	.2806	.3245
	73	.0591	.0785	.0991	.1113	.1337	.1613	.1918	.2215	.2402	.2770	.3205
	74	.0582	.0774	.0977	.1097	.1318	.1591	.1892	.2186	.2371	.2735	.3165
	75	.0574	.0763	.0963	.1082	.1300	.1570	.1867	.2157	.2340	.2701	.3127
	76	.0566	.0753	.0950	.1067	.1282	.1549	.1843	.2130	.2311	.2667	.3090
	77	.0558	.0742	.0937	.1053	.1265	.1529	.1819	.2103	.2282	.2635	.3053
	78	.0551	.0732	.0925	.1039	.1249	.1509	.1796	.2077	.2254	.2603	.3017
	79	.0543	.0722	.0912	.1025	.1233	.1490	.1774	.2051	.2226	.2572	.2982
	80	.0536	.0713	.0901	.1012	.1217	.1471	.1752	.2026	.2199	.2542	.2948
	81	.0529	.0704	.0889	.0999	.1202	.1452	.1730	.2002	.2173	.2512	.2915
	82	.0522	.0695	.0878	.0987	.1187	.1435	.1709	.1978	.2148	.2483	.2882
	83	.0515	.0686	.0867	.0974	.1172	.1417	.1689	.1955	.2123	.2455	.2850
	84	.0509	.0677	.0856	.0962	.1158	.1400	.1669	.1932	.2098	.2427	.2819
	85	.0503	.0669	.0845	.0951	.1144	.1384	.1649	.1910	.2074	.2400	.2789
	86	.0496	.0661	.0835	.0939	.1130	.1367	.1630	.1888	.2051	.2374	.2759
	87	.0490	.0653	.0825	.0928	.1117	.1352	.1612	.1867	.2028	.2348	.2729
	88	.0484	.0645	.0816	.0917	.1104	.1336	.1594	.1846	.2006	.2323	.2701
	89	.0479	.0638	.0806	.0907	.1091	.1321	.1576	.1826	.1984	.2298	.2673
	90	.0473	.0630	.0797	.0896	.1079	.1306	.1558	.1806	.1963	.2274	.2645
	91	.0468	.0623	.0788	.0886	.1067	.1292	.1541	.1786	.1942	.2250	.2618
	92	.0462	.0616	.0779	.0876	.1055	.1278	.1525	.1767	.1921	.2227	.2592
	93	.0457	.0609	.0770	.0866	.1043	.1264	.1508	.1749	.1901	.2204	.2566
	94	.0452	.0602	.0762	.0857	.1032	.1250	.1493	.1731	.1882	.2182	.2541
	95	.0447	.0596	.0753	.0848	.1021	.1237	.1477	.1713	.1863	.2160	.2516
	96	.0442	.0589	.0745	.0839	.1010	.1224	.1462	.1695	.1844	.2138	.2491
	97	.0437	.0583	.0737	.0830	.1000	.1211	.1447	.1678	.1825	.2117	.2468
	98	.0433	.0577	.0730	.0821	.0989	.1199	.1432	.1661	.1807	.2097	.2444
	99	.0428	.0570	.0722	.0812	.0979	.1187	.1418	.1645	.1789	.2077	.2421
	100	.0424	.0565	.0715	.0804	.0969	.1175	.1404	.1629	.1772	.2057	.2398
	101	.0419	.0559	.0707	.0796	.0959	.1163	.1390	.1613	.1755	.2037	.2376
	102	.0415	.0553	.0700	.0788	.0950	.1151	.1376	.1598	.1738	.2018	.2355
	103	.0411	.0547	.0693	.0780	.0940	.1140	.1363	.1582	.1722	.1999	.2333
	104+	4.042*	5.428*	6.924*	7.829*	9.519*	11.67*	14.12*	16.60*	18.21*	21.49*	25.59*

*To approximate p_f, divide the starred entry under f by $(n - 1)$.

**Compute $r^* = n - r$, find p^*_{1-f} for r^* and n, and compute $p_f = 1 - p^*_{1-f}$.

Table 5 (continued)

r	n	.001	.01	.05	.1	.25	f .5	.75	.9	.95	.99	.999
13	14–22	SEE NOTE**										
	23–36	SEE NOTE***										
	37	.1439	.1859	.2285	.2532	.2969	.3486	.4029	.4531	.4834	.5401	.6023
	38	.1395	.1803	.2219	.2459	.2886	.3393	.3926	.4420	.4719	.5280	.5898
	39	.1353	.1751	.2156	.2391	.2808	.3305	.3828	.4314	.4609	.5163	.5777
	40	.1315	.1702	.2097	.2326	.2734	.3221	.3734	.4213	.4503	.5052	.5661
	41	.1278	.1655	.2041	.2265	.2664	.3141	.3645	.4116	.4403	.4945	.5549
	42	.1243	.1611	.1988	.2207	.2598	.3065	.3560	.4024	.4307	.4842	.5441
	43	.1210	.1570	.1938	.2152	.2534	.2992	.3479	.3935	.4214	.4744	.5337
	44	.1179	.1530	.1890	.2099	.2474	.2923	.3402	.3851	.4126	.4649	.5237
	45	.1150	.1492	.1845	.2049	.2417	.2857	.3327	.3770	.4041	.4558	.5140
	46	.1122	.1457	.1801	.2002	.2362	.2794	.3256	.3692	.3960	.4470	.5047
	47	.1095	.1423	.1760	.1956	.2309	.2734	.3188	.3618	.3881	.4385	.4957
	48	.1069	.1390	.1720	.1913	.2259	.2676	.3123	.3546	.3806	.4304	.4869
	49	.1045	.1359	.1683	.1871	.2211	.2621	.3061	.3477	.3734	.4225	.4785
	50	.1022	.1329	.1646	.1832	.2165	.2568	.3000	.3411	.3664	.4150	.4704
	51	.1000	.1301	.1612	.1794	.2121	.2517	.2942	.3347	.3597	.4077	.4625
	52	.0978	.1273	.1579	.1757	.2079	.2468	.2887	.3285	.3532	.4006	.4549
	53	.0958	.1247	.1547	.1722	.2038	.2421	.2833	.3226	.3469	.3938	.4475
	53	.0938	.1222	.1516	.1688	.1999	.2375	.2782	.3169	.3409	.3872	.4403
	55	.0919	.1198	.1487	.1656	.1961	.2332	.2732	.3114	.3350	.3808	.4334
	56	.0901	.1175	.1458	.1625	.1925	.2289	.2684	.3060	.3294	.3746	.4267
	57	.0884	.1153	.1431	.1595	.1890	.2249	.2637	.3009	.3239	.3686	.4202
	58	.0867	.1131	.1405	.1566	.1856	.2210	.2592	.2959	.3187	.3628	.4138
	59	.0851	.1111	.1380	.1538	.1824	.2172	.2549	.2911	.3136	.3572	.4077
	60	.0836	.1091	.1355	.1511	.1792	.2135	.2507	.2864	.3086	.3517	.4017
	61	.0821	.1072	.1332	.1485	.1762	.2100	.2467	.2819	.3038	.3464	.3959
	62	.0807	.1053	.1309	.1460	.1733	.2065	.2427	.2775	.2992	.3413	.3903
	63	.0793	.1035	.1287	.1435	.1704	.2032	.2389	.2732	.2946	.3363	.3848
	64	.0779	.1018	.1266	.1412	.1677	.2000	.2352	.2691	.2903	.3315	.3795
	65	.0766	.1001	.1245	.1389	.1650	.1969	.2317	.2651	.2860	.3267	.3743
	66	.0754	.0985	.1226	.1367	.1624	.1939	.2282	.2612	.2819	.3222	.3692
	67	.0742	.0969	.1206	.1346	.1599	.1910	.2248	.2575	.2779	.3177	.3643
	68	.0730	.0954	.1188	.1325	.1575	.1881	.2215	.2538	.2740	.3134	.3595
	69	.0718	.0939	.1170	.1305	.1552	.1854	.2184	.2502	.2702	.3092	.3548
	70	.0707	.0925	.1152	.1286	.1529	.1827	.2153	.2468	.2665	.3051	.3503
	71	.0697	.0911	.1135	.1267	.1507	.1801	.2123	.2434	.2629	.3011	.3458
	72	.0686	.0898	.1118	.1249	.1485	.1776	.2094	.2401	.2594	.2972	.3415
	73	.0676	.0885	.1102	.1231	.1464	.1751	.2065	.2369	.2560	.2934	.3373
	74	.0666	.0872	.1087	.1213	.1444	.1727	.2038	.2338	.2527	.2897	.3332
	75	.0657	.0860	.1072	.1197	.1424	.1704	.2011	.2308	.2495	.2861	.3292
	76	.0648	.0848	.1057	.1180	.1405	.1682	.1985	.2278	.2463	.2825	.3252
	77	.0639	.0836	.1043	.1164	.1387	.1660	.1959	.2250	.2432	.2791	.3214
	78	.0630	.0825	.1029	.1149	.1368	.1638	.1934	.2222	.2402	.2757	.3176
	79	.0621	.0814	.1015	.1134	.1351	.1617	.1910	.2194	.2373	.2725	.3140
	80	.0613	.0803	.1002	.1119	.1333	.1597	.1886	.2168	.2345	.2693	.3104
	81	.0605	.0793	.0989	.1105	.1316	.1577	.1863	.2142	.2317	.2661	.3069
	82	.0597	.0782	.0976	.1091	.1300	.1558	.1841	.2116	.2290	.2631	.3035
	83	.0590	.0772	.0964	.1077	.1284	.1539	.1819	.2091	.2263	.2601	.3001
	84	.0582	.0763	.0952	.1063	.1268	.1520	.1797	.2067	.2237	.2572	.2969
	85	.0575	.0753	.0940	.1051	.1253	.1502	.1776	.2043	.2212	.2543	.2937
	86	.0568	.0744	.0929	.1038	.1238	.1485	.1756	.2020	.2187	.2515	.2905
	87	.0561	.0735	.0918	.1026	.1224	.1467	.1736	.1997	.2163	.2488	.2875
	88	.0554	.0726	.0907	.1014	.1209	.1451	.1716	.1975	.2139	.2461	.2844
	89	.0548	.0718	.0897	.1002	.1196	.1434	.1697	.1954	.2116	.2435	.2815
	90	.0541	.0709	.0886	.0991	.1182	.1418	.1678	.1932	.2093	.2410	.2786
	91	.0535	.0701	.0876	.0980	.1169	.1402	.1660	.1912	.2071	.2384	.2758
	92	.0529	.0693	.0866	.0969	.1156	.1387	.1642	.1891	.2049	.2360	.2730
	93	.0523	.0686	.0857	.0958	.1143	.1372	.1625	.1871	.2028	.2336	.2703
	94	.0517	.0678	.0847	.0947	.1131	.1357	.1608	.1852	.2007	.2312	.2676
	95	.0511	.0670	.0838	.0937	.1119	.1343	.1591	.1833	.1986	.2289	.2650
	96	.0505	.0663	.0829	.0927	.1107	.1329	.1574	.1814	.1966	.2266	.2625
	97	.0500	.0656	.0820	.0917	.1095	.1314	.1558	.1796	.1946	.2244	.2600
	98	.0495	.0649	.0811	.0908	.1084	.1302	.1542	.1778	.1927	.2222	.2575
	99	.0489	.0642	.0803	.0898	.1072	.1288	.1527	.1760	.1908	.2201	.2551
	100	.0484	.0635	.0795	.0889	.1062	.1275	.1512	.1743	.1890	.2180	.2527
	101	.0479	.0629	.0786	.0880	.1051	.1263	.1497	.1726	.1872	.2160	.2504
	102	.0474	.0623	.0778	.0871	.1040	.1250	.1482	.1710	.1854	.2139	.2481
	103	.0469	.0616	.0771	.0862	.1030	.1238	.1468	.1693	.1836	.2119	.2458

***To approximate p_f, compute $m = r/n$ and $s = \sqrt{m(1 - m)/(n + 1)}$, find z_f in Table 3, and compute $p_f \doteq m + z_f s$.

Table 5 (continued)

r	n	f: .001	.01	.05	.1	.25	.5	.75	.9	.95	.99	.999
13	104	.0465	.0610	.0763	.0854	.1020	.1226	.1454	.1678	.1819	.2100	.2436
	105	.0460	.0604	.0755	.0845	.1010	.1214	.1440	.1662	.1802	.2081	.2415
	106	.0456	.0598	.0748	.0837	.1000	.1203	.1427	.1646	.1786	.2062	.2393
	107	.0451	.0592	.0741	.0829	.0991	.1191	.1414	.1631	.1770	.2044	.2372
	108	.0447	.0587	.0734	.0821	.0982	.1180	.1401	.1617	.1754	.2026	.2352
	109	.0442	.0581	.0727	.0814	.0972	.1169	.1388	.1602	.1738	.2008	.2332
	110	.0438	.0575	.0720	.0806	.0963	.1159	.1375	.1588	.1723	.1990	.2312
	111+	4.611*	6.099*	7.690*	8.646*	10.42*	12.67*	15.22*	17.78*	19.44*	22.82*	27.03*
14	15–23	SEE NOTE**										
	24–40	SEE NOTE***										
	41	.1450	.1851	.2255	.2488	.2901	.3389	.3901	.4376	.4663	.5203	.5800
	42	.1410	.1801	.2196	.2424	.2828	.3307	.3810	.4278	.4562	.5096	.5689
	43	.1373	.1754	.2141	.2363	.2759	.3229	.3724	.4184	.4464	.4993	.5581
	44	.1337	.1710	.2088	.2305	.2693	.3154	.3641	.4095	.4371	.4894	.5477
	45	.1303	.1667	.2037	.2250	.2630	.3083	.3562	.4009	.4282	.4799	.5377
	46	.1271	.1627	.1989	.2198	.2571	.3015	.3486	.3927	.4196	.4707	.5281
	47	.1241	.1589	.1943	.2148	.2513	.2950	.3413	.3847	.4113	.4618	.5187
	48	.1212	.1552	.1899	.2100	.2459	.2888	.3343	.3772	.4034	.4533	.5097
	49	.1184	.1517	.1857	.2054	.2406	.2828	.3276	.3698	.3957	.4451	.5009
	50	.1157	.1484	.1817	.2011	.2356	.2770	.3212	.3628	.3884	.4372	.4925
	51	.1132	.1452	.1779	.1969	.2308	.2715	.3150	.3560	.3813	.4295	.4843
	52	.1108	.1422	.1742	.1928	.2262	.2663	.3091	.3495	.3744	.4221	.4764
	53	.1085	.1392	.1707	.1890	.2218	.2612	.3033	.3432	.3678	.4149	.4687
	54	.1062	.1364	.1673	.1853	.2175	.2563	.2978	.3372	.3614	.4080	.4613
	55	.1041	.1337	.1641	.1817	.2134	.2516	.2925	.3313	.3553	.4013	.4540
	56	.1020	.1311	.1609	.1783	.2094	.2470	.2873	.3256	.3493	.3948	.4470
	57	.1001	.1286	.1579	.1750	.2056	.2426	.2824	.3202	.3435	.3886	.4403
	58	.0982	.1262	.1550	.1718	.2020	.2384	.2776	.3149	.3380	.3825	.4337
	59	.0963	.1239	.1522	.1687	.1984	.2343	.2729	.3097	.3326	.3766	.4273
	60	.0946	.1217	.1495	.1658	.1950	.2304	.2685	.3048	.3273	.3708	.4210
	61	.0929	.1195	.1469	.1629	.1917	.2265	.2641	.3000	.3222	.3653	.4150
	62	.0912	.1175	.1444	.1602	.1885	.2228	.2599	.2953	.3173	.3599	.4091
	63	.0897	.1154	.1420	.1575	.1854	.2193	.2558	.2908	.3126	.3547	.4034
	64	.0881	.1135	.1396	.1549	.1824	.2158	.2519	.2864	.3079	.3496	.3978
	65	.0867	.1116	.1374	.1524	.1795	.2125	.2481	.2822	.3034	.3446	.3924
	66	.0852	.1098	.1352	.1500	.1767	.2092	.2444	.2781	.2991	.3398	.3872
	67	.0839	.1081	.1330	.1476	.1740	.2060	.2408	.2741	.2948	.3351	.3820
	68	.0825	.1064	.1310	.1454	.1714	.2030	.2373	.2702	.2907	.3306	.3770
	69	.0812	.1047	.1290	.1432	.1688	.2000	.2339	.2664	.2867	.3261	.3721
	70	.0800	.1031	.1270	.1410	.1663	.1971	.2306	.2627	.2828	.3218	.3674
	71	.0788	.1016	.1252	.1390	.1639	.1943	.2274	.2591	.2790	.3176	.3628
	72	.0776	.1001	.1233	.1370	.1616	.1916	.2242	.2556	.2753	.3135	.3582
	73	.0764	.0986	.1216	.1350	.1593	.1890	.2212	.2522	.2717	.3095	.3538
	74	.0753	.0972	.1198	.1331	.1571	.1864	.2182	.2489	.2681	.3056	.3495
	75	.0743	.0958	.1182	.1313	.1549	.1839	.2154	.2457	.2647	.3018	.3453
	76	.0732	.0945	.1165	.1295	.1529	.1814	.2126	.2426	.2614	.2981	.3412
	77	.0722	.0932	.1149	.1277	.1508	.1791	.2098	.2395	.2581	.2945	.3372
	78	.0712	.0919	.1134	.1260	.1488	.1767	.2072	.2366	.2550	.2910	.3333
	79	.0702	.0907	.1119	.1243	.1469	.1745	.2046	.2336	.2519	.2875	.3295
	80	.0693	.0895	.1104	.1227	.1450	.1723	.2020	.2308	.2488	.2842	.3257
	81	.0684	.0883	.1090	.1212	.1432	.1701	.1996	.2280	.2459	.2809	.3221
	82	.0675	.0872	.1076	.1196	.1414	.1680	.1972	.2253	.2430	.2777	.3185
	83	.0666	.0861	.1063	.1181	.1397	.1660	.1948	.2227	.2402	.2745	.3150
	84	.0658	.0850	.1050	.1167	.1380	.1640	.1925	.2201	.2374	.2714	.3116
	85	.0650	.0840	.1037	.1153	.1363	.1621	.1903	.2176	.2348	.2684	.3082
	86	.0642	.0829	.1024	.1139	.1347	.1602	.1881	.2151	.2321	.2655	.3049
	87	.0634	.0819	.1012	.1125	.1331	.1583	.1859	.2127	.2296	.2626	.3017
	88	.0626	.0809	.1000	.1112	.1315	.1565	.1838	.2104	.2270	.2598	.2986
	89	.0619	.0800	.0988	.1099	.1300	.1547	.1818	.2081	.2246	.2571	.2955
	90	.0611	.0791	.0977	.1086	.1286	.1530	.1798	.2058	.2222	.2544	.2925
	91	.0604	.0781	.0966	.1074	.1271	.1513	.1778	.2036	.2198	.2517	.2895
	92	.0597	.0773	.0955	.1062	.1257	.1496	.1759	.2014	.2175	.2491	.2866
	93	.0590	.0764	.0944	.1050	.1243	.1480	.1740	.1993	.2152	.2466	.2838
	94	.0584	.0755	.0934	.1039	.1230	.1464	.1722	.1972	.2130	.2441	.2810
	95	.0577	.0747	.0923	.1027	.1217	.1449	.1704	.1952	.2109	.2417	.2783
	96	.0571	.0739	.0913	.1016	.1204	.1434	.1686	.1932	.2087	.2393	.2756
	97	.0565	.0731	.0904	.1006	.1191	.1419	.1669	.1913	.2067	.2369	.2730
	98	.0559	.0723	.0894	.0995	.1179	.1404	.1652	.1894	.2046	.2347	.2704

*To approximate p_f, divide the starred entry under f by $(n-1)$.

**Compute $r^* = n - r$, find p^*_{1-f} for r^* and n, and compute $p_f = 1 - p^*_{1-f}$.

Table 5 (continued)

						f						
r	n	.001	.01	.05	.1	.25	.5	.75	.9	.95	.99	.999
14	99	.0553	.0715	.0885	.0985	.1166	.1390	.1636	.1875	.2026	.2324	.2678
	100	.0547	.0708	.0876	.0975	.1154	.1376	.1620	.1857	.2006	.2302	.2654
	101	.0541	.0701	.0867	.0965	.1143	.1362	.1604	.1839	.1987	.2280	.2629
	102	.0536	.0693	.0858	.0955	.1131	.1349	.1588	.1821	.1968	.2259	.2605
	103	.0530	.0686	.0849	.0945	.1120	.1336	.1573	.1804	.1950	.2288	.2582
	104	.0525	.0680	.0841	.0936	.1109	.1323	.1558	.1787	.1932	.2218	.2559
	105	.0519	.0673	.0833	.0927	.1099	.1310	.1543	.1770	.1914	.2197	.2536
	106	.0514	.0666	.0824	.0918	.1088	.1298	.1528	.1754	.1896	.2178	.2514
	107	.0509	.0660	.0816	.0909	.1078	.1285	.1514	.1738	.1879	.2158	.2492
	108	.0504	.0653	.0809	.0900	.1067	.1273	.1500	.1722	.1862	.2139	.2470
	109	.0500	.0647	.0801	.0892	.1057	.1262	.1487	.1706	.1845	.2120	.2449
	110	.0495	.0641	.0793	.0884	.1048	.1250	.1473	.1691	.1829	.2102	.2428
	111	.0490	.0635	.0786	.0875	.1038	.1239	.1460	.1676	.1813	.2084	.2408
	112	.0485	.0629	.0779	.0867	.1029	.1228	.1447	.1662	.1797	.2066	.2388
	113	.0481	.0623	.0772	.0859	.1019	.1217	.1434	.1647	.1782	.2049	.2368
	114	.0477	.0618	.0765	.0852	.1010	.1206	.1422	.1633	.1767	.2031	.2348
	115	.0472	.0612	.0758	.0844	.1001	.1195	.1410	.1619	.1752	.2015	.2329
	116	.0468	.0607	.0751	.0837	.0993	.1185	.1398	.1605	.1737	.1998	.2310
	117	.0464	.0601	.0745	.0829	.0984	.1175	.1386	.1592	.1723	.1982	.2292
	118+	5.195*	6.782*	8.464*	9.470*	11.33*	13.67*	16.31*	18.96*	20.67*	24.14*	28.45*
15	16–24	SEE NOTE**										
	25–45	SEE NOTE***										
	46	.1426	.1802	.2180	.2397	.2781	.3235	.3714	.4159	.4429	.4940	.5509
	47	.1391	.1759	.2129	.2342	.2719	.3166	.3636	.4075	.4342	.4848	.5413
	48	.1359	.1718	.2081	.2290	.2660	.3099	.3562	.3995	.4259	.4759	.5319
	49	.1327	.1680	.2035	.2240	.2603	.3035	.3491	.3918	.4178	.4673	.5229
	50	.1297	.1642	.1991	.2192	.2548	.2973	.3423	.3844	.4101	.4590	.5141
	51	.1269	.1607	.1949	.2146	.2496	.2914	.3357	.3772	.4026	.4510	.5056
	52	.1241	.1573	.1908	.2102	.2446	.2857	.3293	.3703	.3954	.4433	.4974
	53	.1215	.1540	.1870	.2060	.2398	.2803	.3232	.3637	.3885	.4358	.4895
	54	.1190	.1509	.1833	.2019	.2352	.2750	.3174	.3572	.3817	.4286	.4818
	55	.1166	.1479	.1797	.1980	.2308	.2700	.3117	.3511	.3753	.4216	.4743
	56	.1143	.1450	.1762	.1943	.2265	.2651	.3062	.3451	.3690	.4148	.4670
	57	.1121	.1423	.1729	.1907	.2224	.2604	.3009	.3393	.3629	.4082	.4600
	58	.1099	.1396	.1697	.1872	.2184	.2558	.2958	.3337	.3570	.4019	.4531
	59	.1079	.1370	.1667	.1839	.2146	.2514	.2909	.3283	.3513	.3957	.4465
	60	.1059	.1346	.1637	.1806	.2109	.2472	.2861	.3230	.3458	.3897	.4400
	61	.1040	.1322	.1609	.1775	.2073	.2431	.2815	.3180	.3405	.3839	.4337
	62	.1022	.1299	.1581	.1745	.2038	.2391	.2770	.3130	.3353	.3782	.4276
	63	.1004	.1276	.1554	.1716	.2005	.2353	.2727	.3083	.3303	.3727	.4217
	64	.0987	.1255	.1529	.1688	.1972	.2316	.2685	.3036	.3254	.3674	.4159
	65	.0970	.1234	.1504	.1660	.1941	.2280	.2644	.2991	.3206	.3622	.4103
	66	.0954	.1214	.1480	.1634	.1911	.2245	.2605	.2948	.3160	.3572	.4048
	67	.0939	.1195	.1456	.1608	.1881	.2211	.2566	.2905	.3116	.3523	.3994
	68	.0924	.1176	.1434	.1584	.1853	.2178	.2529	.2864	.3072	.3475	.3942
	69	.0909	.1158	.1412	.1560	.1825	.2146	.2493	.2824	.3030	.3429	.3892
	70	.0895	.1140	.1390	.1536	.1798	.2116	.2458	.2785	.2989	.3384	.3842
	71	.0881	.1123	.1370	.1514	.1772	.2085	.2424	.2747	.2949	.3340	.3794
	72	.0868	.1106	.1350	.1492	.1747	.2056	.2390	.2710	.2910	.3297	.3747
	73	.0855	.1090	.1330	.1470	.1722	.2028	.2358	.2674	.2872	.3255	.3701
	74	.0843	.1074	.1311	.1450	.1698	.2000	.2327	.2640	.2835	.3214	.3656
	75	.0831	.1059	.1293	.1429	.1675	.1973	.2296	.2605	.2798	.3174	.3612
	76	.0819	.1044	.1275	.1410	.1653	.1947	.2266	.2572	.2763	.3135	.3570
	77	.0807	.1030	.1258	.1391	.1631	.1922	.2237	.2540	.2729	.3097	.3528
	78	.0796	.1016	.1241	.1372	.1609	.1897	.2209	.2508	.2695	.3060	.3487
	79	.0786	.1002	.1224	.1354	.1588	.1872	.2181	.2478	.2663	.3024	.3447
	80	.0775	.0989	.1208	.1337	.1568	.1849	.2154	.2447	.2631	.2989	.3408
	81	.0765	.0976	.1193	.1319	.1548	.1826	.2128	.2418	.2600	.2954	.3370
	82	.0755	.0963	.1178	.1303	.1529	.1803	.2102	.2389	.2569	.2921	.3333
	83	.0745	.0951	.1163	.1286	.1510	.1782	.2077	.2362	.2540	.2888	.3296
	84	.0736	.0939	.1148	.1271	.1491	.1760	.2052	.2334	.2511	.2855	.3261
	85	.0726	.0927	.1134	.1255	.1473	.1739	.2029	.2307	.2482	.2824	.3226
	86	.0717	.0916	.1120	.1240	.1456	.1719	.2005	.2281	.2454	.2793	.3191
	87	.0709	.0905	.1107	.1225	.1439	.1699	.1982	.2256	.2427	.2763	.3158
	88	.0700	.0894	.1094	.1211	.1422	.1680	.1960	.2231	.2401	.2733	.3125
	89	.0692	.0883	.1081	.1197	.1406	.1661	.1938	.2207	.2375	.2704	.3093
	90	.0683	.0873	.1069	.1183	.1390	.1642	.1917	.2183	.2349	.2676	.3062

***To approximate p_f, compute $m = r/n$ and $s = \sqrt{m(1-m)/(n+1)}$, find z_f in Table 3, and compute $p_f \doteq m + z_f s$.

Table 5 (continued)

r	n	.001	.01	.05	.1	.25	f .5	.75	.9	.95	.99	.999
15	91	.0675	.0863	.1056	.1169	.1374	.1624	.1896	.2159	.2324	.2648	.3031
	92	.0668	.0853	.1044	.1156	.1359	.1606	.1876	.2136	.2300	.2621	.3000
	93	.0660	.0844	.1033	.1143	.1344	.1589	.1856	.2114	.2276	.2595	.2971
	94	.0653	.0834	.1021	.1131	.1329	.1572	.1836	.2092	.2253	.2568	.2942
	95	.0645	.0825	.1010	.1119	.1315	.1555	.1817	.2071	.2230	.2543	.2913
	96	.0638	.0816	.0999	.1107	.1301	.1539	.1798	.2050	.2208	.2518	.2885
	97	.0631	.0807	.0989	.1095	.1287	.1523	.1780	.2029	.2186	.2493	.2858
	98	.0624	.0798	.0978	.1083	.1274	.1507	.1762	.2009	.2164	.2469	.2831
	99	.0618	.0790	.0968	.1072	.1261	.1492	.1744	.1989	.2143	.2446	.2804
	100	.0611	.0782	.0958	.1061	.1248	.1477	.1727	.1969	.2122	.2422	.2778
	101	.0605	.0774	.0948	.1050	.1235	.1462	.1710	.1950	.2102	.2400	.2753
	102	.0599	.0766	.0938	.1040	.1223	.1447	.1693	.1932	.2082	.2377	.2728
	103	.0592	.0758	.0929	.1029	.1211	.1433	.1677	.1913	.2062	.2355	.2703
	104	.0586	.0750	.0920	.1019	.1199	.1419	.1661	.1895	.2043	.2334	.2679
	105	.0581	.0743	.0911	.1009	.1187	.1406	.1645	.1878	.2024	.2313	.2655
	106	.0575	.0736	.0902	.0999	.1176	.1393	.1630	.1860	.2006	.2292	.2632
	107	.0569	.0728	.0893	.0990	.1165	.1379	.1615	.1843	.1987	.2271	.2609
	108	.0564	.0721	.0884	.0980	.1154	.1367	.1600	.1827	.1970	.2251	.2587
	109	.0558	.0714	.0876	.0971	.1143	.1354	.1585	.1810	.1952	.2232	.2565
	110	.0553	.0708	.0868	.0962	.1132	.1342	.1571	.1794	.1935	.2212	.2543
	111	.0548	.0701	.0860	.0953	.1122	.1329	.1557	.1778	.1918	.2193	.2522
	112	.0542	.0695	.0852	.0944	.1112	.1317	.1543	.1763	.1901	.2175	.2500
	113	.0537	.0688	.0844	.0936	.1102	.1306	.1530	.1747	.1885	.2156	.2480
	114	.0533	.0682	.0836	.0927	.1092	.1294	.1516	.1732	.1869	.2138	.2459
	115	.0528	.0676	.0829	.0919	.1082	.1283	.1503	.1718	.1853	.2120	.2439
	116	.0523	.0670	.0822	.0911	.1073	.1272	.1490	.1703	.1838	.2103	.2420
	117	.0518	.0664	.0814	.0903	.1063	.1261	.1478	.1689	.1822	.2086	.2400
	118	.0514	.0658	.0807	.0895	.1053	.1250	.1465	.1675	.1807	.2069	.2381
	119	.0509	.0652	.0800	.0887	.1045	.1240	.1453	.1661	.1792	.2052	.2363
	120	.0505	.0646	.0793	.0880	.1036	.1229	.1441	.1647	.1778	.2036	.2344
	121	.0500	.0641	.0787	.0872	.1028	.1219	.1429	.1634	.1764	.2020	.2326
	122	.0496	.0635	.0780	.0865	.1019	.1209	.1418	.1621	.1750	.2004	.2308
	123	.0492	.0630	.0773	.0858	.1011	.1199	.1406	.1608	.1736	.1988	.2290
	124	.0488	.0625	.0767	.0851	.1002	.1189	.1395	.1595	.1722	.1973	.2273
	125+	5.794*	7.477*	9.246*	10.30*	12.24*	14.67*	17.40*	20.13*	21.89*	25.45*	29.85*
16	17–26	SEE NOTE**										
	27–49	SEE NOTE***										
	50	.1442	.1804	.2167	.2375	.2742	.3176	.3632	.4057	.4316	.4805	.5353
	51	.1410	.1765	.2121	.2325	.2686	.3113	.3562	.3982	.4237	.4722	.5266
	52	.1379	.1728	.2077	.2277	.2632	.3052	.3495	.3909	.4162	.4641	.5181
	53	.1350	.1692	.2035	.2231	.2580	.2994	.3431	.3839	.4089	.4564	.5099
	54	.1322	.1657	.1994	.2187	.2530	.2938	.3368	.3772	.4018	.4488	.5019
	55	.1295	.1624	.1955	.2145	.2482	.2884	.3308	.3706	.3950	.4415	.4941
	56	.1269	.1592	.1918	.2104	.2436	.2831	.3250	.3643	.3885	.4345	.4866
	57	.1244	.1562	.1881	.2065	.2392	.2781	.3194	.3583	.3821	.4276	.4793
	58	.1221	.1533	.1847	.2028	.2349	.2733	.3140	.3524	.3759	.4210	.4722
	59	.1198	.1504	.1813	.1991	.2308	.2686	.3088	.3467	.3699	.4145	.4653
	60	.1176	.1477	.1781	.1956	.2268	.2641	.3037	.3411	.3642	.4083	.4587
	61	.1154	.1451	.1750	.1922	.2229	.2597	.2988	.3358	.3585	.4022	.4521
	62	.1134	.1425	.1720	.1890	.2192	.2554	.2941	.3306	.3531	.3963	.4458
	63	.1114	.1401	.1691	.1858	.2156	.2514	.2895	.3256	.3478	.3906	.4396
	64	.1095	.1377	.1663	.1827	.2121	.2474	.2850	.3207	.3427	.3850	.4337
	65	.1076	.1354	.1635	.1798	.2088	.2435	.2807	.3159	.3377	.3796	.4278
	66	.1059	.1332	.1609	.1769	.2055	.2398	.2765	.3113	.3329	.3744	.4221
	67	.1041	.1311	.1584	.1741	.2023	.2362	.2725	.3069	.3282	.3692	.4166
	68	.1025	.1290	.1559	.1715	.1993	.2327	.2685	.3025	.3236	.3643	.4112
	69	.1008	.1270	.1535	.1689	.1963	.2293	.2647	.2983	.3192	.3594	.4059
	70	.0993	.1250	.1512	.1663	.1934	.2260	.2609	.2942	.3148	.3547	.4008
	71	.0978	.1232	.1489	.1639	.1906	.2228	.2573	.2902	.3106	.3501	.3958
	72	.0963	.1213	.1468	.1615	.1879	.2196	.2538	.2863	.3065	.3456	.3909
	73	.0949	.1196	.1446	.1592	.1852	.2166	.2504	.2825	.3025	.3412	.3861
	74	.0935	.1178	.1426	.1569	.1826	.2136	.2470	.2788	.2986	.3370	.3815
	75	.0921	.1161	.1406	.1547	.1801	.2108	.2438	.2753	.2948	.3328	.3769
	76	.0908	.1145	.1386	.1526	.1777	.2080	.2406	.2718	.2911	.3287	.3725
	77	.0895	.1129	.1367	.1506	.1753	.2053	.2375	.2683	.2875	.3248	.3681
	78	.0883	.1114	.1349	.1485	.1730	.2026	.2345	.2650	.2840	.3209	.3639
	79	.0871	.1099	.1331	.1466	.1708	.2000	.2316	.2618	.2806	.3171	.3597
	80	.0859	.1084	.1314	.1447	.1686	.1975	.2287	.2586	.2772	.3134	.3557

*To approximate p_f, divide the starred entry under f by $(n - 1)$.

**Compute $r^* = n - r$, find p^*_{1-f} for r^* and n, and compute $p_f = 1 - p^*_{1-f}$.

Table 5 (continued)

r	n						f					
		.001	.01	.05	.1	.25	.5	.75	.9	.95	.99	.999
16	81	.0848	.1070	.1297	.1428	.1665	.1950	.2259	.2555	.2739	.3098	.3517
	82	.0837	.1056	.1280	.1410	.1644	.1926	.2232	.2525	.2707	.3063	.3478
	83	.0826	.1043	.1264	.1392	.1623	.1903	.2205	.2495	.2676	.3028	.3441
	84	.0815	.1030	.1248	.1375	.1604	.1880	.2179	.2466	.2646	.2995	.3403
	85	.0805	.1017	.1233	.1358	.1584	.1858	.2154	.2438	.2616	.2962	.3367
	86	.0795	.1004	.1218	.1342	.1565	.1836	.2129	.2411	.2586	.2929	.3331
	87	.0785	.0992	.1203	.1326	.1547	.1815	.2105	.2384	.2558	.2898	.3297
	88	.0776	.0980	.1189	.1310	.1529	.1794	.2081	.2357	.2530	.2867	.3262
	89	.0767	.0969	.1175	.1295	.1511	.1774	.2058	.2332	.2503	.2837	.3229
	90	.0757	.0957	.1161	.1280	.1494	.1754	.2035	.2306	.2476	.2807	.3196
	91	.0749	.0946	.1148	.1266	.1477	.1734	.2013	.2282	.2450	.2778	.3164
	92	.0740	.0935	.1135	.1251	.1461	.1715	.1992	.2258	.2424	.2750	.3133
	93	.0731	.0925	.1122	.1238	.1445	.1697	.1970	.2234	.2399	.2722	.3102
	94	.0723	.0914	.1110	.1224	.1429	.1679	.1950	.2211	.2374	.2695	.3071
	95	.0715	.0904	.1098	.1211	.1414	.1661	.1929	.2188	.2350	.2668	.3042
	96	.0707	.0894	.1086	.1198	.1399	.1643	.1909	.2166	.2327	.2642	.3013
	97	.0699	.0885	.1074	.1185	.1384	.1626	.1890	.2144	.2304	.2616	.2984
	98	.0692	.0875	.1063	.1172	.1370	.1610	.1871	.2123	.2281	.2591	.2956
	99	.0684	.0866	.1052	.1160	.1356	.1593	.1852	.2102	.2259	.2566	.2928
	100	.0677	.0857	.1041	.1148	.1342	.1577	.1834	.2081	.2237	.2542	.2901
	101	.0670	.0848	.1030	.1136	.1328	.1562	.1816	.2061	.2215	.2518	.2875
	102	.0663	.0839	.1020	.1125	.1315	.1546	.1798	.2042	.2194	.2494	.2849
	103	.0656	.0831	.1009	.1114	.1302	.1531	.1781	.2022	.2174	.2471	.2823
	104	.0650	.0822	.0999	.1103	.1289	.1516	.1764	.2003	.2154	.2449	.2798
	105	.0643	.0814	.0989	.1092	.1277	.1502	.1747	.1985	.2134	.2427	.2773
	106	.0637	.0806	.0980	.1081	.1264	.1487	.1731	.1966	.2114	.2405	.2749
	107	.0631	.0798	.0970	.1071	.1252	.1473	.1715	.1948	.2095	.2384	.2725
	108	.0624	.0791	.0961	.1061	.1240	.1460	.1699	.1931	.2076	.2363	.2702
	109	.0618	.0783	.0952	.1051	.1229	.1446	.1684	.1913	.2058	.2342	.2679
	110	.0612	.0776	.0943	.1041	.1217	.1433	.1668	.1896	.2040	.2322	.2656
	111	.0607	.0768	.0934	.1031	.1206	.1420	.1653	.1880	.2022	.2302	.2634
	112	.0601	.0761	.0926	.1022	.1195	.1407	.1639	.1863	.2004	.2282	.2612
	113	.0595	.0754	.0917	.1012	.1185	.1395	.1624	.1847	.1987	.2263	.2590
	114	.0590	.0747	.0909	.1003	.1174	.1382	.1610	.1831	.1970	.2244	.2569
	115	.0584	.0740	.0901	.0994	.1164	.1370	.1596	.1816	.1954	.2225	.2548
	116	.0579	.0734	.0893	.0985	.1153	.1358	.1583	.1800	.1937	.2207	.2528
	117	.0574	.0727	.0885	.0977	.1143	.1347	.1569	.1785	.1921	.2189	.2508
	118	.0569	.0721	.0877	.0968	.1133	.1335	.1556	.1770	.1905	.2171	.2488
	119	.0564	.0715	.0869	.0960	.1124	.1324	.1543	.1756	.1890	.2154	.2468
	120	.0559	.0708	.0862	.0952	.1114	.1313	.1530	.1741	.1874	.2137	.2449
	121	.0554	.0702	.0855	.0944	.1105	.1302	.1518	.1727	.1859	.2120	.2430
	122	.0549	.0696	.0847	.0936	.1096	.1291	.1505	.1713	.1845	.2103	.2411
	123	.0545	.0690	.0840	.0928	.1087	.1281	.1493	.1700	.1830	.2087	.2393
	124	.0540	.0685	.0833	.0920	.1078	.1270	.1481	.1686	.1816	.2071	.2375
	125	.0536	.0679	.0826	.0913	.1069	.1260	.1470	.1673	.1801	.2055	.2357
	126	.0531	.0673	.0820	.0905	.1060	.1250	.1458	.1660	.1788	.2039	.2339
	127	.0527	.0668	.0813	.0898	.1052	.1240	.1447	.1647	.1774	.2024	.2322
	128	.0522	.0662	.0806	.0891	.1043	.1230	.1435	.1635	.1760	.2009	.2305
	129	.0518	.0657	.0800	.0884	.1035	.1221	.1424	.1622	.1747	.1994	.2288
	130	.0514	.0652	.0794	.0877	.1027	.1211	.1413	.1610	.1734	.1979	.2272
	131+	6.405*	8.181*	10.04*	11.14*	13.15*	15.67*	18.49*	21.29*	23.10*	26.74*	31.24*
17	18–27	SEE NOTE**										
	28–54	SEE NOTE***										
	55	.1427	.1772	.2115	.2311	.2658	.3068	.3499	.3901	.4146	.4612	.5136
	56	.1399	.1737	.2075	.2268	.2609	.3012	.3437	.3835	.4077	.4539	.5059
	57	.1371	.1704	.2035	.2225	.2561	.2959	.3378	.3771	.4011	.4467	.4984
	58	.1345	.1671	.1998	.2185	.2515	.2907	.3321	.3709	.3946	.4398	.4910
	59	.1320	.1640	.1961	.2145	.2471	.2857	.3266	.3649	.3884	.4331	.4839
	60	.1295	.1611	.1926	.2107	.2428	.2809	.3212	.3591	.3823	.4266	.4770
	61	.1272	.1582	.1893	.2071	.2387	.2763	.3161	.3535	.3764	.4203	.4703
	62	.1249	.1554	.1860	.2036	.2347	.2718	.3111	.3480	.3707	.4142	.4637
	63	.1227	.1527	.1829	.2001	.2308	.2674	.3062	.3427	.3652	.4082	.4573
	64	.1206	.1501	.1798	.1968	.2271	.2632	.3015	.3376	.3598	.4024	.4511
	65	.1185	.1476	.1769	.1937	.2235	.2591	.2969	.3326	.3546	.3968	.4451
	66	.1166	.1452	.1740	.1906	.2200	.2551	.2925	.3278	.3496	.3913	.4392
	67	.1147	.1429	.1713	.1876	.2166	.2513	.2882	.3231	.3446	.3860	.4335
	68	.1128	.1406	.1686	.1847	.2133	.2475	.2840	.3185	.3398	.3808	.4279

***To approximate p_f, compute $m = r/n$ and $s = \sqrt{m(1-m)/(n+1)}$, find z_f in Table 3, and compute $p_f \doteq m + z_f s$.

Table 5 (continued)

r	n	f: .001	.01	.05	.1	.25	.5	.75	.9	.95	.99	.999
17	69	.1110	.1384	.1660	.1819	.2101	.2439	.2800	.3141	.3352	.3758	.4224
	70	.1093	.1363	.1635	.1791	.2070	.2404	.2760	.3098	.3306	.3708	.4171
	71	.1076	.1342	.1610	.1765	.2040	.2370	.2722	.3056	.3262	.3660	.4119
	72	.1060	.1322	.1587	.1739	.2011	.2337	.2685	.3015	.3219	.3614	.4068
	73	.1044	.1303	.1564	.1714	.1983	.2304	.2648	.2975	.3177	.3568	.4019
	74	.1029	.1284	.1541	.1690	.1955	.2273	.2613	.2936	.3137	.3524	.3971
	75	.1014	.1266	.1520	.1666	.1928	.2242	.2579	.2899	.3097	.3480	.3924
	76	.0999	.1248	.1499	.1644	.1902	.2213	.2545	.2862	.3058	.3438	.3878
	77	.0985	.1231	.1478	.1621	.1877	.2184	.2513	.2826	.3020	.3396	.3833
	78	.0972	.1214	.1458	.1600	.1852	.2155	.2481	.2791	.2983	.3356	.3789
	79	.0958	.1197	.1439	.1578	.1828	.2128	.2450	.2757	.2947	.3317	.3746
	80	.0945	.1181	.1420	.1558	.1804	.2101	.2420	.2723	.2912	.3278	.3704
	81	.0933	.1166	.1402	.1538	.1782	.2075	.2390	.2691	.2878	.3241	.3662
	82	.0921	.1151	.1384	.1518	.1759	.2049	.2361	.2659	.2844	.3204	.3622
	83	.0909	.1136	.1366	.1499	.1737	.2024	.2333	.2628	.2811	.3168	.3583
	84	.0897	.1122	.1349	.1481	.1716	.2000	.2306	.2598	.2779	.3133	.3544
	85	.0886	.1108	.1333	.1463	.1696	.1976	.2279	.2568	.2748	.3098	.3507
	86	.0875	.1094	.1316	.1445	.1675	.1953	.2253	.2539	.2718	.3065	.3470
	87	.0864	.1081	.1300	.1428	.1656	.1931	.2227	.2511	.2688	.3032	.3433
	88	.0853	.1068	.1285	.1411	.1636	.1909	.2202	.2483	.2658	.2999	.3398
	89	.0843	.1055	.1270	.1394	.1618	.1887	.2178	.2456	.2630	.2968	.3363
	90	.0833	.1043	.1255	.1378	.1599	.1866	.2154	.2430	.2602	.2937	.3329
	91	.0823	.1031	.1241	.1363	.1581	.1845	.2130	.2404	.2574	.2907	.3296
	92	.0814	.1019	.1227	.1347	.1564	.1825	.2107	.2378	.2547	.2877	.3263
	93	.0804	.1007	.1213	.1332	.1546	.1805	.2085	.2353	.2521	.2848	.3231
	94	.0795	.0996	.1200	.1318	.1530	.1786	.2063	.2329	.2495	.2819	.3200
	95	.0786	.0985	.1186	.1303	.1513	.1767	.2041	.2305	.2470	.2791	.3169
	96	.0778	.0974	.1173	.1289	.1497	.1748	.2020	.2282	.2445	.2764	.3139
	97	.0769	.0963	.1161	.1275	.1481	.1730	.2000	.2259	.2421	.2737	.3109
	98	.0761	.0953	.1149	.1262	.1466	.1712	.1980	.2236	.2397	.2711	.3080
	99	.0753	.0943	.1137	.1249	.1451	.1695	.1960	.2214	.2374	.2685	.3051
	100	.0745	.0933	.1125	.1236	.1436	.1678	.1940	.2193	.2351	.2659	.3023
	101	.0737	.0923	.1113	.1223	.1421	.1661	.1921	.2172	.2328	.2635	.2995
	102	.0729	.0914	.1102	.1211	.1407	.1645	.1903	.2151	.2306	.2610	.2968
	103	.0722	.0905	.1091	.1199	.1393	.1629	.1884	.2130	.2284	.2586	.2942
	104	.0714	.0895	.1080	.1187	.1379	.1613	.1866	.2110	.2263	.2563	.2916
	105	.0707	.0887	.1069	.1175	.1366	.1598	.1849	.2091	.2242	.2539	.2890
	106	.0700	.0878	.1059	.1164	.1353	.1582	.1831	.2072	.2222	.2517	.2865
	107	.0693	.0869	.1048	.1153	.1340	.1567	.1815	.2053	.2202	.2494	.2840
	108	.0686	.0861	.1038	.1142	.1327	.1553	.1798	.2034	.2182	.2473	.2816
	109	.0680	.0853	.1028	.1131	.1315	.1539	.1781	.2016	.2163	.2451	.2792
	110	.0673	.0844	.1019	.1120	.1303	.1524	.1765	.1998	.2144	.2430	.2768
	111	.0667	.0836	.1009	.1110	.1291	.1511	.1750	.1980	.2125	.2409	.2745
	112	.0661	.0829	.1000	.1100	.1279	.1497	.1734	.1963	.2107	.2389	.2722
	113	.0654	.0821	.0991	.1090	.1268	.1484	.1719	.1946	.2089	.2368	.2700
	114	.0648	.0813	.0982	.1080	.1256	.1471	.1704	.1929	.2071	.2349	.2678
	115	.0642	.0806	.0973	.1070	.1245	.1458	.1689	.1913	.2053	.2329	.2656
	116	.0637	.0799	.0964	.1061	.1234	.1445	.1675	.1897	.2036	.2310	.2635
	117	.0631	.0792	.0956	.1051	.1223	.1433	.1661	.1881	.2019	.2291	.2614
	118	.0625	.0785	.0947	.1042	.1213	.1421	.1647	.1865	.2003	.2273	.2593
	119	.0620	.0778	.0939	.1033	.1203	.1409	.1633	.1850	.1986	.2254	.2573
	120	.0614	.0771	.0931	.1024	.1192	.1397	.1619	.1835	.1970	.2237	.2553
	121	.0609	.0764	.0923	.1016	.1182	.1385	.1606	.1820	.1954	.2219	.2533
	122	.0604	.0758	.0915	.1007	.1172	.1374	.1593	.1805	.1939	.2202	.2513
	123	.0599	.0752	.0908	.0999	.1163	.1362	.1580	.1791	.1924	.2184	.2494
	124	.0594	.0745	.0900	.0990	.1153	.1351	.1568	.1777	.1908	.2168	.2475
	125	.0589	.0739	.1893	.0982	.1144	.1341	.1555	.1763	.1894	.2151	.2457
	126	.0584	.0733	.0885	.0974	.1135	.1330	.1543	.1749	.1879	.2135	.2439
	127	.0579	.0727	.0878	.0966	.1126	.1319	.1531	.1736	.1865	.2119	.2421
	128	.0574	.0721	.0871	.0959	.1117	.1309	.1519	.1722	.1850	.2103	.2403
	129	.0570	.0715	.0864	.0951	.1108	.1299	.1507	.1709	.1836	.2087	.2385
	130	.0565	.0710	.0857	.0944	.1099	.1289	.1496	.1696	.1823	.2072	.2368
	131	.0560	.0704	.0851	.0936	.1091	.1279	.1484	.1684	.1809	.2057	.2351
	132	.0556	.0698	.0844	.0929	.1082	.1269	.1473	.1671	.1796	.2042	.2334
	133	.0552	.0693	.0837	.0922	.1074	.1260	.1462	.1659	.1783	.2027	.2318
	134	.0547	.0688	.0831	.0915	.1066	.1250	.1451	.1647	.1770	.2012	.2301
	135	.0543	.0682	.0825	.0908	.1058	.1241	.1441	.1635	.1757	.1998	.2285

*To approximate p_f, divide the starred entry under f by $(n - 1)$.
**Compute $r^* = n - r$, find p^*_{1-f} for r^* and n, and compute $p_f = 1 - p^*_{1-f}$.

Table 5 (continued)

						f						
r	n	.001	.01	.05	.1	.25	.5	.75	.9	.95	.99	.999
17	136	.0539	.0677	.0818	.0901	.1050	.1232	.1430	.1623	.1744	.1984	.2270
	137	.0535	.0672	.0812	.0894	.1042	.1223	.1420	.1611	.1732	.1970	.2254
	138+	7.028*	8.895*	10.83*	11.98*	14.07*	16.67*	19.57*	22.45*	24.30*	28.03*	32.62*
18	19–28	SEE NOTE**										
	29–60	SEE NOTE***										
	61	.1392	.1715	.2037	.2221	.2545	.2928	.3332	.3710	.3942	.4382	.4881
	62	.1367	.1685	.2002	.2183	.2502	.2881	.3280	.3653	.3882	.4318	.4813
	63	.1343	.1656	.1968	.2146	.2461	.2834	.3229	.3598	.3824	.4257	.4748
	64	.1319	.1628	.1935	.2111	.2421	.2790	.3179	.3544	.3768	.4196	.4683
	65	.1297	.1601	.1903	.2076	.2383	.2746	.3131	.3492	.3714	.4138	.4621
	66	.1275	.1574	.1873	.2043	.2345	.2704	.3084	.3441	.3661	.4081	.4560
	67	.1254	.1549	.1843	.2011	.2309	.2663	.3039	.3392	.3609	.4026	.4501
	68	.1234	.1524	.1814	.1980	.2274	.2624	.2995	.3344	.3559	.3972	.4443
	69	.1214	.1500	.1786	.1950	.2240	.2585	.2952	.3298	.3511	.3919	.4387
	70	.1195	.1477	.1759	.1920	.2207	.2548	.2911	.3253	.3463	.3868	.4332
	71	.1177	.1455	.1733	.1892	.2175	.2512	.2870	.3209	.3417	.3818	.4278
	72	.1159	.1433	.1707	.1864	.2144	.2477	.2831	.3166	.3372	.3769	.4226
	73	.1142	.1412	.1682	.1838	.2114	.2443	.2793	.3124	.3329	.3722	.4175
	74	.1125	.1391	.1658	.1812	.2084	.2409	.2756	.3083	.3286	.3676	.4125
	75	.1109	.1371	.1635	.1786	.2055	.2377	.2719	.3044	.3244	.3631	.4076
	76	.1093	.1352	.1612	.1762	.2028	.2345	.2684	.3005	.3204	.3587	.4028
	77	.1077	.1333	.1590	.1738	.2001	.2315	.2650	.2968	.3164	.3544	.3982
	78	.1062	.1315	.1569	.1715	.1974	.2285	.2616	.2931	.3126	.3502	.3936
	79	.1048	.1297	.1548	.1692	.1948	.2255	.2584	.2895	.3088	.3461	.3892
	80	.1034	.1280	.1527	.1670	.1923	.2227	.2552	.2860	.3051	.3421	.3848
	81	.1020	.1263	.1508	.1648	.1899	.2199	.2521	.2826	.3015	.3381	.3806
	82	.1006	.1247	.1488	.1627	.1875	.2172	.2490	.2793	.2980	.3343	.3764
	83	.0993	.1231	.1469	.1607	.1852	.2146	.2461	.2760	.2946	.3306	.3723
	84	.0981	.1215	.1451	.1587	.1829	.2120	.2432	.2728	.2912	.3269	.3683
	85	.0968	.1200	.1433	.1568	.1807	.2095	.2403	.2697	.2880	.3233	.3644
	86	.0956	.1185	.1416	.1549	.1786	.2070	.2376	.2667	.2848	.3198	.3606
	87	.0944	.1171	.1399	.1530	.1765	.2046	.2349	.2637	.2816	.3164	.3569
	88	.0933	.1157	.1382	.1512	.1744	.2023	.2322	.2608	.2786	.3130	.3532
	89	.0921	.1143	.1366	.1494	.1724	.2000	.2297	.2580	.2756	.3097	.3496
	90	.0910	.1129	.1350	.1477	.1704	.1978	.2271	.2552	.2726	.3065	.3461
	91	.0900	.1116	.1334	.1460	.1685	.1956	.2247	.2525	.2698	.3034	.3426
	92	.0889	.1103	.1319	.1444	.1666	.1934	.2223	.2498	.2669	.3003	.3392
	93	.0879	.1091	.1304	.1428	.1648	.1913	.2199	.2472	.2642	.2972	.3359
	94	.0869	.1079	.1290	.1412	.1630	.1893	.2176	.2446	.2615	.2943	.3326
	95	.0859	.1067	.1276	.1397	.1613	.1873	.2153	.2421	.2588	.2914	.3294
	96	.0850	.1055	.1262	.1382	.1595	.1853	.2131	.2397	.2562	.2885	.3263
	97	.0840	.1043	.1248	.1367	.1579	.1834	.2109	.2373	.2537	.2857	.3232
	98	.0831	.1032	.1235	.1352	.1562	.1815	.2088	.2349	.2512	.2830	.3202
	99	.0822	.1021	.1222	.1338	.1546	.1797	.2067	.2326	.2488	.2803	.3172
	100	.0814	.1010	.1209	.1324	.1530	.1779	.2047	.2303	.2464	.2776	.3143
	101	.0805	.1000	.1197	.1311	.1515	.1761	.2027	.2281	.2440	.2750	.3115
	102	.0797	.0990	.1185	.1298	.1500	.1744	.2007	.2259	.2417	.2725	.3087
	103	.0788	.0980	.1173	.1285	.1485	.1726	.1988	.2238	.2394	.2700	.3059
	104	.0780	.0970	.1161	.1272	.1470	.1710	.1969	.2217	.2372	.2675	.3032
	105	.0773	.0960	.1150	.1259	.1456	.1693	.1950	.2196	.2350	.2651	.3005
	106	.0765	.0950	.1138	.1247	.1442	.1677	.1932	.2176	.2329	.2628	.2979
	107	.0757	.0941	.1127	.1235	.1428	.1662	.1914	.2156	.2308	.2604	.2953
	108	.0750	.0932	.1116	.1223	.1415	.1646	.1896	.2137	.2287	.2582	.2928
	109	.0743	.0923	.1106	.1212	.1401	.1631	.1879	.2118	.2267	.2559	.2903
	110	.0735	.0914	.1095	.1200	.1388	.1616	.1862	.2099	.2247	.2537	.2879
	111	.0728	.0906	.1085	.1189	.1376	.1601	.1846	.2081	.2228	.2515	.2855
	112	.0722	.0897	.1075	.1178	.1363	.1587	.1829	.2062	.2208	.2494	.2831
	113	.0715	.0889	.1065	.1167	.1351	.1573	.1813	.2045	.2189	.2473	.2808
	114	.0708	.0881	.1056	.1157	.1339	.1559	.1797	.2027	.2171	.2452	.2785
	115	.0702	.0873	.1046	.1147	.1327	.1545	.1782	.2010	.2152	.2432	.2762
	116	.0695	.0865	.1037	.1136	.1315	.1532	.1767	.1993	.2134	.2412	.2740
	117	.0689	.0857	.1028	.1126	.1304	.1519	.1752	.1976	.2117	.2393	.2718
	118	.0683	.0850	.1019	.1117	.1293	.1506	.1737	.1960	.2099	.2373	.2697
	119	.0677	.0842	.1010	.1107	.1282	.1493	.1723	.1944	.2082	.2354	.2676
	120	.0671	.0835	.1001	.1097	.1271	.1481	.1708	.1928	.2065	.2336	.2655
	121	.0665	.0828	.0992	.1088	.1260	.1468	.1694	.1912	.2049	.2317	.2635
	122	.0659	.0821	.0984	.1079	.1249	.1456	.1680	.1897	.2033	.2299	.2614
	123	.0654	.0814	.0976	.1070	.1239	.1444	.1667	.1882	.2017	.2281	.2595

***To approximate p_f, compute $m = r/n$ and $s = \sqrt{m(1-m)/(n+1)}$, find z_f in Table 3, and compute $p_f \doteq m + z_f s$.

Table 5 (continued)

r	n	f: .001	.01	.05	.1	.25	.5	.75	.9	.95	.99	.999
18	124	.0648	.0807	.0968	.1061	.1229	.1433	.1654	.1867	.2001	.2264	.2575
	125	.0643	.0800	.0960	.1052	.1219	.1421	.1640	.1852	.1985	.2246	.2556
	126	.0637	.0793	.0952	.1044	.1209	.1410	.1628	.1838	.1970	.2229	.2537
	127	.0632	.0787	.0944	.1035	.1199	.1398	.1615	.1824	.1955	.2212	.2518
	128	.0627	.0781	.0937	.1027	.1190	.1388	.1602	.1810	.1940	.2196	.2500
	129	.0622	.0774	.0929	.1019	.1181	.1377	.1590	.1796	.1925	.2180	.2481
	130	.0617	.0768	.0922	.1011	.1171	.1366	.1578	.1782	.1911	.2164	.2464
	131	.0612	.0762	.0914	.1003	.1162	.1356	.1566	.1769	.1897	.2148	.2446
	132	.0607	.0756	.0907	.0995	.1153	.1345	.1554	.1756	.1883	.2132	.2428
	133	.0602	.0750	.0900	.0988	.1144	.1335	.1542	.1743	.1869	.2117	.2411
	134	.0598	.0744	.0893	.0980	.1136	.1325	.1531	.1730	.1855	.2102	.2394
	135	.0593	.0739	.0887	.0973	.1127	.1315	.1520	.1718	.1842	.2087	.2378
	136	.0588	.0733	.0880	.0965	.1119	.1305	.1509	.1705	.1829	.2072	.2361
	137	.0584	.0727	.0873	.0958	.1111	.1296	.1498	.1693	.1816	.2058	.2345
	138	.0579	.0722	.0867	.0951	.1102	.1286	.1487	.1681	.1803	.2043	.2329
	139	.0575	.0717	.0860	.0944	.1094	.1277	.1476	.1669	.1790	.2029	.2313
	140	.0571	.0711	.0854	.0937	.1086	.1268	.1466	.1657	.1778	.2015	.2298
	141	.0567	.0706	.0848	.0930	.1079	.1259	.1456	.1646	.1766	.2001	.2282
	142	.0562	.0701	.0842	.0923	.1071	.1250	.1445	.1634	.1753	.1988	.2267
	143	.0558	.0696	.0836	.0917	.1063	.1241	.1435	.1623	.1742	.1975	.2252
	144	7.662*	9.616*	11.63*	12.82*	14.99*	17.67*	20.65*	23.61*	25.50*	29.31*	33.99*
19	20–30	SEE NOTE**										
	31–65	SEE NOTE***										
	66	.1387	.1698	.2006	.2182	.2491	.2857	.3243	.3604	.3825	.4247	.4726
	67	.1364	.1671	.1974	.2147	.2453	.2814	.3195	.3552	.3771	.4189	.4665
	68	.1342	.1644	.1943	.2114	.2416	.2772	.3149	.3502	.3719	.4133	.4605
	69	.1321	.1618	.1913	.2082	.2379	.2732	.3104	.3454	.3668	.4079	.4547
	70	.1300	.1593	.1884	.2051	.2344	.2692	.3061	.3407	.3619	.4026	.4491
	71	.1280	.1569	.1856	.2020	.2310	.2654	.3018	.3361	.3571	.3974	.4435
	72	.1260	.1545	.1829	.1991	.2277	.2617	.2977	.3316	.3524	.3924	.4381
	73	.1241	.1523	.1802	.1962	.2245	.2581	.2937	.3272	.3478	.3875	.4328
	74	.1223	.1500	.1776	.1934	.2214	.2546	.2898	.3230	.3434	.3827	.4277
	75	.1205	.1479	.1751	.1907	.2183	.2511	.2860	.3188	.3391	.3780	.4226
	76	.1188	.1458	.1727	.1881	.2154	.2478	.2823	.3148	.3348	.3734	.4177
	77	.1171	.1438	.1703	.1855	.2125	.2446	.2787	.3108	.3307	.3689	.4129
	78	.1155	.1418	.1680	.1830	.2097	.2414	.2751	.3070	.3267	.3646	.4082
	79	.1139	.1399	.1658	.1806	.2069	.2383	.2717	.3033	.3228	.3603	.4036
	80	.1123	.1380	.1636	.1783	.2043	.2353	.2683	.2996	.3189	.3562	.3991
	81	.1108	.1362	.1615	.1760	.2017	.2324	.2651	.2960	.3152	.3521	.3947
	82	.1094	.1344	.1594	.1737	.1992	.2295	.2619	.2926	.3115	.3481	.3904
	83	.1079	.1327	.1574	.1715	.1967	.2267	.2588	.2891	.3079	.3442	.3862
	84	.1066	.1310	.1554	.1694	.1943	.2240	.2557	.2858	.3044	.3404	.3821
	85	.1052	.1293	.1535	.1673	.1919	.2214	.2528	.2826	.3010	.3367	.3780
	86	.1039	.1277	.1516	.1653	.1897	.2188	.2499	.2794	.2977	.3331	.3741
	87	.1026	.1262	.1498	.1633	.1874	.2162	.2470	.2763	.2944	.3295	.3702
	88	.1013	.1247	.1480	.1614	.1852	.2138	.2443	.2732	.2912	.3260	.3664
	89	.1001	.1232	.1462	.1595	.1831	.2113	.2415	.2703	.2881	.3226	.3627
	90	.0989	.1217	.1445	.1577	.1810	.2090	.2389	.2674	.2850	.3192	.3590
	91	.0978	.1203	.1429	.1559	.1790	.2067	.2363	.2645	.2820	.3160	.3555
	92	.0966	.1189	.1413	.1541	.1770	.2044	.2338	.2617	.2791	.3128	.3520
	93	.0955	.1176	.1397	.1524	.1750	.2022	.2313	.2590	.2762	.3096	.3485
	94	.0944	.1162	.1381	.1507	.1731	.2000	.2288	.2563	.2734	.3065	.3452
	95	.0934	.1149	.1366	.1491	.1713	.1979	.2265	.2537	.2706	.3035	.3418
	96	.0923	.1137	.1351	.1474	.1694	.1958	.2241	.2511	.2679	.3005	.3386
	97	.0913	.1124	.1337	.1459	.1676	.1938	.2218	.2486	.2653	.2976	.3354
	98	.0903	.1112	.1322	.1443	.1659	.1918	.2196	.2461	.2427	.2948	.3323
	99	.0893	.1100	.1308	.1428	.1642	.1898	.2174	.2437	.2601	.2920	.3292
	100	.0884	.1089	.1295	.1413	.1625	.1879	.2153	.2414	.2576	.2892	.3262
	101	.0874	.1077	.1281	.1399	.1609	.1861	.2131	.2390	.2551	.2865	.3232
	102	.0865	.1066	.1268	.1384	.1592	.1842	.2111	.2367	.2527	.2839	.3203
	103	.0856	.1055	.1256	.1371	.1577	.1824	.2090	.2345	.2504	.2813	.3175
	104	.0848	.1045	.1243	.1357	.1561	.1807	.2071	.2323	.2481	.2787	.3147
	105	.0839	.1034	.1231	.1344	.1546	.1789	.2051	.2302	.2458	.2762	.3119
	106	.0831	.1024	.1219	.1331	.1531	.1772	.2032	.2280	.2435	.2738	.3092
	107	.0822	.1014	.1207	.1318	.1517	.1756	.2013	.2260	.2413	.2713	.3065
	108	.0814	.1004	.1195	.1305	.1502	.1739	.1995	.2239	.2392	.2690	.3039
	109	.0806	.0994	.1184	.1293	.1488	.1723	.1976	.2219	.2371	.2666	.3014
	110	.0799	.0985	.1173	.1281	.1474	.1707	.1959	.2200	.2350	.2643	.2988

*To approximate p_f, divide the starred entry under f by $(n-1)$.
**Compute $r^* = n - r$, find p^*_{1-f} for r^* and n, and compute $p_f = 1 - p^*_{1-f}$.

Table 5 (continued)

r	n	f: .001	.01	.05	.1	.25	.5	.75	.9	.95	.99	.999
19	111	.0791	.0976	.1162	.1269	.1461	.1692	.1941	.2180	.2329	.2621	.2963
	112	.0784	.0967	.1151	.1257	.1447	.1677	.1924	.2161	.2309	.2599	.2939
	113	.0776	.0958	.1140	.1246	.1434	.1662	.1907	.2143	.2290	.2577	.2915
	114	.0769	.0949	.1130	.1235	.1422	.1647	.1891	.2124	.2270	.2555	.2891
	115	.0762	.0940	.1120	.1223	.1409	.1633	.1874	.2106	.2251	.2534	.2868
	116	.0755	.0932	.1110	.1213	.1397	.1619	.1858	.2088	.2232	.2513	.2845
	117	.0748	.0923	.1100	.1202	.1384	.1605	.1843	.2071	.2214	.2493	.2822
	118	.0742	.0915	.1090	.1191	.1373	.1591	.1827	.2054	.2196	.2473	.2800
	119	.0735	.0907	.1081	.1181	.1361	.1578	.1812	.2037	.2178	.2453	.2778
	120	.0728	.0899	.1072	.1171	.1349	.1564	.1797	.2020	.2160	.2434	.2757
	121	.0722	.0892	.1062	.1161	.1338	.1551	.1782	.2004	.2143	.2415	.2735
	122	.0716	.0884	.1053	.1151	.1327	.1539	.1768	.1988	.2126	.2396	.2714
	123	.0710	.0876	.1044	.1142	.1316	.1526	.1753	.1972	.2109	.2377	.2694
	124	.0704	.0869	.1036	.1132	.1304	.1514	.1739	.1957	.2092	.2359	.2674
	125	.0698	.0862	.1027	.1123	.1294	.1501	.1726	.1941	.2076	.2341	.2654
	126	.0692	.0855	.1019	.1114	.1284	.1489	.1712	.1926	.2060	.2323	.2634
	127	.0686	.0848	.1011	.1105	.1274	.1478	.1699	.1911	.2044	.2306	.2615
	128	.0681	.0841	.1002	.1096	.1263	.1466	.1685	.1897	.2029	.2288	.2595
	129	.0675	.0834	.0994	.1087	.1253	.1455	.1672	.1882	.2014	.2272	.2577
	130	.0670	.0827	.0986	.1079	.1244	.1443	.1660	.1868	.1999	.2255	.2558
	131	.0664	.0821	.0979	.1070	.1234	.1432	.1637	.1854	.1984	.2238	.2540
	132	.0659	.0814	.0971	.1062	.1225	.1421	.1635	.1840	.1969	.2222	.2522
	133	.0654	.0808	.0964	.1054	.1215	.1411	.1623	.1827	.1955	.2206	.2504
	134	.0649	.0802	.0956	.1046	.1206	.1400	.1611	.1813	.1941	.2190	.2486
	135	.0644	.0795	.0949	.1038	.1197	.1390	.1599	.1800	.1927	.2175	.2469
	136	.0639	.0789	.0942	.1030	.1188	.1379	.1587	.1787	.1913	.2160	.2452
	137	.0634	.0783	.0935	.1022	.1179	.1369	.1576	.1774	.1899	.2144	.2435
	138	.0629	.0777	.0928	.1015	.1170	.1359	.1564	.1762	.1886	.2130	.2419
	139	.0624	.0772	.0921	.1007	.1162	.1349	.1553	.1749	.1873	.2115	.2404
	140	.0620	.0766	.0914	.1000	.1153	.1340	.1542	.1737	.1860	.2100	.2386
	141	.0615	.0760	.0907	.0992	.1145	.1330	.1531	.1725	.1847	.2086	.2370
	142	.0611	.0755	.0901	.0985	.1137	.1321	.1520	.1713	.1834	.2072	.2354
	143	.0606	.0749	.0894	.0978	.1129	.1312	.1510	.1701	.1822	.2058	.2339
	144	.0602	.0744	.0888	.0971	.1121	.1302	.1499	.1690	.1809	.2044	.2324
	145	.0597	.0739	.0882	.0964	.1113	.1293	.1489	.1678	.1797	.2031	.2308
	146	.0593	.0733	.0875	.0958	.1105	.1284	.1479	.1667	.1785	.2017	.2294
	147	.0589	.0728	.0869	.0951	.1098	.1276	.1469	.1656	.1773	.2004	.2279
	148	.0585	.0723	.0863	.0944	.1090	.1267	.1459	.1645	.1762	.1991	.2264
	149	.0581	.0718	.0857	.0928	.1083	.1258	.1449	.1634	.1750	.1978	.2250
	150	.0577	.0713	.0851	.0932	.1075	.1250	.1440	.1623	.1739	.1966	.2236
	151+	8.306*	10.35*	12.44*	13.67*	25.92*	18.67*	21.73*	24.76*	26.69*	30.58*	35.35*
20	21–31	SEE NOTE**										
	32–71	SEE NOTE***										
	72	.1363	.1659	.1951	.2118	.2411	.2757	.3122	.3465	.3675	.4076	.4534
	73	.1343	.1635	.1923	.2087	.2377	.2719	.3080	.3419	.3627	.4025	.4480
	74	.1323	.1611	.1895	.2058	.2344	.2682	.3039	.3375	.3581	.3976	.4427
	75	.1304	.1588	.1869	.2029	.2311	.2646	.3000	.3332	.3536	.3927	.4375
	76	.1285	.1565	.1843	.2001	.2280	.2611	.2961	.3290	.3492	.3880	.4324
	77	.1267	.1543	.1817	.1973	.2249	.2577	.2923	.3249	.3449	.3834	.4275
	78	.1249	.1522	.1793	.1947	.2220	.2543	.2886	.3209	.3407	.3789	.4226
	79	.1232	.1501	.1769	.1921	.2191	.2511	.2850	.3169	.3366	.3744	.4179
	80	.1215	.1481	.1745	.1896	.2163	.2479	.2815	.3131	.3326	.3701	.4132
	81	.1199	.1462	.1723	.1872	.2135	.2448	.2781	.3094	.3287	.3659	.4087
	82	.1183	.1442	.1700	.1848	.2108	.2418	.2747	.3058	.3249	.3618	.4043
	83	.1167	.1424	.1679	.1824	.2082	.2389	.2714	.3022	.3212	.3578	.3999
	84	.1152	.1406	.1658	.1802	.2057	.2360	.2683	.2987	.3175	.3538	.3957
	85	.1138	.1388	.1637	.1780	.2032	.2332	.2651	.2953	.3140	.3500	.3915
	86	.1123	.1371	.1617	.1758	.2008	.2305	.2621	.2920	.3105	.3462	.3874
	87	.1109	.1354	.1598	.1737	.1984	.2278	.2591	.2888	.3071	.3425	.3834
	88	.1096	.1338	.1579	.1716	.1961	.2252	.2562	.2856	.3038	.3389	.3795
	89	.1082	.1322	.1560	.1696	.1938	.2227	.2534	.2825	.3005	.3353	.3756
	90	.1069	.1306	.1542	.1677	.1916	.2202	.2506	.2795	.2973	.3319	.3719
	91	.1057	.1291	.1524	.1658	.1895	.2177	.2479	.2765	.2942	.3285	.3682
	92	.1044	.1276	.1507	.1639	.1873	.2153	.2452	.2736	.2911	.3251	.3646
	93	.1032	.1261	.1490	.1621	.1853	.2130	.2426	.2707	.2881	.3219	.3610
	94	.1021	.1247	.1473	.1603	.1833	.2107	.2401	.2679	.2852	.3187	.3575
	95	.1009	.1233	.1457	.1585	.1813	.2085	.2376	.2652	.2823	.3155	.3541

***To approximate p_f, compute $m = r/n$ and $s = \sqrt{m(1-m)/(n+1)}$, find z_f in Table 3, and compute $p_f \doteq m + z_f s$.

Table 5 (continued)

r	n	.001	.01	.05	.1	.25	f .5	.75	.9	.95	.99	.999
20	96	.0998	.1220	.1441	.1568	.1793	.2063	.2351	.2625	.2795	.3124	.3508
	97	.0987	.1206	.1425	.1551	.1775	.2042	.2327	.2599	.2767	.3094	.3475
	98	.0976	.1193	.1410	.1535	.1756	.2021	.2304	.2573	.2740	.3065	.3442
	99	.0965	.1181	.1395	.1519	.1738	.2000	.2281	.2548	.2714	.3036	.3411
	100	.0955	.1168	.1381	.1503	.1720	.1980	.2258	.2523	.2688	.3007	.3380
	101	.0945	.1156	.1367	.1488	.1703	.1960	.2236	.2499	.2662	.2979	.3349
	102	.0935	.1144	.1353	.1472	.1686	.1941	.2214	.2475	.2637	.2952	.3319
	103	.0925	.1132	.1339	.1458	.1669	.1922	.2193	.2452	.2612	.2925	.3289
	104	.0916	.1121	.1326	.1443	.1652	.1903	.2172	.2429	.2588	.2898	.3260
	105	.0907	.1110	.1312	.1429	.1636	.1885	.2152	.2406	.2564	.2872	.3232
	106	.0898	.1099	.1300	.1415	.1621	.1867	.2132	.2384	.2541	.2847	.3204
	107	.0889	.1088	.1287	.1401	.1605	.1850	.2112	.2362	.2518	.2822	.3176
	108	.0880	.1077	.1274	.1388	.1590	.1832	.2093	.2341	.2496	.2797	.3149
	109	.0871	.1067	.1262	.1375	.1575	.1815	.2074	.2320	.2474	.2773	.3123
	110	.0863	.1057	.1250	.1362	.1560	.1799	.2055	.2300	.2452	.2749	.3097
	111	.0855	.1047	.1239	.1349	.1546	.1783	.2037	.2279	.2431	.2725	.3071
	112	.0847	.1037	.1227	.1337	.1532	.1767	.2019	.2260	.2410	.2702	.3046
	113	.0839	.1027	.1216	.1324	.1518	.1751	.2001	.2240	.2389	.2680	.3021
	114	.0831	.1018	.1205	.1313	.1505	.1735	.1984	.2221	.2369	.2657	.2996
	115	.0823	.1008	.1194	.1301	.1491	.1720	.1966	.2202	.2349	.2636	.2972
	116	.0816	.0999	.1183	.1289	.1478	.1705	.1950	.2183	.2329	.2614	.2948
	117	.0808	.0990	.1173	.1278	.1465	.1691	.1933	.2165	.2310	.2593	.2925
	118	.0801	.0982	.1163	.1267	.1453	.1676	.1917	.2147	.2291	.2572	.2902
	119	.0794	.0973	.1152	.1256	.1440	.1662	.1901	.2130	.2273	.2551	.2879
	120	.0787	.0964	.1143	.1245	.1428	.1648	.1885	.2112	.2254	.2531	.2857
	121	.0780	.0956	.1133	.1234	.1416	.1634	.1870	.2095	.2236	.2511	.2835
	122	.0773	.0948	.1123	.1224	.1404	.1621	.1855	.2079	.2218	.2492	.2813
	123	.0767	.0940	.1114	.1213	.1393	.1608	.1840	.2062	.2201	.2472	.2792
	124	.0760	.0932	.1104	.1204	.1381	.1595	.1825	.2046	.2184	.2453	.2771
	125	.0754	.0924	.1095	.1194	.1370	.1582	.1810	.2030	.2167	.2435	.2751
	126	.0748	.0917	.1086	.1184	.1359	.1569	.1796	.2014	.2150	.2416	.2730
	127	.0741	.0909	.1077	.1174	.1348	.1557	.1782	.1998	.2134	.2398	.2710
	128	.0735	.0902	.1069	.1165	.1337	.1545	.1768	.1983	.2117	.2380	.2690
	129	.0729	.0894	.1060	.1156	.1327	.1533	.1755	.1968	.2102	.2363	.2671
	130	.0723	.0887	.1052	.1147	.1316	.1521	.1741	.1953	.2086	.2345	.2652
	131	.0718	.0880	.1043	.1138	.1306	.1509	.1728	.1939	.2070	.2328	.2633
	132	.0712	.0873	.1035	.1129	.1296	.1498	.1715	.1924	.2055	.2311	.2614
	133	.0706	.0866	.1027	.1120	.1286	.1486	.1702	.1910	.2040	.2295	.2596
	134	.0701	.0860	.1019	.1112	.1276	.1475	.1690	.1896	.2025	.2278	.2577
	135	.0695	.0853	.1012	.1103	.1267	.1464	.1677	.1882	.2011	.2262	.2560
	136	.0690	.0846	.1004	.1095	.1257	.1453	.1665	.1869	.1996	.2246	.2542
	137	.0685	.0840	.0996	.1087	.1248	.1443	.1653	.1855	.1982	.2231	.2525
	138	.0679	.0834	.0989	.1079	.1239	.1432	.1641	.1842	.1968	.2215	.2507
	139	.0674	.0827	.0982	.1071	.1230	.1422	.1629	.1829	.1954	.2200	.2490
	140	.0669	.0821	.0974	.1063	.1221	.1412	.1618	.1816	.1941	.2185	.2474
	141	.0664	.0815	.0967	.1055	.1212	.1401	.1607	.1804	.1927	.2170	.2457
	142	.0659	.0809	.0960	.1047	.1203	.1392	.1595	.1791	.1914	.2155	.2441
	143	.0655	.0803	.0953	.1040	.1195	.1382	.1584	.1779	.1901	.2141	.2425
	144	.0650	.0798	.0947	.1032	.1186	.1372	.1573	.1767	.1888	.2127	.2409
	145	.0645	.0792	.0940	.1025	.1178	.1363	.1563	.1755	.1876	.2113	.2393
	146	.0641	.0786	.0933	.1018	.1170	.1353	.1552	.1743	.1863	.2099	.2378
	147	.0636	.0781	.0927	.1011	.1162	.1344	.1541	.1732	.1851	.2085	.2363
	148	.0632	.0775	.0920	.1004	.1154	.1335	.1531	.1720	.1839	.2071	.2348
	149	.0627	.0770	.0914	.0997	.1146	.1326	.1521	.1709	.1827	.2058	.2333
	150	.0623	.0765	.0908	.0990	.1138	.1317	.1511	.1698	.1815	.2045	.2318
	151	.0618	.0759	.0901	.0984	.1131	.1308	.1501	.1687	.1803	.2032	.2304
	152	.0614	.0754	.0895	.0977	.1123	.1300	.1491	.1676	.1791	.2019	.2289
	153	.0610	.0749	.0889	.0970	.1116	.1291	.1481	.1665	.1780	.2006	.2275
	154	.0606	.0744	.0883	.0964	.1108	.1283	.1472	.1654	.1769	.1994	.2261
	155	.0602	.0739	.0878	.0958	.1101	.1274	.1462	.1644	.1758	.1981	.2247
	156	.0598	.0734	.0872	.0951	.1094	.1266	.1453	.1633	.1747	.1969	.2234
	157+	8.958*	11.08*	13.25*	14.53*	16.83*	19.67*	22.81*	25.90*	27.88*	31.85*	36.70*
21	22–33	SEE NOTE**										
	34–77	SEE NOTE***										
	78	.1345	.1627	.1906	.2064	.2343	.2673	.3020	.3346	.3546	.3930	.4369
	79	.1326	.1605	.1880	.2037	.2313	.2638	.2982	.3305	.3504	.3884	.4320
	80	.1308	.1584	.1856	.2010	.2283	.2605	.2946	.3266	.3462	.3840	.4272

*To approximate p_f, divide the starred entry under f by $(n - 1)$.

**Compute $r^* = n - r$, find p^*_{1-f} for r^* and n, and compute $p_f = 1 - p^*_{1-f}$.

Table 5 (continued)

							f					
r	*n*	.001	.01	.05	.1	.25	.5	.75	.9	.95	.99	.999
21	81	.1290	.1563	.1831	.1984	.2254	.2573	.2910	.3227	.3422	.3796	.4225
	82	.1273	.1542	.1808	.1959	.2225	.2541	.2875	.3189	.3382	.3753	.4180
	83	.1257	.1522	.1785	.1934	.2198	.2510	.2841	.3152	.3344	.3712	.4135
	84	.1240	.1503	.1762	.1910	.2171	.2480	.2807	.3116	.3306	.3671	.4091
	85	.1224	.1484	.1740	.1887	.2145	.2451	.2775	.3080	.3269	.3631	.4048
	86	.1209	.1465	.1719	.1864	.2119	.2422	.2743	.3046	.3233	.3592	.4006
	87	.1194	.1447	.1698	.1841	.2094	.2394	.2712	.3012	.3197	.3554	.3965
	88	.1179	.1430	.1678	.1820	.2070	.2366	.2682	.2979	.3163	.3516	.3924
	89	.1165	.1413	.1658	.1798	.2046	.2340	.2652	.2947	.3129	.3480	.3885
	90	.1151	.1396	.1639	.1777	.2022	.2314	.2623	.2915	.3095	.3444	.3846
	91	.1137	.1380	.1620	.1757	.2000	.2288	.2594	.2884	.3063	.3408	.3808
	92	.1124	.1364	.1601	.1737	.1977	.2263	.2567	.2854	.3031	.3374	.3770
	93	.1111	.1348	.1583	.1718	.1956	.2238	.2539	.2824	.3000	.3340	.3734
	94	.1098	.1333	.1566	.1699	.1934	.2214	.2513	.2795	.2970	.3307	.3698
	95	.1086	.1318	.1549	.1680	.1913	.2191	.2486	.2766	.2940	.3274	.3663
	96	.1074	.1304	.1532	.1662	.1893	.2168	.2461	.2738	.2910	.3243	.3628
	97	.1062	.1289	.1515	.1644	.1873	.2145	.2436	.2711	.2882	.3211	.3594
	98	.1050	.1275	.1499	.1627	.1853	.2123	.2411	.2684	.2853	.3181	.3561
	99	.1039	.1262	.1483	.1610	.1834	.2102	.2387	.2658	.2826	.3151	.3528
	100	.1028	.1248	.1468	.1593	.1815	.2081	.2364	.2632	.2799	.3121	.3496
	101	.1017	.1235	.1452	.1577	.1797	.2060	.2340	.2607	.2772	.3092	.3464
	102	.1006	.1222	.1438	.1561	.1779	.2040	.2318	.2582	.2746	.3064	.3433
	103	.0996	.1210	.1423	.1545	.1761	.2020	.2296	.2558	.2720	.3036	.3403
	104	.0985	.1198	.1409	.1530	.1744	.2000	.2274	.2534	.2695	.3008	.3373
	105	.0975	.1186	.1395	.1515	.1727	.1981	.2252	.2510	.2670	.2981	.3344
	106	.0966	.1174	.1381	.1500	.1710	.1962	.2231	.2487	.2646	.2955	.3315
	107	.0956	.1162	.1368	.1485	.1694	.1944	.2211	.2465	.2622	.2929	.3286
	108	.0947	.1151	.1354	.1471	.1678	.1926	.2190	.2442	.2599	.2903	.3258
	109	.0937	.1140	.1342	.1457	.1662	.1908	.2170	.2421	.2576	.2878	.3231
	110	.0928	.1129	.1329	.1443	.1647	.1890	.2151	.2399	.2553	.2853	.3204
	111	.0919	.1118	.1316	.1430	.1632	.1873	.2132	.2378	.2531	.2829	.3177
	112	.0911	.1108	.1304	.1417	.1617	.1856	.2113	.2357	.2510	.2805	.3151
	113	.0902	.1097	.1292	.1404	.1602	.1840	.2094	.2337	.2488	.2782	.3126
	114	.0894	.1087	.1280	.1391	.1588	.1824	.2076	.2317	.2467	.2759	.3100
	115	.0886	.1077	.1269	.1379	.1574	.1808	.2058	.2297	.2446	.2736	.3075
	116	.0877	.1068	.1257	.1366	.1560	.1792	.2041	.2278	.2426	.2714	.3051
	117	.0869	.1058	.1246	.1354	.1546	.1777	.2023	.2259	.2406	.2692	.3027
	118	.0862	.1049	.1235	.1342	.1533	.1761	.2007	.2240	.2386	.2670	.3003
	119	.0854	.1039	.1225	.1331	.1520	.1747	.1990	.2222	.2367	.2649	.2980
	120	.0846	.1030	.1214	.1319	.1507	.1732	.1973	.2204	.2348	.2628	.2957
	121	.0839	.1021	.1204	.1308	.1494	.1718	.1957	.2186	.2329	.2607	.2934
	122	.0832	.1013	.1193	.1297	.1482	.1703	.1941	.2169	.2311	.2587	.2912
	123	.0825	.1004	.1183	.1286	.1470	.1689	.1926	.2152	.2292	.2567	.2890
	124	.0818	.0996	.1173	.1276	.1458	.1676	.1910	.2135	.2274	.2547	.2868
	125	.0811	.0987	.1164	.1265	.1446	.1662	.1895	.2118	.2257	.2528	.2847
	126	.0804	.0979	.1154	.1255	.1434	.1649	.1880	.2101	.2239	.2509	.2826
	127	.0797	.0971	.1145	.1245	.1422	.1636	.1866	.2085	.2222	.2490	.2805
	128	.0791	.0963	.1136	.1235	.1411	.1623	.1851	.2069	.2206	.2471	.2784
	129	.0784	.0955	.1126	.1225	.1400	.1610	.1837	.2054	.2189	.2453	.2764
	130	.0778	.0948	.1118	.1215	.1389	.1598	.1823	.2038	.2173	.2435	.2744
	131	.0772	.0940	.1109	.1206	.1378	.1586	.1809	.2023	.2157	.2418	.2725
	132	.0766	.0933	.1100	.1196	.1368	.1574	.1795	.2008	.2141	.2400	.2706
	133	.0759	.0925	.1091	.1187	.1357	.1562	.1782	.1993	.2125	.2383	.2687
	134	.0754	.0918	.1083	.1178	.1347	.1550	.1769	.1979	.2110	.2366	.2668
	135	.0748	.0911	.1075	.1169	.1337	.1539	.1756	.1964	.2094	.2349	.2649
	136	.0742	.0904	.1067	.1160	.1327	.1527	.1743	.1950	.2080	.2333	.2631
	137	.0736	.0897	.1059	.1151	.1317	.1516	.1730	.1936	.2065	.2316	.2613
	138	.0731	.0891	.1051	.1143	.1307	.1505	.1718	.1922	.2050	.2300	.2595
	139	.0725	.0884	.1043	.1134	.1298	.1494	.1706	.1909	.2036	.2284	.2578
	140	.0720	.0877	.1035	.1126	.1288	.1483	.1694	.1896	.2022	.2269	.2561
	141	.0714	.0871	.1028	.1118	.1279	.1473	.1682	.1882	.2008	.2253	.2544
	142	.0709	.0864	.1020	.1110	.1270	.1462	.1670	.1869	.1994	.2238	.2527
	143	.0704	.0858	.1013	.1102	.1261	.1452	.1658	.1857	.1980	.2223	.2510
	144	.0699	.0852	.1006	.1094	.1252	.1442	.1647	.1844	.1967	.2208	.2494
	145	.0694	.0846	.0998	.1086	.1243	.1432	.1636	.1831	.1954	.2194	.2478
	146	.0689	.0840	.0991	.1079	.1234	.1422	.1625	.1819	.1941	.2179	.2462
	147	.0684	.0834	.0984	.1071	.1226	.1412	.1614	.1807	.1928	.2165	.2446
	148	.0679	.0828	.0978	.1063	.1218	.1403	.1603	.1795	.1915	.2151	.2430

***To approximate p_f, compute $m = r/n$ and $s = \sqrt{m(1-m)/(n+1)}$, find z_f in Table 3, and compute $p_f \doteq m + z_f s$.

Table 5 (continued)

							f					
r	*n*	.001	.01	.05	.1	.25	.5	.75	.9	.95	.99	.999
21	149	.0674	.0822	.0971	.1056	.1209	.1393	.1592	.1783	.1903	.2137	.2415
	150	.0670	.0817	.0964	.1049	.1201	.1384	.1582	.1772	.1890	.2124	.2400
	151	.0665	.0811	.0958	.1042	.1193	.1375	.1571	.1760	.1878	.2110	.2385
	152	.0660	.0806	.0951	.1035	.1185	.1366	.1561	.1749	.1866	.2097	.2370
	153	.0656	.0800	.0945	.1028	.1177	.1357	.1551	.1737	.1854	.2084	.2355
	154	.0651	.0795	.0939	.1021	.1169	.1348	.1541	.1726	.1843	.2071	.2341
	155	.0647	.0789	.0932	.1015	.1162	.1339	.1531	.1715	.1831	.2058	.2327
	156	.0643	.0784	.0926	.1008	.1154	.1331	.1521	.1705	.1820	.2045	.2312
	157	.0638	.0779	.0920	.1001	.1147	.1322	.1512	.1694	.1808	.2033	.2299
	158	.0634	.0774	.0914	.0995	.1139	.1314	.1502	.1683	.1797	.2020	.2285
	159	.0630	.0769	.0908	.0989	.1132	.1305	.1493	.1673	.1786	.2008	.2271
	160	.0626	.0764	.0902	.0982	.1125	.1297	.1483	.1663	.1775	.1996	.2258
	161	.0622	.0759	.0897	.0976	.1118	.1289	.1474	.1653	.1764	.1984	.2244
	162	.0618	.0754	.0891	.0970	.1111	.1281	.1465	.1643	.1754	.1972	.2231
	163	.0614	.0749	.0885	.0964	.1104	.1273	.1456	.1633	.1743	.1960	.2218
	164+	9.619*	11.83*	14.07*	15.38*	17.75*	20.67*	23.88*	27.05*	29.06*	33.10*	38.04*
22	23–34	SEE NOTE**										
	35–84	SEE NOTE***										
	85	.1313	.1581	.1845	.1994	.2248	.2569	.2898	.3207	.3397	.3761	.4180
	86	.1296	.1561	.1822	.1970	.2231	.2539	.2865	.3171	.3359	.3721	.4136
	87	.1280	.1542	.1800	.1946	.2205	.2510	.2832	.3136	.3323	.3681	.4094
	88	.1264	.1523	.1778	.1923	.2179	.2481	.2801	.3101	.3287	.3643	.4052
	89	.1249	.1505	.1757	.1901	.2154	.2453	.2770	.3068	.3252	.3605	.4012
	90	.1234	.1487	.1737	.1879	.2129	.2425	.2739	.3035	.3217	.3568	.3972
	91	.1219	.1470	.1717	.1857	.2105	.2399	.2710	.3003	.3183	.3531	.3932
	92	.1205	.1453	.1697	.1836	.2082	.2372	.2681	.2971	.3150	.3496	.3894
	93	.1191	.1436	.1678	.1816	.2059	.2347	.2652	.2940	.3118	.3461	.3856
	94	.1177	.1420	.1659	.1795	.2036	.2322	.2624	.2910	.3086	.3426	.3819
	95	.1164	.1404	.1641	.1776	.2014	.2297	.2597	.2880	.3055	.3393	.3783
	96	.1151	.1388	.1623	.1757	.1993	.2273	.2570	.2851	.3025	.3360	.3747
	97	.1138	.1373	.1605	.1738	.1972	.2249	.2544	.2823	.2995	.3328	.3712
	98	.1126	.1358	.1588	.1719	.1951	.2226	.2518	.2795	.2966	.3296	.3678
	99	.1113	.1344	.1571	.1701	.1931	.2203	.2493	.2768	.2937	.3265	.3644
	100	.1101	.1329	.1555	.1684	.1911	.2181	.2469	.2741	.2909	.3234	.3611
	101	.1090	.1315	.1539	.1666	.1892	.2160	.2445	.2714	.2881	.3204	.3579
	102	.1078	.1302	.1523	.1649	.1873	.2138	.2421	.2689	.2854	.3175	.3547
	103	.1067	.1288	.1508	.1633	.1854	.2117	.2398	.2663	.2828	.3146	.3515
	104	.1056	.1275	.1493	.1616	.1836	.2097	.2375	.2638	.2802	.3117	.3485
	105	.1045	.1263	.1478	.1601	.1818	.2077	.2352	.2614	.2776	.3090	.3454
	106	.1035	.1250	.1463	.1585	.1800	.2057	.2331	.2590	.2751	.3062	.3425
	107	.1024	.1238	.1449	.1570	.1783	.2038	.2309	.2566	.2726	.3035	.3395
	108	.1014	.1226	.1435	.1555	.1766	.2019	.2288	.2543	.2702	.3009	.3366
	109	.1004	.1214	.1421	.1540	.1750	.2000	.2267	.2521	.2678	.2983	.3338
	110	.0995	.1202	.1408	.1525	.1734	.1982	.2247	.2498	.2654	.2957	.3310
	111	.0985	.1191	.1395	.1511	.1718	.1964	.2227	.2476	.2631	.2932	.3283
	112	.0976	.1179	.1382	.1497	.1702	.1946	.2207	.2455	.2609	.2908	.3256
	113	.0967	.1168	.1369	.1483	.1687	.1929	.2188	.2434	.2587	.2883	.3229
	114	.0958	.1158	.1356	.1470	.1671	.1912	.2169	.2413	.2565	.2859	.3203
	115	.0949	.1147	.1344	.1457	.1657	.1895	.2150	.2393	.2543	.2836	.3178
	116	.0940	.1137	.1332	.1444	.1642	.1879	.2132	.2372	.2522	.2813	.3152
	117	.0931	.1127	.1320	.1431	.1628	.1863	.2114	.2353	.2501	.2790	.3128
	118	.0923	.1117	.1309	.1419	.1614	.1847	.2096	.2333	.2481	.2768	.3103
	119	.0915	.1107	.1297	.1406	.1600	.1831	.2078	.2314	.2461	.2746	.3079
	120	.0907	.1097	.1286	.1394	.1586	.1816	.2061	.2295	.2441	.2724	.3055
	121	.0899	.1088	.1275	.1382	.1573	.1801	.2044	.2277	.2421	.2703	.3032
	122	.0891	.1078	.1264	.1371	.1560	.1786	.2028	.2259	.2402	.2682	.3009
	123	.0883	.1069	.1254	.1359	.1547	.1771	.2012	.2241	.2383	.2661	.2986
	124	.0876	.1060	.1243	.1348	.1534	.1757	.1995	.2223	.2365	.2641	.2964
	125	.0868	.1051	.1233	.1337	.1522	.1743	.1980	.2206	.2346	.2620	.2942
	126	.0861	.1042	.1223	.1326	.1509	.1729	.1964	.2189	.2328	.2601	.2920
	127	.0854	.1034	.1213	.1315	.1497	.1715	.1949	.2172	.2311	.2581	.2899
	128	.0847	.1025	.1203	.1305	.1485	.1702	.1934	.2155	.2293	.2562	.2878
	129	.0840	.1017	.1193	.1294	.1474	.1688	.1919	.2139	.2276	.2543	.2857
	130	.0833	.1009	.1184	.1284	.1461	.1675	.1904	.2123	.2259	.2525	.2836
	131	.0827	.1001	.1174	.1274	.1451	.1662	.1890	.2107	.2242	.2506	.2816
	132	.0820	.0993	.1165	.1264	.1439	.1650	.1876	.2091	.2226	.2488	.2796
	133	.0813	.0985	.1156	.1254	.1428	.1637	.1862	.2076	.2210	.2470	.2777

*To approximate p_f, divide the starred entry under f by $(n - 1)$.

**Compute $r^* = n - r$, find p^*_{1-f} for r^* and n, and compute $p_f = 1 - p^*_{1-f}$.

Table 5 (continued)

r	n	.001	.01	.05	.1	.25	*f* .5	.75	.9	.95	.99	.999
22	134	.0807	.0977	.1147	.1245	.1418	.1625	.1848	.2061	.2194	.2453	.2757
	135	.0801	.0970	.1138	.1235	.1407	.1613	.1834	.2046	.2178	.2435	.2738
	136	.0795	.0962	.1130	.1226	.1396	.1601	.1821	.2031	.2162	.2418	.2719
	137	.0788	.0955	.1121	.1217	.1386	.1589	.1808	.2017	.2147	.2401	.2701
	138	.0782	.0948	.1113	.1207	.1376	.1578	.1795	.2002	.2132	.2384	.2683
	139	.0777	.0941	.1105	.1199	.1366	.1566	.1782	.1988	.2117	.2368	.2665
	140	.0771	.0934	.1096	.1190	.1356	.1555	.1769	.1974	.2102	.2352	.2647
	141	.0765	.0927	.1088	.1181	.1346	.1544	.1757	.1961	.2088	.2336	.2629
	142	.0759	.0920	.1081	.1173	.1336	.1533	.1745	.1947	.2073	.2321	.2612
	143	.0754	.0913	.1073	.1164	.1327	.1522	.1732	.1934	.2059	.2305	.2595
	144	.0748	.0907	.1065	.1156	.1318	.1512	.1721	.1921	.2045	.2290	.2578
	145	.0743	.0900	.1058	.1148	.1308	.1501	.1709	.1908	.2032	.2275	.2561
	146	.0738	.0894	.1050	.1140	.1299	.1491	.1697	.1895	.2018	.2260	.2545
	147	.0732	.0888	.1043	.1132	.1290	.1481	.1686	.1882	.2005	.2245	.2528
	148	.0727	.0881	.1035	.1124	.1281	.1471	.1674	.1870	.1992	.2230	.2512
	149	.0722	.0875	.1028	.1116	.1273	.1461	.1663	.1857	.1979	.2216	.2496
	150	.0717	.0869	.1021	.1109	.1264	.1451	.1652	.1845	.1966	.2202	.2481
	151	.0712	.0863	.1014	.1101	.1256	.1441	.1641	.1833	.1953	.2188	.2465
	152	.0707	.0857	.1007	.1094	.1247	.1432	.1631	.1821	.1941	.2174	.2450
	153	.0702	.0852	.1001	.1086	.1239	.1422	.1620	.1810	.1928	.2160	.2435
	154	.0697	.0846	.0994	.1079	.1231	.1413	.1610	.1798	.1916	.2147	.2420
	155	.0693	.0840	.0987	.1072	.1223	.1404	.1599	.1787	.1904	.2134	.2405
	156	.0688	.0835	.0981	.1065	.1215	.1395	.1589	.1775	.1892	.2121	.2391
	157	.0684	.0829	.0974	.1058	.1207	.1386	.1579	.1764	.1880	.2108	.2376
	158	.0679	.0824	.0968	.1051	.1199	.1377	.1569	.1753	.1869	.2095	.2362
	159	.0675	.0818	.0962	.1044	.1191	.1368	.1559	.1743	.1857	.2082	.2348
	160	.0670	.0813	.0956	.1038	.1184	.1360	.1550	.1732	.1846	.2070	.2334
	161	.0666	.0808	.0950	.1031	.1176	.1351	.1540	.1721	.1835	.2057	.2320
	162	.0662	.0803	.0944	.1025	.1169	.1343	.1531	.1711	.1824	.2045	.2307
	163	.0657	.0797	.0938	.1018	.1162	.1335	.1521	.1701	.1813	.2033	.2293
	164	.0653	.0792	.0932	.1012	.1155	.1327	.1512	.1690	.1802	.2021	.2280
	165	.0649	.0787	.0926	.1006	.1148	.1319	.1503	.1680	.1791	.2009	.2267
	166	.0645	.0783	.0920	.0999	.1141	.1311	.1494	.1670	.1781	.1997	.2254
	167	.0641	.0778	.0915	.0993	.1134	.1303	.1485	.1661	.1770	.1986	.2241
	168	.0637	.0773	.0909	.0987	.1127	.1295	.1476	.1651	.1760	.1975	.2229
	169	.0633	.0768	.0903	.0981	.1120	.1287	.1468	.1641	.1750	.1963	.2216
	170+	10.29*	12.57*	14.89*	16.24*	18.68*	21.67*	24.96*	28.18*	30.24*	34.35*	39.37*
23	24–35	SEE NOTE**										
	36–91	SEE NOTE***										
	92	.1287	.1542	.1793	.1936	.2186	.2482	.2794	.3088	.3269	.3616	.4016
	93	.1272	.1525	.1773	.1914	.2162	.2455	.2765	.3056	.3235	.3580	.3977
	94	.1257	.1507	.1753	.1893	.2138	.2429	.2736	.3025	.3203	.3545	.3939
	95	.1243	.1490	.1734	.1872	.2115	.2403	.2707	.2994	.3170	.3510	.3902
	96	.1229	.1474	.1715	.1852	.2093	.2378	.2679	.2964	.3139	.3476	.3865
	97	.1215	.1458	.1696	.1832	.2070	.2353	.2652	.2934	.3108	.3443	.3830
	98	.1202	.1442	.1678	.1812	.2049	.2329	.2625	.2905	.3078	.3410	.3794
	99	.1189	.1426	.1660	.1793	.2028	.2305	.2599	.2877	.3048	.3378	.3760
	100	.1176	.1411	.1643	.1775	.2007	.2282	.2574	.2849	.3019	.3346	.3725
	101	.1164	.1396	.1626	.1756	.1987	.2259	.2548	.2822	.2990	.3316	.3692
	102	.1151	.1382	.1609	.1738	.1967	.2237	.2524	.2795	.2962	.3285	.3659
	103	.1139	.1368	.1593	.1721	.1947	.2215	.2500	.2768	.2934	.3255	.3627
	104	.1128	.1354	.1577	.1704	.1928	.2194	.2476	.2743	.2907	.3226	.3595
	105	.1116	.1340	.1561	.1687	.1909	.2173	.2453	.2717	.2881	.3197	.3564
	106	.1105	.1327	.1546	.1670	.1891	.2152	.2430	.2692	.2855	.3169	.3533
	107	.1094	.1314	.1531	.1654	.1873	.2132	.2407	.2668	.2829	.3141	.3503
	108	.1083	.1301	.1516	.1638	.1855	.2112	.2384	.2644	.2804	.3114	.3474
	109	.1072	.1288	.1501	.1623	.1837	.2092	.2364	.2620	.2779	.3087	.3444
	110	.1062	.1276	.1487	.1608	.1820	.2073	.2342	.2597	.2755	.3061	.3416
	111	.1052	.1264	.1473	.1593	.1804	.2054	.2321	.2574	.2731	.3035	.3388
	112	.1042	.1252	.1460	.1578	.1787	.2036	.2301	.2552	.2708	.3009	.3360
	113	.1032	.1240	.1446	.1563	.1771	.2018	.2281	.2530	.2685	.2984	.3332
	114	.1022	.1229	.1433	.1549	.1755	.2000	.2261	.2508	.2662	.2959	.3306
	115	.1013	.1217	.1420	.1535	.1740	.1983	.2242	.2487	.2640	.2935	.3279
	116	.1003	.1206	.1407	.1522	.1724	.1965	.2222	.2466	.2618	.2911	.3253
	117	.0994	.1196	.1395	.1508	.1709	.1948	.2204	.2446	.2596	.2888	.3228
	118	.0985	.1185	.1382	.1495	.1694	.1932	.2185	.2426	.2575	.2864	.3202
	119	.0977	.1175	.1370	.1482	.1680	.1916	.2167	.2406	.2554	.2842	.3178
	120	.0968	.1164	.1358	.1469	.1666	.1900	.2149	.2386	.2533	.2819	.3153

***To approximate p_f, compute $m = r/n$ and $s = \sqrt{m(1-m)/(n+1)}$, find z_f in Table 3, and compute $p_f \doteq m + z_f s$.

Table 5 (continued)

r	n	.001	.01	.05	.1	.25	f .5	.75	.9	.95	.99	.999
23	121	.0959	.1154	.1347	.1457	.1652	.1884	.2132	.2367	.2513	.2797	.3129
	122	.0951	.1144	.1335	.1444	.1638	.1868	.2114	.2348	.2493	.2776	.3105
	123	.0943	.1134	.1324	.1432	.1624	.1853	.2097	.2329	.2474	.2754	.3082
	124	.0935	.1125	.1313	.1420	.1611	.1838	.2080	.2311	.2454	.2733	.3059
	125	.0927	.1115	.1302	.1409	.1598	.1823	.2064	.2293	.2436	.2712	.3036
	126	.0919	.1106	.1291	.1397	.1585	.1809	.2048	.2275	.2417	.2692	.3014
	127	.0911	.1097	.1281	.1386	.1572	.1794	.2032	.2258	.2398	.2672	.2992
	128	.0904	.1088	.1271	.1375	.1560	.1780	.2016	.2241	.2380	.2652	.2970
	129	.0897	.1079	.1260	.1364	.1547	.1766	.2001	.2224	.2362	.2632	.2949
	130	.0889	.1071	.1250	.1353	.1535	.1753	.1985	.2207	.2345	.2613	.2928
	131	.0882	.1062	.1240	.1342	.1523	.1739	.1970	.2191	.2328	.2594	.2907
	132	.0875	.1054	.1231	.1332	.1511	.1726	.1955	.2174	.2310	.2576	.2886
	133	.0868	.1045	.1221	.1322	.1500	.1713	.1941	.2158	.2294	.2557	.2866
	134	.0861	.1037	.1212	.1311	.1489	.1700	.1927	.2143	.2277	.2539	.2846
	135	.0855	.1029	.1202	.1301	.1477	.1687	.1912	.2127	.2261	.2521	.2827
	136	.0848	.1021	.1193	.1292	.1466	.1675	.1898	.2112	.2245	.2503	.2807
	137	.0841	.1014	.1184	.1282	.1455	.1663	.1885	.2097	.2229	.2486	.2788
	138	.0835	.1006	.1175	.1272	.1445	.1651	.1871	.2082	.2213	.2469	.2769
	139	.0829	.0998	.1167	.1263	.1434	.1639	.1858	.2067	.2198	.2452	.2751
	140	.0822	.0991	.1158	.1254	.1424	.1627	.1845	.2053	.2182	.2435	.2732
	141	.0816	.0984	.1150	.1245	.1413	.1615	.1832	.2038	.2167	.2419	.2714
	142	.0810	.0976	.1141	.1236	.1403	.1604	.1819	.2024	.2152	.2402	.2696
	143	.0804	.0969	.1133	.1227	.1393	.1593	.1806	.2011	.2138	.2386	.2678
	144	.0798	.0962	.1125	.1218	.1383	.1581	.1794	.1997	.2123	.2370	.2661
	145	.0793	.0955	.1117	.1209	.1374	.1570	.1782	.1983	.2109	.2355	.2644
	146	.0787	.0949	.1109	.1201	.1364	.1560	.1769	.1970	.2095	.2339	.2627
	147	.0781	.0942	.1101	.1193	.1355	.1549	.1758	.1957	.2081	.2324	.2610
	148	.0776	.0935	.1094	.1184	.1345	.1539	.1746	.1944	.2068	.2309	.2594
	149	.0770	.0929	.1086	.1176	.1336	.1528	.1734	.1931	.2054	.2294	.2577
	150	.0765	.0922	.1079	.1168	.1327	.1518	.1723	.1919	.2041	.2280	.2561
	151	.0760	.0916	.1071	.1160	.1318	.1508	.1711	.1906	.2028	.2265	.2545
	152	.0754	.0910	.1064	.1152	.1310	.1498	.1700	.1894	.2015	.2251	.2529
	153	.0749	.0903	.1057	.1145	.1301	.1488	.1689	.1882	.2002	.2237	.2514
	154	.0744	.0897	.1050	.1137	.1292	.1478	.1678	.1870	.1989	.2223	.2498
	155	.0739	.0891	.1043	.1130	.1284	.1469	.1668	.1858	.1977	.2209	.2483
	156	.0734	.0885	.1036	.1122	.1275	.1459	.1657	.1846	.1964	.2195	.2468
	157	.0729	.0880	.1029	.1115	.1267	.1450	.1646	.1835	.1952	.2182	.2453
	158	.0725	.0874	.1022	.1108	.1259	.1441	.1636	.1823	.1940	.2169	.2439
	159	.0720	.0868	.1016	.1100	.1251	.1432	.1626	.1812	.1928	.2156	.2424
	160	.0715	.0863	.1009	.1093	.1243	.1423	.1616	.1801	.1916	.2143	.2410
	161	.0710	.0857	.1003	.1086	.1235	.1414	.1606	.1790	.1905	.2130	.2396
	162	.0706	.0851	.0996	.1080	.1228	.1405	.1596	.1779	.1893	.2117	.2382
	163	.0701	.0846	.0990	.1073	.1220	.1396	.1586	.1768	.1882	.2105	.2368
	164	.0697	.0841	.0984	.1066	.1212	.1388	.1577	.1758	.1871	.2092	.2354
	165	.0692	.0835	.0978	.1060	.1205	.1379	.1567	.1747	.1860	.2080	.2341
	166	.0688	.0830	.0972	.1053	.1198	.1371	.1558	.1737	.1849	.2068	.2328
	167	.0684	.0825	.0966	.1047	.1190	.1363	.1549	.1727	.1838	.2056	.2314
	168	.0680	.0820	.0960	.1040	.1183	.1355	.1539	.1717	.1827	.2045	.2301
	169	.0675	.0815	.0954	.1034	.1176	.1347	.1530	.1707	.1817	.2033	.2288
	170	.0671	.0810	.0948	.1028	.1169	.1339	.1521	.1697	.1806	.2021	.2276
	171	.0667	.0805	.0943	.1022	.1162	.1331	.1513	.1687	.1796	.2010	.2263
	172	.0663	.0800	.0937	.1015	.1155	.1323	.1504	.1677	.1786	.1999	.2251
	173	.0659	.0796	.0931	.1009	.1148	.1315	.1495	.1668	.1776	.1988	.2238
	174	.0655	.0791	.0926	.1004	.1142	.1308	.1487	.1658	.1766	.1977	.2226
	175	.0651	.0786	.0921	.0998	.1135	.1300	.1478	.1649	.1756	.1966	.2214
	176	.0647	.0782	.0915	.0992	.1129	.1293	.1470	.1640	.1746	.1955	.2202
	177+	10.96*	13.33*	15.72*	17.11*	19.61*	22.67*	26.03*	29.32*	31.41*	35.60*	40.70*
24	25–37	SEE NOTE**										
	38–99	SEE NOTE***										
	100	.1252	.1494	.1731	.1866	.2103	.2383	.2678	.2957	.3128	.3458	.3839
	101	.1238	.1478	.1713	.1847	.2082	.2359	.2652	.2928	.3098	.3426	.3804
	102	.1225	.1463	.1696	.1828	.2061	.2336	.2626	.2900	.3069	.3395	.3771
	103	.1213	.1448	.1679	.1810	.2040	.2313	.2601	.2873	.3041	.3364	.3737
	104	.1200	.1433	.1662	.1792	.2020	.2290	.2577	.2846	.3013	.3334	.3705
	105	.1188	.1419	.1645	.1774	.2000	.2268	.2552	.2820	.2985	.3304	.3673
	106	.1176	.1404	.1629	.1756	.1981	.2247	.2529	.2794	.2958	.3275	.3641
	107	.1164	.1390	.1613	.1739	.1962	.2226	.2505	.2769	.2932	.3246	.3610
	108	.1152	.1377	.1597	.1723	.1944	.2205	.2482	.2744	.2906	.3218	.3580

*To approximate p_f, divide the starred entry under f by $(n - 1)$.
**Compute $r^* = n - r$, find p^*_{1-f} for r^* and n, and compute $p_f = 1 - p^*_{1-f}$.

Table 5 (continued)

						f					
r	n	.001	.01	.05	.1	.25	.5	.75	.9	.95	.99	.999
24	109	.1141	.1363	.1582	.1706	.1925	.2185	.2460	.2720	.2880	.3190	.3350
	110	.1130	.1350	.1567	.1690	.1907	.2165	.2438	.2696	.2855	.3163	.3520
	111	.1119	.1338	.1552	.1675	.1890	.2145	.2416	.2672	.2830	.3136	.3491
	112	.1108	.1325	.1538	.1659	.1873	.2126	.2395	.2649	.2806	.3110	.3463
	113	.1098	.1313	.1524	.1644	.1856	.2107	.2374	.2626	.2782	.3084	.3435
	114	.1088	.1300	.1510	.1629	.1839	.2088	.2353	.2604	.2759	.3059	.3407
	115	.1078	.1289	.1496	.1614	.1823	.2070	.2333	.2582	.2736	.3034	.3380
	116	.1068	.1277	.1483	.1600	.1807	.2052	.2313	.2560	.2713	.3009	.3353
	117	.1058	.1265	.1470	.1586	.1791	.2034	.2293	.2539	.2691	.2985	.3327
	118	.1048	.1254	.1457	.1572	.1775	.2017	.2274	.2518	.2669	.2961	.3301
	119	.1039	.1243	.1444	.1558	.1760	.2000	.2255	.2497	.2647	.2937	.3275
	120	.1030	.1232	.1431	.1545	.1745	.1983	.2237	.2477	.2626	.2914	.3250
	121	.1021	.1221	.1419	.1532	.1731	.1967	.2218	.2457	.2605	.2891	.3225
	122	.1012	.1211	.1407	.1519	.1716	.1951	.2200	.2437	.2584	.2869	.3201
	123	.1003	.1201	.1395	.1506	.1702	.1935	.2183	.2418	.2564	.2847	.3177
	124	.0995	.1190	.1383	.1493	.1688	.1919	.2165	.2399	.2544	.2825	.3153
	125	.0986	.1180	.1372	.1481	.1674	.1904	.2148	.2380	.2524	.2804	.3130
	126	.0978	.1171	.1361	.1469	.1661	.1888	.2131	.2362	.2505	.2783	.3107
	127	.0970	.1161	.1350	.1457	.1647	.1873	.2115	.2344	.2486	.2762	.3084
	128	.0962	.1151	.1339	.1445	.1634	.1859	.2098	.2326	.2467	.2741	.3062
	129	.0954	.1142	.1328	.1434	.1621	.1844	.2082	.2308	.2449	.2721	.3040
	130	.0946	.1133	.1317	.1422	.1608	.1830	.2066	.2291	.2430	.2701	.3018
	131	.0938	.1124	.1307	.1411	.1596	.1816	.2051	.2274	.2412	.2682	.2997
	132	.0931	.1115	.1297	.1400	.1584	.1802	.2035	.2257	.2395	.2662	.2976
	133	.0923	.1106	.1287	.1389	.1572	.1788	.2020	.2240	.2377	.2643	.2955
	134	.0916	.1097	.1277	.1379	.1560	.1775	.2005	.2224	.2360	.2625	.2934
	135	.0909	.1089	.1267	.1368	.1548	.1762	.1990	.2208	.2343	.2606	.2914
	136	.0902	.1081	.1257	.1358	.1536	.1749	.1976	.2192	.2326	.2588	.2894
	137	.0895	.1072	.1248	.1348	.1525	.1736	.1962	.2176	.2310	.2570	.2875
	138	.0888	.1064	.1238	.1338	.1514	.1723	.1948	.2161	.2294	.2552	.2855
	139	.0881	.1056	.1229	.1328	.1503	.1711	.1934	.2146	.2278	.2535	.2836
	140	.0875	.1048	.1220	.1318	.1492	.1699	.1920	.2131	.2262	.2517	.2817
	141	.0868	.1041	.1211	.1308	.1481	.1687	.1906	.2116	.2246	.2500	.2798
	142	.0862	.1033	.1202	.1299	.1470	.1675	.1893	.2101	.2231	.2484	.2780
	143	.0856	.1025	.1194	.1290	.1460	.1663	.1880	.2087	.2216	.2467	.2762
	144	.0849	.1018	.1185	.1280	.1449	.1651	.1867	.2073	.2201	.2451	.2744
	145	.0843	.1011	.1177	.1271	.1439	.1640	.1854	.2059	.2186	.2435	.2726
	146	.0837	.1004	.1168	.1262	.1429	.1628	.1842	.2045	.2172	.2419	.2709
	147	.0831	.0996	.1160	.1254	.1419	.1617	.1829	.2031	.2157	.2403	.2691
	148	.0825	.0989	.1152	.1245	.1410	.1606	.1817	.2018	.2143	.2387	.2674
	149	.0819	.0983	.1144	.1236	.1400	.1596	.1805	.2005	.2129	.2372	.2657
	150	.0814	.0976	.1136	.1228	.1391	.1585	.1793	.1992	.2115	.2357	.2641
	151	.0808	.0969	.1128	.1220	.1381	.1574	.1781	.1979	.2102	.2342	.2624
	152	.0802	.0962	.1121	.1211	.1372	.1564	.1770	.1966	.2088	.2327	.2608
	153	.0797	.0956	.1113	.1203	.1363	.1554	.1758	.1953	.2075	.2313	.2592
	154	.0792	.0949	.1106	.1195	.1354	.1544	.1747	.1941	.2062	.2298	.2576
	155	.0786	.0943	.1098	.1187	.1345	.1534	.1736	.1929	.2049	.2284	.2561
	156	.0781	.0937	.1091	.1180	.1336	.1524	.1725	.1917	.2036	.2270	.2545
	157	.0776	.0931	.1084	.1172	.1328	.1514	.1714	.1905	.2024	.2256	.2530
	158	.0771	.0924	.1077	.1164	.1319	.1504	.1703	.1893	.2011	.2242	.2515
	159	.0765	.0918	.1070	.1157	.1311	.1495	.1692	.1881	.1999	.2229	.2500
	160	.0760	.0912	.1063	.1149	.1302	.1485	.1682	.1870	.1987	.2216	.2485
	161	.0756	.0907	.1056	.1142	.1294	.1476	.1671	.1858	.1975	.2202	.2471
	162	.0751	.0901	.1050	.1135	.1286	.1467	.1661	.1847	.1963	.2189	.2456
	163	.0746	.0895	.1043	.1128	.1278	.1458	.1651	.1836	.1951	.2176	.2442
	164	.0741	.0889	.1037	.1121	.1270	.1449	.1641	.1825	.1939	.2164	.2428
	165	.0736	.0884	.1030	.1114	.1262	.1440	.1631	.1814	.1928	.2151	.2414
	166	.0732	.0878	.1024	.1107	.1255	.1431	.1621	.1803	.1917	.2139	.2400
	167	.0727	.0873	.1017	.1100	.1247	.1423	.1612	.1793	.1905	.2126	.2387
	168	.0723	.0867	.1011	.1093	.1239	.1414	.1602	.1782	.1894	.2114	.2373
	169	.0718	.0862	.1005	.1087	.1232	.1406	.1593	.1772	.1883	.2102	.2360
	170	.0714	.0857	.0999	.1080	.1225	.1398	.1584	.1762	.1873	.2090	.2347
	171	.0709	.0852	.0993	.1074	.1217	.1389	.1574	.1751	.1862	.2078	.2334
	172	.0705	.0847	.0987	.1067	.1210	.1381	.1565	.1741	.1851	.2067	.2321
	173	.0701	.0842	.0981	.1061	.1203	.1373	.1556	.1732	.1841	.2055	.2308
	174	.0697	.0837	.0975	.1055	.1196	.1365	.1547	.1722	.1831	.2044	.2296
	175	.0693	.0832	.0970	.1049	.1189	.1358	.1539	.1712	.1820	.2033	.2283

***To approximate p_f, compute $m = r/n$ and $s = \sqrt{m(1 - m)/(n + 1)}$, find z_f in Table 3, and compute $p_f \doteq m + z_f s$.

Table 5 (continued)

r	n	.001	.01	.05	.1	.25	f .5	.75	.9	.95	.99	.999
24	176	.0688	.0827	.0964	.1043	.1182	.1350	.1530	.1702	.1810	.2021	.2271
	177	.0684	.0822	.0958	.1037	.1176	.1342	.1521	.1693	.1800	.2010	.2259
	178	.0680	.0817	.0953	.1031	.1169	.1335	.1513	.1684	.1790	.2000	.2247
	179	.0676	.0812	.0947	.1025	.1162	.1327	.1504	.1674	.1781	.1989	.2235
	180	.0673	.0808	.0942	.1019	.1156	.1320	.1496	.1665	.1771	.1978	.2223
	181	.0669	.0803	.0937	.1013	.1149	.1312	.1488	.1656	.1761	.1968	.2211
	182	.0665	.0799	.0931	.1008	.1143	.1305	.1480	.1647	.1752	.1957	.2200
	183+	11.65*	14.09*	16.55*	17.97*	20.54*	23.67*	27.10*	30.45*	32.59*	36.84*	42.02*
25	26–38	SEE NOTE**										
	39–107	SEE NOTE***										
	108	.1223	.1454	.1679	.1807	.2032	.2298	.2579	.2844	.3007	.3321	.3685
	109	.1211	.1439	.1663	.1790	.2013	.2277	.2556	.2818	.2980	.3293	.3654
	110	.1199	.1426	.1648	.1773	.1995	.2256	.2533	.2794	.2955	.3265	.3624
	111	.1187	.1412	.1632	.1757	.1976	.2236	.2510	.2769	.2929	.3237	.3594
	112	.1176	.1399	.1617	.1741	.1958	.2216	.2488	.2745	.2904	.3210	.3565
	113	.1165	.1386	.1602	.1725	.1941	.2196	.2466	.2722	.2879	.3183	.3536
	114	.1154	.1373	.1587	.1709	.1923	.2177	.2445	.2698	.2855	.3157	.3508
	115	.1143	.1360	.1573	.1694	.1906	.2157	.2424	.2676	.2831	.3131	.3480
	116	.1133	.1348	.1559	.1678	.1889	.2139	.2403	.2653	.2808	.3106	.3452
	117	.1122	.1336	.1545	.1664	.1873	.2120	.2383	.2631	.2784	.3081	.3425
	118	.1112	.1324	.1531	.1649	.1857	.2102	.2363	.2609	.2762	.3056	.3399
	119	.1102	.1312	.1518	.1635	.1841	.2085	.2343	.2588	.2739	.3032	.3372
	120	.1092	.1300	.1505	.1621	.1825	.2067	.2324	.2567	.2717	.3008	.3346
	121	.1083	.1289	.1492	.1607	.1810	.2050	.2305	.2546	.2696	.2985	.3321
	122	.1073	.1278	.1479	.1593	.1794	.2033	.2286	.2526	.2675	.2962	.3296
	123	.1064	.1267	.1467	.1580	.1780	.2016	.2268	.2506	.2654	.2939	.3271
	124	.1055	.1256	.1454	.1567	.1765	.2000	.2250	.2486	.2633	.2917	.3247
	125	.1046	.1246	.1442	.1554	.1751	.1984	.2232	.2467	.2613	.2895	.3223
	126	.1037	.1235	.1430	.1541	.1736	.1968	.2215	.2448	.2593	.2873	.3199
	127	.1029	.1225	.1419	.1528	.1722	.1953	.2197	.2429	.2573	.2851	.3176
	128	.1020	.1215	.1407	.1516	.1709	.1937	.2180	.2411	.2553	.2830	.3153
	129	.1012	.1205	.1396	.1504	.1695	.1922	.2163	.2392	.2534	.2810	.3130
	130	.1003	.1196	.1385	.1492	.1682	.1907	.2147	.2375	.2516	.2789	.3108
	131	.0995	.1186	.1374	.1480	.1669	.1893	.1231	.2357	.2497	.2769	.3086
	132	.0987	.1177	.1363	.1469	.1656	.1878	.2115	.2339	.2479	.2749	.3064
	133	.0979	.1167	.1352	.1457	.1643	.1864	.2099	.2322	.2461	.2729	.3043
	134	.0972	.1158	.1342	.1446	.1631	.1850	.2084	.2305	.2443	.2710	.3022
	135	.0964	.1149	.1332	.1435	.1618	.1836	.2068	.2289	.2425	.2691	.3001
	136	.0957	.1140	.1321	.1424	.1606	.1823	.2053	.2272	.2408	.2672	.2981
	137	.0949	.1132	.1311	.1414	.1594	.1809	.2038	.2256	.2391	.2653	.2960
	138	.0942	.1123	.1302	.1403	.1583	.1796	.2024	.2240	.2374	.2635	.2940
	139	.0935	.1115	.1292	.1393	.1571	.1783	.2009	.2224	.2358	.2617	.2921
	140	.0928	.1106	.1282	.1383	.1560	.1770	.1995	.2209	.2341	.2599	.2901
	141	.0921	.1098	.1273	.1372	.1548	.1758	.1981	.2193	.2325	.2582	.2882
	142	.0914	.1090	.1264	.1363	.1537	.1745	.1967	.2178	.2309	.2564	.2863
	143	.0907	.1082	.1255	.1353	.1526	.1733	.1954	.2163	.2294	.2547	.2844
	144	.0901	.1074	.1246	.1343	.1516	.1721	.1940	.2149	.2278	.2430	.2826
	145	.0894	.1067	.1237	.1334	.1505	.1709	.1927	.2134	.2263	.2514	.2808
	146	.0888	.1059	.1228	.1324	.1494	.1697	.1914	.2120	.2248	.2497	.2790
	147	.0881	.1051	.1219	.1315	.1484	.1686	.1901	.2106	.2233	.2481	.2772
	148	.0875	.1044	.1211	.1306	.1474	.1674	.1888	.2092	.2219	.2465	.2754
	149	.0869	.1037	.1202	.1297	.1464	.1663	.1876	.2078	.2204	.2449	.2737
	150	.0863	.1030	.1194	.1288	.1454	.1652	.1863	.2065	.2190	.2434	.2720
	151	.0857	.1023	.1186	.1279	.1444	.1641	.1851	.2051	.2176	.2418	.2703
	152	.0851	.1016	.1178	.1271	.1435	.1630	.1839	.2038	.2162	.2403	.2686
	153	.0845	.1009	.1170	.1262	.1425	.1619	.1827	.2025	.2148	.2388	.2670
	154	.0839	.1002	.1162	.1254	.1416	.1609	.1815	.2012	.2135	.2373	.2654
	155	.0834	.0995	.1154	.1245	.1406	.1598	.1804	.1999	.2121	.2359	.2638
	156	.0828	.0988	.1147	.1237	.1397	.1588	.1792	.1987	.2108	.2344	.2622
	157	.0823	.0982	.1139	.1229	.1388	.1578	.1781	.1974	.2095	.2330	.2606
	158	.0817	.0975	.1132	.1221	.1379	.1568	.1770	.1962	.2082	.2316	.2590
	159	.0812	.0969	.1125	.1213	.1370	.1558	.1759	.1950	.2069	.2302	.2575
	160	.0806	.0963	.1117	.1206	.1362	.1548	.1748	.1938	.2057	.2288	.2560
	161	.0801	.0957	.1110	.1198	.1353	.1539	.1737	.1926	.2044	.2274	.2545
	162	.0796	.0950	.1103	.1190	.1345	.1529	.1726	.1915	.2032	.2261	.2530
	163	.0791	.0944	.1096	.1183	.1336	.1520	.1716	.1903	.2020	.2248	.2516
	164	.0786	.0938	.1089	.1175	.1328	.1510	.1705	.1892	.2008	.2234	.2501
	165	.0781	.0933	.1083	.1168	.1320	.1501	.1695	.1880	.1996	.2221	.2487

*To approximate p_f, divide the starred entry under f by $(n - 1)$.
**Compute $r^* = n - r$, find p^*_{1-f} for r^* and n, and compute $p_f = 1 - p^*_{1-f}$.

Table 5 (continued)

							f					
r	*n*	.001	.01	.05	.1	.25	.5	.75	.9	.95	.99	.999
25	166	.0776	.0927	.1076	.1161	.1312	.1492	.1685	.1869	.1984	.2209	.2473
	167	.0771	.0921	.1069	.1154	.1304	.1483	.1675	.1858	.1973	.2196	.2459
	168	.0766	.0915	.1063	.1147	.1296	.1474	.1665	.1847	.1961	.2183	.2445
	169	.0762	.0910	.1056	.1140	.1288	.1465	.1655	.1837	.1950	.2171	.2431
	170	.0757	.0904	.1050	.1133	.1280	.1457	.1646	.1826	.1939	.2159	.2418
	171	.0752	.0899	.1043	.1126	.1273	.1448	.1636	.1816	.1928	.2146	.2404
	172	.0748	.0893	.1037	.1120	.1265	.1440	.1627	.1805	.1917	.2134	.2391
	173	.0743	.0888	.1031	.1113	.1258	.1431	.1617	.1795	.1906	.2123	.2378
	174	.0739	.0883	.1025	.1106	.1251	.1423	.1608	.1785	.1895	.2111	.2365
	175	.0734	.0877	.1019	.1100	.1243	.1415	.1599	.1775	.1885	.2099	.2352
	176	.0730	.0872	.1013	.1094	.1236	.1407	.1590	.1765	.1874	.2088	.2340
	177	.0726	.0867	.1007	.1087	.1229	.1399	.1581	.1755	.1864	.2076	.2327
	178	.0721	.0862	.1001	.1081	.1222	.1391	.1572	.1745	.1854	.2065	.2315
	179	.0717	.0857	.0996	.1075	.1215	.1383	.1563	.1736	.1844	.2054	.2302
	180	.0713	.0852	.0990	.1069	.1208	.1376	.1555	.1726	.1834	.2043	.2290
	181	.0709	.0847	.0984	.1063	.1202	.1368	.1546	.1717	.1824	.2032	.2278
	182	.0705	.0843	.0979	.1057	.1195	.1360	.1538	.1708	.1814	.2021	.2266
	183	.0701	.0838	.0973	.1051	.1188	.1353	.1530	.1699	.1804	.2011	.2255
	184	.0697	.0833	.0968	.1045	.1182	.1345	.1521	.1689	.1795	.2000	.2243
	185	.0693	.0828	.0963	.1039	.1175	.1338	.1513	.1681	.1785	.1990	.2232
	186	.0689	.0824	.0957	.1034	.1169	.1331	.1505	.1672	.1776	.1979	.2220
	187	.0685	.0819	.0952	.1028	.1163	.1324	.1497	.1663	.1766	.1969	.2209
	188	.0682	.0815	.0947	.1022	.1156	.1317	.1489	.1654	.1757	.1959	.2198
	189	.0678	.0810	.0942	.1017	.1150	.1310	.1481	.1646	.1748	.1949	.2187
	190+	12.34*	14.85*	17.38*	18.84*	21.47*	24.67*	28.17*	31.58*	33.75*	38.08*	43.33*
26	27–39	SEE NOTE**										
	40–115	SEE NOTE***										
	116	.1198	.1419	.1635	.1757	.1972	.2225	.2494	.2746	.2902	.3203	.3551
	117	.1187	.1406	.1621	.1742	.1955	.2206	.2472	.2723	.2878	.3177	.3523
	118	.1177	.1394	.1606	.1726	.1938	.2188	.2452	.2701	.2855	.3152	.3496
	119	.1166	.1382	.1592	.1711	.1921	.2169	.2431	.2679	.2832	.3127	.3469
	120	.1156	.1369	.1578	.1697	.1905	.2151	.2411	.2657	.2809	.3102	.3442
	121	.1146	.1357	.1565	.1682	.1889	.2133	.2392	.2636	.2787	.3078	.3416
	122	.1136	.1346	.1551	.1668	.1873	.2115	.2372	.2615	.2765	.3054	.3390
	123	.1126	.1334	.1538	.1654	.1857	.2098	.2353	.2594	.2743	.3031	.3365
	124	.1116	.1323	.1525	.1640	.1842	.2081	.2334	.2574	.2722	.3008	.3340
	125	.1107	.1312	.1513	.1627	.1827	.2064	.2316	.2554	.2701	.2985	.3315
	126	.1097	.1301	.1500	.1613	.1812	.2048	.2298	.2534	.2680	.2962	.3291
	127	.1088	.1290	.1488	.1600	.1798	.2032	.2280	.2514	.2660	.2940	.3267
	128	.1079	.1280	.1476	.1587	.1783	.2016	.2262	.2495	.2639	.2919	.3243
	129	.1070	.1269	.1464	.1575	.1769	.2000	.2245	.2476	.2620	.2897	.3220
	130	.1061	.1259	.1452	.1562	.1756	.1985	.2228	.2458	.2600	.2876	.3197
	131	.1053	.1249	.1441	.1550	.1742	.1969	.2211	.2440	.2581	.2855	.3175
	132	.1044	.1239	.1430	.1538	.1728	.1954	.2194	.2422	.2562	.2835	.3152
	133	.1036	.1229	.1418	.1526	.1715	.1940	.2178	.2404	.2544	.2815	.3130
	134	.1028	.1219	.1407	.1514	.1702	.1925	.2162	.2386	.2525	.2795	.3109
	135	.1020	.1210	.1397	.1503	.1689	.1911	.2146	.2369	.2507	.2775	.3087
	136	.1012	.1201	.1386	.1491	.1677	.1897	.2130	.2352	.2489	.2756	.3066
	137	.1004	.1192	.1375	.1480	.1664	.1883	.2115	.2335	.2472	.2737	.3046
	138	.0996	.1182	.1364	.1469	.1652	.1869	.2100	.2319	.2454	.2718	.3025
	139	.0989	.1174	.1355	.1458	.1640	.1855	.2085	.2302	.2437	.2699	.3005
	140	.0981	.1165	.1345	.1447	.1628	.1842	.2070	.2286	.2420	.2681	.2985
	141	.0974	.1156	.1335	.1437	.1616	.1829	.2056	.2271	.2404	.2663	.2965
	142	.0967	.1148	.1325	.1426	.1605	.1816	.2041	.2255	.2387	.2645	.2946
	143	.0960	.1139	.1316	.1416	.1593	.1803	.2027	.2240	.2371	.2627	.2926
	144	.0953	.1131	.1306	.1406	.1582	.1791	.2013	.2224	.2355	.2610	.2908
	145	.0946	.1123	.1297	.1396	.1571	.1778	.1999	.2209	.2340	.2593	.2889
	146	.0939	.1115	.1288	.1386	.1560	.1766	.1986	.2195	.2324	.2576	.2870
	147	.0932	.1107	.1279	.1377	.1549	.1754	.1972	.2180	.2309	.2559	.2852
	148	.0926	.1099	.1270	.1367	.1538	.1742	.1959	.2165	.2294	.2543	.2834
	149	.0919	.1091	.1261	.1358	.1528	.1730	.1946	.2151	.2279	.2526	.2816
	150	.0913	.1084	.1252	.1348	.1518	.1719	.1933	.2137	.2264	.2510	.2799
	151	.0906	.1076	.1244	.1339	.1507	.1707	.1921	.2123	.2249	.2494	.2781
	152	.0900	.1069	.1235	.1330	.1497	.1696	.1908	.2110	.2235	.2479	.2764
	153	.0894	.1062	.1227	.1321	.1487	.1685	.1896	.2096	.2221	.2463	.2747
	154	.0888	.1055	.1219	.1312	.1477	.1674	.1884	.2083	.2207	.2448	.2731
	155	.0882	.1048	.1211	.1304	.1468	.1663	.1871	.2070	.2193	.2433	.2714
	156	.0876	.1041	.1203	.1295	.1458	.1652	.1860	.2057	.2179	.2418	.2698
	157	.0870	.1034	.1195	.1287	.1449	.1642	.1848	.2044	.2166	.2403	.2682

***To approximate p_f, compute $m = r/n$ and $s = \sqrt{m(1-m)/(n+1)}$, find z_f in Table 3, and compute $p_f \doteq m + z_f s$.

Table 5 (continued)

		f										
r	*n*	.001	.01	.05	.1	.25	.5	.75	.9	.95	.99	.999
26	158	.0864	.1027	.1187	.1278	.1440	.1631	.1836	.2031	.2152	.2389	.2666
	159	.0858	.1020	.1179	.1270	.1430	.1621	.1825	.2019	.2139	.2374	.2650
	160	.0853	.1014	.1172	.1262	.1421	.1611	.1813	.2006	.2126	.2360	.2634
	161	.0847	.1007	.1164	.1254	.1412	.1601	.1802	.1994	.2113	.2346	.2619
	162	.0842	.1001	.1157	.1246	.1403	.1591	.1791	.1982	.2101	.2332	.2604
	163	.0836	.0994	.1150	.1238	.1395	.1581	.1780	.1970	.2088	.2318	.2589
	164	.0831	.0988	.1142	.1230	.1386	.1571	.1770	.1958	.2076	.2305	.2574
	165	.0826	.0982	.1135	.1223	.1378	.1562	.1759	.1947	.2064	.2291	.2559
	166	.0821	.0975	.1128	.1215	.1369	.1552	.1748	.1935	.2051	.2278	.2545
	167	.0815	.0969	.1121	.1208	.1361	.1543	.1738	.1924	.2039	.2265	.2530
	168	.0810	.0963	.1114	.1200	.1353	.1534	.1728	.1913	.2028	.2252	.2516
	169	.0805	.0958	.1108	.1193	.1344	.1525	.1718	.1901	.2016	.2239	.2502
	170	.0800	.0952	.1101	.1186	.1336	.1516	.1708	.1891	.2004	.2227	.2488
	171	.0795	.0946	.1094	.1179	.1328	.1507	.1698	.1880	.1993	.2214	.2474
	172	.0791	.0940	.1088	.1172	.1321	.1498	.1688	.1869	.1982	.2202	.2461
	173	.0786	.0935	.1081	.1165	.1313	.1489	.1678	.1858	.1971	.2190	.2447
	174	.0781	.0929	.1075	.1158	.1305	.1481	.1669	.1848	.1959	.2177	.2434
	175	.0776	.0924	.1069	.1151	.1298	.1472	.1659	.1837	.1949	.2165	.2421
	176	.0772	.0918	.1062	.1145	.1290	.1464	.1650	.1827	.1938	.2154	.2408
	177	.0767	.0913	.1056	.1138	.1283	.1456	.1641	.1817	.1927	.2142	.2395
	178	.0763	.0907	.1050	.1131	.1276	.1447	.1631	.1807	.1917	.2130	.2382
	179	.0758	.0902	.1044	.1125	.1268	.1439	.1622	.1797	.1906	.2119	.2370
	180	.0754	.0897	.1038	.1119	.1261	.1431	.1613	.1787	.1896	.2108	.2357
	181	.0750	.0892	.1032	.1112	.1254	.1423	.1605	.1778	.1886	.2096	.2345
	182	.0745	.0887	.1026	.1106	.1247	.1415	.1596	.1768	.1875	.2085	.2333
	183	.0741	.0882	.1021	.1100	.1240	.1408	.1587	.1759	.1865	.2074	.2320
	184	.0737	.0877	.1015	.1094	.1233	.1400	.1579	.1749	.1856	.2063	.2309
	185	.0733	.0872	.1009	.1088	.1227	.1392	.1570	.1740	.1846	.2053	.2297
	186	.0729	.0867	.1004	.1082	.1220	.1385	.1562	.1731	.1836	.2042	.2285
	187	.0725	.0862	.0998	.1076	.1213	.1377	.1553	.1722	.1826	.2031	.2273
	188	.0721	.0858	.0993	.1070	.1207	.1370	.1545	.1713	.1817	.2021	.2262
	189	.0717	.0853	.0987	.1064	.1200	.1363	.1537	.1704	.1808	.2011	.2250
	190	.0713	.0848	.0982	.1059	.1194	.1356	.1529	.1695	.1798	.2000	.2239
	191	.0709	.0844	.0977	.1053	.1188	.1349	.1521	.1686	.1789	.1990	.2228
	192	.0705	.0839	.0972	.1047	.1181	.1341	.1513	.1677	.1780	.1980	.2217
	193	.0701	.0835	.0966	.1042	.1175	.1335	.1505	.1669	.1771	.1970	.2206
	194	.0697	.0830	.0961	.1036	.1169	.1328	.1498	.1660	.1762	.1961	.2195
	195	.0694	.0826	.0956	.1031	.1163	.1321	.1490	.1652	.1753	.1951	.2185
	196+	13.03*	15.62*	18.22*	19.72*	22.40*	25.67*	29.23*	32.71*	34.92*	39.31*	44.64*
27	28–41	SEE NOTE**										
	42–124	SEE NOTE***										
	125	.1168	.1378	.1584	.1700	.1904	.2145	.2400	.2640	.2788	.3075	.3407
	126	.1158	.1367	.1571	.1686	.1889	.2128	.2381	.2619	.2767	.3052	.3382
	127	.1148	.1355	.1558	.1672	.1873	.2111	.2362	.2599	.2746	.3029	.3357
	128	.1139	.1344	.1545	.1659	.1858	.2094	.2344	.2580	.2725	.3007	.3333
	129	.1129	.1333	.1533	.1645	.1844	.2078	.2326	.2560	.2705	.2985	.3309
	130	.1120	.1323	.1520	.1632	.1829	.2062	.2308	.2541	.2685	.2963	.3286
	131	.1111	.1312	.1508	.1619	.1815	.2046	.2291	.2522	.2665	.2941	.3263
	132	.1102	.1302	.1496	.1607	.1801	.2031	.2274	.2503	.2645	.2920	.3240
	133	.1093	.1291	.1485	.1594	.1787	.2015	.2257	.2485	.2626	.2900	.3217
	134	.1085	.1281	.1473	.1582	.1774	.2000	.2240	.2467	.2607	.2879	.3195
	135	.1076	.1271	.1462	.1570	.1760	.1985	.2224	.2449	.2589	.2859	.3172
	136	.1068	.1261	.1451	.1558	.1747	.1970	.2207	.2432	.2570	.2839	.3152
	137	.1059	.1252	.1440	.1546	.1734	.1956	.2192	.2414	.2552	.2819	.3130
	138	.1051	.1242	.1429	.1535	.1721	.1942	.2176	.2397	2534	.2800	.3109
	139	.1043	.1233	.1418	.1524	.1709	.1928	.2160	.2380	.2517	.2781	.3088
	140	.1035	.1224	.1408	.1512	.1696	.1914	.2145	.2364	.2499	.2762	.3068
	141	.1028	.1215	.1398	.1501	.1684	.1900	.2130	.2347	.2482	.2743	.3048
	142	.1020	.1206	.1387	.1490	.1672	.1887	.2115	.2331	.2465	.2725	.3028
	143	.1013	.1197	.1377	.1480	.1660	.1874	.2100	.2315	.2448	.2707	.3008
	144	.1005	.1188	.1367	.1469	.1648	.1861	.2086	.2300	.2432	.2689	.2989
	145	.0998	.1180	.1358	.1459	.1637	.1848	.2072	.2284	.2416	.2671	.2969
	146	.0991	.1171	.1348	.1448	.1625	.1835	.2058	.2269	.2400	.2654	.2950
	147	.0983	.1163	.1339	.1438	.1614	.1822	.2044	.2254	.2384	.2637	.2932
	148	.0976	.1155	.1329	.1428	.1603	.1810	.2030	.2239	.2368	.2620	.2913
	149	.0970	.1147	.1320	.1418	.1592	.1798	.2017	.2224	.2353	.2603	.2895
	150	.0963	.1139	.1311	.1409	.1581	.1786	.2003	.2210	.2338	.2586	.2877

*To approximate p_f, divide the starred entry under f by $(n - 1)$.
**Compute $r^* = n - r$, find p^*_{1-f} for r^* and n, and compute $p_f = 1 - p^*_{1-f}$.

Table 5 (continued)

r	n	f: .001	.01	.05	.1	.25	.5	.75	.9	.95	.99	.999
27	151	.0956	.1131	.1302	.1399	.1571	.1774	.1990	.2195	.2323	.2570	.2859
	152	.0949	.1123	.1293	.1390	.1560	.1762	.1977	.2181	.2308	.2554	.2841
	153	.0943	.1115	.1284	.1380	.1550	.1751	.1964	.2167	.2293	.2538	.2824
	154	.0936	.1108	.1276	.1371	.1539	.1739	.1952	.2154	.2279	.2522	.2807
	155	.0930	.1100	.1267	.1362	.1529	.1728	.1939	.2140	.2264	.2507	.2790
	156	.0924	.1093	.1259	.1353	.1519	.1717	.1927	.2126	.2250	.2491	.2773
	157	.0918	.1086	.1251	.1344	.1510	.1706	.1915	.2113	.2236	.2476	.2757
	158	.0912	.1079	.1243	.1336	.1500	.1695	.1903	.2100	.2223	.2461	.2740
	159	.0906	.1072	.1234	.1327	.1490	.1684	.1891	.2087	.2209	.2446	.2724
	160	.0900	.1065	.1226	.1318	.1481	.1674	.1879	.2074	.2196	.2432	.2708
	161	.0894	.1058	.1219	.1310	.1471	.1663	.1868	.2062	.2182	.2417	.2692
	162	.0888	.1051	.1211	.1302	.1462	.1653	.1856	.2049	.2169	.2403	.2677
	163	.0882	.1044	.1203	.1294	.1453	.1643	.1845	.2037	.2156	.2389	.2661
	164	.0877	.1038	.1196	.1286	.1444	.1633	.1834	.2025	.2144	.2375	.2646
	165	.0871	.1031	.1188	.1278	.1435	.1623	.1823	.2013	.2131	.2361	.2631
	166	.0866	.1025	.1181	.1270	.1427	.1613	.1812	.2001	.2118	.2347	.2616
	167	.0860	.1018	.1174	.1262	.1418	.1603	.1801	.1989	.2106	.2334	.2601
	168	.0855	.1012	.1166	.1254	.1409	.1594	.1790	.1978	.2094	.2321	.2587
	169	.0849	.1006	.1159	.1247	.1401	.1584	.1780	.1966	.2082	.2307	.2572
	170	.0844	.1000	.1152	.1239	.1392	.1575	.1769	.1955	.2070	.2294	.2558
	171	.0839	.0994	.1145	.1232	.1384	.1566	.1759	.1944	.2058	.2881	.2544
	172	.0834	.0988	.1138	.1224	.1376	.1556	.1749	.1932	.2046	.2269	.2530
	173	.0829	.0982	.1132	.1217	.1368	.1547	.1739	.1921	.2035	.2256	.2516
	174	.0824	.0976	.1125	.1210	.1360	.1538	.1729	.1911	.2024	.2244	.2502
	175	.0819	.0970	.1118	.1203	.1352	.1530	.1719	.1900	.2012	.2231	.2489
	176	.0814	.0964	.1112	.1196	.1344	.1521	.1710	.1889	.2001	.2219	.2476
	177	.0809	.0959	.1105	.1189	.1337	.1512	.1700	.1879	.1990	.2207	.2462
	178	.0805	.0953	.1099	.1182	.1329	.1504	.1691	.1868	.1979	.2195	.2449
	179	.0800	.0948	.1093	.1175	.1321	.1495	.1681	.1858	.1968	.2183	.2436
	180	.0795	.0942	.1087	.1169	.1314	.1487	.1672	.1848	.1958	.2172	.2423
	181	.0791	.0937	.1080	.1162	.1307	.1479	.1663	.1838	.1947	.2160	.2411
	182	.0786	.0931	.1074	.1156	.1299	.1471	.1654	.1828	.1937	.2149	.2398
	183	.0782	.0926	.1068	.1149	.1292	.1463	.1645	.1818	.1926	.2137	.2386
	184	.0777	.0921	.1062	.1143	.1285	.1455	.1636	.1809	.1916	.2126	.2374
	185	.0773	.0916	.1056	.1136	.1278	.1447	.1627	.1799	.1906	.2115	.2361
	186	.0769	.0911	.1051	.1130	.1271	.1439	.1618	.1789	.1896	.2104	.2349
	187	.0764	.0906	.1045	.1124	.1264	.1431	.1610	.1780	.1886	.2093	.2337
	188	.0760	.0901	.1039	.1118	.1257	.1424	.1601	.1771	.1876	.2083	.2326
	189	.0756	.0896	.1033	.1112	.1251	.1416	.1593	.1762	.1867	.2072	.2314
	190	.0752	.0891	.1028	.1106	.1244	.1408	.1585	.1752	.1857	.2062	.2302
	191	.0748	.0886	.1022	.1100	.1237	.1401	.1576	.1743	.1848	.2051	.2291
	192	.0744	.0881	.1017	.1094	.1231	.1394	.1568	.1735	.1838	.2041	.2280
	193	.0740	.0877	.1012	.1088	.1224	.1387	.1560	.1726	.1829	.2031	.2268
	194	.0736	.0872	.1006	.1083	.1218	.1379	.1552	.1717	.1820	.2020	.2257
	195	.0732	.0867	.1001	.1077	.1212	.1372	.1544	.1708	.1811	.2010	.2246
	196	.0728	.0863	.0996	.1071	.1205	.1365	.1536	.1700	.1802	.2001	.2235
	197	.0724	.0858	.0990	.1066	.1199	.1358	.1529	.1691	.1793	.1991	.2225
	198	.0720	.0854	.0985	.1060	.1193	.1351	.1521	.1683	.1784	.1981	.2214
	199	.0716	.0849	.0980	.1055	.1187	.1345	.1513	.1674	.1775	.1971	.2203
	200	.0713	.0845	.0975	.1050	.1181	.1338	.1506	.1666	.1766	.1962	.2193
	201	.0709	.0841	.0970	.1044	.1175	.1331	.1498	.1658	.1758	.1952	.2182
	202	.0705	.0836	.0965	.1039	.1169	.1325	.1491	.1650	.1749	.1943	.2172
	203+	13.73*	16.40*	19.06*	20.59*	23.34*	26.67*	30.30*	33.84*	36.08*	40.53*	45.94*
28	29–42	SEE NOTE**										
	43–134	SEE NOTE***										
	135	.1133	.1333	.1528	.1638	.1831	.2060	.2301	.2529	.2670	.2942	.3258
	136	.1124	.1323	.1516	.1625	.1818	.2044	.2284	.2511	.2651	.2922	.3236
	137	.1115	.1312	.1505	.1613	.1804	.2029	.2268	.2493	.2632	.2901	.3214
	138	.1107	.1302	.1492	.1601	.1791	.2015	.2252	.2475	.2614	.2882	.3193
	139	.1098	.1293	.1482	.1589	.1778	.2000	.2236	.2458	.2596	.2862	.3171
	140	.1090	.1283	.1471	.1578	.1765	.1986	.2220	.2441	.2578	.2842	.3150
	141	.1082	.1273	.1460	.1566	.1752	.1972	.2204	.2424	.2560	.2823	.3130
	142	.1074	.1264	.1450	.1555	.1739	.1958	.2189	.2407	.2543	.2804	.3109
	143	.1066	.1255	.1439	.1543	.1727	.1944	.2174	.2391	.2525	.2786	.3089
	144	.1058	.1246	.1429	.1532	.1715	.1930	.2159	.2375	.2508	.2767	.3069
	145	.1050	.1237	.1419	.1522	.1703	.1917	.2144	.2359	.2492	.2749	.3049
	146	.1043	.1228	.1409	.1511	.1691	.1904	.2129	.2343	.2475	.2731	.3030
	147	.1035	.1219	.1399	.1500	.1679	.1891	.2115	.2327	.2459	.2714	.3011

***To approximate p_f, compute $m = r/n$ and $s = \sqrt{m(1-m)/(n+1)}$, find z_f in Table 3, and compute $p_f \doteq m + z_f s$.

Table 5 (continued)

						f						
r	n	.001	.01	.05	.1	.25	.5	.75	.9	.95	.99	.999
28	148	.1028	.1210	.1389	.1490	.1668	.1878	.2101	.2312	.2443	.2696	.2992
	149	.1021	.1202	.1379	.1480	.1656	.1865	.2087	.2297	.2427	.2679	.2973
	150	.1013	.1194	.1370	.1469	.1645	.1853	.2073	.2282	.2411	.2662	.2955
	151	.1006	.1185	.1360	.1459	.1634	.1840	.2060	.2267	.2396	.2645	.2936
	152	.0999	.1177	.1351	.1450	.1623	.1828	.2046	.2253	.2380	.2629	.2918
	153	.0993	.1169	.1342	.1440	.1612	.1816	.2033	.2238	.2365	.2612	.2900
	154	.0986	.1161	.1333	.1430	.1602	.1804	.2020	.2224	.2350	.2596	.2883
	155	.0979	.1154	.1324	.1421	.1591	.1793	.2007	.2210	.1336	.2580	.2865
	156	.0972	.1146	.1315	.1411	.1581	.1781	.1994	.2196	.2321	.2564	.2848
	157	.0966	.1138	.1307	.1402	.1570	.1770	.1982	.2182	.2307	.2549	.2831
	158	.0960	.1131	.1298	.1393	.1560	.1759	.1969	.2169	.2293	.2533	.2814
	159	.0953	.1123	.1290	.1384	.1550	.1747	.1957	.2155	.2279	.2518	.2798
	160	.0947	.1116	.1281	.1375	.1541	.1736	.1945	.2142	.2265	.2503	.2781
	161	.0941	.1109	.1273	.1366	.1531	.1726	.1933	.2129	.2251	.2488	.2765
	162	.0935	.1102	.1265	.1358	.1521	.1715	.1921	.2116	.2238	.2473	.2749
	163	.0929	.1095	.1257	.1349	.1512	.1704	.1909	.2104	.2224	.2459	.2733
	164	.0923	.1088	.1249	.1341	.1502	.1694	.1898	.2091	.2211	.2445	.2718
	165	.0917	.1081	.1241	.1333	.1493	.1684	.1886	.2079	.2198	.2430	.2702
	166	.0911	.1074	.1234	.1324	.1484	.1673	.1875	.2066	.2185	.2416	.2687
	167	.0905	.1067	.1226	.1316	.1475	.1663	.1864	.2054	.2173	.2402	.2672
	168	.0900	.1061	.1219	.1308	.1466	.1653	.1853	.2042	.2160	.2389	.2657
	169	.0894	.1054	.1211	.1300	.1457	.1644	.1842	.2030	.2148	.2375	.2642
	170	.0889	.1048	.1204	.1292	.1449	.1634	.1831	.2019	.2135	.2362	.2627
	171	.0883	.1041	.1197	.1285	.1440	.1624	.1821	.2007	.2123	.2349	.2613
	172	.0878	.1035	.1189	.1277	.1431	.1615	.1810	.1996	.2111	.2335	.2599
	173	.0872	.1029	.1182	.1269	.1423	.1605	.1800	.1984	.2099	.2322	.2584
	174	.0867	.1023	.1175	.1262	.1415	.1596	.1790	.1973	.2087	.2310	.2570
	175	.0862	.1017	.1168	.1255	.1407	.1587	.1779	.1962	.2076	.2297	.2557
	176	.0857	.1011	.1162	.1247	.1398	.1578	.1769	.1951	.2064	.2284	.2543
	177	.0852	.1005	.1155	.1240	.1390	.1569	.1759	.1940	.2053	.2272	.2529
	178	.0847	.0999	.1148	.1233	.1383	.1560	.1750	.1930	.2042	.2260	.2516
	179	.0842	.0993	.1142	.1226	.1375	.1551	.1740	.1919	.2031	.2248	.2503
	180	.0837	.0987	.1135	.1219	.1367	.1543	.1730	.1909	.2020	.2236	.2489
	181	.0832	.0982	.1129	.1212	.1359	.1534	.1721	.1898	.2009	.2224	.2476
	182	.0827	.0976	.1122	.1205	.1352	.1526	.1711	.1888	.1998	.2212	.2464
	183	.0823	.0971	.1116	.1199	.1344	.1517	.1702	.1878	.1987	.2200	.2451
	184	.0818	.0965	.1110	.1192	.1337	.1509	.1693	.1868	.1977	.2189	.2438
	185	.0813	.0960	.1104	.1185	.1329	.1501	.1684	.1858	.1966	.2178	.2426
	186	.0809	.0955	.1098	.1179	.1322	.1493	.1675	.1848	.1956	.2166	.2413
	187	.0804	.0949	.1092	.1172	.1315	.1485	.1666	.1838	.1946	.2155	.2401
	188	.0800	.0944	.1086	.1166	.1308	.1477	.1657	.1829	.1936	.2144	.2389
	189	.0795	.0939	.1080	.1160	.1301	.1469	.1648	.1819	.1926	.2133	.2377
	190	.0791	.0934	.1074	.1153	.1294	.1461	.1640	.1810	.1916	.2122	.2365
	191	.0787	.0929	.1068	.1147	.1287	.1454	.1631	.1801	.1906	.2112	.2353
	192	.0783	.0924	.1062	.1141	.1280	.1446	.1623	.1791	.1896	.2101	.2342
	193	.0778	.0919	.1057	.1135	.1274	.1439	.1615	.1782	.1887	.2090	.2330
	194	.0774	.0914	.1051	.1129	.1267	.1431	.1606	.1773	.1877	.2080	.2319
	195	.0770	.0909	.1046	.1123	.1260	.1424	.1598	.1764	.1868	.2070	.2308
	196	.0766	.0904	.1040	.1117	.1254	.1416	.1590	.1755	.1859	.2060	.2296
	197	.0762	.0899	.1035	.1112	.1247	.1409	.1582	.1747	.1849	.2050	.2285
	198	.0758	.0895	.1029	.1106	.1241	.1402	.1574	.1738	.1840	.2040	.2274
	199	.0754	.0890	.1024	.1100	.1235	.1395	.1566	.1729	.1831	.2030	.2264
	200	.0750	.0886	.1019	.1095	.1229	.1388	.1558	.1721	.1822	.2020	.2253
	201	.0746	.0881	.1014	.1089	.1222	.1381	.1551	.1713	.1813	.2010	.2242
	202	.0742	.0877	.1008	.1084	.1216	.1374	.1543	.1704	.1805	.2001	.2232
	203	.0738	.0872	.1003	.1078	.1210	.1367	.1536	.1696	.1796	.1991	.2221
	204	.0735	.0868	.0998	.1073	.1204	.1361	.1528	.1688	.1787	.1982	.2211
	205	.0731	.0863	.0993	.1067	.1198	.1354	.1521	.1680	.1779	.1972	.2200
	206	.0727	.0859	.0988	.1062	.1192	.1347	.1513	.1672	.1770	.1963	.2190
	207	.0724	.0855	.0984	.1057	.1186	.1341	.1506	.1664	.1762	.1954	.2180
	208	.0720	.0850	.0979	.1052	.1181	.1334	.1499	.1656	.1754	.1945	.2170
	209+	14.44*	17.17*	19.90*	21.47*	24.27*	27.67*	31.36*	34.96*	37.23*	41.76*	47.23*
29	30–43	SEE NOTE**										
	44–145	SEE NOTE***										
	146	.1095	.1285	.1469	.1574	.1757	.1973	.2201	.2417	.2550	.2809	.3109
	147	.1088	.1276	.1459	.1563	.1744	.1959	.2186	.2401	.2534	.2790	.3089
	148	.1080	.1267	.1449	.1552	.1732	.1946	.2172	.2385	.2517	.2772	.3070
	149	.1072	.1258	.1439	.1541	.1721	.1933	.2157	.2369	.2501	.2755	.3051
	150	.1065	.1249	.1429	.1530	.1709	.1920	.2143	.2354	.2484	.2737	.3032

*To approximate p_f, divide the starred entry under f by $(n - 1)$.

**Compute $r^* = n - r$, find p^*_{1-f} for r^* and n, and compute $p_f = 1 - p^*_{1-f}$.

Table 5 (continued)

r	n	.001	.01	.05	.1	.25	f .5	.75	.9	.95	.99	.999
29	151	.1057	.1240	.1419	.1520	.1697	.1907	.2129	.2339	.2469	.2720	.3013
	152	.1050	.1232	.1409	.1510	.1686	.1894	.2115	.2324	.2453	.2703	.2995
	153	.1043	.1223	.1400	.1500	.1675	.1882	.2101	.2309	.2437	.2686	.2976
	154	.1035	.1215	.1390	.1490	.1664	.1870	.2088	.2294	.2422	.2670	.2958
	155	.1028	.1207	.1381	.1480	.1653	.1857	.2074	.2280	.2407	.2653	.2940
	156	.1021	.1199	.1372	.1470	.1642	.1846	.2061	.2265	.2392	.2637	.2923
	157	.1015	.1191	.1363	.1460	.1631	.1834	.2048	.2251	.2377	.2621	.2905
	158	.1008	.1183	.1354	.1451	.1621	.1822	.2035	.2237	.2362	.2605	.2888
	159	.1001	.1175	.1345	.1441	.1611	.1811	.2023	.2224	.2348	.2589	.2871
	160	.0995	.1168	.1337	.1432	.1600	.1799	.2010	.2210	.2334	.2574	.2854
	161	.0988	.1160	.1328	.1423	.1590	.1788	.1998	.2197	.2320	.2559	.2838
	162	.0982	.1153	.1320	.1414	.1580	.1777	.1986	.2183	.2306	.2544	.2821
	163	.0975	.1145	.1311	.1405	.1570	.1766	.1974	.2170	.2292	.2529	.2805
	164	.0969	.1138	.1303	.1396	.1561	.1755	.1962	.2157	.2278	.2514	.2789
	165	.0963	.1131	.1295	.1388	.1551	.1744	.1950	.2144	.2265	.2499	.2773
	166	.0957	.1124	.1287	.1379	.1542	.1734	.1938	.2132	.2252	.2485	.2758
	167	.0951	.1117	.1279	.1371	.1532	.1723	.1927	.2119	.2239	.2471	.2742
	168	.0945	.1110	.1271	.1362	.1523	.1713	.1915	.2107	.2226	.2457	.2727
	169	.0939	.1103	.1263	.1354	.1514	.1703	.1904	.2095	.2213	.2443	.2711
	170	.0933	.1096	.1256	.1346	.1505	.1693	.1893	.2083	.2200	.2429	.2696
	171	.0928	.1090	.1248	.1338	.1496	.1683	.1882	.2071	.2188	.2415	.2682
	172	.0922	.1083	.1241	.1330	.1487	.1673	.1871	.2059	.2175	.2402	.2667
	173	.0916	.1077	.1233	.1322	.1478	.1663	.1860	.2047	.2163	.2389	.2652
	174	.0911	.1070	.1226	.1314	.1470	.1654	.1850	.2036	.2151	.2375	.2638
	175	.0905	.1064	.1219	.1307	.1461	.1644	.1839	.2024	.2139	.2362	.2624
	176	.0900	.1058	.1212	.1299	.1453	.1635	.1829	.2013	.2127	.2349	.2610
	177	.0895	.1051	.1205	.1291	.1444	.1626	.1819	.2002	.2116	.2337	.2596
	178	.0889	.1045	.1198	.1284	.1436	.1617	.1809	.1991	.2104	.2324	.2582
	179	.0884	.1039	.1191	.1277	.1428	.1608	.1799	.1980	.2093	.2312	.2569
	180	.0879	.1033	.1184	.1269	.1420	.1599	.1789	.1969	.2081	.2299	.2555
	181	.0874	.1027	.1177	.1262	.1412	.1590	.1779	.1958	.2070	.2287	.2542
	182	.0869	.1021	.1171	.1255	.1404	.1581	.1769	.1948	.2059	.2275	.2529
	183	.0864	.1016	.1164	.1248	.1396	.1572	.1760	.1937	.2048	.2263	.2515
	184	.0859	.1010	.1157	.1241	.1389	.1564	.1750	.1927	.2037	.2251	.2503
	185	.0854	.1004	.1151	.1234	.1381	.1555	.1741	.1917	.2026	.2240	.2490
	186	.0849	.0999	.1145	.1228	.1374	.1547	.1731	.1907	.2016	.2228	.2477
	187	.0845	.0993	.1138	.1221	.1366	.1538	.1722	.1897	.2005	.2216	.2465
	188	.0840	.0988	.1132	.1214	.1359	.1530	.1713	.1887	.1995	.2205	.2452
	189	.0835	.0982	.1126	.1208	.1351	.1522	.1704	.1877	.1985	.2194	.2440
	190	.0831	.0977	.1120	.1201	.1344	.1514	.1695	.1867	.1974	.2183	.2428
	191	.0826	.0972	.1114	.1195	.1337	.1506	.1686	.1858	.1964	.2172	.2416
	192	.0822	.0966	.1108	.1188	.1330	.1498	.1678	.1848	.1954	.2161	.2404
	193	.0817	.0961	.1102	.1182	.1323	.1490	.1669	.1839	.1944	.2150	.2392
	194	.0813	.0956	.1096	.1176	.1316	.1483	.1660	.1829	.1935	.2139	.2380
	195	.0809	.0951	.1090	.1170	.1309	.1475	.1652	.1820	.1925	.2129	.2369
	196	.0804	.0946	.1085	.1164	.1303	.1468	.1644	.1811	.1915	.2118	.2357
	197	.0800	.0941	.1079	.1158	.1296	.1460	.1635	.1802	.1906	.2108	.2346
	198	.0796	.0936	.1074	.1152	.1289	.1453	.1627	.1793	.1896	.2098	.2335
	199	.0792	.0931	.1068	.1146	.1283	.1445	.1619	.1784	.1887	.2088	.2323
	200	.0788	.0926	.1063	.1140	.1276	.1438	.1611	.1776	.1878	.2077	.2312
	201	.0783	.0922	.1057	.1134	.1270	.1431	.1603	.1767	.1869	.2068	.2301
	202	.0779	.0917	.1052	.1128	.1263	.1424	.1595	.1758	.1860	.2058	.2291
	203	.0775	.0912	.1046	.1123	.1257	.1417	.1587	.1750	.1851	.2048	.2280
	204	.0772	.0908	.1041	.1117	.1251	.1410	.1580	.1741	.1842	.2038	.2269
	205	.0768	.0903	.1036	.1111	.1245	.1403	.1572	.1733	.1833	.2029	.2259
	206	.0764	.0899	.1031	.1106	.1239	.1396	.1564	.1725	.1824	.2019	.2248
	207	.0760	.0894	.1026	.1100	.1232	.1389	.1557	.1716	.1816	.2010	.2238
	208	.0756	.0890	.1021	.1095	.1226	.1383	.1549	.1708	.1807	.2000	.2228
	209	.0752	.0885	.1016	.1090	.1221	.1376	.1542	.1700	.1799	.1991	.2218
	210	.0749	.0881	.1011	.1084	.1215	.1369	.1535	.1692	.1790	.1982	.2207
	211	.0745	.0877	.1006	.1079	.1209	.1363	.1528	.1684	.1782	.1973	.2197
	212	.0741	.0872	.1001	.1074	.1203	.1356	.1520	.1677	.1774	.1964	.2188
	213	.0738	.0868	.0996	.1069	.1197	.1350	.1513	.1669	.1766	.1955	.2178
	214	.0734	.0864	.0991	.1064	.1192	.1344	.1506	.1661	.1758	.1946	.2168
	215+	15.15*	17.96*	20.75*	22.35*	25.21*	28.67*	32.43*	36.08*	38.39*	42.98*	48.52*
30	31–44	SEE NOTE**										
	45–156	SEE NOTE***										
	157	.1064	.1244	.1420	.1519	.1692	.1898	.2115	.2320	.2447	.2693	.2979
	158	.1057	.1236	.1410	.1509	.1682	.1886	.2102	.2306	.2432	.2677	.2962

***To approximate p_f, compute $m = r/n$ and $s = \sqrt{m(1 - m)/(n + 1)}$, find z_f in Table 3, and compute $p_f \doteq m + z_f s$.

Table 5 (continued)

r	n	.001	.01	.05	.1	.25	f .5	.75	.9	.95	.99	.999
30	159	.1050	.1228	.1401	.1499	.1671	.1874	.2089	.2292	.2417	.2660	.2944
	160	.1043	.1220	.1392	.1489	.1660	.1862	.2076	.2278	.2402	.2645	.2927
	161	.1036	.1212	.1383	.1480	.1650	.1850	.2063	.2264	.2388	.2629	.2910
	162	.1029	.1204	.1374	.1470	.1639	.1839	.2050	.2250	.2374	.2613	.2893
	163	.1023	.1196	.1366	.1461	.1629	.1828	.2038	.2237	.2360	.2598	.2876
	164	.1016	.1189	.1357	.1452	.1619	.1816	.2025	.2223	.2346	.2583	.2860
	165	.1010	.1181	.1349	.1443	.1609	.1805	.2013	.2210	.2332	.2568	.2844
	166	.1003	.1174	.1340	.1434	.1599	.1794	.2001	.2197	.2318	.2553	.2828
	167	.0997	.1167	.1332	.1425	.1590	.1784	.1989	.2184	.2305	.2539	.2812
	168	.0991	.1159	.1324	.1417	.1580	.1773	.1978	.2171	.2291	.2524	.2796
	169	.0984	.1152	.1316	.1408	.1570	.1762	.1966	.2159	.2278	.2510	.2781
	170	.0978	.1145	.1308	.1400	.1561	.1752	.1955	.2146	.2265	.2496	.2765
	171	.0972	.1138	.1300	.1391	.1552	.1742	.1943	.2134	.2252	.2482	.2750
	172	.0966	.1131	.1292	.1383	.1543	.1732	.1932	.2122	.2240	.2468	.2735
	173	.0961	.1124	.1284	.1375	.1534	.1722	.1921	.2110	.2227	.2454	.2720
	174	.0955	.1118	.1277	.1367	.1525	.1712	.1910	.2098	.2214	.2441	.2705
	175	.0949	.1111	.1269	.1359	.1516	.1702	.1899	.2086	.2202	.2427	.2691
	176	.0943	.1105	.1262	.1351	.1507	.1692	.1888	.2075	.2190	.2414	.2676
	177	.0938	.1098	.1254	.1343	.1498	.1682	.1878	.2063	.2178	.2401	.2662
	178	.0932	.1092	.1247	.1335	.1490	.1673	.1867	.2052	.2166	.2388	.2648
	179	.0927	.1085	.1240	.1327	.1481	.1664	.1859	.2041	.2154	.2375	.2634
	180	.0921	.1079	.1233	.1320	.1473	.1654	.1847	.2029	.2143	.2363	.2620
	181	.0916	.1073	.1226	.1312	.1465	.1645	.1837	.2018	.2131	.2350	.2607
	182	.0911	.1067	.1219	.1305	.1457	.1636	.1827	.2008	.2120	.2338	.2593
	183	.0906	.1061	.1212	.1298	.1449	.1627	.1817	.1997	.2108	.2325	.2580
	184	.0901	.1055	.1205	.1291	.1441	.1618	.1807	.1986	.2097	.2313	.2567
	185	.0895	.1049	.1199	.1283	.1433	.1609	.1797	.1976	.2086	.2301	.2553
	186	.0890	.1043	.1192	.1276	.1425	.1601	.1788	.1965	.2075	.2289	.2540
	187	.0885	.1037	.1185	.1269	.1417	.1592	.1778	.1955	.2064	.2278	.2528
	188	.0881	.1032	.1179	.1263	.1410	.1584	.1769	.1945	.2054	.2266	.2515
	189	.0876	.1026	.1173	.1256	.1402	.1575	.1760	.1935	.2043	.2254	.2502
	190	.0871	.1020	.1166	.1249	.1394	.1567	.1750	.1925	.2033	.2243	.2490
	191	.0866	.1015	.1160	.1242	.1387	.1559	.1741	.1915	.2022	.2232	.2477
	192	.0861	.1009	.1154	.1236	.1380	.1551	.1732	.1905	.2012	.2221	.2465
	193	.0857	.1004	.1148	.1229	.1373	.1542	.1723	.1895	.2002	.2209	.2453
	194	.0852	.0999	.1142	.1223	.1365	.1535	.1715	.1886	.1992	.2199	.2441
	195	.0848	.0993	.1136	.1216	.1358	.1527	.1706	.1876	.1982	.2188	.2429
	196	.0843	.0988	.1130	.1210	.1351	.1519	.1697	.1867	.1972	.2177	.2418
	197	.0839	.0983	.1124	.1204	.1344	.1511	.1689	.1857	.1962	.2166	.2406
	198	.0834	.0978	.1118	.1197	.1337	.1503	.1680	.1848	.1952	.2146	.2394
	199	.0830	.0973	.1112	.1191	.1331	.1496	.1672	.1839	.1943	.2145	.2383
	200	.0825	.0968	.1106	.1185	.1324	.1488	.1663	.1830	.1933	.2135	.2372
	201	.0821	.0963	.1101	.1179	.1317	.1481	.1655	.1821	.1924	.2125	.2360
	202	.0817	.0958	.1095	.1173	.1311	.1474	.1647	.1812	.1915	.2115	.2349
	203	.0813	.0953	.1090	.1167	.1304	.1466	.1639	.1803	.1905	.2104	.2338
	204	.0809	.0948	.1084	.1161	.1298	.1459	.1631	.1795	.1896	.2095	.2327
	205	.0805	.0943	.1079	.1156	.1291	.1452	.1623	.1786	.1887	.2085	.2317
	206	.0800	.0938	.1073	.1150	.1285	.1445	.1615	.1778	.1878	.2075	.2306
	207	.0796	.0934	.1068	.1144	.1279	.1438	.1608	.1769	.1869	.2065	.2295
	208	.0792	.0929	.1063	.1139	.1272	.1431	.1600	.1761	.1861	.2056	.2285
	209	.0789	.0925	.1058	.1133	.1266	.1424	.1592	.1752	.1852	.2046	.2274
	210	.0785	.0920	.1052	.1128	.1260	.1417	.1585	.1744	.1843	.2037	.2264
	211	.0781	.0916	.1047	.1122	.1254	.1410	.1577	.1736	.1835	.2027	.2254
	212	.0777	.0911	.1042	.1117	.1248	.1404	.1570	.1728	.1826	.2018	.2244
	213	.0773	.0907	.1037	.1111	.1242	.1397	.1563	.1720	.1818	.2009	.2234
	214	.0769	.0902	.1032	.1106	.1236	.1391	.1555	.1712	.1810	.2000	.2224
	215	.0766	.0898	.1027	.1101	.1230	.1384	.1548	.1704	.1801	.1991	.2214
	216	.0762	.0894	.1023	.1096	.1225	.1378	.1541	.1697	.1793	.1982	.2204
	217	.0758	.0889	.1018	.1091	.1219	.1371	.1534	.1689	.1785	.1973	.2195
	218	.0755	.0885	.1013	.1085	.1213	.1365	.1527	.1681	.1777	.1964	.2185
	219	.0751	.0881	.1008	.1080	.1208	.1359	.1520	.1674	.1769	.1956	.2175
	220	.0748	.0877	.1004	.1075	.1202	.1353	.1513	.1666	.1761	.1947	.2166
	221	.0744	.0873	.0999	.1070	.1197	.1346	.1506	.1659	.1754	.1939	.2157
	222+	15.87*	18.74*	21.59*	23.23*	26.15*	29.67*	33.49*	37.20*	39.54*	44.19	49.80*
31	32–46	SEE NOTE**										
	47–169	SEE NOTE***										
	170	.1024	.1194	.1360	.1453	.1617	.1811	.2016	.2210	.2330	.2562	.2834

*To approximate p_f, divide the starred entry under f by $(n - 1)$.
**Compute $r^* = n - r$, find p^*_{1-f} for r^* and n, and compute $p_f = 1 - p^*_{1-f}$.

Table 5 (continued)

r	n	.001	.01	.05	.1	.25	f .5	.75	.9	.95	.99	.999
31	171	.1018	.1187	.1352	.1445	.1608	.1800	.2004	.2197	.2317	.2548	.2818
	172	.1011	.1180	.1344	.1436	.1598	.1790	.1993	.2185	.2303	.2534	.2803
	173	.1005	.1173	.1336	.1427	.1589	.1780	.1981	.2172	.2291	.2520	.2787
	174	.0999	.1166	.1328	.1419	.1580	.1769	.1970	.2160	.2278	.2506	.2772
	175	.0993	.1159	.1320	.1411	.1571	.1759	.1959	.2148	.2265	.2492	.2757
	176	.0987	.1152	.1312	.1403	.1561	.1749	.1948	.2136	.2253	.2479	.2743
	177	.0981	.1145	.1304	.1394	.1552	.1739	.1937	.2124	.2240	.2465	.2728
	178	.0976	.1138	.1297	.1386	.1544	.1729	.1926	.2113	.2228	.2452	.2714
	179	.0970	.1132	.1289	.1378	.1535	.1720	.1916	.2101	.2216	.2439	.2699
	180	.0964	.1125	.1282	.1371	.1526	.1710	.1905	.2090	.2204	.2426	.2685
	181	.0959	.1119	.1274	.1363	.1518	.1701	.1895	.2078	.2192	.2413	.2671
	182	.0953	.1112	.1268	.1355	.1509	.1691	.1884	.2067	.2180	.2400	.2657
	183	.0948	.1106	.1260	.1348	.1501	.1682	.1874	.2056	.2169	.2388	.2644
	184	.0942	.1100	.1253	.1340	.1493	.1673	.1864	.2045	.2157	.2375	.2630
	185	.0937	.1094	.1246	.1333	.1484	.1664	.1854	.2034	.2146	.2363	.2617
	186	.0932	.1088	.1240	.1325	.1476	.1655	.1844	.2023	.2135	.2351	.2603
	187	.0926	.1082	.1233	.1318	.1468	.1646	.1834	.2013	.2123	.2339	.2590
	188	.0921	.1076	.1226	.1311	.1460	.1637	.1825	.2002	.2112	.2327	.2577
	189	.0916	.1070	.1219	.1304	.1453	.1628	.1815	.1992	.2102	.2315	.2564
	190	.0911	.1064	.1213	.1297	.1445	.1620	.1806	.1982	.2091	.2303	.2552
	191	.0906	.1058	.1206	.1290	.1437	.1611	.1796	.1971	.2080	.2291	.2539
	192	.0901	.1052	.1200	.1283	.1430	.1603	.1787	.1961	.2070	.2280	.2527
	193	.0896	.1047	.1193	.1276	.1422	.1594	.1778	.1951	.2059	.2269	.2514
	194	.0892	.1041	.1187	.1270	.1415	.1586	.1769	.1941	.2049	.2257	.2502
	195	.0887	.1036	.1181	.1263	.1407	.1578	.1760	.1932	.2039	.2246	.2490
	196	.0882	.1030	.1175	.1256	.1400	.1570	.1751	.1922	.2028	.2235	.2478
	197	.0877	.1025	.1169	.1250	.1393	.1562	.1742	.1912	.2018	.2224	.2466
	198	.0873	.1019	.1162	.1243	.1386	.1554	.1733	.1903	.2008	.2213	.2454
	199	.0868	.1014	.1156	.1237	.1379	.1546	.1724	.1894	.1999	.2203	.2442
	200	.0864	.1009	.1151	.1231	.1372	.1538	.1716	.1884	.1989	.2192	.2431
	201	.0859	.1004	.1145	.1224	.1365	.1531	.1707	.1875	.1979	.2182	.2419
	202	.0855	.0999	.1139	.1218	.1358	.1523	.1699	.1866	.1970	.2171	.2408
	203	.0850	.0994	.1133	.1212	.1351	.1516	.1691	.1857	.1960	.2161	.2397
	204	.0846	.0988	.1127	.1206	.1344	.1508	.1682	.1848	.1951	.2151	.2385
	205	.0842	.0984	.1122	.1200	.1338	.1501	.1674	.1839	.1941	.2141	.2374
	206	.0837	.0979	.1116	.1194	.1331	.1494	.1666	.1830	.1932	.2131	.2363
	207	.0833	.0974	.1111	.1188	.1325	.1486	.1658	.1822	.1923	.2121	.2353
	208	.0829	.0969	.1105	.1182	.1318	.1479	.1650	.1813	.1914	.2111	.2342
	209	.0825	.0964	.1100	.1177	.1312	.1472	.1643	.1804	.1905	.2101	.2331
	210	.0821	.0959	.1094	.1171	.1305	.1465	.1635	.1796	.1896	.2091	.2321
	211	.0817	.0955	.1089	.1165	.1299	.1458	.1627	.1788	.1887	.2082	.2310
	212	.0813	.0950	.1084	.1160	.1293	.1451	.1619	.1779	.1879	.2072	.2300
	213	.0809	.0945	.1079	.1154	.1287	.1444	.1612	.1771	.1870	.2063	.2290
	214	.0805	.0941	.1073	.1149	.1281	.1438	.1604	.1763	.1862	.2054	.2279
	215	.0801	.0936	.1068	.1143	.1275	.1431	.1597	.1755	.1853	.2044	.2269
	216	.0797	.0932	.1063	.1138	.1269	.1424	.1590	.1747	.1845	.2035	.2259
	217	.0793	.0927	.1058	.1132	.1263	.1418	.1582	.1739	.1836	.2026	.2249
	218	.0790	.0923	.1053	.1127	.1257	.1411	.1575	.1731	.1828	.2017	.2239
	219	.0786	.0919	.1048	.1122	.1251	.1405	.1568	.1723	.1820	.2008	.2230
	220	.0782	.0914	.1043	.1117	.1245	.1398	.1561	.1716	.1812	.1999	.2220
	221	.0778	.0910	.1039	.1111	.1240	.1392	.1554	.1708	.1804	.1991	.2210
	222	.0775	.0906	.1034	.1106	.1234	.1386	.1547	.1700	.1796	.1982	.2201
	223	.0771	.0902	.1029	.1101	.1228	.1379	.1540	.1693	.1788	.1973	.2192
	224	.0768	.0898	.1024	.1096	.1223	.1373	.1533	.1685	.1780	.1965	.2182
	225	.0764	.0893	.1020	.1091	.1217	.1367	.1526	.1678	.1772	.1956	.2173
	226	.0761	.0889	.1015	.1086	.1212	.1361	.1520	.1671	.1765	.1948	.2164
	227	.0757	.0885	.1011	.1082	.1207	.1355	.1513	.1664	.1757	.1940	.2155
	228+	16.59*	19.53*	22.44*	24.11*	27.09*	30.67*	34.55*	38.32*	40.69*	45.40*	51.08*
32	33–47	SEE NOTE**										
	48–182	SEE NOTE***										
	183	.0990	.1152	.1309	.1398	.1553	.1737	.1931	.2115	.2229	.2450	.2707
	184	.0984	.1145	.1302	.1390	.1545	.1727	.1921	.2104	.2217	.2437	.2694
	185	.0979	.1139	.1294	.1382	.1536	.1718	.1910	.2093	.2205	.2424	.2680
	186	.0973	.1133	.1287	.1375	.1528	.1709	.1900	.2082	.2194	.2412	.2666
	187	.0968	.1126	.1280	.1367	.1520	.1699	.1890	.2071	.2182	.2399	.2653
	188	.0962	.1120	.1273	.1360	.1511	.1690	.1880	.2060	.2171	.2387	.2639
	189	.0957	.1114	.1266	.1352	.1503	.1681	.1870	.2049	.2160	.2375	.2626
	190	.0952	.1108	.1259	.1345	.1495	.1673	.1861	.2039	.2149	.2363	.2613

***To approximate p_f, compute $m = r/n$ and $s = \sqrt{m(1-m)/(n+1)}$, find z_f in Table 3, and compute $p_f \doteq m + z_f s$.

Table 5 (continued)

							f					
r	*n*	.001	.01	.05	.1	.25	.5	.75	.9	.95	.99	.999
32	191	.0947	.1102	.1253	.1338	.1487	.1663	.1851	.2028	.2138	.2351	.2600
	192	.0941	.1096	.1246	.1331	.1479	.1655	.1841	.2018	.2127	.2339	.2587
	193	.0936	.1090	.1239	.1324	.1472	.1646	.1832	.2007	.2116	.2328	.2575
	194	.0931	.1084	.1233	.1317	.1464	.1638	.1823	.1997	.2106	.2316	.2562
	195	.0926	.1078	.1226	.1310	.1456	.1630	.1813	.1987	.2095	.2305	.2550
	196	.0921	.1073	.1220	.1303	.1449	.1621	.1804	.1977	.2085	.2293	.2537
	197	.0917	.1067	.1213	.1296	.1441	.1613	.1795	.1967	.2074	.2282	.2525
	198	.0912	.1061	.1207	.1289	.1434	.1605	.1786	.1958	.2064	.2271	.2513
	199	.0907	.1056	.1201	.1283	.1427	.1597	.1777	.1948	.2054	.2260	.2501
	200	.0902	.1050	.1195	.1276	.1419	.1589	.1768	.1938	.2044	.2249	.2489
	201	.0897	.1045	.1189	.1270	.1412	.1581	.1760	.1929	.2034	.2238	.2478
	202	.0893	.1040	.1183	.1263	.1405	.1573	.1751	.1920	.2024	.2228	.2466
	203	.0888	.1034	.1177	.1257	.1398	.1565	.1742	.1910	.2014	.2217	.2455
	204	.0884	.1029	.1171	.1251	.1391	.1557	.1734	.1901	.2005	.2207	.2443
	205	.0879	.1024	.1165	.1244	.1384	.1550	.1725	.1892	.1995	.2196	.2432
	206	.0875	.1019	.1159	.1238	.1378	.1542	.1717	.1883	.1986	.2186	.2421
	207	.0870	.1014	.1153	.1232	.1371	.1535	.1709	.1874	.1976	.2176	.2409
	208	.0866	.1009	.1148	.1226	.1364	.1527	.1701	.1865	.1967	.2166	.2398
	209	.0862	.1004	.1142	.1220	.1358	.1520	.1693	.1856	.1958	.2156	.2388
	210	.0857	.0999	.1136	.1214	.1351	.1513	.1685	.1848	.1949	.2146	.2377
	211	.0853	.0994	.1131	.1208	.1344	.1506	.1677	.1839	.1940	.2136	.2366
	212	.0849	.0989	.1125	.1203	.1338	.1498	.1669	.1831	.1931	.2126	.2355
	213	.0845	.0984	.1120	.1197	.1332	.1491	.1661	.1822	.1922	.2117	.2345
	214	.0841	.0980	.1115	.1191	.1325	.1484	.1653	.1814	.1913	.2107	.2335
	215	.0837	.0975	.1109	.1185	.1319	.1477	.1646	.1805	.1905	.2098	.2324
	216	.0833	.0970	.1104	.1180	.1313	.1471	.1638	.1797	.1896	.2088	.2314
	217	.0829	.0966	.1099	.1174	.1307	.1464	.1631	.1789	.1887	.2079	.2304
	218	.0825	.0961	.1094	.1169	.1301	.1457	.1623	.1781	.1879	.2070	.2294
	219	.0821	.0956	.1089	.1163	.1295	.1450	.1616	.1773	.1871	.2061	.2284
	220	.0817	.0952	.1084	.1158	.1289	.1444	.1609	.1765	.1862	.2052	.2274
	221	.0813	.0948	.1079	.1153	.1283	.1437	.1601	.1757	.1854	.2043	.2264
	222	.0809	.0943	.1074	.1147	.1277	.1431	.1594	.1749	.1846	.2034	.2254
	223	.0806	.0939	.1069	.1142	.1271	.1424	.1587	.1742	.1838	.2025	.2245
	224	.0802	.0934	.1064	.1137	.1266	.1418	.1580	.1734	.1830	.2016	.2235
	225	.0798	.0930	.1059	.1132	.1260	.1412	.1573	.1726	.1822	.2007	.2226
	226	.0794	.0926	.1054	.1127	.1254	.1405	.1566	.1719	.1814	.1999	.2216
	227	.0791	.0922	.1049	.1122	.1249	.1399	.1559	.1711	.1806	.1990	.2207
	228	.0787	.0918	.1045	.1117	.1243	.1393	.1553	.1704	.1798	.1982	.2198
	229	.0784	.0913	.1040	.1112	.1238	.1387	.1546	.1697	.1791	.1973	.2189
	230	.0780	.0909	.1035	.1107	.1232	.1381	.1539	.1689	.1783	.1965	.2179
	231	.0777	.0905	.1031	.1102	.1227	.1375	.1532	.1682	.1775	.1957	.2170
	232	.0773	.0901	.1026	.1097	.1221	.1369	.1526	.1675	.1768	.1949	.2162
	233	.0770	.0897	.1022	.1092	.1216	.1363	.1519	.1668	.1761	.1941	.2153
	234+	17.32*	20.32*	23.30*	25.00*	28.03*	31.67*	35.61*	39.43*	41.84*	46.61*	52.36*
33	34–48	SEE NOTE**										
	49–196	SEE NOTE***										
	197	.0956	.1109	.1259	.1343	.1490	.1664	.1848	.2022	.2130	.2340	.2585
	198	.0951	.1104	.1252	.1336	.1482	.1655	.1839	.2012	.2120	.2328	.2572
	199	.0946	.1098	.1246	.1329	.1475	.1647	.1830	.2002	.2109	.2317	.2560
	200	.0941	.1092	.1239	.1322	.1467	.1639	.1821	.1992	.2099	.2306	.2548
	201	.0936	.1087	.1233	.1315	.1460	.1631	.1812	.1983	.2089	.2295	.2536
	202	.0931	.1081	.1227	.1309	.1453	.1623	.1803	.1973	.2079	.2284	.2524
	203	.0926	.1076	.1220	.1302	.1445	.1615	.1794	.1964	.2069	.2273	.2512
	204	.0922	.1070	.1214	.1295	.1438	.1607	.1785	.1954	.2059	.2262	.2501
	205	.0917	.1065	.1208	.1289	I1431	.1599	.1776	.1945	.2049	.2252	.2489
	206	.0912	.1059	.1202	.1283	.1424	.1591	.1768	.1935	.2039	.2241	.2477
	207	.0908	.1054	.1196	.1276	.1417	.1583	.1759	.1926	.2030	.2231	.2466
	208	.0903	.1049	.1190	.1270	.1410	.1576	.1751	.1917	.2020	.2220	.2455
	209	.0899	.1044	.1184	.1264	.1403	.1568	.1743	.1908	.2011	.2210	.2444
	210	.0894	.1038	.1179	.1258	.1397	.1561	.1734	.1899	.2001	.2200	.2433
	211	.0890	.1033	.1173	.1252	.1390	.1553	.1726	.1890	.1992	.2190	.2422
	212	.0885	.1028	.1167	.1246	.1383	.1546	.1718	.1882	.1983	.2180	.2411
	213	.0881	.1023	.1162	.1240	.1377	.1538	.1710	.1873	.1974	.2170	.2400
	214	.0877	.1018	.1156	.1234	.1370	.1531	.1702	.1864	.1965	.2160	.2389
	215	.0873	.1014	.1151	.1228	.1364	.1524	.1694	.1856	.1956	.2151	.2379
	216	.0868	.1009	.1145	.1222	.1357	.1517	.1687	.1847	.1947	.2141	.2368
	217	.0864	.1004	.1140	.1216	.1351	.1510	.1679	.1839	.1938	.2132	.2358
	218	.0860	.0999	.1134	.1211	.1345	.1503	.1671	.1831	.1930	.2122	.2348

*To approximate p_f, divide the starred entry under f by $(n - 1)$.
**Compute $r^* = n - r$, find p^*_{1-f} for r^* and n, and compute $p_f = 1 - p^*_{1-f}$.

Table 5 (continued)

r	n	.001	.01	.05	.1	.25	f .5	.75	.9	.95	.99	.999
33	219	.0856	.0994	.1129	.1205	.1338	.1496	.1664	.1822	.1921	.2113	.2338
	220	.0852	.0990	.1124	.1199	.1332	.1489	.1656	.1814	.1913	.2103	.2327
	221	.0848	.0985	.1119	.1194	.1326	.1483	.1649	.1806	.1904	.2094	.2317
	222	.0844	.0981	.1113	.1188	.1320	.1476	.1641	.1798	.1896	.2085	.2307
	223	.0840	.0976	.1108	.1183	.1314	.1469	.1634	.1790	.1887	.2076	.2298
	224	.0836	.0972	.1103	.1178	.1308	.1463	.1627	.1782	.1879	.2067	.2288
	225	.0832	.0967	.1098	.1172	.1302	.1456	.1620	.1775	.1871	.2058	.2278
	226	.0828	.0963	.1093	.1167	.1296	.1450	.1613	.1767	.1863	.2049	.2269
	227	.0825	.0958	.1088	.1162	.1291	.1443	.1605	.1759	.1855	.2041	.2259
	228	.0821	.0954	.1083	.1156	.1285	.1437	.1598	.1752	.1847	.2032	.2250
	229	.0817	.0950	.1079	.1151	.1279	.1431	.1592	.1744	.1839	.2023	.2240
	230	.0814	.0945	.1074	.1146	.1274	.1424	.1585	.1737	.1831	.2015	.2231
	231	.0810	.0941	.1069	.1141	.1268	.1418	.1578	.1729	.1823	.2007	.2222
	232	.0806	.0937	.1064	.1136	.1263	.1512	.1571	.1722	.1816	.1998	.2213
	233	.0803	.0933	.1060	.1131	.1257	.1406	.1564	.1715	.1808	.1990	.2203
	234	.0799	.0929	.1055	.1126	.1252	.1400	.1558	.1707	.1800	.1982	.2194
	235	.0796	.0925	.1050	.1121	.1246	.1394	.1551	.1700	.1793	.1973	.2186
	236	.0792	.0921	.1046	.1117	.1241	.1388	.1545	.1693	.1786	.1965	.2177
	237	.0789	.0917	.1041	.1112	.1236	.1382	.1538	.1686	.1778	.1957	.2168
	238	.0785	.0913	.1037	.1107	.1230	.1376	.1532	.1679	.1771	.1949	.2159
	239	.0782	.0909	.1032	.1102	.1225	.1371	.1525	.1672	.1764	.1941	.2151
	240+	18.05*	21.12*	24.25*	25.89*	28.97*	32.67*	36.67*	40.54*	42.98*	47.81*	53.63*
34	35–50	SEE NOTE**										
	51–212	SEE NOTE***										
	213	.0918	.1063	.1203	.1283	.1422	.1586	.1759	.1924	.2026	.2223	.2455
	214	.0913	.1058	.1198	.1276	.1415	.1578	.1751	.1915	.2016	.2213	.2444
	215	.0909	.1052	.1192	.1270	.1408	.1571	.1743	.1906	.2007	.2204	.2433
	216	.0904	.1047	.1186	.1264	.1402	.1563	.1735	.1897	.1998	.2194	.2423
	217	.0900	.1042	.1181	.1258	.1395	.1556	.1727	.1889	.1989	.2184	.2412
	218	.0896	.1037	.1175	.1253	.1389	.1549	.1719	.1880	.1980	.2174	.2402
	219	.0891	.1033	.1169	.1247	.1382	.1542	.1711	.1872	.1971	.2165	.2391
	220	.0887	.1028	.1164	.1241	.1376	.1535	.1704	.1864	.1963	.2155	.2381
	221	.0883	.1023	.1159	.1235	.1369	.1528	.1696	.1855	.1954	.2146	.2371
	222	.0879	.1018	.1153	.1229	.1363	.1521	.1688	.1847	.1945	.2136	.2360
	223	.0875	.1013	.1148	.1224	.1357	.1514	.1681	.1839	.1937	.2127	.2350
	224	.0871	.1009	.1143	.1218	.1351	.1507	.1674	.1831	.1928	.2118	.2340
	225	.0867	.1004	.1138	.1213	.1345	.1501	.1666	.1823	.1920	.2109	.2330
	226	.0863	.1000	.1132	.1207	.1339	.1494	.1659	.1815	.1912	.2100	.2321
	227	.0859	.0995	.1127	.1202	.1333	.1488	.1652	.1807	.1903	.2091	.2311
	228	.0855	.0991	.1122	.1197	.1327	.1481	.1644	.1799	.1895	.2082	.2301
	229	.0851	.0986	.1117	.1191	.1321	.1474	.1637	.1791	.1887	.2073	.2292
	230	.0847	.0982	.1112	.1186	.1315	.1468	.1630	.1784	.1879	.2065	.2282
	231	.0843	.0977	.1107	.1181	.1309	.1462	.1623	.1776	.1871	.2056	.2273
	232	.0840	.0973	.1102	.1175	.1304	.1455	.1616	.1769	.1863	.2047	.2263
	233	.0836	.0969	.1098	.1170	.1298	.1449	.1609	.1761	.1855	.2039	.2254
	234	.0832	.0964	.1093	.1165	.1293	.1443	.1602	.1754	.1848	.2030	.2245
	235	.0828	.0960	.1088	.1160	.1287	.1437	.1596	.1746	.1840	.2022	.2236
	236	.0825	.0956	.1083	.1155	.1281	.1431	.1589	.1739	.1832	.2014	.2227
	237	.0821	.0952	.1079	.1150	.1276	.1425	.1582	.1732	.1825	.2006	.2218
	238	.0818	.0948	.1074	.1145	.1271	.1419	.1576	.1725	.1817	.1997	.2209
	239	.0814	.0944	.1069	.1140	.1265	.1413	.1569	.1718	.1810	.1989	.2200
	240	.0811	.0940	.1065	.1136	.1260	.1407	.1563	.1711	.1802	.1981	.2191
	241	.0807	.0936	.1060	.1131	.1255	.1401	.1556	.1704	.1795	.1973	.2183
	242	.0804	.0932	.1056	.1126	.1249	.1395	.1550	.1697	.1788	.1965	.2174
	243	.0800	.0928	.1051	.1121	.1244	.1389	.1544	.1690	.1781	.1958	.2165
	244	.0797	.0924	.1047	.1117	.1239	.1384	.1537	.1683	.1774	.1950	.2157
	245	.0793	.0920	.1043	.1112	.1234	.1378	.1531	.1676	.1766	.1942	.2148
	246	.0790	.0916	.1038	.1107	.1229	.1372	.1525	.1670	.1759	.1934	.2140
	247+	18.78*	21.92*	25.01*	26.77*	29.91*	33.67*	37.73*	41.65*	44.12*	49.01*	54.90*
35	36–51	SEE NOTE**										
	52–230	SEE NOTE***										
	231	.0877	.1014	.1146	.1220	.1351	.1505	.1668	.1823	.1919	.2105	.2323
	232	.0873	.1009	.1141	.1215	.1345	.1499	.1661	.1815	.1911	.2096	.2314
	233	.0869	.1005	.1136	.1210	.1339	.1492	.1654	.1808	.1903	.2088	.2304
	234	.0865	.1000	.1131	.1204	.1333	.1486	.1647	.1800	.1895	.2079	.2295
	235	.0862	.0996	.1126	.1199	.1328	.1479	.1640	.1792	.1887	.2071	.2286
	236	.0858	.0991	.1121	.1194	.1322	.1473	.1633	.1785	.1879	.2062	.2277
	237	.0854	.0987	.1116	.1189	.1316	.1467	.1626	.1778	.1871	.2054	.2267

***To approximate p_f, compute $m = r/n$ and $s = \sqrt{m(1 - m)/(n + 1)}$, find z_f in Table 3, and compute $p_f \doteq m + z_f s$.

Table 5 (continued)

							f					
r	*n*	.001	.01	.05	.1	.25	.5	.75	.9	.95	.99	.999
35	238	.0850	.0983	.1111	.1184	.1311	.1461	.1620	.1770	.1864	.2045	.2258
	239	.0847	.0979	.1107	.1179	.1305	.1455	.1613	.1763	.1856	.2037	.2249
	240	.0843	.0974	.1102	.1174	.1300	.1448	.1606	.1756	.1848	.2029	.2240
	241	.0839	.0970	.1097	.1169	.1294	.1442	.1600	.1749	.1841	.2021	.2231
	242	.0836	.0966	.1093	.1164	.1289	.1436	.1593	.1741	.1833	.2013	.2223
	243	.0832	.0962	.1088	.1159	.1283	.1431	.1587	.1734	.1826	.2005	.2214
	244	.0829	.0958	.1083	.1154	.1278	.1425	.1580	.1727	.1819	.1997	.2205
	245	.0825	.0954	.1079	.1149	.1273	.1419	.1574	.1720	.1812	.1989	.2197
	246	.0822	.0950	.1074	.1145	.1268	.1413	.1567	.1714	.1804	.1981	.2188
	247	.0818	.0946	.1070	.1140	.1262	.1407	.1561	.1707	.1797	.1973	.2180
	248	.0815	.0942	.1065	.1135	.1257	.1402	.1555	.1700	.1790	.1965	.2171
	249	.0811	.0938	.1061	.1130	.1252	.1396	.1549	.1693	.1783	.1958	.2163
	250	.0808	.0934	.1057	.1126	.1247	.1390	.1542	.1687	.1776	.1950	.2155
	251	.0805	.0930	.1052	.1121	.1242	.1385	.1536	.1680	.1769	.1943	.2146
	252	.0801	.0927	.1048	.1117	.1237	.1379	.1530	.1673	.1762	.1935	.2138
	253+	19.52*	22.72*	25.87*	27.66*	30.85*	34.67*	38.79*	42.76*	45.27*	50.21*	56.16*
36	37–52	SEE NOTE**										
	53–249	SEE NOTE***										
	250	.0839	.0968	.1092	.1162	.1285	.1431	.1584	.1730	.1820	.1996	.2202
	251	.0836	.0964	.1088	.1158	.1280	.1425	.1578	.1723	.1813	.1988	.2193
	252	.0832	.0960	.1083	.1153	.1275	.1419	.1572	.1716	.1806	.1980	.2185
	253	.0829	.0956	.1079	.1148	.1270	.1413	.1566	.1710	.1799	.1973	.2177
	254	.0826	.0952	.1075	.1144	.1265	.1408	.1559	.1703	.1792	.1965	.2168
	255	.0822	.0948	.1070	.1139	.1260	.1402	.1553	.1696	.1785	.1958	.2160
	256	.0819	.0944	.1066	.1135	.1255	.1397	.1547	.1690	.1778	.1950	.2152
	257	.0816	.0941	.1062	.1130	.1250	.1391	.1541	.1683	.1772	.1943	.2144
	258	.0812	.0937	.1058	.1126	.1245	.1386	.1535	.1677	.1765	.1936	.2136
	259+	20.26*	23.53*	26.73*	28.56*	31.79*	35.67*	29.85*	43.87*	46.40*	51.41*	57.42*
37	38–53	SEE NOTE**										
	54–264	SEE NOTE***										
	265+	21.00*	24.33*	27.59*	29.45*	32.74*	36.67*	40.90*	44.98*	47.54*	52.60*	58.67*
38	39–55	SEE NOTE**										
	56–271	SEE NOTE***										
	272+	21.75*	25.14*	28.46*	30.34*	33.68*	37.67*	41.96*	46.08*	48.68*	53.79*	59.93*
39	40–56	SEE NOTE**										
	57–277	SEE NOTE***										
	278+	22.51*	25.96*	29.33*	31.24*	34.63*	38.67*	43.01*	47.19*	49.81*	54.98*	61.17*
40	41–57	SEE NOTE**										
	58–283	SEE NOTE***										
	284+	23.26*	26.77*	30.20*	32.14*	35.57*	39.67*	44.07*	48.29*	50.94*	56.16*	62.42*
41	42–58	SEE NOTE**										
	59–289	SEE NOTE***										
	290+	24.02*	27.59*	31.07*	33.04*	36.52*	40.67*	45.12*	49.39*	52.07*	57.35*	63.66*
42	43–59	SEE NOTE**										
	60–295	SEE NOTE***										
	296+	24.78*	28.41*	31.94*	33.94*	37.47*	41.67*	46.17*	50.49*	53.20*	58.53*	64.90*
43	44–61	SEE NOTE**										
	62–302	SEE NOTE***										
	303+	25.54*	29.23*	32.81*	34.84*	38.41*	42.67*	47.22*	51.59*	54.32*	59.71*	66.14*
44	45–62	SEE NOTE**										
	63–308	SEE NOTE***										
	309+	26.31*	30.05*	33.69*	35.74*	39.36*	43.67*	48.27*	52.69*	55.45*	60.88*	67.37*
45	46–63	SEE NOTE**										
	64–314	SEE NOTE***										
	315+	27.08*	30.88*	34.56*	36.65*	40.31*	44.67*	49.32*	53.78*	56.57*	62.06*	68.60*
46	47–64	SEE NOTE**										
	65–320	SEE NOTE***										
	321+	27.85*	31.70*	35.44*	37.55*	41.26*	45.67*	50.38*	54.88*	57.69*	63.23*	69.83*
47	48–66	SEE NOTE**										
	67–326	SEE NOTE***										
	327+	28.62*	32.53*	36.32*	38.46*	42.21*	46.67*	51.42*	55.97*	58.82*	64.40*	71.06*
48	49–67	SEE NOTE**										
	68–332	SEE NOTE***										
	333+	29.40*	33.36*	37.20*	39.36*	43.16*	47.67*	52.47*	57.07*	59.94*	65.57*	72.28*

*To approximate p_f, divide the starred entry under f by $(n - 1)$.
**Compute $r^* = n - r$, find p^*_{1-f} for r^* and n, and compute $p_f = 1 - p^*_{1-f}$.

Table 5 (continued)

							f					
r	n	.001	.01	.05	.1	.25	.5	.75	.9	.95	.99	.999
49	50–68	SEE NOTE**										
	69–338	SEE NOTE***										
	339+	30.18*	34.20*	38.08*	40.27*	44.11*	48.67*	53.52*	58.16*	61.05*	66.74*	73.51*
50	51–69	SEE NOTE**										
	70–345	SEE NOTE***										
	346+	30.96*	35.03*	38.96*	41.18*	45.07*	49.67*	54.57*	59.25*	62.17*	67.90*	74.71*
51	52–70	SEE NOTE**										
	71–351	SEE NOTE***										
	352+	31.74*	35.87*	39.85*	42.09*	46.02*	50.67*	55.62*	60.34*	63.29*	69.07*	75.94*
52	53–72	SEE NOTE**										
	73–357	SEE NOTE***										
	358+	32.53*	36.71*	40.73*	43.00*	46.97*	51.67*	56.67*	61.43*	64.40*	70.23*	77.16*
53	54–73	SEE NOTE**										
	74–363	SEE NOTE***										
	364+	33.31*	37.55*	41.62*	43.91*	47.93*	52.67*	57.71*	62.52*	65.52*	71.39*	78.37*
54	55–74	SEE NOTE**										
	75–369	SEE NOTE***										
	370+	34.10*	38.39*	42.51*	44.82*	48.88*	53.67*	58.76*	63.61*	66.63*	72.55*	79.58*
55	56–75	SEE NOTE**										
	76–375	SEE NOTE***										
	376+	34.89*	39.23*	43.40*	45.74*	49.83*	54.67*	59.80*	64.69*	67.74*	73.71*	80.79*
56	57–76	SEE NOTE**										
	77–381	SEE NOTE***										
	382+	35.69*	40.07*	44.29*	46.65*	50.79*	55.67*	60.85*	65.78*	68.85*	74.86*	82.00*
57	58–78	SEE NOTE**										
	79–387	SEE NOTE***										
	388+	36.48*	40.92*	45.18*	47.56*	51.74*	56.67*	61.89*	66.86*	69.96*	76.02*	83.20*
58	59–79	SEE NOTE**										
	80–393	SEE NOTE***										
	394+	37.28*	41.76*	46.07*	48.48*	52.70*	57.67*	62.94*	67.95*	71.07*	77.17*	84.41*
59	60–80	SEE NOTE**										
	81–399	SEE NOTE***										
	400+	38.08*	42.61*	46.96*	49.40*	53.65*	58.67*	63.98*	69.03*	72.18*	78.32*	85.61*
60	61–81	SEE NOTE**										
	82–405	SEE NOTE***										
	406+	38.88*	43.46*	47.85*	50.31*	54.61*	59.67*	65.03*	70.12*	73.28*	79.48*	86.81*
61	62–82	SEE NOTE**										
	83–411	SEE NOTE***										
	412+	39.68*	44.31*	48.75*	51.23*	55.57*	60.67*	66.07*	71.20*	74.39*	80.62*	88.01*
62	63–83	SEE NOTE**										
	84–417	SEE NOTE***										
	418+	40.48*	45.16*	49.64*	52.15*	56.52*	61.67*	67.11*	72.28*	75.49*	81.77*	89.20*
63	64–85	SEE NOTE**										
	86–423	SEE NOTE***										
	424+	41.29*	46.02*	50.54*	53.07*	57.48*	62.67*	68.16*	73.36*	76.60*	82.92*	90.40*
64	65–86	SEE NOTE**										
	87–429	SEE NOTE***										
	430+	42.09*	46.87*	51.43*	53.99*	58.44*	63.67*	69.20*	74.44*	77.70*	84.07*	91.59*
65	66–87	SEE NOTE**										
	88–435	SEE NOTE***										
	436+	42.90*	47.73*	52.33*	54.91*	59.40*	64.67*	70.24*	75.52*	78.80*	85.21*	92.79*
66	67–88	SEE NOTE**										
	89–442	SEE NOTE***										
	443+	43.71*	48.58*	53.23*	55.83*	60.35*	65.67*	71.28*	76.60*	79.91*	86.36*	93.98*
67	68–89	SEE NOTE**										
	90–447	SEE NOTE***										
	448+	44.52*	49.44*	54.13*	56.75*	61.31*	66.67*	72.32*	77.68*	81.01*	87.50*	95.17*
68	69–90	SEE NOTE**										
	91–454	SEE NOTE***										
	455+	45.33*	50.30*	55.03*	57.67*	62.27*	67.67*	73.37*	78.76*	82.11*	88.64*	96.35*
69	70–92	SEE NOTE**										
	93–460	SEE NOTE***										
	461+	46.15*	51.16*	55.93*	58.59*	63.23*	68.67*	74.41*	79.84*	83.21*	89.78*	97.54*

***To approximate p_f, compute $m = r/n$ and $s = \sqrt{m(1-m)/(n+1)}$, find z_f in Table 3, and compute $p_f \doteq m + z_f s$.

Table 5 (continued)

r	n	.001	.01	.05	.1	.25	*f* .5	.75	.9	.95	.99	.999
70	71-93	SEE NOTE**										
	94-466	SEE NOTE***										
	467+	46.96*	52.02*	56.83*	59.51*	64.19*	69.67*	75.45*	80.91*	84.31*	90.92*	98.73*
71	72-94	SEE NOTE**										
	95-472	SEE NOTE***										
	473+	47.78*	52.88*	57.73*	60.44*	65.15*	70.67*	76.49*	81.99*	85.40*	92.06*	99.91*
72	73-95	SEE NOTE**										
	96-478	SEE NOTE***										
	479+	48.60*	53.74*	58.63*	61.36*	66.11*	71.67*	77.53*	83.07*	86.50*	93.20*	101.1*
73	74-96	SEE NOTE**										
	97-484	SEE NOTE***										
	485+	49.42*	54.60*	59.54*	62.29*	67.07*	72.67*	78.57*	84.14*	87.60*	94.33*	102.3*
74	75-97	SEE NOTE**										
	98-490	SEE NOTE***										
	491+	50.24*	55.47*	60.44*	63.21*	68.03*	73.67*	79.61*	85.22*	88.70*	95.47*	103.5*
75	76-98	SEE NOTE**										
	99-496	SEE NOTE***										
	497+	51.06*	56.33*	61.35*	64.14*	68.99*	74.67*	80.65*	86.29*	89.79*	96.60*	104.6*
76	77-100	SEE NOTE**										
	101-502	SEE NOTE***										
	503+	51.88*	57.20*	62.25*	65.06*	69.95*	75.67*	81.68*	87.36*	90.89*	97.74*	105.8*
77	78-101	SEE NOTE**										
	102-508	SEE NOTE***										
	509+	52.70*	58.07*	63.16*	65.99*	70.91*	76.67*	82.72*	88.44*	91.98*	98.87*	107.0*
78	79-102	SEE NOTE**										
	103-513	SEE NOTE***										
	514+	53.53*	58.93*	64.06*	66.92*	71.88*	77.67*	83.76*	89.51*	93.07*	100.0*	108.2*
79	80-103	SEE NOTE**										
	104-519	SEE NOTE***										
	520+	54.35*	59.80*	64.97*	67.84*	72.84*	78.67*	84.80*	90.58*	94.17*	101.1*	109.3*
80	81-104	SEE NOTE**										
	105-525	SEE NOTE***										
	526+	55.18*	60.67*	65.88*	68.77*	73.80*	79.67*	85.84*	91.66*	95.26*	102.3*	110.5*
81	82-105	SEE NOTE**										
	106-531	SEE NOTE***										
	532+	56.01*	61.54*	66.79*	69.70*	74.76*	80.67*	86.88*	92.73*	96.35*	103.4*	111.7*
82	83-106	SEE NOTE**										
	107-537	SEE NOTE***										
	538+	56.84*	62.41*	67.69*	70.63*	75.72*	81.67*	87.91*	93.80*	97.44*	104.5*	112.9*
83	84-108	SEE NOTE**										
	109-543	SEE NOTE***										
	544+	57.67*	63.29*	68.60*	71.56*	76.69*	82.67*	88.95*	94.87*	98.53*	105.7*	114.0*
84	85-109	SEE NOTE**										
	110-549	SEE NOTE***										
	550+	58.50*	64.16*	69.51*	72.49*	77.65*	83.67*	89.99*	95.94*	99.62*	106.8*	115.2*
85	86-110	SEE NOTE**										
	111-555	SEE NOTE***										
	556+	59.33*	65.03*	70.42*	73.42*	78.61*	84.67*	91.02*	97.01*	100.7*	107.9*	116.4*
86	87-111	SEE NOTE**										
	112-561	SEE NOTE***										
	562+	60.16*	65.91*	71.34*	74.35*	79.58*	85.67*	92.06*	98.08*	101.8*	109.0*	117.5*
87	88-112	SEE NOTE**										
	113-567	SEE NOTE***										
	568+	61.00*	66.78*	72.25*	75.28*	80.54*	86.67*	93.10*	99.15*	102.9*	110.2*	118.7*
88	89-113	SEE NOTE**										
	114-573	SEE NOTE***										
	574+	61.83*	67.66*	73.16*	76.21*	81.50*	87.67*	94.13*	100.2*	104.0*	111.3*	119.9*
89	90-114	SEE NOTE**										
	115-579	SEE NOTE***										
	580+	62.67*	68.53*	74.07*	77.14*	82.47*	88.67*	95.17*	101.3*	105.1*	112.4*	121.0*
90	91-116	SEE NOTE**										
	117-585	SEE NOTE***										
	586+	63.51*	69.41*	74.98*	78.08*	83.43*	89.67*	96.20*	102.4*	106.2*	113.5*	122.2*

*To approximate p_f, divide the starred entry under f by $(n - 1)$.

**Compute $r^* = n - r$, find p^*_{1-f} for r^* and n, and compute $p_f = 1 - p^*_{1-f}$.

Table 5 (continued)

r	n	.001	.01	.05	.1	.25	f .5	.75	.9	.95	.99	.999
91	92–117	SEE NOTE**										
	118–591	SEE NOTE***										
	592+	64.34*	70.29*	75.90*	79.01*	84.40*	90.67*	97.24*	103.4*	107.2*	114.7*	123.3*
92	93–118	SEE NOTE**										
	119–597	SEE NOTE***										
	598+	65.18*	71.17*	76.81*	79.94*	85.36*	91.67*	98.28*	104.5*	108.3*	115.8*	124.5*
93	94–119	SEE NOTE**										
	120–603	SEE NOTE**										
	604+	66.02*	72.05*	77.73*	80.88*	86.33*	92.67*	99.31*	105.6*	109.4*	116.9*	125.7*
94	95–120	SEE NOTE**										
	121–609	SEE NOTE***										
	610+	66.86*	72.92*	78.64*	81.81*	87.29*	93.67*	100.3*	106.6*	110.5*	118.0*	126.8*
95	96–121	SEE NOTE**										
	122–615	SEE NOTE***										
	616+	67.70*	73.81*	79.56*	82.74*	88.26*	94.67*	101.4*	107.7*	111.6*	119.1*	128.0*
96	97–122	SEE NOTE**										
	123–620	SEE NOTE***										
	621+	68.55*	74.69*	80.47*	83.68*	89.22*	95.67*	102.4*	108.8*	112.7*	120.3*	129.1*
97	98–123	SEE NOTE**										
	124–626	SEE NOTE***										
	627+	69.39*	75.57*	81.39*	84.61*	90.19*	96.67*	103.4*	109.8*	113.7*	121.4*	130.3*
98	99–125	SEE NOTE**										
	126–632	SEE NOTE***										
	633+	70.23*	76.45*	82.30*	85.55*	91.15*	97.67*	104.5*	110.9*	114.8*	122.5*	131.5*
99	100–126	SEE NOTE**										
	127–638	SEE NOTE***										
	639+	71.08*	77.33*	83.22*	86.48*	92.12*	98.67*	105.5*	111.9*	115.9*	123.6*	132.6*
100	101–127	SEE NOTE**										
	128–644	SEE NOTE***										
	645+	71.92*	78.22*	84.14*	87.42*	93.09*	99.67*	106.6*	113.0*	117.0*	124.7*	133.8*

$101 \leqslant r \leqslant 357$

$n/r < 1.28$	SEE NOTE**
$n/r \geqslant 1.28$	SEE NOTE***

$r > 357$

$n - r < 100$	SEE NOTE**
$n - r \geqslant 100$	SEE NOTE***

***To approximate p_f, compute $m = r/n$ and $s = \sqrt{m(1-m)/(n+1)}$, find z_f in Table 3, and compute $p_f \doteq m + z_f s$.

These Tables of "Beta Cumulative Functions" appear in *Analysis of Decisions under Uncertainty* by Robert Schlaifer, published by McGraw-Hill Book Company, Inc., 1969. They are reproduced here by specific permission of the copyright holder, The President and Fellows of Harvard College.

Table 6. Unit Normal Linear Loss Integral

This table gives values of the unit normal linear loss integral,

$$L_N(D) = \int_D^\infty (\theta - D) f_N(\theta)\, d\theta = \int_D^\infty (\theta - D) \frac{1}{\sqrt{2\pi}} e^{-\theta^2/2}\, d\theta,$$

for $\qquad D = 0(.01)4.99.$

To find $L_N(D)$ for $D < 0$, take $L_N(D) = -D + L_N(-D)$.

Examples:

$$L_N(3.57) = .0_4 4417 = .00004417$$

and $\qquad L_N(-3.57) = 3.57 + L_N(3.57) = 3.57004417.$

These Tables of "Unit Normal Loss Integral" appear in *Probability and Statistics for Business Decisions* by Robert Schlaifer, published by McGraw-Hill Book Company, Inc., 1959. They are reproduced here by specific permission of the copyright holder, The President and Fellows of Harvard College.

Table 6

D	.00	.01	.02	.03	.04	.05	.06	.07	.08	.09
.0	.3989	.3940	.3890	.3841	.3793	.3744	.3697	.3649	.3602	.3556
.1	.3509	.3464	.3418	.3373	.3328	.3284	.3240	.3197	.3154	.3111
.2	.3069	.3027	.2986	.2944	.2904	.2863	.2824	.2784	.2745	.2706
.3	.2668	.2630	.2592	.2555	.2518	.2481	.2445	.2409	.2374	.2339
.4	.2304	.2270	.2236	.2203	.2169	.2137	.2104	.2072	.2040	.2009
.5	.1978	.1947	.1917	.1887	.1857	.1828	.1799	.1771	.1742	.1714
.6	.1687	.1659	.1633	.1606	.1580	.1554	.1528	.1503	.1478	.1453
.7	.1429	.1405	.1381	.1358	.1334	.1312	.1289	.1267	.1245	.1223
.8	.1202	.1181	.1160	.1140	.1120	.1100	.1080	.1061	.1042	.1023
.9	.1004	.09860	.09680	.09503	.09328	.09156	.08986	.08819	.08654	.08491
1.0	.08332	.08174	.08019	.07866	.07716	.07568	.07422	.07279	.07138	.06999
1.1	.06862	.06727	.06595	.06465	.06336	.06210	.06086	.05964	.05844	.05726
1.2	.05610	.05496	.05384	.05274	.05165	.05059	.04954	.04851	.04750	.04650
1.3	.04553	.04457	.04363	.04270	.04179	.04090	.04002	.03916	.03831	.03748
1.4	.03667	.03587	.03508	.03431	.03356	.03281	.03208	.03137	.03067	.02998
1.5	.02931	.02865	.02800	.02736	.02674	.02612	.02552	.02494	.02436	.02380
1.6	.02324	.02270	.02217	.02165	.02114	.02064	.02015	.01967	.01920	.01874
1.7	.01829	.01785	.01742	.01699	.01658	.01617	.01578	.01539	.01501	.01464
1.8	.01428	.01392	.01357	.01323	.01290	.01257	.01226	.01195	.01164	.01134
1.9	.01105	.01077	.01049	.01022	$.0^{2}9957$	$.0^{2}9698$	$.0^{2}9445$	$.0^{2}9198$	$.0^{2}8957$	$.0^{2}8721$
2.0	$.0^{2}8491$	$.0^{2}8266$	$.0^{2}8046$	$.0^{2}7832$	$.0^{2}7623$	$.0^{2}7418$	$.0^{2}7219$	$.0^{2}7024$	$.0^{2}6835$	$.0^{2}6649$
2.1	$.0^{2}6468$	$.0^{2}6292$	$.0^{2}6120$	$.0^{2}5952$	$.0^{2}5788$	$.0^{2}5628$	$.0^{2}5472$	$.0^{2}5320$	$.0^{2}5172$	$.0^{2}5028$
2.2	$.0^{2}4887$	$.0^{2}4750$	$.0^{2}4616$	$.0^{2}4486$	$.0^{2}4358$	$.0^{2}4235$	$.0^{2}4114$	$.0^{2}3996$	$.0^{2}3882$	$.0^{2}3770$
2.3	$.0^{2}3662$	$.0^{2}3556$	$.0^{2}3453$	$.0^{2}3352$	$.0^{2}3255$	$.0^{2}3159$	$.0^{2}3067$	$.0^{2}2977$	$.0^{2}2889$	$.0^{2}2804$
2.4	$.0^{2}2720$	$.0^{2}2640$	$.0^{2}2561$	$.0^{2}2484$	$.0^{2}2410$	$.0^{2}2337$	$.0^{2}2267$	$.0^{2}2199$	$.0^{2}2132$	$.0^{2}2067$
2.5	$.0^{2}2004$	$.0^{2}1943$	$.0^{2}1883$	$.0^{2}1826$	$.0^{2}1769$	$.0^{2}1715$	$.0^{2}1662$	$.0^{2}1610$	$.0^{2}1560$	$.0^{2}1511$
2.6	$.0^{2}1464$	$.0^{2}1418$	$.0^{2}1373$	$.0^{2}1330$	$.0^{2}1288$	$.0^{2}1247$	$.0^{2}1207$	$.0^{2}1169$	$.0^{2}1132$	$.0^{2}1095$

Table 6 (continued)

D	.00	.01	.02	.03	.04	.05	.06	.07	.08	.09
2.7	$.0^{2}1060$	$.0^{2}1026$	$.0^{3}9928$	$.0^{3}9607$	$.0^{3}9295$	$.0^{3}8992$	$.0^{3}8699$	$.0^{3}8414$	$.0^{3}8138$	$.0^{3}7870$
2.8	$.0^{3}7611$	$.0^{3}7359$	$.0^{3}7115$	$.0^{3}6879$	$.0^{3}6650$	$.0^{3}6428$	$.0^{3}6213$	$.0^{3}6004$	$.0^{3}5802$	$.0^{3}5606$
2.9	$.0^{3}5417$	$.0^{3}5233$	$.0^{3}5055$	$.0^{3}4883$	$.0^{3}4716$	$.0^{3}4555$	$.0^{3}4398$	$.0^{3}4247$	$.0^{3}4101$	$.0^{3}3959$
3.0	$.0^{3}3822$	$.0^{3}3689$	$.0^{3}3560$	$.0^{3}3436$	$.0^{3}3316$	$.0^{3}3199$	$.0^{3}3087$	$.0^{3}2978$	$.0^{3}2873$	$.0^{3}2771$
3.1	$.0^{3}2673$	$.0^{3}2577$	$.0^{3}2485$	$.0^{3}2396$	$.0^{3}2311$	$.0^{3}2227$	$.0^{3}2147$	$.0^{3}2070$	$.0^{3}1995$	$.0^{3}1922$
3.2	$.0^{3}1852$	$.0^{3}1785$	$.0^{3}1720$	$.0^{3}1657$	$.0^{3}1596$	$.0^{3}1537$	$.0^{3}1480$	$.0^{3}1426$	$.0^{3}1373$	$.0^{3}1322$
3.3	$.0^{3}1273$	$.0^{3}1225$	$.0^{3}1179$	$.0^{3}1135$	$.0^{3}1093$	$.0^{3}1051$	$.0^{3}1012$	$.0^{4}9734$	$.0^{4}9365$	$.0^{4}9009$
3.4	$.0^{4}8666$	$.0^{4}8335$	$.0^{4}8016$	$.0^{4}7709$	$.0^{4}7413$	$.0^{4}7127$	$.0^{4}6852$	$.0^{4}6587$	$.0^{4}6331$	$.0^{4}6085$
3.5	$.0^{4}5848$	$.0^{4}5620$	$.0^{4}5400$	$.0^{4}5188$	$.0^{4}4984$	$.0^{4}4788$	$.0^{4}4599$	$.0^{4}4417$	$.0^{4}4242$	$.0^{4}4073$
3.6	$.0^{4}3911$	$.0^{4}3755$	$.0^{4}3605$	$.0^{4}3460$	$.0^{4}3321$	$.0^{4}3188$	$.0^{4}3059$	$.0^{4}2935$	$.0^{4}2816$	$.0^{4}2702$
3.7	$.0^{4}2592$	$.0^{4}2486$	$.0^{4}2385$	$.0^{4}2287$	$.0^{4}2193$	$.0^{4}2103$	$.0^{4}2016$	$.0^{4}1933$	$.0^{4}1853$	$.0^{4}1776$
3.8	$.0^{4}1702$	$.0^{4}1632$	$.0^{4}1563$	$.0^{4}1498$	$.0^{4}1435$	$.0^{4}1375$	$.0^{4}1317$	$.0^{4}1262$	$.0^{4}1208$	$.0^{4}1157$
3.9	$.0^{4}1108$	$.0^{4}1061$	$.0^{4}1016$	$.0^{5}9723$	$.0^{5}9307$	$.0^{5}8908$	$.0^{5}8525$	$.0^{5}8158$	$.0^{5}7806$	$.0^{5}7469$
4.0	$.0^{5}7145$	$.0^{5}6835$	$.0^{5}6538$	$.0^{5}6253$	$.0^{5}5980$	$.0^{5}5718$	$.0^{5}5468$	$.0^{5}5227$	$.0^{5}4997$	$.0^{5}4777$
4.1	$.0^{5}4566$	$.0^{5}4364$	$.0^{5}4170$	$.0^{5}3985$	$.0^{5}3807$	$.0^{5}3637$	$.0^{5}3475$	$.0^{5}3319$	$.0^{5}3170$	$.0^{5}3027$
4.2	$.0^{5}2891$	$.0^{5}2760$	$.0^{5}2635$	$.0^{5}2516$	$.0^{5}2402$	$.0^{5}2292$	$.0^{5}2188$	$.0^{5}2088$	$.0^{5}1992$	$.0^{5}1901$
4.3	$.0^{5}1814$	$.0^{5}1730$	$.0^{5}1650$	$.0^{5}1574$	$.0^{5}1501$	$.0^{5}1431$	$.0^{5}1365$	$.0^{5}1301$	$.0^{5}1241$	$.0^{5}1183$
4.4	$.0^{5}1127$	$.0^{5}1074$	$.0^{5}1024$	$.0^{6}9756$	$.0^{6}9296$	$.0^{6}8857$	$.0^{6}8437$	$.0^{6}8037$	$.0^{6}7655$	$.0^{6}7290$
4.5	$.0^{6}6942$	$.0^{6}6610$	$.0^{6}6294$	$.0^{6}5992$	.0 5704	$.0^{6}5429$	$.0^{6}5167$	$.0^{6}4917$	$.0^{6}4679$	$.0^{6}4452$
4.6	$.0^{6}4236$	$.0^{6}4029$	$.0^{6}3833$	$.0^{6}3645$	$.0^{6}3467$	$.0^{6}3297$	$.0^{6}3135$	$.0^{6}2981$	$.0^{6}2834$	$.0^{6}2694$
4.7	$.0^{6}2560$	$.0^{6}2433$	$.0^{6}2313$	$.0^{6}2197$	$.0^{6}2088$	$.0^{6}1984$	$.0^{6}1884$	$.0^{6}1790$	$.0^{6}1700$	$.0^{6}1615$
4.8	$.0^{6}1533$	$.0^{6}1456$	$.0^{6}1382$	$.0^{6}1312$	$.0^{6}1246$	$.0^{6}1182$	$.0^{6}1122$	$.0^{6}1065$	$.0^{6}1011$	$.0^{7}9588$
4.9	$.0^{7}9096$	$.0^{7}8629$	$.0^{7}8185$	$.0^{7}7763$	$.0^{7}7362$	$.0^{7}6982$	$.0^{7}6620$	$.0^{7}6276$	$.0^{7}5950$	$.0^{7}5640$

Table 7. Random Digits

10	09	73	25	33	76	52	01	35	86	34	67	35	48	76	80	95	90	91	17	39	29	27	49	45
37	54	20	48	05	64	89	47	42	96	24	80	52	40	37	20	63	61	04	02	00	82	29	16	65
08	42	26	89	53	19	64	50	93	03	23	20	90	25	60	15	95	33	47	64	35	08	03	36	06
99	01	90	25	29	09	37	67	07	15	38	31	13	11	65	88	67	67	43	97	04	43	62	76	59
12	80	79	99	70	80	15	73	61	47	64	03	23	66	53	98	95	11	68	77	12	17	17	68	33
66	06	57	47	17	34	07	27	68	50	36	69	73	61	70	65	81	33	98	85	11	19	92	91	70
31	06	01	08	05	45	57	18	24	06	35	30	34	26	14	86	79	90	74	39	23	40	30	97	32
85	26	97	76	02	02	05	16	56	92	68	66	57	48	18	73	05	38	52	47	18	62	38	85	79
63	57	33	21	35	05	32	54	70	48	90	55	35	75	48	28	46	82	87	09	83	49	12	56	24
73	79	64	57	53	03	52	96	47	78	35	80	83	42	82	60	93	52	03	44	35	27	38	84	35
98	52	01	77	67	14	90	56	86	07	22	10	94	05	58	60	97	09	34	33	50	50	07	39	98
11	80	50	54	31	39	80	82	77	32	50	72	56	82	48	29	40	52	42	01	52	77	56	78	51
83	45	29	96	34	06	28	89	80	83	13	74	67	00	78	18	47	54	06	10	68	71	17	78	17
88	68	54	02	00	86	50	75	84	01	36	76	66	79	51	90	36	47	64	93	29	60	91	10	62
99	59	46	73	48	87	51	76	49	69	91	82	60	89	28	93	78	56	13	68	23	47	83	41	13
65	48	11	76	74	17	46	85	09	50	58	04	77	69	74	73	03	95	71	86	40	21	81	65	44
80	12	43	56	35	17	72	70	80	15	45	31	82	23	74	21	11	57	82	53	14	38	55	37	63
74	35	09	98	17	77	40	27	72	14	43	23	60	02	10	45	52	16	42	37	96	28	60	26	55
69	91	62	68	03	66	25	22	91	48	36	93	68	72	03	76	62	11	39	90	94	40	05	64	18
09	89	32	05	05	14	22	56	85	14	46	42	75	67	88	96	29	77	88	22	54	38	21	45	98
91	49	91	45	23	68	47	92	76	86	46	16	28	35	54	94	75	08	99	23	37	08	92	00	48
80	33	69	45	98	26	94	03	68	58	70	29	73	41	35	53	14	03	33	40	42	05	08	23	41
44	10	48	19	49	85	15	74	79	54	32	97	92	65	75	57	60	04	08	81	22	22	20	64	13
12	55	07	37	42	11	10	00	20	40	12	86	07	46	97	96	64	48	94	39	28	70	72	58	15
63	60	64	93	29	16	50	53	44	84	40	21	95	25	63	43	65	17	70	82	07	20	73	17	90
61	19	69	04	46	26	45	74	77	74	51	92	43	37	29	65	39	45	95	93	42	58	26	05	27
15	47	44	52	66	95	27	07	99	53	59	36	78	38	48	82	39	61	01	18	33	21	15	94	66
94	55	72	85	73	67	89	75	43	87	54	62	24	44	31	91	19	04	25	92	92	92	74	59	73
42	48	11	62	13	97	34	40	87	21	16	86	84	87	67	03	07	11	20	29	25	70	14	66	70
23	52	37	83	17	73	20	88	98	37	68	93	59	14	16	26	25	22	96	63	05	52	28	25	62
04	49	35	24	94	75	24	63	38	24	45	86	25	10	25	61	96	27	93	35	65	33	71	24	72
00	54	99	76	54	64	05	18	81	59	96	11	96	38	96	54	69	28	23	91	23	28	72	95	29
35	96	31	53	07	26	89	80	93	54	33	35	13	54	62	77	97	45	00	24	90	10	33	93	33
59	80	80	83	91	45	42	72	68	42	83	60	94	97	00	13	02	12	48	92	78	56	52	01	06
46	05	88	52	36	01	39	09	22	86	77	28	14	40	77	93	91	08	36	47	70	61	74	29	41
32	17	90	05	97	87	37	92	52	41	05	56	70	70	07	86	74	31	71	57	85	39	41	18	38
69	23	46	14	06	20	11	74	52	04	15	95	66	00	00	18	74	39	24	23	97	11	89	63	38
19	56	54	14	30	01	75	87	53	79	40	41	92	15	85	66	67	43	68	06	84	96	28	52	07
45	15	51	49	38	19	47	60	72	46	43	66	79	45	43	59	04	79	00	33	20	82	66	95	41
94	86	43	19	94	36	16	81	08	51	34	88	88	15	53	01	54	03	54	56	05	01	45	11	76
98	08	62	48	26	45	24	02	84	04	44	99	90	88	96	39	09	47	34	07	35	44	13	18	80
33	18	51	62	32	41	94	15	09	49	89	43	54	85	81	88	69	54	19	94	37	54	87	30	43
80	95	10	04	06	96	38	27	07	74	20	15	12	33	87	25	01	62	52	98	94	62	46	11	71
79	75	24	91	40	71	96	12	82	96	69	86	10	25	91	74	85	22	05	39	00	38	75	95	79
18	63	33	25	37	98	14	50	65	71	31	01	02	46	74	05	45	56	14	27	77	93	89	19	36
74	02	94	39	02	77	55	73	22	70	97	79	01	71	19	52	52	75	80	21	80	81	45	17	48
54	17	84	56	11	80	99	33	71	43	05	33	51	29	69	56	12	71	92	55	36	04	09	03	24
11	66	44	98	83	52	07	98	48	27	59	38	17	15	39	09	97	33	34	40	88	46	12	33	56
48	32	47	79	28	31	24	96	47	10	02	29	53	68	70	32	30	75	75	46	15	02	00	99	94
69	07	49	41	38	87	63	79	19	76	35	58	40	44	01	10	51	82	16	15	01	84	87	69	38

This table is reproduced by permission from the Rand Corporation, *A Million Random Digits*. New York: The Free Press, 1975.

Table 8. Factorials of Integers

n	$n!$	n	$n!$
1	1	**26**	4.03291×10^{26}
2	2	**27**	1.08889×10^{28}
3	6	**28**	3.04888×10^{29}
4	24	**29**	8.84176×10^{30}
5	120	**30**	2.65253×10^{32}
6	720	**31**	8.22284×10^{33}
7	5040	**32**	2.63131×10^{35}
8	40320	**33**	8.68332×10^{36}
9	362880	**34**	2.95233×10^{38}
10	3.62880×10^{6}	**35**	1.03331×10^{40}
11	3.99168×10^{7}	**36**	3.71993×10^{41}
12	4.79002×10^{8}	**37**	1.37638×10^{43}
13	6.22702×10^{9}	**38**	5.23023×10^{44}
14	8.71783×10^{10}	**39**	2.03979×10^{46}
15	1.30767×10^{12}	**40**	8.15915×10^{47}
16	2.09228×10^{13}	**41**	3.34525×10^{49}
17	3.55687×10^{14}	**42**	1.40501×10^{51}
18	6.40327×10^{15}	**43**	6.04153×10^{52}
19	1.21645×10^{17}	**44**	2.65827×10^{54}
20	2.43290×10^{18}	**45**	1.19622×10^{56}
21	5.10909×10^{19}	**46**	5.50262×10^{57}
22	1.12400×10^{21}	**47**	2.58623×10^{59}
23	2.58520×10^{22}	**48**	1.24139×10^{61}
24	6.20448×10^{23}	**49**	6.08282×10^{62}
25	1.55112×10^{25}	**50**	3.04141×10^{64}

Table 9. Powers and Roots

N	N^2	$\sqrt{N}$	$\sqrt{10N}$
1	1	1.00 000	3.16 228
2	4	1.41 421	4.47 214
3	9	1.73 205	5.47 723
4	16	2.00 000	6.32 456
5	25	2.23 607	7.07 107
6	36	2.44 949	7.74 597
7	49	2.64 575	8.36 660
8	64	2.82 843	8.94 427
9	81	3.00 000	9.48 683
10	100	3.16 228	10.00 00
11	121	3.31 662	10.48 81
12	144	3.46 410	10.95 45
13	169	3.60 555	11.40 18
14	196	3.74 166	11.83 22
15	225	3.87 298	12.24 74
16	256	4.00 000	12.64 91
17	289	4.12 311	13.03 84
18	324	4.24 264	13.41 64
19	361	4.35 890	13.78 40
20	400	4.47 214	14.14 21
21	441	4.58 258	14.49 14
22	484	4.69 042	14.83 24
23	529	4.79 583	15.16 58
24	576	4.89 898	15.49 19
25	625	5.00 000	15.81 14
26	676	5.09 902	16.12 45
27	729	5.19 615	16.43 17
28	784	5.29 150	16.73 32
29	841	5.38 516	17.02 94
30	900	5.47 723	17.32 05
31	961	5.56 776	17.60 68
32	1 024	5.65 685	17.88 85
33	1 089	5.74 456	18.16 59
34	1 156	5.83 095	18.43 91
35	1 225	5.91 608	18.70 83
36	1 296	6.00 000	18.97 37
37	1 369	6.08 276	19.23 54
38	1 444	6.16 441	19.49 36
39	1 521	6.24 500	19.74 84
40	1 600	6.32 456	20.00 00
41	1 681	6.40 312	20.24 85
42	1 764	6.48 074	20.49 39
43	1 849	6.55 744	20.73 64
44	1 936	6.63 325	20.97 62
45	2 025	6.70 820	21.21 32
46	2 116	6.78 233	21.44 76
47	2 209	6.85 565	21.67 95
48	2 304	6.92 820	21.90 89
49	2 401	7.00 000	22.13 59
50	2 500	7.07 107	22.36 07
N	N^2	$\sqrt{N}$	$\sqrt{10N}$

N	N^2	$\sqrt{N}$	$\sqrt{10N}$
50	2 500	7.07 107	22.36 07
51	2 601	7.14 143	22.58 32
52	2 704	7.21 110	22.80 35
53	2 809	7.28 011	23.02 17
54	2 916	7.34 847	23.23 79
55	3 025	7.41 620	23.45 21
56	3 136	7.48 331	23.66 43
57	3 249	7.54 983	23.87 47
58	3 364	7.61 577	24.08 32
59	3 481	7.68 115	24.28 99
60	3 600	7.74 597	24.49 49
61	3 721	7.81 025	24.69 82
62	3 844	7.87 401	24.89 98
63	3 969	7.93 725	25.09 98
64	4 096	8.00 000	25.29 82
65	4 225	8.06 226	25.49 51
66	4 356	8.12 404	25.69 05
67	4 489	8.18 535	25.88 44
68	4 624	8.24 621	26.07 68
69	4 761	8.30 662	26.26 79
70	4 900	8.36 660	26.45 75
71	5 041	8.42 615	26.64 58
72	5 184	8.48 528	26.83 28
73	5 329	8.54 400	27.01 85
74	5 476	8.60 233	27.20 29
75	5 625	8.66 025	27.38 61
76	5 776	8.71 780	27.56 81
77	5 929	8.77 496	27.74 89
78	6 084	8.83 176	27.92 85
79	6 241	8.88 819	28.10 69
80	6 400	8.94 427	28.28 43
81	6 561	9.00 000	28.46 05
82	6 724	9.05 539	28.63 56
83	6 889	9.11 043	28.80 97
84	7 056	9.16 515	28.98 28
85	7 225	9.21 954	29.15 48
86	7 396	9.27 362	29.32 58
87	7 569	9.32 738	29.49 58
88	7 744	9.38 083	29.66 48
89	7 921	9.43 398	29.83 29
90	8 100	9.48 683	30.00 00
91	8 281	9.53 939	30.16 62
92	8 464	9.59 166	30.33 15
93	8 649	9.64 365	30.49 59
94	8 836	9.69 536	30.65 94
95	9 025	9.74 679	30.82 21
96	9 216	9.79 796	30.98 39
97	9 409	9.84 886	31.14 48
98	9 604	9.89 949	31.30 50
99	9 801	9.94 987	31.46 43
100	10 000	10.00 000	31.62 28
N	N^2	$\sqrt{N}$	$\sqrt{10N}$

Table 9 (continued)

N	N^2	$\sqrt{N}$	$\sqrt{10N}$
100	10 000	10.00 00	31.62 28
101	10 201	10.04 99	31.78 05
102	10 404	10.09 95	31.93 74
103	10 609	10.14 89	32.09 36
104	10 816	10.19 80	32.24 90
105	11 025	10.24 70	32.40 37
106	11 236	10.29 56	32.55 76
107	11 449	10.34 41	32.71 09
108	11 664	10.39 23	32.86 34
109	11 881	10.44 03	33.01 51
110	12 100	10.48 81	33.16 62
111	12 321	10.53 57	33.31 67
112	12 544	10.58 30	33.46 64
113	12 769	10.63 01	33.61 55
114	12 996	10.67 71	33.76 39
115	13 225	10.72 38	33.91 16
116	13 456	10.77 03	34.05 88
117	13 689	10.81 67	34.20 53
118	13 924	10.86 28	34.35 11
119	14 161	10.90 87	34.49 64
120	14 400	10.95 45	34.64 10
121	14 641	11.00 00	34.78 51
122	14 884	11.04 54	34.92 85
123	15 129	11.09 05	35.07 14
124	15 376	11.13 55	35.21 36
125	15 625	11.18 03	35.35 53
126	15 876	11.22 50	35.49 65
127	16 129	11.26 94	35.63 71
128	16 384	11.31 37	35.77 71
129	16 641	11.35 78	35.91 66
130	16 900	11.40 18	36.05 55
131	17 161	11.44 55	36.19 39
132	17 424	11.48 91	36.33 18
133	17 689	11.53 26	36.46 92
134	17 956	11.57 58	36.60 60
135	18 225	11.61 90	36.74 23
136	18 496	11.66 19	36.87 82
137	18 769	11.70 47	37.01 35
138	19 044	11.74 73	37.14 84
139	19 321	11.78 98	37.28 27
140	19 600	11.83 22	37.41 66
141	19 881	11.87 43	37.55 00
142	20 164	11.91 64	37.68 29
143	20 449	11.95 83	37.81 53
144	20 736	12.00 00	37.94 73
145	21 025	12.04 16	38.07 89
146	21 316	12.08 30	38.20 99
147	21 609	12.12 44	38.34 06
148	21 904	12.16 55	38.47 08
149	22 201	12.20 66	38.60 05
150	22 500	12.24 74	38.72 98
N	N^2	$\sqrt{N}$	$\sqrt{10N}$

N	N^2	$\sqrt{N}$	$\sqrt{10N}$
150	22 500	12.24 74	38.72 98
151	22 801	12.28 82	38.85 87
152	23 104	12.32 88	38.98 72
153	23 409	12.36 93	39.11 52
154	23 716	12.40 97	39.24 28
155	24 025	12.44 99	39.37 00
156	24 336	12.49 00	39.49 68
157	24 649	12.53 00	39.62 32
158	24 964	12.56 98	39.74 92
159	25 281	12.60 95	39.87 48
160	25 600	12.64 91	40.00 00
161	25 921	12.68 86	40.12 48
162	26 244	12.72 79	40.24 92
163	26 569	12.76 71	40.37 33
164	26 896	12.80 62	40.49 69
165	27 225	12.84 52	40.62 02
166	27 556	12.88 41	40.74 31
167	27 889	12.92 28	40.86 56
168	28 224	12.96 15	40.98 78
169	28 561	13.00 00	41.10 96
170	28 900	13.03 84	41.23 11
171	29 241	13.07 67	41.35 21
172	29 584	13.11 49	41.47 29
173	29 929	13.15 29	41.59 33
174	30 276	13.19 09	41.71 33
175	30 625	13.22 88	41.83 30
176	30 976	13.26 65	41.95 24
177	31 329	13.30 41	42.07 14
178	31 684	13.34 17	42.19 00
179	32 041	13.37 91	42.30 84
180	32 400	13.41 64	42.42 64
181	32 761	13.45 36	42.54 41
182	33 124	13.49 07	42.66 15
183	33 489	13.52 77	42.77 85
184	33 856	13.56 47	42.89 52
185	34 225	13.60 15	43.01 16
186	34 596	13.63 82	43.12 77
187	34 969	13.67 48	43.24 35
188	35 344	13.71 13	43.35 90
189	35 721	13.74 77	43.47 41
190	36 100	13.78 40	43.58 90
191	36 481	13.82 03	43.70 35
192	36 864	13.85 64	43.81 78
193	37 249	13.89 24	43.93 18
194	37 636	13.92 84	44.04 54
195	38 025	13.96 42	44.15 88
196	38 416	14.00 00	44.27 19
197	38 809	14.03 57	44.38 47
198	39 204	14.07 12	44.49 72
199	39 601	14.10 67	44.60 94
200	40 000	14.14 21	44.72 14
N	N^2	$\sqrt{N}$	$\sqrt{10N}$

Table 9 (continued)

N	N^2	$\sqrt{N}$	$\sqrt{10N}$
200	40 000	14.14 21	44.72 14
201	40 401	14.17 74	44.83 30
202	40 804	14.21 27	44.94 44
203	41 209	14.24 78	45.05 55
204	41 616	14.28 29	45.16 64
205	42 025	14.31 78	45.27 69
206	42 436	14.35 27	45.38 72
207	42 849	14.38 75	45.49 73
208	43 264	14.42 22	45.60 70
209	43 681	14.45 68	45.71 65
210	44 100	14.49 14	45.82 58
211	44 521	14.52 58	45.93 47
212	44 944	14.56 02	46.04 35
213	45 369	14.59 45	46.15 19
214	45 796	14.62 87	46.26 01
215	46 225	14.66 29	46.36 81
216	46 656	14.69 69	46.47 58
217	47 089	14.73 09	46.58 33
218	47 524	14.76 48	46.69 05
219	47 961	14.79 86	46.79 74
220	48 400	14.83 24	46.90 42
221	48 841	14.86 61	47.01 06
222	49 284	14.89 97	47.11 69
223	49 729	14.93 32	47.22 29
224	50 176	14.96 66	47.32 86
225	50 625	15.00 00	47.43 42
226	51 076	15.03 33	47.53 95
227	51 529	15.06 65	47.64 45
228	51 984	15.09 97	47.74 93
229	52 441	15.13 27	47.85 39
230	52 900	15.16 58	47.95 83
231	53 361	15.19 87	48.06 25
232	53 824	15.23 15	48.16 64
233	54 289	15.26 43	48.27 01
234	54 756	15.29 71	48.37 35
235	55 225	15.32 97	48.47 68
236	55 696	15.36 23	48.57 98
237	56 169	15.39 48	48.68 26
238	56 644	15.42 72	48.78 52
239	57 121	15.45 96	48.88 76
240	57 600	15.49 19	48.98 98
241	58 081	15.52 42	49.09 18
242	58 564	15.55 63	49.19 35
243	59 049	15.58 85	49.29 50
244	59 536	15.62 05	49.39 64
245	60 025	15.65 25	49.49 75
246	60 516	15.68 44	49.59 84
247	61 009	15.71 62	49.69 91
248	61 504	15.74 80	49.79 96
249	62 001	15.77 97	49.89 99
250	62 500	15.81 14	50.00 00
N	N^2	$\sqrt{N}$	$\sqrt{10N}$

N	N^2	$\sqrt{N}$	$\sqrt{10N}$
250	62 500	15.81 14	50.00 00
251	63 001	15.84 30	50.09 99
252	63 504	15.87 45	50.19 96
253	64 009	15.90 60	50.29 91
254	64 516	15.93 74	50.39 84
255	65 025	15.96 87	50.49 75
256	65 536	16.00 00	50.59 64
257	66 049	16.03 12	50.69 52
258	66 564	16.06 24	50.79 37
259	67 081	16.09 35	50.89 20
260	67 600	16.12 45	50.99 02
261	68 121	16.15 55	51.08 82
262	68 644	16.18 64	51.18 59
263	69 169	16.21 73	51.28 35
264	69 696	16.24 81	51.38 09
265	70 225	16.27 88	51.47 82
266	70 756	16.30 95	51.57 52
267	71 289	16.34 01	51.67 20
268	71 824	16.37 07	51.76 87
269	72 361	16.40 12	51.86 52
270	72 900	16.43 17	51.96 15
271	73 441	16.46 21	52.05 77
272	73 984	16.49 24	52.15 36
273	74 529	16.52 27	52.24 94
274	75 076	16.55 29	52.34 50
275	75 625	16.58 31	52.44 04
276	76 176	16.61 32	52.53 57
277	76 729	16.64 33	52.63 08
278	77 284	16.67 33	52.72 57
279	77 841	16.70 33	52.82 05
280	78 400	16.73 32	52.91 50
281	78 961	16.76 31	53.00 94
282	79 524	16.79 29	53.10 37
283	80 089	16.82 26	53.19 77
284	80 656	16.85 23	53.29 17
285	81 225	16.88 19	53.38 54
286	81 796	16.91 15	53.47 90
287	82 369	16.94 11	53.57 24
288	82 944	16.97 06	53.66 56
289	83 521	17.00 00	53.75 87
290	84 100	17.02 94	53.85 16
291	84 681	17.05 87	53.94 44
292	85 264	17.08 80	54.03 70
293	85 849	17.11 72	54.12 95
294	86 436	17.14 64	54.22 18
295	87 025	17.17 56	54.31 39
296	87 616	17.20 47	54.40 59
297	88 209	17.23 37	54.49 77
298	88 804	17.26 27	54.58 94
299	89 401	17.29 16	54.68 09
300	90 000	17.32 05	54.77 23
N	N^2	$\sqrt{N}$	$\sqrt{10N}$

Table 9 (continued)

N	N^2	$\sqrt{N}$	$\sqrt{10N}$
300	90 000	17.32 05	54.77 23
301	90 601	17.34 94	54.86 35
302	91 204	17.37 81	54.95 45
303	91 809	17.40 69	55.04 54
304	92 416	17.43 56	55.13 62
305	93 025	17.46 42	55.22 68
306	93 636	17.49 29	55.31 73
307	94 249	17.52 14	55.40 76
308	94 864	17.54 99	55.49 77
309	95 481	17.57 84	55.58 78
310	96 100	17.60 68	55.67 76
311	96 721	17.63 52	55.76 74
312	97 344	17.66 35	55.85 70
313	97 969	17.69 18	55.94 64
314	98 596	17.72 00	56.03 57
315	99 225	17.74 82	56.12 49
316	99 856	17.77 64	56.21 39
317	100 489	17.80 45	56.30 28
318	101 124	17.83 26	56.39 15
319	101 761	17.86 06	56.48 01
320	102 400	17.88 85	56.56 85
321	103 041	17.91 65	56.65 69
322	103 684	17.94 44	56.74 50
323	104 329	17.97 22	56.83 31
324	104 976	18.00 00	56.92 10
325	105 625	18.02 78	57.00 88
326	106 276	18.05 55	57.09 64
327	106 929	18.08 31	57.18 39
328	107 584	18.11 08	57.27 13
329	108 241	18.13 84	57.35 85
330	108 900	18.16 59	57.44 56
331	109 561	18.19 34	57.53 26
332	110 224	18.22 09	57.61 94
333	110 889	18.24 83	57.70 62
334	111 556	18.27 57	57.79 27
335	112 225	18.30 30	57.87 92
336	112 896	18.33 03	57.96 55
337	113 569	18.35 76	58.05 17
338	114 244	18.38 48	58.13 78
339	114 921	18.41 20	58.22 37
340	115 600	18 43 91	58.30 95
341	116 281	18.46 62	58.39 52
342	116 964	18.49 32	58.48 08
343	117 649	18.52 03	58.56 62
344	118 336	18.54 72	58.65 15
345	119 025	18.57 42	58.73 67
346	119 716	18.60 11	58.82 18
347	120 409	18.62 79	58.90 67
348	121 104	18.65 48	58.99 15
349	121 801	18.68 15	59.07 62
350	122 500	18.70 83	59.16 08
N	N^2	$\sqrt{N}$	$\sqrt{10N}$

N	N^2	$\sqrt{N}$	$\sqrt{10N}$
350	122 500	18.70 83	59.16 08
351	123 201	18.73 50	59.24 53
352	123 904	18.76 17	59.32 96
353	124 609	18.78 83	59.41 38
354	125 316	18.81 49	59.49 79
355	126 025	18.84 14	59.58 19
356	126 736	18.86 80	59.66 57
357	127 449	18.89 44	59.74 95
358	128 164	18.92 09	59.83 31
359	128 881	18.94 73	59.91 66
360	129 600	18.97 37	60.00 00
361	130 321	19.00 00	60.08 33
362	131 044	19.02 63	60.16 64
363	131 769	19.05 26	60.24 95
364	132 496	19.07 88	60.33 24
365	133 225	19.10 50	60.41 52
366	133 956	19.13 11	60.49 79
367	134 689	19.15 72	60.58 05
368	135 424	19.18 33	60.66 30
369	136 161	19.20 94	60.74 54
370	136 900	19.23 54	60.82 76
371	137 641	19.26 14	60.90 98
372	138 384	19.28 73	60.99 18
373	139 129	19.31 32	61.07 37
374	139 876	19.33 91	61.15 55
375	140 625	19.36 49	61.23 72
376	141 376	19.39 07	61.31 88
377	142 129	19.41 65	61.40 03
378	142 884	19.44 22	61.48 17
379	143 641	19.46 79	61.56 30
380	144 400	19.49 36	61.64 41
381	145 161	19.51 92	61.72 52
382	145 924	19.54 48	61.80 61
383	146 689	19.57 04	61.88 70
384	147 456	19.59 59	61.96 77
385	148 225	19.62 14	62.04 84
386	148 996	19.64 69	62.12 89
387	149 769	19.67 23	62.20 93
388	150 544	19.69 77	62.28 96
389	151 321	19.72 31	62.36 99
390	152 100	19.74 84	62.45 00
391	152 881	19.77 37	62.53 00
392	153 664	19.79 90	62.60 99
393	154 449	19.82 42	62.68 97
394	155 236	19.84 94	62.76 94
395	156 025	19.87 46	62.84 90
396	156 816	19.89 97	62.92 85
397	157 609	19.92 49	63.00 79
398	158 404	19.94 99	63.08 72
399	159 201	19.97 50	63.16 64
400	160 000	20.00 00	63.24 56
N	N^2	$\sqrt{N}$	$\sqrt{10N}$

Table 9 (continued)

N	N^2	$\sqrt{N}$	$\sqrt{10N}$
400	160 000	20.00 00	63.24 56
401	160 801	20.02 50	63.32 46
402	161 604	20.04 99	63.40 35
403	162 409	20.07 49	63.48 23
404	163 216	20.09 98	63.56 10
405	164 025	20.12 46	63.63 96
406	164 836	20.14 94	63.71 81
407	165 649	20.17 42	63.79 66
408	166 464	20.19 90	63.87 49
409	167 281	20.22 37	63.95 31
410	168 100	20.24 85	64.03 12
411	168 921	20.27 31	64.10 93
412	169 744	20.29 78	64.18 72
413	170 569	20.32 24	64.26 51
414	171 396	20.34 70	64.34 28
415	172 225	20.37 15	64.42 05
416	173 056	20.39 61	64.49 81
417	173 889	20.42 06	64.57 55
418	174 724	20.44 50	64.65 29
419	175 561	20.46 95	64.73 02
420	176 400	20.49 39	64.80 74
421	177 241	20.51 83	64.88 45
422	178 084	20.54 26	64.96 15
423	178 929	20.56 70	65.03 85
424	179 776	20.59 13	65.11 53
425	180 625	20.61 55	65.19 20
426	181 476	20.63 98	65.26 87
427	182 329	20.66 40	65.34 52
428	183 184	20.68 82	65.42 17
429	184 041	20.71 23	65.49 81
430	184 900	20.73 64	65.57 44
431	185 761	20.76 05	65.65 06
432	186 624	20.78 46	65.72 67
433	187 489	20.80 87	65.80 27
434	188 356	20.83 27	65.87 87
435	189 225	20.85 67	65.95 45
436	190 096	20.88 06	66.03 03
437	190 969	20.90 45	66.10 60
438	191 844	20.92 84	66.18 16
439	192 721	20.95 23	66.25 71
440	193 600	20.97 62	66.33 25
441	194 481	21.00 00	66.40 78
442	195 364	21.02 38	66.48 31
443	196 249	21.04 76	66.55 82
444	197 136	21.07 13	66.63 33
445	198 025	21.09 50	66.70 83
446	198 916	21.11 87	66.78 32
447	199 809	21.14 24	66.85 81
448	200 704	21.16 60	66.93 28
449	201 601	21.18 96	67.00 75
450	202 500	21.21 32	67.08 20
N	N^2	$\sqrt{N}$	$\sqrt{10N}$

N	N^2	$\sqrt{N}$	$\sqrt{10N}$
450	202 500	21.21 32	67.08 20
451	203 401	21.23 68	67.15 65
452	204 304	21.26 03	67.23 09
453	205 209	21.28 38	67.30 53
454	206 116	21.30 73	67.37 95
455	207 025	21.33 07	67.45 37
456	207 936	21.35 42	67.52 78
457	208 849	21.37 76	67.60 18
458	209 764	21.40 09	67.67 57
459	210 681	21.42 43	67.74 95
460	211 600	21.44 76	67.82 33
461	212 521	21.47 09	67.89 70
462	213 444	21.49 42	67.97 06
463	214 369	21.51 74	68.04 41
464	215 296	21.54 07	68.11 75
465	216 225	21.56 39	68.19 09
466	217 156	21.58 70	68.26 42
467	218 089	21.61 02	68.33 74
468	219 024	21.63 33	68.41 05
469	219 961	21.65 64	68.48 36
470	220 900	21.67 95	68.55 65
471	221 841	21.70 25	68.62 94
472	222 784	21.72 56	68.70 23
473	223 729	21.74 86	68.77 50
474	224 676	21.77 15	68.84 77
475	225 625	21.79 45	68.92 02
476	226 576	21.81 74	68.99 28
477	227 529	21.84 03	69.06 52
478	228 484	21.86 32	69.13 75
479	229 441	21.88 61	69.20 98
480	230 400	21.90 89	69.28 20
481	231 361	21.93 17	69.35 42
482	232 324	21.95 45	69.42 62
483	233 289	21.97 73	69.49 82
484	234 256	22.00 00	69.57 01
485	235 225	22.02 27	69.64 19
486	236 196	22.04 54	69.71 37
487	237 169	22.06 81	69.78 54
488	238 144	22.09 07	69.85 70
489	239 121	22.11 33	69.92 85
490	240 100	22.13 59	70.00 00
491	241 081	22.15 85	70.07 14
492	242 064	22.18 11	70.14 27
493	243 049	22.20 36	70.21 40
494	244 036	22.22 61	70.28 51
495	245 025	22.24 86	70.35 62
496	246 016	22.27 11	70.42 73
497	247 009	22.29 35	70.49 82
498	248 004	22.31 59	70.56 91
499	249 001	22.33 83	70.63 99
500	250 000	22.36 07	70.71 07
N	N^2	$\sqrt{N}$	$\sqrt{10N}$

Table 9 (continued)

N	N^2	$\sqrt{N}$	$\sqrt{10N}$
500	250 000	22.36 07	70.71 07
501	251 001	22.38 30	70.78 14
502	252 004	22.40 54	70.85 20
503	253 009	22.42 77	70.92 25
504	254 016	22.44 99	70.99 30
505	255 025	22.47 22	71.06 34
506	256 036	22.49 44	71.13 37
507	257 049	22.51 67	71.20 39
508	258 064	22.53 89	71.27 41
509	259 081	22.56 10	71.34 42
510	260 100	22.58 32	71.41 43
511	261 121	22.60 53	71.48 43
512	262 144	22.62 74	71.55 42
513	263 169	22.64 95	71.62 40
514	264 196	22.67 16	71.69 38
515	265 225	22.69 36	71.76 35
516	266 256	22.71 56	71.83 31
517	267 289	22.73 76	71.90 27
518	268 324	22.75 96	71.97 22
519	269 361	22.78 16	72.04 17
520	270 400	22.80 35	72.11 10
521	271 441	22.82 54	72.18 03
522	272 484	22.84 73	72.24 96
523	273 529	22.86 92	72.31 87
524	274 576	22.89 10	72.38 78
525	275 625	22.91 29	72.45 69
526	276 676	22.93 47	72.52 59
527	277 729	22.95 65	72.59 48
528	278 784	22.97 83	72.66 36
529	279 841	23.00 00	72.73 24
530	280 900	23.02 17	72.80 11
531	281 961	23.04 34	72.86 97
532	283 024	23.06 51	72.93 83
533	284 089	23.08 68	73.00 68
534	285 156	23.10 84	73.07 53
535	286 225	23.13 01	73.14 37
536	287 296	23.15 17	73.21 20
537	288 369	23.17 33	73.28 03
538	289 444	23.19 48	73.34 85
539	290 521	23.21 64	73.41 66
540	291 600	23.23 79	73.48 47
541	292 681	23.25 94	73.55 27
542	293 764	23.28 09	73.62 06
543	294 849	23.30 24	73.68 85
544	295 936	23.32 38	73.75 64
545	297 025	23.34 52	73.82 41
546	298 116	23.36 66	73.89 18
547	299 209	23.38 80	73.95 94
548	300 304	23.40 94	74.02 70
549	301 401	23.43 07	74.09 45
550	302 500	23.45 21	74.16 20
N	N^2	$\sqrt{N}$	$\sqrt{10N}$

N	N^2	$\sqrt{N}$	$\sqrt{10N}$
550	302 500	23.45 21	74.16 20
551	303 601	23.47 34	74.22 94
552	304 704	23.49 47	74.29 67
553	305 809	23.51 60	74.36 40
554	306 916	23.53 72	74.43 12
555	308 025	23.55 84	74.49 83
556	309 136	23.57 97	74.56 54
557	310 249	23.60 08	74.63 24
558	311 364	23.62 20	74.69 94
559	312 481	23.64 32	74.76 63
560	313 600	23.66 43	74.83 31
561	314 721	23.68 54	74.89 99
562	315 844	23.70 65	74.96 67
563	316 969	23.72 76	75.03 33
564	318 096	23.74 87	75.09 99
565	319 225	23.76 97	75.16 65
566	320 356	23.79 08	75.23 30
567	321 489	23.81 18	75.29 94
568	322 624	23.83 28	75.36 58
569	323 761	23.85 37	75.43 21
570	324 900	23.87 47	75.49 83
571	326 041	23.89 56	75.56 45
572	327 184	23.91 65	75.63 07
573	328 329	23.93 74	75.69 68
574	329 476	23.95 83	75.76 28
575	330 625	23.97 92	75.82 88
576	331 776	24.00 00	75.89 47
577	332 929	24.02 08	75.96 05
578	334 084	24.04 16	76.02 63
579	335 241	24.06 24	76.09 20
580	336 400	24.08 32	76.15 77
581	337 561	24.10 39	76.22 34
582	338 724	24.12 47	76.28 89
583	339 889	24.14 54	76.35 44
584	341 056	24.16 61	76.41 99
585	342 225	24.18 68	76.48 53
586	343 396	24.20 74	76.55 06
587	344 569	24.22 81	76.61 59
588	345 744	24.24 87	76.68 12
589	346 921	24.26 93	76.74 63
590	348 100	24.28 99	76.81 15
591	349 281	24.31 05	76.87 65
592	350 464	24.33 11	76.94 15
593	351 649	24.35 16	77.00 65
594	352 836	24.37 21	77.07 14
595	354 025	24.39 26	77.13 62
596	355 216	24.41 31	77.20 10
597	356 409	24.43 36	77.26 58
598	357 604	24.45 40	77.33 05
599	358 801	24.47 45	77.39 51
600	360 000	24.49 49	77.45 97
N	N^2	$\sqrt{N}$	$\sqrt{10N}$

Table 9 (continued)

N	N^2	$\sqrt{N}$	$\sqrt{10N}$
600	360 000	24.49 49	77.45 97
601	361 201	24.51 53	77.52 42
602	362 404	24.53 57	77.58 87
603	363 609	24.55 61	77.65 31
604	364 816	24.57 64	77.71 74
605	366 025	24.59 67	77.78 17
606	367 236	24.61 71	77.84 60
607	368 449	24.63 74	77.91 02
608	369 664	24.65 77	77.97 44
609	370 881	24.67 79	78.03 85
610	372 100	24.69 82	78.10 25
611	373 321	24.71 84	78.16 65
612	374 544	24.73 86	78.23 04
613	375 769	24.75 88	78.29 43
614	376 996	24.77 90	78.35 82
615	378 225	24.79 92	78.42 19
616	379 456	24.81 93	78.48 57
617	380 689	24.83 95	78.54 93
618	381 924	24.85 96	78.61 30
619	383 161	24.87 97	78.67 66
620	384 400	24.89 98	78.74 01
621	385 641	24.91 99	78.80 36
622	386 884	24.93 99	78.86 70
623	388 129	24.96 00	78.93 03
624	389 376	24.98 00	78.99 37
625	390 625	25.00 00	79.05 69
626	391 876	25.02 00	79.12 02
627	393 129	25.04 00	79.18 33
628	394 384	25.05 99	79.24 65
629	395 641	25.07 99	79.30 95
630	396 900	25.09 98	79.37 25
631	398 161	25.11 97	79.43 55
632	399 424	25.13 96	79.49 84
633	400 689	25.15 95	79.56 13
634	401 956	25.17 94	79.62 41
635	403 225	25.19 92	79.68 69
636	404 496	25.21 90	79.74 96
637	405 769	25.23 89	79.81 23
638	407 044	25.25 87	79.87 49
639	408 321	25.27 84	79.93 75
640	409 600	25.29 82	80.00 00
641	410 881	25.31 80	80.06 25
642	412 164	25.33 77	80.12 49
643	413 449	25.35 74	80.18 73
644	414 736	25.37 72	80.24 96
645	416 025	25.39 69	80.31 19
646	417 316	25.41 65	80.37 41
647	418 609	25.43 62	80.43 63
648	419 904	25.45 58	80.49 84
649	421 201	25.47 55	80.56 05
650	422 500	25.49 51	80.62 26
N	N^2	$\sqrt{N}$	$\sqrt{10N}$

N	N^2	$\sqrt{N}$	$\sqrt{10N}$
650	422 500	25.49 51	80.62 26
651	423 801	25.51 47	80.68 46
652	425 104	25.53 43	80.74 65
653	426 409	25.55 39	80.80 84
654	427 716	25.57 34	80.87 03
655	429 025	25.59 30	80.93 21
656	430 336	25.61 25	80.99 38
657	431 649	25.63 20	81.05 55
658	432 964	25.65 15	81.11 72
659	434 281	25.67 10	81.17 88
660	435 600	25.69 05	81.24 04
661	436 921	25.70 99	81.30 19
662	438 244	25.72 94	81.36 34
663	439 569	25.74 88	81.42 48
664	440 896	25.76 82	81.48 62
665	442 225	25.78 76	81.54 75
666	443 556	25.80 70	81.60 88
667	444 889	25.82 63	81.67 01
668	446 224	25.84 57	81.73 13
669	447 561	25.86 50	81.79 24
670	448 900	25.88 44	81.85 35
671	450 241	25.90 37	81.91 46
672	451 584	25.92 30	81.97 56
673	452 929	25.94 22	82.03 66
674	454 276	25.96 15	82.09 75
675	455 625	25.98 08	82.15 84
676	456 976	26.00 00	82.21 92
677	458 329	26.01 92	82.28 00
678	459 684	26.03 84	82.34 08
679	461 041	26.05 76	82.40 15
680	462 400	26.07 68	82.46 21
681	463 761	26.09 60	82.52 27
682	465 124	26.11 51	82.58 33
683	466 489	26.13 43	82.64 38
684	467 856	26.15 34	82.70 43
685	469 225	26.17 25	82.76 47
686	470 596	26.19 16	82.82 51
687	471 969	26.21 07	82.88 55
688	473 344	26.22 98	82.94 58
689	474 721	26.24 88	83.00 60
690	476 100	26.26 79	83.06 62
691	477 481	26.28 69	83.12 64
692	478 864	26.30 59	83.18 65
693	480 249	26.32 49	83.24 66
694	481 636	26.34 39	83.30 67
695	483 025	26.36 29	83.36 67
696	484 416	26.38 18	83.42 66
697	485 809	26.40 08	83.48 65
698	487 204	26.41 97	83.54 64
699	488 601	26.43 86	83.60 62
700	490 000	26.45 75	83.66 60
N	N^2	$\sqrt{N}$	$\sqrt{10N}$

Table 9 (continued)

N	N^2	$\sqrt{N}$	$\sqrt{10N}$
700	490 000	26.45 75	83.66 60
701	491 401	26.47 64	83.72 57
702	492 804	26.49 53	83.78 54
703	494 209	26.51 41	83.84 51
704	495 616	26.53 30	83.90 47
705	497 025	26.55 18	83.96 43
706	498 436	26.57 07	84.02 38
707	499 849	26.58 95	84.08 33
708	501 264	26.60 83	84.14 27
709	502 681	26.62 71	84.20 21
710	504 100	26.64 58	84.26 15
711	505 521	26.66 46	84.32 08
712	506 944	26.68 33	84.38 01
713	508 369	26.70 21	84.43 93
714	509 796	26.72 08	84.49 85
715	511 225	26.73 95	84.55 77
716	512 656	26.75 82	84.61 68
717	514 089	26.77 69	84.67 59
718	515 524	26.79 55	84.73 49
719	516 961	26.81 42	84.79 39
720	518 400	26.83 28	84.85 28
721	519 841	26.85 14	84.91 17
722	521 284	26.87 01	84.97 06
723	522 729	26.88 87	85.02 94
724	524 176	26.90 72	85.08 82
725	525 625	26.92 58	85.14 69
726	527 076	26.94 44	85.20 56
727	528 529	26.96 29	85.26 43
728	529 984	26.98 15	85.32 29
729	531 441	27.00 00	85.38 15
730	532 900	27.01 85	85.44 00
731	534 361	27.03 70	85.49 85
732	535 824	27.05 55	85.55 70
733	537 289	27.07 40	85.61 54
734	538 756	27.09 24	85.67 38
735	540 225	27.11 09	85.73 21
736	541 696	27.12 93	85.79 04
737	543 169	27.14 77	85.84 87
738	544 644	27.16 62	85.90 69
739	546 121	27.18 46	85.96 51
740	547 600	27.20 29	86.02 33
741	549 081	27.22 13	86.08 14
742	550 564	27.23 97	86.13 94
743	552 049	27.25 80	86.19 74
744	553 536	27.27 64	86.25 54
745	555 025	27.29 47	86.31 34
746	556 516	27.31 30	86.37 13
747	558 009	27.33 13	86.42 92
748	559 504	27.34 96	86.48 70
749	561 001	27.36 79	86.54 48
750	562 500	27.38 61	86.60 25
N	N^2	$\sqrt{N}$	$\sqrt{10N}$

N	N^2	$\sqrt{N}$	$\sqrt{10N}$
750	562 500	27.38 61	86.60 25
751	564 001	27.40 44	86.66 03
752	565 504	27.42 26	86.71 79
753	567 009	27.44 08	86.77 56
754	568 516	27.45 91	86.83 32
755	570 025	27.47 73	86.89 07
756	571 536	27.49 55	86.94 83
757	573 049	27.51 36	87.00 57
758	574 564	27.53 18	87.06 32
759	576 081	27.55 00	87.12 06
760	577 600	27.56 81	87.17 80
761	579 121	27.58 62	87.23 53
762	580 644	27.60 43	87.29 26
763	582 169	27.62 25	87.34 99
764	583 696	27.64 05	87.40 71
765	585 225	27.65 86	87.46 43
766	586 756	27.67 67	87.52 14
767	588 289	27.69 48	87.57 85
768	589 824	27.71 28	87.63 56
769	591 361	27.73 08	87.69 26
770	592 900	27.74 89	87.74 96
771	594 441	27.76 69	87.80 66
772	595 984	27.78 49	87.86 35
773	597 529	27.80 29	87.92 04
774	599 076	27.82 09	87.97 73
775	600 625	27.83 88	88.03 41
776	602 176	27.85 68	88.09 09
777	603 729	27.87 47	88.14 76
778	605 284	27.89 27	88.20 43
779	606 841	27.91 06	88.26 10
780	608 400	27.92 85	88.31 76
781	609 961	27.94 64	88.37 42
782	611 524	27.96 43	88.43 08
783	613 089	27.98 21	88.48 73
784	614 656	28.00 00	88.54 38
785	616 225	28.01 79	88.60 02
786	617 796	28.03 57	88.65 66
787	619 369	28.05 35	88.71 30
788	620 944	28.07 13	88.76 94
789	622 521	28.08 91	88.82 57
790	624 100	28.10 69	88.88 19
791	625 681	28.12 47	88.93 82
792	627 264	28.14 25	88.99 44
793	628 849	28.16 03	89.05 05
794	630 436	28.17 80	89.10 67
795	632 025	28.19 57	89.16 28
796	633 616	28.21 35	89.21 88
797	635 209	28.23 12	89.27 49
798	636 804	28.24 89	89.33 08
799	638 401	28.26 66	89.38 68
800	640 000	28.28 43	89.44 27
N	N^2	$\sqrt{N}$	$\sqrt{10N}$

Table 9 (continued)

N	N^2	$\sqrt{N}$	$\sqrt{10N}$
800	640 000	28.28 43	89.44 27
801	641 601	28.30 19	89.49 86
802	643 204	28.31 96	89.55 45
803	644 809	28.33 73	89.61 03
804	646 416	28.35 49	89.66 60
805	648 025	28.37 25	89.72 18
806	649 636	28.39 01	89.77 75
807	651 249	28.40 77	89.83 32
808	652 864	28.42 53	89.88 88
809	654 481	28.44 29	89.94 44
810	656 100	28.46 05	90.00 00
811	657 721	28.47 81	90.05 55
812	659 344	28.49 56	90.11 10
813	660 969	28.51 32	90.16 65
814	662 596	28.53 07	90.22 19
815	664 225	28.54 82	90.27 74
816	665 856	28.56 57	90.33 27
817	667 489	28.58 32	90.38 81
818	669 124	28.60 07	90.44 34
819	670 761	28.61 82	90.49 86
820	672 400	28.63 56	90.55 39
821	674 041	28.65 31	90.60 91
822	675 684	28.67 05	90.66 42
823	677 329	28.68 80	90.71 93
824	678 976	28.70 54	90.77 44
825	680 625	28.72 28	90.82 95
826	682 276	28.74 02	90.88 45
827	683 929	28.75 76	90.93 95
828	685 584	28.77 50	90.99 45
829	687 241	28.79 24	91.04 94
830	688 900	28.80 97	91.10 43
831	690 561	28.82 71	91.15 92
832	692 224	28.84 44	91.21 40
833	693 889	28.86 17	91.26 88
834	695 556	28.87 91	91.32 36
835	697 225	28.89 64	91.37 83
836	698 896	28.91 37	91.43 30
837	700 569	28.93 10	91.48 77
838	702 244	28.94 82	91.54 23
839	703 921	28.96 55	91.59 69
840	705 600	28.98 28	91.65 15
841	707 281	29.00 00	91.70 61
842	708 964	29.01 72	91.76 06
843	710 649	29.03 45	91.81 50
844	712 336	29.05 17	91.86 95
845	714 025	29.06 89	91.92 39
846	715 716	29.08 61	91.97 83
847	717 409	29.10 33	92.03 26
848	719 104	29.12 04	92.08 69
849	720 801	29.13 76	92.14 12
850	722 500	29.15 48	92.19 54
N	N^2	$\sqrt{N}$	$\sqrt{10N}$

N	N^2	$\sqrt{N}$	$\sqrt{10N}$
850	722 500	29.15 48	92.19 54
851	724 201	29.17 19	92.24 97
852	725 904	29.18 90	92.30 38
853	727 609	29.20 62	92.35 80
854	729 316	29.22 33	92.41 21
855	731 025	29.24 04	92.46 62
856	732 736	29.25 75	92.52 03
857	734 449	29.27 46	92.57 43
858	736 164	29.29 16	92.62 83
859	737 881	29.30 87	92.68 23
860	739 600	29.32 58	92.73 62
861	741 321	29.34 28	92.79 01
862	743 044	29.35 98	92.84 40
863	744 769	29.37 69	92.89 78
864	746 496	29.39 39	92.95 16
865	748 225	29.41 09	93.00 54
866	749 956	29.42 79	93.05 91
867	751 689	29.44 49	93.11 28
868	753 424	29.46 18	93.16 65
869	755 161	29.47 88	93.22 02
870	756 900	29.49 58	93.27 38
871	758 641	29.51 27	93.32 74
872	760 384	29.52 96	93.38 09
873	762 129	29.54 66	93.43 45
874	763 876	29.56 35	93.48 80
875	765 625	29.58 04	93.54 14
876	767 376	29.59 73	93.59 49
877	769 129	29.61 42	93.64 83
878	770 884	29.63 11	93.70 17
879	772 641	29.64 79	93.75 50
880	774 400	29.66 48	93.80 83
881	776 161	29.68 16	93.86 16
882	777 924	29.69 85	93.91 49
883	779 689	29.71 53	93.96 81
884	781 456	29.73 21	94.02 13
885	783 225	29.74 89	94.07 44
886	784 996	29.76 58	94.12 76
887	786 769	29.78 25	94.18 07
888	788 544	29.79 93	94.23 38
889	790 321	29.81 61	94.28 68
890	792 100	29.83 29	94.33 98
891	793 881	29.84 96	94.39 28
892	795 664	29.86 64	94.44 58
893	797 449	29.88 31	94.49 87
894	799 236	29.89 98	94.55 16
895	801 025	29.91 66	94.60 44
896	802 816	29.93 33	94.65 73
897	804 609	29.95 00	94.71 01
898	806 404	29.96 66	94.76 29
899	808 201	29.98 33	94.81 56
900	810 000	30.00 00	94.86 83
N	N^2	$\sqrt{N}$	$\sqrt{10N}$

Table 9 (continued)

N	N^2	$\sqrt{N}$	$\sqrt{10N}$
900	810 000	30.00 00	94.86 83
901	811 801	30.01 67	94.92 10
902	813 604	30.03 33	94.97 37
903	815 409	30.05 00	95.02 63
904	817 216	30.06 66	95.07 89
905	819 025	30.08 32	95.13 15
906	820 836	30.09 98	95.18 40
907	822 649	30.11 64	95.23 65
908	824 464	30.13 30	95.28 90
909	826 281	30.14 96	95.34 15
910	828 100	30.16 62	95.39 39
911	829 921	30.18 28	95.44 63
912	831 744	30.19 93	95.49 87
913	833 569	30.21 59	95.55 10
914	835 396	30.23 24	95.60 33
915	837 225	30.24 90	95.65 56
916	839 056	30.26 55	95.70 79
917	840 889	30.28 20	95.76 01
918	842 724	30.29 85	95.81 23
919	844 561	30.31 50	95.86 45
920	846 400	30.33 15	95.91 66
921	848 241	30.34 80	95.96 87
922	850 084	30.36 45	96.02 08
923	851 929	30.38 09	96.07 29
924	853 776	30.39 74	96.12 49
925	855 625	30.41 38	96.17 69
926	857 476	30.43 02	96.22 89
927	859 329	30.44 67	96.28 08
928	861 184	30.46 31	96.33 28
929	863 041	30.47 95	96.38 46
930	864 900	30.49 59	96.43 65
931	866 761	30.51 23	96.48 83
932	868 624	30.52 87	96.54 01
933	870 489	30.54 50	96.59 19
934	872 356	30.56 14	96.64 37
935	874 225	30.57 78	96.69 54
936	876 096	30.59 41	96.74 71
937	877 969	30.61 05	96.79 88
938	879 844	30.62 68	96.85 04
939	881 721	30.64 31	96.90 20
940	883 600	30.65 94	96.95 36
941	885 481	30.67 57	97.00 52
942	887 364	30.69 20	97.05 67
943	889 249	30.70 83	97.10 82
944	891 136	30.72 46	97.15 97
945	893 025	30.74 09	97.21 11
946	894 916	30.75 71	97.26 25
947	896 809	30.77 34	97.31 39
948	898 704	30.78 96	97.36 53
949	900 601	30.80 58	97.41 66
950	902 500	30.82 21	97.46 79
N	N^2	$\sqrt{N}$	$\sqrt{10N}$

N	N^2	$\sqrt{N}$	$\sqrt{10N}$
950	902 500	30.82 21	97.46 79
951	904 401	30.83 83	97.51 92
952	906 304	30.85 45	97.57 05
953	908 209	30.87 07	97.62 17
954	910 116	30.88 69	97.67 29
955	912 025	30.90 31	97.72 41
956	913 936	30.91 92	97.77 53
957	915 849	30.93 54	97.82 64
958	917 764	30.95 16	97.87 75
959	919 681	30.96 77	97.92 85
960	921 600	30.98 39	97.97 96
961	923 521	31.00 00	98.03 06
962	925 444	31.01 61	98.08 16
963	927 369	31.03 22	98.13 26
964	929 296	31.04 83	98.18 35
965	931 225	31.06 44	98.23 44
966	933 156	31.08 05	98.28 53
967	935 089	31.09 66	98.33 62
968	937 024	31.11 27	98.38 70
969	938 961	31.12 88	98.43 78
970	940 900	31.14 48	98.48 86
971	942 841	31.16 09	98.53 93
972	944 784	31.17 69	98.59 01
973	946 729	31.19 29	98.64 08
974	948 676	31.20 90	98.69 14
975	950 625	31.22 50	98.74 21
976	952 576	[illegible].24 10	98.79 27
977	954 529	31.25 70	98.84 33
978	956 484	31.27 30	98.89 39
979	958 441	31.28 90	98.94 44
980	960 400	31.30 50	98.99 49
981	962 361	31.32 09	99.04 54
982	964 324	31.33 69	99.09 59
983	966 289	31.35 28	99.14 64
984	968 256	31.36 88	99.19 68
985	970 225	31.38 47	99.24 72
986	972 196	31.40 06	99.29 75
987	974 169	31.41 66	99.34 79
988	976 144	31.43 25	99.39 82
989	978 121	31.44 84	99.44 85
990	980 100	31.46 43	99.49 87
991	982 081	31.48 02	99.54 90
992	984 064	31.49 60	99.59 92
993	986 049	31.51 19	99.64 94
994	988 036	31.52 78	99.69 95
995	990 025	31.54 36	99.74 97
996	992 016	31.55 95	99.79 98
997	994 009	31.57 53	99.84 99
998	996 004	31.59 11	99.89 99
999	998 001	31.60 70	99.95 00
1000	1000 000	31.62 28	100.00 00
N	N^2	$\sqrt{N}$	$\sqrt{10N}$

BIBLIOGRAPHY

Aitchison, J. (1964), "Bayesian Tolerance Regions," *Journal of the Royal Statistical Society B, 26*, 161–175.

Aitchison, J. (1970a), *Choice against Chance: An Introduction to Statistical Decision Theory*. Reading, Mass.: Addison-Wesley.

Aitchison, J. (1970b), "Statistical Problems of Treatment Allocation," *Journal of the Royal Statistical Society A, 133*, 206–238 (with discussion).

Aitchison, J., and Dunsmore, I.R. (1968), "Linear-Loss Interval Estimation of Location and Scale Parameters," *Biometrika, 55*, 141–148.

Alderfer, C.P., and Bierman, H. (1970), "Choices with Risk: Beyond the Mean and Variance," *Journal of Business, 43*, 341–353.

Allais, M. (1953), "Le Comportement De L'Homme Rationnel Devant Le Risque: Critique Des Postulats et Axiomes De L'Ecole Americaine," *Econometrica, 21*, 503–546.

Altham, P.M.E. (1969), "Exact Bayesian Analysis of a 2 × 2 Contingency Table, and Fisher's 'Exact' Significance Test," *Journal of the Royal Statistical Society B, 31*, 261–269.

Ando, A., and Kaufman, G.M. (1965), "Bayesian Analysis of the Independent Multinormal Process—Neither Mean nor Precision Known," *Journal of the American Statistical Association, 60*, 347–358.

Anscombe, F.J. (1963), "Sequential Medical Trials," *Journal of the American Statistical Association, 58*, 365–383.

Anscombe, F.J. (1964), "Some Remarks on Bayesian Statistics," in M.W. Shelly and G.L. Bryan (eds.), *Human Judgments and Optimality*. New York: Wiley, 155–177.

Anscombe, F.J., and Aumann, R.J. (1963), "A Definition of Subjective Probability," *Annals of Mathematical Statistics, 34*, 199–205.

Antelman, G.R. (1965), "Insensitivity to Non-Optimal Design in Bayesian Decision Theory," *Journal of the American Statistical Association, 60*, 584–601.

Aoki, M. (1967), *Optimization of Stochastic Systems*. New York: Academic Press.

Arrow, K.J. (1951a) "Alternative Approaches to the Theory of Choice in Risk-Taking Situations," *Econometrica, 19*, 404–437.

Arrow, K.J. (1951b), *Social Choice and Individual Values*. New York: Wiley.

Arrow, K.J. (1958), "Utilities, Attitudes, Choices: A Review Note," *Econometrica, 26*, 1–23.

Arrow, K.J. (1965), *Aspects of the Theory of Risk Bearing*. Helsinki: Yrjo Jahnssonin Saatio.

Arrow, K.J. (1971), *Essays in the Theory of Risk Bearing.* Chicago: Markham.

Barnard, G.A. (1967a), "The Bayesian Controversy in Statistical Inference," *Journal of the Institute of Actuaries, 93*, 229–269 (with discussion).

Barnard, G.A. (1967b), "The Use of the Likelihood Function in Statistical Practice," in J. Neyman (ed.), *Proceedings of the Fifth Berkeley Symposium on Mathematical Statistics and Probability, 1*, 27–40. Berkeley: University of California Press.

Barnard, G.A., Jenkins, G.M., and Winsten, C.B. (1962), "Likelihood Inference and Time Series," *Journal of the Royal Statistical Society A, 125*, 321–372 (with discussion).

Baron, D.P. (1970), "Price Uncertainty, Utility, and Industry Equilibrium in Pure Competition," *International Economic Review, 11*, 463–480.

Bartholomew, D.J. (1965), "A Comparison of Some Bayesian and Frequentist Inferences," *Biometrika, 52*, 19–35.

Bartlett, M.S. (1965), "R.A. Fisher and the Last Fifty Years of Statistical Methodology," *Journal of the American Statistical Association, 60*, 395–409.

Bather, J.A. (1963), "Control Charts and the Minimization of Costs," *Journal of the Royal Statistical Society B, 25*, 49–80 (with discussion).

Bayes, T., "An Essay towards Solving A Problem in the Doctrine of Chance," *Philosophical Transactions of the Royal Society, 53* (1763), 370–418. Reproduced with biography of Bayes in G.A. Barnard, "Studies in the History of Probability and Statistics: IX," *Biometrika, 45*, (1958), 293–315.

Becker, G.M., De Groot, M.H., and Marschak, J. (1963), "Stochastic Models of Choice Behavior," *Behavioral Science, 8*, 41–55.

Becker, G.M., De Groot, M.H., and Marschak, J. (1964), "Measuring Utility by a Single-Response Sequential Method," *Behavioral Science, 9*, 226–232.

Becker, G.M., and McClintock, C.G. (1967), "Value: Behavioral Decision Theory," *Annual Review of Psychology, 18*, 239–286.

Becker, S.W., and Brownson, F.O. (1964), "What Price Ambiguity? Or the Role of Ambiguity in Decision Making," *Journal of Political Economy, 72*, 62–73.

Bellman, R.E. (1957), *Dynamic Programming.* Princeton: Princeton University Press.

Bellman, R.E., and Dreyfus, S.E. (1962), *Applied Dynamic Programming.* Princeton: Princeton University Press.

Berkson, J. (1938), "Some Difficulties of Interpretation Encountered in the Application of the Chi-Square Test," *Journal of the American Statistical Association, 33*, 526–536.

Bernoulli, D. (1938), "Specimen Theoriae Novae De Mensura Sortis," *Commentarii Academide Scientiarum Imperialis Petropolitanae, 5*, 175–192, translated by L. Sommer, *Econometrica, 22*, (1954), 23–36.

Bers, L. (1969), *Calculus.* New York: Holt, Rinehart and Winston.

Betaque, N.E., and Gorry, A. (1971), "Automating Judgmental Decision Making for a Serious Medical Problem," *Management Science B, 17*, 421–434.

Bhattacharya, S.K. (1967), "Bayesian Approach to Life Testing and Reliability Estimation," *Journal of the American Statistical Association, 62*, 48–62.

Bierman, H., and Hausman, W.H. (1970), "The Credit Granting Decision,"

Management Science B, 16, 519–532.

Birnbaum, A. (1962a), "Another View of the Foundations of Statistics," *The American Statistician, 16,* 17–21.

Birnbaum, A. (1962b), "On the Foundations of Statistical Inference," *Journal of the American Statistical Association, 33,* 526–536.

Birnbaum, A. (1968), "Likelihood," *International Encyclopedia of the Social Sciences, 9,* 299–301.

Birnbaum, A. (1969), "Concepts of Statistical Evidence," in S. Morgenbesser, P. Suppes, and M. White (eds.), *Philosophy, Science, and Method: Essays in Honor of Ernest Nagel.* New York: St. Martin's Press, 112–143.

Birnbaum, A. (1970), "On Durbin's Modified Principle of Conditionality," *Journal of the American Statistical Association, 65,* 402–403.

Blackman, A.W. (1971), "The Use of Bayesian Techniques in Delphi Forecasts," *Technological Forecasting and Social Change, 2,* 261–268.

Blackwell, D., and Girshick, M.A. (1954), *Theory of Games and Statistical Decisions.* New York: Wiley.

Borch, K. (1963), "A Note on Utility and Attitudes to Risk," *Management Science, 9,* 697–700.

Borch, K. (1966), "A Utility Function Derived from a Survival Game," *Management Science B, 12,* 287–295.

Borch, K. (1967), "The Theory of Risk," *Journal of the Royal Statistical Society B, 29,* 432–467 (with discussion).

Borch, K. (1968a), "The Allais Paradox: A Comment," *Behavioral Science, 13,* 488–489.

Borch, K. (1968b), *The Economics of Uncertainty.* Princeton: Princeton University Press.

Borch, K., and Mossin, J. (eds.) (1968), *Risk and Uncertainty.* London: Macmillan.

Box, G.E.P., and Draper, N.R. (1965), "The Bayesian Estimation of Common Parameters from Several Responses," *Biometrika, 52,* 355–365.

Box, G.E.P., and Tiao, G.C. (1962), "A Further Look at Robustness via Bayes' Theorem," *Biometrika, 49,* 419–432.

Box, G.E.P., and Tiao, G.C. (1964), "A Bayesian Approach to the Importance of Assumptions Applied to the Comparison of Variances," *Biometrika, 51,* 153–167.

Bracken, J., and Soland, R.M. (1969), "Statistical Decision Models for Brokering," *Management Science, 15,* 619–625.

Brender, D.M. (1968a), "The Prediction and Measurement of System Availability: A Bayesian Treatment," *IEEE Transactions on Reliability, R-17,* 127–138.

Brender, D.M. (1968b), "The Bayesian Assessment of System Availability: Advanced Applications and Techniques," *IEEE Transactions on Reliability, R-17,* 138–147.

Bross, I. (1953), *Design for Decision.* New York: Macmillan.

Bross, I.D.J. (1961), "Statistical Dogma: A Challenge," *The American Statistician, 15,* 14–15.

Brown, R.V. (1969), *Research and the Credibility of Estimates.* Boston: Division of Research, Harvard Business School.

Brown, R.V. (1970), "Do Managers Find Decision Theory Useful?," *Harvard Business Review, 48*, 78–89.

Buzzell, R.D., Cox, D.F., and Brown, R.V. (1969), *Marketing Research and Information Systems.* New York: McGraw-Hill.

Campling, G.E.G. (1968), "Serial Sampling Acceptance Schemes for Large Batches of Items Where the Mean Quality Has a Normal Prior Distribution," *Biometrika, 55*, 393–399.

Carnap, R. (1950), *Logical Foundations of Probability.* Chicago: University of Chicago Press.

Chambers, M.L. (1970), "A Simple Problem with Strikingly Different Frequentist and Bayesian Solutions," *Journal of the Royal Statistical Society B, 32*, 278–282.

Chernoff, H., and Moses, L.E. (1959), *Elementary Decision Theory.* New York: Wiley.

Chetty, V.K. (1968), "Bayesian Analysis of Haavelmo's Models," *Econometrica, 36*, 582–602.

Chipman, J.S. (1960), "The Foundations of Utility," *Econometrica, 28*, 193–224.

Christenson, C. (1965), *Strategic Aspects of Bidding for Corporate Securities.* Boston: Division of Research, Harvard Business School.

Clelland, R.C., De Cani, J.S., Brown, F.E., Bursk, J.P., and Murray, D.S. (1966), *Basic Statistics with Business Applications.* New York: Wiley.

Cook, W. H. (1968), "Decision Analysis for Product Development," *IEEE Transactions on Systems Science and Cybernetics, SSC-4*, 342–354.

Cornfield, J. (1966a), "A Bayesian Test of Some Classical Hypotheses, with Applications to Sequential Clinical Trials," *Journal of the American Statistical Association, 61*, 577–594.

Cornfield, J. (1966b), "Sequential Trials, Sequential Analysis, and the Likelihood Principle," *The American Statistician, 20*, 18–23.

Cornfield, J. (1967), "Bayes Theorem," *Review of the International Statistical Institute, 35*, 34–49.

Cornfield, J. (1969), "The Bayesian Outlook and Its Applications," *Biometrics, 25*, 617–657 (with discussion).

Cramér, H. (1946), *Mathematical Methods of Statistics.* Princeton: Princeton University Press.

Cyert, R.M., and De Groot, M.H. (1970), "Bayesian Analysis and Duopoly Theory," *Journal of Political Economy, 78*, 1168–1184.

Dantzig, G.B. (1963), *Linear Programming and Extensions.* Princeton: Princeton University Press.

Davidson, D., Suppes, P., and Siegel, S. (1957), *Decision Making: An Experimental Approach.* Stanford: Stanford University Press.

Debreu, G. (1959), *Theory of Value.* New York: Wiley.

de Finetti, B. (1937), "La Prévision: Ses Lois Logiques, Ses Sources Subjectives," *Annales De L'Institut Henri Poincaré, 7*, Translated by H.E. Kyburg ("Foresight: Its Logical Laws, Its Subjective Sources") in Kyburg and Smokler (1964).

de Finetti, B. (1961), "The Bayesian Approach to the Rejection of Outliers," in *Proceedings of the Fourth Berkeley Symposium on Mathematical Statistics and Probability, 1*, 199–210. Berkeley: University of California Press.

de Finetti, B. (1962), "Does It Make Sense to Speak of 'Good Probability Appraisers'?," in I.J. Good (ed.), *The Scientist Speculates: An Anthology of Partly-Baked Ideas.* New York: Basic Books.

de Finetti, B. (1965), "Methods for Discriminating Levels of Partial Knowledge Concerning a Test Item," *British Journal of Mathematical and Statistical Psychology, 18*, 87–123.

de Finetti, B. (1968), "Probability: Interpretations," in *International Encyclopedia of the Social Sciences.* New York: Macmillan, *12*, 496–504.

de Finetti, B. (1969), "Initial Probabilities: A Prerequisite for Any Valid Induction," *Synthese, 20*, 2–24 (with discussion).

de Finetti, B. (1970), "Logical Foundations and Measurement of Subjective Probability," *Acta Psychologica, 34*, 129–145.

De Groot, M.H. (1963), "Some Comments on the Experimental Measurement of Utility," *Behavioral Science, 8*, 146–149.

De Groot, M.H. (1968), "Some Problems of Optimal Stopping," *Journal of the Royal Statistical Society B, 30*, 108–122.

De Groot, M.H. (1970), *Optimal Statistical Decisions.* New York: McGraw-Hill.

De Groot, M.H., and Starr, N. (1969), "Optimal Two-Stage Stratified Sampling," *Annals of Mathematical Statistics, 40*, 575–582.

Dempster, A.P. (1963), "On Direct Probabilities," *Journal of the Royal Statistical Society B, 25*, 100–110.

Dempster, A.P. (1968), "A Generalization of Bayesian Inference," *Journal of the Royal Statistical Society B, 30*, 205–247 (with discussion).

Dickey, J.M. (1971), "The Weighted Likelihood Ratio, Linear Hypotheses on Normal Location Parameters," *Annals of Mathematical Statistics, 42*, 204–223.

Dickey, J.M., and Lientz, B.P. (1970), "The Weighted Likelihood Ratio, Sharp Hypotheses About Chances, the Order of a Markov Chain," *Annals of Mathematical Statistics, 41*, 214–226.

Dolbear, F.T. (1963), "Individual Choice Under Uncertainty—An Experimental Study," *Yale Economic Essays, 3*, 418–469.

Doob, J.L. (1953), *Stochastic Processes.* New York: Wiley.

Draper, N.R., and Guttman, I. (1968), "Some Bayesian Stratified Two-Phase Sampling Results," *Biometrika, 55*, 131–139.

Draper, N.R., and Hunter, W.G. (1967), "The Use of Prior Distributions in the Design of Experiments for Parameter Estimation in Non-Linear Situations," *Biometrika, 54*, 147–153.

Du Charme, W.M. (1970), "Response Bias Explanation of Conservative Human Inference," *Journal of Experimental Psychology, 85*, 66–74.

Dunsmore, I.R. (1966), "A Bayesian Approach to Classification," *Journal of the Royal Statistical Society B, 28*, 568–577.

Dunsmore, I.R. (1968), "A Bayesian Approach to Calibration," *Journal of the Royal Statistical Society B, 30*, 396–405.

Dunsmore, I.R. (1969), "Regulation and Optimization," *Journal of the Royal Statistical Society B, 31*, 160–170.

Durbin, J. (1970), "On Birnbaum's Theorem on the Relation between Sufficiency, Conditionality, and Likelihood," *Journal of the American Statistical Association, 65*, 395–398.

Easterling, R.G., and Weeks, D.L. (1970), "An Accuracy Criterion for Bayesian

Tolerance Intervals," *Journal of the Royal Statistical Society B*, *32*, 236–240.

Edwards, W. (1954), "The Theory of Decision Making," *Psychological Bulletin*, *51*, 380–417.

Edwards, W. (1961), "Behavioral Decision Theory," *Annual Review of Psychology*, *12*, 473–498.

Edwards, W. (1962), "Dynamic Decision Theory and Probabilistic Information Processing," *IEEE Transactions on Human Factors in Electronics*, *4*, 59–73.

Edwards, W. (1965), "Tactical Note on the Relation between Scientific and Statistical Hypotheses," *Psychological Bulletin*, *63*, 400–402.

Edwards, W. (1966), "Nonconservative Probabilistic Information Processing Systems," Report ESD-TR-66-404, Engineering Psychology Laboratory, University of Michigan, Ann Arbor.

Edwards, W. (1968), "Conservatism in Human Information Processing," in B. Kleinmuntz (ed.), *Formal Representation of Human Judgment.* New York: Wiley.

Edwards, W. (1969), *A Bibliography of Research on Behavioral Decision Processes to 1968.* Ann Arbor: Engineering Psychology Laboratory, University of Michigan.

Edwards, W., Lindman, H., and Savage, L.J. (1963), "Bayesian Statistical Inference for Psychological Research," *Psychological Review*, *70*, 193–242.

Edwards, W., Phillips, L.D., Hays, W.L., and Goodman, B.C. (1968), "Probabilistic Information Processing Systems: Design and Evaluation," *IEEE Transactions on Systems Science and Cybernetics*, *SSC-4*, 248–265.

Edwards, W., and Tversky, A. (eds.) (1967), *Decision Making.* Baltimore: Penguin.

Eisenberg, E., and Gale, D. (1959), "Consensus of Subjective Probabilities: The Pari-Mutuel Method," *Annals of Mathematical Statistics*, *30*, 165–168.

Ellsberg, D. (1954), "Classical and Current Notions of 'Measurable Utility,'" *Economic Journal*, *64*, 528–556.

Ellsberg, D. (1961), "Risk, Ambiguity, and the Savage Axioms," *Quarterly Journal of Economics*, *75*, 643–669.

Epstein, E.S. (1962), "A Bayesian Approach to Decision Making in Applied Meteorology," *Journal of Applied Meteorology*, *1*, 169–177.

Ericson, W.A. (1965), "Optimum Stratified Sampling Using Prior Information," *Journal of the American Statistical Association*, *60*, 750–771.

Ericson, W.A. (1967a), "On the Economic Choice of Experiment Sizes for Decisions Regarding Certain Linear Combinations," *Journal of the Royal Statistical Society B*, *29*, 503–512.

Ericson, W.A. (1967b), "Optimal Sample Design with Nonresponse," *Journal of the American Statistical Association*, *62*, 63–78.

Ericson, W.A. (1968), "Optimum Allocation in Stratified and Multistage Samples Using Prior Information," *Journal of the American Statistical Association*, *63*, 964–983.

Ericson, W.A. (1969a), "A Note on the Posterior Mean of a Population Mean," *Journal of the Royal Statistical Society B*, *31*, 332–334.

Ericson, W.A. (1969b), "Subjective Bayesian Models in Sampling Finite Populations," *Journal of the Royal Statistical Society B*, *31*, 195–233 (with discussion).

Ericson, W.A. (1969c), "Subjective Bayesian Models in Sampling Finite Populations: Stratification," in N.L. Johnson and H. Smith, Jr. (eds.), *New Developments in Survey Sampling*, 326–357. New York: Wiley.

Ericson, W.A. (1970), "On the Posterior Mean and Variance of a Population Mean," *Journal of the American Statistical Association, 65*, 649–652.

Fama, E.F. (1970), "Multiperiod Consumption-Investment Decisions," *American Economic Review, 60*, 163–174.

Farrar, D.W. (1962), *The Investment Decision under Uncertainty*. Englewood Cliffs, N.J.: Prentice-Hall.

Feller, W. (1968), *An Introduction to Probability Theory and Its Applications*, Vol. 1, 3rd Ed. New York: Wiley.

Fellner, W. (1961), "Distortion of Subjective Probabilities as a Reaction to Uncertainty," *Quarterly Journal of Economics, 75*, 670–689.

Fellner, W. (1965), *Probability and Profit: A Study of Economic Behavior Along Bayesian Lines*. Homewood, Ill.: Irwin.

Ferguson, T.S. (1967), *Mathematical Statistics: A Decision-Theoretic Approach*. New York: Academic Press.

Fishburn, P.C. (1964), *Decision and Value Theory*. New York: Wiley.

Fishburn, P.C. (1966a), "Decision under Uncertainty: An Introductory Exposition," *Journal of Industrial Engineering, 17*, 341–353.

Fishburn, P.C. (1966b), "On the Prospects of a Useful Unified Theory of Value for Engineering," *IEEE Transactions on Systems Science and Cybernetics, SSC-2*, 27–35.

Fishburn, P.C. (1967), "Methods of Estimating Additive Utilities," *Management Science, 13*, 435–453.

Fishburn, P.C. (1968), "Utility Theory," *Management Science, 14*, 335–378.

Fishburn, P.C. (1970), *Utility Theory for Decision Making*. New York: Wiley.

Fishburn, P.C., Murphy, A.H., and Isaacs, H.H. (1968), "Sensitivity of Decisions to Probability Estimation Errors: A Reexamination," *Operations Research, 16*, 254–267.

Fisher, R.A. (1958), *Statistical Methods for Research Workers*, 13th Ed. Edinburgh: Oliver and Boyd.

Fisher, R.A. (1959), *Statistical Methods and Scientific Inference*, 2nd Ed. Edinburgh: Oliver and Boyd.

Fisher, R.A. (1960), *The Design of Experiments*, 7th Ed. Edinburgh: Oliver and Boyd.

Forester, J. (1968), *Statistical Selection of Business Strategies*. Homewood, Ill.: Irwin.

Frank, R.E., and Green, P.E. (1967), *Quantitative Methods in Marketing*. Englewood Cliffs, N.J.: Prentice-Hall.

Fraser, D.A.S. (1963), "On the Sufficiency and Likelihood Principles," *Journal of the American Statistical Association, 58*, 641–647.

Freund, J.E. (1962), *Mathematical Statistics*. Englewood Cliffs, N.J.: Prentice-Hall.

Friedman, L. (1956), "A Competitive-Bidding Strategy," *Operations Research, 4*, 104–112.

Friedman, M., and Savage, L.J. (1948), "The Utility Analysis of Choices Involving Risk," *Journal of Political Economy, 56*, 279–304.

Friedman, M., and Savage, L.J. (1952), "The Expected-Utility Hypothesis and the Measurability of Utility," *Journal of Political Economy, 60*, 463–474.

Ginsberg, A.S. (1970), "Decision Analysis in Clinical Patient Management," *Proceedings of the Second Conference on the Diagnostic Process.* New York: Academic Press.

Ginsberg, A.S., and Offensend, F.L. (1968), "An Application of Decision Theory to a Medical Diagnosis-Treatment Problem," *IEEE Transactions on Systems Science and Cybernetics, SSC-4*, 355–362.

Godambe, V.P., and Sprott, D.A. (eds.) (1971), *Foundations of Statistical Inference.* Toronto: Holt, Rinehart and Winston of Canada.

Goldberg, S. (1960), *Probability: An Introduction.* Englewood Cliffs, N.J.: Prentice-Hall.

Good, I.J. (1950), *Probability and the Weighing of Evidence.* London: Griffin.

Good, I.J. (1952), "Rational Decisions," *Journal of the Royal Statistical Society B, 14*, 107–114.

Good, I.J. (1959), "Kinds of Probability," *Science, 129*, 443–447.

Good, I.J. (1962), "How Rational Should a Manager Be?," *Management Science, 8*, 383–393.

Good, I.J. (1965), *The Estimation of Probabilities: An Essay on Modern Bayesian Methods.* Cambridge: M.I.T. Press.

Good, I.J. (1967), "A Bayesian Significance Test for Multinomial Distributions," *Journal of the Royal Statistical Society B, 29*, 399–431 (with discussion).

Good, I.J. (1969), "A Subjective Evaluation of Bode's Law and an 'Objective' Test for Approximate Numerical Rationality," *Journal of the American Statistical Association, 64*, 23–66 (with discussion).

Grayson, C.J., Jr. (1960), *Decisions under Uncertainty: Drilling Decisions by Oil and Gas Operators.* Boston: Division of Research, Harvard Business School.

Green, P.E. (1962), "Bayesian Decision Theory in Advertising," *Journal of Advertising Research, 2*, 33–41.

Green, P.E. (1963), "Bayesian Decision Theory in Pricing Strategy," *Journal of Marketing, 27*, 5–14.

Green, P.E., and Frank, R.E. (1966), "Bayesian Statistics and Marketing Research," *Applied Statistics, 15*, 173–190.

Green, P.E., Halbert, M.H., and Minas, F.S. (1964), "An Experiment in Information Buying," *Journal of Advertising Research, 4*, 17–23.

Green, P.E., Halbert, M.H., and Robinson, P.J. (1965), "An Experiment in Probability Estimation," *Journal of Marketing Research, 2*, 266–273.

Gustafson, D.H., Edwards, W., Phillips, L.D., and Slack, W.V. (1969), "Subjective Probabilities in Medical Diagnosis," *IEEE Transactions on Man-Machine Systems, MMS-10*, 61–65.

Guttman, I. (1970), *Statistical Tolerance Regions, Classical and Bayesian.* London: Griffin.

Hacking, I. (1965), *Logic of Statistical Inference.* Cambridge: Cambridge University Press.

Hadley, G. (1961), *Linear Algebra.* Reading, Mass.: Addison-Wesley.

Hadley, G. (1962), *Linear Programming.* Reading, Mass.: Addison-Wesley.

Hadley G. (1964), *Nonlinear and Dynamic Programming.* Reading, Mass.: Addison-Wesley.

Hadley, G. (1967), *Introduction to Probability and Statistical Decision Theory.* San Francisco: Holden-Day.

Hakansson, N.H. (1969), "An Induced Theory of Accounting under Risk," *Accounting Review, 44*, 495–514.

Hakansson, N.H. (1970a), "Friedman-Savage Utility Functions Consistent with Risk Aversion," *Quarterly Journal of Economics, 84*, 472–487.

Hakansson, N.H. (1970b), "Optimal Investment and Consumption Strategies under Risk for a Class of Utility Functions," *Econometrica, 38*, 587–607.

Hakansson, N.H. (1971a), "Capital Growth and the Mean-Variance Approach to Portfolio Selection," *Journal of Financial and Quantitative Analysis, 6*, 517–557.

Hakansson, N.H. (1971b), "Optimal Entrepreneurial Decisions in a Completely Stochastic Environment," *Management Science, 17*, 427–449.

Halter, A.N., and Dean, G.W. (1971), *Decisions under Uncertainty.* Cincinnati: South-Western.

Hammond, J.S. (1967), "Better Decisions with Preference Theory," *Harvard Business Review, 45*, Nov.–Dec., 123–141.

Hanoch, G., and Levy, H. (1970), "Efficient Portfolio Selection with Quadratic and Cubic Utility," *Journal of Business, 43*, 181–189.

Harnett, D.L. (1970), *Statistical Methods.* Reading, Mass.: Addison-Wesley.

Harsanyi, J.C. (1967), "Games with Incomplete Information Played by 'Bayesian' Players. Part I, The Basic Model," *Management Science, 14*, 159–182.

Harsanyi, J.C. (1968a), "Games with Incomplete Information Played by 'Bayesian' Players. Part II, Bayesian Equilibrium Points," *Management Science, 14*, 320–334.

Harsanyi, J.C. (1968b), "Games with Incomplete Information Played by 'Bayesian' Players. Part III, The Basic Probability Distribution of the Game," *Management Science, 14*, 486–502.

Hartigan, J.A. (1967), "The Likelihood and Invariance Principles," *Journal of the Royal Statistical Society B, 29*, 533–539.

Hartigan, J.A. (1969), "Linear Bayesian Methods," *Journal of the Royal Statistical Society B, 31*, 446–454.

Hartley, H.O. (1963), "In Dr. Bayes' Consulting Room," *The American Statistician, 17*, 22–24.

Hausman, W.H., and White, W.L. (1968), "Theory of Option Strategy under Risk Aversion," *Journal of Financial and Quantitative Analysis, 3*, 343–358.

Hausner, M. (1954), "Multidimensional Utilities," in R.M. Thrall *et al, Decision Processes.* New York: Wiley.

Hayes, R.H. (1969a), "Optimal Strategies for Divestiture," *Operations Research, 17*, 242–310.

Hayes, R.H. (1969b), "Qualitative Insights from Quantitative Methods," *Harvard Business Review, 47*, July–August, 108–117.

Hayes, R.H. (1969c), "Statistical Estimation Problems in Inventory Control," *Management Science, 15*, 686–701.

Hays, W.L. (1963), *Statistics for Psychologists.* New York: Holt, Rinehart and Winston.

Hays, W.L., and Winkler, R.L. (1970), *Statistics: Probability, Inference, and Decision*, 2 Vols. New York: Holt, Rinehart and Winston.

Helmer, O. (1966), *Social Technology*. New York: Basic Books.

Hershman, R.L., and Levine, J.R. (1970), "Deviations from Optimum Information-Purchase Strategies in Human Decision-Making," *Organizational Behavior and Human Performance*, *5*, 313–329.

Hertz, D.B. (1964), "Risk Analysis in Capital Investment," *Harvard Business Review* *42*, January–February, 95–106.

Hertz, D.B. (1968), "Investment Policies that Pay Off," *Harvard Business Review*, *46*, January–February, 96–108.

Hespos, R.F., and Strassmann, P.A. (1965), "Stochastic Decision Trees for the Analysis of Investment Decisions," *Management Science B*, *11*, 244–259.

Hildreth, C. (1963), "Bayesian Statisticians and Remote Clients," *Econometrica*, *31*, 422–438.

Hill, B.M. (1963), "The Three-Parameter Lognormal Distribution and Bayesian Analysis of a Point-Source Epidemic," *Journal of the American Statistical Association*, *58*, 72–84.

Hill, B.M. (1967), "Correlated Errors in the Random Model," *Journal of the American Statistical Association*, *62*, 1387–1400.

Hill, B.M. (1968), "Posterior Distribution of Percentiles: Bayes' Theorem for Sampling from a Population," *Journal of the American Statistical Association*, *63*, 677–691.

Hill, B.M. (1969), "Foundations for the Theory of Least Squares," *Journal of the Royal Statistical Society B*, *31*, 89–97.

Hill, B.M. (1970), "Zipf's Law and Prior Distributions," *Journal of the American Statistical Association*, *65*, 1220–1232.

Hillier, F.S., and Lieberman, G.J. (1967), *Introduction to Operations Research*. San Francisco: Holden-Day.

Hirschleifer, J. (1961), "The Bayesian Approach to Statistical Decision: An Exposition," *Journal of Business*, 471–489.

Hoadley, A.B. (1969), "The Compound Multinomial Distribution and Bayesian Analysis of Categorical Data from Finite Populations," *Journal of the American Statistical Association*, *64*, 216–229.

Hoadley, A.B. (1970), "A Bayesian Look at Inverse Linear Regression," *Journal of the American Statistical Association*, *65*, 356–369.

Hodges, J.L., and Lehmann, E.L. (1970), *Basic Concepts of Probability and Statistics*, 2nd Ed. San Francisco: Holden-Day.

Hoel, P.G. (1962), *Introduction to Mathematical Statistics*, 3rd Ed. New York: Wiley.

Horowitz, I. (1970), *Decision Making and the Theory of the Firm*. New York: Holt, Rinehart and Winston.

Horowitz, I. (1972), *An Introduction to Quantitative Business Analysis*, 2nd Ed. New York: McGraw-Hill.

Howard, R.A. (1960), *Dynamic Programming and Markov Processes*. Cambridge, Mass.: M.I.T. Press.

Howard, R.A. (1965a), "Bayesian Decision Models for System Engineering," *IEEE Transactions*, *SSC-1*, 36–40.

Howard, R.A. (1965b), "Dynamic Inference," *Operations Research, 13*, 712–733.

Howard, R.A. (1966a), "Decision Analysis: Applied Decision Theory," *Proceedings of the Fourth International Conference on Operational Research.* Boston: International Federation of Operational Research Societies.

Howard, R.A. (1966b), "Information Value Theory," *IEEE Transactions on Systems Science and Cybernetics, SSC-2*, 22–26.

Howard, R.A. (1967), "Value of Information Lotteries," *IEEE Transactions on Systems Science and Cybernetics, SSC-3*, 54–60.

Howard, R.A. (1968), "The Foundations of Decision Analysis," *IEEE Transactions on Systems Science and Cybernetics, SSC-4*, 211–219.

Howard, R.A. (1971), "Proximal Decision Analysis," *Management Science, 17*, 507–541.

Howell, W.C., Gettys, C.F., and Martin, D.W. (1971), "On the Allocation of Inference Functions in Decision Systems," *Organizational Behavior and Human Performance, 6*, 132–149.

Hurst, E.G., Jr. (1968), "Bayesian Autoregressive Time Series Analysis," *IEEE Transactions on Systems Science and Cybernetics, SSC-4*, 317–324.

Isaacs, H.H. (1963), "Sensitivity of Decisions to Probability Estimation Errors," *Operations Research, 11*, 536–552.

Jamison, D. (1969), "Conjoint Measurement of Time Preference and Utility." Santa Monica: Rand, RM-6029-PR.

Jeffrey, R.C. (1965), *The Logic of Decision.* New York: McGraw-Hill.

Jeffreys, H. (1961), *Theory of Probability*, 3rd Ed. Oxford: Clarendon Press.

Johnson, N., and Kotz, S. (1969), *Distributions in Statistics: Discrete Distributions.* Boston: Houghton Mifflin.

Johnson, N., and Kotz, S. (1970), *Distributions in Statistics: Continuous Univariate Distributions*, 2 Vols. Boston: Houghton Mifflin.

Kalymon, B.A. (1971), "Estimation Risk in the Portfolio Selection Model," *Journal of Financial and Quantitative Analysis, 6*, 559–582.

Kaplan, R.J., and Newman, J.R. (1966), "Studies in Probabilistic Information Processing," *IEEE Transactions on Human Factors in Electronics, HFE-7*, 49–63.

Karlin, S. (1966), *A First Course in Stochastic Processes.* New York: Academic Press.

Kaufman, G.M. (1963a), "Sequential Investment Analysis under Uncertainty," *Journal of Business, 36*, 39–64.

Kaufman, G.M. (1963b), *Statistical Decision and Related Techniques in Oil and Gas Exploration.* Englewood Cliffs, N.J.: Prentice-Hall.

Kaufman, G.M. (1968), "Optimal Sample Size in Two-Action Problems When the Sample Observations Are Lognormal and the Precision Is Known," *Journal of the American Statistical Association, 63*, 653–659.

Keeney, R.L. (1968a), "Evaluating Multidimensional Situations Using a Quasi-Separable Utility Function," *IEEE Transactions on Man-Machine Systems, MMS-9*, 25–28.

Keeney, R.L. (1968b), "Quasi-Separable Utility Functions," *Naval Research Logistics Quarterly, 15*, 551–566.

Keeney, R.L. (1969), "Multidimensional Utility Functions: Theory, Assessment, and Application," *Technical Report No. 43*, Operations Research Center, Massachusetts Institute of Technology.

Kendall, M.G. (1966), "Statistical Inference in the Light of the Theory of the Electronic Computer," *Review of the International Statistical Institute, 34*, 1–12.

Kendall, M.G. (1968), "On the Future of Statistics—A Second Look," *Journal of the Royal Statistical Society A, 131*, 182–204 (with discussion).

Kendall, M., and Stuart, A. (1958, 1961, 1966), *The Advanced Theory of Statistics*, 3 Vols. London: Griffin.

Keynes, J.M. (1921), *A Treatise on Probability*. London: Macmillan.

Kogan, N., and Wallach, M.A. (1964), *Risk Taking: A Study in Cognition and Personality*. New York: Holt, Rinehart and Winston.

Kolmogorov, A.N. (1933), *Grundbegriffe Der Wahrscheinlichkeitsrechnung*. Berlin: Springer. Translated, Edited by N. Morrison (1950) as *Foundations of the Theory of Probability*. New York: Chelsea.

Kyburg, H.E. (1966), "Probability and Decision," *Philosophy of Science, 33*, 250–261.

Kyburg, H.E., and Smokler, H.E. (eds.) (1964), *Studies in Subjective Probability*. New York: Wiley.

Laplace, P.S. (1820), *Essai Philosophique Sur Les Probabilités*, Translated by F.W. Truscott and F.L. Emory (1951), *A Philosophical Essay on Probabilities*. New York: Dover.

LaValle, I.H. (1967), "A Bayesian Approach to an Individual Player's Choice of Bid in Competitive Sealed Auctions," *Management Science, 13*, 584–597.

LaValle, I.H. (1968), "On Cash Equivalents and Information Evaluation in Decisions under Uncertainty," *Journal of the American Statistical Association, 63*, 252–290.

LaValle, I.H. (1970), *An Introduction to Probability, Decision, and Inference*. New York: Holt, Rinehart and Winston.

Lee, T.C., Judge, G.G., and Zellner, A. (1968), "Maximum Likelihood and Bayesian Estimation of Transition Probabilities," *Journal of the American Statistical Association, 63*, 1162–1179.

Lehmann, E.L. (1959), *Testing Statistical Hypotheses*. New York: Wiley.

Levi, I. (1967), *Gambling with Truth*. New York: Alfred A. Knopf.

Levy, H., and Hanoch, G. (1970), "Relative Effectiveness of Efficiency Criteria for Portfolio Selection," *Journal of Financial and Quantitative Analysis, 5*, 63–76.

Lindgren, B.W. (1968), *Statistical Theory*, 2nd ed. New York: Macmillan.

Lindley, D.V. (1953), "Statistical Inference," *Journal of the Royal Statistical Society B, 11*, 30–76.

Lindley, D.V. (1957), "A Statistical Paradox," *Biometrika, 44*, 187–192.

Lindley, D.V. (1961a), "Dynamic Programming and Decision Theory," *Applied Statistics, 10*, 39–51.

Lindley, D.V. (1961b), "The Use of Prior Probability Distributions in Statistical Inference and Decision," in J. Neyman (ed.), *Proceedings of the Fourth Berkeley Symposium on Mathematical Statistics and Probability, 1*, 453–468. Berkeley: University of California Press.

Lindley, D.V. (1964), "The Bayesian Analysis of Contingency Tables," *Annals of Mathematical Statistics, 35*, 1622–1643.

Lindley, D.V. (1965), *Introduction to Probability and Statistics from a Bayesian Viewpoint*, 2 Vols., Cambridge: Cambridge University Press.

Lindley, D.V. (1968), "The Choice of Variables in Multiple Regression," *Journal of the Royal Statistical Society B, 30*, 31–66.

Lindley, D.V. (1971), *Making Decisions*. New York: Wiley.

Lindley, D.V., and Barnett, B.N. (1965), "Sequential Sampling: Two Decision Problems with Linear Losses for Binomial and Normal Variables," *Biometrika, 52*, 507–532.

Lindley, D.V., and El-Sayyad, G.M. (1968), "The Bayesian Estimation of a Linear Functional Relationship," *Journal of the Royal Statistical Society B*, 190–202.

Lippman, S.A. (1971), *Elements of Probability and Statistics*. New York: Holt, Rinehart and Winston.

Luce, R.D., and Raiffa, H. (1955), *Games and Decisions*. New York: Wiley.

Luce, R.D., and Suppes, P. (1965), "Preference, Utility and Subjective Probability," in R.D. Luce, R.R. Bush, and E. Galanter (eds.), *Handbook of Mathematical Psychology, 3*. New York: Wiley.

Lusted, L.B. (1968), *Introduction to Medical Decision Making*. Springfield, Ill.: Thomas.

McCall, J.J. (1965), "The Economics of Information and Optimal Stopping Rules," *Journal of Business, 38*, 300–317.

MacCrimmon, K.R. (1968), "Descriptive and Normative Implications of the Decision-Theory Postulates," in Borch and Mossin (1968), 3–23.

Machol, R.E. (ed.) (1960), *Information and Decision Processes*. New York: McGraw-Hill.

Machol, R.E., and Gray, P. (eds.) (1962), *Recent Developments in Information and Decision Processes*. New York: Macmillan.

Magee, J.F. (1964a), "Decision Trees for Decision Making," *Harvard Business Review, 42*, July–August, 126–138.

Magee, J.F. (1964b), "How to Use Decision Trees in Capital Investment," *Harvard Business Review, 42*, September–October, 79–96.

Mao, J.C.T., and Särndal, C.E. (1966), "A Decision Theory Approach to Portfolio Selection," *Management Science B, 12*, 323–333.

Maritz, J.S. (1970), *Empirical Bayes Methods*. London: Methuen.

Markowitz, H. (1952), "The Utility of Wealth," *Journal of Political Economy, 60*, 151–158.

Markowitz, H.M. (1959), *Portfolio Selection*. New York: Wiley.

Marschak, J. (1950), "Rational Behavior, Uncertain Prospects, and Measurable Utility," *Econometrica, 18*, 111–141.

Marschak, J. (1964), "Actual versus Consistent Decision Behavior," *Behavioral Science, 9*, 103–110.

Martin, E.W. (1969), *Mathematics for Decision Making*, 2 Vols. Homewood, Ill.: Irwin.

Martin, J.J. (1967), *Bayesian Decision Problems and Markov Chains*. New York: Wiley.

Massy, W.F., Montgomery, D.G., and Morrison, D.G. (1970), *Stochastic Models of Buying Behavior*. Cambridge: M.I.T. Press.

Matheson, J.E. (1969), "Decision Analysis Practice: Examples and Insights,"

Proceedings of the Fifth International Conference of Operational Research, 677–691.

Meehl, P.E. (1954), *Clinical versus Statistical Prediction: A Theoretical Analysis and a Review of the Evidence.* Minneapolis: University of Minnesota Press.

Menezes, C.F., and Hanson, D.L. (1970), "On the Theory of Risk Aversion," *International Economic Review, 11,* 481–487.

Mercer, A., and Russell, J.I.T. (1969), "Recurrent Competitive Bidding," *Operational Research Quarterly, 20,* 209–221.

Meyer, R.F., and Pratt, J.W. (1968), "The Consistent Assessment and Fairing of Preference Functions," *IEEE Transactions on Systems Science and Cybernetics, SSC-4,* 270–278.

Mills, E.S. (1959), "Uncertainty and Price Theory," *Quarterly Journal of Economics, 73,* 116–130.

Mood, A.M., and Graybill, F.A. (1963), *Introduction to the Theory of Statistics.* New York: McGraw-Hill.

Moore, P.G. (1966), "The Bayesian Approach to Statistics," *Journal of the Institute of Actuaries, 92,* 326–339.

Morrison, D.G. (1967), "On the Consistency of Preferences in Allais' Paradox," *Behavioral Science, 12,* 373–383.

Morris, W.T. (1968), *Management Science: A Bayesian Introduction.* Englewood Cliffs, N.J.: Prentice-Hall.

Mosteller, F., and Nogee, P. (1951), "An Experimental Measurement of Utility," *Journal of Political Economy, 59,* 371–404.

Mosteller, F., Rourke, R.E.K., and Thomas, G.B. (1970), *Probability with Statistical Applications,* 2nd ed. Reading, Mass.: Addison-Wesley.

Mosteller, F., and Wallace, D.L. (1964), *Inference and Disputed Authorship: The Federalist.* Reading, Mass.: Addision-Wesley.

Murphy, A.H. (1966), "A Note on the Utility of Probability Predictions and the Probability Score in the Cost-Loss Ratio Decision Situation," *Journal of Applied Meteorology, 5,* 534–537.

Murphy, A.H. (1969), "Measures of the Utility of Probabilistic Predictions in Cost-Loss Ratio Decision Situations in Which Knowledge of the Cost-Loss Ratios is Incomplete," *Journal of Applied Meteorology, 8,* 863–873.

Murphy, A.H., and Winkler, R.L. (1970), "Scoring Rules in Probability Assessment and Evaluation," *Acta Psychologica, 34,* 273–286.

Murray, G.R., and Silver, E.A. (1966), "A Bayesian Analysis of the Style Goods Inventory Problem," *Management Science, 12,* 785–797.

Nelson, R.R., and Winter, S.G. (1964), "A Case Study in the Economics of Information and Coordination: The Weather Forecasting System," *Quarterly Journal of Economics, 78,* 420–441.

Nemhauser, G.L. (1966), *Introduction to Dynamic Programming.* New York: Wiley.

Neyman, J. (1952), *Lectures and Conferences on Mathematical Statistics and Probability,* 2nd ed. Washington: U.S. Department of Agriculture.

Neyman, J. (1962), "Two Breakthroughs in the Theory of Statistical Decision Making," *Review of the International Statistical Institute, 30,* 11–27.

Neyman, J., and Pearson, E.S. (1928), "On the Use and Interpretation of

Certain Test Criteria for Purposes of Statistical Inference," *Biometrika*, *20*, 175–240, 263–294.

Neyman, J., and Pearson, E.S. (1933), "On the Problem of the Most Efficient Tests of Statistical Hypotheses," *Philosophical Transactions of the Royal Society, London* (Series A), *231*, 289–337.

Nogee, P., and Lieberman, B. (1960), "The Auction Value of Certain Risky Situations," *Journal of Psychology*, *49*, 167–169.

North, D.W. (1968), "A Tutorial Introduction to Decision Theory," *IEEE Transactions on Systems Science and Cybernetics*, *SSC-4*, 200–210.

Novick, M.R. (1969), "Multiparameter Bayesian Indifference Procedure," *Journal of the Royal Statistical Society B*, *31*, 29–64 (with discussion).

Novick, M.R., and Grizzle, J.E. (1965), "A Bayesian Approach to the Analysis of Data from Clinical Trials," *Journal of the American Statistical Association*, *60*, 81–96.

Novick, M.R., and Hall, W.J. (1965), "A Bayesian Indifference Procedure," *Journal of the American Statistical Association*, *60*, 1104–1117.

Page, A.N. (1968), *Utility Theory: A Book of Readings*. New York: Wiley.

Pankoff, L.D., and Roberts, H.V. (1968), "Bayesian Synthesis of Clinical and Statistical Prediction," *Psychological Bulletin*, *70*, 762–773.

Parzen, E. (1960), *Modern Probability Theory and Its Applications*. New York: Wiley.

Parzen, E. (1962), *Stochastic Processes*. San Francisco: Holden-Day.

Patil, G.P., and Joshi, S.W. (1968), *A Dictionary and Bibliography of Discrete Distributions*. Edinburgh: Oliver and Boyd.

Pearson, E.S. (1925), "Bayes' Theorem, Examined in the Light of Experimental Sampling," *Biometrika*, *17*, 388–442.

Peterson, C.R., and Beach, L.R. (1967), "Man as an Intuitive Statistician," *Psychological Bulletin*, *68*, 29–46.

Pfanzagl, J. (1963), "Sampling Procedures Based on Prior Distributions and Costs," *Technometrics*, *5*, 47–61.

Phillips, L., and Edwards, W. (1966), "Conservatism in a Simple Probability Inference Task," *Journal of Experimental Psychology*, *72*, 346–354.

Phillips, L.D., Hays, W.L., and Edwards, W. (1966), "Conservatism in Complex Probabilistic Inference," *IEEE Transactions on Human Factors in Electronics*, *7*, 7–18.

Pierce, D.A., and Folks, J.L. (1969), "Sensitivity of Bayes Procedures to the Prior Distribution," *Operations Research*, *17*, 344–350.

Plackett, R.L. (1966), "Current Trends in Statistical Inference," *Journal of the Royal Statistical Society A*, *129*, 249–267.

Popper, K. (1959), *The Logic of Scientific Discovery*. London: Hutchinson.

Pratt, J.W. (1964), "Risk Aversion in the Small and in the Large," *Econometrica*, *32*, 122–136.

Pratt, J.W. (1965), "Bayesian Interpretation of Standard Inference Statements," *Journal of the Royal Statistical Society B*, *27*, 169–203 (with discussion).

Pratt, J.W., Raiffa, H., and Schlaifer, R. (1964), "The Foundations of Decision under Uncertainty: An Elementary Exposition," *Journal of the American Statistical Association*, *59*, 353–375.

Pratt, J.W., Raiffa, H., and Schlaifer, R. (1965), *Introduction to Statistical Decision Theory*, preliminary ed. New York: McGraw-Hill.

Press, S.J. (1971), *Applied Multivariate Analysis Including Bayesian Techniques*. New York: Holt, Rinehart and Winston.

Preston, M.G., and Baratta, P. (1948), "An Experimental Study of the Auction Value of an Uncertain Outcome," *American Journal of Psychology, 61*, 183–193.

Pritsker, A.A.B. (1965), "A Decision Theory Approach to Enrollment Prediction," *Journal of Industrial Engineering, 16*, 164–170.

Raiffa, H. (1961), "Risk, Ambiguity, and the Savage Axioms: Comment," *Quarterly Journal of Economics, 75*, 690–694.

Raiffa, H. (1968), *Decision Analysis*. Reading, Mass.: Addison-Wesley.

Raiffa, H. (1969), "Preferences for Multi-Attributed Alternatives." Santa Monica, Cal.: Rand Memorandum RM-5868-DOT/RC.

Raiffa, H., and Schlaifer, R. (1961), *Applied Statistical Decision Theory*. Boston: Division of Research, Harvard Business School.

Ramsey, F.P. (1931), "Truth and Probability," in F.P. Ramsey, *The Foundations of Mathematics and Other Logical Essays*. New York: Harcourt. [Reprinted in Kyburg and Smokler (1964).]

Rapoport, A. (1964), "Sequential Decision-Making in a Computer-Controlled Task," *Journal of Mathematical Psychology, 1*, 351–374.

Rapoport, A. (1967), "Dynamic Programming Models for Multistage Decision-Making Tasks," *Journal of Mathematical Psychology, 4*, 48–71.

Rapoport, A. (1968), "Choice Behavior in a Markovian Decision Task," *Journal of Mathematical Psychology, 5*, 163–181.

Rapoport, A., and Tversky, A. (1970), "Choice Behavior in an Optional Stopping Task," *Organizational Behavior and Human Performance, 5*, 105–120.

Rappaport, A. (1967), "Sensitivity Analysis in Decision Making," *Accounting Review, 42*, 441–456.

Robbins, H. (1955), "An Empirical Bayes Approach to Statistics," *Proceedings of the Third Berkeley Symposium on Mathematical Statistics and Probability, 1*, 157–164.

Robbins, H. (1964), "The Empirical Bayes Approach to Statistical Decision Problems," *Annals of Mathematical Statistics, 35*, 1–20.

Roberts, H.V. (1960), "The New Business Statistics," *Journal of Business, 33*, 21–30.

Roberts, H.V. (1963a), "Bayesian Statistics in Marketing," *Journal of Marketing, 27*, 1–4.

Roberts, H.V. (1963b), "Risk, Ambiguity, and the Savage Axioms: Comment," *Quarterly Journal of Economics, 77*, 327–335.

Roberts, H.V. (1965), "Probabilistic Prediction," *Journal of the American Statistical Association, 60*, 50–62.

Roberts, H.V. (1966a), "Statistical Dogma: One Response to a Challenge," *The American Statistician, 20*, 25–27.

Roberts, H.V. (1966b), *Statistical Inference and Decision*. Chicago: Graduate School of Business, University of Chicago.

Roberts, H.V. (1967), "Informative Stopping Rules about Population Size," *Journal of the American Statistical Association, 62*, 763–775.

Rothkopf, M.H. (1969), "A Model of Rational Competitive Bidding," *Management Science, 15*, 362–373.

Royden, H.L., Suppes, P., and Walsh, K. (1959), "A Model for the Experimental Measurement of the Utility of Gambling," *Behavioral Science, 4*, 11–18.

Salmon, W.C. (1966), *The Foundations of Scientific Inference.* Pittsburgh: University of Pittsburgh Press.

Samuelson, P. (1952), "Probability, Utility, and the Independence Axiom," *Econometrica, 20*, 670–678.

Sandmo, A. (1970), "The Effect of Uncertainty on Saving Decisions," *Review of Economic Studies, 37*, 353–360.

Sandmo, A. (1971), "On the Theory of the Competitive Firm under Price Uncertainty," *American Economic Review, 61*, 65–73.

Savage, L.J. (1951), "The Theory of Statistical Decision," *Journal of the American Statistical Association, 46*, 56–67.

Savage, L.J. (1954), *The Foundations of Statistics.* New York: Wiley.

Savage, L.J. (1961), "The Foundations of Statistics Reconsidered," in J. Neyman (ed.), *Proceedings of the Fourth Berkeley Symposium on Mathematical Statistics and Probability, 1.* Berkeley: University of California Press.

Savage, L.J. (1970a), "Comments on a Weakened Principle of Conditionality," *Journal of the American Statistical Association, 65*, 399–401.

Savage, L.J. (1970b), "Diagnosis and the Bayesian Viewpoint," *Proceedings of the Second Conference on the Diagnostic Process.* New York: Academic Press.

Savage, L.J. (1970c), "Reading Suggestions for the Foundations of Statistics," *The American Statistician, 24*, 23–27.

Savage, L.J. (1971), "The Elicitation of Personal Probabilities and Expectations," *Journal of the American Statistical Association, 66*, in press.

Savage, L.J., *et al.* (1962), *The Foundations of Statistical Inference.* London: Methuen.

Sawaragi, Y., Sunahara, Y., and Nakamizo, T. (1967), *Statistical Decision Theory in Adaptive Control Systems.* New York: Academic Press.

Sawyer, J. (1966), "Measurement *and* Prediction, Clinical *and* Statistical," *Psychological Bulletin, 66*, 178–200.

Schlaifer, R. (1959), *Probability and Statistics for Business Decisions.* New York: McGraw-Hill.

Schlaifer, R. (1968), *Manual of Cases on Decision under Uncertainty with Analyses and Teaching Notes* [for use in conjunction with Schlaifer (1969)]. New York: McGraw-Hill.

Schlaifer, R. (1969), *Analysis of Decisions under Uncertainty.* New York: McGraw-Hill.

Schleifer, A. (1969), "Two-Stage Normal Sampling in Two-Action Problems with Linear Economics," *Journal of the American Statistical Association, 64*, 1504–1541.

Schmitt, S.A. (1969), *Measuring Uncertainty: An Elementary Introduction to Bayesian Statistics.* Reading, Mass.: Addison-Wesley.

Schrenk, L.P. (1969), "Aiding the Decision Maker—A Decision Process Model," *IEEE Transactions on Man-Machine Systems, MMS-10*, 204–218.

Seal, H.L. (1969), *Stochastic Theory of a Risk Business.* New York: Wiley.

Searle, S.R., and Hausman, W.H. (1970), *Matrix Algebra for Business and Economics.* New York: Wiley.

Sharpe, W.F. (1970), *Portfolio Theory and Capital Markets.* New York: McGraw-Hill.

Shelly, M.W., and Bryan, G.L. (eds.) (1964), *Human Judgments and Optimality.* New York: Wiley.

Sheridan, T.B. (1970), "On How Often the Supervisor Should Sample," *IEEE Transactions on Systems Science and Cybernetics, SSC-6,* 140–145.

Silver, E.A. (1965), "Bayesian Determination of the Reorder Point of a Slow Moving Item," *Operations Research, 13,* 989–997.

Smallwood, R.D. (1968), "A Decision Analysis of Model Selection," *IEEE Transactions on Systems Science and Cybernetics, SSC-4,* 333–342.

Smith, C.A.B. (1961), "Consistency in Statistical Inference and Decision," *Journal of the Royal Statistical Society B, 23,* 1–37.

Smith, C.A.B. (1965), "Personal Probability and Statistical Analysis," *Journal of the Royal Statistical Society B, 27,* 469–489.

Smith, V.L. (1969), "Measuring Nonmonetary Utilities in Uncertain Choices: The Ellsberg Urn," *Quarterly Journal of Economics, 83,* 324–329.

Soland, R.M. (1968), "Bayesian Analysis of the Weibull Process with Unknown Scale Parameter and Its Application to Acceptance Sampling," *IEEE Transactions on Reliability, R-17,* 84–90.

Soland, R.M. (1969), "Bayesian Analysis of the Weibull Process with Unknown Scale and Shape Parameters," *IEEE Transactions on Reliability, R-18,* 181–184.

Solomon, H., and Zacks, S. (1970), "Optimal Design of Sampling from Finite Populations: A Critical Review and Indication of New Research Areas," *Journal of the American Statistical Association, 65,* 653–677.

Sorenson, J.E. (1969), "Bayesian Analysis in Auditing," *Accounting Review, 44,* 555–561.

Spetzler, C.S. (1968), "The Development of a Corporate Risk Policy for Capital Investment Decisions," *IEEE Transactions on Systems Science and Cybernetics, SSC-4,* 279–300.

Spivey, W.A., and Thrall, R.M. (1970), *Linear Optimization.* New York: Holt, Rinehart and Winston.

Sprott, D.A., and Kalbfleisch, J.G. (1965), "Use of the Likelihood Function in Inference," *Psychological Bulletin, 64,* 15–22.

Staël von Holstein, C.-A.S. (1970a), *Assessment and Evaluation of Subjective Probability Distributions.* Stockholm: Economic Research Institute, Stockholm School of Economics.

Staël von Holstein, C.-A.S. (1970b), "Measurement of Subjective Probability," *Acta Psychologica, 34,* 146–159.

Stimson, D.H. (1969), "Utility Measurement in Public Health Decision Making," *Management Science B, 16,* 17–30.

Stone, M. (1969), "The Role of Experimental Randomization in Bayesian Statistics: Finite Sampling and Two Bayesians," *Biometrika, 56,* 681–683.

Suppes, P. (1956), "The Role of Subjective Probability and Utility in Decision-

Making," *Proceedings of the Third Berkeley Symposium on Mathematical Statistics and Probability*, *5*, 61–73. Berkeley: University of California Press.

Suppes, P. (1961), "Behavioristic Foundations of Utility," *Econometrica*, *29*, 186–202.

Suppes, P. (1969), *Studies in the Methodology and Foundations of Science*. Dordrecht, Holland: D. Reidel Publishing Co.

Suppes, P., and Winet, M. (1955), "An Axiomatization of Utility Based on the Notion of Utility Differences," *Management Science*, *1*, 259–270.

Suppes, P., and Walsh, K. (1959), "A Non-Linear Model for the Experimental Measurement of Utility," *Behavioral Science*, *4*, 204–211.

Swalm, R.O. (1966), "Utility Theory—Insights into Risk Taking," *Harvard Business Review*, 123–136.

Thatcher, A.R. (1964), "Relationships between Bayesian and Confidence Limits for Predictions," *Journal of the Royal Statistical Society B*, *26*, 176–192.

Thrall, R.M., Coombs, C.H., and Davis, R.L. (eds.) (1954), *Decision Processes*. New York: Wiley.

Tiao, G.C., and Zellner, A. (1964), "Bayes' Theorem and the Use of Prior Knowledge in Regression Analysis," *Biometrika*, *51*, 219–230.

Tisdell, C.A. (1968), *The Theory of Price Uncertainty, Production, and Profit*. Princeton: Princeton University Press.

Toda, M., and Shuford, E.H. (1965), "Utility, Induced Utilities, and Small Worlds," *Behavioral Science*, *10*, 238–254.

Tracy, J.A. (1969), "Bayesian Statistical Models in Auditing," *Accounting Review*, *44*, 90–98.

Tribus, M. (1969), *Rational Descriptions, Decisions, and Designs*. New York: Pergamon.

Tukey, J.W. (1960), "Conclusions vs. Decisions," *Technometrics*, *2*, 423–433.

Tukey, J.W. (1962), "The Future of Data Analysis," *Annals of Mathematical Statistics*, *33*, 1–67.

Tversky, A. (1969), "Intransitivity of Preferences," *Psychological Review*, *76*, 31–48.

Tversky, A., and Edwards, W. (1966), "Information versus Reward in Binary Choices," *Journal of Experimental Psychology*, *71*, 680–683.

von Neumann, J., and Morgenstern, O. (1947), *Theory of Games and Economic Behavior*, 2nd ed. Princeton: Princeton University Press.

Wagner, H.M. (1969), *Principles of Operations Research*. Englewood Cliffs, N.J.: Prentice-Hall.

Wald, A. (1947), *Sequential Analysis*. New York: Wiley.

Wald, A. (1950), *Statistical Decision Functions*. New York: Wiley.

Wallis, W.A., and Roberts, H.V. (1956), *Statistics: A New Approach*. Glencoe, Ill.: Free Press.

Watts, D.G. (ed.) (1968), *The Future of Statistics*. New York: Academic Press.

Weiss, L. (1961), *Statistical Decision Theory*. New York: McGraw-Hill.

Wendt, D. (1969), "Value of Information for Decisions," *Journal of Mathematical Psychology*, *6*, 430–443.

Wetherill, G.B. (1961), "Bayesian Sequential Analysis," *Biometrika*, *48*, 281–292.

Wetherill, G.B. (1966), *Sequential Methods in Statistics*. New York: Wiley.

Wetherill, G.B., and Campling, G.E.G. (1966), "The Decision Theory Approach to Sampling Inspection," *Journal of the Royal Statistical Society, 28,* 381–416 (with discussion).

White, L.S. (1967), "Bayes Markovian Decision Models for a Multiperiod Reject Allowance Problem," *Operations Research, 15,* 857–865.

Wilks, S.S. (1962), *Mathematical Statistics.* New York: Wiley.

Williams, J. (1954), *The Compleat Strategyst.* New York: McGraw-Hill.

Wilson, R.B. (1967), "Competitive Bidding with Asymmetric Information," *Management Science, 13,* 816–820.

Wilson, R.B. (1968a), "Decision Analysis in a Corporation," *IEEE Transactions on Systems Science and Cybernetics, SSC-4,* 220–226.

Wilson, R.B. (1968b), "On the Theory of Syndicates," *Econometrica, 36,* 119–132.

Wilson, R.B. (1969), "Investment Analysis under Uncertainty," *Management Science B, 15,* 650–664.

Winkler, R.L. (1967a), "The Assessment of Prior Distributions in Bayesian Analysis," *Journal of the American Statistical Association, 62,* 776–800.

Winkler, R.L. (1967b), "The Quantification of Judgment: Some Methodological Suggestions," *Journal of the American Statistical Association, 62,* 1105–1120.

Winkler, R.L. (1968), "The Consensus of Subjective Probability Distributions," *Management Science, 15,* 61–75.

Winkler, R.L. (1969), "Scoring Rules and the Evaluation of Probability Assessors," *Journal of the American Statistical Association, 64,* 1073–1078.

Winkler, R.L. (1971), "Probabilistic Prediction: Some Experimental Results," *Journal of the American Statistical Association, 66,* in press.

Winkler, R.L., and Murphy, A.H. (1968a), "Evaluation of Subjective Precipitation Probability Forecasts," *Proceedings of the First Conference on Statistical Meteorology,* American Meteorological Society, 148–157.

Winkler, R.L., and Murphy, A.H. (1968b), "'Good' Probability Assessors," *Journal of Applied Meteorology, 7,* 751–768.

Wolfowitz, J. (1962), "Bayesian Inference and Axioms of Consistent Decision," *Econometrica, 30,* 470–479.

Woods, D.H. (1966), "Improving Estimates that Involve Uncertainty," *Harvard Business Review, 44,* July–August, 91–98.

Ying, C.C. (1967), "Learning by Doing—An Adaptive Approach to Multiperiod Decisions," *Operations Research, 15,* 797–812.

Ying, C.C. (1969), "A Model of Adaptive Team Decision," *Operations Research, 17,* 800–811.

Zeckhauser, R., and Keeler, E. (1970), "Another Type of Risk Aversion," *Econometrica, 38,* 661–665.

Zellner, A. (1970), "Estimation of Regression Relationships Containing Unobservable Independent Variables," *International Economic Review, 11,* 441–454.

Zellner, A., and Chetty, V.K. (1965), "Prediction and Decision Problems in Regression Models from the Bayesian Point of View," *Journal of the American Statistical Association, 60,* 608–616.

Zellner, A., and Tiao, G.C. (1964), "Bayesian Analysis of the Regression Model with Autocorrelated Errors," *Journal of the American Statistical Association, 59,* 763–778.

ANSWERS TO SELECTED EXERCISES

CHAPTER 2

3. 2^n.
6. (b) 1/6 (e) 13/18 (g) 1/12.
8. 25/52, 1/2.
9. 125/216.
17. (a) 2/3 (c) 3/10.
18. (b) 1 to 4 (c) 7 to 1.
41. (b) 3/7, 12/13, 4/7, 1/13 (e) No.
43. $P(E_1,E_2,E_3) = P(E_1)P(E_2 \mid E_1)P(E_3 \mid E_1,E_2)$.
44. 2/3.
45. 0, .24, .4.
48. 8/23, 5/23, 10/23.
50. 343/370.
51. 5/29.

CHAPTER 3

1. (c) .5 (e) .9 (g) 0.
2. (b) 0 (c) $E(\tilde{x}) = 7/4$, $V(\tilde{x}) = 3/16$.
4. (b) 9/13 (d) 8/13.
5. (a) 1.8 (c) 5.56 (f) 0, 1.
6. (a) $E(\tilde{x}) = 8205$, $V(\tilde{x}) = 445{,}475$ (b) $E(\tilde{z}) = 3{,}025$, $V(\tilde{z}) = 11{,}136{,}875$.
11. The values of $\tilde{z}$ are -1.81, -1.06, -0.31, 0.44, 1.19, 1.94, and 2.69.
12. $5.18.
14. (b) Yes (e) 1.5, 2.4, 3.9, 1.2 (f) 0.45, 0.84, 1.29, 10.56.

17. (a) $P(\tilde{x} = 1 \mid \tilde{y} = 1) = .83$, $P(\tilde{x} = 2 \mid \tilde{y} = 1) = .17$
(d) $P(\tilde{y} = 1 \mid \tilde{x} = 2) = .042$, $P(\tilde{y} = 2 \mid \tilde{x} = 2) = .333$,
$P(\tilde{y} = 3 \mid \tilde{x} = 2) = .500$, $P(\tilde{y} = 4 \mid \tilde{x} = 2) = .125$
(e) 1.17, 1.40, 2.862, 2.708.
21. .0584, .0781.
23. .1933, .3844, .3097, .1126.
25. $P(\tilde{p} = .4) = 2/15$, $P(\tilde{p} = .5) = 10/15$, $P(\tilde{p} = .6) = 3/15$.
26. The posterior probabilities are .06990, .10744, .15713, .31012, .28831, and .06710, respectively, and the posterior mean and variance are .1568 and .000699.
31. .0067, .0681, .1251, 10.
33. .109, .561, .330.
34. The posterior probabilities are $P(\tilde{\lambda} = 10) = .071$, $P(\tilde{\lambda} = 11) = .210$, $P(\tilde{\lambda} = 12) = .470$, and $P(\tilde{\lambda} = 13) = .249$.
38. $P(\tilde{\lambda} = 50) = .00$, $P(\tilde{\lambda} = 75) = .14$, $P(\tilde{\lambda} = 100) = .86$.
41. $P(\tilde{\lambda} = 4) = .3244$, $P(\tilde{\lambda} = 5) = .5027$, $P(\tilde{\lambda} = 6) = .1729$.
47. From the hypergeometric, .881; from the binomial approximation, .821.
48. .0324, .0220.
49. .0353, .0014.
50. .000685, .0154.
53. $P(\tilde{\theta} = 2) = 1/19$, $P(\tilde{\theta} = 1) = 6/19$, $P(\tilde{\theta} = 0) = 12/19$.
57. $P(\tilde{r} = 0) = .03194$, $P(\tilde{r} = 1) = .10406$, $P(\tilde{r} = 2) = .17550$,
$P(\tilde{r} = 3) = .20824$, $P(\tilde{r} = 4) = .19391$, $P(\tilde{r} = 5) = .14558$,
$P(\tilde{r} = 6) = .08661$, $P(\tilde{r} = 7) = .03905$, $P(\tilde{r} = 8) = .01240$,
$P(\tilde{r} = 9) = .00247$, $P(\tilde{r} = 10) = .00023$.
59. Some predictive probabilities are $P(\tilde{r} = 0) = .0487$, $P(\tilde{r} = 5) = .1121$, and $P(\tilde{r} = 9) = .0059$.

CHAPTER 4

1. (a) 6 (c) 5/32 (d) 1 if $x \geq 1$, $3x^2 - 2x^3$ if $0 < x < 1$, 0 if $x \leq 0$
(e) 1/2, 3/10, 1/20.
2. (a) $x/2$ if $0 \leq x < 2$, 0 elsewhere (d) -1, 2.
3. 7/36, 1/9.
6. (a) 1/8 (c) $(x + 1)/4$ if $0 \leq x \leq 2$, 0 elsewhere
(d) $(2x + 1)/6$ if $0 \leq x \leq 2$, 0 elsewhere (f) 7/6, 7/6.
9. (b) $f(\theta \mid y) = 229(\theta + .10)^2$ if $-.10 \leq \theta \leq .10$, $459(\theta + .10)(.20 - \theta)$ if $.10 \leq \theta \leq .12$, $1261(.20 - \theta)^2$ if $.12 \leq \theta \leq .20$, 0 elsewhere.
13. $r' = 10$, $n' = 15$.
14. $r' = 15$, $n' = 75$.
16. $r' = 2$, $n' = 4$, $r'' = 4$, $n'' = 10$.
20. (a) .5328, .6915, .248 (b) $c = 7.56$ (c) -3.56, 0.32, 4.76, 12.32.
21. 109.1, 10.5.
24. .969, .831.

27. The posterior probabilities are $P(\tilde{\mu} = 109.4) = .0734$, $P(\tilde{\mu} = 109.7) = .2572$, $P(\tilde{\mu} = 110.0) = .5063$, $P(\tilde{\mu} = 110.3) = .1412$, and $P(\tilde{\mu} = 110.6) = .0219$.
28. normal, $m'' = 109.73$, $\sigma''^2 = .267$.
30. normal, $m' = 50$, $\sigma'^2 = 16.67$.
33. normal, $m'' = 51.4$, $\sigma''^2 = 8.97$; $P(\tilde{\mu} \geq 50) = .681$.
34. normal, $m'' = 967$, $\sigma''^2 = 1667$.
36. (c) normal-gamma, $m'' = 967$, $n'' = 30$, $v'' = 55{,}111$, $d'' = 21$.
43. beta, $r' = 450$, $n' = 1000$.
44. $\lambda = .55$.
45. gamma, $r'' = 34.5$, $t'' = 10$, $E(\tilde{\lambda}) = 3.45$, $V(\tilde{\lambda}) = .345$.
55. normal, $E(\tilde{m}) = 50$, $V(\tilde{m}) = 39$; $P(\tilde{m} \geq 55) = .212$.
56. beta-binomial, $n = 6$, $r' = 4$, $n' = 10$; $E(\tilde{r}) = 2.4$, $V(\tilde{r}) = 2.09$.
57. $P(\tilde{r} = r) = 1/(n + 1)$ for $r = 0, 1, \ldots, n$.

CHAPTER 5

2. (a) Action 3
4. (d)

	DEMAND				
	100	*150*	*200*	*250*	*300*
Order 100	200	175	150	125	100
Order 200	0	300	600	575	550
Order 300	−150	150	450	750	1050

7.

	PERCENTAGE CHANGE IN MARKET SHARE			
	0	*10%*	*20%*	*30%*
Advertise	−200,000	−50,000	100,000	250,000
Don't Advertise	0	0	0	0

8. (a) Action 2 (b) Action 4 (c) Action 4.
9. Order 100, order 300, order 300.
12. Action 2.
14. $ER(1) = 26.0$, $ER(2) = 48.0$, $ER(4) = 39.5$, $EL(1) = 90.5$, $EL(2) = 68.5$, $EL(4) = 77.0$.

15. Action 2.
16. ER(Order 100) = 2900/19, ER(Order 200) = 8200/19, ER(Order 300) = 7950/19.
18. ER(advertise) = −1,800,000/21, ER(Don't advertise) = 0.
22. (b) P(Adverse weather) > C/L.
24. ER(Hire) = −131.5, ER(Don't hire) = 0; don't hire.
25. ER(Plan 1) = 9957.9, ER(Plan 2) = 8500, ER(Plan 3) = 11,017.5.
26. (c) ER(Adjust) = 2400, ER(Don't adjust) = 2438.
29. (a) Action 2 (d) Action 2.
31. (a) Don't drill (c) Drill (e) Drill (f) Don't drill.
35. (b) EU(Bet) = 2160, EU(No bet) = 1900 (c) EU(Bet) = 9840, EU(No bet) = 9900.
37. (b) For the 6 utility functions, the optimal actions are bet, don't bet, either, don't bet, bet, and don't bet, respectively.
38. (d) Yes.
39. EU(Don't invest) = .50, EU($1000 in one investment) = .40, EU($500 in one investment) = .49, EU($500 in each investment) = .52.
41. Approximate expected utilities: EU(Advertise) = .08, EU(Don't advertise) = .12.
43. (b) For the 6 utility functions, the risk premiums are −2.67, 2.72, 0, 2.62, −59, and 22.75, respectively.
51. ER(Advertise) = 100,857, ER(Don't advertise) = 0.
56. ER(Order 0) = −91.25, ER(Order 100) = 142.50, ER(Order 200) = 381.25, ER(Order 300) = 328.75, ER(Order 400) = 95.00.

CHAPTER 6

2. In Exercise 14, EVPI = 68.5; in Exercise 15, EVPI = 0.70.
3. $121.05.
4. EVPI = 50,000.
5. (b) 1.95 (d) 1.40.
7. For $\tilde{\lambda}$, EVPI = 751.5.
9. (a) 14.38 (b) 0 (c) 12.22 (d) 0.
12. (a) 0 (b) 0 (c) 2571.
13. VPI (Oil) = 0, VPI (No Oil) = 40,000, EVPI = 28,000.
15. EVPI = 3.8.
16. (a) VSI = 3.231 (b) VSI = 0 (c) EVSI = .84.
17. (a) 1.09 (c) 2.05.
18. (b) $3.20 (c) $0.80 (d) $1.90.
19. (b) $4.80 (c) $1.20 (d) $2.84.
20. (b) 4,000 (c) ENGS = 1196.
22. (a) 0 (b) 14,286 (c) 5,079.
23. 27,466.
24. (a) $150,000 (b) $52,000 (c) $69,000.
25. ENGS (Use A *and* B) = $62,300.

26. ENGS (Sequential plan) = \$73,900.
27. (b) .34.
34. (a) 20 (c) 5 (d) 0 (e) 5.726 (g) 3.956.
35. (c) a_1 (d) beta distribution, $r'' = 8$ and $n'' = 39$, a_2.
36. ENGS(1) = 1.83, ENGS(5) = 3.35, ENGS(10) = 3.32.
39. (b) A (c) VPI = 0.
40. Drill with 100% interest.
44. (a) $n = 6$ (b) $n = 10$.
46. $E'U(a_1) = 11{,}750$, $E'U(a_2) = 11{,}433$, $E''U(a_1) = 12{,}167$, $E''U(a_2) = 12{,}254$.
47. (b) a_2 (c) $E''U(a_3) = 11{,}899$.
48. $E'U(a_1) = 207{,}359$, $E'U(a_2) = 262{,}891$.
49. $E'R(a_1) = 4$, $E'R(a_2) = 3.14$, $E'R(a_3) = 3.38$.

CHAPTER 7

9. (b) $\sigma^2, \sigma^2/4, 3\sigma^2/10, 4\sigma^2/9$ (c) 0, 0, 0, $\mu/3$.
10. (d) (108.96, 111.04) (e) (107.73, 110.67) (f) (108.88, 110.58).
11. (a) (46.5, 61.5) (c) (49.9, 52.9).
12. (b) (.1875, .5300) (d) (.2910, .5020).
14. 5 houses; EVPI = \$2,000.
15. 9,140; EVPI = \$218.
16. 43,300.
17. .1032.
19. 674.
24. .10.
26. .1875 for the posterior distribution, .1688 for the prior distribution.
27. 109.73.
29. (109.48, 109.98).
30. (a) 5.667 (b) .225 (c) 1.275 (d) $P''(H_1) = .56$.
31. (a) Poisson (d) 13/1126 (e) .005.
33. (a) .25 (b) 1.45 (c) .36.
35. $\Omega' = 1$, $\Omega'' = 2.33$.
36. $n \geq 66$, $n \geq 714$.
39. (a) .842 (c) .046 (e) .984 (g) .842.
40. $P''(\tilde{\mu} \geq 20) = .305$, $P''(19 < \tilde{\mu} < 21) = .281$.
41. 0; 10.6.
42. (b) .3625 (d) .325.
43. (a) normal, $m'' = -66$ and $\sigma''^2 = 13.33$ (b) (−69.65, −62.35) (c) .864.
44. (a) .136, .813, .051.
47. $\Omega'' \times L_{I/II} = 3.825$; do not market.
48. B.
49. $\Omega'' \times L_{I/II} = .50$; accept H_2.
50. (a) .2552 (b) .6284 (c) $\Omega'' \times L_{I/II} = .4189$.

INDEX